Lecture Notes in Computer Sc

Commenced Publication in 1973
Founding and Former Series Editors:
Gerhard Goos, Juris Hartmanis, and Jan van Leeuwen

Frank Dehne   Marina Gavrilova
Jörg-Rüdiger Sack   Csaba D. Tóth (Eds.)

Algorithms and Data Structures

11th International Symposium, WADS 2009
Banff, Canada, August 21-23, 2009
Proceedings

 Springer

Volume Editors

Frank Dehne
Carleton University, Ottawa, ON, Canada
E-mail: frank@dehne.net

Marina Gavrilova
University of Calgary, Calgary, AB, Canada
E-mail: mgavrilo@ucalgary.ca

Jörg-Rüdiger Sack
Carleton University, Ottawa, ON, Canada
E-mail: sack@scs.carleton.ca

Csaba D. Tóth
University of Calgary, Calgary, AB, Canada
E-mail: cdtoth@math.ucalgary.ca

Library of Congress Control Number: 2009931940

CR Subject Classification (1998): F.2, E.1, G.2, I.3.5, G.1

LNCS Sublibrary: SL 1 – Theoretical Computer Science and General Issues

ISSN   0302-9743
ISBN-10   3-642-03366-0 Springer Berlin Heidelberg New York
ISBN-13   978-3-642-03366-7 Springer Berlin Heidelberg New York

springer.com

© Springer-Verlag Berlin Heidelberg 2009
Printed in Germany

Typesetting: Camera-ready by author, data conversion by Scientific Publishing Services, Chennai, India
Printed on acid-free paper   SPIN: 12721795   06/3180   5 4 3 2 1 0

Preface

This volume contains the papers presented at the 11th Algorithms and Data Structures Symposium, WADS 2009 (formerly Workshop on Algorithms and Data Structures), held during August 21–23, 2009 in Banff, Alberta, Canada. WADS alternates with the Scandinavian Workshop on Algorithms Theory (SWAT), continuing the tradition of SWAT and WADS starting with SWAT 1988 and WADS 1989.

In response to the call for papers, 126 papers were submitted. From these submissions, the Program Committee selected 49 papers for presentation at WADS 2009. In addition, invited lectures were given by the following distinguished researchers: Erik Demaine, Richard Karp, and Christos Papadimitriou.

On behalf of the Program Committee, we would like to express our appreciation to the invited speakers, reviewers and all authors who submitted papers.

May 2009

Frank Dehne
Marina Gavrilova
Jörg-Rüdiger Sack
Csaba D. Tóth

Conference Organization

Program Chairs

Frank Dehne
Marina Gavrilova
Joerg-Ruediger Sack
Csaba D. Toth

Program Committee

Sergei Bereg
Allan Borodin
Gerth Stolting Brodal
Timothy Chan
Mark de Berg
Frank Devai
Matt Duckham
Rolf Fagerberg
Randolph Franklin
Joachim Gudmundsson
Susanne Hambrusch
Rolf Klein
Mike Langston
Ming Li
Friedhelm Meyer auf der Heide
Ian Munro
Jon Rokne
Shashi Shekhar
Bettina Speckmann
Paul Spirakis
Gabor Tardos
Jeff Vitter
Frances F. Yao

Local Organization

Marina Gavrilova
Csaba D. Toth
Jon Rokne
Maruf Monwar
Kushan Ahmadian

Table of Contents

On the Power of the
Semi-Separated Pair Decomposition

Mohammad Ali Abam[1,*], Paz Carmi[2,**], Mohammad Farshi[3,**],
and Michiel Smid[3,**]

[1] MADALGO Center, Aarhus University, Denmark
abam@madalgo.au.dk
[2] Ben-Gurion University of the Negev, Israel
paz@cg.scs.carleton.ca
[3] School of Computer Science, Carleton University, Ottawa, ON, K1S 5B6, Canada
mfarshi@cg.scs.carleton.ca, michiel@scs.carleton.ca

Abstract. A Semi-Separated Pair Decomposition (SSPD), with parameter $s > 1$, of a set $S \subset \mathbb{R}^d$ is a set $\{(A_i, B_i)\}$ of pairs of subsets of S such that for each i, there are balls D_{A_i} and D_{B_i} containing A_i and B_i respectively such that $d(D_{A_i}, D_{B_i}) \geq s \cdot \min(\mathrm{radius}(D_{A_i}), \mathrm{radius}(D_{B_i}))$, and for any two points $p, q \in S$ there is a unique index i such that $p \in A_i$ and $q \in B_i$ or vice-versa. In this paper, we use the SSPD to obtain the following results: First, we consider the construction of geometric t-spanners in the context of imprecise points and we prove that any set $S \subset \mathbb{R}^d$ of n imprecise points, modeled as pairwise disjoint balls, admits a t-spanner with $\mathcal{O}(n \log n/(t-1)^d)$ edges which can be computed in $\mathcal{O}(n \log n/(t-1)^d)$ time. If all balls have the same radius, the number of edges reduces to $\mathcal{O}(n/(t-1)^d)$. Secondly, for a set of n points in the plane, we design a query data structure for half-plane closest-pair queries that can be built in $\mathcal{O}(n^2 \log^2 n)$ time using $\mathcal{O}(n \log n)$ space and answers a query in $\mathcal{O}(n^{1/2+\varepsilon})$ time, for any $\varepsilon > 0$. By reducing the pre-processing time to $\mathcal{O}(n^{1+\varepsilon})$ and using $\mathcal{O}(n \log^2 n)$ space, the query can be answered in $\mathcal{O}(n^{3/4+\varepsilon})$ time. Moreover, we improve the preprocessing time of an existing axis-parallel rectangle closest-pair query data structure from quadratic to near-linear. Finally, we revisit some previously studied problems, namely spanners for complete k-partite graphs and low-diameter spanners, and show how to use the SSPD to obtain simple algorithms for these problems.

1 Introduction

Background. The Well-Separated Pair Decomposition (WSPD) introduced by Callahan and Kosaraju [1] has found numerous applications in proximity problems [2, Chapter 10]. A WSPD for a point set $S \subset \mathbb{R}^d$ with respect to a constant $s > 1$ is a set of pairs $\{(A_i, B_i)\}_i$ where (i) $A_i, B_i \subset S$, (ii) A_i and B_i

* MAA was supported by the MADALGO Center for Massive Data Algorithmics, a Center of the Danish National Research Foundation.
** PC, MF and MS were supported by NSERC of Canada.

F. Dehne et al. (Eds.): WADS 2009, LNCS 5664, pp. 1–12, 2009.

are s-well-separated, i.e., there are balls D_{A_i} and D_{B_i} containing A_i and B_i, respectively, such that $d(D_{A_i}, D_{B_i}) \geq s \cdot \max(\text{radius}(D_{A_i}), \text{radius}(D_{B_i}))$, and (iii) for any two points $p, q \in S$ there is a unique index i such that $p \in A_i$ and $q \in B_i$ or vice-versa. Callahan and Kosaraju showed that a WSPD containing $\mathcal{O}(s^d n)$ pairs can be constructed in $\mathcal{O}(s^d n + n \log n)$ time. Although they showed that $\sum \min(|A_i|, |B_i|) = \mathcal{O}(n \log n)$, the summation $\sum(|A_i| + |B_i|)$, the so-called *weight* of the WSPD, can be $\Theta(n^2)$. This disadvantage led Varadarajan [3] to define the Semi-Separated Pair Decomposition (SSPD).

An SSPD is defined as a WSPD, except for the condition (ii) which is relaxed to the requirement that A_i and B_i are s-semi-separated, i.e., there are balls D_{A_i} and D_{B_i} containing A_i and B_i, respectively, such that $d(D_{A_i}, D_{B_i}) \geq s \cdot \min(\text{radius}(D_{A_i}), \text{radius}(D_{B_i}))$. Varadarajan [3] showed how to compute an SSPD of weight $\mathcal{O}(n \log^4 n)$ for a set of n points in the plane in $\mathcal{O}(n \log^5 n)$ time and used the decomposition to solve the min-cost perfect-matching problem. Recently, Abam *et al.* [4] presented an algorithm which improves the construction time to $\mathcal{O}(n \log n)$ and the weight to $\mathcal{O}(n \log n)$; in [5], the same bounds were obtained in $\mathbb{R}^d$. It follows from results by Hansel [6] that any SSPD of any set of n points has weight $\Omega(n \log n)$—see Bollobás and Scott [7] as well.

Abam *et al.* [4] used the SSPD to compute a region fault-tolerant t-spanner in $\mathbb{R}^2$, which is a geometric t-spanner, as defined next, and remains a t-spanner after everything inside a half-plane fault region is removed from the spanner.

Let $G = (S, E)$ be a geometric graph on a set S of n points in $\mathbb{R}^d$. That is, G is an edge-weighted graph where the weight of an edge $(p, q) \in E$ is equal to $|pq|$, the Euclidean distance between p and q. The distance in G between two points p and q, denoted by $d_G(p, q)$, is defined as the length of a shortest (that is, minimum-weight) path between p and q in G. The graph G is called a (geometric) t-*spanner*, for some $t \geq 1$, if for any two points $p, q \in S$ we have $d_G(p, q) \leq t \cdot |pq|$. We define a t-*path* between p and q to be any path between p and q having length at most $t \cdot |pq|$. Geometric spanners have received a lot of attention in the past few years—see the book by Narasimhan and Smid [2] for more details.

Our results. In this paper, we present more applications of the SSPD and show how powerful the SSPD can be:

(i) We consider geometric t-spanners in the context of imprecise points. We model each imprecise point as a ball which specifies the possible location of the point. For a set of n pairwise disjoint imprecise points in $\mathbb{R}^d$, for a constant d, we compute a geometric t-spanner with $\mathcal{O}(n \log n / (t - 1)^d)$ edges such that regardless of the position of each point in its associated ball, it remains a t-spanner. Moreover, we improve the number of edges to $\mathcal{O}(n / (t - 1)^d)$ if the associated balls have the same radius.

(ii) We present a query data structure for the half-plane closest-pair query problem that uses $\mathcal{O}(n \log^2 n)$ space and can be computed in $\mathcal{O}(n^{1+\varepsilon})$ time and answers a query in $\mathcal{O}(n^{3/4+\varepsilon})$ time, where $\varepsilon > 0$ is an arbitrary constant. By increasing the pre-processing time to $\mathcal{O}(n^2 \log^2 n)$ and using $\mathcal{O}(n \log n)$ space, we achieve $\mathcal{O}(n^{1/2+\varepsilon})$ query time. We also improve the pre-processing time of the axis-parallel rectangle closest-pair query data structure of [8] from quadratic to near-linear.

(iii) We revisit some previously studied problems, specifically spanners for k-partite graphs [9] and low-diameter spanners [10,11], and show how to use the SSPD to obtain simple algorithms for these problems. Here, we just emphasize on the simplicity of the algorithms; we do not improve the existing results.

2 Spanners for Imprecise Points

Computational geometers traditionally assume that input data, such as points, are precise. However, in the real-world, the input comes from measuring devices which are subject to finite precision. Therefore, the input data given to an algorithm is imprecise and running the algorithm on the input may lead to incorrect output. One solution is to design algorithms that explicitly compute with imprecise data which can be modeled in different ways. One possible model, for data that consists of points, is to consider each point as a region. This region represents all possible locations where the point might be. Given a collection of such imprecise points, one can then ask questions about these points. What is their convex hull? What is their Voronoi diagram/Delaunay triangulation? These questions were recently studied— see [12,13]— and here we consider one more interesting question: "Is it possible to design a t-spanner with few edges for imprecise points such that regardless of the positions of the points in their associated regions, it remains a t-spanner?". In this section we answer this question affirmatively using the SSPD.

We model each imprecise input point p_i as a ball D_i in $\mathbb{R}^d$ and we assume that the balls D_i are pairwise disjoint. Indeed, the input is a set of n pairwise disjoint balls $D_i = (c_i, r_i)$, where c_i and r_i are the center and the radius of D_i.

2.1 Balls with Similar Sizes

We first consider the case when all balls are unit-balls. We can easily extend the results to the case when all balls have similar sizes. Our spanner construction is based on the WSPD approach [14] which works as follows. It computes a WSPD of the point set with respect to a constant s, and then for each pair (A, B) in the WSPD, it adds an edge between an arbitrary point from A and an arbitrary point from B. Choosing an appropriate value s based on t leads us to a t-spanner.

The above construction is applicable to an imprecise point set if we are able to construct a WSPD of the imprecise point set, i.e., regardless of the positions of the points in their associated balls, the pairs in the decomposition remain s-well-separated. The following lemma states that it is possible to obtain a WSPD of imprecise points using a WSPD of the center points.

Lemma 1. *If a WSPD of the center points $\{c_i\}_i$ with respect to $s' := 2s + 2$ is available, then we can obtain a WSPD of the points $\{p_i\}_i$ with respect to s.*

Proof. Let (A, B) be an s'-well-separated pair in the WSPD of the points $\{c_i\}_i$. Let A' (B') be the set containing the points $p_i \in D_i$ corresponding to the points $c_i \in A$ $(c_i \in B)$. We will show that (A', B') is an s-well-separated pair.

Since (A, B) is an s'-well-separated pair for $s' = 2s + 2$, there are two balls D_A and D_B containing the points in A and B, respectively, and $d(D_A, D_B) \geq (2s + 2) \cdot \max(\text{radius}(D_A), \text{radius}(D_B))$. If $\max(\text{radius}(D_A), \text{radius}(D_A)) < 1$, then the disjointness of the balls D_i implies that A and B are singletons, which implies that (A', B') is an s-well-separated pair. Otherwise, let D'_A (D'_B) be a ball with radius $\text{radius}(D_A) + 1$ $(\text{radius}(D_B) + 1)$ co-centered with D_A (D_B). Since $|p_i c_i| \leq 1$, from $c_i \in D_A$ we can conclude that $p_i \in D'_A$. The same property holds for B. Therefore D'_A and D'_B contain all points in A' and B', respectively, and it is easy to prove that $d(D'_A, D'_B) \geq s \cdot \max(\text{radius}(D'_A), \text{radius}(D'_B))$. ☐

Theorem 1. *For any set of n imprecise points in $\mathbb{R}^d$ modeled as pairwise disjoint balls with similar sizes and any $t > 1$, there is a t-spanner with $\mathcal{O}(n/(t-1)^d)$ edges which can be computed in $\mathcal{O}(n/(t-1)^d + n \log n)$ time.*

2.2   Balls with Arbitrary Sizes

When the sizes of the balls vary greatly, we cannot simply construct a WSPD of the points p_i using a WSPD of the center points c_i. Hence, a more sophisticated approach is needed. As we will see, the SSPD comes handy here. The overall idea is to construct an SSPD of the points $\{p_i\}_i$ using an SSPD of the points $\{c_i\}_i$ and then construct a t-spanner using the SSPD of the points $\{p_i\}_i$.

Lemma 2. *If an SSPD of the center points $\{c_i\}_i$ with respect to $s' := 3s + 3$ is available, we can obtain an SSPD of the points $\{p_i\}_i$ with respect to s.*

Proof. Let (A, B) be an s'-semi-separated pair in the SSPD of the points $\{c_i\}_i$. Let A' (B') be the set containing the points $p_i \in D_i$ corresponding to the points $c_i \in A$ $(c_i \in B)$. Since (A, B) is an s'-semi-separated pair, there are two balls D_A and D_B containing all points in A and B, respectively, such that $d(D_A, D_B) \geq s' \cdot \min(\text{radius}(D_A), \text{radius}(D_B))$. Without loss of generality, assume that $\text{radius}(D_A) \leq \text{radius}(D_B)$. If $\text{radius}(D_i) \geq 2 \cdot \text{radius}(D_A)$, the disjointness of the balls implies that A, and as a consequence A', is a singleton and therefore (A', B') is an s-semi-separated pair.

Otherwise, assume that for any point $c_i \in A$, $\text{radius}(D_i) < 2 \cdot \text{radius}(D_A)$. Therefore, every point p_i corresponding to the point $c_i \in A$ must lie in the ball co-centered with D_A and having radius $3 \cdot \text{radius}(D_A)$. Let D be the ball co-centered with D_A and radius $s' \cdot \text{radius}(D_A)$. Using a packing argument, it can be shown that the number of points $c_i \in B$ whose associated balls intersect D and have radius greater than $\text{radius}(D_A)$ is bounded by a constant. For each such point c_i, $(A', \{p_i\})$ is an s-semi-separated pair. For the remaining points c_i, the corresponding point p_i is at least $(s' - 1) \cdot \text{radius}(D_A)$ away from D_A. This implies that these points are at least $(s' - 3) \cdot \text{radius}(D_A)$ away from the points in A'; the latter points are inside a ball with radius $3 \cdot \text{radius}(D_A)$. This all together implies that these points and A' are s-semi-separated, because $(s' - 3)/3 = s$.

Note that each pair (A, B) produces a constant number of pairs each of which has linear size based on A and B and therefore the weight of the generated SSPD remains $\mathcal{O}(n \log n)$.   ⊡

Our spanner construction is as follows. First we use Lemma 2 to compute an SSPD $\mathcal{S}$ of the points $\{p_i\}_i$ with respect to $4/(t-1)$. Then, for each pair $(A, B) \in \mathcal{S}$, assuming radius$(D_A) \leq$ radius(D_B), we select an arbitrary point from A and connect it by an edge to every other point in $A \cup B$. The number of edges added to the spanner is at most $\sum_{(A,B) \in \mathcal{S}} (|A| + |B|)$ which is $\mathcal{O}(n \log n)$ based on the property of the SSPD. We claim that this gives a t-spanner. To prove this, let p and q be two arbitrary points. There is a pair $(A, B) \in \mathcal{S}$ such that $p \in A$ and $q \in B$ or vice-versa. Assume that radius$(D_A) \leq$ radius(D_B), $p \in A$, and $q \in B$. Based on our construction, both p and q are connected to a point w in A—note that w can be p. Therefore the length of the path between p and q in the graph is at most $|pw| + |wq|$, which can be bounded as follows:

$$|pw| + |wq| \leq 2|pw| + |pq| \leq 4 \text{ radius}(D_A) + |pq| \leq (t-1)|pq| + |pq| \leq t \cdot |pq|.$$

This shows that the path is a t-path between p and q.

Theorem 2. *For any set of n imprecise points in $\mathbb{R}^d$ modeled as pairwise disjoint balls and any $t > 1$, there is a t-spanner with $\mathcal{O}(n \log n/(t-1)^d)$ edges which can be computed in $\mathcal{O}(n \log n/(t-1)^d)$ time.*

3   Range Closest-Pair Query

The range searching problem is a well-studied problem in computational geometry. In such a problem, we are given a set of geometric objects, such as points or line segments, and want to pre-process the set into a data structure such that we can report the objects in a query region quickly—see the survey by Agarwal and Erickson [15]. However, in several applications, we need more information about the objects in the query area, for example the closest pair or the proximity of these objects. For this kind of queries, a so-called *aggregation function* can be defined to satisfy the property we are looking for. This *range-aggregate* query problem has been studied in recent years in both the computational geometry [16] and the database communities [17].

The range-aggregate query problem for the case when ranges are axis-parallel rectangles and the aggregation function is the closest pair, was first considered by Shan *et al.* [18]. They proposed an algorithm and showed that it works well in practice, but no theoretical bound was provided. Later Gupta [19] gave a data structure with constant query time using $\mathcal{O}(n)$ space for points in $\mathbb{R}$. For points in the plane, their structure answers a query in $\mathcal{O}(\log^3 n)$ time and uses $\mathcal{O}(n^2 \log^3 n)$ space. Later, Sharathkumar and Gupta [20] improved the space in the 2D case to $\mathcal{O}(n \log^3 n)$ while guaranteeing the same query time. Recently, Gupta *et al.* [8] improved the query time to $\mathcal{O}(\log^2 n)$ using $\mathcal{O}(n \log^5 n)$ space. It is unknown whether the data structures in [8,20] can be built in sub-quadratic time.

In this section, we first present a data structure for range closest-pair query problem when ranges are half-planes. Then, we show how to modify Gupta *et al.*'s data structure in [8] such that it can be built in near-linear time without affecting the query time and space bound.

3.1   Half-Plane Closest-Pair Query

Let S be a set of n points in the plane. We first start investigating which pairs of points can be a closest pair for some half-plane. Let G be the graph with vertex set S where p and q are connected if and only if (p, q) is a closest pair in $S \cap h$ for some half-plane h. The following lemma states that the number of such closest pairs is $\mathcal{O}(n)$, even though the number of "different" half-planes is $\Theta(n^2)$.

Lemma 3. *The graph G defined above is plane.*

Proof. For the sake of contradiction, assume that (p, q) and (r, s) properly intersect, where (p, q) and (r, s) are the closest pairs inside the half-planes h_1 and h_2, respectively. It is easy to see that h_1 contains at least one of the points r and s. Assume that r is inside h_1. Since (p, q) is the closest pair inside h_1, $|pr|$ and $|qr|$ are at least $|pq|$. The same argument holds for (r, s). Under the assumption that p is in h_2, we can conclude that $|pr|$ and $|ps|$ are at least $|rs|$. This all together implies that $|pq| + |rs| \leq |ps| + |rq|$. On the other hand, $|pq| + |rs| > |ps| + |rq|$, since (p, q) and (r, s) properly intersect. This contradiction implies that the graph G is plane. $\boxdot$

We describe our data structure under the assumption that G is available to us. Later, we will explain how to construct G. We construct a half-plane segment-reporting data structure for the edges of G, which is a multi-level partition tree—see [21, Section 16.2]. This data structure stores n segments not sharing any endpoint in such a way that the segments inside the half-plane query can be reported as the union of $\mathcal{O}(n^{1/2+\varepsilon})$ disjoint canonical subsets. The data structure uses $\mathcal{O}(n \log n)$ space and can be constructed in $\mathcal{O}(n^{1+\varepsilon})$ time. We also pre-compute the closest pair for each canonical subset of nodes in the second level of the tree to be able to report the closest pair without visiting all the edges in the query region.

The assumption that the segments are not sharing any endpoint can be relaxed to the assumption that each endpoint can be adjacent to at most a constant number of segments. Indeed, by such an assumption, the size of the associated partition tree with a node v in the first level is still proportional to $|S(v)|$, where $S(v)$ is the set of points stored at the subtree rooted at v. This is the key property in the analysis of the space and time complexity. Unfortunately, this assumption does not hold in our graph G, as we can simply find a configuration of n points such that the maximum degree in G is $\Theta(n)$. To make it work, we first make the graph G directed such that the out-degree of each vertex is constant. This can be performed as follows. We select a set S_1 of $n/2$ vertices whose degrees are at most 12—this is always possible, since the degree sum in any plane graph is at most 6 times the number of vertices. For each edge (p, q), if both p and q are in S_1 we give this edge an arbitrary direction. If one of them is in S_1, say p, we make the edge (p, q) directed in the direction $\overrightarrow{pq}$. We remove every directed edge as well as the vertices in S_1 from the graph and recurse on the remaining graph. At the end, we have a directed graph $\overrightarrow{G}$ such that the out-degree of each node is at most 12.

Now, given the graph $\overrightarrow{G}$, for each node v in the first level of the multi-level partition tree, we look at the edges going out from $S(v)$. Since the number of such edges is proportional to $|S(v)|$, the same query-time bound can be obtained.

One easy way of computing G is to compute the closest pair for all $2 \cdot \binom{n}{2}$ possible half-planes which takes $\mathcal{O}(n^3 \log n)$ time. Unfortunately it seems difficult to compute G in near-linear time. Hence, we introduce a new graph G' with $\mathcal{O}(n \log n)$ edges which contains G as a subgraph and can be computed in near-linear time. To define G', we use the convex region fault-tolerant t-spanner of the points, as introduced by Abam et al. [4]. This graph has the property that after removing all vertices and edges which are inside a convex fault region, what remains is a t-spanner of the complete graph minus the vertices and edges in the convex fault region. They used an SSPD to construct a convex region fault-tolerant t-spanner containing $\mathcal{O}(n \log n)$ edges in $\mathcal{O}(n \log^2 n)$ time. When the fault regions are half-planes, what remains from the graph is a t-spanner of the remaining points due to the fact that the line segment between any two points outside the half-plane fault region does not touch the fault region. Since any t-spanner, for $t < 2$, contains the closest pair as an edge, we set G' to be a region fault-tolerant t-spanner for some $t < 2$.

There are two possibilities of using G': (i) use G' instead of G in the above construction and (ii) use G' to compute G faster. Next we look at each of them more precisely.

(i) Using G' instead of G in our structure will obviously affect the asymptotic complexity of the space bound by a factor of $\mathcal{O}(\log n)$. Moreover, since we cannot make G' directed such that the out-degree of each vertex is bounded, we are unable to obtain the same query-time bound. We can show, however, that the query time is $\mathcal{O}(n^{3/4+\varepsilon})$: Searching in the first level of the multi-level partition tree boils down to $\mathcal{O}(n^{1/2+\varepsilon})$ associated partition trees which are disjoint and whose total size is $\mathcal{O}(n \log n)$. If x is the size of one of the associated trees, searching in the associated tree takes $\mathcal{O}(x^{1/2+\varepsilon})$ time. By the Cauchy-Schwarz inequality, we know that $\sum_{i=1}^{m} \sqrt{x_i}/m \leq \sqrt{\sum_{i=1}^{m} x_i/m}$. Therefore, the total search costs $\mathcal{O}(n^{3/4+\varepsilon})$—note that $m = \mathcal{O}(n^{1/2+\varepsilon})$ and $\sum_{i=1}^{m} x_i = \mathcal{O}(n \log n)$.

(ii) We can construct G from G' as follows. We sort all edges of G' by their lengths and process them in ascending order. Initially, we set G to be the graph on the point set whose edge set is empty. Let e be the edge of G' to be processed. We check in linear time whether it intersects any of the current edges of G. If so, we ignore e. Otherwise, we perform two rotational sweeps around the endpoints of e, in $\mathcal{O}(n \log n)$ time, to see whether there is a half-plane containing e which does not contain any edge in the current graph G. If so, e is inserted into G, otherwise, we ignore e. Since we process $\mathcal{O}(n \log n)$ edges, each of which takes $\mathcal{O}(n \log n)$ time, the total construction time is $\mathcal{O}(n^2 \log^2 n)$.

Theorem 3. *Let S be a set of n points in the plane. For any $\varepsilon > 0$, there is a data structure for S*

(i) of size $\mathcal{O}(n \log^2 n)$ that can be constructed in $\mathcal{O}(n^{1+\varepsilon})$ time and answers a half-plane closest-pair query in $\mathcal{O}(n^{3/4+\varepsilon})$ time; or

(ii) of size $\mathcal{O}(n \log n)$ that can be constructed in $\mathcal{O}(n^2 \log^2 n)$ time and answers a half-plane closest-pair query in $\mathcal{O}(n^{1/2+\varepsilon})$ time.

3.2   Axis-Parallel Rectangle Closest-Pair Query

We now consider the axis-parallel rectangle closest-pair query. As mentioned above, Gupta *et al.* [8] presented a data structure of size $\mathcal{O}(n \log^5 n)$ and query time $\mathcal{O}(\log^2 n)$. It is unknown whether their structure can be built in subquadratic time. Their data structure works as follows: They first construct a data structure to answer closest-pair queries for two-sided queries (vertical/horizontal strips and quadrants). To do that, they pre-compute a graph G with vertex set S whose edges are closest pair for some two-sided region. They show that G has linear size for quadrants and $\mathcal{O}(n \log n)$ size for vertical/horizontal strips; however, it is unknown how to compute G quickly. For three- and four-sided queries, they use the data structure for two-sided queries together with some additional information which can be computed in near-linear time. Therefore, the time-consuming ingredient of their structure is computing the graph G.

As in the previous section, we introduce a graph G' which has $\mathcal{O}(n \log n)$ edges, including all edges of G. We use G' instead of G. The graph G' indeed is a kind of t-spanner which we call *local t-spanner*.

A geometric t-spanner G is an F-local spanner, for a region F in the plane, if the part of G which is completely inside F is a t-spanner of the points inside F. For a family $\mathcal{F}$ of regions, we call a graph G an $\mathcal{F}$-local t-spanner, if for any region $F \in \mathcal{F}$ the graph G is an F-local t-spanner. As an example, any convex region fault-tolerant t-spanner is an $\mathcal{H}$-local t-spanner, where $\mathcal{H}$ is the family of half-planes. We will show that there are $\mathcal{F}$-local t-spanners with $\mathcal{O}(n \log n)$ edges, when $\mathcal{F}$ is the family of all axis-parallel two-sided regions in the plane. To this end, we set G' to be an $\mathcal{F}$-local t-spanner for some $t < 2$ which therefore contains the closest pair for every possible query region.

Theorem 4. *A set S of n points in the plane can be stored in a structure of size $\mathcal{O}(n \log^5 n)$ such that for any axis-parallel query rectangle Q, the closest pair in $S \cap Q$ can be reported in $\mathcal{O}(\log^2 n)$ time. Moreover, the structure can be built in $\mathcal{O}(n \log^5 n)$ time.*

Local t-spanner. In this section, we construct $\mathcal{F}$-local t-spanners with $\mathcal{O}(n \log n)$ edges, when $\mathcal{F}$ is the family of all axis-parallel two-sided regions in the plane. Due to similarity, we just consider the family $\mathcal{VS}$ of vertical strips and the family $\mathcal{NE}$ of north-east quadrants. Our construction is based on the region fault-tolerant t-spanner [4]. To re-use the approach in [4], we construct the graph such that the following property holds for every s-semi-separated pair (A, B) (assuming that radius$(D_A) \leq$ radius(D_B)).

(I) For every region $R \in \mathcal{F}$ such that $R \cap A \neq \emptyset$ and $R \cap B \neq \emptyset$, the point in $R \cap B$ that is closest to the center of D_A is connected to a point in $R \cap A$.

This property is equivalent to Lemma 3.2 of [4] and following a similar argument as in the proof of Lemma 3.3 of the same paper, proves that any graph satisfying this property is an $\mathcal{F}$-local t-spanner.

We will show how to satisfy property (I) using $\mathcal{O}(|A|+|B|)$ edges when regions are vertical strips and north-east quadrants. Therefore, this gives us an $\mathcal{F}$-local t-spanner which contains $\mathcal{O}(n \log n)$ edges.

Vertical strips. We first sort the points in A based on their x-coordinates. Then, for each point $b \in B$, we find two consecutive points $a, a' \in A$ surrounding b on the x-axis. We then connect b to both a and a'.

Lemma 4. *The above connecting schema satisfies property (I) and uses $\mathcal{O}(|A| + |B|)$ edges and can be performed in $\mathcal{O}((|A| + |B|) \log |A|)$ time.*

Proof. Assume that an arbitrary region $R \in \mathcal{VS}$ contains at least one point from each subset A and B. Let $a_1, \ldots, a_k$ be the sorted list of points in A, based on their x-coordinates. Let $b \in B \cap R$ be the point that is closest to the center of D_A. Our schema connects b to a_i and a_{i+1} for some i. If R does not contain a_i or a_{i+1}, then $R \cap A$ must be empty which is not true by assumption—note that since R is a vertical strip, it contains a contiguous subsequence of the sorted list. Therefore, the point b must be connected to a point of A.

Since the above schema just needs a sorted list of $|A|$, and it makes $|B|$ binary searches in this list, it can be performed in $\mathcal{O}((|A| + |B|) \log |A|)$ time. ☐

North-east quadrants. For a point $p = (p_x, p_y)$, let NE(p) be the north-east quadrant with apex at p. More precisely, NE$(p) = [p_x, +\infty) \times [p_y, +\infty)$. Similarly we define NW$(p) = (-\infty, p_x] \times [p_y, +\infty)$ and SE$(p) = [p_x, +\infty) \times (-\infty, p_y)$. The connecting schema is as follows:

(1) We connect every point $a \in A$ to the point in NE$(a) \cap B$, if it exists, that is closest to the center of D_A.
(2) We connect each point $b \in B$ to an arbitrary point in NE$(b) \cap A$, to the highest point in SE$(b) \cap A$ and to the rightmost point in NW$(b) \cap A$, if they exist.

Lemma 5. *The above connecting schema satisfies property (I) and uses $\mathcal{O}(|A| + |B|)$ edges and can be performed in $\mathcal{O}((|A| + |B|) \log^2(|A| + |B|))$ time.*

Proof. Let $R \in \mathcal{NE}$ be an arbitrary north-east quadrant containing at least one point of each subset A and B, and let $b \in R \cap B$ be the point that is closest to the center of D_A. If there exists a point of $R \cap A$ in NE(b), SE(b), or NW(b), our schema guarantees that b is connected to one of points in $R \cap A$. If this is not the case, then for every $a \in R \cap A$, the point b must be in NE(a) which then, by the first step of our schema, guarantees that a is connected to b.

To perform the above schema, we need two 2-dimensional range trees T_A and T_B for the points in A and B, respectively. We perform $|A|$ searches in T_B and $3|B|$ searches in T_A which in total can be done in $\mathcal{O}((|A| + |B|) \log^2(|A| + |B|))$ time. ⊟

4 SSPD Makes Life Easier

4.1 Spanners for Complete k-Partite Graphs

Bose *et al.* [9] introduced the following problem: Given a complete k-partite graph K on a set of n points in $\mathbb{R}^d$, compute a sparse spanner of the graph K. They presented an algorithm running in $\mathcal{O}(n \log n)$ time that computes a $(5+\varepsilon)$-spanner of K with $\mathcal{O}(n)$ edges. They also gave an algorithm of $\mathcal{O}(n \log n)$ time complexity which computes a $(3 + \varepsilon)$-spanner of K with $\mathcal{O}(n \log n)$ edges. This algorithm is based on a WSPD of the points and a bit involved. They also showed that every t-spanner of K for $t < 3$ must contain $\Omega(n \log n)$ edges.

We present a simpler algorithm, using the SSPD, to compute a $(3 + \varepsilon)$-spanner of K with $\mathcal{O}(n \log n)$ edges in $\mathcal{O}(n \log n)$ time. We first present the algorithm when K is a complete bipartite graph; at the end, we describe how to extend it to any k-partite graph. To this end, assume that we are given a complete bipartite graph of n red and blue points.

We first compute an SSPD of the point set with respect to $s = 6/\varepsilon$, no matter what the color of the points is. Consider a pair (A, B) in the SSPD. There exist two disjoint balls D_A and D_B containing A and B, resp., such that $d(D_A, D_B) \geq s \cdot \min(\text{radius}(D_A), \text{radius}(D_B))$. Assume that $\text{radius}(D_A) \leq \text{radius}(D_B)$. We choose a red and a blue representative point in A, denoted by $\text{rep}_r(A)$ and $\text{rep}_b(A)$, resp., if they exist. We also choose red and blue representative points in B, denoted by $\text{rep}_r(B)$ and $\text{rep}_b(B)$, which are the red and the blue points in B that are closest to A. Then we connect $\text{rep}_r(A)$ to all blue points in B and $\text{rep}_b(A)$ to all red points in B. We apply the same procedure for the representative points in B.

Consider a pair (x, y) of points, where x is red and y is blue and assume that (A, B) is the pair in the SSPD such that $x \in A$ and $y \in B$. Assume that $\text{radius}(D_A) \leq \text{radius}(D_B)$. Our algorithm connects x to $\text{rep}_b(B)$, $\text{rep}_b(B)$ to $\text{rep}_r(A)$, and $\text{rep}_r(A)$ to y. Let Π be this 3-hop path between x and y. For ease of presentation let $z = \text{rep}_r(A)$ and $w = \text{rep}_b(B)$, and let o and r be the center and the radius of D_A. We have:

$$\begin{aligned}
|\Pi| &\leq |xw| + |wz| + |zy| \\
&\leq r + |wo| + r + |ow| + 2r + |xy| = 4r + 2|wo| + |xy| \\
&\leq 4r + 2|yo| + |xy| \\
&\leq 4r + 2(|xy| + r) + |xy| = 6r + 3|xy| \\
&\leq 6|xy|/s + 3|xy| = (3 + 6/s)|xy| = (3 + \varepsilon)|xy|.
\end{aligned}$$

Extending the results to k-partite complete graphs is simple. We choose a representative point for any component for each color and we connect each representative to all the other points whose colors are different form the representative. This gives a $(3 + \varepsilon)$-spanner of size $\mathcal{O}(kn \log n)$.

4.2   Low-Diameter Spanners

The diameter of a t-spanner is the minimum integer Δ such that for any pair of points, there exists a t-path between them in the t-spanner containing at most Δ links. Spanners with low diameter are desirable to many applications like ad hoc networks where in order to quickly get a packet to the receiver it must pass through few stations. There are several t-spanners with $\mathcal{O}(\log n)$ diameter and $\mathcal{O}(n)$ edges; for example see [11,22,23]. Moreover, Arya *et al.* [10] presented an algorithm for constructing a t-spanner of diameter 2 which contains $\mathcal{O}(n \log n)$ edges. They also showed that a t-spanner with a constant diameter cannot have a linear number of edges.

t-spanner with diameter $\mathcal{O}(\log n)$. We present one more t-spanner with $\mathcal{O}(n)$ edges and diameter $\mathcal{O}(\log n)$. Our construction is the same as that of the region fault-tolerant t-spanner given in [4], except that instead of using $\mathcal{O}(|A| + |B|)$ edges to connect a pair (A, B) of the SSPD, we use $\mathcal{O}(1/(t-1))$ edges. For a pair (A, B) in the SSPD, assuming that $\mathrm{radius}(D_A) \leq \mathrm{radius}(D_B)$, we draw k cones with apex at the center of D_A. Let B_i be the set of points of B inside ith cone. We connect the point of B_i that is closest to A to an arbitrary point of A. Since the SSPD constructed in [4] is based on BAR tree [24] which has a depth of $\mathcal{O}(\log n)$, it is straightforward to see that the diameter of the new t-spanner is $\mathcal{O}(\log n)$.

t-spanner with diameter 2. Computing t-spanner of diameter 2 is even simpler using the SSPD. We compute an SSPD of the points with respect to $4/(t-1)$. Then for each pair (A, B) in the SSPD, assuming $\mathrm{radius}(D_A) \leq \mathrm{radius}(D_B)$, we choose an arbitrary point p in A and connect all the points in $A \cup B \setminus \{p\}$ to p. This gives us a spanner with $\mathcal{O}(n \log n)$, because of the SSPD property.

Let p and q be two arbitrary points. There is a pair (A, B) in the SSPD such that $p \in A$ and $q \in B$ or vice-versa. Assume that $\mathrm{radius}(D_A) \leq \mathrm{radius}(D_B)$. Based on our construction, both p and q are connected to a point w in A. Since

$$|pw| + |wq| \leq 2|pw| + |pq| \leq 4\,\mathrm{radius}(D_A) + |pq| \leq (t-1)|pq| + |pq| \leq t \cdot |pq|,$$

it follows that the spanner has diameter 2.

References

1. Callahan, P.B., Kosaraju, S.R.: A decomposition of multidimensional point sets with applications to k-nearest-neighbors and n-body potential fields. Journal of the ACM 42, 67–90 (1995)
2. Narasimhan, G., Smid, M.: Geometric spanner networks. Cambridge University Press, Cambridge (2007)
3. Varadarajan, K.R.: A divide-and-conquer algorithm for min-cost perfect matching in the plane. In: FOCS 1998, pp. 320–331 (1998)
4. Abam, M.A., de Berg, M., Farshi, M., Gudmundsson, J.: Region-fault tolerant geometric spanners. In: SODA 2007, pp. 1–10 (2007)
5. Abam, M.A., de Berg, M., Farshi, M., Gudmundsson, J., Smid, M.: Geometric spanners for weighted point sets (manuscript) (2009)

6. Hansel, G.: Nombre minimal de contacts de fermeture nécessaires pour réaliser une fonction booléenne symétrique de n variables. C. R. Acad. Sci. Paris 258, 6037–6040 (1964); Russian transl., Kibern. Sb. (Nov. Ser.) 5, 47–52 (1968)
7. Bollobás, B., Scott, A.: On separating systems. European Journal of Combinatorics 28(4), 1068–1071 (2007)
8. Gupta, P., Janardan, R., Kumar, Y., Smid, M.: Data structures for range-aggregate extent queries. In: CCCG 2008, pp. 7–10 (2008)
9. Bose, P., Carmi, P., Courture, M., Maheshvari, A., Morin, P., Smid, M.: Spanners of complete k-partite geometric graphs. In: Laber, E.S., Bornstein, C., Nogueira, L.T., Faria, L. (eds.) LATIN 2008. LNCS, vol. 4957, pp. 170–181. Springer, Heidelberg (2008)
10. Arya, S., Das, G., Mount, D.M., Salowe, J.S., Smid, M.: Euclidean spanners: short, thin, and lanky. In: STOC 1995, pp. 489–498 (1995)
11. Arya, S., Mount, D.M., Smid, M.: Randomized and deterministic algorithms for geometric spanners of small diameter. In: FOCS 1994, pp. 703–712 (1994)
12. Löffler, M., Snoeyink, J.: Delaunay triangulations of imprecise points in linear time after preprocessing. In: SCG 2008, pp. 298–304 (2008)
13. van Kreveld, M., Löffler, M.: Approximating largest convex hulls for imprecise points. Journal of Discrete Algorithms 6(4), 583–594 (2008)
14. Callahan, P.B., Kosaraju, S.R.: Faster algorithms for some geometric graph problems in higher dimensions. In: SODA 1993, pp. 291–300 (1993)
15. Agarwal, P.K., Erickson, J.: Geometric range searching and its relatives. In: Advances in Discrete and Computational Geometry, pp. 1–56 (1999)
16. Nievergelt, J., Widmayer, P.: Spatial data structures: Concepts and design choices. In: Handbook of Computational Geometry, pp. 725–764 (2000)
17. Tao, Y., Papadias, D.: Range aggregate processing in spatial databases. IEEE Transactions on Knowledge and Data Engineering 16(12), 1555–1570 (2004)
18. Shan, J., Zhang, D., Salzberg, B.: On spatial-range closest-pair query. In: Hadzilacos, T., Manolopoulos, Y., Roddick, J., Theodoridis, Y. (eds.) SSTD 2003. LNCS, vol. 2750, pp. 252–269. Springer, Heidelberg (2003)
19. Gupta, P.: Range-aggregate query problems involving geometric aggregation operations. Nordic Journal of Computing 13(4), 294–308 (2006)
20. Sharathkumar, R., Gupta, P.: Range-aggregate proximity queries. Technical Report IIIT/TR/2007/80, IIIT Hyderabad (2007)
21. de Berg, M., Cheong, O., van Kreveld, M., Overmars, M.: Computational Geometry: Algorithms and Applications, 3rd edn. Springer, Heidelberg (2008)
22. Arya, S., Mount, D.M., Smid, M.: Dynamic algorithms for geometric spanners of small diameter: Randomized solutions. Computational Geometry: Theory and Applications 13(2), 91–107 (1999)
23. Bose, P., Gudmundsson, J., Morin, P.: Ordered theta graphs. Computational Geometry: Theory and Applications 28, 11–18 (2004)
24. Duncan, C.A., Goodrich, M.T., Kobourov, S.: Balanced aspect ratio trees: Combining the advances of k-d trees and octrees. J. of Algorithms 38, 303–333 (2001)

Plane Graphs with Parity Constraints[*]

Oswin Aichholzer[1], Thomas Hackl[1], Michael Hoffmann[2], Alexander Pilz[1], Günter Rote[3], Bettina Speckmann[4], and Birgit Vogtenhuber[1]

[1] Institute for Software Technology, Graz University of Technology, Austria
{oaich,thackl,bvogt}@ist.tugraz.at, alexander.pilz@student.tugraz.at
[2] Institute for Theoretical Computer Science, ETH Zürich, Switzerland
hoffmann@inf.ethz.ch
[3] Institut für Informatik, FU Berlin, Germany
rote@inf.fu-berlin.de
[4] Dep. of Mathematics and Computer Science, TU Eindhoven, The Netherlands
speckman@win.tue.nl

Abstract. Let S be a set of n points in general position in the plane. Together with S we are given a set of parity constraints, that is, every point of S is labeled either even or odd. A graph G on S satisfies the parity constraint of a point $p \in S$, if the parity of the degree of p in G matches its label. In this paper we study how well various classes of planar graphs can satisfy arbitrary parity constraints. Specifically, we show that we can always find a plane tree, a two-connected outerplanar graph, or a pointed pseudo-triangulation which satisfy all but at most three parity constraints. With triangulations we can satisfy about 2/3 of all parity constraints. In contrast, for a given simple polygon H with polygonal holes on S, we show that it is NP-complete to decide whether there exists a triangulation of H that satisfies all parity constraints.

1 Introduction

Computing a simple graph that meets a given *degree sequence* is a classical problem in graph theory and theoretical computer science, dating back to the work of Erdös and Gallai [6]. A degree sequence is a vector $d = (d_1, \ldots, d_n)$ of n positive numbers. It is *realizable*, iff there exists a simple graph whose nodes have precisely this sequence of degrees. Erdös and Gallai gave necessary and sufficient conditions for a degree sequence to be realizable, and several algorithms have been developed that generate a corresponding abstract graph.

An extension of this problem prescribes not only a degree sequence d, but also gives a set $S \subset \mathbb{R}^2$ of n points in general position, where $p_i \in S$ is assigned

[*] This research was initiated during the Fifth European Pseudo-Triangulation Research Week in Ratsch a.d. Weinstraße, Austria, 2008. Research of O. Aichholzer, T. Hackl, and B. Vogtenhuber supported by the FWF [Austrian Fonds zur Förderung der Wissenschaftlichen Forschung] under grant S9205-N12, NFN Industrial Geometry. Research by B. Speckmann supported by the Netherlands Organisation for Scientific Research (NWO) under project no. 639.022.707.

F. Dehne et al. (Eds.): WADS 2009, LNCS 5664, pp. 13–24, 2009.
© Springer-Verlag Berlin Heidelberg 2009

degree d_i. It is well known that a degree sequence d is realizable as a tree if and only if $\sum_{i=1}^{n} d_i = 2n - 2$. Tamura and Tamura [11] extended this result to plane (straight line) spanning trees, giving an $O(n^2 \log n)$ time embedding algorithm, which in turn was improved by Bose et al. [4] to optimal $O(n \log n)$ time.

In this paper we study a relaxation of this problem, where we replace exact degrees with degree parity: odd or even. Although parity constrains are significantly weaker than actual degree constrains, they still characterize certain (classes of) graphs. For example, Eulerian graphs are exactly those connected graphs where all vertices have even degree, and a classical theorem of Whitney states that a maximal planar graph is 3-colorable iff all vertices have even degree. A given graph might satisfy only a subset of the parity constraints. So we study how well various classes of planar graphs can satisfy arbitrary parity constraints.

Definitions and notation. Let $S \subset \mathbb{R}^2$ be a set of n points in general position. We denote the convex hull of S by $CH(S)$. The points of S have parity constraints, that is, every point of S is labeled either *even* or *odd*; for ease of explanation we refer to even and odd points. We denote by n_e and n_o the number of even and odd points in S, respectively. Throughout the paper an even point is depicted by ⊖, an odd point by ⦶, and a point that can be either by ⊕. A graph G on S makes a point $p \in S$ *happy*, if the parity of $\deg_G(p)$ matches its label. If p is not happy, then it is *unhappy*. Throughout the paper a happy point is depicted by ○, an unhappy point by ●, and a point that can be either by ◐.

Results. Clearly, not every arbitrary set of parity constraints can be fulfilled. For example, in any graph the number of odd-degree vertices is even. Hence, the number of unhappy vertices has the same parity as n_o. For the class of plane trees, the aforementioned results on degree sequences immediately imply:

Theorem 1. *On every point set $S \subset \mathbb{R}^2$ with parity constraints, there exists a plane spanning tree that makes (i) all but two points happy if $n_o = 0$, (ii) all but one point happy if n_o is odd, and (iii) all points happy if $n_o \geq 2$ is even.*

We show that we can always find a two-connected outerplanar graph (which is a Hamiltonian cycle with additional edges in the interior, Theorem 2) and a pointed pseudo-triangulation (Theorem 3), which satisfy all but at most three parity constraints. For triangulations (Theorem 4), we can satisfy about $2/3$ of the parity constraints. Our proofs are based on simple inductive constructions, but sometimes involve elaborate case distinctions. We also argue that for triangulations the number of unhappy vertices might grow linearly in n. Finally, in Section 5 we show that if we are given a simple polygon H with polygonal holes on S, it is NP-complete to decide whether there exists a triangulation of H that satisfies all parity constraints.

Related work. Many different types of degree restrictions for geometric graphs have been studied. For example, for a given set $S \subset \mathbb{R}^2$ of n points, are there planar graphs on S for which the maximum vertex degree is bounded? There clearly is a path, and hence a spanning tree, of maximum degree at most two. Furthermore, there is always a pointed pseudo-triangulation of maximum degree five [8], although there are point sets where every triangulation must have a vertex of degree $n - 1$. Another related question is the following: we are given a set $S \subset \mathbb{R}^2$ of n points, together with a planar graph G on n vertices. Is there a plane straight-line embedding of G on S? Outerplanar graphs are the largest class of planar graphs for which this is always possible, in particular, Bose [3] showed how to compute such an embedding in $O(n \log^2 n)$ time.

One motivation for our work on parity restrictions stems from a bi-colored variation of a problem stated by Erdős and Szekeres in 1935: Is there a number $f^{\mathrm{ES}}(k)$ such that any set $S \subset \mathbb{R}^2$ of at least $f^{\mathrm{ES}}(k)$ bi-colored points in general position has a monochromatic subset of k points that form an empty convex k-gon (that is, a k-gon that does not contain any points of S in its interior)? It has been shown recently [1] that every bi-colored point set of at least 5044 points contains an empty (not necessarily convex) monochromatic quadrilateral. The proof uses, among others, a result that for any point set there exists a triangulation where at least half of the points have odd parity. Any increase in the guaranteed number of odd parity points translates into a lower minimum number of points required in the above statement. More specifically, Theorem 4 below shows that the above result holds for any set of at least 2760 points.

2   Outerplanar Graphs

After trees as minimally connected graphs, a natural next step is to consider two-connected graphs. In particular, outerplanar graphs generalize trees both in terms of connectivity and with respect to treewidth. In this section we consider two-connected outerplanar graphs, which are the same as outerplanar graphs with a unique Hamiltonian cycle [5], in other words, simple polygons augmented with a set of pairwise non-crossing diagonals.

The following simple construction makes all but at most three points happy. Pick an arbitrary point p. Set $p_1 = p$ and denote by $p_2, \ldots, p_n$ the sequence of points from S, as encountered by a counterclockwise radial sweep around p, starting from some suitable direction (if p is on $\mathrm{CH}(S)$ towards its counterclockwise neighbor). The outerplanar graph G consists of the closed polygonal chain $P = (p_1, \ldots, p_n)$ plus an edge between p and every odd point in $p_3, \ldots, p_{n-1}$. All points are happy, with the possible exception of p, p_2, and p_n. The figure below shows an example of a point set S with parity constraints and an outerplanar graph on S such that all but two points are happy.

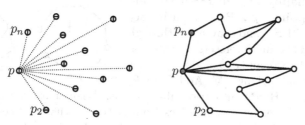

Theorem 2. *For every set $S \subset \mathbb{R}^2$ of n points with parity constraints, there exists an outerplanar graph on S that makes all but at most three points happy.*

3 Pointed Pseudo-triangulations

Pseudo-triangulations are related to triangulations and use *pseudo-triangles* in addition to triangles. A pseudo-triangle is a simple polygon with exactly three interior angles smaller than π. A pseudo-triangulation is called *pointed* if every vertex p has one incident region whose angle at p is greater than π. In the following we describe a recursive construction for a pointed pseudo-triangulation $\mathcal{P}$ on S that makes all but at most three points of S happy.

At any time in our construction we have only one recursive sub-problem to consider. This subproblem consists of a point set S^* whose convex hull edges have already been added to $\mathcal{P}$. The current set $\mathcal{P}$ is a pointed set of edges that subdivides the exterior of $\mathrm{CH}(S^*)$ into pseudo-triangles such that all points outside $\mathrm{CH}(S^*)$ are happy. $\mathcal{P}$ contains no edges inside $\mathrm{CH}(S^*)$. We say that S* is *hopeful* if at least one point on $\mathrm{CH}(S^*)$ is made happy by the current version of $\mathcal{P}$. Otherwise, we say that S^* is *unhappy*.

We initialize our construction by setting $S^* = S$ and adding $\mathrm{CH}(S)$ to $\mathcal{P}$. Now we distinguish four cases.

(1) **S^* is hopeful.** Let v be a point on $\mathrm{CH}(S^*)$ that is currently happy, let p and q be its neighbors, and let S' be the (possibly empty) set of points from S that lie in the interior of the triangle $\triangle_{qvp}$. Then $\mathrm{CH}(S' \cup \{p,q\})$ without the edge 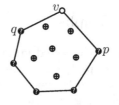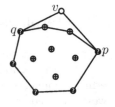 pq defines a convex chain C from p to q, in a way that C and v together form a pseudo-triangle. (If $S' = \emptyset$, then $C = pq$.) Remove v from consideration by adding C to $\mathcal{P}$. If $|S^*| \geq 5$, recurse on $S^* \setminus \{v\}$. Otherwise, there are at most three unhappy points.

(2) **S^* is unhappy and has no interior points.** Choose one point p on $\mathrm{CH}(S^*)$ and triangulate $\mathrm{CH}(S^*)$ by adding edges from p. There are at most three unhappy points, namely p and its two neighbors.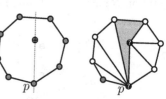

(3) **S^* is unhappy and has exactly one interior point, p_i.** Pick an arbitrary point p on $\mathrm{CH}(S^*)$ and draw a line through p and p_i. This line intersects exactly one edge e of $\mathrm{CH}(S^*)$, and e, p, and p_i together define a pseudo-triangle ∇. Add ∇ to $\mathcal{P}$, 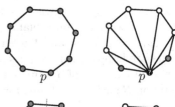 which splits $\mathrm{CH}(S^*)$ into two sub-polygons. Triangulate the sub-polygon which contains p_i by adding edges from p_i to all other vertices, except to its neighbors. Similarly, triangulate the other sub-polygon by adding edges from p. There are at most three unhappy points: p, p_i, and a neighbor of p.

(4) S^* is unhappy and has more than one interior point. Let S_i be the set of interior points. First add the edges of $\text{CH}(S_i)$ to $\mathcal{P}$. Then connect each point on $\text{CH}(S^*)$ tangentially to $\text{CH}(S_i)$ in clock-   wise direction, thereby creating a "lens shutter" pattern. Each point on $\text{CH}(S^*)$ is now happy. If $|S_i| > 3$, then recurse on S_i. Otherwise, there are at most three unhappy points.

Theorem 3. *For every point set $S \subset \mathbb{R}^2$ with parity constraints, there exists a pointed pseudo-triangulation on S that makes all but at most three points of S happy.*

4   Triangulations

The final class of planar graphs which we consider are triangulations. If the point set S lies in convex position, then all pseudo-triangulations of S are in fact triangulations. Thus we obtain the following as a consequence of Theorem 3:

Corollary 1. *For every point set $S \subset \mathbb{R}^2$ in convex position with parity constraints, and any three points p, q, r that are consecutive along $\text{CH}(S)$, there exists a triangulation on S that makes all points of S happy, with the possible exception of p, q, and r.*

The following simple observation will prove to be useful.

Observation 1. *For every set $S \subset \mathbb{R}^2$ of four points in convex position with parity constraints and every $p \in S$ there exists a triangulation on S that makes at least two of the points from $S \setminus \{p\}$ happy.*

For point sets of small cardinality we can investigate the number of happy vertices with the help of the order type data base [2]. For any set of 11 points with parity constraints we can always find a triangulation which makes at least 7 vertices happy. This immediately implies that there is always a triangulation that makes at least $7n/11 \approx 0.63n$ vertices happy.

The figure below shows a double circle for 10 points with parity constraints, such that at most 5 points can be made happy. This is in fact the only point configuration for $n = 10$ (out of 14 309 547) with this property.

Based on the double circle we have been able to construct large exam-   ples with a repeating parity pattern (starting at an extreme vertex) $\sigma = \langle (ee(oe)^3ee(oe)^7ee(oe)^5)^3 \rangle$ of length 108, where e denotes even, and o odd parity. It can be shown by inductive arguments that for such configurations for any triangulation we get at least $n/108 + 2$ unhappy vertices. Triangulating the interior of the double circle is equivalent to triangulating a simple polygon, as the inner vertices are connected by *unavoid-able edges*, that is, edges that have to be in any triangulation of the set. Hence, all base cases (over 46000) for the required induction can be checked using dynamic programming, see the full version of the paper and [10] for details. Open Problem 1 in [1] asks which is the maximum constant c such that for any point set there always exists a triangulation where $cn - o(n)$ points have odd degree. While for the question as stated we still believe that $c = 1$ is possible, the above construction shows that for general parity constraints we have $c \leq \frac{107}{108}$.

Theorem 4. *For every set $S \subset \mathbb{R}^2$ of n points with parity constraints, there exists a triangulation on S that makes at least $\lceil 2(n-1)/3 \rceil - 6$ points of S happy.*

Proof. Pick an arbitrary point p on $\mathrm{CH}(S)$, set $p_1 = p$, and denote by $p_2, \dots, p_n$ the sequence of points from S, as encountered by a counterclockwise radial sweep around p. Consider the closed polygonal chain $P = (p_1, \dots, p_n)$ and observe that P describes the boundary of a simple polygon (Fig. 1). With $\angle pqr$ denote the counterclockwise angle between the edges pq and qr around q. A point p_i, $2 \leq i < n$, is *reflex* if the interior angle of P at p_i is reflex, that is, $\angle p_{i-1}p_ip_{i+1} > \pi$; otherwise, p_i is *convex*. Thus, p_1, p_2, and p_n are convex.

We construct a triangulation T on S as follows. As a start, we take the edges of $\mathrm{CH}(S)$ and all edges of P, and denote the resulting graph by T_0. If P is convex then T_0 forms a convex polygon. Otherwise $\mathrm{CH}(S)$ is partitioned into two or more faces by the edges of P. Thinking of p as a light source and of P as opaque, we call the face of T_0 that contains p the *light face* and the other faces of T_0 *dark faces*. Dark faces are shown gray in figures.

In a next step, we insert further edges to ensure that all faces are convex. The light face is made convex by adding all edges pp_i where p_i is reflex. Hence the light face of T_0 might be split into a number of faces, all of which we refer to as light faces in the following. We partition the dark faces into convex faces as follows. First, we add all edges to connect the subsequence of P that consists of all convex points by a polygonal path. Note that some of those edges may be edges of P or $\mathrm{CH}(S)$ and, hence, already be present. Next, we triangulate those dark faces that are not convex. For now, let us say that these faces are triangulated arbitrarily. Later, we add a little twist.

Our construction is based on choosing particular triangulations for those faces that share at least two consecutive edges with P. Let us refer to these faces as

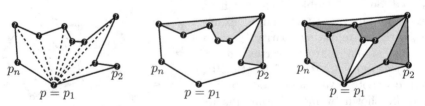

Fig. 1. The simple polygon bounded by P, the initial graph T_0 (with dark faces shown gray), and the graph T_1 in which all faces are convex (interesting light and dark faces shown light gray and dark gray, respectively)

interesting, while the remaining ones are called *uninteresting*. The interesting faces can be ordered linearly along P, such that any two successive faces share exactly one edge. We denote this order by $f_1, \ldots, f_m$. Note that f_i is light for i odd and dark for i even, and that both f_1 and f_m are light. Also observe that p is a vertex of every light face; therefore, any interesting light face other than f_1 and f_m has at least four vertices and all uninteresting light faces are triangles. On the dark side, however, there may be both interesting triangles and uninteresting faces with more than three vertices. Similar to above, we triangulate all uninteresting dark faces, for now, arbitrarily (a little twist will come later). We denote the resulting graph by T_1.

As a final step, we triangulate the interesting faces $f_1, \ldots, f_m$ of T_1 in this order to obtain a triangulation on S with the desired happiness ratio. We always treat a light face f_i and the following dark face f_{i+1} together. The vertices that do not occur in any of the remaining faces are *removed*, and the goal is to choose a local triangulation for f_i and f_{i+1} that makes a large fraction of those vertices happy. The progress is measured by the *happiness ratio* h/t, if h vertices among t removed vertices are happy. Note that these ratios are similar to fractions. But in order to determine the collective happiness ratio of two successive steps, the corresponding ratios have to be added component-wise. In that view, for instance, $2/2$ is different from $3/3$.

We say that some set of points can be made happy "using a face f", if f can be triangulated—for instance using Corollary 1 or Observation 1—such that all these points are happy. Two vertices are *aligned*, if either both are currently happy or both are currently unhappy. Two vertices that are not aligned are *contrary*. Denote the boundary of a face f by ∂f, and let $\partial f_i = (p, p_j, \ldots, p_k)$, for some $k \geq j + 2$, and $\partial f_{i+1} = (p_{k-1}, \ldots, p_r)$, for some $r \geq k + 1$.

After treating f_i and f_{i+1}, we have removed all vertices up to, but not including, the last two vertices p_{r-1} and p_r of f_{i+1}, which coincide with the first two vertices of the next face f_{i+2}. Sometimes, the treatment of f_i and f_{i+1} leaves the freedom to vary the parity of the vertex p_{r-1} while maintaining the desired happiness ratio as well as the parity of p_r. This means that the future treatment of f_{i+2} and f_{i+3} does not need to take care of the parity of p_{r-1}. By adjusting the triangulation of f_i and f_{i+1} we can always guarantee that p_{r-1} is happy.

Therefore, we distinguish two different settings regarding the treatment of a face pair: no choice (the default setting with no additional help from outside) and 1st choice (we can flip the parity of the first vertex p_j of the face and, thus, always make it happy).

No choice. We distinguish cases according to the number of vertices in f_i.

(1.1) $k \geq j + 3$, that is, f_i has at least five vertices. Then $p_j, \ldots, p_{k-2}$ can be made happy using f_i, and $p_{k-1}, \ldots, p_{r-3}$ can be made happy using f_{i+1}. Out of the $r - j - 1$ points removed, at least $(k - 2 - j + 1) + (r - 3 - (k - 1) + 1) = r - j - 2$ are happy. As $r - j \geq 4$, this yields a happiness ratio of at least $2/3$. The figure to the right shows the case $r = k + 1$ as an example.

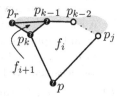

(1.2) $k = j + 2$, that is, f_i is a convex quadrilateral. We distinguish subcases according to the number of vertices in f_{i+1}.

(1.2.1) $r \geq j + 4$, that is, f_{i+1} has at least four vertices. Using f_{i+1}, all of $p_{j+3}, \ldots, p_{r-2}$ can be made happy. Then at least two out of $p_j, \ldots, p_{j+2}$ can be made happy using f_i. Overall, at least $r - 2 - (j + 3) + 1 + 2 = r - j - 2$ out of $r - j - 1$ removed points are happy. As $r - j \geq 4$, the happiness ratio is at least $2/3$.

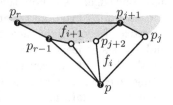

(1.2.2) $r = j + 3$, that is, f_{i+1} is a triangle. If both p_j and p_{j+1} can be made happy using f_i, the happiness ratio is $2/2$. Otherwise, regardless of how f_i is trian-

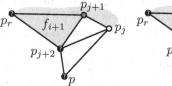

gulated exactly one of p_j and p_{j+1} is happy, see the figure to the right. This yields a ratio of $1/2$ and 1st choice for f_{i+2}.

First choice. Denote by f' the other (than f_i) face incident to the edge $p_j p_{j+1}$ in the current graph. As all of $f_1, \ldots, f_{i-1}$ are triangulated already, f' is a triangle whose third vertex (other than p_j and p_{j+1}) we denote by p'. Recall that in the 1st choice setting we assume that, regardless of how f_i is triangulated, p_j can be made happy. More precisely, we assume the following in a 1st choice scenario with a face pair f_i, f_{i+1} to be triangulated: By adjusting the triangulations of $f_1, \ldots, f_{i-1}$, we can synchronously flip the parity of both p_j and p', such that

   (C1) All faces $f_i, f_{i+1}, \ldots, f_m$ as well as f' remain unchanged,
   (C2) the degree of all of $p_{j+1}, \ldots, p_n$ remains unchanged, and
   (C3) the number of happy vertices among $p_2, \ldots, p_{j-1}$ does not decrease.

Observe that these conditions hold after Case 1.2.2. Using this 1st *choice flip*, we may suppose that p' is happy. Then by (C3) the number of happy vertices among $\{p_2, \ldots, p_{j-1}\} \setminus \{p'\}$ does not decrease, in case we do the 1st choice flip (again) when processing f_i, f_{i+1}. We distinguish cases according to the number of vertices in f_i.

(2.1) $k \geq j + 3$, that is, f_i has at least five vertices. Then $p_{j+1}, \ldots, p_{k-1}$ can be made happy using f_i. If f_{i+1} is a triangle (as shown in the figure to the right), this yields a ratio of at least $3/3$. Otherwise ($r \geq k + 2$), apart from keeping p_{k-1} happy, f_{i+1} can be used to make all of $p_k, \ldots, p_{r-3}$ happy. At least $r - j - 2$ out of $r - j - 1$ vertices removed are happy, for a happiness ratio of at least $3/4$.

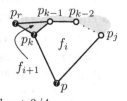

(2.2) $k = j + 2$, that is, f_i is a convex quadrilateral. We distinguish subcases according to the size of f_{i+1}.

(2.2.1) $r \geq j + 5$, that is, f_{i+1} has at least five vertices. Triangulate f_i arbitrarily and use f_{i+1} to make all of $p_{j+1}, \ldots, p_{r-3}$ happy. At least $r - j - 2$ out of $r - j - 1$ vertices removed are happy, for a happiness ratio of at least $3/4$.

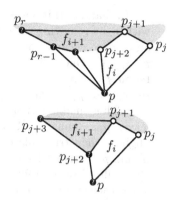

(2.2.2) $r = j + 3$, that is, f_{i+1} is a triangle. Use f_i to make p_{j+1} happy for a perfect ratio of $2/2$.

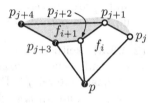

(2.2.3) $r = j + 4$, that is, f_{i+1} is a convex quadrilateral. If p_{j+1} and p_{j+2} are aligned, then triangulating f_i arbitrarily makes them contrary. Using f_{i+1} both can be made happy, for a perfect $3/3$ ratio overall. Thus, suppose that p_{j+1} and p_{j+2} are contrary. We make a further case distinction according to the position of p_j with respect to f_{i+1}.

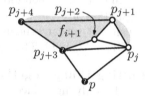

(2.2.3.1) $\angle p_{j+3}p_{j+2}p_j \leq \pi$, that is, p, p_j, p_{j+2}, p_{j+3} form a convex quadrilateral. Add edge $p_j p_{j+2}$ and exchange edge $p p_{j+2}$ with edge $p_j p_{j+3}$. In this way, p_{j+1} and p_{j+2} remain contrary. Hence, both p_{j+1} and p_{j+2} can be made happy using f_{i+1}, for a perfect ratio of $3/3$ overall.

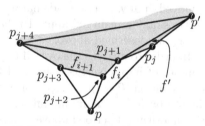

(2.2.3.2) $\angle p_j p_{j+1}p_{j+3} \leq \pi$, that is, the points $p_j, p_{j+4}, p_{j+3}, p_{j+1}$ form a convex quadrilateral. To conquer this case we need $p' p_{j+4}$ to be an edge of T_1. In order to ensure this, we apply the before mentioned little twist: before triangulating the non-convex dark faces, we scan through the sequence of dark faces for

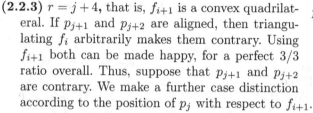

configurations of points like in this case. Call a dark quadrilateral f_i with $\partial f_i = (p_{j+1}, \ldots, p_{j+4})$ *delicate* if $\angle p_j p_{j+1}p_{j+3} \leq \pi$. For every delicate dark quadrilateral f_i in $f_4, f_6, \ldots, f_{m-1}$ such that f_{i-2} is not delicate, add the edge $p_{j+4}p_h$, where p_h is the first vertex of f_{i-2}. Observe that this is possible as $p_h, \ldots, p_{j+1}, p_{j+3}, p_{j+4}$ form a convex polygon f^*: $p_h, \ldots, p_{j+1}$ and $p_{j+1}, p_{j+3}, p_{j+4}$ form convex chains being vertices of f_{i-2} and f_i, respectively, and p_{j+1} is a convex vertex of f^* because $\angle p_j p_{j+1}p_{j+3} \leq \pi$. Then we triangulate the remaining non-convex and the uninteresting dark faces arbitrarily to get T_1.

To handle this case we join f_{i+1} with f' by removing the edges $p_{j+1}p_{j+4}$ and $p'p_{j+1}$ and adding the edge $p_{j+3}p_{j+1}$, which yields a convex pentagon $f^* = p_{j+4}, p_{j+3}, p_{j+1}, p_j, p'$. Observe that p_{j+1} and p_{j+2} are aligned now. Thus, making p_{j+2} happy using f_i leaves p_{j+1} unhappy. If p' and p_j are

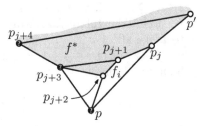

aligned, then triangulate f^* using a star from p', making p_{j+1} happy. As p' and p_j remain aligned, both can be made happy—possibly using the 1st choice flip—for a perfect 3/3 ratio. If, on the other hand, p' and p_j are contrary, then triangulate f^* using a star from p_{j+4}, making p_{j+1} happy. Now p' and p_j are aligned and both can made happy—possibly using the 1st choice flip—for a perfect 3/3 ratio.

(2.2.3.3) Neither of the previous two cases occurs and, thus, $p_j, p_{j+1}, p_{j+3}, p_{j+2}$ form a convex quadrilateral f^*. Remove $p_{j+1}p_{j+2}$ and add $p_{j+1}p_{j+3}$ and p_jp_{j+2}. Note that p_j is happy because of 1st choice for f_i, and p_{j+1} and p_{j+2} are still contrary. Therefore, independent of the trian-

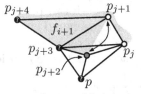

gulation of f^*, at least two vertices out of p_j, p_{j+1}, p_{j+2} are happy. Moreover, using f^* we can synchronously flip the parity of both p_{j+1} and p_{j+3} such that (C1)–(C3) hold. This gives us a ratio of 2/3 and 1st choice for f_{i+2}.

Putting things together. Recall that the first face f_1 and the last face f_m are the only light faces that may be triangles. In case that f_1 is a triangle, we just accept that p_2 may stay unhappy, and using f_2 the remaining vertices removed, if any, can be made happy. Similarly, from the last face f_m up to three vertices may remain unhappy. To the remaining faces $f_3, \ldots, f_{m-1}$ we apply the algorithm described above.

In order to analyze the overall happiness ratio, denote by $h_0(n)$ the minimum number of happy vertices obtained by applying the algorithm described above to a sequence $P = (p_1, \ldots, p_n)$ of $n \geq 3$ points in a no choice scenario. Similarly, denote by $h_1(n)$ the minimum number of happy vertices obtained by applying the algorithm described above to a sequence $P = (p_1, \ldots, p_n)$ of $n \geq 3$ points in a 1st choice scenario. From the case analysis given above we deduce the following recursive bounds.

a) $h_0(n) = 0$ and $h_1(n) = 1$, for $n \leq 4$.
b) $h_0(n) \geq \min\{2 + h_0(n-3), 1 + h_1(n-2)\}$.
c) $h_1(n) \geq \min\{3 + h_0(n-4), 2 + h_0(n-2), 2 + h_1(n-3)\}$.

By induction on n we can show that $h_0(n) \geq \lceil (2n-8)/3 \rceil$ and $h_1(n) \geq \lceil (2n-7)/3 \rceil$. Taking the at most four unhappy vertices from f_1 and f_m into account yields the claimed overall happiness ratio. $\square$

5   Triangulating Polygons with Holes

Theorem 5. *Let H be a polygon with holes and with parity constraints on the vertices. It is NP-complete to decide whether there exists a triangulation of H such that all vertices of H are happy.*

Proof. Following Jansen [7], we use a restricted version of the NP-complete *planar 3-SAT* problem [9], in which each clause contains at most three literals and each variable occurs in at most three clauses.

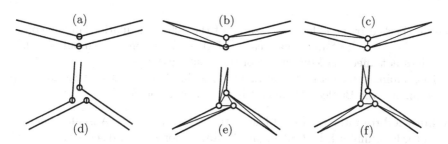

Fig. 2. Wire (a) that transfers TRUE (b), and FALSE (c). The short edge between the two vertices is in every triangulation. A variable (d) in TRUE (e) and FALSE (f) state.

The *edges* of the planar formula are represented by *wires* (Fig. 2(a)–(c)), narrow corridors which can be triangulated in two possible ways, and thereby transmit information between their ends. Negation can easily be achieved by swapping the labels of a single vertex pair in a wire. The construction of a *variable* (Fig. 2(d)–(f)) ensures that all wires emanating from it carry the same state, that is, their diagonals are oriented in the same direction.

To check clauses we use an OR-gate with two inputs and one output wire which we build by cascading two OR-gates and fixing the output of the second gate to true (Fig. 3(b)). The OR-gate is a convex 9-gon with three attached wires, and a *don't-care loop* (Fig. 3(a)) attached to the two top-most vertices. This loop has two triangulations and gives more freedom for the two vertices to which it is attached: they must have an even number of incident diagonals *in total*.

Fig. 4 shows triangulations for the four possible input configurations, where the output is FALSE iff both inputs are false. We have to ensure that the configuration where both inputs are FALSE and the output is TRUE is infeasible. This

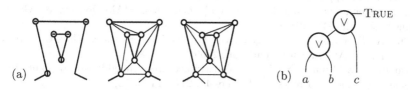

Fig. 3. A don't-care loop (a), checking a clause $a \lor b \lor c$ by joining two OR-gates (b)

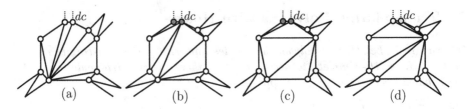

Fig. 4. An OR-gate with inputs FALSE, FALSE (a), TRUE, FALSE (b), FALSE, TRUE (c), and TRUE, TRUE (d). The two inputs are at the lower side and the output is at the upper right side. A don't-care loop dc is attached to the two top-most vertices.

can be checked by an exhaustive search of the 429 triangulations of the convex 9-gon. (The output of an OR-gate can be FALSE even if only one input is FALSE; this does not affect the correctness of the clause gadget.)

To combine the constructed elements to a simple polygon H with holes representing a given Boolean formula ϕ is now straightforward. $\square$

Acknowledgments. We would like to thank W. Aigner, F. Aurenhammer, M. Demuth, E. Mumford, D. Orden, and P. Ramos for fruitful discussions.

References

1. Aichholzer, O., Hackl, T., Huemer, C., Hurtado, F., Vogtenhuber, B.: Large bichromatic point sets admit empty monochromatic 4-gons (submitted) (2008)
2. Aichholzer, O., Krasser, H.: The point set order type data base: A collection of applications and results. In: Proc. 13th Canadian Conference on Computational Geometry, Waterloo, Ontario, Canada, pp. 17–20 (2001)
3. Bose, P.: On embedding an outer-planar graph in a point set. Computational Geometry: Theory and Applications 23(3), 303–312 (2002)
4. Bose, P., McAllister, M., Snoeyink, J.: Optimal algorithms to embed trees in a point set. Journal of Graph Algorithms and Applications 1(2), 1–15 (1997)
5. Colbourn, C., Booth, K.: Linear time automorphism algorithms for trees, interval graphs, and planar graphs. SIAM Journal on Computing 10(1), 203–225 (1981)
6. Erdös, P., Gallai, T.: Graphs with prescribed degree of vertices. Mat. Lapok 11, 264–274 (1960)
7. Jansen, K.: One strike against the min-max degree triangulation problem. Computational Geometry: Theory and Applications 3(2), 107–120 (1993)
8. Kettner, L., Kirkpatrick, D., Mantler, A., Snoeyink, J., Speckmann, B., Takeuchi, F.: Tight degree bounds for pseudo-triangulations of points. Computational Geometry: Theory and Applications 25(1&2), 1–12 (2003)
9. Lichtenstein, D.: Planar formulae and their uses. SIAM Journal on Computing 11(2), 329–343 (1982)
10. Pilz, A.: Parity properties of geometric graphs. Master's thesis, Graz University of Technology, Austria (in preparation, 2009)
11. Tamura, A., Tamura, Y.: Degree constrained tree embedding into points in the plane. Information Processing Letters 44, 211–214 (1992)

Straight-Line Rectangular Drawings of Clustered Graphs*

Patrizio Angelini[1], Fabrizio Frati[1], and Michael Kaufmann[2]

[1] Dipartimento di Informatica e Automazione – Roma Tre University
{angelini,frati}@dia.uniroma3.it
[2] Wilhelm-Schickard-Institut für Informatik – Universität Tübingen, Germany
mk@informatik.uni-tuebingen.de

Abstract. We show that every c-planar clustered graph admits a straight-line c-planar drawing in which each cluster is represented by an axis-parallel rectangle, thus solving a problem posed by Eades, Feng, Lin, and Nagamochi [*Algorithmica, 2006*].

1 Introduction

A *clustered graph* is a pair (G, T), where G is a graph, called *underlying graph*, and T is a rooted tree, called *inclusion tree*, such that the leaves of T are the vertices of G. Each internal node ν of T corresponds to the subset of vertices of G, called *cluster*, that are the leaves of the subtree of T rooted at ν.

Clustered graphs are widely used in applications where it is needed at the same time to represent relationships between entities and to group entities with semantic affinities. For example, in the Internet network, links among routers give rise to a graph; geographically close routers are grouped into areas, which in turn are grouped into Autonomous Systems.

Visualizing clustered graphs turns out to be a difficult problem, due to the simultaneous need for a readable drawing of the underlying structure and for a good rendering of the recursive clustering relationship. As for the visualization of graphs, the most important aesthetic criterion for a drawing of a clustered graph to be "nice" is commonly regarded to be the *planarity*, which however needs a refinement in order to take into account the clustering structure.

A *drawing* of a clustered graph $C(G, T)$ consists of a drawing of G (vertices are points in the plane and edges are Jordan curves between their endvertices) and of a representation of each node μ of T as a simple closed region containing all and only the vertices of μ. In what follows, when we say "cluster", we refer both to a set of vertices and to the region representing the cluster in a drawing, the meaning being clear from the context. A drawing has an *edge crossing* if two

* Fabrizio Frati was partially supported by the Italian Ministry of Research, Grant RBIP06BZW8, project "Advanced tracking system in intermodal freight transportation". Michael Kaufmann was partially supported by the German Research Foundation (DFG), Grant KA812/13-1, project "Scalable visual analytics".

F. Dehne et al. (Eds.): WADS 2009, LNCS 5664, pp. 25–36, 2009.

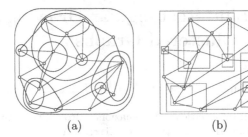

(a)   (b)

Fig. 1. (a) A clustered graph C. (b) A straight-line rectangular drawing of C.

edges of G cross, an *edge-region crossing* if an edge crosses a cluster boundary more than once, and a *region-region crossing* if two cluster boundaries cross. A drawing is *c-planar* if it has no edge crossing, no edge-region crossing, and no region-region crossing. A clustered graph is c-planar if it has a c-planar drawing.

Given a clustered graph, testing whether it admits a c-planar drawing is a problem of unknown complexity, perhaps the most studied problem in the Graph Drawing community during the last ten years (see, e.g., [12,10,4,14,13,2,15,3]).

Suppose that a c-planar clustered graph C is given together with a c-planar embedding, that is, together with an equivalence class of c-planar drawings of C, where two c-planar drawings are equivalent if they have the same order of the edges incident to each vertex and the same order of the edges incident to each cluster. How can the graph be drawn? Such a problem has been intensively studied in the literature and a number of papers have been presented for constructing c-planar drawings of clustered graphs within many drawing conventions.

Eades *et al.* [7] show how to construct $O(n^2)$-area c-planar orthogonal and poly-line drawings of c-planar clustered graphs with clusters drawn as axis-parallel rectangles. Di Battista *et al.* [5] give algorithms and show bounds for constructing small-area drawings of c-planar clustered trees within several drawing styles. The strongest result in the area is perhaps the one of Eades *et al.* [6]. Namely, the authors present an algorithm for constructing c-planar straight-line drawings of c-planar clustered graphs in which each cluster is drawn as a convex region (see also [17]). Such an algorithm requires, in general, exponential area. However, such a bound is asymptotically optimal in the worst case [11].

In this paper we address a problem posed by Eades *et al.* [8,10,6]: Does every c-planar clustered graph admit a *straight-line rectangular drawing*, i.e., a c-planar straight-line drawing in which each cluster is an axis-parallel rectangle (see Fig. 1)? Eades *et al.* observe how pleasant and readable straight-line rectangular drawings are; however, they argue that their algorithm [6] for constructing c-planar straight-line convex drawings cannot be modified to obtain straight-line rectangular drawings without introducing edge-region crossings.

We show that every c-planar clustered graph has a straight-line rectangular drawing. We obtain such a result as a corollary of a stronger theorem stating that a straight-line rectangular drawing of a c-planar clustered graph exists for an arbitrary *convex-separated* drawing of its outer face, that is, a drawing satisfying some properties of convexity and of visibility among vertices and clusters.

Such a stronger result is proved by means of an inductive algorithm reminiscent of Fary's drawing algorithm for planar graphs [9]. Namely, the algorithm consists of three inductive cases. Each case considers a clustered graph C and performs an operation (removal of a cluster, split of the graph in correspondence of a separating 3-cycle, contraction of an edge) turning C into a smaller clustered graph C', for which a straight-line rectangular drawing can be inductively constructed. Then, such a drawing can be easily augmented to a straight-line rectangular drawing of C. The algorithm is described more in detail in Sect. 5.

When none of the three inductive cases applies, every cluster contains a vertex incident to the outer face. We call *outerclustered graph* a clustered graph satisfying this property. We prove that every outerclustered graph admits a straight-line rectangular drawing even when a convex-separated drawing of its outer face is arbitrarily fixed, thus providing a base case for the above inductive algorithm for general clustered graphs. In order to draw an outerclustered graph C, we split it into three *linearly-ordered outerclustered graphs* (an even more restricted family of clustered graphs), we separately draw such graphs, and we compose the obtained drawings to get a drawing of C. How to split C and how to compose the drawings of the obtained linearly-ordered outerclustered graphs into a drawing of C are described in Sect. 4.

A linearly-ordered outerclustered graph is an outerclustered graph in which all the vertices of the underlying graph belong to a path in the inclusion tree. A drawing algorithm is provided for constructing a straight-line rectangular drawing of any linearly-ordered outerclustered graph $C(G, T)$ for an arbitrary convex-separated drawing of its outer face. Such an inductive algorithm finds a subgraph of G (a path plus an edge) that splits G into smaller linearly-ordered outerclustered graphs and draws such a subgraph so that the outer faces of the smaller linearly-ordered outerclustered graphs are convex-separated, thus allowing the induction to go through. Such an algorithm is presented in Sect. 3.

Omitted and sketched proofs can be found in the full version of the paper [1].

2 Preliminaries

Let $C(G, T)$ be a clustered graph. An edge (u, v) of G is *incident to a cluster* μ of T if u belongs to μ and v does not. Let $\sigma(u_1, u_2, \ldots, u_k)$ be the *smallest cluster* of T containing vertices $u_1, u_2, \ldots, u_k$ of G, i.e., the node of T containing all of $u_1, u_2, \ldots, u_k$ and such that none of its children in T, if any, contains all of $u_1, u_2, \ldots, u_k$. A cluster is *minimal* if it contains no other cluster. A cluster μ is an ancestor (descendant) of a cluster ν if μ is an ancestor (descendant) of ν in T. C is *c-connected* if each cluster induces a connected subgraph of G.

A *straight-line rectangular drawing* of a clustered graph is a c-planar drawing such that each edge is a straight-line segment and each cluster is an axis-parallel rectangle. From now on, "clustered graph" will always mean c-planar clustered graph, and "drawing" will always mean straight-line rectangular drawing.

A clustered graph $C(G, T)$ is *maximal* if G is a maximal planar graph. In order to prove that every clustered graph admits a straight-line rectangular drawing,

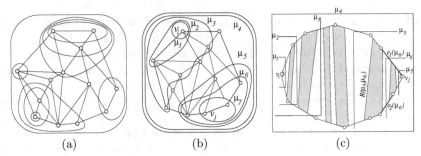

Fig. 2. (a) An outerclustered graph. (b) A linearly-ordered outerclustered graph. (c) A convex-separated drawing of the outer face of a linearly-ordered outerclustered graph.

it suffices to consider maximal clustered graphs. Namely, every non-maximal c-planar clustered graph $C(G, T)$ can be augmented to a maximal c-planar clustered graph by adding dummy edges to G [16]. Further, by the results of Feng *et al.* [12], every maximal c-planar clustered graph is c-connected. From now on, we assume that the embedding (that is, the order of the edges incident to each vertex) and the outer face of any graph G is fixed in advance. We denote by $o(G)$ the outer face of G. A clustered graph $C(G, T)$ is *internally-triangulated* if every internal face of G is delimited by a 3-cycle.

Let $C(G, T)$ be a clustered graph and let f be any face of G. Denote by $C_f(G_f, T_f)$ the clustered graph such that G_f is the cycle delimiting f, and such that T_f is obtained from T by removing the clusters not containing any vertex incident to f. The *outer face of* $C(G, T)$ is the clustered graph $C_{o(G)}(G_{o(G)}, T_{o(G)})$, simply denoted by C_o. In Sect.s 3, 4, and 5, we prove that a drawing of a clustered graph can be constructed given an arbitrary drawing of its outer face satisfying some geometric properties to be described below. Then, a straight-line rectangular drawing $\Gamma(C)$ of C *completes* a straight-line rectangular drawing $\Gamma(C_o)$ of C_o if the part of $\Gamma(C)$ representing C_o coincides with $\Gamma(C_o)$.

We now introduce the following class of clustered graphs, whose study is essential to achieve the main result of this paper. A c-planar clustered graph $C(G, T)$ is an *outerclustered graph* if (see Fig. 2.a): (O1) every cluster contains at least one vertex incident to $o(G)$; (O2) the boundary of every cluster μ of T that contains some but not all the vertices incident to $o(G)$ intersects $o(G)$ exactly twice, namely it intersects exactly two edges $e_1(\mu)$ and $e_2(\mu)$ incident to $o(G)$; and (O3) every edge (u, v) with $\sigma(u) = \sigma(v)$ is incident to $o(G)$.

The following class of outerclustered graphs is used as a base case in the algorithm for drawing outerclustered graphs. An internally-triangulated biconnected outerclustered graph $C(G, T)$ is *linearly-ordered* if a sequence $\mu_1, \mu_2, \ldots, \mu_k$ of clusters in T and an index $1 \leq h \leq k$ exist such that (see Fig. 2.b): (LO1) for each vertex v of G, $\sigma(v) = \mu_i$, for some index $1 \leq i \leq k$; (LO2) let v_i and v_j be any two vertices incident to $o(G)$ such that $\sigma(v_i) = \mu_1$ and $\sigma(v_j) = \mu_k$; then, $o(G)$ is delimited by two *monotone paths* $\mathcal{P}_1 = (v_i, v_{i+1}, \ldots, v_{j-1}, v_j)$ and $\mathcal{P}_2 = (v_i, v_{i-1}, \ldots, v_{j+1}, v_j)$, i.e., paths such that, if $\sigma(v_t) = \mu_a$ and $\sigma(v_{t+1}) = \mu_b$, then $a \leq b$ if $(v_t, v_{t+1}) \in \mathcal{P}_1$ and $b \leq a$ if $(v_t, v_{t+1}) \in \mathcal{P}_2$; and (LO3) μ_{i+1} is the parent of μ_i, for each $1 \leq i < h$, and μ_{i+1} is a child of μ_i, for each $h \leq i < k$.

In Sect. 3 we prove that a drawing of a linearly-ordered outerclustered graph $C(G,T)$ exists completing an arbitrary *convex-separated drawing* $\Gamma(C_o)$ of C_o, that is, a straight-line rectangular drawing such that (see Fig. 2.c):

- CS1: the polygon P representing $o(G)$ is convex;
- CS2: there exist two vertices v_i and v_j such that $\sigma(v_i) = \mu_1$, $\sigma(v_j) = \mu_k$, and the angles incident to v_i and v_j in P are strictly less than $180°$; and
- CS3: for every pair of clusters μ and ν such that μ is the parent of ν in T and such that μ is not an ancestor of the smallest cluster containing all the vertices of $o(G)$, there exists a convex region $R(\mu, \nu)$ such that: (i) $R(\mu, \nu)$ is entirely contained inside $\mu \cap (P \cup int(P))$, where $int(P)$ denotes the interior of polygon P; (ii) for any cluster $\mu' \neq \mu$ and any child ν' of μ', $R(\mu, \nu)$ intersects neither $R(\mu', \nu')$ nor the boundary of μ'; (iii) $R(\mu, \nu) \cap P$ consists of two polygonal lines $l_1(\mu, \nu)$ and $l_2(\mu, \nu)$ such that $l_1(\mu, \nu)$ belongs to the polygonal line representing $\mathcal{P}_1$ in $\Gamma(C_o)$ and $l_2(\mu, \nu)$ belongs to the polygonal line representing $\mathcal{P}_2$ in $\Gamma(C_o)$; further, at least one endpoint of $l_1(\mu, \nu)$ (resp. of $l_2(\mu, \nu)$) lies on $e_1(\nu)$ (resp. on $e_2(\nu)$).

Let $C(G,T)$ be a linearly-ordered outerclustered graph with outer face $o(G)$ delimited by cycle $\mathcal{C} = (v_i, v_{i+1}, \ldots, v_{j-1}, v_j, v_{j+1}, \ldots, v_{i-1}, v_i)$. Let (v_x, v_y) be a chord of $\mathcal{C}$. Consider the clustered graphs $C^1(G^1, T^1)$ and $C^2(G^2, T^2)$ such that G^1 (resp. G^2) is the subgraph of G induced by the vertices incident to and internal to cycle $\mathcal{C}^1 = (v_x, v_{x+1}, \ldots, v_{y-1}, v_y, v_x)$ (resp. incident to and internal to cycle $\mathcal{C}^2 = (v_y, v_{y+1}, \ldots, v_{x-1}, v_x, v_y)$), and such that T^1 (resp. T^2) is the subtree of T induced by the clusters containing vertices of G^1 (resp. of G^2).

Lemma 1. $C^1(G^1, T^1)$ and $C^2(G^2, T^2)$ are linearly-ordered outerclustered graphs.

Let Γ be any convex-separated drawing of C_o. Suppose that v_x and v_y are not collinear with any vertex of $o(G)$. Let Γ_1 and Γ_2 be the drawings of C_o^1 and C_o^2 obtained by drawing (v_x, v_y) in Γ as a straight-line segment.

Lemma 2. Γ_1 and Γ_2 are convex-separated drawings.

When dealing with outerclustered and clustered graphs, it is sufficient to consider underlying graphs with triangular outer faces. Then, let $C(G,T)$ be a clustered graph such that G is a 3-cycle (u, v, z). Denote by $e_1(\mu)$ and $e_2(\mu)$ the edges of G incident to a cluster μ of T not containing all the vertices of G. A straight-line rectangular drawing $\Gamma(C)$ of C is a *triangular-convex-separated drawing* if, for every pair of clusters μ and ν such that μ is the parent of ν in T and such that μ is not an ancestor of $\sigma(u, v, z)$, there exists a convex region $R(\mu, \nu)$ such that: (i) $R(\mu, \nu)$ is entirely contained inside $\mu \cap (P \cup int(P))$, where P is the triangle representing G in $\Gamma(C)$; (ii) for any cluster $\mu' \neq \mu$ and any child ν' of μ', $R(\mu, \nu)$ intersects neither $R(\mu', \nu')$ nor the boundary of μ'; (iii) $R(\mu, \nu) \cap P$ consists of two polygonal lines $l_1(\mu, \nu)$ and $l_2(\mu, \nu)$ such that at least one endpoint of $l_1(\mu, \nu)$ (resp. of $l_2(\mu, \nu)$) belongs to $e_1(\nu)$ (resp. to $e_2(\nu)$).

We observe the following relationship between convex-separated drawings and triangular-convex-separated drawings.

Lemma 3. *Let $C(G,T)$ be a linearly-ordered maximal outerclustered graph. Then, a triangular-convex-separated drawing of C_o is a convex-separated drawing of C_o.*

Finally, we define a class of drawings in which the properties of convexity and visibility among vertices and clusters are imposed on all the internal faces rather than on the outer face. Let $C(G,T)$ be an internally-triangulated clustered graph. A drawing $\Gamma(C)$ of C is an *internally-convex-separated drawing* if, for every internal face f of G, the part $\Gamma(C_f)$ of $\Gamma(C)$ representing C_f is a triangular-convex-separated drawing.

3   Drawing Linearly-Ordered Outerclustered Graphs

In this section we show how to construct an internally-convex-separated drawing of any linearly-ordered outerclustered graph C for an arbitrary convex-separated drawing of the outer face C_o of C. This is done by means of an inductive algorithm that uses the following lemma as the main tool:

Lemma 4. *Let $C(G,T)$ be an internally-triangulated triconnected outerclustered graph. Suppose that C is linearly-ordered according to a sequence $\mu_1, \mu_2, \ldots, \mu_k$ of clusters of T. Let v_i and v_j be any two vertices such that $\sigma(v_i) = \mu_1$ and $\sigma(v_j) = \mu_k$. Let V_1 (resp. V_2) be the set of vertices between v_i and v_j (resp. between v_j and v_i) in the clockwise order of the vertices around $o(G)$. Then, if $V_1 \neq \emptyset$, there exists a path $\mathcal{P}_u = (u_1, u_2, \ldots, u_r)$ such that (see Fig. 3.a):*

- *P1: u_1 and u_r belong to $V_2 \cup \{v_i, v_j\}$;*
- *P2: u_i is an internal vertex of G, for each $2 \leq i \leq r - 1$;*
- *P3: if $\sigma(u_i) = \mu_{j_1}$ and $\sigma(u_{i+1}) = \mu_{j_2}$, then $j_1 < j_2$, for each $1 \leq i \leq r - 1$;*
- *P4: there exists exactly one vertex u_x, where $2 \leq x \leq r - 1$, that is adjacent to at least one vertex v_x in V_1;*
- *P5: there exist no chords among the vertices of path $(u_1, u_2, \ldots, u_x)$ and no chords among the vertices of path $(u_x, u_{x+1}, \ldots, u_r)$.*

A lemma similar to Lemma 4 can be proved in which V_1 replaces V_2 and vice versa (see also [1]). We now present the main theorem of this section.

Theorem 1. *Let $C(G,T)$ be a linearly-ordered internally-triangulated triconnected outerclustered graph. For every convex-separated drawing $\Gamma(C_o)$ of C_o, there exists an internally-convex-separated drawing $\Gamma(C)$ of C completing $\Gamma(C_o)$.*

Proof sketch: The proof consists of an inductive drawing algorithm that distinguishes two cases. In the first case, the vertices of $V_1 \cup \{v_i, v_j\}$ are not all collinear. Then, Lemma 4 applies and a path $\mathcal{P}_u = (u_1, u_2, \ldots, u_x, \ldots, u_r)$ is found. The three internal faces in the graph $o(G) \cup \mathcal{P}_u \cup (u_x, v_x)$ are the outer faces of the underlying graphs G_1, G_2, and G_3 of three clustered graphs C_1, C_2, and C_3, respectively, whose inclusion trees are the subtrees of T induced by the clusters containing vertices of G_1, G_2, and G_3, respectively. Then, C_1, C_2, and C_3 can be proved to be linearly-ordered outerclustered graphs. Further,

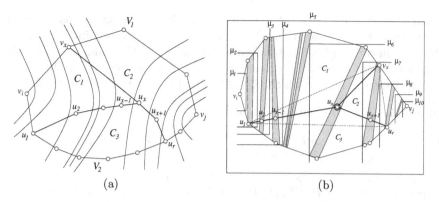

Fig. 3. (a) A path $\mathcal{P}_u$ satisfying Properties P1–P5. (b) Drawing $\mathcal{P}_u$ in $\Gamma(C_o)$.

path $\mathcal{P}_u$ can be suitably drawn (see Fig. 3.b) in $\Gamma(C_o)$, so that the outer faces of C_1, C_2, and C_3 are represented by convex-separated drawings. By Lemma 1, possible chords between the vertices of the outer faces of G_1, G_2, and G_3 split C_1, C_2, and C_3 into smaller triconnected linearly-ordered outerclustered graphs; by Lemma 2, drawing such chords as straight-line segments splits the drawings of the outer faces of C_1, C_2, and C_3 into convex-separated drawings of the obtained linearly-ordered outerclustered graphs, thus allowing the induction to go through. In the second case, the vertices of $V_2 \cup \{v_i, v_j\}$ are not all collinear. Then, a path $\mathcal{P}_u$ is found, with its endvertices in $V_1 \cup \{v_i, v_j\}$ and with a vertex u_x adjacent to a vertex v_x in V_2. $\mathcal{P}_u$ and (u_x, v_x) split C into smaller linearly-ordered outerclustered graphs C_1, C_2, and C_3; further, suitable drawings of $\mathcal{P}_u$ and (u_x, v_x) split $\Gamma(C_o)$ into convex-separated drawings of the outer faces of C_1, C_2, and C_3. Since $\Gamma(C_o)$ is a convex-separated drawing, one of the two cases applies, otherwise the polygon representing $o(G)$ would not be convex.   □

4   Drawing Outerclustered Graphs

In this section we generalize from linearly-ordered outerclustered graphs to general outerclustered graphs. In order to show that any maximal outerclustered graph has an internally-convex-separated drawing completing an arbitrary triangular-convex-separated drawing of its outer face, we show how to reduce the problem of drawing an outerclustered graph to the one of drawing some linearly-ordered outerclustered graphs.

Consider a maximal outerclustered graph $C(G, T)$ and let u, v, and z be the vertices incident to $o(G)$. Suppose that G has internal vertices, that the smallest containing clusters of u, v, and z are three distinct clusters, that is, $\sigma(u) \neq \sigma(v)$, $\sigma(u) \neq \sigma(z)$, and $\sigma(v) \neq \sigma(z)$, and that, if there exists a cluster containing exactly two vertices incident to $o(G)$, then such vertices are u and v. The following lemma aims at finding three (or two) paths $\mathcal{P}_u$, $\mathcal{P}_v$, and $\mathcal{P}_z$ (resp. $\mathcal{P}_u$ and $\mathcal{P}_v$), all starting from the same internal vertex u_1 of G (resp. from z and ending at u, v, and z (resp. at u and v), respectively. The paths, that share no vertex other than u_1 (resp. than z), are such that their insertion into C_o splits C into three linearly-ordered outerclustered graphs $C_{u,v}$, $C_{u,z}$, and $C_{v,z}$. In order to perform such a split, the three paths have to be monotone for the cluster sequences with respect to which $C_{u,v}$, $C_{u,z}$, and $C_{v,z}$ are linearly-ordered.

Lemma 5. *One of the following holds:*

1. *There exist three paths* $\mathcal{P}_u = (u_1, u_2, \ldots, u_U)$, $\mathcal{P}_v = (v_1, v_2, \ldots, v_V)$, *and* $\mathcal{P}_z = (z_1, z_2, \ldots, z_Z)$ *such that (see Fig. 4.a):*
   (a) *$u_U = u$, $v_V = v$, $z_Z = z$, and $u_1 = v_1 = z_1$;*
   (b) *the vertices of $\mathcal{P}_u \setminus \{u_1\}$, $\mathcal{P}_v \setminus \{v_1\}$, and $\mathcal{P}_z \setminus \{z_1\}$ are distinct;*
   (c) *each of paths $\mathcal{P}_u \setminus \{u_1\}$, $\mathcal{P}_v \setminus \{v_1\}$, and $\mathcal{P}_z$ has no chords;*
   (d) *$\sigma(u_i)$ does not contain neither v nor z, for each $2 \leq i \leq U$; $\sigma(v_i)$ does not contain neither u nor z, for each $2 \leq i \leq V$; $\sigma(z_i)$ does not contain neither u nor v, for each $Z^* \leq i \leq Z$, where Z^* is an index such that $1 \leq Z^* \leq Z$;*
   (e) *$\sigma(u_{i+1})$ is a descendant of $\sigma(u_i)$, for each $2 \leq i \leq U - 1$; $\sigma(v_{i+1})$ is a descendant of $\sigma(v_i)$, for each $2 \leq i \leq V - 1$;*
   (f) *$\sigma(z_1)$ is either a cluster containing z and not containing u and v (then $Z^* = 1$ and $\sigma(z_{i+1})$ is a descendant of $\sigma(z_i)$, for each $1 \leq i \leq Z - 1$), or is $\sigma(u, v, z)$ (then $Z^* = 2$ and $\sigma(z_{i+1})$ is a descendant of $\sigma(z_i)$, for each $1 \leq i \leq Z - 1$), or a cluster not containing z and containing u and v. In the latter case $Z^* \geq 2$, $\sigma(z_{i+1})$ is an ancestor of $\sigma(z_i)$, for each $1 \leq i \leq Z^* - 2$, $\sigma(z_{i+1})$ is a descendant of $\sigma(z_i)$, for each $Z^* \leq i \leq Z - 1$, and either $\sigma(z_{Z^*})$ is a descendant of $\sigma(z_{Z^*-1})$ (if $\sigma(z_{Z^*-1}) = \sigma(u, v, z)$), or $\sigma(z_{Z^*})$ is not comparable with $\sigma(z_{Z^*-1})$ (if $\sigma(z_{Z^*-1})$ contains u and v and does not contain z).*
   (g) *G contains an internal face having incident vertices u_1, u_2, and v_2.*
2. *There exist two paths* $\mathcal{P}_u = (u_1, u_2, \ldots, u_U)$ *and* $\mathcal{P}_v = (v_1, v_2, \ldots, v_V)$ *such that:*
   (a) *$u_U = u$, $v_V = v$, and $u_1 = v_1 = z$;*
   (b) *the vertices of $\mathcal{P}_u \setminus \{u_1\}$ and $\mathcal{P}_v \setminus \{v_1\}$ are distinct;*
   (c) *each of paths $\mathcal{P}_u \setminus \{u_1\}$ and $\mathcal{P}_v \setminus \{v_1\}$ has no chords;*
   (d) *$\sigma(u_i)$ does not contain neither v nor z, for each $2 \leq i \leq U$; $\sigma(v_i)$ does not contain neither u nor z, for each $2 \leq i \leq V$;*
   (e) *$\sigma(u_{i+1})$ is a descendant of $\sigma(u_i)$, for each $2 \leq i \leq U - 1$; $\sigma(v_{i+1})$ is a descendant of $\sigma(v_i)$, for each $2 \leq i \leq V - 1$;*
   (f) *G contains an internal face having incident vertices u_2, v_2, and z.*

Suppose that Condition 1 of Lemma 5 holds. Denote by $C_{u,v}$, by $C_{u,z}$, and by $C_{v,z}$ the clustered graphs whose underlying graphs $G_{u,v}$, $G_{u,z}$, and $G_{v,z}$ are the subgraphs of G induced by the vertices incident to and internal to cycles $\mathcal{C}_{u,v} \equiv (u, v) \cup (\mathcal{P}_u \setminus \{u_1\}) \cup (u_2, v_2) \cup (\mathcal{P}_v \setminus \{v_1\})$, $\mathcal{C}_{u,z} \equiv (u, z) \cup \mathcal{P}_u \cup \mathcal{P}_z$, and $\mathcal{C}_{v,z} \equiv (v, z) \cup \mathcal{P}_v \cup \mathcal{P}_z$, and whose inclusion trees $T_{u,v}$, $T_{u,z}$, and $T_{v,z}$ are the subtrees of T induced by the clusters containing vertices of $G_{u,v}$, $G_{u,z}$, and $G_{v,z}$.

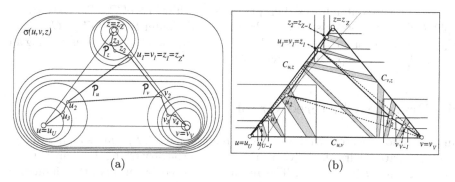

Fig. 4. (a) Paths $\mathcal{P}_u$, $\mathcal{P}_v$, and $\mathcal{P}_z$. (b) Drawing $\mathcal{P}_u$, $\mathcal{P}_v$, and $\mathcal{P}_z$ in $\Gamma(C_o)$.

Lemma 6. $C_{u,v}$, $C_{u,z}$, and $C_{v,z}$, are linearly-ordered outerclustered graphs.

Suppose that Condition 2 of Lemma 5 holds. Denote by $C_{u,v}$, by $C_{u,z}$, and by $C_{v,z}$ the clustered graphs whose underlying graphs $G_{u,v}$, $G_{u,z}$, and $G_{v,z}$ are the subgraphs of G induced by the vertices incident to and internal to cycles $\mathcal{C}_{u,v} \equiv (u,v) \cup (\mathcal{P}_u \setminus \{u_1\}) \cup (u_2,v_2) \cup (\mathcal{P}_v \setminus \{v_1\})$, $\mathcal{C}_{u,z} \equiv (u,z) \cup \mathcal{P}_u$, and $\mathcal{C}_{v,z} \equiv (v,z) \cup \mathcal{P}_v$, and whose inclusion trees $T_{u,v}$, $T_{u,z}$, and $T_{v,z}$ are the subtrees of T induced by the clusters containing vertices of $G_{u,v}$, $G_{u,z}$, and $G_{v,z}$.

Lemma 7. $C_{u,v}$, $C_{u,z}$, and $C_{v,z}$, are linearly-ordered outerclustered graphs.

We are now ready to exhibit the main theorem of this section.

Theorem 2. *Let $C(G,T)$ be a maximal outerclustered graph. Then, for every triangular-convex-separated drawing $\Gamma(C_o)$ of C_o, there exists an internally-convex-separated drawing $\Gamma(C)$ of C completing $\Gamma(C_o)$.*

Proof sketch: If $\sigma(u) = \sigma(v)$, or $\sigma(u) = \sigma(z)$, or $\sigma(v) = \sigma(z)$, then C can be proved to be a linearly-ordered outerclustered graph and the theorem follows from Theorem 1 and Lemma 3. Otherwise, $\sigma(u) \neq \sigma(v)$, $\sigma(u) \neq \sigma(z)$, and $\sigma(v) \neq \sigma(z)$. Then, Lemma 5 applies, and either three paths $\mathcal{P}_u = (u_1, u_2, \ldots, u_U)$, $\mathcal{P}_v = (v_1, v_2, \ldots, v_V)$, and $\mathcal{P}_z = (z_1, z_2, \ldots, z_Z)$ are found satisfying Condition 1 of Lemma 5, or two paths $\mathcal{P}_u = (u_1, u_2, \ldots, u_U)$ and $\mathcal{P}_v = (v_1, v_2, \ldots, v_V)$ are found satisfying Condition 2 of Lemma 5. In the first case, the three internal faces different from (u_1, u_2, v_2) in the graph $o(G) \cup \mathcal{P}_u \cup \mathcal{P}_v \cup \mathcal{P}_z \cup (u_2, v_2)$ are the outer faces of $G_{u,v}$, $G_{u,z}$, and $G_{v,z}$, respectively. By Lemma 6, $C_{u,v}$, $C_{u,z}$, and $C_{v,z}$ are linearly-ordered outerclustered graphs. Further, paths $\mathcal{P}_u$, $\mathcal{P}_v$, $\mathcal{P}_z$, and edge (u_2, v_2) can be suitably drawn in $\Gamma(C_o)$ so that the outer faces of $C_{u,v}$, $C_{u,z}$, and $C_{v,z}$ are represented by convex-separated drawings (see Fig. 4.b). Then, Theorem 1 applies to draw $C_{u,v}$, $C_{u,z}$, and $C_{v,z}$, thus constructing an internally-convex-separated drawing of C. In the second case, analogously as in the first case, paths $\mathcal{P}_u$, $\mathcal{P}_v$, and edge (u_2, v_2) split C into three clustered graphs $C_{u,v}$, $C_{u,z}$, and $C_{v,z}$ that, by Lemma 7, are linearly-ordered outerclustered graphs. Suitable drawings of $\mathcal{P}_u$, $\mathcal{P}_v$, and (u_2, v_2) split $\Gamma(C_o)$ into convex-separated drawings of the outer faces of $C_{u,v}$, $C_{u,z}$, and $C_{v,z}$. Again, Theorem 1 applies to draw $C_{u,v}$, $C_{u,z}$, and $C_{v,z}$, thus constructing an internally-convex-separated drawing of C.   □

5   Drawing Clustered Graphs

In this section we prove that every clustered graph $C(G, T)$ admits an internally-convex-separated drawing $\Gamma(C)$ completing an arbitrary triangular-convex-separated drawing $\Gamma(C_o)$ of C_o. The result is achieved by means of an inductive algorithm, where the induction is on the number of vertices of G plus the number of clusters in T. In the base case, C is an outerclustered graph and the statement follows from Theorem 2. Consider any maximal clustered graph $C(G, T)$.

  Case 1: There exists a minimal cluster μ containing exactly one vertex v internal to G and containing no vertex incident to $o(G)$. Remove μ from T obtaining a clustered graph $C'(G, T')$. Observe that C_o and C'_o are the same graph. The number of vertices plus the number of clusters in C' is one less than in C. Hence, the inductive hypothesis applies and there exists an internally-convex-separated drawing $\Gamma(C')$ of C' completing an arbitrary triangular-convex-separated drawing $\Gamma(C_o)$ of C_o. In $\Gamma(C')$ a small disk D can be drawn centered at v, not intersecting the boundary of any cluster, not containing any vertex of G different from v, and intersecting only the edges incident to v. For each edge e_i incident to v, choose two points p_i^1 and p_i^2 inside D, where p_i^1 is closer to v than p_i^2. Insert a drawing of μ in $\Gamma(C')$ as a rectangle containing v and contained inside the polygon $(p_1^1, p_2^1, \ldots, p_k^1, p_1^1)$, thus obtaining a drawing $\Gamma(C)$ that can be proved to be an internally-convex-separated drawing of C. In particular, for each face f_i of G incident to edges e_i and e_{i+1}, the quadrilateral having p_i^1, p_i^2, p_{i+1}^1, and p_{i+1}^2 as vertices satisfies the property of a triangular-convex-separated drawing.

  Case 2: There exists a separating 3-cycle (u', v', z') in G. Let $C^1(G^1, T^1)$ $(C^2(G^2, T^2))$ be the clustered graph defined as follows. G^1 (resp. G^2) is the subgraph of G induced by u', v', z', and by the vertices outside (u', v', z') (resp. by u', v', z', and by the vertices inside (u', v', z')). T^1 (resp. T^2) is the subtree of T whose clusters contain vertices of G^1 (resp. of G^2). Observe that C_o and C_o^1 are the same graph. Since (u', v', z') is a separating 3-cycle, the number of vertices plus the number of clusters in each of C^1 and C^2 is strictly less than in C. Hence, the inductive hypothesis applies and there exists an internally-convex-separated drawing $\Gamma(C^1)$ of C^1 completing an arbitrary triangular-convex-separated drawing $\Gamma(C_o)$ of C_o. Cycle (u', v', z') is a face f of G^1. Then, the drawing $\Gamma(C_f)$ of C_f in $\Gamma(C^1)$ is a triangular-convex-separated drawing. Observe that C_f and C_o^2 are the same graph. Hence, the inductive hypothesis applies again and an internally-convex-separated drawing $\Gamma(C^2)$ can be constructed completing $\Gamma(C_o^2)$. Plugging $\Gamma(C^2)$ in $\Gamma(C^1)$ provides a drawing $\Gamma(C)$ of C, that can be proved to be an internally-convex-separated drawing of C.

Case 3: There exists no separating 3-cycle, and there exist two adjacent vertices u' and v' such that $\sigma(u') = \sigma(v')$ and such that they are not both external. Suppose that G contains two adjacent vertices u' and v' such that $\sigma(u') = \sigma(v')$ and such that u' is internal, and suppose that there exists no separating 3-cycle in G. Since G is maximal, u' and v' have exactly two common neighbors z'_1 and z'_2. Contract edge (u', v') to a vertex w', that is, replace vertices u' and v' with a vertex w' connected to all the vertices u' and v' are connected to. Vertex w' belongs to $\sigma(u')$ and to all the ancestors of $\sigma(u')$ in T'. The resulting clustered graph $C'(G', T')$ is easily shown to be a maximal c-planar clustered graph. In particular, the absence of separating 3-cycles in G guarantees that G' is simple and maximal. Observe that C_o and C'_o are the same graph. Hence, the inductive hypothesis applies and there exists an internally-convex-separated drawing $\Gamma(C')$ of C' completing an arbitrary triangular-convex-separated drawing $\Gamma(C_o)$ of C_o. Then consider a small disk D centered at w' and consider any line l from w' to an interior point of the segment between z'_1 and z'_2. Replace w' with u' and v' so that such vertices lie on l and inside D. Connect u' and v' to their neighbors, obtaining a drawing $\Gamma(C)$ of C, that can be proved to be an internally-convex-separated drawing of C.

It remains to observe that, if none of Cases 1, 2, and 3 applies, then C is an outerclustered graph. Hence, we get the following:

Theorem 3. *Let $C(G, T)$ be a maximal c-planar clustered graph. Then, for every triangular-convex-separated drawing $\Gamma(C_o)$ of C_o, there exists an internally-convex-separated drawing $\Gamma(C)$ of C completing $\Gamma(C_o)$.*

6 Conclusions

In this paper we have shown that every c-planar clustered graph admits a c-planar straight-line rectangular drawing. Actually, the algorithms we proposed do not exploit at all the fact that clusters are drawn as rectangles. The only property that must be satisfied by each region representing a cluster for the algorithm to work is that an edge incident to the cluster must cross its boundary exactly once. Hence, the algorithm we proposed can be modified in order to construct a c-planar straight-line drawing of a given clustered graph for an arbitrary assignment of convex shapes to the clusters (more generally, even *star-shaped polygons* are feasible, that is, polygons that have a set of points, called *kernel*, from which it is possible to draw edges towards all the vertices of the polygon without crossing its sides).

The algorithm we described in this paper uses real coordinates, hence it requires exponential area to be implemented in a system with a finite resolution rule. However, this drawback is unavoidable, since there exist clustered graphs requiring exponential area in any straight-line drawing in which clusters are represented by convex regions, as proved by Feng *et al.* [11]. We believe worth of interest the problem of determining whether clustered graphs whose hierarchy is *flat*, i.e., all clusters different from the root do not contain smaller clusters, admit straight-line convex drawings and straight-line rectangular drawings in polynomial area.

References

1. Angelini, P., Frati, F., Kaufmann, M.: Straight-line rectangular drawings of clustered graphs. Tech. Report RT-DIA-144-2009, Dept. of Computer Sci., Roma Tre Univ. (2009),
   http://web.dia.uniroma3.it/ricerca/rapporti/rt/2009-144.pdf
2. Cornelsen, S., Wagner, D.: Completely connected clustered graphs. J. Discr. Alg. 4(2), 313–323 (2006)
3. Cortese, P.F., Di Battista, G., Frati, F., Patrignani, M., Pizzonia, M.: C-planarity of c-connected clustered graphs. J. Graph Alg. Appl. 12(2), 225–262 (2008)
4. Dahlhaus, E.: A linear time algorithm to recognize clustered planar graphs and its parallelization. In: Lucchesi, C.L., Moura, A.V. (eds.) LATIN 1998. LNCS, vol. 1380, pp. 239–248. Springer, Heidelberg (1998)
5. Di Battista, G., Drovandi, G., Frati, F.: How to draw a clustered tree. In: Dehne, F., Sack, J.-R., Zeh, N. (eds.) WADS 2007. LNCS, vol. 4619, pp. 89–101. Springer, Heidelberg (2007)
6. Eades, P., Feng, Q., Lin, X., Nagamochi, H.: Straight-line drawing algorithms for hierarchical graphs and clustered graphs. Algorithmica 44(1), 1–32 (2006)
7. Eades, P., Feng, Q., Nagamochi, H.: Drawing clustered graphs on an orthogonal grid. J. Graph Alg. Appl. 3(4), 3–29 (1999)
8. Eades, P., Feng, Q.W., Lin, X.: Straight-line drawing algorithms for hierarchical graphs and clustered graphs. In: North, S.C. (ed.) GD 1996. LNCS, vol. 1190, pp. 113–128. Springer, Heidelberg (1997)
9. Fary, I.: On straight line representations of planar graphs. Acta. Sci. Math. 11, 229–233 (1948)
10. Feng, Q.: Algorithms for Drawing Clustered Graphs. PhD thesis, The University of Newcastle, Australia (1997)
11. Feng, Q., Cohen, R.F., Eades, P.: How to draw a planar clustered graph. In: Li, M., Du, D.-Z. (eds.) COCOON 1995. LNCS, vol. 959, pp. 21–30. Springer, Heidelberg (1995)
12. Feng, Q., Cohen, R.F., Eades, P.: Planarity for clustered graphs. In: Spirakis, P.G. (ed.) ESA 1995. LNCS, vol. 979, pp. 213–226. Springer, Heidelberg (1995)
13. Goodrich, M.T., Lueker, G.S., Sun, J.Z.: C-planarity of extrovert clustered graphs. In: Healy, P., Nikolov, N.S. (eds.) GD 2005. LNCS, vol. 3843, pp. 211–222. Springer, Heidelberg (2006)
14. Gutwenger, C., Jünger, M., Leipert, S., Mutzel, P., Percan, M., Weiskircher, R.: Advances in c-planarity testing of clustered graphs. In: Goodrich, M.T., Kobourov, S.G. (eds.) GD 2002. LNCS, vol. 2528, pp. 220–235. Springer, Heidelberg (2002)
15. Jelínková, E., Kára, J., Kratochvíl, J., Pergel, M., Suchý, O., Vyskočil, T.: Clustered planarity: small clusters in Eulerian graphs. In: Hong, S.-H., Nishizeki, T., Quan, W. (eds.) GD 2007. LNCS, vol. 4875, pp. 303–314. Springer, Heidelberg (2008)
16. Jünger, M., Leipert, S., Percan, M.: Triangulating clustered graphs. Technical report, Zentrum für Angewandte Informatik Köln (December 2002)
17. Nagamochi, H., Kuroya, K.: Drawing c-planar biconnected clustered graphs. Discr. Appl. Math. 155(9), 1155–1174 (2007)

Online Priority Steiner Tree Problems

Spyros Angelopoulos

Max-Planck-Institut für Informatik
Campus E1 4, Saarbrücken 66123, Germany

Abstract. A central issue in the design of modern communication networks is the provision of Quality-of-Service (QoS) guarantees at the presence of heterogeneous users. For instance, in QoS multicasting, a source needs to efficiently transmit a message to a set of receivers, each requiring support at a different QoS level (e.g., bandwidth). This can be formulated as the *Priority Steiner tree* problem: Here, each link of the underlying network is associated with a priority value (namely the QoS level it can support) as well as a cost value. The objective is to find a tree of minimum cost that spans all receivers and the source, such that the path from the source to any given receiver can support the QoS level requested by the said receiver. The problem has been studied from the point of view of approximation algorithms.

In this paper we introduce and address the on-line variant of the problem, which models the situation in which receivers join the multicast group dynamically. Our main technical result is a tight bound on the competitive ratio of $\Theta\left(\min\left\{b\log\frac{k}{b}, k\right\}\right)$ (when $k > b$), and $\Theta(k)$ (when $k \leq b$), where b is the total number of different priority values and k is the total number of receivers. The bound holds for undirected graphs, and for both deterministic and randomized algorithms. For the latter class, the techniques of Alon *et al.* [*Trans. on Algorithms* 2005] yield a $O(\log k \log m)$-competitive randomized algorithm, where m is the number of edges in the graph. Last, we study the competitiveness of online algorithms assuming directed graphs; in particular, we consider directed graphs of bounded edge-cost asymmetry.

1 Introduction

Problem statement and motivation. The provision of Quality-of-Service (QoS) guarantees is amongst the most important design considerations in modern telecommunication networks. Today's networks are routinely used to route vast amounts of traffic, ranging from high-definition video to voice and text. However, reliable data dissemination remains a complex issue, due to the heterogeneity of both the network and the end users. More specifically, some types of traffic may require more stringent QoS guarantees than others: for instance, high-definition video requires significantly higher bandwidth than text. On the other hand, the Internet itself consists of highly heterogeneous subnetworks, each capable of supporting only certain levels of QoS.

As a concrete application, consider multicast communication in an environment comprised by heterogeneous users. A multicast group consists of a source

F. Dehne et al. (Eds.): WADS 2009, LNCS 5664, pp. 37–48, 2009.

which disseminates data to a number of receivers, i.e., the members of the group. These members may vary significantly in their characteristics (such as the bandwidth of the end-connection or their computational power), which implies that they may require vastly different QoS guarantees in terms of the delivered traffic. The objective is to deliver information to all members of the group, while meeting the requirements of each individual member. Furthermore, this dissemination must be efficient in terms of utilization of network links.

Multicast communication in networks is often modeled by means of *Steiner tree problems* (for an in-depth study of the interplay between Steiner trees and network multicasting, the interested reader is referred to [1]). In terms of QoS multicasting, Charikar, Naor and Schieber [2] introduced formulations based on the *Priority Steiner tree problem* (or PST for brevity). In particular, the underlying network is represented by a graph $G = (V, E)$ (which unless specified, is assumed to be undirected). Let $r \in V$ denote the source of the multicast group (which we also call *root*). We let $K \subseteq V$ denote the set of receivers of the multicast group, also called *terminals*.

In addition, the different QoS levels that can be supported by the network are modeled by b integral *priorities* $1 \ldots b$, where b is the highest priority and 1 the lowest. Every edge $e \in E$ is associated with its own *priority value* $p(e)$, which reflects the level of QoS capabilities of the corresponding link (e.g., the bandwidth of the link) as well as with a cost value $c(e)$ which reflects the cost incurred when including the link in the multicast tree. Last, every terminal $t \in K$ is associated with a priority $p(t) \in [1, b]$, which describes the QoS level it requires. A feasible Steiner tree T in this model is a tree rooted in r that spans K, and is such that for every terminal $t \in K$, the priority of *each* edge in the path from r to t in T is at least $p(t)$. The interpretation of this requirement is that terminal t should be able to receive traffic at a level of QoS at least as good as $p(t)$. The objective of the problem is to identify a feasible tree of smallest cost, where the cost of the tree is defined as the total edge-cost of the tree.

A generalization of this problem follows along the lines of the Generalized Steiner tree problem (see eg. [3]). Here, the set K consists of k *pairs* of vertices $(s_1, t_1), \ldots (s_k, t_k)$. Each pair (s_i, t_i) is associated with a priority level $p_i \in [1, b]$. The objective is to identify a feasible Steiner *forest* of smallest cost, such that for all $i \in [1, k]$ the path connecting s_i to p_i in the forest has priority at least p_i[1]. We refer to this problem as the *Generalized Priority Steiner* problem (or GPS).

In this work we address priority Steiner tree problems from the point of view of *online* algorithms, in that the set K of terminals in PST (or pairs of terminals in GPS) is not known to the algorithm in advance, but instead is revealed as a sequence of *requests*. This models the situation in which subscribers to a multicast group are not predetermined, but rather issue dynamic requests to join a group. Thus it is not surprising that many variants of Steiner tree problems have been studied extensively in the on-line setting (see also Section 1). For PST, when a new request for a terminal t is issued, the algorithm must guarantee a path from r to t, of priority at least $p(t)$. Likewise, for GPS, at request (s_i, t_i),

---

[1] The priority of a path is defined as the smallest priority among its edges.

the algorithm must guarantee a path connecting s_i and t_i of priority at least p_i. For both problems, the graph G is assumed to be known to the algorithm. In terms of performance analysis, we apply the standard framework of competitive analysis (see, e.g., [4]). More precisely, the competitive ratio of an algorithm is defined as the supremum (over all request sequences and input graphs) of the ratio of the cost of the solution produced by the algorithm over the optimal off-line cost assuming complete knowledge of the request sequence.

The definition of PST in [2] assumes an underlying undirected graph, however a directed graph is a more accurate and realistic representation of a real network. Indeed, a typical communication network consists of links asymmetric in the quality of service they provide. This motivates the definition of the *edge asymmetry* α (or simply asymmetry) of a directed graph G, originally due to Ramanathan [5], as the maximum ratio of the cost of antiparallel links in G. More formally, let A denote the set of pairs of vertices in V such that if the pair u, v is in A, then either $(v, u) \in E$ or $(u, v) \in E$ (i.e, there is an edge from u to v or an edge from v to u or both). Then the edge asymmetry is defined as

$$\alpha = \max_{\{v,u\} \in A} \frac{c(v, u)}{c(u, v)}$$

According to this measure, undirected graphs are graphs of asymmetry $\alpha = 1$, whereas directed graphs in which there is at least one pair of vertices v, u such that $(v, u) \in E$, but $(u, v) \notin E$ are graphs with unbounded asymmetry ($\alpha = \infty$). Between these extreme cases, graphs of small asymmetry model networks relatively homogeneous networks in terms of the cost of antiparallel links.

The objective is to address the efficiency of an algorithm (in our case, the competitiveness of algorithms for PST) assuming that the underlying graph has bounded asymmetry α. More specifically, for PST, we need to maintain (on-line) an arborescence rooted at r that spans all requests, such that the directed path from the root to each terminal has priority at least the priority of the terminal.

Related Work. The priority Steiner tree problem was introduced by Charikar, Naor and Schieber [2]. In their work, they provided $O(\min\{\log k, b\})$ approximation algorithms for both PST and GPS. The question on whether PST could be approximated within a constant factor was left open in [2], and sparked considerable interest in this problem, until Chuzoy et al. [6] showed a lower bound of $\Omega(\log \log n)$ (under the complexity assumption that NP has slightly superpolynomial time deterministic algorithms). Interestingly, this is one of few problems for which a log log inapproximability result is currently the best known.

Steiner tree problems have been extensively studied from the point of view of online algorithms. For graphs of either constant or unbounded asymmetry, the competitive ratio is tight. For the former class, Imase and Waxman [7] showed that a simple greedy algorithm is optimal and achieves competitive ratio $\Theta(\log k)$. Berman and Coulston [8] extended the result to the Generalized Steiner problem by providing a more sophisticated algorithm. The performance of the greedy algorithm for online Steiner Trees and its generalizations has also been studied by Awerbuch *et al.* [3] and Westbrook and Yan [9]. For the on-line Steiner Tree in the Euclidean plane, the best known lower bound on the competitive ratio is $\Omega(\log k/\log\log k)$ due to Alon and Azar [10]. Westbrook and Yan [11] showed that in directed graphs (of unbounded asymmetry), the competitive ratio can be as bad as $\Omega(k)$, which is trivially matched by a naive algorithm that serves each request by buying a least-cost path from the root to the requested terminal.

The first study of the online Steiner tree problem in graphs of bounded asymme-try is due to Faloutsos *et al.* [12] and continued with work of this author [13,14]. Cur-rently, the best competitive ratio is $O\left(\min\left\{\max\left\{\alpha\frac{\log k}{\log\alpha},\alpha\frac{\log k}{\log\log k}\right\},k\right\}\right)$ and is achieved by a simple greedy algorithm. The known lower bounds are $\Omega\left(\min\left\{\alpha\frac{\log}{\log\alpha},k\right\}\right)$ [12] and $\Omega\left(\min\left\{\alpha\frac{\log}{\log\log k},k^{1-\epsilon}\right\}\right)$ [13], where ϵ is any arbitrarily small constant.

Summary of our results. The main technical result of this paper is a tight bound on the competitive ratio of deterministic algorithms for both PST and GPS equal to $\Theta\left(\min\left\{b\log\frac{k}{b},k\right\}\right)$ (if $k>b$), and $\Theta(k)$, otherwise (c.f. Theorem 1 and Theorem 2). The bound extends to randomized algorithms (Theorem 3). For the latter class of algorithms, the techniques introduced by Alon *et al.* [15] yield a near-optimal $O(\log k\log m)$-competitive randomized algorithm, where m is the number of edges in the graph (c.f Theorem 4). Thus, when k is comparable to m, there exist efficient online algorithms regardless of the number of priority levels.

Last, we study the competitiveness of algorithms for PST assuming directed graphs. If antiparallel links have the same costs, but their priorities can differ by as little as one, it is easy to show a tight bound of $\Theta(k)$ on the competitive ratio. Hence we focus on the case in which antiparallel links have the same priority, but their costs may vary. In particular, we consider directed graphs of bounded edge-cost asymmetry α. For this case, we derive an upper bound of $O\left(\min\left\{\max\left\{\alpha b\frac{\log(k/b)}{\log\alpha},\alpha b\frac{\log(k/b)}{\log\log(k/b)}\right\},k\right\}\right)$, and a corresponding lower bound of $\Omega\left(\min\left\{\alpha b\frac{\log(k/b)}{\log\alpha},k^{1-\epsilon}\right\}\right)$ (where ϵ is any arbitrarily small constant).

2   Tight Bounds for Online PST and GPS in Undirected Graphs

We begin by showing the following lower bound for undirected graphs:

Theorem 1. *The competitive ratio of every deterministic online algorithm for online PST and GPS is* $\Omega\left(\min\{b\log\frac{k}{b},k\}\right)$.

Proof. We will prove the bound for online PST, then the result carries over to online GPS. We distinguish two cases, based on whether $k\geq b$ or not. Depending on the case, the adversary will present a suitable input graph G and a request sequence σ of k terminals (the latter is derived by means of a game between the algorithm and the adversary).

Case 1: $k > b$. We will describe the construction of G in b consecutive *phases*. Each phase, in turn, will be defined in terms of additions of appropriate vertices and edges, which occurs in consecutive *rounds*.

• *Construction of the adversarial graph.*

We begin with the definition of some auxiliary constructions. Let $T_1 = \{v_1, \ldots, v_l\}$ and $T_2 = \{v_1', \ldots, v_l'\}$ be two disjoint sets of l vertices each (we should think of T_1 as lying higher than T_2). For every $i \leq l$, we say that the *distance* between v_i and v_i' is equal to 1. We call index i the i-th *column*, and require that l is a power of 2, and that it is large compared to k (e.g., at least 2^k). For a column $i \leq l$, we say that we *insert a vertex w at depth $d < 1$ in column i* if we introduce a new vertex w at distance d from vertex v_i and distance $1 - d$ from vertex v_i'. We denote by E the set of all l columns.

On the set E of columns, we define a construction $B(E, \beta)$ called *block* which appropriately inserts vertices and edges in $\log(k/b) - 1$ consecutive *rounds*[2]; here β is an integral parameter, with $\beta \in [1, b]$. The rounds are defined inductively as follows: In round 1, l vertices $w^{1,1} \ldots, w^{1,l}$ are inserted in each column of E, at depth equal to $1/2$. We partition the l vertices in two groups, $S^{1,1}$ and $S^{1,2}$, consisting of vertices $\{w^{1,1}, \ldots w^{1,l/2}\}$, and $\{w^{1,l/2+1}, \ldots w^{1,l}\}$, respectively. We will refer to groups of w vertices as *s-sets*. For each one of the two *s*-sets add a corresponding vertex, namely we add vertices $u^{1,1}$ and $u^{1,2}$. For every vertex $w \in S^{1,1}$ (resp. $w \in S^{1,2}$) we add the edge $(w, u^{1,1})$ (resp. $(w, u^{1,2})$), which has zero cost and priority equal to β.

We now describe the j-th round in the construction of $B(E, \beta)$, assuming that rounds $1, \ldots, j-1$ have been defined. Let d denote the size of the *s*-set of smallest cardinality that was inserted in round $j - 1$. For every $i \in [1, 2^j - 1]$, we insert l vertices at depth $i/2^j$, one for each column in E, *unless* some other w vertex has been inserted at this same depth in a previous round, in which case we do not perform any additional insertion. This has the effect that 2^{j-1} additional *layers* of vertices are inserted in round j. We call the i-th deepest layer of vertices that is inserted in round j the i-*th layer of round j*, and denote by $w^{j,i,1} \ldots w^{j,i,l}$ its vertices. We partition the l w-vertices of the i-th layer of round j in $(2^i l)/d$ *s*-sets (from left to right), all of the same size $d/2^i$: In particular, the *s*-set $S^{j,i,q}$ (in words, the q-th *s*-set, from left to right, of the i-th layer in round j) consists of $d/2^i$ vertices $\{w^{j,i,\frac{(q-1)d}{2^i}+1}, \ldots w^{j,i,\frac{qd}{2^i}}\}$. For each such *s*-set of the form $S^{j,i,q}$, we add a vertex $u^{j,i,q}$, and for every vertex $w \in S^{j,i,q}$ we add the edge $(w, u^{j,i,q})$ of zero cost and priority β. The *depth* of a u-vertex is defined as the depth of any of the w-vertices in its corresponding *s*-set, and its *height* as the quantity (1-depth). This completes the definition of round j. Figure 1 illustrates an example.

Finally, we turn $B(E, \beta)$ into a well-defined graph by inserting edges between any two pairs of consecutive vertices in all columns. More precisely, for each pair of consecutive vertices w, w' of every column i, we add an edge (w, w') (which we call *vertical* edge) of cost equal to the distance between w and w' in the column

---

[2] Throughout the paper we omit floors and ceilings since they do not affect the asymptotic behavior of the algorithms.

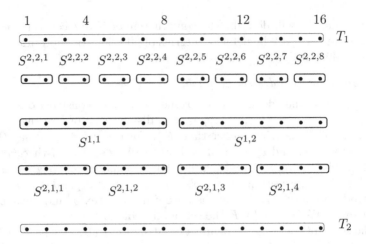

Fig. 1. An illustration of $B(E, \beta)$, assuming a set E of 16 columns, and two rounds. For simplicity, the figure illustrates only the partition of w-vertices into s-sets (i.e., we omit the u-vertices). In the figure, $S^{1,1}$ and $S^{1,2}$ are the s-sets of the first round. The sets $S^{2,1,1} \ldots S^{2,1,4}$ are the s-sets at the first layer of round 2, and $S^{2,2,1} \ldots S^{2,2,8}$ are the s-sets at the second layer of round 2.

and priority equal to b. We use the same notation $B(E, \beta)$ to refer to this graph, when clear from context.

We say that an s-set *crosses* a certain set of columns if and only if the set of columns in which the vertices of S lie intersects the set of columns in question. Two s-sets cross each other iff the intersection of the sets of columns crossed by each one is non-empty. Note that there is a 1-1 correspondence between s-sets and u-vertices, which means that several properties/definitions pertaining to s-sets carry over to the corresponding u vertices (e.g., we will say that two u vertices cross if their s-sets cross).

The adversarial graph G is defined by performing a series of block constructions, in b consecutive *phases*. Let T_1, T_2 and E be as defined earlier. Phase 1 consists only of the insertion of block $B(E, 1)$. Suppose that phase $p - 1 < b$ has been defined, we will define phase p. Let d denote the smallest cardinality among s-sets inserted[3] in phase $p - 1$. This induces a partition of the columns in E in d sets, where the i-th set in this partition, say set E_i, consists of columns $(i - 1)d + 1, \ldots id$. Phase p consists of inserting d blocks, namely blocks $B(E_i, p)$, for all $i \in [1, d]$ (see Fig. 2 for an illustration).

One subtle point has to do with w-vertices of the same column, which are at the same depth but belong to different phases. Say w_1 and w_2 are such vertices, added in phases j_1 and j_2, respectively. For every such pair, there is an edge (w_1, w_2) in G of zero weight and priority equal to $\min\{j_1, j_2\}$. In words, this means that using this edge is free, however it can be of use only if we route traffic at a level at most the minimum of the two corresponding phases.

---

[3] In other words, d is the cardinality of s-sets in the last (i.e., highest) layer inserted in the last round of phase $p - 1$.

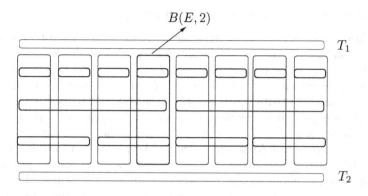

Fig. 2. An illustration of the insertion of blocks $B(E, 2)$ in the creation of G. In this figure, $B(E, 1)$ consists of 2 rounds, similar to Figure 1, i.e., three layers are inserted in total. Each of the blocks $B(E, 2)$ is represented by a rectangle; its edges and vertices are omitted due to its complexity. There are 8 such blocks inserted, each crossing the columns of the highest S-set of $B(E, 1)$.

This essentially completes the construction of G. We also add a root vertex r and edges (r, v), for all $v \in T_1$, as well as edges (t, v'), for all $v' \in T_w$ (where t is a new vertex). These edges are assigned zero cost and priority b.

Before proceeding to the description of the game between the algorithm and the adversary, we provide some alternative classification of vertices added in round j of phase p, which will make the description of the game easier. Consider the collection of deepest s-sets inserted during round j of phase p. This collection induces a partition of the set of all columns E into disjoint sets $E_p^{j,1}, E_p^{j,2}, \ldots$, from left to right: every s-set inserted during round j of phase p will cross only one of the sets of columns $E_p^{j,1}, E_p^{j,2}, \ldots$. We then say that an s-set (or a corresponding u-vertex) inserted during round j of phase p that crosses edges of the set $E_p^{j,i}$ belongs in *component* $C_p^{j,i}$. This provides a convenient way to identify s-sets and u-vertices in this construction. Namely, we use the notation $C_p^{j,i}(level, pos)$ to describe the s-set which belongs in component $C_p^{j,i}$, is in the *level*-th deepest layer among layers of the component $C_p^{j,i}$ and is also the *pos*-th s-set (from left to right) in this component. We call the s-set $C_p^{j,i}(1, 1)$ the s-*root* of component $C_p^{j,i}$ (by definition, the unique deepest s-set of the component).

The purpose of introducing this notation is to define a *parent/child* relation among s-sets in the same component. From construction, every s-set in layer L of $C_p^{j,i}$ (say S) crosses exactly two other s-sets in the same component which are inserted at layer $L + 1$ (say S_1 and S_2, from left to right). The only exception is the highest layer of the component, which does not have any children. We say that S_1 (resp. S_2) is the left (resp. right) *child* of S in this component.

We emphasize that the definitions that relate s-sets apply also to u-vertices, due to their 1-1 correspondence.

• *The game between the algorithm and the adversary*

We show how to construct an adversarial sequence σ of at most k requests (the actual number of requests will be $k - b + 1$, but this does not affect the asymptotic analysis). The request sequence σ is constructed in b *phases*, and each phase will request a u-vertex which was added in the corresponding phase in the construction of G (with the exception of the very first request, which is for vertex t). Every requested vertex in phase $p \leq b$ will be assigned a priority equal to the phase index, namely p. For every such vertex u, the online algorithm must establish a *connection path* from r to u denoted by $path(u)$ (of priority at least $p(u)$), possibly buying new edges. We can assume, without loss of generality, that the algorithm provides a single connection path for each requested terminal (if more than one such paths exist, the adversary can declare one arbitrarily as $path(u)$, and ignore all others. Similar arguments are used in [12] and [13]).

All phases consist of $\log(k/b) - 1$ rounds, with the exception of phase 1, which consists of $\log(k/b)$ rounds (with round 0 an "initialization" round). To give some intuition about the game, we first describe the actions of the the adversary on the first three rounds of phase 1, then provide a more formal iterative description of the game. Every request in phase 1 is assigned priority equal to 1. In round 0, vertex t is requested, and the connection path $path(t)$ chosen by the algorithm will cross exactly one of the sets $S^{1,1}$ and $S^{1,2}$ (which are also the s-roots of the trivial components $C_1^{1,1}$ and $C_1^{1,2}$). Round 1 consists of a single request to a u-vertex determined as follows: if $path(t)$ crosses the s-root of $C_1^{1,1}$ (resp. $C_1^{1,2}$), then the adversary requests the root of $C_1^{1,2}$ (resp. $C_1^{1,1}$). At this point, no matter what the connection paths chosen so far, there is a component $C_1^{2,x}$, with $x \in [1,4]$, such that no columns of $C_1^{2,x}$ are crossed by existing connection paths[4]. Round 2 then begins, during which the adversary will request one u vertex per layer of $C_1^{2,x}$. In the first request of the round, the adversary requests the root of $C_1^{2,x}$. The connection path for the latter will cross exactly one of its children: if it crosses the left child, then the next request will be to the right child and vice versa. In total 2 requests to u-vertices are made in this round.

For a formal description of the algorithm/adversary game, we will first need a straightforward proposition concerning u-vertices.

Proposition 1. *Let u be a u-vertex requested by the adversary which belongs to component $C_p^{j,x}$, for some round j and index x. Suppose that u satisfies the precondition that prior to its request in σ, no connection path for previously requested vertices crosses u. Then the following hold:*

(a) *If u is the highest u-vertex in $C_p^{j,x}$, (i.e., u has no children in $C_p^{j,x}$) and $j < \log(k/b) - 1$, then after $path(u)$ is established, there is a component $C_p^{j+1,y}$ whose root does not cross any connection paths for all previous requests.*

(b) *If u is not the highest u-vertex in $C_p^{j,x}$ (i.e., u has children in $C_p^{j,x}$) then after $path(u)$ is established, there is a child of u in $C_p^{j,x}$ that is not crossed by any connection paths for all previous requests.*

---

[4] We will say that a connection path for a terminal crosses an s-set (or a corresponding u-vertex) if it contains (vertical) edges in a column that is crossed by the s-set.

(c) *If u is the highest u-vertex in $C_p^{j,x}$, and $j = \log(k/b) - 1$, then after $path(u)$ is established, there is a component $C_{p+1}^{1,y}$ whose root does not cross any connection paths of previous requests.*

The game proceeds as follows: First, suppose that rounds $1, \ldots j$ of phase p have been defined (with $j < \log(k/b) - 1$), we will now describe the requests of round $j + 1$ of phase p. Let s denote the s-set of the highest u-vertex requested in round j. Then, from Proposition 1(a), there is a component, namely $C_p^{j+1,y}$ for some index y such that no connection path established so far crosses the s-root of $C_p^{j+1,y}$. Round $j + 1$ begins with the adversary requesting the u-root of this component. The connection path for this request will cross only one of its children in the component, and the child that is not crossed will be the 2nd terminal to be requested in this round. Round j of this phase proceeds with requesting u-vertices in an upwards fashion, one u-vertex per layer of $C_p^{j+1,y}$: when a vertex is requested, the next request is the unique child of the vertex that is not crossed by any connection paths up to that point (as Proposition 1(b) suggests). The round continues until the highest vertex in component $C_p^{j+1,y}$ is requested, i.e., continues with 2^j requests in total.

It remains to argue how to generate the first request for any phase $p > 1$. Note that the last request of phase $p - 1$ is to a u-vertex which is highest in some component $C_{p-1}^{\log(k/b)-1,x}$ for some index value x. Property 1(c) guarantees that there is a component $C_p^{1,y}$, for some index y, whose root does not cross any connection paths of previous requests. Then the first request in phase p (indeed the only request of the first round in phase p) is to the u-root of $C_p^{1,y}$.

• *Analysis*

We first observe that the adversarial sequence of requests σ is such that for any request u, there exists a column in u that crosses all previously requested u-vertices in σ (this is easy to show by means of a straightforward induction). This implies that all requested u-vertices in σ share a common column, say i. Then, an offline solution suffices to buy: i) all vertical edges of column i (for a total cost equal to 1); ii) edges from a w vertex in column i to a requested u-vertex (zero cost); iii) edges of the form (r, v_i), and (t, v_i') (at zero cost). Hence the optimal offline cost is at most 1.

On the other hand, consider the requests which belong in round j of phase p. These requests are part of a component $C_p^{j,x}$ for some index x. Due to the actions of the adversary, as well as Proposition 1(b), every time a request is issued for a u-vertex in $C_p^{j,x}$ other than the root of this component the algorithm must pay cost at least $1/2^j$. Since there are 2^{j-1} requests in σ that belong to $C_p^{j,x}$, the algorithm pays a cost which is at least $1/2 - 1/2^j$ during the round. Since there are $\log(k/b) - 1$ rounds and b phases, the overall cost is $\Omega(b \log \frac{k}{b})$.

Case 2: $k \leq b$. This case is easier to show, and its proof is omitted.

The theorem follows by combining the results of the two cases.   □

We now show a simple algorithm for the problem that is asymptotically optimal. The algorithm applies not only to PST but also to its generalization, namely GPS (as defined in Section 1). We use the online algorithm for the (plain) Generalized Steiner problem in undirected graphs due to Berman and Coulston [8], which we denote by BC. In particular, the algorithm maintains b Steiner forests, $F_1, \ldots F_b$, one for each priority value, in an online fashion (the forests may not be edge-disjoint, i.e., F_i may share edges with F_j). When a request $r_i = (s_i, t_i, b_i)$ is issued, the algorithm will assign r_i to forest F_{b_i}, by running the BC algorithm for (s_i, t_i) (ignoring the priority values). We denote the resulting algorithm by PRBC.

Theorem 2. *The competitive ratio of* PRBC *is* $O\left(\min\left\{b\log\frac{k}{b}, k\right\}\right)$.

Proof. For any fixed integer $j \leq b$, let G_j denote the subgraph of G induced by all edges e of priority at least j. For any sequence of requests $R = r_1, \ldots r_k$, partition R into subsequences $R_1, \ldots, R_b$ such that R_j consists of all requests in R of priority equal to j. Let (G_j, R_j) denote an instance of the (plain) online Generalized Steiner Problem on graph G_j and request sequence R_j: here we ignore the priority of edges and requests, and our objective is to minimize the cost of the Steiner forest for the sequence R_j without concerns about priorities. Let OPT_j denote the cost of the optimal Steiner forest for the above instance. Since the algorithm of Berman and Coulston is $O(\log|R_j|)$-competitive, it follows that $c(R_j) = O(\log|R_j|)OPT_j = O(\log|R_j|)OPT$ (this follows from the fact that $OPT_j \leq OPT$). Here, we denote by $c(R_j)$ the cost paid by PRBC on request set R_j (which is the same as the cost of BC for instance (G_j, R_j)), and by OPT the cost of the optimal offline solution to the problem.

Therefore, the total cost of PRBC on sequence R is bounded by $c(R) = O\left(\sum_{j=1}^{b}\log|R_j|OPT\right) = O\left(\log\prod_{j=1}^{b}|R_j|\right)OPT$, which is maximized when $|R_j| = \frac{k}{b}$ (for $k > b$.) Hence $c(R) = O\left(b\log\frac{k}{b} \cdot OPT\right)$.

In addition, $c(R) \leq k \cdot OPT$. Therefore, $c(R) = O\left(\min\left\{b\log\frac{k}{b}, k\right\}\right)OPT$, when $k > b$, and $c(R) = O(k)OPT$, otherwise. $\square$

3 Extensions: Randomized Algorithms and Directed Graphs

3.1 Randomized Algorithms

We begin by showing that the lower bound of Theorem 1 holds even when considering randomized algorithms against an oblivious adversary.

Theorem 3. *The competitive ratio of every randomized online algorithm against the oblivious adversary for online PST and GPS is* $\Omega\left(\min\{b\log\frac{k}{b}, k\}\right)$.

Proof sketch. We resort to Yao's principle [16]. We use the same adversarial graph G as in the proof of Theorem 1. The distribution for the case $k > b$ is motivated by the request sequence in the deterministic case: more specifically, for a given round of a phase, the next terminal to be requested is chosen, uniformly at random (i.e. with probability $1/2$) among the two children of the last requested

terminal. This ensures that any deterministic algorithm will pay, on average, a
cost at least half the cost determined in the analysis of Theorem 1, whereas the
optimal average cost remains as in the same analysis. A similar (and simpler)
argument can be made for the case $k \leq b$.   □

Even though Theorem 1 and Theorem 3 are tight, in the case where b is large,
namely when $b \in \Omega(k)$, the competitive ratio of any deterministic or randomized
algorithm is disappointingly bad (i.e., $\Omega(k)$). However, observe that the adversarial graph G requires at least an exponential number of vertices (and edges)
in the number of requested terminals k. Thus the following question arises: if
the number of edges is comparable to the number of terminals, can we achieve
better competitive ratios? To answer this, we can use the framework of Alon et
$al.$ [15], which addresses online algorithms for broad classes of (edge-weighted)
connectivity and cut problems. In particular, we can obtain the following:

Theorem 4. *There exists a randomized algorithm for GPS and PST with competitive ratio $O(\log k \log m)$, where m is the number of edges in the graph.*

3.2   Directed Graphs

So far in this paper we assumed that the input graph G is undirected. However,
the competitiveness of PST changes dramatically when G is directed. Specifically,
since the online Steiner tree problem is $\Omega(k)$-competitive for directed graphs [11],
the same lower bound carries over to online PST. More importantly, it is not
difficult to show that the same bound applies even in graphs in which antiparallel
edges have the same cost, as long as their priorities can differ by as little as one.
    We thus focus on graphs of bounded edge-cost asymmetry, denoted by α,
assuming that antiparallel links have the same priority. In other words, the
"directedness" of a graph manifests itself in terms of edge costs, and not in
terms of edge priorities. In this model, we can show the following:

Theorem 5. *The competitive ratio of the greedy algorithm for PST is*

$$O \left(\min \left\{ \max \left\{ \alpha b \frac{\log(k/b)}{\log \alpha}, \alpha b \frac{\log(k/b)}{\log \log(k/b)} \right\}, k \right\} \right).$$

*Furthermore, the competitive ratio of every deterministic (or randomized) algo-
rithm is $\Omega \left(\min \left\{ \alpha b \frac{\log(k/b)}{\log \alpha}, k^{1-\epsilon} \right\} \right)$, for arbitrarily small constant ϵ.*

Proof sketch. For the upper bound, we use a classification of terminals according
to their priority (similar to the proof of Theorem 2), in combination with the
analysis of the greedy algorithm for the asymmetric Steiner tree problem of [14].
We also make use of the concavity of the functions $\log(x)$ and $\log(x)/\log \log(x)$
so as to upper-bound the overall contribution of terminals.
    For the lower bound, we combine ideas from Theorem 1 and the lower bounds
of [12] and [13]. The adversarial sequence will force any algorithm to pay a cost
of $\Omega \left(\min \left\{ \alpha \frac{\log(k/b)}{\log \alpha}, (k/b)^{1-\epsilon} \right\} \right)$, in each of a total number of b phases.   □

References

1. Oliveira, C.A.S., Pardalos, P.M.: A survey of combinatorial optimization problems in multicast routing. Computers and Operations Research 32(8), 1953–1981 (2005)
2. Charikar, M., Naor, J., Schieber, B.: Resource optimization in QoS multicast routing of real-time multimedia. IEEE/ACM Transactions on Networking 12(2), 340–348 (2004)
3. Awerbuch, B., Azar, Y., Bartal, Y.: On-line generalized Steiner problem. Theoretical Computer Science 324(2–3), 313–324 (2004)
4. Borodin, A., El-Yaniv, R.: Online computation and competitive analysis. Cambridge University Press, Cambridge (1998)
5. Ramanathan, S.: Multicast tree generation in networks with asymmetric links. IEEE/ACM Transactions on Networking 4(4), 558–568 (1996)
6. Chuzhoy, J., Gupta, A., Naor, J., Sinha, A.: On the approximability of some network design problems. Transactions on Algorithms 4(2) (2008)
7. Imase, M., Waxman, B.: The dynamic Steiner tree problem. SIAM Journal on Discrte Mathematics 4(3), 369–384 (1991)
8. Berman, P., Coulston, C.: Online algorithms for Steiner tree problems. In: Proceedings of the 39th Symposium on the Theory of Computing (STOC), pp. 344–353 (1997)
9. Westbrook, J., Yan, D.C.K.: The performance of greedy algorithms for the on-line Steiner tree and related problems. Mathematical Systems Theory 28(5), 451–468 (1995)
10. Alon, N., Azar, Y.: On-line Steiner trees in the Euclidean plane. Discrete and Computational Geometry 10, 113–121 (1993)
11. Westbrook, J., Yan, D.C.K.: Linear bounds for on-line Steiner problems. Information Processing Letters 55(2), 59–63 (1995)
12. Faloutsos, M., Pankaj, R., Sevcik, K.C.: The effect of asymmetry on the on-line multicast routing problem. International Journal of Foundations of Computer Science 13(6), 889–910 (2002)
13. Angelopoulos, S.: Improved bounds for the online Steiner tree problem in graphs of bounded edge-asymmetry. In: Proceedings of the 18th Annual Symposium on Discrete Algorithms (SODA), pp. 248–257 (2007)
14. Angelopoulos, S.: A near-tight bound for the online steiner tree problem in graphs of bounded asymmetry. In: Halperin, D., Mehlhorn, K. (eds.) Esa 2008. LNCS, vol. 5193, pp. 76–87. Springer, Heidelberg (2008)
15. Alon, N., Awerbuch, B., Azar, Y., Buchbinder, N., Naor, J.: A general approach to online network optimization problems. In: Proceedings of the 15th Symposium on Discrete Algorithms (SODA), pp. 570–579 (2005)
16. Yao, A.: Probabilistic computations: Toward a unified measure of complexity. In: Proceedings of the 18th IEEE Symposium on Foundations of Computer Science (FOCS), pp. 222–227 (1977)

Connect the Dot:
Computing Feed-Links with Minimum Dilation*

Boris Aronov[1], Kevin Buchin[2], Maike Buchin[3], Marc van Kreveld[3],
Maarten Löffler[3], Jun Luo[4], Rodrigo I. Silveira[3], and Bettina Speckmann[2]

[1] Dep. Computer Science and Engineering, Polytechnic Institute of NYU, USA
aronov@poly.edu
[2] Dep. of Mathematics and Computer Science, TU Eindhoven, The Netherlands
{kbuchin,speckman}@win.tue.nl
[3] Dep. of Information and Computing Sciences, Utrecht University, The Netherlands
{maike,marc,loffler,rodrigo}@cs.uu.nl
[4] Shenzhen Institute of Advanced Technology, Chinese Academy of Sciences, China
jun.luo@sub.siat.ac.cn

Abstract. A *feed-link* is an artificial connection from a given location
p to a real-world network. It is most commonly added to an incomplete
network to improve the results of network analysis, by making p part
of the network. The feed-link has to be "reasonable", hence we use the
concept of dilation to determine the quality of a connection.

We consider the following abstract problem: Given a simple polygon
P with n vertices and a point p inside, determine a point q on P such that
adding a feedlink $\overline{pq}$ minimizes the maximum dilation of any point on P.
Here the *dilation* of a point r on P is the ratio of the shortest route from
r over P and $\overline{pq}$ to p, to the Euclidean distance from r to p. We solve this
problem in $O(\lambda_7(n) \log n)$ time, where $\lambda_7(n)$ is the slightly superlinear
maximum length of a Davenport-Schinzel sequence of order 7. We also
show that for convex polygons, two feed-links are always sufficient and
sometimes necessary to realize constant dilation, and that k feed-links
lead to a dilation of $1 + O(1/k)$. For (α, β)-covered polygons, a constant
number of feed-links suffices to realize constant dilation.

1 Introduction

Network analysis is a type of geographical analysis on real-world networks, such
as road, subway, or river networks. Many facility location problems involve network analysis. For example, when a location for a new hospital needs to be
chosen, a feasibility study typically includes values that state how many people

* This research has been supported by the Netherlands Organisation for Scientific Re-
search (NWO) under BRICKS/FOCUS grant number 642.065.503, under the project
GOGO, and under project no. 639.022.707. B. Aronov has been partially supported
by a grant from the U.S.-Israel Binational Science Foundation, by NSA MSP Grant
H98230-06-1-0016, and NSF Grant CCF-08-30691. M. Buchin is supported by the
German Research Foundation (DFG) under grant number BU 2419/1-1.

F. Dehne et al. (Eds.): WADS 2009, LNCS 5664, pp. 49–60, 2009.
© Springer-Verlag Berlin Heidelberg 2009

would have their travel time to the nearest hospital decreased to below 30 minutes due to the new hospital location. In a more global study of connectivity, one may analyze how many households are reachable within 45 minutes from a fire station. In this case, the households are typically aggregated by municipality or postal-code region, and the centroid of this region is taken as the representative point. This representative point might not lie on the road network. It might even be far removed from it, since nation-wide connectivity studies seldomly use detailed network data for their analysis. A similar situation occurs when the quality of the network data is not very high. In developing countries, data sets are often incomplete due to omissions in the digitization process, or due to lack of regular updates. In both cases a network study must be executed that involves a set of locations that are not connected to the network in the available data.

A workable solution in such cases is to connect the given locations to the known road network by *feed-links*. A feed-link is an artificial connection between a location and the known network that is "reasonable", that is, it is conceivable that such a connection exists in the real world [2,6]. A road network forms an embedded, mostly planar graph. Hence a location that does not lie on the network, lies inside some face of this graph. Such a face can be represented by a simple polygon. A feed-link is then a connection from the given location to the boundary of the simple polygon.

When computing feed-links we need to be able to judge their quality. That is, we have to assess if a particular connection could possibly exist in reality. To do this, we use the concept of *dilation*, also known as *stretch factor* or *crow flight conversion coefficient*. People in general do not like detours, so a connection that causes as little detour as possible, is more likely to be "real". Given an embedded plane graph, the dilation of two points p and q on the graph is the ratio of their distance within the graph to their Euclidean distance. The concept of dilation is commonly used in computational geometry for the construction of *spanners*: a t-spanner is a graph defined on a set of points such that the dilation between any two points is at most t, see [7,10,13,14,15].

In this paper we consider a single point p inside a simple polygon, whose boundary we denote by P. We solve the problem of placing one feed-link between p and P so that the maximum dilation over all points on P to p is minimized. We assume that a feed-link is a straight-line connection between p and exactly one point q on P. We allow the feed-link $\overline{pq}$ to intersect P in more points, see Fig. 1 (left), but assume that it is not possible to "hop on" the feed-link at any such point other than q (the white points in the figure provide no access to the feed-link). Fig. 1 (middle) shows that the feed-link yielding minimum dilation may intersect P in a point other than q. One could also choose to disallow feed-links that intersect the outside of P, or to use geodesic shortest paths inside P as feed-links, and measure the dilation of any point on P with respect to its geodesic distance to p. We also study the problem of connecting several feed-links to p to bound the dilation. Then any point on P uses exactly one of the feed-links to reach p over the network. Fig. 1 (right) shows that $n/2$ feed-links may be necessary to bound the dilation by a constant, if P has n vertices.

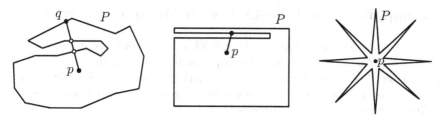

Fig. 1. A feed-link that intersects P gives no access to the feed-link other than q (left). A minimum dilation feed-link may intersect P in the interior of the feed-link (middle). Simple polygons may require many feed-links to achieve constant dilation (right).

In a recent paper [2] we showed how to compute the dilation of a polygon when a collection of feed-links to a point inside is given. We also gave heuristic algorithms to place one or more feed-links and compared them experimentally on generated polygons. The simple heuristic for one feed-link that connects p to the closest point on P is a factor-2 approximation for the optimal feed-link placement. We also studied the problem of placing as few feed-links as possible to realize a specified dilation. A simple incremental algorithm exists that uses at most one more feed-link than the minimum possible.

Results. In Section 2 we give an efficient algorithm to compute an optimal feed-link. For a simple polygon with n vertices, our algorithm runs in $O(\lambda_7(n) \log n)$ time, where $\lambda_7(n)$ is the maximum length of a Davenport-Schinzel sequence of order 7, which is only slightly superlinear [1,16]. If we are interested in the dilation with respect to only m fixed points on P, the running time reduces to $O(n + m \log m)$. Furthermore, we give a $(1 + \varepsilon)$-approximation algorithm for the general problem that runs in $O(n + (1/\varepsilon) \log(1/\varepsilon))$ time, for any $\varepsilon > 0$. The results in this section also hold with geodesic dilation and feed-links, or with feed-links that are not allowed to intersect the outside of P.

In Section 3.1 we show that for any convex polygon and any point inside, two feed-links are sufficient and sometimes necessary to achieve constant dilation. In this case the dilation is at most $3 + \sqrt{3}$. There are convex polygons where no two feed-links can realize a dilation better than $2 + \sqrt{3}$. We also show that we can realize a dilation of $1 + O(1/k)$ with k feed-links. Finally, in Section 3.2 we show that for (α, β)-covered polygons [8] (a class of realistic polygons), a constant number of feed-links suffices to obtain constant dilation. This result does not hold for most other classes of realistic polygons.

Notation. P denotes the boundary of a convex or simple polygon, and p is a point inside it. For two points a and b on P, $P[a, b]$ denotes the portion of P from a clockwise to b, its length is denoted by $\mu(a, b)$. Furthermore, $\mu(P)$ denotes the length (perimeter) of P. The Euclidean distance between two points p and q is denoted by $|pq|$. For two points q and r on P, the dilation of point r when the feed-link is $\overline{pq}$ is denoted by $\delta_q(r)$. For an edge e, $\delta_q(e)$ denotes the maximum dilation of any point on e when the feed-link is $\overline{pq}$.

2   Computing One Feed-Link with Minimum Dilation

Let $v_0, \ldots, v_{n-1}$ be the vertices of P and let p be a point inside P. We seek a point q on P such that the feed-link $\overline{pq}$ minimizes the maximum dilation to any point on P. We first consider the restricted case of minimizing the dilation only for m given points on P. Then we solve the general case. In both cases, the feed-link may connect to any point on P.

Let r be a point on P and let r' be the point opposite r, that is, the distance along P between r and r' is exactly $\mu(r, r') = \mu(r', r) = \mu(P)/2$. For any given location of q, r has a specific dilation. We study the change in dilation of r as q moves along P. If $q \in P[r', r]$, then the graph distance between p and r is $|pq| + \mu(q, r)$, otherwise it is $|pq| + \mu(r, q)$.

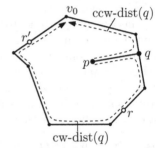

We choose a fixed point v_0 on P and define two functions cw-dist(q) and ccw-dist(q) that measure the distance from p to v_0 via the feed-link $\overline{pq}$ and then from q either clockwise or counterclockwise along P, see Fig. 2. The dilation $\delta_q(r)$ of r can be expressed using either cw-dist(q) or ccw-dist(q), depending on the order in which v_0, q, r, and r' appear along P. In particular, we distinguish four cases that follow from the six possible clockwise orders of v_0, q, r, and r':

Fig. 2.   cw-dist(q) and ccw-dist(q); shown is case 1 with order $v_0 q r r'$

1. If the clockwise boundary order is $v_0 q r r'$ or $v_0 r' q r$, then the dilation is
   $\delta_q(r) = (\text{cw-dist}(q) - \mu(r, v_0)) / |rp|$.
2. If the clockwise boundary order is $v_0 r r' q$, then the dilation is
   $\delta_q(r) = (\text{cw-dist}(q) + \mu(v_0, r)) / |rp|$.
3. If the clockwise boundary order is $v_0 q r' r$, then the dilation is
   $\delta_q(r) = (\text{ccw-dist}(q) + \mu(r, v_0)) / |rp|$.
4. If the clockwise boundary order is $v_0 r q r'$ or $v_0 r' r q$, then the dilation is
   $\delta_q(r) = (\text{ccw-dist}(q) - \mu(v_0, r)) / |rp|$.

As q moves along P in clockwise direction, starting from v_0, three of the cases above apply consecutively. Either we have $v_0 q r r' \rightarrow v_0 r q r' \rightarrow v_0 r r' q$, or $v_0 q r' r \rightarrow v_0 r' q r \rightarrow v_0 r' r q$. We parameterize the location of q both by cw-dist(q) and ccw-dist(q). This has the useful effect that the dilation $\delta_q(r)$ of r is a linear function on the intervals where it is defined (see Fig. 3). In particular, for a fixed point r, $\delta_q(r)$ consists of three linear pieces. Note that we cannot combine the two graphs into one, because the parameterizations of the location of q by cw-dist(q) and ccw-dist(q) are not linearly related. This follows from the fact that cw-dist$(q)+$ ccw-dist$(q) = \mu(P) + 2 \cdot |pq|$.

We now solve the restricted case of minimizing the dilation only for m given points on P. For each point r we determine the line segments in the two graphs that give the dilation of r as a function of cw-dist(q) and ccw-dist(q). These line segments can be found in $O(n + m)$ time in total. Next, we compute the upper

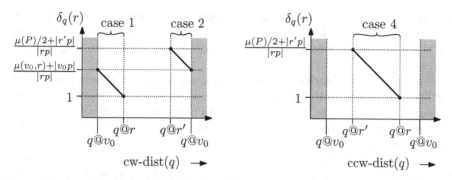

Fig. 3. Two graphs showing the dilation of a point r as a function of cw-dist(q) (left) and ccw-dist(q) (right); $q@r$ indicates "q is at position r"

envelope of the line segments in each of the two graphs. This takes $O(m \log m)$ time using the algorithm of Hershberger [12], and results in two upper envelopes with complexity $O(m \cdot \alpha(m))$. Finally, we scan the two envelopes simultaneously, one from left to right and the other from right to left, taking the maximum of the corresponding positions on the two upper envelopes, and recording the lowest value encountered. This is the optimal position of q.

To implement the scan, we first add the vertices of P to the two envelopes. Since we need to compute the intersection points of the two envelopes we must unify their parameterizations. Consider the locations of q that fall within an interval I which is determined by two envelope edges e_1 and e_2. Since cw-dist$(q) = -$ccw-dist$(q) + 2 \cdot |pq| + \mu(P)$, the line segment of one envelope restricted to I becomes a hyperbolic arc in the parametrization of the other envelope. Hence e_1 and e_2 can intersect at most twice in a unified parametrization, and the scan takes time linear in the sum of the complexities of the two envelopes.

Theorem 1. *Given the boundary P of a simple polygon with n vertices, a point p inside P, and a set S of m points on P, we can compute the feed-link (which might connect to any point on P) that minimizes the maximum dilation from p to any point in S in $O(n + m \log m)$ time.*

Next we extend our algorithm to minimize the dilation over all points on P. Let $r_e(q)$ denote the point with the maximum dilation on a given edge e of P. Instead of considering the graphs of the dilation for a set of fixed points, we consider the graphs for the points $r_e(q)$ for all edges of P. The positions of $r_e(q)$ change with q. The graphs of the dilation do not consist of line segments anymore, but of more complex functions, which, however, intersect at most six times per pair, as we prove in the full paper. As a consequence, we can compute their upper envelope in $O(\lambda_7(n) \log n)$ time [12], where $\lambda_7(n)$ is the maximum length of a Davenport-Schinzel sequence of order 7, which is slightly superlinear [1,16].

Theorem 2. *Given the boundary P of a simple polygon with n vertices and a point p inside P, we can compute the feed-link that minimizes the maximum dilation from p to any point on P in $O(\lambda_7(n) \log n)$ time.*

Note that our algorithms ignore the degenerate case where p lies on a line supporting an edge e of P. In this case cw-dist(q) and ccw-dist(q) are both constant on e. This is in fact easy to handle, as we describe below when discussing geodesic dilation.

We can adapt our algorithms to not allow feed-links that intersect the exterior of P. We first compute the visibility polygon $V(p)$ of p with respect to P. The vertices of $V(p)$ partition the edges of P into parts that are allowed to contain q and parts that are not. The number of parts is $O(n)$ in total, and they can be computed in $O(n)$ time.

We compute the upper envelopes exactly as before. Before we start scanning the two envelopes, we add the vertices of P *and also* the vertices of the visibility polygon to the two envelopes. The envelopes now have the property that between two consecutive vertices, a feed-link is allowed everywhere or nowhere. During the scan, we keep the maximum of the dilation functions and record the lowest value that is allowed. The time complexity of our algorithms does not change.

We can also adapt our algorithms to use geodesic feed-links and geodesic shortest distances. In this case the feed-link is a geodesic shortest path between p and q, and the dilation of a point r on P is defined as the ratio of the graph distance between r and p (necessarily via q) and the geodesic shortest path between r and p.

By computing the shortest path tree of p inside P, we obtain the geodesic shortest distances of p to every vertex of P, and hence can partition P into $O(n)$ parts, such that the first vertex on a geodesic shortest path to p is the same (this first vertex can also be p itself) [11].

When we use cw-dist(q) and ccw-dist(q) to represent the location of q, we use the length of the geodesic from q to p instead of $|pq|$, plus the clockwise or counterclockwise distance to v_0. But now a value of cw-dist(q) or ccw-dist(q) does not necessarily represent a unique position of q anymore: when q traverses an edge of P and the geodesic from q to p is along this edge in the opposite direction, cw-dist(q) and ccw-dist(q) do not change in value. However, it is sufficient to consider only the location of q that gives the shortest feed-link (if any such feed-link is optimal, then the shortest one is optimal too). All other adaptations to the algorithms are straightforward, and we obtain the same time complexities.

3   Number of Feed-Links vs. Dilation

In this section we study how many feed-links are needed to achieve constant dilation. We immediately observe that there are simple polygons that need $n/2$ feed-links to achieve constant dilation, see Fig. 1. For convex polygons, we establish that two feed-links are necessary and sufficient to obtain constant dilation. For realistic ("fat") simple polygons, there are several definitions one can use to capture realism ([4,5,8,9,17] and others). Most of these definitions are not sufficient to guarantee constant dilation with a constant number of feed-links. However, for (α, β)-covered polygons [8] we can show that a constant number of feed-links suffices for constant dilation.

3.1   Convex Polygons

Let P be the boundary of a convex polygon and let p be a point inside P. We explore how many feed-links are necessary and sufficient to guarantee constant dilation for all points on P.

One feed-link is not sufficient to guarantee constant dilation. Consider a rectangle with width w and height $h < w$, and let p be its center. One of the points in the middle of the long sides will have dilation greater than $2w/h$, which can become arbitrarily large. Hence two feed-links may be necessary.

Two feed-links are also sufficient to guarantee constant dilation for all points on P. In fact we argue that we can always choose two feed-links such that the dilation is at most $3 + \sqrt{3} \approx 4.73$. This bound is not far from the optimum, since an equilateral triangle with p placed in the center has dilation at least $2 + \sqrt{3} \approx 3.73$ for any two feed-links. To see that, observe that one of the sides of the equilateral triangle does not have a feed-link attached to it (or only at a vertex), which causes the middle of that side to have dilation at least $2 + \sqrt{3}$.

Let q be the closest point to p on P. We choose $\overline{pq}$ as the first feed-link and argue that the dilation is now constant for all points in some part of P which includes q. Then we show how to place the second feed-link to guarantee constant dilation for the remaining part of P. Consider the smallest equilateral triangle Δ that contains P and which is oriented in such a way, that one of its edges contains q. Let e_0 be the edge of Δ containing q, and let e_1 and e_2 be the other edges, in clockwise order from e_0 (see Fig. 4). By construction, each edge of Δ is in contact with P. Let t_1 be a point of P in contact with e_1, and let t_2 be a point of P in contact with e_2.

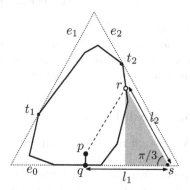

Fig. 4. The smallest equilateral triangle that contains P

Lemma 1. *For any point* $r \in P[t_2, t_1]$, $\delta_q(r) \leq 3 + \sqrt{3}$.

We prove Lemma 1 by arguing that $\mu(r, q) \leq l_1 + l_2$. The details can be found in the full paper. The second feed-link connects p to the point q' on $P[t_1, t_2]$ closest to p. Lemma 2 can be proven with similar arguments as Lemma 1.

Lemma 2. *For any point* $r \in P[t_1, t_2]$, $\delta_{q'}(r) \leq 3 + \sqrt{3}$.

These two lemmas jointly imply

Theorem 3. *Given the boundary P of a convex polygon and a point p inside it, two feed-links from p to P are sufficient to achieve a dilation of $3 + \sqrt{3}$.*

We now consider the general setting of placing k feed-links, where k is a constant. We prove that placing the feed-links at an equal angular distance of $\eta = 2\pi/k$ guarantees a dilation of $1 + O(1/k)$. To simplify the argument we choose $k \geq 6$ (the result for smaller k immediately follows from the result for two feed-links). Our proof uses the following lemma.

Lemma 3. *Let q_1 and q_2 be two points on the boundary P of a convex polygon such that the angle $\angle q_1 p q_2 = \eta \leq \pi/3$. Then for all points $r \in P[q_1, q_2]$, we have $\delta(r)/\max(\delta(q_1), \delta(q_2)) \leq 1 + \eta$.*

Note that $\overline{pq_1}$ and $\overline{pq_2}$ need not be feed-links in Lemma 3. The lemma implies that for $\eta = 2\pi/k$ we obtain the following result.

Theorem 4. *Given the boundary P of a convex polygon and a point p inside it, k feed-links from p to P are sufficient to achieve a dilation of $1 + O(1/k)$.*

Approximation algorithm for convex polygons. We can use Lemma 3 to obtain a linear-time $(1 + \varepsilon)$-approximation algorithm to place one feed-link optimally. We measure dilation only at $2\pi/\varepsilon$ points on P, and hence the running time of the approximation algorithm is $O(n + (1/\varepsilon)\log(1/\varepsilon))$ by Theorem 1. The points at which we measure the dilation are chosen on P such that the angle between two consecutive points measured at p is ε. Since Lemma 3 bounds the dilation between two consecutive points, the theorem follows.

Theorem 5. *For any $\varepsilon > 0$, given the boundary P of a convex polygon with n vertices and a point p inside it, we can compute a feed-link that approximately minimizes the maximum dilation from p to any point on P within a factor $1 + \varepsilon$ in $O(n + (1/\varepsilon)\log(1/\varepsilon))$ time.*

3.2   Realistic Polygons

A constant number of feed-links should guarantee constant dilation for *realistic* polygons. Therefore, we define a simple polygon to be *feed-link realistic* if there are two constants $\delta > 1$ and $c \geq 1$, such that there exist c feed-links that achieve a dilation of at most δ for any point on its boundary. Many different definitions of realistic polygons exist in the literature. We show that most of them do not imply feed-link realism. However, we also argue that polygons that are (α, β)-covered [8] are feed-link realistic.

Consider the left polygon in Fig. 5. At least c feed-links are required to obtain a dilation smaller than δ, if the number of prongs is c and their length is at least

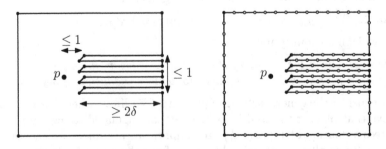

Fig. 5. A β-fat polygon (left) and an adaptation (right) that require many feed-links

δ times larger than the distance of their leftmost vertex to p. No feed-link can give a dilation at most δ for the leftmost vertex of more than one dent. However, the polygon is β-fat [5]. Definitions that depend on the spacing between the vertices or edge-vertex distances will also not give feed-link realism, because the left polygon in Fig. 5 can be turned into a realistic polygon according to such definitions. We simply add extra vertices on the edges to get the right polygon: it has edge lengths that differ by a factor of at most 2, it has no vertex close to an edge in relation to the length of that edge, and it has no sharp angles. The extra vertices obviously have no effect on the dilation. This shows that definitions like low density (of the edges) [17], unclutteredness (of the edges) [4,5], locality [9], and another fatness definition [18] cannot imply feed-link realism.

(α, β)-covered polygons. For an angle ϕ and a distance d, a (ϕ, d)-*triangle* is a triangle with all angles at least ϕ and all edge lengths at least d. Let P be the boundary of a simple polygon, let $\mathrm{diam}(P)$ be the diameter of P, and let $0 < \alpha < \pi/3$ and $0 < \beta < 1$ be two constants. P is (α, β)-*covered* if for each point on P, an $(\alpha, \beta \cdot \mathrm{diam}(P))$-triangle exists with a vertex at that point, whose interior is completely inside P [8]. Furthermore, P is (α, β)-*immersed* if for each point on P there is such a triangle completely inside P and one completely outside P. For ease of description, we assume that $\mathrm{diam}(P) = 1$.

We use a result by Bose and Dujmović [3] that bounds the perimeter of P as a function of α and β.

Lemma 4. *The perimeter of an (α, β)-covered polygon P is at most $\frac{c}{\beta \sin \alpha}$, for some absolute constant $c > 0$.*

Also, we need a technical lemma that states that if the distance between two points on P is short enough, then it is proportional to the Euclidean distance.

Lemma 5. *If p and q can see each other on the inside of an (α, β)-covered polygon P and $\mu(p, q) < \beta$, then $\mu(p, q) < f(\alpha) \cdot |pq|$, where $f(\alpha) \leq \frac{2\pi}{\alpha \sin \frac{1}{4}\alpha}$.*

When P is (α, β)-immersed, each point on the boundary has an empty (α, β)-triangle outside P as well as inside P. This implies that the lemma also holds for two points p and q that can see each other on the *outside* of the polygon.

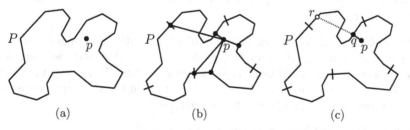

Fig. 6. (a) A polygon P that is (α, β)-immersed. (b) A feed-link to the closest point on each boundary piece of length β. (c) The dilation of r is constant, because the boundary distance between r and q is bounded by their Euclidean distance.

Theorem 6. *When P is (α, β)-immersed, we can place $\frac{c}{\beta^2 \sin \alpha}$ feed-links such that the dilation of every point on P is at most $1 + \frac{4\pi}{\alpha \sin \frac{1}{4}\alpha}$.*

Proof. We give a constructive proof. Given an (α, β)-immersed polygon and a point p inside it, we split P into pieces of length β. By Lemma 4 there are only $\frac{c}{\beta^2 \sin \alpha}$ pieces. On each piece, we place a feed-link to the closest point to p. Fig. 6(b) shows the resulting feed-links in an example.

For each point r on P, we show that the dilation is constant. Consider the piece of P containing r and the point q that is the closest point to p on that piece, as in Fig. 6(c). The segment $\overline{qr}$ may intersect P in a number of points. For each pair of consecutive intersection points, they can see each other either inside or outside P. Since P is (α, β)-immersed, Lemma 5 applies to each pair, and hence $\mu(q, r) \leq f(\alpha) \cdot |qr|$. Also, we know that $|pq| \leq |pr|$. We conclude that the dilation is bounded by

$$
\begin{aligned}
\delta_q(r) = \frac{|pq| + \mu(q, r)}{|pr|} &\leq \frac{|pq| + f(\alpha)|qr|}{|pr|} \\
&\leq \frac{|pq| + f(\alpha)(|pr| + |pq|)}{|pr|} \leq \frac{|pr| + f(\alpha)(|pr| + |pr|)}{|pr|} = 1 + 2f(\alpha).
\end{aligned}
$$

$\square$

When P is (α, β)-covered but not (α, β)-immersed, the proof no longer works since there can be two points that see each other outside the polygon, in which case Lemma 5 does not hold. However, we can still prove that (α, β)-covered polygons are feed-link realistic, albeit with a different dependence on α and β.

Let $C = \frac{4\pi c}{\beta^2 \alpha \sin \alpha \sin \frac{1}{2}\alpha}$ be a constant (depending on α and β). We incrementally place feed-links until the dilation is at most C everywhere. In particular, after placing the first i feed-links, consider the set of points on the boundary of P that have dilation worse than C. If q_{i+1} is the point of this set that is closest to p, then we let the next feed-link be $\overline{pq_{i+1}}$.

We now need to prove that this results in a constant number of feed-links. So, say we placed k feed-links this way, and let their points be $q_1 \ldots q_k$. Obviously, we have $|pq_i| \leq |pq_j|$ if $i < j$.

Lemma 6. *Unless $k = 1$, all points q_i are inside the circle D centered at p of radius $R = \frac{1}{2}\beta \sin \frac{1}{2}\alpha$.*

Inside the circle D, there cannot be edges of P of length β or longer. So, each point q_i has an empty (α, β)-triangle t_i with one corner at q_i and the other two corners outside D. Fig. 7(a) illustrates the situation, where the grey part is inside P. Let d_i be the direction of the bisector of t_i at q_i. In the full paper we prove that two directions d_i and d_j differ by at least $\frac{1}{2}\alpha$.

Lemma 7. *The angle between d_i and d_j is at least $\frac{1}{2}\alpha$.*

Theorem 7. *Given the boundary P of an (α, β)-covered polygon and a point p inside it, $\frac{4\pi}{\alpha}$ feed-links are sufficient to achieve a dilation of $\frac{4\pi c}{\beta^2 \alpha \sin \alpha \sin \frac{1}{2}\alpha}$.*

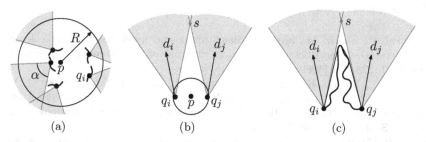

(a)   (b)   (c)

Fig. 7. (a) A circle around p of radius R contains points q_i such that $\overline{pq_i}$ is a feed-link. (b) If the angle between two bisecting directions d_i and d_j is small, the (α, β)-triangles intersect in s. (c) The boundary length between q_i and q_j cannot be too long.

Proof. We place feed-links incrementally as described, until all points on P have dilation at most C. By Lemma 7 there cannot be more than $\frac{4\pi}{\alpha}$ feed-links, because otherwise some pair q_i and q_j would have (α, β)-triangles with directions d_i and d_j whose angle is smaller than $\frac{1}{2}\alpha$.   □

4   Conclusions

We presented an efficient algorithm to compute an optimal feed-link for the boundary of a simple polygon and a point inside it. Furthermore, we showed that two feed-links are sometimes necessary and always sufficient to guarantee constant dilation for convex polygons; by placing k feed-links, we can even guarantee a dilation of at most $1 + O(1/k)$. Finally, we considered the number of feed-links necessary for realistic polygons, and proved that (α, β)-covered polygons require only a constant number of feed-links for constant dilation. For other definitions of realistic polygons such a result does provably not hold.

It is open whether the optimal feed-link can be placed in $O(n \log n)$ time or even faster. It is also open whether a linear-time, $(1+\varepsilon)$-approximation algorithm exists for computing an optimal feed-link in a simple polygon (we proved this only for convex polygons).

A number of interesting and challenging extensions of our work are possible. Firstly, the optimal placement of more than one feed-link seems difficult. Secondly, we did not consider the situation where several points lie inside P and need to be connected via feed-links. Here we may or may not want to allow one feed-link to connect to another feed-link. Thirdly, assume we are given an incomplete road network N and several locations, which might fall into different faces of the graph induced by N. How should we place feed-links optimally?

Acknowledgements. We thank Tom de Jong for introducing us to this problem and Mark de Berg for pointing us to the results in [3].

References

1. Agarwal, P., Sharir, M., Shor, P.: Sharp upper and lower bounds on the length of general Davenport-Schinzel sequences. Journal of Combinatorial Theory, Series A 52, 228–274 (1989)
2. Aronov, B., Buchin, K., Buchin, M., Jansen, B., de Jong, T., van Kreveld, M., Löffler, M., Luo, J., Silveira, R.I., Speckmann, B.: Feed-links for network extensions. In: Proc. 16th International Symposium on Advances in Geographic Information Systems, pp. 308–316 (2008)
3. Bose, P., Dujmović, V.: A note on the perimeter of (α, β)-covered objects (manuscript, 2009)
4. de Berg, M.: Linear size binary space partitions for uncluttered scenes. Algorithmica 28, 353–366 (2000)
5. de Berg, M., van der Stappen, A., Vleugels, J., Katz, M.: Realistic input models for geometric algorithms. Algorithmica 34, 81–97 (2002)
6. de Jong, T., Tillema, T.: Transport network extensions for accessibility analysis in geographic information systems. In: Proc. AfricaGIS (2005)
7. Ebbers-Baumann, A., Grüne, A., Klein, R.: The geometric dilation of finite point sets. Algorithmica 44, 137–149 (2006)
8. Efrat, A.: The complexity of the union of (α, β)-covered objects. SIAM Journal on Computing 34, 775–787 (2005)
9. Erickson, J.: Local polyhedra and geometric graphs. Computational Geometry: Theory and Applications 31, 101–125 (2005)
10. Farshi, M., Giannopoulos, P., Gudmundsson, J.: Finding the best shortcut in a geometric network. In: Proc. 21st Symposium on Computational Geometry, pp. 327–335 (2005)
11. Guibas, L.J., Hershberger, J., Leven, D., Sharir, M., Tarjan, R.E.: Linear-time algorithms for visibility and shortest path problems inside triangulated simple polygons. Algorithmica 2, 209–233 (1987)
12. Hershberger, J.: Finding the upper envelope of n line segments in $O(n \log n)$ time. Information Processing Letters 33, 169–174 (1989)
13. Langerman, S., Morin, P., Soss, M.: Computing the maximum detour and spanning ratio of planar paths, trees, and cycles. In: Alt, H., Ferreira, A. (eds.) STACS 2002. LNCS, vol. 2285, pp. 250–261. Springer, Heidelberg (2002)
14. Narasimhan, G., Smid, M.: Approximating the stretch factor of Euclidean graphs. SIAM Journal on Computing 30, 978–989 (2000)
15. Narasimhan, G., Smid, M.: Geometric Spanner Networks. Cambridge University Press, Cambridge (2007)
16. Sharir, M., Agarwal, P.: Davenport-Schinzel Sequences and Their Geometric Applications. Cambridge University Press, Cambridge (1995)
17. van der Stappen, A.F., Overmars, M.H., de Berg, M., Vleugels, J.: Motion planning in environments with low obstacle density. Discrete & Computational Geometry 20, 561–587 (1998)
18. van Kreveld, M.: On fat partitioning, fat covering and the union size of polygons. Computational Geometry: Theory and Applications 9, 197–210 (1998)

Minimal Locked Trees

Brad Ballinger[1], David Charlton[2], Erik D. Demaine[3,*], Martin L. Demaine[3],
John Iacono[4], Ching-Hao Liu[5,**], and Sheung-Hung Poon[5,**]

[1] Davis School for Independent Study, 526 B Street, Davis, CA 95616, USA
ballingerbrad@yahoo.com
[2] Boston University Computer Science, 111 Cummington Street,
Boston, MA 02135, USA
dcharlton@gmail.com
[3] MIT Computer Science and Artificial Intelligence Laboratory, 32 Vassar Street,
Cambridge, MA 02139, USA
{edemaine,mdemaine}@mit.edu
[4] Department of Computer Science and Engineering, Polytechnic Institute of NYU,
5 MetroTech Center, Brooklyn NY 11201, USA
http://john.poly.edu
[5] Department of Computer Science, National Tsing Hua University,
Hsinchu, Taiwan
chinghao.liu@gmail.com, spoon@cs.nthu.edu.tw

Abstract. Locked tree linkages have been known to exist in the plane
since 1998, but it is still open whether they have a polynomial-time characterization. This paper examines the properties needed for planar trees
to lock, with a focus on finding the smallest locked trees according to
different measures of complexity, and suggests some new avenues of research for the problem of algorithmic characterization. First we present
a locked linear tree with only eight edges. In contrast, the smallest previous locked tree has 15 edges. We further show minimality by proving
that every locked linear tree has at least eight edges. We also show that
a six-edge tree can interlock with a four-edge chain, which is the first
locking result for individually unlocked trees. Next we present several
new examples of locked trees with varying minimality results. Finally,
we provide counterexamples to two conjectures of [12], [13] by showing
the existence of two new types of locked tree: a locked orthogonal tree
(all edges horizontal and vertical) and a locked equilateral tree (all edges
unit length).

1 Introduction

A *locked tree* is a tree graph (linkage) embedded in the plane that is unable
to reconfigure to some other configuration if we treat the edges as rigid bars
that cannot intersect each other. The idea of locked trees goes back to 1997, in

---

* Partially supported by NSF CAREER award CCF-0347776.
** Supported by grant 97-2221-E-007-054-MY3 of the National Science Council (NSC),
Taiwan, R.O.C.

F. Dehne et al. (Eds.): WADS 2009, LNCS 5664, pp. 61–73, 2009.
© Springer-Verlag Berlin Heidelberg 2009

the context of an origami problem [8]. Only four main families of locked trees have been discovered so far. The first two locked trees, shown in Figure 1(a–b), were discovered soon after in 1998 [3]. In 2000, it was established that locked trees must have vertices of degree more than 2 (the Carpenter's Rule Theorem) [6,14]. The third locked tree, shown in Figure 1c, shows that this result is tight: a single degree-3 vertex suffices to lock a tree [5]. The fourth locked tree, shown in Figure 1d, modified the first locked tree to reduce its graph diameter to 4, which is the smallest of any locked tree [12].

All four trees have a similar structure: they arrange repeated pieces in a cycle so that no piece can individually squeeze and so that no piece can individually expand without squeezing the other pieces (which in turn is impossible). Do all locked trees have this structure? This paper aims to find minimal examples of locked trees, with the goal of finding the "heart" of being locked. In particular we find smaller locked trees that lack the cyclic structure of previous examples.

It seems difficult to characterize locked trees. Toward this goal, some types of trees are easy to prove locked via recent algorithmic tools [5,4], and we use this theory extensively here. On the other hand, deciding whether a tree link-age can be transformed from one configuration to another is PSPACE-complete [2]. However, this hardness result says nothing about the special case of testing whether a tree is locked. In the sections that follow, we describe several new examples and counterexamples in locked trees, and suggest ways in which they may hint at deeper results in the associated algorithmic theory.

Our results. We discover several new families of locked trees with several previously unobtained properties. We also introduce a new general category of locked tree, the *linear* locked tree, which in addition to being important for the study of locked linkages also provides an interesting special case for the algorithmic characterization of lockedness.

First, in Section 3, we present a locked tree with only eight edges. In contrast, the smallest previous locked tree is Figure 1a with 15 edges. Our tree is also the only locked tree other than Figure 1c that has just one degree-3 vertex (and the other degrees at most 2). Therefore we improve the number of edges in the smallest such tree from 21 to eight.

Our tree has the additional property that it is *linear*: its vertices lie (roughly) along a line (see the full definition in Section 3). In Section 4, we prove that

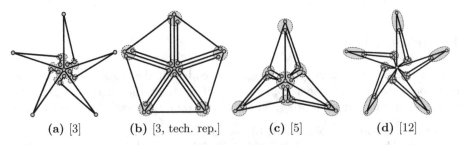

(a) [3]   **(b)** [3, tech. rep.]   **(c)** [5]   **(d)** [12]

Fig. 1. All previous families of locked trees. Shaded regions are tighter than drawn

all linear locked trees have at least eight edges, establishing minimality of our eight-edge locked tree. We conjecture further that all locked trees have at least eight edges, though this problem remains open.

We also show in Section 4 that all linear locked trees have diameter at least 5. In Section 5, we find a linear locked tree of precisely this diameter, using nine edges, and further show that this is the smallest number of edges possible for a diameter-5 linear locked tree. In contrast, the (nonlinear) locked tree in Figure 1d has diameter 4, while no locked trees have diameter 3 [12].

Next we consider interlocked trees, in the spirit of interlocked 3D chains [9,10,11]. In Section 6, we show that though diameter 3 trees cannot lock, they can interlock. (In contrast, any number of diameter-2 trees cannot interlock, as they are star-shaped [7,15,10].) As a consequence, caterpillar trees, which generalize diameter-3 trees, can lock. Additionally, we prove for the first time that smaller trees suffice for interlocking: a six-edge tree can interlock with a four-edge chain.

Finally we solve two conjectures about the existence of locked trees with particular properties. On the easier side, we show in Section 7 that certain linear locked trees, such as our eight-edge locked tree, can be transformed to obtain locked orthogonal trees. Such trees were previously conjectured not to exist [13] because all examples in Figure 1 critically use angles strictly less than 90°.

Our technically most challenging result is the design of a locked equilateral tree, where every edge has the same length. The hexagonal analog of Figure 1b is tantalizingly close to this goal, as the edges can have lengths arbitrarily close to each other. But if the tree is to not overlap itself, the lengths cannot be made equal. For this reason, equilateral locked trees were conjectured not to exist [12]. Nonetheless, in Section 8, we find one. This result is quite challenging because previous algorithmic frameworks were unable to analyze the lockedness of trees with fixed edge lengths. Specifically, where previous locked trees were very tightly locked (within an arbitrarily small constant), our locked equilateral tree has fairly large positive gaps between edges, forcing us to carefully compute the freedom of motion instead of simply using topological limiting arguments.

2   Terminology

A *(planar) linkage* is a simple graph together with an assignment of a nonnegative real length to each edge and a combinatorial planar embedding (clockwise order of edges around each vertex and which edges form the outer face). A *configuration* of a linkage is a (possibly self-intersecting) straight-line drawing of that graph in the plane, respecting the linkage's combinatorial embedding, such that the Euclidean distance between adjacent nodes equals the length assigned to their shared edge.

We are primarily interested in *nontouching* configurations, that is, configurations in which no edges intersect each other except at a shared vertex. The set of all such configurations is called the *configuration space* of the linkage. A *motion* of a nontouching configuration C is a continuous path in the configuration space beginning at C. A configuration of a tree linkage can be *flattened*

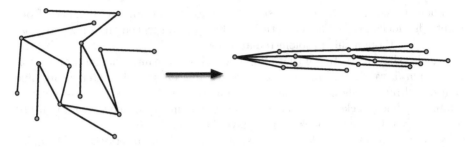

Fig. 2. Flattening a linkage: the initial tree (left) can be continuously transformed into the "flat" tree on the right, with all edges trailing off in the same direction from the root

if it has a motion transforming it as in Figure 2 so that all edges are trailing off in the same direction from an arbitrarily chosen root node. Otherwise, it is *unflattenable*. (Which node is chosen as the root does not affect the definition; see [3].) We say a tree configuration is *locked* if it is unflattenable, and a tree linkage is locked if it has a locked configuration.

To analyze nontouching configurations it is helpful to also consider *self-touching* configurations, where edges may overlap as long as they do not cross each other. This complicates the definitions, because edges can share the same geometric location. Care is also needed in generalizing the definition of a motion, because two geometrically identical configurations may have different sets of valid motions depending on the combinatorial ordering of the edges. A full discussion of these details is beyond our scope, so we rely on the formalization and results of [5], [4] and [1]. The reader who wishes for the intuition behind this theory can think of a self-touching configuration as a convergent sequence of nontouching configurations, but for the formal definitions see the references.

A self-touching configuration is *rigid* if it has no nonrigid motion. A configuration is *locked within ε* if no motion can change the position of any vertex by a distance of more than ε (modulo equivalence by rigid motions). A configuration C is *strongly locked* if, for any $\varepsilon > 0$, for any sufficiently small perturbation of C's vertices (respecting the original combinatorial relations between edges—see [5]), the resulting perturbed configuration is locked within ε. This property trivially implies unflattenability, and thus also that the underlying linkage is locked. Note that strongly locked configurations must be rigid and thus self-touching.

3   Minimal Locked Linear Tree

In this section we describe a new locked tree that is edge-minimal within an important class of locked trees, and is conjectured to be edge-minimal among all locked trees. Namely, a *linear configuration* is a (usually self-touching) configuration of a linkage in which all vertices lie on a single line. A *locked linear tree* is a tree linkage having an unflattenable linear configuration.

Note that our primary interest is still in nontouching configurations, but the existence of a linear configuration has implications for the general configuration space of the tree. Specifically, we make extensive use of the following lemma:

Lemma 1 (Theorem 8.1 from [5]). *Any rigid self-touching configuration is strongly locked.*

Because our proofs proceed by showing our linear trees rigid, this result implies that they remain locked even when the parameters are modified slightly to allow a nontouching configuration.

Consider the self-touching tree in Figure 3a. The linear geometry of this tree is a straight vertical line with only three distinct vertices at the top, center and bottom, but it is shown "pulled apart" to ease exposition. We claim this tree is rigid and thus, by Lemma 1, strongly locked. To show rigidity, we use two lemmas from [4]:

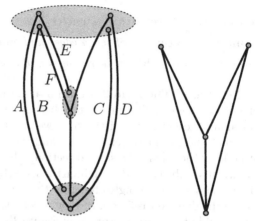

(a) Eight-bar locked linear tree. Shaded regions are tighter than drawn. All edges are straight lines, but shown "pulled apart".

(b) Reduced version of the tree after applying Lemma 2 and Lemma 3.

Fig. 3. The fewest-edge locked linear tree

Lemma 2 (Rule 1 from [4]). *If a bar b is collocated with another bar b′ of equal length, and two other bars incident to b′ on each end form angles less than 90° on the same side as b, then any motion must keep b collocated with b′ for some positive time.*

Lemma 3 (Rule 2 from [4]). *If a bar b is collocated with an incident bar b′ of the same length whose other incident bar b″ forms a convex angle with b′ surrounding b, then any motion must keep b collocated with b′ for some positive time.*

Theorem 1. *The tree in Figure 3a is strongly locked.*

Proof: By Lemma 2, edges A and B must be collocated for positive time under any continuous motion, as must edges C and D. With these identifications, Lemma 3 shows that edges E and F must also remain collocated. We conclude that for positive time, the tree is equivalent to Figure 3b, which is trivially rigid. Therefore, the original tree is rigid and, by Lemma 1, strongly locked.   □

4   Unfolding Linear Trees of Seven Edges

In Section 3, we presented a linear locked tree with eight edges. Now we will show that this is minimal: linear trees with at most seven edges can always be flattened. Because the full proof requires an extensive case analysis, we defer this to the full paper, and here present a sketch of how our arguments exploit the linearity of a tree.

Theorem 2. *A linear tree of diameter 4 can always be flattened.*

Lemma 4. *A linear tree of seven edges and diameter 5 can always be flattened.*

Lemma 5. *A linear tree of seven edges and diameter 6 can always be flattened.*

Proof sketch of Theorem 2, Lemma 4, Lemma 5: Because the tree's initial configuration lies on a line, many steps become simpler: first, if there are any loose edges along the perimeter of the tree, we can immediately straighten these. In Theorem 2, the tree has a center node, and we can then pivot all subtrees around that node so they lie in the same direction. This allows us to sequentially rotate out individual subtrees and straighten them one by one (a case analysis shows that if distinct subtrees are tangled together they can be safely pulled apart).

When the diameter is 5 or 6, the key observation is that the constraints do not allow a double-triangle structure as in Figure 3b. Specifically, case analysis shows the center edge cannot be formed, and thus the bounding quadrilateral can be expanded. When this quadrilateral becomes convex, the tree pulls apart easily. □

Because it was already shown in [6] that a seven-edge, diameter-7 tree (i.e., a 7-chain) cannot lock, combining these results immediately gives us the following:

Theorem 3. *A linear tree of at most seven edges can always be flattened.*

We thus conclude that the linear locked tree in Figure 3a has the fewest possible edges.

5   Additional Locked Linear Trees

Theorem 4. *The trees in Figure 4 are strongly locked.*

Like Theorem 1, all these theorems are proven by repeatedly applying Lemmas 2 and 3 until the configuration simplifies to Figure 3b, after which Lemma 1 applies.

By a slight extension to the results of Section 4, we can prove the minimality of Figure 4a in a second sense, by showing that any diameter-5 linear locked tree requires at least nine edges:

Theorem 5. *A linear tree of 8 edges and of diameter 5 can always be flattened.*

This claim is nearly implicit in the proof of Lemma 4; see the full paper.

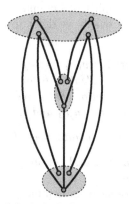 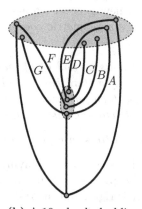 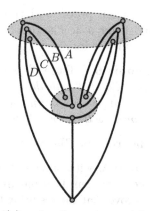

(a) A locked tree having nine edges and the lowest diameter (5) of any possible locked linear tree.

(b) A 10-edge locked linear tree with a somewhat different structure. Edge labels appear to the right of their respective edge.

(c) Another symmetric locked linear tree, this time with 11 edges. Edge labels appear to the right of their respective edge.

Fig. 4. Additional locked linear trees

6 Interlocked Trees

6.1 Diameter-3 Interlocked Trees

In this section we describe a set of eight interlocked trees of diameter 3 (although four of the "trees" are in fact 2-chains). Because diameter-2 trees cannot interlock (as they are star-shaped [7,15,10]), this example is tight. Because diameter-3 trees cannot lock, this is also the first example showing that the required diameter for interlocked (planar) trees is strictly below that of locked trees.

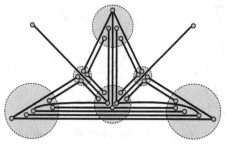   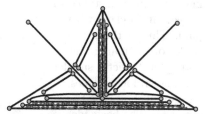

(a) Interlocked configuration (shaded regions are self-touching or very close).

(b) Identifications obtained from Lemma 2 and Lemma 3 (darkened areas indicate edges glued together).

Fig. 5. Eight interlocked diameter-3 trees

For our proof, we introduce a new general lemma in the spirit of Lemma 2 and Lemma 3. The proof requires a geometric computation which we defer to the full version.

Lemma 6 ("Rule 3"). *If endpoints v_1 and v_3 of incident bars v_1v_2 and v_2v_3 are collocated with the endpoints of a third bar b, and bars incident to b form acute angles containing v_1 and v_3, then for positive time, any motion that moves v_1 or v_3 with respect to b must strictly increase the distance between v_2 and b.*

Theorem 6. *The eight diameter-3 trees in Figure 5a are strongly (inter)locked.*

Proof: As with previous examples we begin by applying Lemma 2 and Lemma 3. The edge identifications from this process are shown in Figure 5b. It is enough to prove that the resulting figure is rigid, and the rest will follow from Lemma 1.

Now, observe that the 2-chains inside each of the four regions of the figure satisfy the requirements of Lemma 6, and that therefore the long diagonal edges are rigid: any rotation on their part would decrease the angular space allocated to some region, and push the center of the corresponding 2-chain closer to the opposing edge, contradicting the lemma. But then Lemma 6 implies the 2-chains themselves are glued in place for positive time.

The preceding leaves only the four loose edges around the outside. But because the 2-chains glue to their base vertices and to the long diagonals, Lemma 3 now applies, so these edges too are locked in place.   □

6.2   Six-Edge Interlocked Tree

Here we describe a simple transformation that applies to many locked linear trees, yielding a smaller tree interlocked with a chain. Applying this transformation to Figure 4a, we obtain the smallest known instance of a tree interlocked with a chain. This is the first example of a planar interlocking tree strictly smaller than known locked trees.

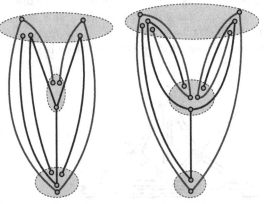

(a) Interlocked version of Figure 4a.   (b) Interlocked version of Figure 4c.

Fig. 6. Interlocked variations of our locked trees

Figure 6 shows the transformation. The basic idea is to disconnect a subtree at one end of the tree, replacing the connection with an extra edge that serves the same purpose, that is, such that the edge also constrains the subtree to remain adjacent to the same node. In Figure 6a, this gives us a 6-edge tree interlocked with a 4-edge chain, the best known.

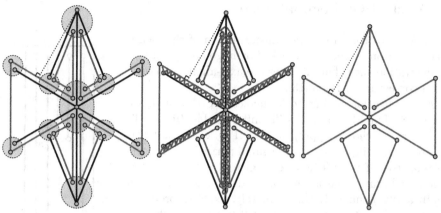

(a) Interlocked caterpillar (black) and two 9-chains.

(b) Identifications obtained from Lemma 2.

(c) Simplified linkage after identifications and removal of extraneous bars.

Fig. 7. A locked caterpillar

Theorem 7. *The configurations in Figure 6a and Figure 6b are interlocked.*

As with their locked predecessors, successive applications of Lemma 2 and Lemma 3 suffice to prove rigidity of the configurations; see the full paper.

6.3   (Inter)Locked Caterpillar

Here we describe an interesting example originally inspired by the search for the diameter-3 interlocked trees of Section 6.1. A *caterpillar graph* is a graph where removal of all leaf vertices and their incident edges results in a path. Because every vertex is at most one edge away from the central chain, the leaf vertices form the "legs" of the central chain "body" of the caterpillar. The intuition is of a graph that is locally low-diameter, or almost chain-like. Caterpillars provide a natural intermediate structure between collections of diameter-3 graphs and the full power of diameter 4, which was already known to lock.

Because a caterpillar can take the place of any number of diameter-3 graphs, we can implicitly obtain a locked caterpillar directly from Figure 5a. However, in this section we describe a much simpler structure, and one that can be realized as the interlocking of a single ten-edge caterpillar and two 9-chains (or one 22-chain). We can also produce a single locking (not interlocking) caterpillar by merging the separate chains into the main body of the caterpillar.

Theorem 8. *The configuration in Figure 7a is rigid, and therefore strongly locked.*

This claim follows from successive applications of Lemma 2, Lemma 3 and Lemma 6, similar to Theorem 6.

7 Locked Orthogonal Tree

We now show that a simple transformation of the locked
tree in Section 3 produces a locked orthogonal tree (a tree
configuration such that all edges are axis-aligned), resolving a conjecture of Poon [13].

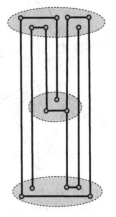

A modification to Figure 3a makes it orthogonal: see
Figure 8. This diagram is still unflattenable (if the dimensions are chosen appropriately). The key is that this
diagram can still be viewed as a small perturbation of the
original tree if we add a zero-length edge to Figure 3a
wherever Figure 8 has a horizontal edge, and thus we can
again apply Lemma 1. Unfortunately, existing proofs of
Lemma 1 do not work when the self-touching configuration has zero-length edges. It is a straightforward but technical matter to extend the lemma in this way. We defer
the formal details to the full version.

Fig. 8. Orthogonal
version of Figure 3a

8 Locked Equilateral Tree

In [13], Poon conjectured that an equilateral tree (a tree linkage all of whose
edges are equal length) could not lock. We provide a counterexample, shown in
Figure 9a. This follows the "pinwheel" style of previous locked trees (Figure 1).
The difference is that all previous locked trees (including the other examples in
this paper) select their edge lengths so as to obtain an infinitesimally tight fit,
whereas with unit edges we are limited to explicit numerical constraints.

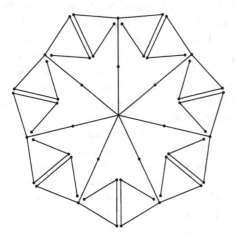

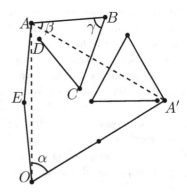

(a) The locked unit tree, having seven arms
of radius 2.

(b) Close-up of two adjacent arms.
Roman letters refer to vertices,
Greek to angles.

Fig. 9. A locked unit tree

Theorem 9. *The tree in Figure 9a is locked.*

Proof sketch: To prove lockedness in the absence of machinery derived from rigidity theory, we consider the angles and vertices labelled in Figure 9b. We claim that, under any continuous motion, the following invariants hold:

$$\sqrt{3} \leq \|A - A'\| \leq \sqrt{3} + 0.025 \quad (1) \qquad \left|\beta - \tfrac{\pi}{6}\right| \leq 0.078\pi \quad (4)$$

$$1.94 \leq \|A - O\| \quad (2) \qquad \tfrac{\pi}{3} \leq \gamma \leq \pi\left(\tfrac{1}{3} + 0.02\right) \quad (5)$$

$$0.2850989\pi \leq \alpha \leq 0.28941\pi \quad (3) \qquad \|C - AE\| \leq 0.386 \quad (6)$$

We do so by showing that, if these inequalities hold for some valid configuration, then they actually hold strictly, that is, every instance of $\leq$ above is actually $<$. Thus, these properties are preserved by any continuous motion. Due to space constraints, we give here the proofs for Equation 1 and Equation 3.

Consider Equation 3. The minimal α is attained when A and A' are at minimum distance from each other and maximum distance from the center vertex O. The latter is trivially 2, and the former is $\sqrt{3}$ by Equation 1. The angle so obtained is $2\arcsin(\frac{\sqrt{3}}{2\cdot2}) > 0.2850989\pi$, as required. On the other hand, there are seven arms in the tree, and by the preceding that leaves $< 2\pi - 7(0.2850989)\pi$ of angular free space, and even if one pair of arms uses all of it, we still have $\alpha < 2\pi - 6(0.2850989)\pi < 0.28941\pi$, so Equation 3 holds strictly.

Now consider the distance $\|A - A'\|$ between two adjacent arms. If we look at the line from A to A', (4) and (5) show that it must pass through edge BC (and, by symmetry, that of the neighbor arm). The distance from A to BC is least when the three vertices form an equilateral triangle, in which case it is $\sqrt{3}/2$. Because this is true for both arms, and because the tree is not self-touching, the true distance between A and A' must be strictly greater than $\sqrt{3}$. On the other hand, by (3) the maximum angular distance between A and A' is 0.28941π. Given this, $\|A - A'\|$ is maximized when both vertices are at distance 2 from the center (it could be higher if one of the vertices approached the center very closely, but (2) prevents this; see below). In this case, the distance between them is $2 \cdot 2\sin(0.28941\pi/2) < \sqrt{3} + 0.25$, proving Equation 1 strict. $\qquad\square$

9   Open Problems

The results of this paper open up several new questions, both in terms of their optimality and in exploring the newly discovered class of locked linear trees.

Figure 3a has eight edges. Is this the smallest possible for a locked (not necessarily linear) tree? We conjecture yes. We believe that a proof along the general outline of Theorem 3 may work, but the case analysis must be arranged more carefully in a general tree.

The orthogonal tree in Figure 8 has 14 edges. Is this minimal? We suspect so. A possible path to proving this conjecture is to show that any smaller orthogonal tree can be projected down to a locked linear tree with fewer than eight edges, contradicting Theorem 3.

Stefan Langerman proposed the idea of interlocked trees by asking whether multiple diameter-3 trees could interlock, which we have shown in Section 6 to be the case. However, this leaves minimality open: four diameter-3 trees can interlock with four 2-chains. Can this be done with fewer trees? Fewer edges?

Our results suggest some more general algorithmic questions. All of our locked trees were reduced to the two triangles of Figure 3b by repeated applications of Lemmas 2 and 3. This may not be a coincidence. Can every rigid linear tree can be reduced to a set of connected triangles by applying these lemmas, or simple extensions of them? In particular, we believe that rigidity in linear trees is a purely combinatorial (rather than geometric) property. Even more generally, is there an efficient algorithm to decide rigidity of linear trees? We suspect so.

In linear trees, there may also be a closer connection between rigidity and lockedness than is (known to be) true in general. If we start with a locked linear tree, and extend its loose edges until they are tightly constrained (and hence possibly satisfy the preconditions for Lemma 2 and Lemma 3), does the resulting graph have a rigid subtree? This is true for all the examples we are aware of, and may provide a starting point for an algorithmic characterization. Analysis of linear trees seems much more feasible than the general case. Is there an efficient algorithm to decide lockedness?

Acknowledgments. This work was initiated during a series of open-problem sessions for an MIT class on Geometric Folding Algorithms (6.885 in Fall 2007), and continued at the 23rd Bellairs Winter Workshop on Computational Geometry organized by Godfried Toussaint (February 2008). We thank the other participants of those sessions and workshops for providing a productive and inspiring environment. In particular, we thank Stefan Langerman for posing the problem of interlocked trees.

References

1. Abbott, T.G., Demaine, E.D., Gassend, B.: A generalized carpenter's rule theorem for self-touching linkages (preprint) (December 2007)
2. Alt, H., Knauer, C., Rote, G., Whitesides, S.: On the complexity of the linkage reconfiguration problem. In: Towards a Theory of Geometric Graphs. Contemporary Mathematics, vol. 342, pp. 1–14. AMS (2004)
3. Biedl, T., Demaine, E.D., Demaine, M.L., Lazard, S., Lubiw, A., O'Rourke, J., Robbins, S., Streinu, I., Toussaint, G., Whitesides, S.: A note on reconfiguring tree linkages: Trees can lock. Discrete Applied Mathematics 117(1–3), 293–297 (2002); The full paper is Technical Report SOCS-00.7, School of Computer Science, McGill University (September 2000) (Originally appeared at CCCG 1998)
4. Connelly, R., Demaine, E.D., Demaine, M.L., Fekete, S., Langerman, S., Mitchell, J.S.B., Ribó, A., Rote, G.: Locked and unlocked chains of planar shapes. In: Proceedings of the 22nd Annual ACM Symposium on Computational Geometry, Sedona, Arizona, June 2006, pp. 61–70 (2006)
5. Connelly, R., Demaine, E.D., Rote, G.: Infinitesimally locked self-touching linkages with applications to locked trees. In: Physical Knots: Knotting, Linking, and Folding of Geometric Objects in $\mathbb{R}^3$, pp. 287–311. AMS (2002)

6. Connelly, R., Demaine, E.D., Rote, G.: Straightening polygonal arcs and convexifying polygonal cycles. Discrete & Computational Geometry 30(2), 205–239 (2003)
7. Dawson, R.: On removing a ball without disturbing the others. Mathematics Magazine 57(1), 27–30 (1984)
8. Demaine, E.D., Demaine, M.L.: Computing extreme origami bases. Technical Report CS-97-22, Dept. of Computer Science, University of Waterloo (May 1997)
9. Demaine, E.D., Langerman, S., O'Rourke, J., Snoeyink, J.: Interlocked open linkages with few joints. In: Proceedings of the 18th Annual ACM Symposium on Computational Geometry, Barcelona, Spain, June 2002, pp. 189–198 (2002)
10. Demaine, E.D., Langerman, S., O'Rourke, J., Snoeyink, J.: Interlocked open and closed linkages with few joints. Computational Geometry: Theory and Applications 26(1), 37–45 (2003)
11. Glass, J., Lu, B., O'Rourke, J., Zhong, J.K.: A 2-chain can interlock with an open 11-chain. Geombinatorics 15(4), 166–176 (2006)
12. Poon, S.-H.: On straightening low-diameter unit trees. In: Healy, P., Nikolov, N.S. (eds.) GD 2005. LNCS, vol. 3843, pp. 519–521. Springer, Heidelberg (2006)
13. Poon, S.-H.: On unfolding lattice polygons/trees and diameter-4 trees. In: Proceedings of the 12th Annual International Computing and Combinatorics Conference, pp. 186–195 (2006)
14. Streinu, I.: Pseudo-triangulations, rigidity and motion planning. Discrete & Computational Geometry 34(4), 587–635 (2005)
15. Toussaint, G.T.: Movable separability of sets. In: Computational Geometry, pp. 335–375. North-Holland, Amsterdam (1985)

Approximating Transitive Reductions for Directed Networks

Piotr Berman[1], Bhaskar DasGupta[2], and Marek Karpinski[3]

[1] Pennsylvania State University, University Park, PA 16802, USA
berman@cse.psu.edu
Research partially done while visiting Dept. of Computer Science, University of Bonn
and supported by DFG grant Bo 56/174-1
[2] University of Illinois at Chicago, Chicago, IL 60607-7053, USA
dasgupta@cs.uic.edu
Supported by NSF grants DBI-0543365, IIS-0612044 and IIS-0346973
[3] University of Bonn, 53117 Bonn, Germany
marek@cs.uni-bonn.de
Supported in part by DFG grants, Procope grant 31022, and Hausdorff Center
research grant EXC59-1

Abstract. We consider *minimum equivalent digraph* problem, its *maximum optimization variant* and some *non-trivial extensions* of these two types of problems motivated by biological and social network applications. We provide $\frac{3}{2}$-approximation algorithms for *all the minimization problems* and 2-approximation algorithms for *all the maximization problems* using appropriate primal-dual polytopes. We also show lower bounds on the integrality gap of the polytope to provide some intuition on the final limit of such approaches. Furthermore, we provide APX-hardness result for all those problems *even if the length of all simple cycles is bounded by* 5.

1 Introduction

Finding an *equivalent digraph* is a classical computational problem (cf. [13]). The statement of the basic problem is simple. For a digraph $G = (V, E)$, we use the notation $u \xrightarrow{E} v$ to indicate that E contains a path from u to v and the *transitive closure* of E is the relation $u \xrightarrow{E} v$ over all pairs of vertices of V. Then, the digraph (V, A) is an equivalent digraph for $G = (V, E)$ if **(a)** $A \subseteq E$ and **(b)** transitive closures of A and E are the same. To formulate the above as an optimization problem, besides the definition of a valid solution we need an objective function. Two versions are considered:

- MIN-ED, in which we minimize $|A|$, and
- MAX-ED, in which we maximize $|E - A|$.

If we skip condition **(a)** we obtain the *transitive reduction* problem which was optimally solved in polynomial time by Aho *et al.* [1]. These names are a bit confusing because one would expect a *reduction* to be a subset and an *equivalent set* to be unrestricted, but transitive reduction was first discussed when the name

F. Dehne et al. (Eds.): WADS 2009, LNCS 5664, pp. 74–85, 2009.

minimum equivalent digraph was already introduced [13]. This could motivate renaming the equivalent digraph as a *strong transitive reduction* [14].

Further applications in biological and social networks have recently introduced the following *non-trivial* extensions of the above basic versions of the problems. Below we introduce these extensions, leaving discussions about their motivations in Section 1.3 and in the references [2,3,5,8].

The first extension is the case when we specify a subset $D \subset E$ of edges which have to be present in every valid solution. It is not difficult to see that this requirement may change the nature of an optimal solution. We call this problem as MIN-TR$_1$ or MAX-TR$_1$ depending on whether we wish to minimize $|A|$ or maximize $|E - A|$, respectively.

A further generalization can be obtained when each edge e has a *character* $\ell(e) \in \mathbb{Z}_2$, where edge characters define the character of a path as the sum modulo 2. In a valid solution we want to have paths with every character that is possible in the full set of edges. This concept than be applied to any group, but our method works only for $\mathbb{Z}_p$ where p is prime. Formally,

① $\ell : E \mapsto \mathbb{Z}_p$;
② a path $P = (u_0, u_1, \ldots, u_k)$ has character $\ell(P) = \sum_{i=1}^{k} \ell(u_{i-1}, u_i) \pmod{p}$;
③ $Closure_\ell(E) = \{(u, v, q) : \exists P \text{ in } E \text{ from } u \text{ to } v \text{ such that } \ell(P) = q\}$;

Then we generalize the notion of "preserving the transitive closure" as follows: (V, A) is a *p-ary transitive reduction* of $G = (V, E)$ with a required subset D if $D \subseteq A \subseteq E$ and $Closure_\ell(A) = Closure_\ell(E)$.

Our two objective functions, namely minimizing $|A|$ or maximizing $|E - A|$, define the two optimization problems MIN-TR$_p$ and MAX-TR$_p$, respectively.

For readers convenience, we indicate the relationships for the various versions below where $A \prec B$ indicates that problem B is a proper generalization of problem A:

$$\text{MIN-ED} \prec \text{MIN-TR}_1 \prec \text{MIN-TR}_p$$
$$\text{MAX-ED} \prec \text{MAX-TR}_1 \prec \text{MAX-TR}_p$$

1.1 Related Earlier Results

The initial work on the minimum equivalent digraph by Moyles and Thomson [13] described an efficient reduction to the case of strongly connected graphs and an exact exponential time algorithm for the latter.

Several approximation algorithms for MIN-ED have been described in the literature, most notably by Khuller *et al.* [11] with an approximation ratio of $1.617 + \varepsilon$ and by Vetta [15] with a claimed approximation ratio of $\frac{3}{2}$. The latter result seems to have some gaps in a correctness proof.

Albert *et al.* [2] showed how to convert an algorithm for MIN-ED with approximation ratio r to an algorithm for MIN-TR$_1$ with approximation ratio $3 - 2/r$. They have also proved a 2-approximation for MIN-TR$_p$. Other heuristics for these problems were investigated in [3,8].

On the hardness side, Papadimitriou [14] indicated that the strong transitive reduction is NP-hard, Khuller *et al.* proved it formally and also showed its APX-hardness. Motivated by their *cycle contraction* method in [11], they were interested in the complexity of the problem when there is an upper bound γ on the cycle length; in [10] they showed that MIN-ED is solvable in polynomial time if $\gamma = 3$, NP-hard if $\gamma = 5$ and MAX-SNP-hard if $\gamma = 17$.

Finally, Frederickson and JàJà [7] provides a 2-approximation for a weighted generalization of MIN-ED based on the works of [6,9].

1.2   Results in This Paper

Table 1 summarizes our results. We briefly discuss the results below.

We first show a 1.5-approximation algorithm for MIN-ED that can be extended for MIN-TR$_1$. Our approach is inspired by the work of Vetta [15], but our combinatorial approach makes a more explicit use of the primal-dual formulation of Edmonds and Karp, and this makes it much easier to justify edge selections within the promised approximation ratio.

Next, we show how to modify that algorithm to provide a 1.5-approximation for MIN-TR$_1$. Notice that *one cannot* use a method for MIN-ED as a "black box" because we need to control which edges we keep and which we delete.

We then design a 2-approximation algorithm MAX-TR$_1$. Simple greedy algorithms that provides a constant approximation for MIN-ED, such as delete an unnecessary edge as long as one exists, would not provide any bounded approximation at all since it is easy to provide an example of MAX-ED instance with n nodes and $2n - 2$ edges in which greedy removes only one edge, and the optimum solution removes $n - 2$ edges. Other known algorithms for MIN-ED are not much better in the worst case when applied to MAX-ED.

Next, we show that for a prime p we can transform a solution of MIN-TR$_1$/MAX-TR$_1$ to a solution of MIN-TR$_p$/MAX-TR$_p$ by a single edge insertion per strongly connected component, thereby obtaining 1.5-approximation

Table 1. Summary of our results. The parameter γ indicates the maximum cycle length and the parameter p is prime. **Our results in a particular row holds for all problems in that row.** We also provide a lower bound of $4/3$ and $3/2$ on the integrality gap of the polytope used in our algorithms for MIN-ED and MAX-ED, respectively (not mentioned in the table).

Problem names	Our results	Previous best (if any)	
		Result	Ref
MIN-ED,MIN-TR$_1$, MIN-TR$_p$	1.5-approx. MAX-SNP-hard for $\gamma = 5$	1.5-approx. for MIN-ED	[15]
		1.78-approx. for MIN-TR$_1$	[2]
		$2 + o(1)$-approx. for MIN-TR$_p$	[2]
		MAX-SNP-hard for $\gamma = 17$	[10]
		NP-hard for $\gamma = 5$	[10]
MAX-ED,MAX-TR$_1$, MAX-TR$_p$	2-approx. MAX-SNP-hard for $\gamma = 5$	NP-hard for $\gamma = 5$	[10]

for MIN-TR$_p$ and a 2-approximation for MAX-TR$_p$ (we can compensate for an insertion of a single edge, so we do not add a $o(1)$ to the approximation ratio).

Finally, We provide an approximation hardness proof for MIN-ED and MAX-ED when γ, the maximum cycle length, is 5. This leaves unresolved only the case of $\gamma = 4$.

1.3 Some Motivations and Applications

Application of MIN-ED*: Connectivity Requirements in Computer Networks.*
Khuller *et al.* [10] indicated applications of MIN-ED to design of computer networks that satisfy given connectivity requirements. With preexisting sets of connections, this application motivates MIN-TR$_1$ (cf. [12]).

Application of MIN-TR$_1$*: Social Network Analysis and Visualization.*
MIN-TR$_1$ can be applied to social network analysis and visualization. For example, Dubois and Cécile [5] applies MIN-TR$_1$ to the publicly available (and famous) social network built upon interaction data from email boxes of Enron corporation to study useful properties (such as scale-freeness) of such networks as well as help in the visualization process. The approach employed in [5] is the straightforward greedy approach which, as we have discussed, has inferior performance, both for MIN-TR$_1$ and MAX-TR$_1$.

Application of MIN-TR$_2$*: Inferring Biological Signal Transduction Networks.*
In the study of biological signal transduction networks two types of interactions are considered. For example, nodes can represent genes and an edge (u, v) means that gene u *regulates* gene v. Without going into biological details, *regulates* may mean two different things: when u is *expressed*, *i.e.* molecules of the protein coded by u are created, the expression of v can be *repressed* or *promoted*. A path in this network is an indirect interaction, and promoting a repressor represses, while repressing a repressor *promotes*. Moreover, for certain interactions we have direct evidence, so an instance description includes set $D \subset E$ of edges which have to be present in every valid solution. The MIN-TR$_2$ problem allows to determine the sparsest graph consistent with experimental observations; it is a key part of the network synthesis software described in [3,8] and downloadable from http://www.cs.uic.edu/~dasgupta/network-synthesis/.

2 Overview of Our Algorithmic Techniques

Moyles and Thompson [13] showed that MIN-ED can be reduced in linear time to the case when the input graph (V, E) is strongly connected, therefore we will assume that $G = (V, E)$ is already strongly connected. In Section 2.5 we will use a similar result obtained for MIN-TR$_p$ and MAX-TR$_p$ obtained in [2]. We use the following additional notations.

- $G = (V, E)$ is the input digraph;
- $\iota(U) = \{(u, v) \in E : u \notin U \ \& \ v \in U\}$;
- $o(U) = \{(u, v) \in E : u \in U \ \& \ v \notin U\}$;

- $scc_A(u)$ is the strongly connected component containing vertex u in the digraph (V, A);
- $T[u]$ is the node set of the subtree with root u (of a rooted tree T).

A starting point for our approximation algorithms for both MIN-TR$_1$ and MAX-TR$_1$ is a certain polytope for them as described below.

2.1   A Primal-Dual LP Relaxation for MIN-TR$_1$ and MAX-TR$_1$

The *minimum cost rooted out-arborescence*[1] problem is defined as follows. We are given a weighted digraph $G = (V, E)$ with a cost function $c : E \to \mathbb{R}_+$ and root node $r \in V$. A valid solution is $A \subseteq E$ such that in (V, A) there is a path from r to every other node and we need to minimize $\sum_{e \in A} c(e)$. The following exponential-size LP formulation for this was provided by Edmonds and Karp. Let $x = (\ldots, x_e, \ldots) \in \{0,1\}^{|E|}$ be the 0-1 selection vector of edges with $x_e = 1$ if the edge e being selected and $x_e = 0$ otherwise. Abusing notations slightly, let $\iota(U) \in \{0,1\}^{|E|}$ also denote the 0-1 indicator vector for the edges in $\iota(U)$. Then, the LP formulation is:

(primal **P1**)

minimize $c \cdot x$ subject to

$x \geq 0$

$\iota(U) \cdot x \geq 1$ for all U s.t. $\varnothing \subset U \subset V$ and $r \notin U$ (1)

Edmonds [6] and Karp [9] showed that the above LP always has an integral optimal solution and that we can find it in polynomial-time.

From now on, by a *requirement* we mean a set of edges R that any valid solution must intersect; in particular, it means that the LP formulation has the constraint $Rx \geq 1$. We modify **P1** to an LP formulation for MIN-ED by setting $c = 1$ in (1) and removing "and $r \notin U$" from the condition (so we have a requirement for every non-empty $\iota(U)$). The dual program of this LP can be constructed by having a vector y that has a coordinate y_U for every $\varnothing \subset U \subset V$; both the primal and the dual is written down below for clarity:

(primal **P2**)   (dual **D2**)

minimize $1 \cdot x$ subject to   **maximize** $1 \cdot y$ subject to

$x \geq 0$ $y \geq 0$

$\iota(U) \cdot x \geq 1$ for all U s.t. $\varnothing \subset U \subset V$ $\sum_{e \in \iota(U)} y_U \iota(U) \leq 1$ for all $e \in E$

We can change **P2** into the LP formulation for MAX-ED by replacing the objective to "maximize $1 \cdot (1 - x)$". and the dual is changed accordingly to reflect this change. Finally, we can extend **P2** to an LP formulation for MIN-TR$_1$ or MAX-TR$_1$ by adding one-edge requirements $\{e\}$ (and thus inequality $x_e \geq 1$) for each $e \in D$ where D is the set of edges that have to be present in a valid solution. *Abusing notations slightly, we will denote all these polytopes by* **P2** *when it is clear from the context.*

---

[1] The corresponding in-arborescence problem must have a path from every node to r.

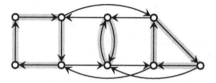 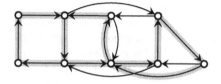

Fig. 1. The left panel shows an example of a lower bound solution L, the right panel shows an example of an eventual solution

2.2 Using the Polytope to Approximate MIN-TR₁

For MIN-TR$_1$, our goal is to prove the following theorem.

Theorem 1. *There is a polynomial time algorithm for* MIN-TR$_1$ *that produces a solution with at most* $1.5\,OPT - 1$ *edges, where* OPT *is the number of edges in an optimum solution.*

We mention the key ideas in the proof in the next few subsections.

A Combinatorial Lower Bound L for MIN-TR$_1$. We will form a set of edges L that satisfies $|L| \leq OPT$ by solving an LP **P3** derived from **P2** by keeping a subset of requirements (hence, with the optimum that is not larger) and which has an integer solution (hence, it corresponds to a set of edges L). We form an **P3** by keeping only those requirements $Rx \geq 1$ of **P2** that for some node u satisfy $R \subseteq \iota(u)$ or $R \subseteq o(u)$. To find the requirements of **P3** efficiently, for each $u \in V$ we find strongly connected components of $V - \{u\}$. Then,

(a) for every source component C have requirement $\iota(C) \subset o(u)$;
(b) for every sink component C have requirement $o(C) \subset \iota(u)$;
(c) if we have one edge requirement $\{e\} \subset R$ we remove R.

After **(c)** the requirements of **P3** contained in a particular $\iota(u)$ or $o(u)$ are pairwise disjoint, hence requirements of **P3** form a bipartite graph in which connections have the form of shared edges. If we have m requirements, a minimum solution can be formed by finding a maximum matching in this graph, say of size a, and then greedily adding $m - 2a$ edges. See Figure 1 for an illustration of calculation of L.

Converting L to a Valid Solution of MIN-TR$_1$. We will convert L into a valid solution. In a nutshell, we divide L into strongly connected components of (V, L) which we will call *objects*. We merge objects into larger strongly connected components, using amortized analysis to attribute each edge of the resulting solution to one or two objects. To prove that we use at most $1.5|L| - 1$ edges, an object with a edges of L can be responsible for at most $1.5a$ edges, and in one case, the "root object", for at most a edges.

Starting Point: the DFS. One can find an equivalent digraph using *depth first search* starting at any root node r. Because we operate in a strongly connected

```
DFS(u)
{   COUNTER ←COUNTER+1
    NUMBER[u] ←LOWDONE[u] ←LOWCANDO[u] ←COUNTER
    for each edge (u, v)     // scan the adjacency list of u
        ifNUMBER[v] = 0
            INSERT(T, (u, v))     // (u, v) is a tree edge
            DFS(v)
            ifLOWDONE[u] > LOWDONE[v]
                LOWDONE[u] ←LOWDONE[v]
            ifLOWCANDO[u] > LOWCANDO[v]
                LOWCANDO[u] ←LOWCANDO[v]
                LOWEDGE[u] ←LOWEDGE[v]
        elseifLOWCANDO[u] > NUMBER[v]
            LOWCANDO[u] ←NUMBER[v]
            LOWEDGE[u] ← (u, v)
    // the final check: do we need another back edge?
    ifLOWDONE[u] = NUMBER[u] and u ≠ r
        INSERT(B,LOWEDGE[u])     // LOWEDGE[u] is a back edge
        LOWDONE[u] ←LOWCANDO[u]
}

T ← B ← ∅
for every node u
    NUMBER[u] ← 0
COUNTER ← 0
DFS(r)
```

Fig. 2. DFS for finding an equivalent digraph of a strongly connected graph

graph, only one root call of the depth first search is required. This algorithm (see Fig. 2) mimics Tarjan's algorithm for finding strongly connected components and biconnected components. As usual for depth first search, the algorithm forms a spanning tree T in which we have an edge (u, v) if and only if DFS(u) made a call DFS(v). The invariant is

(A) if DFS(u) made a call DFS(v) and DFS(v) terminated then
$$T[v] \subset scc_{T \cup B}(u).$$

(A) implies that $(V, T \cup B)$ is strongly connected when DFS(r) terminates. Moreover, in any depth first search the arguments of calls that already have started and have not terminated yet form a simple path starting at the root. By (A), every node already visited is, in $(V, T \cup B)$, strongly connected to an ancestor who has not terminated. Thus, (A) implies that the strongly connected components of $(V, T \cup B)$ form a simple path. This justifies our convention of using the term *back edge* for all non-tree edges.

To prove the invariant, we first observe that when DFS(u) terminates then LOWCANDO[u] is the lowest number of an end of an edge that starts in $T[u]$.

Application of (A) to each child of v shows that $T[v] \subset scc_{T \cup B}(v)$ when we perform the final check of DFS(v).

If the condition of the final check is false, we already have a B edge from $T[v]$ to an ancestor of u, and thus we have a path from v to u in $T \cup B$. Otherwise, we attempt to insert such an edge. If LowCanDo$[v]$ is "not good enough" then there is no path from $T[v]$ to u, a contradiction with the assumption that the graph is strongly connected.

The actual algorithm is based on the above DFS, *but we also need to alter the set of selected edges in some cases.*

An Overview of the Amortized Scheme

Objects, Credits, Debits. The initial solution L to **P3** is divided into *objects*, namely the strongly connected components of (V, L). L-edges are either inside objects, or between objects. We allocate L-edges to objects, and give 1.5 for each. In turn, an object has to pay for solution edges that connect it, for a T-edge that enters this object and for a B-edge that connects it to an ancestor. Each solution edge costs 1. Some objects have enough money to pay for all L-edges inside, so they become strongly connected, and two more edges of the solution, to enter and to exit. We call them *rich*. Other objects are *poor* and we have to handle them somehow.

Allocation of L-Edges to Objects

- L-edge inside object A: allocate to A;
- from object A: call the first L-edge primary, and the rest secondary;
  - primary L-edge $A \to B$, $|A| = 1$: 1.5 to A;
  - primary L-edge $A \to B$, $|A| > 1$: 1 to A, and 0.5 to B;
  - secondary L-edge $A \to B$ (while there is a primary L-edge $A \to C$): if $|A| > 1$, 1.5 to B, otherwise 0.5 to each of A, B and C.

When is an Object A rich?

1. A is the root object, no payment for incoming and returning edges;
2. $|A| \geq 4$: it needs at most L-edges inside, plus two edges, and it has at least $0.5|A|$ for these two edges;
3. if $|A| > 1$ and an L-edge exits A: it needs at most L-edges inside, plus two edges, and it has at least $(1 + 0.5|A|)$ for these two edges;
4. if $|A| = 1, 3$ and a secondary L-edge enters A;
5. if $|A| = 1, 3$ and a primary L-edge enters A from some D where $|D| > 1$.

To discuss a poor object A, we call it a *path node*, *digons* or a *triangles* when $|A| = 1, 2,$ or 3 respectively.

Guiding. DFS For a rich object A, we decide at once to use L-edges inside A in our solution, and we consider it in DFS as a single node, with combined adjacency list. This makes point **(1)** below moot. Otherwise, the preferences are in the order: **(1)** L-edges inside the same object; **(2)** primary L-edges; **(3)** other edges.

The analysis of the balance of poor objects for enough credits is somewhat complicated, especially since we desire to extend the same approach from MIN-ED to MIN-TR$_1$. The details are available in the full version of the paper.

2.3   Using the Polytope to Approximate MAX-TR$_1$

Theorem 2. *There is a polynomial time algorithm for* MAX-TR$_1$ *that produces a solution set of edges H with $|E - H| \geq \frac{1}{2}OPT + 1$, where $OPT = |E - H|$ if H is an optimum solution.*

Proof. (In the proof, we add in parenthesis the parts needed to prove $0.5OPT + 1$ bound rather than $0.5OPT$.) First, we determine the *necessary* edges: e is necessary if $e \in D$ or $\{e\} = \iota(S)$ for some node set S. (If there are any cycles of necessary edges, we replace them with single nodes.)

We give a cost of 0 to the necessary edges and a cost of 1 for the remaining ones. We set $x_e = 1$ if e is a necessary edge and $x_e = 0.5$ otherwise. This is a valid solution for the fractional relaxation of the problem as defined in **P1**.

Now, pick any node r. (Make sure that no necessary edges enter r.) Consider the out-arborescence problem with r as the root. Obviously, edges of cost 0 can be used in every solution. An optimum (integral) out-arborescence T can be computed in polynomial time by the greedy heuristic in [9]; this algorithm also provides a set of cuts that forms a dual solution.

Suppose that $m + 1$ edges of cost 1 are *not* included in T, then no solution can delete more than m edges (because to the cuts collected by the greedy algorithm we can add $\iota(r)$). Let us reduce the cost of edges in T to 0. Our fractional solution is still valid for the in-arborescence, so we can find the in-arborescence with at most $(m + 1)/2$ edges that still have cost 1. Thus we delete at least $(m + 1)/2$ edges, while the upper bound is m.

To assure deletion of at least $k/2 + 1$ edges, where k is the optimum number, we can try in every possible way one initial deletion. If the first deletion is correct, subsequently the optimum is $k - 1$ and our method finds at least $(k - 1 + 1)/2$ deletions, so together we have at least $k/2 + 1$.   ❏

2.4   Some Limitations of the Polytope

We also show an inherent limitation of our approach by showing an integrality gap of the LP relaxation of the polytope **P2** for MIN-TR$_1$ and MAX-TR$_1$.

Lemma 1. *The LP formulation P2 for* MIN-ED *and* MAX-ED *has an integrality gap of at least $4/3$ and $3/2$, respectively.*

To prove the lemma, we use a graph with $2n + 2$ nodes and $4n + 2$ edges; Fig. 3 shows an example with $n = 5$. This graph has no cycles of five or more edges while every cut has at least 2 incoming and 2 outgoing

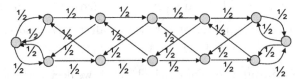

Fig. 3. A graph for the integrality gap of the polytope. The fractional solution is indicated.

edges. For MIN-ED, one could show that the optimal fractional and integral solutions of the polytope **P2** are $2n + 1$ and $(8n + 8)/3$, respectively; the claim for MAX-ED also follows from these bounds.

2.5 Approximating MIN-TR$_p$ and MAX-TR$_p$ for Prime p

We will show how to transform our approximation algorithms for MIN-TR$_1$ and MIN-TR$_1$ into approximation algorithms for MIN-TR$_p$ and MAX-TR$_p$ with ratios 1.5 and 2 respectively. In a nutshell, we can reduce the approximation in the general case the case of a strongly connected graph, and in a strongly connected graph we will show that a solution to MIN-TR$_1$ (MAX-TR$_1$) can be transformed into a solution to MIN-TR$_p$ (MAX-TR$_p$) by adding a single edge, and in polynomial time we can find that edge.

In turn, when we run approximation algorithms within strongly connected components, we obtain the respective ratio even if we add one extra edge.

Let G be the input graph. The following proposition says that it suffices to restrict our attention to strongly connected components of G^2. (One should note that the algorithm implicit in this Proposition runs in time proportional to p.)

Proposition 3. [2] *Suppose that we can compute a ρ-approximation of TR_p on each strongly connected component of G for some $\rho > 1$. Then, we can also compute a ρ-approximation of TR_p on G.*

The following characterization of scc's of G appear in [2].

Lemma 2. [2] *Every strongly connected component $U \subset V$ is one of the following two types:*

(Multiple Parity Component) $\{q : (u,v,q) \in Closure_\ell(E(U))\} = \mathbb{Z}_p$ *for any two vertices $u, v \in U$;*
(Single Parity Component) $|\{q : (u,v,q) \in Closure_\ell(E(U))\}| = 1$ *for any two vertices $u, v \in U$.*

Based on the above lemma, we can use the following approach. Consider an instance (V, E, ℓ, D) of MIN-TR$_p$. For every strongly connected component $U \subset V$ we consider an induced instance of MIN-TR$_1$, $(U, E(U), D \cap U)$. We find an approximate solution A_U that contains an out-arborescence T_U with root r. We label each node $u \in U$ with $\ell(u) = \ell(P_u)$ where P_u is the unique path in T_U from r to u.

Now for every $(u,v) \in E(U)$ we check if $\ell(v) = \ell(u) + \ell(u,v)$ **mod** p.

If this is true for every $e \in E(U)$ then U is a single parity component. Otherwise, we pick a single edge (u,v) violating the test and we insert it to A_U, thus assuring that (U, A_U) becomes a multiple parity component.

2.6 Inapproximability of MIN-ED and MAX-ED

Theorem 4. *Let γ be the length of the longest cycle in the given graph. Then, both 5-MIN-ED and 5-MAX-ED are MAX-SNP-hard even if $\gamma = 5$.*

---

[2] The authors in [2] prove their result only for MIN-TR$_p$, but the proofs work for MAX-TR$_p$ as well.

Proof. We will use a single approximation reduction that reduces 2REG-MAX-SAT to MIN-ED and MAX-ED with $\gamma = 5$.

In MAX-SAT problem the input is a set S of disjunctions of literals, a valid solution is an assignment of truth values (a mapping from variables to $\{0, 1\}$), and the objective function is the number of clauses in S that are satisfied. 2REG-MAX-SAT is MAX-SAT restricted to sets of clauses in which every variable x occurs *exactly four times* (of course, if it occurs at all), twice as literal x and twice as literal $\bar{x}$. This problem is MAX-SNP hard even if we impose another constraint, namely that each clause has *exactly three* literals [4].

Consider an instance S of 2REG-MAX-SAT with n variables and m clauses. We construct a graph with $1 + 6n + m$ nodes and $14n + m$ edges. One node is h, *the hub*. For each clause c we have node c. For each variable x we have a gadget G_x with 6 nodes, two switch nodes labeled x, two nodes that are occurrences of literal x and two nodes that are occurrences of literal $\bar{x}$.

We have the following edges: (h, x^*) for every switch node, (c, h) for every clause node, (l, c) for every occurrence l of a literal in clause c, while each node gadget is connected with 8 edges as shown in Fig. 4.

We show that

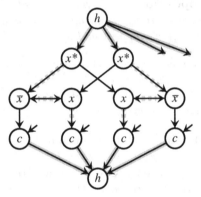

① if we can satisfy k clauses, then we have a solution of MIN-ED with $8n + 2m - k$ nodes, which is also a solution of MAX-ED that deletes $6n - m + k$ edges;

② if we have a solution of MIN-ED with $8n + 2m - k$ edges, we can show a solution of 2REG-MAX-SAT that satisfies k clauses.

Fig. 4. Illustration of our reduction. Marked edges are necessary. Dash-marked edges show set A_x that we can interpret it as x =true. If some i clause nodes are not reached (*i.e.*, the corresponding clause is not satisfied) then we need to add k extra edges. Thus, k unsatisfied clauses correspond to $8n + m + k$ edges being used ($6n - k$ deleted) and k satisfied clauses correspond to $8n + 2m - k$ edges being used ($6n + m - k$ deleted).

To show ①, we take a truth assignment and form an edge set as follows: include all edges from h to switch nodes ($2n$ edges) and from clauses to h (m edges). For a variable x assigned as true pick set A_x of 6 edges forming two paths of the form $(x^*, \bar{x}, x, c)$, where c is the clause where literal x occurs, and if x is assigned false, we pick set $A_{\bar{x}}$ of edges from the paths of the form $(x^*, x, \bar{x}, c)$ ($6n$ edges). At this point, the only nodes that are not on cycles including h are nodes of unsatisfied clauses, so for each unsatisfied clause c we pick one of its literal occurrences, l and add edge (l, c) ($m - k$ edges).

The proof of ② will appear in the full version.   ❑

Remark 1. Berman *et al.* [4] have a randomized construction of 2REG-MAX-SAT instances with $90n$ variables and $176n$ clauses for which it is NP-hard to tell if we can leave at most εn clauses unsatisfied or at least $(1 - \varepsilon)n$. The above construction converts it to graphs with $(14 \times 90 + 176)$ edges in which it is NP-hard to tell if we need at least $(8 \times 90 + 176 + 1 - \varepsilon)n$ edges or at most $(8 \times 90 + 176 + \varepsilon)n$, which gives an inapproximability bound of $1 + 1/896$ for MIN-ED and $1 + 1/539$ for MAX-ED.

Acknowledgments. The authors thank Samir Khuller for useful discussions.

References

1. Aho, A., Garey, M.R., Ullman, J.D.: The transitive reduction of a directed graph. SIAM Journal of Computing 1(2), 131–137 (1972)
2. Albert, R., DasGupta, B., Dondi, R., Sontag, E.: Inferring (Biological) Signal Transduction Networks via Transitive Reductions of Directed Graphs. Algorithmica 51(2), 129–159 (2008)
3. Albert, R., DasGupta, B., Dondi, R., Kachalo, S., Sontag, E., Zelikovsky, A., Westbrooks, K.: A Novel Method for Signal Transduction Network Inference from Indirect Experimental Evidence. Journal of Computational Biology 14(7), 927–949 (2007)
4. Berman, P., Karpinski, M., Scott, A.D.: Approximation Hardness of Short Symmetric Instances of MAX-3SAT, Electronic Colloquium on Computational Complexity, Report TR03-049 (2003),
   http://eccc.hpi-web.de/eccc-reports/2003/TR03-049/index.html
5. Dubois, V., Bothorel, C.: Transitive reduction for social network analysis and visualization. In: IEEE/WIC/ACM International Conference on Web Intelligence, pp. 128–131 (2005)
6. Edmonds, J.: Optimum Branchings. In: Dantzig, G.B., Veinott Jr., A.F. (eds.) Mathematics and the Decision Sciences. Amer. Math. Soc. Lectures Appl. Math., vol. 11, pp. 335–345 (1968)
7. Frederickson, G.N., JàJà, J.: Approximation algorithms for several graph augmentation problems. SIAM Journal of Computing 10(2), 270–283 (1981)
8. Kachalo, S., Zhang, R., Sontag, E., Albert, R., DasGupta, B.: NET-SYNTHESIS: A software for synthesis, inference and simplification of signal transduction networks. Bioinformatics 24(2), 293–295 (2008)
9. Karp, R.M.: A simple derivation of Edmonds' algorithm for optimum branching. Networks 1, 265–272 (1972)
10. Khuller, S., Raghavachari, B., Young, N.: Approximating the minimum equivalent digraph. SIAM Journal of Computing 24(4), 859–872 (1995)
11. Khuller, S., Raghavachari, B., Young, N.: On strongly connected digraphs with bounded cycle length. Discrete Applied Mathematics 69(3), 281–289 (1996)
12. Khuller, S., Raghavachari, B., Zhu, A.: A uniform framework for approximating weighted connectivity problems. In: 19th Annual ACM-SIAM Symposium on Discrete Algorithms, pp. 937–938 (1999)
13. Moyles, D.M., Thompson, G.L.: Finding a minimum equivalent of a digraph. JACM 16(3), 455–460 (1969)
14. Papadimitriou, C.: Computational Complexity, p. 212. Addison-Wesley, Reading (1994)
15. Vetta, A.: Approximating the minimum strongly connected subgraph via a matching lower bound. In: 12th ACM-SIAM Symposium on Discrete Algorithms, pp. 417–426 (2001)

1.25-Approximation Algorithm for Steiner Tree Problem with Distances 1 and 2

Piotr Berman[1,*], Marek Karpinski[2,**], and Alexander Zelikovsky[3]

[1] Department of Computer Science & Engineering, Pennsylvania State University,
University Park, PA 16802
berman@cse.psu.edu
[2] Department of Computer Science, University of Bonn, 53117 Bonn
marek@cs.uni-bonn.de
[3] Department of Computer Science, Georgia State University, Atlanta, GA 30303
alexz@cs.gsu.edu

Abstract. Given a connected graph $G = (V, E)$ with nonnegative costs on edges, $c : E \rightarrow \mathcal{R}^+$, and a subset of terminal nodes $R \subset V$, the Steiner tree problem asks for the minimum cost subgraph of G spanning R. The Steiner Tree Problem with distances 1 and 2 (i.e., when the cost of any edge is either 1 or 2) has been investigated for long time since it is MAX SNP-hard and admits better approximations than the general problem. We give a 1.25 approximation algorithm for the Steiner Tree Problem with distances 1 and 2, improving on the previously best known ratio of 1.279.

1 Introduction

Given a connected graph $G = (V, E)$ with nonnegative costs on edges, $c : E \rightarrow \mathcal{R}^+$, and a subset of terminal nodes $R \subset V$, the Steiner tree problem asks for the minimum cost subgraph of G spanning R. This is a well-known MAX SNP-hard problem with a long history of efforts to improve the approximation ratio achievable in polynomial time. For several decades the best known approximation algorithm has been the minimum spanning tree heuristic (MST heuristic) which reduces the original Steiner tree problem to the minimum spanning tree problem in the graph induced by the set of terminal nodes R. The best up-to-date ratio is 1.55 [3].

This paper is focused on a special case of the general Steiner tree problem when the cost of any edge between different vertices is either 1 or 2 (usually referred to as the Steiner Tree Problem with distances 1 and 2, or STP[1,2]) which is also MAX SNP-hard. The Rayward-Smith heuristic [2] has been the first algorithm with a better approximation ratio than the MST-heuristic for STP[1,2]. Bern and Plassman [1] have shown that its ratio is $\frac{4}{3}$ while the ratio of MST-heuristic is 2. The previously best approximation ratio for this problem is 1.279 which is achieved by polynomial-time

---

* Research partially done while visiting Department of Computer Science, University of Bonn and supported by DFG grant Bo 56/174-1.
** Supported in part by DFG grants, Procope grant 31022, and the Hausdorff Center grant EXC59-1.

F. Dehne et al. (Eds.): WADS 2009, LNCS 5664, pp. 86–97, 2009.

approximation scheme of so called loss-contracting algorithms, the same scheme that achieves the best known ratio of 1.55 in the general case [3]. For a decade it remained open if there exists an algorithm (different from the loss-contracting algorithm applicable in the general case) which can have a better approximation ratio.

In this paper, we give a new 1.25-approximation algorithm for STP[1,2]. Unlike previously best 1.279-approximation, which is a PTAS and achieves the ratio in the limit [3], the proposed Seven-Phase algorithm is a single polynomial-time algorithm with runtime $O(|V|^{3.5})$.[1] The main idea behind the new algorithm is considering only Steiner full components of certain type, so called comets which are generalization of stars. In the proof of the approximation ratio we compare the found solution with the optimal Steiner tree whose only Steiner full components are comets and repeatedly associate several types of potential functions with different elements of Steiner trees.

The rest of the paper is organized as follows. Section 2 introduce several standard definitions related to Steiner trees. In Section 3 we give a strengthen proof of $\frac{4}{3}$-approximation ratio for Rayward-Smith heuristic based on the two types of potentials. Section 4 describes the Seven-Phase algorithm and the worst-case performance. Finally, Section 5 concludes with the proof of the $\frac{5}{4}$-approximation ratio of the Seven-Phase algorithm for STP[1,2].

2 Definitions and Notation

A metric with distances 1 and 2 can be represented as a graph, so edges are pairs in distance 1 and non-edges are pairs in distance 2.

The problem instance of STP[1,2] is a graph $G = (V, E)$ that defines a metric in this way, and a set $R \subset V$ of *terminal nodes*. A valid solution is a set of unordered node pairs S, i.e., a *Steiner tree*, such that R is contained in a connected component of (V, S). We minimize $|S \cap E| + 2|S - E|$.

We will further use the following standard notations. Let S be a Steiner tree, each node of S that is not a terminal will be called *Steiner node*. The edges of S can be split into *Steiner full components*, i.e., maximal connected subsets of S in which the degree of each terminal node is at most 1. For brevity, we will also refer to a Steiner full components as an S-comp.

Let T^* be an optimal Steiner tree and let $T = T^* \cap E$ be its *Steiner skeleton* consisting of its edges (cost-1 connections. We can assume that all cost-2 connections in T^* (as well as in any Steiner tree) are only between terminals, therefore, S-comps of T^* are connected (i.e., have only cost-1 connections).

Each full Steiner component has gain and loss (see [3]) that for STP[1,2] can be defined as follows. The *gain* of an S-comp K equals $2k - 2$ (cost of connecting all k terminals of K with cost-2 connections) minus $cost(K)$. The *loss* of an S-comp is the cost of connection of all its Steiner points to the terminals. The loss symbolizes by how much addition of an S-comp can deviate solution from the optimum. Let *gloss* of an S-comp S be the ratio of gain over loss, i.e., $gloss(S) = \frac{gain(S)}{loss(S)}$. The loss-contracting algorithm (LCA) from [3] greedily processes S-comps with the largest gloss.

---

[1] A slight modification of the algorithm (omitted for brevity) reduces the runtime to $O(|V|^3)$.

A simplest example of an S-comp is an *s-star* consisting of a Steiner node c, called the center, s terminal nodes $t_1, \ldots, t_s$ and edges $(c, t_1), \ldots, (c, t_s)$. If $s < 3$ we say that the star is *degenerate*, and *proper* otherwise. One can easily see that the gloss of an s-star is $s - 2$ since its gain is $s - 2$ and and its loss is 1.

We now generalize a notion of a star. A (a, b) *comet* (sometimes referred as b-comet) is an S-comp consisting of a center c connected to b terminals and a non-terminals called *fork nodes*; each fork node is connected only to c and two terminals, those three nodes and two edges form a *fork*. Pictorially, a comet is like a star with trailing tail consisting of forks (see Fig. 1). It is easy to see that the gloss of an (a, b)-comet is $gloss(S) = \frac{a+b-2}{a+1}$. It is somewhat less obvious that

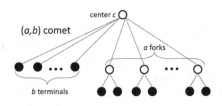

Fig. 1. An (a,b) comet with center c connected to b terminals and a forks

Lemma 1. *A comet S with the maximum $gloss(S)$ can be found in time $O(|V|^{2.5})$.*

Proof. First, observe that if $b \geq 3$, the gloss of a (a, b)-comet is not greater than the gloss of a b-star, so we can remove all the forks. If $b \leq 2$ the situation is opposite: the gloss grows with the number of forks. However, gloss grows if we add a terminal and remove a fork. Thus we can find the best comet as follows. Try every possible center c and find b_c, the number of terminals to which it can be connected. If $b_c > 2$ for some c, select c with the largest b_c. Otherwise, for each c form the following graph G_c: for every non-terminal u that is connected to c and to two terminals v, w (not directly connected to c) add an edge $\{v, w\}$ to E_c. The maximum number of forks with the center c is the size of maximum matching which can be found in $O(n^{1.5})$ time since $|E_c| < n$.   □

All recent approximation algorithms restrict the type of S-comps. LCA approximates the optimal k-restricted Steiner tree, i.e., the one whose S-comps has at most k terminals. It is shown that the approximation ratio of LCA converges to 1.279 with $k \to \infty$ [3]. The Rayward-Smith heuristic (RSH) approximates the optimal *star-restricted* Steiner tree whose S-comps are stars [2]. In the next section we will give our proof of $\frac{4}{3}$-approximation ratio for RSH. In Section 4, we describe the new Seven-Phase algorithm approximating the optimal *comet-restricted* Steiner tree whose S-comps are comets.

When we analyze an algorithm, we view its selections as transformations of the input instance, so after each phase we have a partial solution and a residual instance. We formalize these notions as follows. A partition Π of V induces a graph $(\Pi, E(\Pi))$ where $(A, B) \in E(\Pi)$ if $(u, v) \in E$ for some $u \in A, v \in B$). We say that (u, v) is a representative of (A, B). Similarly, Π induces the set of terminals $R_\Pi = \{A \in \Pi : A \cap R \neq \varnothing\}$. In our algorithms, we augment initially empty solution F. Edge set F defines partition $\Pi(F)$ into connected components of (V, F). In a step, we identify a connected set A in the induced graph $(\Pi(F), E(\Pi(F)))$ and we augment F with representatives of edges that form a spanning tree of A. We will call it *collapsing A*, because A will become a single node of $(\Pi(F), E(\Pi(F)))$.

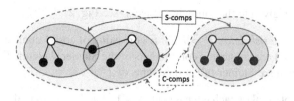

Fig. 2. Three S-comps and two C-comps of a Steiner skeleton. Terminal nodes are black and Steiner nodes are white.

3   A Potential-Based Analysis of the Rayward-Smith Heuristic

In this section we introduce a new way of analyzing greedy heuristics for STP[1,2], applying it to RSH. The potential-based analysis method will allow to tighten the performance analysis of Bern and Plassman [1] (see Theorem 1). We have reformulated the RSH as follows.

1. **Preprocessing:** Collapse all edges (cost-1 connections) between terminals.
2. **Greedy collapsing of s-stars, $s \geq 4$:** Find an s-star S with the largest possible s. If $s \leq 3$, then exit. Collapse S.
3. **Greedy collapsing of 3-stars:** Find an s-star S with the largest possible s. If $s = 2$, then exit. Collapse S.
4. **Finishing:** Connect the remaining terminals with non-edges.

In the rest of the section we are going to prove the following

Theorem 1. *Let T_{RS} be the Steiner tree given by RSH. Then $cost(T_{RS}) \leq cost(T^*) + \frac{1}{3}cost(T)$, where T^* is the optimal tree and T is its skeleton*

If Preprocessing is nontrivial, then it can only improve the approximation ratio since all such edges can be forced into the optimal solution. So further we assume that any two terminals are within distance 2 from each other. In the analysis of Steps 2-4, we update the following three values after each iteration:

CA = the cost of edges collapsed so far, initially, $CA = 0$;
CR = the cost of a certain solution T' where all collapsed edges have cost 0, initially, $T' = T^*$, the optimum solution, and $CR = cost(T^*)$.
P = the sum of potentials distributed among objects, which will be defined later.

We will define the potential satisfying the following conditions:

(a) initially, $P < cost(T^*)/3$, at each stage, $P \geq 0$;
(b) after collapse of each star S, the decrease in *full_cost* (which is equal to $CR + P$) is at least $cost(S)$, i.e., $\Delta CA = cost(S) \leq \Delta full\_cost$;
(c) the final Steiner skeleton is empty.

The potential is given to the following objects (see Fig. 2):

- edges of a *Steiner skeleton* $T = T^* \cap E$ consisting of cost-1 connections of T^*;
- *C-comps* which are connected components of the Steiner skeleton T;
- *S-comps* which are Steiner full components of T^*.

Initially, the potential of each edge e is $p(e) = 1/3$ and the potential of each C-comp and S-comp is zero. The total edge potential is denoted PE and the total C-comps and S-comps potential is denoted PC.

We will now modify the Steiner skeleton without increasing the full cost such that the resulting Steiner tree will be star-restricted, i.e., each S-comp will be a either a proper star or non-edge. This star-restricted Steiner tree is constructed by repeatedly applying the following two steps.

Path step. Let Steiner skeleton T contain a Steiner point v of degree 2. We remove two edges incident to v from T adding a non-edge (cost-2 connection) to T^*. The potential for the both resulting C-comps is set to 0. We then repeatedly remove all degree-1 Steiner points that may appear.

Bridge Step. Let each Steiner node have degree at lest 3 and let e be a *bridge*, i.e., an edge $e = (u, v)$ between Steiner points. We remove this edge from T (adding a non-edge to T^*) thus splitting a connected component C into C_0 and C_1. Each new C-comp has at least two edges since u and v originally have degree at least 3. We set $p(C_0) = p(C)$ and $p(C_1) = -2/3$.

Lemma 2. *There exists a star-restricted Steiner tree T_B^* and a potential p for its skeleton T_B such that*

- *(i) each edge $e \in T_B$ has $p(e) = 1/3$;*
- *(ii) each C-comp C has $p(C) \geq -2/3$ and each trivial C-comp (with at most one edge) has $p(C) = 0$;*
- *(iii) each S-comp S has $p(S) = 0$;*

with full cost not larger than in the optimal tree, full_cost$(T_B^*) \leq$ full_cost(T^*).

Proof. After Path Step, the cost does not increase, edge potential PE is decreased by $2/3$ due to removal of 2 edges, while we may increase PC by at most $2/3$: we start with one C-comp C with $p(C) \geq -2/3$ and we end with two C-comps with potential 0; thus we do not increase *full_cost*. After Bridge Step, the cost is increased by 1, the total edge potential is decreased by $1/3$ and the total C-comp potential is decreased by $2/3$ resulting in unchanged *full_cost*.   □

Lemma 3. *After greedy collapsing of an s-star S, $s > 3$, conditions (i)-(iii) are satisfied and cost$(S) \leq \Delta$ full_cost(T^*).*

Proof. Let terminals of S be in a C-comps. To break cycles created in T^* when we collapse S, we replace $s-1$ connections, of which $a-1$ are cost-2 connections between different C-comps and $s - a$ edges within C-comps. Note that deletion of edges within a C-comp C decreases degree of its Steiner point and if all terminals of C are in S, then the Steiner point becomes a leaf and the edge connecting it to the terminal is removed from the optimal solution. Let k be the number of nontrivial C-comps with all terminals in S. The decrement in the cost is at least $\Delta CR \geq 2(a-1)+s-a+k = s+a+k-2$ and the decrement of the edge potential is $\Delta PE \geq \frac{s-a+k}{3}$.

Collapsing of S replaces a different C-comps with a single C-comp. If the resulting C-comp is nontrivial or one of C-comps intersecting S is trivial then the number of nontrivial C-comps is reduced by at most $a-1$ and the C-comp potential is incremented by at most $\frac{2}{3}(a-1)$, i.e., $\Delta PC \geq -\frac{2}{3}(a-1)$, and

$$\begin{aligned} \Delta full\_cost &= \Delta CR + \Delta PE + \Delta PC \\ &\geq s+a+k-2 + \frac{s-a+k}{3} - \frac{2}{3}(a-1) \\ &\geq s + \frac{1}{3}(s-4) \\ &\geq cost(S) \end{aligned}$$

Otherwise, $\Delta PC \geq -\frac{2}{3}a$ and there exits at least one nontrivial C-comp with all terminals in S, i. e., $k \geq 1$. Similarly,

$$\Delta full\_cost \geq s+a+k-2 + \frac{s-a+k}{3} - \frac{2}{3}a \geq cost(S) \qquad \square$$

When collapsing of s-stars with $s > 3$ is completed, we redistribute potential between C-comps and S-comps by increasing potential of each nontrivial C-comp by $\frac{1}{6}$ bringing it to $-\frac{1}{2}$ and decreasing potential of one of its S-comps by $\frac{1}{6}$. This will replace conditions (ii)-(iii) with

(ii') each C-comp C has $p(C) \geq -1/2$ and each trivial C-comp (with at most one edge) has $p(C) = 0$;

(iii') each S-comp S has $p(S) \geq -\frac{1}{6}$;

Lemma 4. *After greedy collapsing of any 3-star S, conditions (i)-(ii')-(iii') are satisfied and $cost(S) \leq \Delta full\_cost(T^*)$.*

Proof. Suppose that the terminals of the selected star belong to 3 different C-comps. We remove two cost-2 connections from T^*, hence $\Delta CR = 4$, and we replace three C-comps with one, hence $\Delta PC \geq -2/2$, so we have $cost(S) = 3$ and $\Delta full\_cost \geq 4-1$.

Suppose that the terminals of the selected star belong to 2 different C-comps. $\Delta CR = 3$ because we remove one cost-2 connection from T^* and one edge from an S-comp. This S-comp becomes a 2-star, hence we remove it from T using a Path Step, so together we remove 3 edges from T and $\Delta PE = 1$.

One S-comp disappears, so $\Delta PS = -1/6$. Because we collapse two C-comps into one, $\Delta PC = -1/2$.

If the terminals of the selected star belong to a single C-comp and we remove 2 edges from a single S-comp, we also remove the third edge of this S-comp and $\Delta CR = -3$, while $\Delta PE = -1$, $\Delta PS = 1/6$, and if its C-comp degenerates to a single node, we have $\Delta PC = 1/2$ (otherwise, zero). Thus the balance $\Delta full\_cost - cost(S)$ is at least $1/3$.

Finally, if the terminals of the selected star belong to a single C-comp and we remove 2 edges from two S-comps, we have $\Delta CR = -2$. Because we apply Path Steps to those two S-comps, $\Delta PE = -2$. while $\Delta PS = 1/3$ and $\Delta PC \leq 1/2$. Thus the balance is at least $1/6$.   □

Proof of Theorem 1. We initialized $CA = 0$, $CR = opt$ and $P \leq opt/3$. The sum of costs of all collapsed stars is at most $\Delta full\_cost(T^*)$ and when we finish star collapsing we have $P = 0$ and $T = \varnothing$. At this point, we can connect the partial solution using exactly the same number of cost-2 connections as we have in T^* so we obtain a solution with cost at most $full\_cost(T^*)$.

4   Seven-Phase Approximation Algorithm

We will formulate a new approximation algorithm and analyze its runtime and give a worst-case example.

The new *Seven-Phase Algorithm* proceeds as follows:

---

1. Collapse all edges between terminals.
2. Repeatedly, collapse an s-stars with the largest s, $s > 4$.
3. Collapse all s-stars with $s = 4$.
4. Find T_3, a maximum size set of 3-stars.
5. Repeatedly, whenever possible replace each 3-star from T_3 with a (1,3)-comet and collapse it. Collapse 3-stars remaining in T_3.
6. Repeatedly, collapse a comet with the maximum gloss.
7. Connect the remaining terminals with non-edges.

In the following section we will prove the following

Theorem 2. *The Seven-Phase Algorithm has the approximation ratio of* $5/4$ *for the Steiner Tree Problem in metrics with distances 1 and 2.*

The tightness of the $\frac{5}{4}$-ratio follows from the following example (see Figure 3). Let $X_1, ..., X_k$ be the set of nonintersecting 3-stars, each X_i with a center c_i, such that there is an edge between c_i and c_{i+1}, $i = 1, ..., k - 1$. Thus there is a Steiner tree of length $4k - 1$. Also let there exists a star X_i' for each X_i with a center c_i' instead of c_i

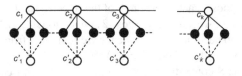

Fig. 3. Solid edges are in optimal solution and dashed edges are in approximate solution

but with the same terminals. Let Step 4 find X_i''s rather than X_i's. Then Step 5 cannot guarantee to find existing $(3, 1)$ comets since one such comet replaces two 3-stars. Finally, the cost of the approximate solution will be 5k -2.

The runtime of this algorithm is dominated by the runtime for Steps 4 and 6. Step 4 can be performed in time $O(|V|^{3.5})$ [4] and step 6 requires $O(|V|^{2.5})$ per iteration (see Lemma 1). Therefore, the total runtime is $O(|V|^{3.5})$.

5   Proof of the 5/4-Approximation Ratio

In the analysis, we use similar potential as in Section 3, assigned to objects defined by the skeleton T_R of the restriced solution T_R^*: edges, S-comps and C-comps. Now we start with $p(e) = 1/4$ (and zero potential for component objects).

5.1   Analysis of Phases 1 and 2

Collapsing edges between terminals has balance $\Delta full\_cost - cost(e) = 1 + 1/4 - 1 > 0$.

Collapsing an s-star S for $s > 4$ removes $s - 1$ edges from T, so the balance is $\Delta full\_cost - cost(S) \geq (s-1)5/4 - s = (s-5)/4 \geq 0$.

5.2   Preliminary to the Analysis of Phases 3-6

As in Section 3, we will modify the Steiner skeleton T of the optimal tree without increasing the full cost such that the resulting Steiner tree T_R^* will be comet-restricted, i.e., each nontrivial S-comp of T_R is either an s-star, $s > 2$ or an (a, b)-comet, $a + b > 2$. The Path Step is the same and the Bridge Step is altered – we remove an edge between Steiner nodes splitting a C-comp C into C_0 and C_1 only if both C_i's have at least 3 nodes. We set $p(C_0) = p(C)$ and $p(C_1) = -3/4$.

Repeatedly applying Path Steps and altered Bridge Steps to the optimum tree T^* we obtain a comet-restricted Steiner tree for which the following analogue of Lemma 2 holds.

Lemma 5. *There exists a comet-restricted Steiner tree T_R^* and a potential p for its skeleton T_R such that each edge $e \in T_R$ has $p(e) = 1/3$, each C-comp C has $p(C)$ either 0 or $-3/4$, each S-comp has $p(S) = 0$ and with full cost not larger than in the optimal tree, $full\_cost(T_B^*) \leq full\_cost(T^*)$.*

5.3   Analysis of Phase 3

Now we discuss the phase of selecting 4-stars. We change the potential distribution by setting $p(C) = -2/3$ for each C-comp C that had potential $-3/4$, and we compensate by setting, for one of its S-comps, say, S, $p(S) = -1/12$.

When we select a 4-star S, we remove 3 connections from T_R^*. With each such connection e' we associate $cost(e') = cost(S)/3 = 4/3$ and the corresponding components of $\Delta full\_cost$ to compute the balance.

When we remove a non-edge from T_R^* we have $\Delta CR = 2$. We also coalesce two C-comps, so $\Delta PC \geq -2/3$. Edges and S-comps are not affected, so we get balance $2 - 2/3 - 4/3 \geq 0$.

When we remove an edge from a fork (i.e. incident to a fork node of a comet), we can apply Path Step and remove two more edges from T_R. Thus we have $\Delta CR = 1$,

$\Delta PE = 3/4$, and balance at least $1 + 3/4 - 4/3$. The balance is even better when we remove two edges from the same fork with two connections, because in that case ΔCR is better; in T_R we erase three edges of a fork, rather then erasing one and replacing two with a non-edge. Thus we have $\Delta CR = 3$ and $\Delta PE = 3/4$ and balance $3 + 3/4 - 8/3$.

When we remove an edge from a 3-star or a "borderline" comet, like $(2, 1)$-comet or $(3, 0)$-comet, the reasoning is similar to the fork case. We have significant surplus. We also eliminate a negative potential of the star, but the surplus is so big we will not calculate it here.

The cases that remain is removing edges from stars, or edges that connect terminals with centers of comets. Without changing the potential of the affected S-comp we would have a deficit: $\Delta CR = 1$ and $\Delta PE = 1/4$, for the "preliminary" balance of $1 + 1/4 - 4/3 = 1/12$. Surplus of other connections can give non-negative balance, but if not, we obtain zero balance by decreasing the potential of the affected S-comp by $1/12$.

This process has the following invariants:

(a) the sum of the potentials of S-comps of a C-comp, and of that C-comp, is a multiple of $1/4$.
(b) a s-star or a s-comet has potential at least $-(5 - s)/12$.

Invariant (a) follows from the fact that cost change is integral, and the other potentials that change are edge potentials, each $1/4$. Moreover, we coalesce a group of C-comps if we charge more than one. (A careful reasoning would consider consequences of breaking a C-comp by applying a Path Step).

Invariant (b) is clearly true at the start of the process. Then when we remove an edge from an s star we subtract 1 from s and $1/12$ from the potential, and the invariant is preserved.

5.4 Preliminary to the Analysis of Phases 4-5

When Phase 3 is over, we reorganize the potential in a manner more appropriate for considering 3-stars. We increase a potential of a C-comp from $-2/3$ to $-1/2$, and we decrease the potential of one or two of its S-comps. In the same time, we want to have the following potential for S-comps:

$$p_4(S) = \begin{cases} -\frac{1}{4} & \text{if } S \text{ is a 3-star or a 3-comet} \\ -\frac{3-s}{4} & \text{if } S \text{ is an } s\text{-comet}, s < 3 \end{cases}$$

Note that before the decrease, a 3-star or a 3-comet has potential at least $-2/12$, so it can aborb $1/12$. Similarly, a 1-comet had potential at least $-4/12$, so it can absorb $2/12$, and the balance of 0 comets is even better. We would have a problem if we have only 2-comets that have the minimum potential of $1/4$ and perhaps only one exception in the form of a 3-star/3-comet with the minimum potential or 2-comet with potential only $1/12$ above the minimum. With no exceptions, the sum of potentials of S-comps and the C-comp before the distribution is a multiple of $1/4$ plus $p(C) = -2/3$, and this violates invariant (b). With one exception, that sum is a multiple of $1/4$ plus $-2/3 + 1/6$ and this also violates that invariant.

Before phase 6 we need a different distribution of potential. In that phase we do not have 3-stars, the best gloss is below 1. The following values of potential are sufficiently high for the analysis:

$$
p_6(S) = \begin{cases} -\frac{7}{12} & \text{if } S \text{ is a 2-comet} \\[2mm] -1 & \text{if } S \text{ is a 1-comet} \\[2mm] -\frac{29}{20} & \text{if } S \text{ is a 0-comet} \end{cases}
$$

Moreover, we will have potential zero for C-comps except for C-comps that consist of one S-comp only; for such C-comp C we can have $p_6(C) = -1/3$ if C is a 2-comet and $p_6(C) = -1/4$ if C is an 1-comet.

5.5   Analysis of Phases 4-5

In phase 4 we insert a maximum set of 3-stars to the solution, and this selection can be modified in phase 5. Note that we can obtain a set of 3-stars from the T_R by taking all 3-stars and stripping forks from 3-comets.

If a selected 3-star has terminals in three C-comps, we can collapse it, $\Delta PE = \Delta PS = 0$, $\Delta CA = 3$, $\Delta CR = -4$ and $\Delta PC = -1$, so this analysis step has balance zero. And we still can find at least as many 3-stars as we have 3-stars and 3-comets in T_R.

We will consider connected component created by the inserted 3-stars together with C-comps of T_R. When we consider selection of a 3-star, we view it as a pair of connections e_0, e_1, each with $cost(e_i) = 3/2$, we insert such a connection to T_R^* and break the cycle by removing a connection, a non-edge or an edge from an S-comp S. This changes the potential of S; the first change we account by comparing $p_4(S)$ with $p_6(S')$, where S' is S with one connection removed. When a subsequent such change is considered, we compare $p_6(S)$ with $p_6(S')$. In that accounting we need to collect some surplus, so we will be able to cover increases of the potential of C-comps (from $-1/2$ to 0, with some critically important exceptions).

The following cases analyze the impact of a single new connection introduced in phase 4, followed by a deletion of a non-edge or a connection within S-comp S, which "annihilates" S or changes it into S'. We number the cases as follows: the first digit describes the type of S-comp from which a connection is removed, 2 for 3-star, 3, 4, 5, 6 for 3-, 2-, 1- and 0- comet; the second digit indicates the initial or a subsequent removal, and the third digit, with values 1 and 2, indicates the outcome of the removal: 1 for annihilation (when the S-comp ceases to be cheaper that cost-2 connections), 2 for the decrease in the number of terminals and 3 for a removal of a fork.

Case 1: the new connection causes a deletion of a non-edge. This entails $\Delta CR = -2$ and $\Delta PC = 1/2$, so we have balance zero.

Case 2: the new connection causes a deletion of a connection in a 3-star. This entails $\Delta CR = -1$, $\Delta PE = -3/4$ and $\Delta PS = 1/4$ for the balance of $-3/2 + 1 + 3/4 - 1/4 = 0$. (Second deletion, if any, will be treated as Case 1.)

Case 3: the new connection causes the first deletion of a connection in a 3-comet S. Because no 3-star survives phase 4, we can assume that this is a connection from the center to a terminal, which alters $(a, 3)$-comet S into $(a, 2)$-comet S'. We have $\Delta CR = 1$, $\Delta PE = 1/4$, $\Delta PS = p_4(S) - p_6(S') = 4/12$ for the balance of $-3/2 + 1 + 1/4 + 4/12 = 1/12$. (Second deletion will be treated as Case 4.2).

Case 4.1: the new connection causes the first deletion of a connection in a $(a, 2)$-comet.

Case 4.1.1: S is "annihilated", which implies that S is a $(1,2)$-comet. We have $\Delta CR = 1$, $\Delta PE = 5/4$, $\Delta PS = p_4(S) = -1/4$ for the balance of $-3/2 + 1 + 5/4 - 1/4 = 1/2$.

Case 4.1.2: the deletion is from the center to a terminal, so S' is a $(a, 1)$-comet. We have $\Delta CR = 1$, $\Delta PE = 1/4$, $\Delta PS = p_4(S) - p_6(S') = 3/4$ for the balance of $-3/2 + 1 + 1/4 + 3/4 = 1/2$.

Case 4.1.3: the deletion is in a fork, so S' is an $(a-1, 2)$-comet. This entails $\Delta CR = 1$, $\Delta PE = 3/4$ and $\Delta PS = 4/12$ for the balance of $-3/2 + 1 + 3/4 + 4/12 = 7/12$.

Case 4.2: the new connection causes a subsequent deletion in what is now a $(a, 2)$-comet. We have similar cases, but with balance lower by $4/12$, as our starting potential of S is lower; thus the balance of 4.2.1, 4.2.2, 4.2.3 is $1/6$, $1/6$ and $1/4$ respectively.

Case 5.1: the new connection causes the first deletion of a connection in a $(a, 1)$-comet.

Case 5.1.1: S is "annihilated", which implies that S is a $(2,1)$-comet. We have $\Delta CR = 1$, $\Delta PE = 7/4$, $\Delta PS = p_4(S) = -1/2$ for the balance of $-3/2 + 1 + 7/4 - 1/2 = 3/4$.

Case 5.1.2: the deletion is from the center to a terminal, so S' is a $(a, 0)$-comet. We have $\Delta CR = 1$, $\Delta PE = 1/4$, $\Delta PS = p_4(S) - p_6(S') = 19/20$ for the balance of $-3/2 + 1 + 1/4 + 19/20 = 7/10$.

Case 5.1.3: the deletion is in a fork, so S' is an $(a-1, 1)$-comet. This entails $\Delta CR = 1$, $\Delta PE = 3/4$ and $\Delta PS = 1/2$ for the balance of $-3/2 + 1 + 3/4 + 1/2 = 3/4$.

Case 5.2: the new connection causes a subsequent deletion in what is now a $(a, 1)$-comet. We have similar cases, but with balance lower by $1/2$, as our starting potential of S is lower; thus the balance of 5.2.1, 5.2.2, 5.2.3 is $1/4$, $1/5$ and $1/4$ respectively.

Case 6: the new connection causes a deletion in a 0-comet. The calculation is similar to the previous cases except that the balance is even more favorable, except for the case of subsequent deletion that removes a fork, the balance is the same $1/4$ for 2-, 1- and 0-stars.

As we see, no deletion has negative balance. Now we can make the final accounting that includes the potential of C-comps. We have to pay for "annihilating" the negative potential, which amounts to $1/2$ per affected C-comp.

Coalescing C-comps increases the potential, but it is accounted for, because this is Case 1. Removing connections when we annihilate forks or non-viable comets may sometimes decrease the potential (when we separate a C-comp that consists of a single S-comp, 1- or 2-comet), and we disregard it. Thus we need to account for $1/2$ in each of the resulting C-comps.

If the "history" of a resulting C-comp C involves a deletion in a pre-existing 2-, 1- or 0-comet, we had Case 4.1, 5.1 or 6.1 with balance at least $1/2$. If C includes a comet S in which we had no deletions, we have surplus balance $p_4(S) - p_6(S) \geq 1/3$. Thus we have no problem if there are two such comets.

Now suppose that C contains a comet S' that was obtained from a 3-comet S. If S was affected by one deletion only, in Phase 5 we can replace the triple that caused that with S; this would lead to a net gain of two "cheap" connections that provide a gain of $1/2$ each (or less, if they cause subsequent removals in pre-existing 2-, 1-, or 0-comets), thus a gain of 1. Because inserting a (1,3)-comet makes 4 connections, greedily we can find 1/4-th o them (or more, if they caused removals in pre-existing 2-, 1- or 0-comets, rather than in the "correct" 3-comets). Therefore with a 3-comet we associate a gain of at least 1/4. (We left out the case when a 3-comet is annihilated by two removals and it annihilates its C-comp; if this is done by two 3-stars, they coalesce with some other C-comps and thus Case 1 pays for that, and if it was done by a single 3-star, we still have the ability of replacing it with S.)

If C contains two comets, then each can provide at least 1/4 toward the increase of $p(C)$. The same holds if it contains a single 0-comet (we have larger drop $p_4(S) - p_6(S)$). If C contains one larger comet, then the drop is reduced by at least 1/4.

What remains is to consider former C-comps that consisted solely of 3-stars and all these 3-stars were annihilated. Observe that because Phase 4 uses an exact algorithm for maximizing the number of 3-stars, we found at least as many stars as were annihilated, so we did not increase the solution cost (while the sum of potentials that vanished in the process could not be negative, positive PE always dominates negative potentials of S-comps and C-comps).

5.6 Analysis of Phase 6

Basically, when we contract a comet with $i+1$ terminals, we have to delete i connections from T_R, and the accounting is like in Cases 4.2, 5.2 and 6.2 of the previous sections, except that we have a higher ΔCA, which was in that calculation assumed to be $3/2$.

If we remove from a 2-comet, then the portion of the cost per connection is at most $5/3 = 3/2 + 1/6$, so it suffices that in the respective subcases of Case 4.2 we had balance at least $1/6$. Similarly, if we delete from a 1-comet, the cost per connection is at most $7/4 = 3/1/4$ and it suffices that we had a balance of $1/4$ in Case 5.2. For 0-stars, the cost is at most $9/5 = 3/2 + 3/10$, and when we anihilate, we have such balance in Case 6.2.1. When we remove a fork without annihilation, then we have at least (4,1)-comet and the cost is at most $12/7 < 3/2 + 1/4$, and the balance of the fork removal is $1/4$.

Thus we proved Theorem 2.

References

1. Bern, M., Plassmann, P.: The Steiner problem with edge lengths 1 and 2. Information Processing letters 32, 171–176 (1989)
2. Rayward-Smith, V.J.: The computation of nearly minimal Steiner trees in graphs. Internat. J. Math. Educ. Sci. Tech. 14, 15–23 (1983)
3. Robins, G., Zelikovsky, A.: Tighter Bounds for Graph Steiner Tree Approximation. SIAM Journal on Discrete Mathematics 19(1), 122–134 (2005); Preliminary version appeared in Proc. SODA 2000, pp. 770–779
4. Gabow, H.N., Stallmann, M.: An augmenting path algorithm for linear matroid parity. Combinatorica 6, 123–150 (1986)

Succinct Orthogonal Range Search Structures on a Grid with Applications to Text Indexing*

Prosenjit Bose[1], Meng He[2], Anil Maheshwari[1], and Pat Morin[1]

[1] School of Computer Science, Carleton University, Canada
{jit,anil,morin}@scs.carleton.ca
[2] Cheriton School of Computer Science, University of Waterloo, Canada
mhe@uwaterloo.ca

Abstract. We present a succinct representation of a set of n points on an $n \times n$ grid using $n \lg n + o(n \lg n)$ bits[1] to support orthogonal range counting in $O(\lg n / \lg \lg n)$ time, and range reporting in $O(k \lg n / \lg \lg n)$ time, where k is the size of the output. This achieves an improvement on query time by a factor of $\lg \lg n$ upon the previous result of Mäkinen and Navarro [1], while using essentially the information-theoretic minimum space. Our data structure not only can be used as a key component in solutions to the general orthogonal range search problem to save storage cost, but also has applications in text indexing. In particular, we apply it to improve two previous space-efficient text indexes that support substring search [2] and position-restricted substring search [1]. We also use it to extend previous results on succinct representations of sequences of small integers, and to design succinct data structures supporting certain types of orthogonal range query in the plane.

1   Introduction

The two-dimensional *orthogonal range search* problem is a fundamental problem in computational geometry. In this problem, we store a set, N, of points in a data structure so that given a query rectangle R, information about the points in R can be retrieved efficiently. There are two common types of queries: *orthogonal range counting* queries and *orthogonal range reporting* queries. An orthogonal range counting query returns the number of points in $N \cap R$, and an orthogonal range reporting query returns these points. The orthogonal range search problem has applications in many areas of computer science, including databases and computer graphics, and thus has been studied extensively [3,4,5,6,7]. Many trade-offs for this problem have been achieved. For example, for the two-dimensional range reporting query, there are data structures achieving the optimal $O(\lg n + k)$ query time using $O(n \lg^\epsilon n)$ words of space, where k is the size of the output and $0 < \epsilon < 1$ [6], and structures of linear space that answer queries in $O(\lg n + k \lg^\epsilon n)$ time [4]. See [7] for a recent survey.

---

* This work was supported by NSERC of Canada. The work was done when the second author was in School of Computer Science, Carleton University, Canada.

[1] $\lg n$ denotes $\log_2 n$.

F. Dehne et al. (Eds.): WADS 2009, LNCS 5664, pp. 98–109, 2009.
© Springer-Verlag Berlin Heidelberg 2009

In this paper, we mainly study the orthogonal range search problem in two-dimensional *rank space*, i.e. on an $n \times n$ grid, where n is the size of the point set. The general orthogonal range search problem in which the points are real numbers can be reduced to this problem using a standard approach [3]. Thus, solutions to the general range search problem are often based on range search structures in rank space [3,6]. For example, one key component for the data structure achieving the optimal $O(\lg n + k)$ query time for orthogonal range reporting by Alstrup *et al.* [6] is a data structure supporting orthogonal range reporting in rank space in $O(\lg \lg n + k)$ time using $O(n \lg^{\epsilon} n)$ words of space.

More recently, the orthogonal range search problem on an $n \times n$ grid was studied to design *succinct data structures*, and in particular succinct text indexes. Succinct data structures provide solutions to reduce the storage cost of modern applications that process huge amounts of data, such as textual data in databases and on the World Wide Web, geometric data in GIS systems, and genomic data in bioinformatics applications. They were first proposed by Jacobson [8] to encode bit vectors, (unlabeled) trees and planar graphs using space close to the information-theoretic lower bound, while supporting efficient navigation operations in them. For example, Jacobson showed how to represent a tree on n nodes using $2n + o(n)$ bits, so that the parent and the children of a node can be efficiently located. The obvious approach uses $3n$ words, which is about 96 times as much as the space required for the succinct representation on a 64-bit machine. This approach was also successfully applied to various other abstract data types, including dictionaries [9], strings [10,11,12], binary relations [11,12] and labeled trees [13,14,11,12]. For orthogonal range search in rank space, Mäkinen and Navarro [1] designed a succinct data structure that encodes the point set using $n \lg n + o(n \lg n)$ bits to support orthogonal range counting in $O(\lg n)$ time and range reporting in $O(k \lg n)$ time. This space cost is close to the information-theoretic minimum, but their structure requires that there does not exist two points in the set with the same coordinate in one of the two dimensions.

The succinct range search structure mentioned in the previous paragraph was further used to design space-efficient text indexes. Mäkinen and Navarro [1] initially designed this structure for the problem of *position-restricted substring search*. The goal is to construct an index for a text string T of length n such that given a query substring P of length m and a range $[i..j]$ of positions in T, the occurrences of P in this range can be reported efficiently. They showed how to reduce the problem to orthogonal range search on an $n \times n$ grid, and designed a text index of $3n \lg n + o(n \lg n)$ bits to support position-restricted substring search in $O(m + \text{occ} \lg n)$ time, where occ is the number of occurrences of P in T. Chien *et al.* [2] considered the problem of indexing a text string to support the general *substring search* that reports the occurrences of a query pattern P in a text string T. This is a fundamental problem in computer science. They designed a succinct text index using $O(n \lg \sigma)$ bits, where σ is the alphabet size, to support substring search in $O(m + \lg n(\lg_{\sigma} n + \text{occ} \lg n))$ time. One key data structure in their solution is the same succinct range search structure [1].

1.1   Our Results

In this paper, we design succinct data structures for orthogonal range search on an $n \times n$ grid. Our range search structure is an improvement upon that of Mäkinen and Navarro [1], and we use it to improve previous results on designing space-efficient text indexes for substring search [2] and position-restricted substring search [1]. We also apply our structure to extend the previous result on representing a sequence of small integers succinctly [15], as well as a restricted version of orthogonal range search in which the query range is defined by two points in the point set [16]. More precisely, we present the following results, among which the first one is our main result and the rest are its applications:

1. A succinct data structure that encodes a point set, N, of n points in an $n \times n$ grid using $n \lg n + o(n \lg n)$ bits to support orthogonal range counting in $O(\lg n / \lg \lg n)$ time, and orthogonal range reporting in $O(k \lg n / \lg \lg n)$ time, where k is the size of the output. Compared to the succinct structure of Mäkinen and Navarro [1], this data structure achieves an improvement on query time by a factor of $\lg \lg n$, while still using space close to the information-theoretic minimum. Another improvement is that our structure does not require each point to have a distinct x-coordinate or y-coordinate.

2. A succinct text index of $O(n \lg \sigma)$ bits for a text string T of length n over an alphabet of size σ that supports substring search in $O(m + \lg n (\lg_\sigma n + \text{occ} \lg n) / \lg \lg n)$ time, where m is the length of the query substring, and occ is the number of its occurrences in T. This provides faster query support than the structure of Chien et al. [2] while using the same amounts of space.

3. A text index of $3n \lg n + o(n \lg n)$ bits that supports position-restricted substring search in $O(m + \text{occ} \lg n / \lg \lg n)$ time. This improves the query time of the index of Mäkinen and Navarro [1] using the same amount of space.

4. A succinct data structure that encodes a sequence, S, of n numbers in $[1..s]$, where $s = \texttt{polylog}(n)$, in $n H_0(S) + o(n)$ bits[2] to support the following query in constant time: given a range, $[p_1..p_2]$, of positions in S and a range, $[v_1..v_2]$, of values, compute the number of entries in $S[p_1..p_2]$ whose values are in the range $[v_1..v_2]$. These entries can also be reported in constant time per entry. This extends the result of Ferragina et al. [15] on the same input data to support more operations.

5. A space-efficient data structure that encodes a point set N in the plane in $cn + n \lg n + o(n \lg n)$ bits, where c is the number of bits required to encode the coordinate pair of a point, to provide $O(\lg n / \lg \lg n)$-time support for a restricted version of orthogonal range counting in which the query rectangle is defined by two points in N. The points in the query range can be reported in $O(k \lg n / \lg \lg n)$ time.

All our results are under the word RAM model of $\Theta(\lg n)$-bit word size.

---

[2] $H_0(S)$ is the zeroth-order empirical entropy of S, defined as $\sum_{i=1}^{s} (p_i \log_2 \frac{1}{p_i})$, where p_i is the frequency of the occurrence of i in S, and $0 \log_2 0$ is interpreted as 0.

2   Preliminaries

Bit vectors. A key structure for many succinct data structures and for our research is a bit vector $B[1..n]$ that supports rank and select operations. For $\alpha \in \{0, 1\}$, the operator $\text{rank}_B(\alpha, x)$ returns the number of occurrences of α in $B[1..x]$, and $\text{select}_B(\alpha, r)$ returns the position of the rth occurrence of α in B. Lemma 1 addresses the problem of succinct representations of bit vectors.

Lemma 1 ([8,17]). *A bit vector $B[1..n]$ with v 1s can be represented using $n + o(n)$ bits to support the access to each bit, rank and select in $O(1)$ time.*

Sequences of small numbers. The rank/select operations can also be performed on a sequence, S, of n integers in $[1..s]$. To define $\text{rank}_S(\alpha, x)$ and $\text{select}_S(\alpha, r)$, we simply let $\alpha \in \{1, 2, \cdots, s\}$ to extend the definitions of these operations on bit vectors. Ferragina *et al.* [15] proved the following lemma:

Lemma 2 ([15]). *A sequence, S, of n numbers in $[1..s]$, where $2 \leq s \leq \sqrt{n}$, can be represented using $nH_0(S) + O(s(n \lg \lg n)/\log_s n)$ bits to support the access of each number, rank and select in $O(1)$ time.*

3   Succinct Range Search Structures on a Grid

In this section, we design a succinct data structure that supports orthogonal range search in rank space. We first design a structure for a narrow grid (more precisely, an $n \times O(\lg^\epsilon n)$ grid) in Section 3.1 with the restriction that each point has a distinct x-coordinate. Based on this structure, we further design structures for an $n \times n$ grid in Section 3.2 without any similar restrictions.

3.1   Orthogonal Range Search on an $n \times O(\lg^\epsilon n)$ Grid

We first consider range counting on a narrow grid. We make use of the well-known fact that the orthogonal range counting problem can be reduced to *dominance counting* queries. A point whose coordinates are (x_1, y_1) *dominates* another point (x_2, y_2) if $x_1 \geq x_2$ and $y_1 \geq y_2$, and a dominance counting query computes the number of points dominating the query point.

Lemma 3. *Let N be a set of points from the universe $M = [1..n] \times [1..t]$, where $n = |N|$ and $t = O(\lg^\epsilon n)$ for any constant ϵ such that $0 < \epsilon < 1$. If each point in N has a distinct x-coordinate, the set N can be represented using $n\lceil \lg t \rceil + o(n)$ bits to support orthogonal range counting in $O(1)$ time.*

Proof. As each point in N has a distinct x-coordinate, we can store the coordinates as a sequence $S[1..n]$, in which $S[i]$ stores the y-coordinate of the point whose x-coordinate is i. Thus S occupies $n\lceil \lg t \rceil$ bits, and it suffices to show how to construct auxiliary data structures of $o(n)$ bits to support dominance counting (recall that dominance counting can be used to support range counting).

We first partition the universe M into regions called *blocks* of size $\lceil \lg^2 n \rceil \times t$ by dividing the first dimension into ranges of size $\lceil \lg^2 n \rceil$. More precisely, the ith block of M under this partition is $L_i = [(i-1)\lceil \lg^2 n \rceil + 1..i\lceil \lg^2 n \rceil] \times [1..t]$. We assume that n is divisible by $\lceil \lg^2 n \rceil$ for simplicity.

For each block L_i, we further partition it into *subblocks* of size $\lceil \lg^\lambda n \rceil \times t$ by dividing the first dimension into ranges of size $\lceil \lg^\lambda n \rceil$, where λ is a constant such that $\epsilon < \lambda < 1$. Under this partition, the jth subblock of L_i is $L_{i,j} = [(i-1)\lceil \lg^2 n \rceil + (j-1)\lceil \lg^\lambda n \rceil + 1..(i-1)\lceil \lg^2 n \rceil + j\lceil \lg^\lambda n \rceil] \times [1..t]$. For simplicity, we assume that $\lceil \lg^2 n \rceil$ is divisible by $\lceil \lg^\lambda n \rceil$.

We construct the following auxiliary data structures:

- A two-dimensional array $A[1..n/\lceil \lg^2 n \rceil, 1..t]$, in which $A[i,j]$ stores the number of points in N dominating the coordinate pair $(i\lceil \lg^2 n \rceil, j)$;
- A two-dimensional array $B[1..n/\lceil \lg^\lambda n \rceil, 1..t]$, in which $B[i,j]$ stores the number of points in N dominating the coordinate pair $(i\lceil \lg^\lambda n \rceil, j)$ in the block that contains this coordinate pair;
- A table C that stores for each possible set of $\lceil \lg^\lambda n \rceil$ points in the universe $[1..\lceil \lg^\lambda n \rceil] \times [1..t]$ (each point in this set has a distinct x-coordinate), every integer i in $[1..\lceil \lg^\lambda n \rceil]$ and every integer j in $[1..t]$, the number of points in this set that dominates the coordinate pair (i,j).

We now analyze the space costs of the above data structures. A occupies $n/\lceil \lg^2 n \rceil \times t \times \lceil \lg n \rceil = O(n/\lg^{1-\epsilon} n) = o(n)$ bits. As there are $\lceil \lg^2 n \rceil$ points inside each block, each entry of B can be stored in $O(\lg \lg n)$ bits. Therefore, B occupies $n/\lceil \lg^\lambda n \rceil \times t \times O(\lg \lg n) = O(n \lg \lg n / \lg^{\lambda-\epsilon} n) = o(n)$ bits. To compute the space cost of C, we first count the number, b, of possible $\lceil \lg^\lambda n \rceil$-point set in the universe $[1..\lceil \lg^\lambda n \rceil] \times [1..t]$, where each point in this set has a distinct x-coordinate. We have $b = t^{\lceil \lg^\lambda n \rceil} = 2^{\lceil \lg^\lambda n \rceil \lg t}$. Let $f = \lceil \lg^\lambda n \rceil \lg t$. Then $f = O(\lg^\lambda n \lg \lg n) = o(\lg n)$. By the definition of order notation, there exists a constant n_0 such that $f < \frac{1}{2} \lg n$ for any $n > n_0$. As $b = 2^f$, we have $b < 2^{\frac{1}{2} \lg n} = \sqrt{n}$ when $n > n_0$. Therefore, when $n > n_0$, the space cost of C in bits is less than $\sqrt{n} \times \lceil \lg^\lambda n \rceil \times t$. Thus, the space cost of C is $O(\sqrt{n} \lg^{\lambda+\epsilon} n) = o(n)$ bits. Hence the auxiliary data structures occupy $O(n \lg \lg n / \lg^{\lambda-\epsilon} n) = o(n)$ bits in total.

With the above data structures, we can support dominance counting. Let (u,v) be the coordinates of the query point q. Let L_i and $L_{i,j}$ be the block and subblock that contain q, respectively. The result is the sum of the following three values: k_1, the number of points in blocks $L_{i+1}, L_{i+2}, \cdots$ that dominate q; k_2, the number of points in subblocks $L_{i,j+1}, L_{i,j+2}, \cdots, L_{i,v}$ that dominate q, where v is the number of subblocks in block L_i; and k_3, the number of points in subblock $L_{i,j}$ that dominate q. By the definitions of the data structures we constructed, we have $k_1 = A[i,y]$ and $k_2 = B[(i-1) \times \lceil \lg^2 n \rceil / \lceil \lg^\lambda n \rceil + j, y]$. To compute k_3, we first compute the coordinates of q inside block $L_{i,j}$ by treating $L_{i,j}$ as a universe of size $\lceil \lg^\lambda n \rceil \times t$, and get the encoding of the subsequence of S that corresponds to points inside $L_{i,j}$. With these we can perform table lookup on C to compute k_3 in constant time. $\qquad \square$

We next show how to support range reporting.

Lemma 4. *Let N be a set of points from the universe $M = [1..n] \times [1..t]$, where $n = |N|$ and $t = O(\lg^\epsilon n)$ for any constant ϵ such that $0 < \epsilon < 1$. If each point in N has a distinct x-coordinate, the set N can be represented using $n\lceil \lg t \rceil + o(n)$ bits to support orthogonal range reporting in $O(k)$ time, where k is the size of the output.*

Proof. As with the proof of Lemma 3, we encode N as the string S, and divide M into blocks and subblocks. Based on this, we design auxiliary data structures to support orthogonal range reporting. We answer a query in two steps. Given a query rectangle R, we first compute the set, Y, of y-coordinates of the output in $O(k')$ time, where k' is the number of distinct y-coordinates of the points in the output. Then, for each y-coordinate, v, in Y, we compute the points in R whose y-coordinate is v (we spend $O(1)$ time on each such point).

We first show how to compute Y in $O(k')$ time. We construct the following auxiliary data structures:

- A two-dimensional array $D[1..n/\lceil \lg^2 n \rceil, 1..\lceil \lg n \rceil]$. Each entry, $D[i, j]$, stores a bit vector of length t whose l^{th} bit is 1 iff there is at least one point (from the set N) in blocks $L_i, L_{i+1}, \cdots, L_{i+2^j-1}$ whose y-coordinate is l;
- A two-dimensional array $E_i[1..w][1..\lceil \lg w \rceil]$ for each block L_i, where $w = \lceil \lg^2 n \rceil / \lceil \lg^\lambda n \rceil$ (i.e. the maximum number of subblocks in a given block). Each entry, $E_i[j, u]$, stores a bit vector of length t whose l^{th} bit is 1 iff there is at least one point (from the set N) in subblocks $L_{i,j}, L_{i,j+1}, \cdots, L_{i,j+2^u-1}$ whose y-coordinate is l;
- A table F which stores for every possible set of $\lceil \lg^\lambda n \rceil$ point in the universe $[1..\lceil \lg^\lambda n \rceil] \times [1..t]$ (each point in this set has a distinct x-coordinate), every pair of integers i and j in $[1..\lceil \lg^\lambda n \rceil]$, a bit vector of length t whose l^{th} bit is 1 iff there is at least one point from this set whose x-coordinate is between (and including) i and j and whose y-coordinate is l.

To analyze the space cost, we have that D occupies $O(n/\lg^2 n \times \lg n \times t) = O(n/\lg^{1-\epsilon} n)$ bits. As there are $n/\lceil \lg^\lambda n \rceil$ subblocks in total, all the E_i's occupy $O(n/\lg^\lambda n \times \lg \lg n \times \lg^\epsilon n) = O(n \lg \lg n/\lg^{\lambda-\epsilon})$ bits. Similarly to the analysis in the proof of Lemma 3, we have F occupies $O(\sqrt{n} \times \lceil \lg^\lambda n \rceil \times \lceil \lg^\lambda n \rceil \times \lceil \lg^\epsilon n \rceil)$ bits. Therefore, these data structures occupy $O(n \lg \lg n/\lg^{\lambda-\epsilon}) = o(n)$ bits.

To use the above data structures to compute Y, let $R = [x_1..x_2] \times [y_1..y_2]$ be the query rectangle. We first show how to compute a bit vector Z of length t, where $Z[i] = 1$ iff there is a point from N whose y-coordinate is i and whose x-coordinates are between (and including) x_1 and x_2. Let $L_{a,b}$ and $L_{c,d}$ be the two subblocks whose ranges of x-coordinates contain x_1 and x_2, respectively. Assume that $a < c$ (the case in which $a = c$ can be handled similarly). Then Z is the result of bitwise OR operation on the the following five bit vectors of length t:

- Z_1, where $Z_1[i] = 1$ iff there is a point in blocks $L_{a+1}, L_{a+2}, \cdots, L_{c-1}$ whose y-coordinate is i;
- Z_2, where $Z_2[i] = 1$ iff there is a point in subblocks $L_{a,b+1}, L_{a,b+2}, \cdots, L_{a,q}$ whose y-coordinate is i (let $L_{a,q}$ be the last subblock in block L_a);

- Z_3, where $Z_3[i] = 1$ iff there is a point in subblocks $L_{c,1}, L_{c,2}, \cdots, L_{c,d-1}$ whose y-coordinate is i;
- Z_4, where $Z_4[i] = 1$ iff there is a point in subblock $L_{a,b}$ that is in the query rectangle and whose y-coordinate is i;
- Z_5, where $Z_4[i] = 1$ iff there is a point in subblock $L_{a,b}$ that is in the query rectangle and whose y-coordinate is i.

To compute Z_1, we first observe that the corresponding range of indexes of blocks is $[a+1..c-1] = [a+1, a+2^g] \cup [c-2^g, c-1]$, where $g = \lfloor \lg(c-a-2) \rfloor$ (similar ideas were used by Bender and Farach-Colton to support range minimum queries [18]). Hence Z_1 is the result of bitwise OR operation on the bit vectors stored in $D[a+1, g]$ and $D[c-2^g, g]$. Z_2 and Z_3 can be computed in a similar way using E_a and E_c. Z_4 and Z_5 can be computed by performing table lookups on F in constant time. Therefore, Z can be computed in constant time.

To compute Y using Z in $O(k')$ time, it suffices to support rank and select operations on Y in constant time. As Y is of size $t = O(\lg^\epsilon n)$, this can be achieved by precomputing a table of $o(n)$ bits [17].

To further report the points in R, we observe that we store the coordinates as a sequence, S, of numbers in $[1..t]$. The data structures of Ferragina *et al.* [15] designed for Lemma 2 has two parts: a compressed encoding of the sequence of $nH_0(S)$ bits and an auxiliary data structure of $O(s(n \lg \lg n)/\lg_s n)$ bits. Their data structures still work if we replace the first part by the uncompressed version of the original sequence. Thus we can construct the auxiliary data structures in Lemma 2 to support rank and select on S in constant time. As $t = O(\lg^\epsilon n)$, these data structures occupy $O(n(\lg \lg n)^2/\lg^{1-\epsilon} n) = o(n)$ bits. For each y-coordinate, v, in Y, the set of the points in R whose y-coordinates are equal to v can be computed by performing rank and select operations on S, which takes constant time per point in the output. $\qquad\Box$

As Lemma 3 and Lemma 4 both encode and store the coordinates in the same sequence and build auxiliary structures of $o(n)$ bits, we can combine them:

Lemma 5. *Let N be a set of points from the universe $M = [1..n] \times [1..t]$, where $n = |N|$ and $t = O(\lg^\epsilon n)$ for any constant ϵ such that $0 < \epsilon < 1$. If each point in N has a distinct x-coordinate, the set N can be represented using $n\lceil \lg t \rceil + o(n)$ bits to support orthogonal range counting in $O(1)$ time, and orthogonal range reporting in $O(k)$ time, where k is the size of the output.*

3.2 Orthogonal Range Search on an $n \times n$ Grid

To design succinct data structures for range search in rank space, we first consider range counting with the restriction that each point has a distinct x-coordinate.

Lemma 6. *Let N be a set of points from the universe $M = [1..n] \times [1..n]$, where $n = |N|$. If each point in N has a distinct x-coordinate, the set N can be represented using $n \lg n + o(n \lg n)$ bits to support orthogonal range counting in $O(\lg n/\lg \lg n)$ time.*

Proof. The main idea is to combine the techniques of Lemma 3 with the generalized wavelet tree structures proposed by Ferragina *et al.* [15] to design a representation of N, based on which we design algorithms to support range counting.

We construct our structure recursively; at each level, we construct an orthogonal range counting structure over a set of points whose y-coordinates are in the range $[1..t]$ using Lemma 3, where $t = O(\lg^\epsilon n)$ for any constant ϵ such that $0 < \epsilon < 1$. At the first (i.e. top) level, we consider a conceptual point set N_1 from the universe $M_1 = [1..n] \times [1..t]$. N_1 can be obtained by dividing the range of y-coordinates in the universe M into t ranges of the same size, and there is a point (a, b) in N_1 iff there is a point in N whose x-coordinate is a, and whose y-coordinate is in the ith range. More precisely, if there is a point (x, y) in N, then there is a point $(x, \lfloor y/(n/t) \rfloor)$ in N_1. We then construct an orthogonal range counting structure, C_1 for N_1 using Lemma 3. Note that when we use the approach of Lemma 3 to construct C_1, we store the set N_1 as a sequence S_1 over alphabet $[t]$ in which $S_1[i]$ stores the y-coordinate of the point whose x-coordinate is i. The approach of Lemma 2 can be used here to construct an auxiliary structure of $o(n)$ bits to support rank/select operations on S_1 in constant time. The space cost of the data structures constructed for the first level is clearly $n\lceil \lg t \rceil + o(n)$ bits.

At the second level, we consider t conceptual point sets $N_{2,1}, N_{2,2}, \cdots, N_{2,t}$. The set $N_{2,i}$ is from the universe $M_{2,i} = [1..n_{2,i}] \times [1..t]$, which corresponds to the ith range of y-coordinates of M for the level above (i.e. the first level), and $n_{2,i}$ is the number of points in N whose y-coordinates are in this range. We further divide this range into t subranges of the same size, such that the point (x, y) is in $N_{2,i}$ iff there is a point in N_1 whose x-coordinate is $\text{select}_{S_1}(i, x)$, and whose y-coordinate is in the yth subrange of $M_{2,i}$. Note that $\sum_{i=1}^{t} n_{2,i} = n$. Thus, if we combine all the universes $M_{2,1}, M_{2,2}, \cdots, M_{2,t}$ such that the universe $M_{2,i-1}$ is adjacent and to the left of $M_{2,i}$, we can get a universe $M_2 = [1..n] \times [1..t]$. We also transform the coordinates of the points in $N_{2,1}, N_{2,2}, \cdots, N_{2,t}$ into coordinates in the universe M_2, and denote the set that contains all these (n) points N_2. We construct an orthogonal range counting structure, C_2, for N_2 using Lemma 3 (same as S_1, S_2 denotes the corresponding string). We can count the total number of points in the sets $N_{2,1}, N_{2,2}, \cdots, N_{2,j-1}$ in constant time for any given j by performing range counting on C_1. Thus, we can determine the range of x-coordinates in C_2 that correspond to the points in $C_{2,i}$ in constant time, which allows us to use C_2 to answer orthogonal range queries on each set $C_{2,i}$ in constant time. We also construct the auxiliary structures of Lemma 2 to support rank/select operations on C_2 in constant time, which can be further used to support rank/select operations on each substring of S_2 that corresponds to the set $C_{2,i}$. The total space of the data structures constructed for the second level is thus $n\lceil \lg t \rceil + o(n)$ bits.

We continue the above process recursively, and at each level l, we construct t point sets for each point set considered at level $l - 1$. Figure 1 illustrates the hierarchy of our structure. This structures use $n\lceil \lg t \rceil + o(n)$ bits for each level

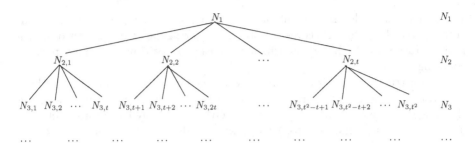

Fig. 1. The hierarchy of the data structures in Lemma 6

to support orthogonal range counting in each point set at this level, as well as rank/select operations on the substrings of S_l corresponding to each set. Note that the substring of S_l and the sub-universe of M_l that correspond to any set at this level can be located in a top-down traversal, performing range counting at each level in constant time until we reach level l. We continue this process until we can no longer divide a point set into t subsets (i.e. the y-coordinates of the points in this set are from a range of size t in M). Thus our structures have $\log_t n$ levels, and the set of data structures for each level occupy $n\lceil \lg t\rceil + o(n)$ bits. Therefore, the overall space of our structures is $n \lg n + o(n \lg n)$ bits.

We now design a recursive algorithm to support orthogonal range counting using our data structures. Let $R = [x_1..x_2] \times [y_1..y_2]$ be the query rectangle. We consider the case in which $y_2 - y_1 \geq n/t$. Let $z_1 = \lceil y_1/(n/t)\rceil$ and $z_2 = \lfloor y_2/(n/t)\rfloor$. Then R can be partitioned into three query rectangles: $R_1 = [x_1..x_2] \times [y_1..z_1(n/t)]$, $R_2 = [x_1..x_2] \times [z_1(n/t) + 1..z_2(n/t)]$ and $R_3 = [x_1..x_2] \times [z_2(n/t) + 1..y_2]$. The result is the sum of the numbers, r_1, r_2 and r_3, of points in R_1, R_2 and R_3, respectively. We observe that r_2 can be computed by performing an orthogonal range counting query over the structure C_1, using $[x_1..x_2] \times [z_1..z_2 - 1]$ as the query rectangle. Thus we need only compute r_1 (the computation of r_3 is similar). Note that R_2 is the maximum sub-rectangle of R whose range of y-coordinates starts with a multiple of n/t and whose width is divisible by n/t, and we use C_1 to compute the number of points in it. Using the same strategy, we can compute the maximum sub-rectangle of R_1 whose range of y-coordinates starts with a multiple of n/t^2 and whose width is divisible by n/t^2, and we use C_{2,z_1-1} to compute this result. To perform this query on C_{2,z_1-1}, we need to scale down the range of x-coordinates of R_1 to $[\mathbf{rank}_{S_1}(z_1 - 1, x_1)..\mathbf{rank}_{S_1}(z_1 - 1, x_2)]$. The number of points in the remaining part of R_1 (note that they are all above this sub-rectangle) can be computed in a recursive fashion using the same approach. Thus r_1 can be computed by performing a top-down traversal to at most the bottom level, and we require constant time per level. Therefore, r_1 can be computed in $O(\log_t n) = O(\lg n/ \lg\lg n)$ time. Hence we can compute r in $O(\lg n/ \lg\lg n)$ time. The case in which $y_2 - y_1 < n/t$ can be handled similarly. □

We now remove the restriction that each point has a distinct x-coordinate:

Lemma 7. *Let N be a set of points from the universe $M = [1..n] \times [1..n]$, where $n = |N|$. N can be represented using $n \lg n + o(n \lg n)$ bits to support orthogonal range counting in $O(\lg n / \lg \lg n)$ time.*

Proof. We construct a point set N' whose points have distinct x-coordinates as follows: Sort the points in N in increasing order using their x-coordinates as the primary key and their y-coordinates as the secondary key. If the ith point in this order has y-coordinate y_i, we add the point (i, y_i) into N'. We also construct a bit vector C to encode the number of points having the same x-coordinates. More precisely, $C = 10^{k_1} 10^{k_2} \cdots 10^{k_n}$, where k_j is the number of points whose x-coordinates are j.

We represent N' using Lemma 6 with $n \lg n + o(n \lg n)$ bits. There are n 1s and n 0s in C, so we can represent C in $2n + o(n)$ bits using Lemma 1 to support rank/select operations. The overall space cost of our data structures is $n \lg n + o(n \lg n)$ bits.

To answer an orthogonal range query, let $R = [x_1..x_2] \times [y_1..y_2]$ be the query rectangle. Consider the points from N that is in R. We observe that the points in N' corresponding to them are in the rectangle $R' = [\text{rank}_C(0, \text{select}_C(1, x_1)) + 1..\text{rank}_C(0, \text{select}_C(1, x_2+1))] \times [y_1..y_2]$. Thus we need only perform an orthogonal range counting query on N' using R' as the query rectangle. □

Lemma 8. *Let N be a set of points from the universe $M = [1..n] \times [1..n]$, where $n = |N|$. N can be represented using $n \lg n + o(n \lg n)$ bits to support orthogonal range reporting in $O(k \lg n / \lg \lg n)$ time, where k is the size of the output.*

Proof. We only consider the case in which each point has a distinct x-coordinate; the approach in Lemma 7 can be used to extend this to the more general case.

We use the approach of Lemma 6 to construct a hierarchy of structures. The only difference is that at each level i, we use Lemma 4 to construct C_i to support range reporting on N_i. The same algorithm can be used to support orthogonal range reporting; at each level we report a set of points in the answer. The challenge here is how to get the original coordinates of each point reported at the ith level. Let (x, y) be the coordinates of a point, v, reported in the set $N_{j,k}$. Then the set $N_{j-1,\lceil k/t \rceil}$ contains v in the level above. Note that in previous steps, we have computed the number, u, of points in the sets $N_{j-1,1}, N_{j-1,2}, \cdots, N_{j-1,\lceil k/t \rceil - 1}$. The x-coordinate of the point in $N_{j,k}$ corresponding to v is $\text{select}_{S_{j-1}}(k, x + \text{rank}_{S_{j-1}}(k, u)) - u)$. Using this approach, we can go up one level at a time, until we reach the top level, where we get the original x-coordinate of v. Thus the x-coordinate of v can be computed in $O(\lg n / \lg \lg n)$ time. To retrieve the y-coordinate of v, we use the fact that each successive level (after level i) divides the range in N corresponding to each y-coordinate at the level above into t ranges of the same size. Thus, by going down our hierarchy of structures until reaching the bottom level, we can compute the original y-coordinate of v. This can be performed in $O(\lg n / \lg \lg n)$ time. □

Combing Lemma 7 and Lemma 8, we have our main result:

Theorem 1. *Let N be a set of points from the universe $M = [1..n] \times [1..n]$, where $n = |N|$. N can be represented using $n \lg n + o(n \lg n)$ bits to support orthogonal range counting in $O(\lg n / \lg \lg n)$ time, and orthogonal range reporting in $O(k \lg n / \lg \lg n)$ time, where k is the size of the output.*

4   Applications

Substring search. The succinct text index of Chien *et al.* [2] uses the succinct orthogonal range search structure of Mäkinen and Navarro [1]. Thus, we can speed up substring search by using our structure in Theorem 1:

Theorem 2. *A text string T of length n over an alphabet of size σ can be encoded in $O(n \lg \sigma)$ bits to support substring search in $O(m + \lg n(\lg_\sigma n + \text{occ} \lg n) / \lg \lg n)$ time, where m is the length of the query substring, and occ is the number of its occurrences in T.*

Position-restricted substring search. As Mäkinen and Navarro [1] designed a text index that supports position-restricted substring search by reducing this problem to orthogonal range search on a grid, we can improve their result by applying Theorem 1:

Theorem 3. *Given a text string T of length n over an alphabet of size σ, there is an index of $O(3n \lg n)$ bits that supports position-restricted range search in $O(m + \text{occ}(\lg n) / \lg \lg n)$ time, where m is the length of the query substring, and occ is the number of its occurrences in T.*

Sequences of small numbers. Lemma 2 is interesting only if $s = o(\lg n / \lg \lg n)$ because otherwise, the second term in its space bound becomes a dominating term. Thus, Ferragina *et al.* [15] designed another approach to encode a sequence, S, of n integers bounded by $s = \texttt{polylog}(n)$ in $nH_0(S) + o(n)$ bits to support rank/select operations in constant time. We can further extend their representation to support one more operation: retrieving the entries in any given subsequence of S whose values are in a given range. This is equivalent to the problem of supporting range search on an $n \times t$ grid where $t = \texttt{polylog}(n)$ (each point has a distinct x-coordinate). If we apply the techniques in Section 3.2 to this problem, we only build a constant number of levels of structures. The approach in [15] can also be applied here to achieve compression. Thus:

Theorem 4. *A sequence, S, of n numbers in $[1..s]$, where $s = \texttt{polylog}(n)$, can be encoded in $nH_0(S) + o(n)$ bits such that given a range, $[p_1..p_2]$, of positions in S and a range, $[v_1..v_2]$, of values, the number of entries in $S[p_1..p_2]$ whose values are in the range $[v_1..v_2]$ can be computed in constant time. These entries can be listed in $O(k)$ time, where k is the size of the output. The access to each number, \texttt{rank} and \texttt{select} operations can also be supported in $O(1)$ time.*

A restricted version of orthogonal range search. We consider a restricted version of range search (a weaker operation was proposed by Bauernöppel *et al.* [16]) and we have the following theorem (we omit the proof):

Theorem 5. *A point set N in the plane can be encoded in $cn + n \lg n + o(n \lg n)$ bits, where $n = |N|$ and c is the number of bits required to encode the coordinate pair of each point, to support orthogonal range counting in $O(\lg n / \lg \lg n)$ time and orthogonal range reporting in $O(k \lg n / \lg \lg n)$ time (k is the size of the output) if the query rectangle is defined by two points in N.*

References

1. Mäkinen, V., Navarro, G.: Rank and select revisited and extended. Theor. Comput. Sci. 387, 332–347 (2007)
2. Chien, Y.F., Hon, W.K., Shah, R., Vitter, J.S.: Geometric burrows-wheeler transform: Linking range searching and text indexing. In: DCC, pp. 252–261 (2008)
3. Gabow, H.N., Bentley, J.L., Tarjan, R.E.: Scaling and related techniques for geometry problems. In: STOC, pp. 135–143 (1984)
4. Chazelle, B.: A functional approach to data structures and its use in multidimensional searching. SIAM Journal on Computing 17(3), 427–462 (1988)
5. Overmars, M.H.: Efficient data structures for range searching on a grid. Journal of Algorithms 9(2), 254–275 (1988)
6. Alstrup, S., Brodal, G.S., Rauhe, T.: New data structures for orthogonal range searching. In: FOCS, pp. 198–207 (2000)
7. Nekrich, Y.: Orthogonal range searching in linear and almost-linear space. Computational Geometry: Theory and Applications 42(4), 342–351 (2009)
8. Jacobson, G.: Space-efficient static trees and graphs. In: FOCS, pp. 549–554 (1989)
9. Raman, R., Raman, V., Satti, S.R.: Succinct indexable dictionaries with applications to encoding k-ary trees, prefix sums and multisets. ACM Transactions on Algorithms 3(4), 43 (2007)
10. Grossi, R., Gupta, A., Vitter, J.S.: High-order entropy-compressed text indexes. In: SODA, pp. 841–850 (2003)
11. Barbay, J., Golynski, A., Munro, J.I., Rao, S.S.: Adaptive searching in succinctly encoded binary relations and tree-structured documents. Theoretical Computer Science 387(3), 284–297 (2007)
12. Barbay, J., He, M., Munro, J.I., Rao, S.S.: Succinct indexes for strings, binary relations and multi-labeled trees. In: SODA, pp. 680–689 (2007)
13. Geary, R.F., Raman, R., Raman, V.: Succinct ordinal trees with level-ancestor queries. ACM Transactions on Algorithms 2(4), 510–534 (2006)
14. Ferragina, P., Luccio, F., Manzini, G., Muthukrishnan, S.: Structuring labeled trees for optimal succinctness, and beyond. In: FOCS, pp. 184–196 (2005)
15. Ferragina, P., Manzini, G., Mäkinen, V., Navarro, G.: Compressed representations of sequences and full-text indexes. ACM Trans. Alg. 3(2), 20 (2007)
16. Bauernöppel, F., Kranakis, E., Krizanc, D., Maheshwari, A., Sack, J.R., Urrutia, J.: Planar stage graphs: Characterizations and applications. Theoretical Computer Science 175(2), 239–255 (1997)
17. Clark, D.R., Munro, J.I.: Efficient suffix trees on secondary storage. In: SODA, pp. 383–391 (1996)
18. Bender, M.A., Farach-Colton, M.: The LCA problem revisited. In: Gonnet, G.H., Viola, A. (eds.) LATIN 2000. LNCS, vol. 1776, pp. 88–94. Springer, Heidelberg (2000)

A Distribution-Sensitive Dictionary with Low Space Overhead*

Prosenjit Bose, John Howat, and Pat Morin

School of Computer Science, Carleton University
1125 Colonel By Dr., Ottawa, Ontario, Canada, K1S 5B6
{jit,jhowat,morin}@scs.carleton.ca

Abstract. The time required for a sequence of operations on a data structure is usually measured in terms of the worst possible such sequence. This, however, is often an overestimate of the actual time required. *Distribution-sensitive* data structures attempt to take advantage of underlying patterns in a sequence of operations in order to reduce time complexity, since access patterns are non-random in many applications. Unfortunately, many of the distribution-sensitive structures in the literature require a great deal of space overhead in the form of pointers. We present a dictionary data structure that makes use of both randomization and existing space-efficient data structures to yield very low space overhead while maintaining distribution sensitivity in the expected sense.

1   Introduction

For the *dictionary problem*, we would like to efficiently support the operations of INSERT, DELETE and SEARCH over some totally ordered universe. There exist many such data structures: AVL trees [1], red-black trees [7] and splay trees [11], for instance. Splay trees are of particular interest because they are *distribution-sensitive*, that is, the time required for certain operations can be measured in terms of the distribution of those operations. In particular, splay trees have the *working set property*, which means that the time required to search for an element is logarithmic in the number of distinct accesses since that element was last searched for. Splay trees are not the only dictionary to provide the working set property; the working set structure [8], the unified structure [2] and a variant of the skip list [3] also have it.

Unfortunately, such dictionaries often require a significant amount of space overhead. Indeed, this is a problem with data structures in general. Space overhead often takes the form of pointers: a binary search tree, for instance, might have three pointers per node in the tree: one to the parent and one to each child. If this is the case and we assume that pointers and keys have the same size, then it is easy to see that 3/4 of the storage used by the binary search tree consists of pointers. This seems to be wasteful, since we are really interested in the data itself and would rather not invest such a large fraction of space in overhead.

---

* This research was partially supported by NSERC and MRI.

F. Dehne et al. (Eds.): WADS 2009, LNCS 5664, pp. 110–118, 2009.

To remedy this situation, there has been a great deal of research in the area of *implicit* data structures. An implicit data structure uses only the space required to hold the data itself (in addition to only a constant number of words, each of size $O(\log n)$ bits). Implicit dictionaries are a particularly well-studied problem [4,6,10].

Our goal is to combine notions of distribution sensitivity with ideas from implicit dictionaries to yield a distribution-sensitive dictionary with low space overhead.

1.1   Our Results

In this paper, we present a dictionary data structure with worst-case insertion and deletion times $O(\log n)$ and expected search time $O(\log t(x))$, where x is the key being searched for and $t(x)$ is the number of distinct queries made since x was last searched for, or n if x has not yet been searched for. The space overhead required for this data structure is $O(\log \log n)$, *i.e.*, $O(\log \log n)$ additional words of memory (each of size $O(\log n)$ bits) are required aside from the data itself. Current data structures that can match this query time include the splay tree [11] (in the amortized sense) and the working set structure [8] (in the worst case), although these require $3n$ and $5n$ pointers respectively, assuming three pointers per node (one parent pointer and two child pointers). We also show how to modify this structure (and by extension the working set structure [8]) to support predecessor queries in time logarithmic in the working set number of the predecessor.

The rest of the paper is organized in the following way. Section 2 briefly summarizes the working set structure [8] and shows how to modify it to reduce its space overhead. Section 3 shows how to modify the new dictionary to support more useful queries with additional–but sublinear–overhead. These modifications are also applicable to the working set structure [8] and make both data structures considerably more useful. Section 4 concludes with possible directions for future research.

2   Modifying the Working Set Structure

In this section, we describe the data structure. We will begin by briefly summarizing the working set structure [8]. We then show how to use randomization to remove the queues from the working set structure, and finally how to shrink the size of the trees in the working set structure.

2.1   The Working Set Structure

The working set structure [8] consists of k balanced binary search trees $T_1, \ldots, T_k$ and k queues $Q_1, \ldots, Q_k$. Each queue has precisely the same elements as its corresponding tree, and the size of T_i and Q_i is 2^{2^i}, except for T_k and Q_k which simply contain the remaining elements. Therefore, since there are n elements,

$k = O(\log\log n)$. The structure is manipulated with a *shift* operation in the following manner. A shift from i to j is performed by dequeuing an element from Q_i and removing the corresponding element from T_i. The removed element is then inserted into the next tree and queue (where "next" refers to the tree closer to T_j), and the process is repeated until we reach T_j and Q_j. In this manner, the oldest elements are removed from the trees every time. The result of a shift is that the size of T_i and Q_i has decreased by one and the size of T_j and Q_j has increased by one.

Insertions are made by performing a usual dictionary insertion into T_1 and Q_1, and then shifting from the first index to the last index. Such a shift makes room for the newly inserted element in the first tree and queue by moving the oldest element in each tree and queue down one index. Deletions are accomplished by searching for the element and deleting it from the tree (and queue) it was found in, and then shifting from the last index to the index the element was found at. Such a shift fills in the gap created by the removed element by bringing elements up from further down the data structure. Finally, a search is performed by searching successively in $T_1, T_2, \ldots, T_k$ until the element is found. This element is then removed from the tree and queue in which it was found and inserted into T_1 and Q_1 in the manner described previously. By performing this shift, we ensure that elements searched for recently are towards the front of the data structure and will therefore by found quickly on subsequent searches.

The working set structure was shown by Iacono [8] to have insertion and deletion costs of $O(\log n)$ and a search cost of $O(\log t(x))$, where x is the key being searched for and $t(x)$ is the number of distinct queries made since x was last searched for, or n if x has not yet been searched for. To see that the search cost is $O(\log t(x))$, consider that if x is found in T_i, it must have been dequeued from Q_{i-1} at some point. If this is the case, then $2^{2^{i-1}}$ accesses to elements other than x have taken place since the last access to x, and therefore $t(x) \geq 2^{2^{i-1}}$. Since the search time for x is dominated by the search time in the tree it was found in, the cost is $O\left(\log 2^{2^i}\right) = O(\log t(x))$.

2.2   Removing the Queues

Here we present a simple use of randomization to remove the queues from the working set structure. Rather than relying on the queue to inform the shifting procedure of the oldest element in the tree, we simply pick a random element in the tree and treat it exactly as we would the dequeued element. Lemma 1 shows that we still maintain the working set property in the expected sense.

Lemma 1. *The expected search cost in the randomized working set structure is* $O(\log t(x))$.

Proof. Fix an element x and let $t = t(x)$ denote the number of distinct accesses since x was last accessed. Suppose that x is in T_i and a sequence of accesses occurs during which t distinct accesses occur. Since T_i has size 2^{2^i}, the probability that x is not removed from T_i during these accesses is at least

$$\Pr\{x \text{ not removed from } T_i \text{ after } t \text{ accesses}\} \geq \left(1 - \frac{1}{2^{2^i}}\right)^t$$

$$= \left(1 - \frac{1}{2^{2^i}}\right)^{\frac{t2^{2^i}}{2^{2^i}}}$$

$$\geq \left(\frac{1}{4}\right)^{\frac{t}{2^{2^i}}}$$

Now, if $x \in T_i$, it must have been selected for removal from T_{i-1} at some point. The probability that x is in at least the i-th tree is therefore at most

$$\Pr\{x \in T_j \text{ for } j \geq i\} \leq 1 - \left(\frac{1}{4}\right)^{\frac{t}{2^{2^{i-1}}}}$$

An upper bound on the expectation $E[S]$ of the search cost S is therefore

$$E[S] = \sum_{i=1}^{k} O(\log|T_i|) \times \Pr\{x \in T_i\}$$

$$\leq \sum_{i=1}^{k} O(\log|T_i|) \times \Pr\{x \in T_j \text{ for } j \geq i\}$$

$$\leq \sum_{i=1}^{k} O\left(\log\left(2^{2^i}\right)\right)\left(1 - \left(\frac{1}{4}\right)^{\frac{t}{2^{2^{i-1}}}}\right)$$

$$= \sum_{i=1}^{k} O(2^i)\left(1 - \left(\frac{1}{4}\right)^{\frac{t}{2^{2^{i-1}}}}\right)$$

$$= \sum_{i=1}^{\lfloor \log\log t \rfloor} O(2^i)\left(1 - \left(\frac{1}{4}\right)^{\frac{t}{2^{2^{i-1}}}}\right) + \sum_{i=1+\lfloor \log\log t \rfloor}^{k} O(2^i)\left(1 - \left(\frac{1}{4}\right)^{\frac{t}{2^{2^{i-1}}}}\right)$$

$$= O(\log t) + \sum_{i=1}^{k-\lfloor \log\log t \rfloor} O\left(2^{i+\lfloor \log\log t \rfloor}\right)\left(1 - \left(\frac{1}{4}\right)^{\frac{t}{2^{2^{(i+\lfloor \log\log t \rfloor)-1}}}}\right)$$

$$= O(\log t) + O(\log t)\sum_{i=1}^{k-\lfloor \log\log t \rfloor} O(2^i)\left(1 - \left(\frac{1}{4}\right)^{\frac{t}{2^{2^{(i+\lfloor \log\log t \rfloor)-1}}}}\right)$$

$$\leq O(\log t) + O(\log t)\sum_{i=1}^{k-\lfloor \log\log t \rfloor} O(2^i)\left(1 - \left(\frac{1}{4}\right)^{\frac{t}{2^{2^i \log t}}}\right)$$

$$= O(\log t) + O(\log t)\sum_{i=1}^{k-\lfloor \log\log t \rfloor} O(2^i)\left(1 - \left(\frac{1}{4}\right)^{\frac{1}{t^{2^i-1}}}\right)$$

It thus suffices to show that the remaining sum is $O(1)$. We will assume that $t \geq 2$, since otherwise x can be in at most the second tree and can therefore be found in $O(1)$ time. Considering only the remaining sum, we have

$$\sum_{i=1}^{k-\lfloor \log\log t\rfloor} O(2^i)\left(1-\left(\frac{1}{4}\right)^{\frac{1}{t^{2^i}-1}}\right) \le \sum_{i=1}^{k-\lfloor \log\log t\rfloor} O(2^i)\left(1-\left(\frac{1}{4}\right)^{\frac{1}{2^{2^i}-1}}\right)$$

$$\le \sum_{i=1}^{k-\lfloor \log\log t\rfloor} O(2^i)\left(1-\left(\frac{1}{16}\right)^{\frac{1}{2^{2^i}}}\right)$$

$$\le \sum_{i=1}^{\infty} O(2^i)\left(1-\left(\frac{1}{16}\right)^{\frac{1}{2^{2^i}}}\right)$$

All that remains to show is that this infinite sum is bounded by a decreasing geometric series (and is therefore constant.) The ratio of consecutive terms is

$$\lim_{i\to\infty} \frac{2^{i+1}\left(1-\left(\frac{1}{16}\right)^{\frac{1}{2^{2^{i+1}}}}\right)}{2^i\left(1-\left(\frac{1}{16}\right)^{\frac{1}{2^{2^i}}}\right)} = 2\lim_{i\to\infty} \frac{\left(1-\left(\frac{1}{16}\right)^{\frac{1}{2^{2^{i+1}}}}\right)}{\left(1-\left(\frac{1}{16}\right)^{\frac{1}{2^{2^i}}}\right)}$$

If we substitute $u=\frac{1}{2^{2^i}}$, we find that $\frac{1}{2^{2^{i+1}}}=u^2$. Observe that as $i\to\infty$, we have $u\to 0$. Thus

$$2\lim_{i\to\infty} \frac{\left(1-\left(\frac{1}{16}\right)^{\frac{1}{2^{2^{i+1}}}}\right)}{\left(1-\left(\frac{1}{16}\right)^{\frac{1}{2^{2^i}}}\right)} = 2\lim_{u\to 0} \frac{\left(1-\left(\frac{1}{16}\right)^{u^2}\right)}{\left(1-\left(\frac{1}{16}\right)^{u}\right)} = 2\lim_{u\to 0} \frac{8\left(\frac{1}{16}\right)^{u^2}u\ln 2}{4\left(\frac{1}{16}\right)^{u}\ln 2} = 0$$

Therefore, the ratio of consecutive terms is $o(1)$ and the series is therefore bounded by a decreasing geometric series. An expected search time of $O(\log t(x))$ follows.

At this point, we have seen how to eliminate the queues from the structure at a cost of an *expected* search cost. In the next section, we will show how to further reduce space overhead by shrinking the size of the trees.

2.3   Shrinking the Trees

Another source of space overhead in the working set structure is that of the trees. As mentioned before, many pointers are required to support a binary search tree. Instead, we will borrow some ideas from the study of implicit data structures. Observe that there is nothing special about the trees used in the working set structure: they are simply dictionary data structures that support logarithmic time queries and update operations. In particular, we do not rely on the fact that they are trees. Therefore, we can replace these trees with one of the many implicit dictionary data structures in the literature (see, *e.g.*, [4,5,6,10].) The dictionary of Franceschini and Grossi [5] provides a worst-case optimal implicit dictionary with access costs $O(\log n)$, and so we will employ these results.[1]

---

[1] It is useful to note that any dictionary that offers polylogarithmic access times will yield the same results: the access cost for each operation in our data structure will be the maximum of the access costs for the substructure, since the shifting operation consists of searches, insertions and deletions in the substructures.

Unfortunately, the resulting data structure is not implicit in the strict sense. Since each substructure can use $O(1)$ words of size $O(\log n)$ bits and we have $O(\log \log n)$ such substructures, the data structure as a whole could use as much as $O(\log \log n)$ words of size $O(\log n)$ bits each. Nevertheless, this is a significant improvement over the $O(n)$ additional words used by the traditional data structures. We have

Theorem 1. *There exists a dictionary data structure that stores only the data required for its elements in addition to $O(\log \log n)$ words of size $O(\log n)$ bits each. This dictionary supports insertions and deletions in worst-case $O(\log n)$ time and searches in expected $O(\log t(x))$ time, where $t(x)$ is the working set number of the query x.*

3   Further Modifications

In this section, we describe a simple modification to the data structure outlined in Section 2 that makes searches more useful. This improvement comes at the cost of additional space overhead.

Until now, we have implicitly assumed that searches in our data structure are successful. If they are not, then we will end up searching in each substructure at a total cost of $O(\log n)$ and returning nothing. Unfortunately, this is not very useful.[2] Typically, a dictionary will return the largest element in the dictionary that is smaller than the element searched for or the smallest element larger than the element searched for. Such predecessor and successor queries are a very important feature of comparison-based data structures: without them, one could simply use hashing to achieve $O(1)$ time operations. Predecessor queries are simple to implement in binary search trees since we can simply examine where we "fell off" the tree. This trick will not work in our data structure, however, since we have many such substructures and we will have to know when to stop.

Our goal is thus the following. Given a search key x, we would (as before) like to return x in time $O(\log t(x))$ if x is in the data structure. If x is not in the data structure, we would like to like to return $pred(x)$ in time $O(\log t(pred(x)))$, where $pred(x)$ denotes the predecessor of x.

To accomplish this, we will augment our data structure with some pointers. In particular, every item in the data structure will have a pointer to its successor. During an insertion, each substructure will be searched for the inserted element for a total cost of $O(\log n)$ and the smallest successor in each substructure will be recorded. The smallest such successor is clearly the new element's successor in the whole structure. Therefore, the cost of insertion remains $O(\log n)$. Similarly, during a deletion only the predecessor of the deleted element will need to have its successor pointer updated and thus the total cost of deletion remains $O(\log n)$.

During a search for x, we proceed as before. Consider searching in any particular substructure i. If the result of the search in substructure i is in fact x,

---

[2] Note that this is also true of the original working set structure [8]. The modifications described here are also applicable to it.

then the analysis is exactly the same as before and we can return x in time $O(\log t(x))$. Otherwise, we won't find x in substructure i. In this case, we search substructure i for the predecessor (in that substructure) of x.[3] Denote this element by $pred_i(x)$. Since every element knows its successor in the structure as a whole, we can determine $succ(pred_i(x))$. If $succ(pred_i(x)) = x$, then we know that x is indeed in the structure and thus the query can be completed as before. If $succ(pred_i(x)) < x$, then we know that there is still an element smaller than x but larger than what we have seen, and so we continue searching for x. Finally, if $succ(pred_i(x)) > x$, then we have reached the largest element less than or equal to x, and so our search stops.

In any case, after we find x or $pred(x)$, we shift the element we returned to the first substructure as usual. The analysis of the time complexity of this search algorithm is exactly the same as before; we are essentially changing the element we are searching for during the search. The search time for the substructure we stop in dominates the cost of the search and since we return x if we found it or $pred(x)$ otherwise, the search time is $O(\log t(x))$ if x is in the dictionary and $O(\log t(pred(x)))$ otherwise.

Of course, this augmentation incurs some additional space overhead. In particular, we now require n pointers, resulting in a space overhead of $O(n)$. While the queues are now gone, we still have one pointer for each element in the dictionary. To fix this, observe that we can leave out the pointers for the last few trees and simply do a brute-force search at the cost of slightly higher time complexity. Suppose we leave the pointers out of the last j trees: T_{k-j+1} to T_k, where k represents the index of the last tree, as before. Therefore, each element x in the trees $T_1, \ldots, T_{k-j}$ has a pointer to $succ(x)$. Now, suppose we are searching in the data structure and get to T_{k-j+1}. At this point, we may need to search all remaining trees, since if we do not find the key we are looking for, we have no way of knowing when we have found its predecessor. Consider the following lemma.

Lemma 2. *Let $0 \le j \le k$. A predecessor search for x takes expected time $O\big(2^j \log t(x)\big)$ if x is in the dictionary and $O\big(2^j \log t(pred(x))\big)$ otherwise.*

Proof. Assume the search reaches T_{k-j+1}, since otherwise our previous analyses apply. We therefore have $t(x) \ge 2^{2^{k-j}}$, since at least $2^{2^{k-j}}$ operations have taken place.[4] As before, the search time is bounded by the search time in T_k. Since T_k has size at most 2^{2^k}, we have an expected search time of

$$O\left(\log 2^{2^k}\right) = O\left(\log 2^{2^{k-j} 2^j}\right) = O\left(2^j \log 2^{2^{k-j}}\right) \le O\big(2^j \log t(x)\big)$$

This analysis applies as well when x is not in the dictionary. In this case, the search can be accomplished in time $O\big(2^j \log t(pred(x))\big)$.

---

[3] Here we are assuming that substructures support predecessor queries in the same time required for searching. This is not a strong assumption, since any comparison-based dictionary must compare x to $succ(x)$ and $pred(x)$ during an unsuccessful search for x.

[4] This follows from Lemma 1, which is why the results here hold in the expected sense.

One further consideration is that once an element is found in the last j trees, we need to find its successor so that it knows where it is once it is shifted to the front. However, this is straightforward because we have already examined all substructures in the data structure and so we can make a second pass. It remains to consider how much space we have saved using this scheme. Since each tree has size the square of the previous, by leaving out the last j trees, the total number of extra pointers used is $O\left(n^{1/2^j}\right)$. We therefore have

Theorem 2. *Let $0 \leq j \leq k$. There exists a dictionary that stores only the data required for its elements in addition to $O\left(n^{1/2^j}\right)$ words of size $O(\log n)$ bits each. This dictionary supports insertions and deletions in worst-case $O(\log n)$ time and searches in expected $O\left(2^j \log t(x)\right)$ time if x is found in the dictionary and expected $O\left(2^j \log t(pred(x))\right)$ time otherwise (in which case $pred(x)$ is returned).*

Observe that $j \leq k = O(\log \log n)$, and so while the dependence on j is exponential, it is still quite small relative to n. In particular, take $j = 1$ to get

Corollary 1. *There exists a dictionary that stores only the data required for its elements in addition to $O(\sqrt{n})$ words of size $O(\log n)$ bits each. This dictionary supports insertions and deletions in worst-case $O(\log n)$ time and searches in expected $O(\log t(x))$ time if x is found in the dictionary. If x is not in the dictionary, $pred(x)$ is returned in expected $O(\log t(pred(x)))$ time.*

4   Conclusion

We have seen how to modify the Iacono's working set structure [8] in several ways. To become more space efficient, we can remove the queues and use randomization to shift elements, while replacing the underlying binary search trees with implicit dictionaries. To support more useful search queries, we can sacrifice some space overhead to maintain information about some portion of the elements in the dictionary in order to support returning the predecessor of any otherwise unsuccessful search queries. All such modifications maintain the working set property in an expected sense.

4.1   Future Work

The modifications described in this paper leave open a few directions for research.

The idea of relying on the properties of the substructures (in this case, implicitness) proved fruitful. A natural question to ask, then, is what other substructure properties can carry over to the dictionary as a whole in a useful way? Other substructures could result in a combination of the working set property and some other useful properties.   .

In this paper, we concerned ourselves with the working set property. There are other types of distribution sensitivity, such as the *dynamic finger property*, which means that query time is logarithmic in the rank difference between successive queries. A sorted array, for example, has the dynamic finger property (assuming we keep a pointer to the result of the previous query) but does not support efficient updates. One could also consider a notion complementary to the idea of the working set property, namely the *queueish property* [9], wherein query time is logarithmic in the number of items *not* accessed since the query item was last accessed. Are there implicit dictionaries that provide either of these properties? Could we provide any of these properties (or some analogue of them) for other types of data structures?

Finally, it would be of interest to see if a data structure that does not rely on randomization is possible, in order to guarantee a *worst case* time complexity of $O(\log t(x))$ instead of an expected one.

References

1. Adelson-Velskii, G.M., Landis, E.M.: An algorithm for the organization of information. Soviet Math. Doklady 3, 1259–1263 (1962)
2. Badoiu, M., Cole, R., Demaine, E.D., Iacono, J.: A unified access bound on comparison-based dynamic dictionaries. Theoretical Computer Science 382(2), 86–96 (2007)
3. Bose, P., Douieb, K., Langerman, S.: Dynamic optimality for skip lists and B-trees. In: SODA 2008: Proceedings of the 19th Annual ACM-SIAM Symposium on Discrete Algorithms, pp. 1106–1114 (2008)
4. Franceschini, G., Grossi, R.: Implicit dictionaries supporting searches and amortized updates in $O(\log n \log \log n)$ time. In: SODA 2003: Proceedings of the 14th Annual ACM-SIAM Symposium on Discrete Algorithms, pp. 670–678 (2003)
5. Franceschini, G., Grossi, R.: Optimal worst-case operations for implicit cache-oblivious search trees. In: Dehne, F., Sack, J.-R., Smid, M. (eds.) WADS 2003. LNCS, vol. 2748, pp. 114–126. Springer, Heidelberg (2003)
6. Franceschini, G., Munro, J.I.: Implicit dictionaries with $O(1)$ modifications per update and fast search. In: SODA 2006: Proceedings of the 17th Annual ACM-SIAM Symposium on Discrete Algorithms, pp. 404–413 (2006)
7. Guibas, L.J., Sedgewick, R.: A dichromatic framework for balanced trees. In: FOCS 1978: Proceedings of the 19th Annual IEEE Symposium on Foundations of Computer Science, pp. 8–21 (1978)
8. Iacono, J.: Alternatives to splay trees with $O(\log n)$ worst-case access times. In: SODA 2001: Proceedings of the 12th Annual ACM-SIAM Symposium on Discrete Algorithms, pp. 516–522 (2001)
9. Iacono, J., Langerman, S.: Queaps. Algorithmica 42(1), 49–56 (2005)
10. Ian Munro, J.: An implicit data structure supporting insertion, deletion, and search in $O(\log^2 n)$ time. J. Comput. Syst. Sci. 33(1), 66–74 (1986)
11. Sleator, D.D., Tarjan, R.E.: Self-adjusting binary search trees. J. ACM 32(3), 652–686 (1985)

A Comparison of
Performance Measures for Online Algorithms*

Joan Boyar[1], Sandy Irani[2], and Kim S. Larsen[1]

[1] Department of Mathematics and Computer Science, University of Southern
Denmark, Campusvej 55, DK-5230 Odense M, Denmark
{joan,kslarsen}@imada.sdu.dk
[2] Department of Computer Science, University of California, Irvine, CA 92697, USA
irani@ics.uci.edu

Abstract. This paper provides a systematic study of several proposed
measures for online algorithms in the context of a specific problem,
namely, the two server problem on three colinear points. Even though the
problem is simple, it encapsulates a core challenge in online algorithms
which is to balance greediness and adaptability. We examine Competitive Analysis, the Max/Max Ratio, the Random Order Ratio, Bijective
Analysis and Relative Worst Order Analysis, and determine how these
measures compare the Greedy Algorithm and Lazy Double Coverage,
commonly studied algorithms in the context of server problems. We find
that by the Max/Max Ratio and Bijective Analysis, Greedy is the better
algorithm. Under the other measures, Lazy Double Coverage is better,
though Relative Worst Order Analysis indicates that Greedy is sometimes better. Our results also provide the first proof of optimality of an
algorithm under Relative Worst Order Analysis.

1 Introduction

Since its introduction by Sleator and Tarjan in 1985 [16], Competitive Analysis has been the most widely used method for evaluating online algorithms. A
problem is said to be *online* if the input to the problem is given a piece at a
time, and the algorithm must commit to parts of the solution over time before
the entire input is revealed to the algorithm. *Competitive Analysis* evaluates an
online algorithm in comparison to the optimal offline algorithm which receives
the input in its entirety in advance and has unlimited computational power in
determining a solution. Informally speaking, we look at the worst-case input
which maximizes the ratio of the cost of the online algorithm for that input
to the cost of the optimal offline algorithm on that same input. The maximum
ratio achieved is called the *Competitive Ratio*. Thus, we factor out the inherent

---

* The work of Boyar and Larsen was supported in part by the Danish Natural Science
Research Council. Part of this work was carried out while these authors were visiting
the University of California, Irvine. The work of Irani was supported in part by NSF
Grant CCR-0514082.

F. Dehne et al. (Eds.): WADS 2009, LNCS 5664, pp. 119–130, 2009.

difficulty of a particular input (for which the offline algorithm is penalized along with the online algorithm) and measure what is lost in making decisions with partial information.

Despite the popularity of Competitive Analysis, researchers have been well aware of its deficiencies and have been seeking better alternatives almost since the time that it came into wide use. (See [9] for a recent survey.) Many of the problems with Competitive Analysis stem from the fact that it is a worst case measure and fails to examine the performance of algorithms on instances that would be expected in a particular application. It has also been observed that Competitive Analysis sometimes fails to distinguish between algorithms which have very different performance in practice and intuitively differ in quality.

Over the years, researchers have devised alternatives to Competitive Analysis, each designed to address one or all of its shortcomings. There are exceptions, but it is fair to say that many alternatives are application-specific, and very often, these papers only present a direct comparison between a new measure and Competitive Analysis.

This paper is a study of several generally-applicable alternative measures for evaluating online algorithms that have been suggested in the literature. We perform this comparison in the context of a particular problem: the 2-server problem on the line with three possible request points, nick-named here the *baby server problem*. Investigating simple k-servers problems to shed light on new ideas has also been done in [2], for instance.

We concentrate on two algorithms (GREEDY and LAZY DOUBLE COVERAGE (LDC) [8]) and four different analysis techniques (measures): Bijective Analysis, the Max/Max Ratio, Random Order Ratio and Relative Worst Order Analysis.

In investigating the baby server problem, we find that according to some quality measures for online algorithms, GREEDY is better than LDC, whereas for others, LDC is better than GREEDY.

The ones that conclude that LDC is best are focused on a worst-case sequence for the ratio of an algorithm's cost compared to OPT. In the case of GREEDY and LDC, this conclusion makes use of the fact that there exists a family of sequences for which GREEDY's cost is unboundedly larger than the cost of OPT, whereas LDC's cost is always at most a factor two larger than the cost of OPT.

On the other hand, the measures that conclude that GREEDY is best compare two algorithms based on the multiset of costs stemming from the set of all sequences of a fixed length. In the case of GREEDY and LDC, this makes use of the fact that for any fixed n, both the maximum as well as the average cost of LDC over all sequences of length n are greater than the corresponding values for GREEDY.

Using Relative Worst Order Analysis a more nuanced result is obtained, concluding that LDC can be a factor at most two worse than GREEDY, while GREEDY can be unboundedly worse than LDC.

All omitted proofs may be found in the full version of the paper [7].

2   Preliminaries

2.1   The Server Problem

Server problems [4] have been the objects of many studies. In its full generality, one assumes that some number k of servers are available in some metric space. Then a sequence of requests must be treated. A request is simply a point in the metric space, and a k-server algorithm must move servers in response to the request to ensure that at least one server is placed on the request point. A cost is associated with any move of a server (this is usually the distance moved in the given metric space), and the objective is to minimize total cost. The initial configuration (location of servers) may or may not be a part of the problem formulation.

In investigating the strengths and weaknesses of the various measures for the quality of online algorithms, we define the simplest possible nontrivial server problem:

Definition 1. *The* baby server problem *is a 2-server problem on the line with three possible request points A, B, and C, in that order from left to right, with distance one between A and B and distance $d > 1$ between B and C. The cost of moving a server is defined to be the distance it is moved. We assume that initially the two servers are placed on A and C.*

All results in the paper pertain to this problem. Even though the problem is simple, it contains a core k-server problem of balancing greediness and adaptability, and this simple set-up is sufficient to show the non-competitiveness of GREEDY with respect to Competitive Analysis [4].

2.2   Server Algorithms

First, we define some relevant properties of server algorithms:

Definition 2. *A server algorithm is called*

- noncrossing *if servers never change their relative position on the line.*
- lazy *[15] if it never moves more than one server in response to a request and it does not move any servers if the requested point is already occupied by a server.*

A server algorithm fulfilling both these properties is called compliant.

Given an algorithm, $\mathbb{A}$, we define the algorithm *lazy* $\mathbb{A}$, $\mathcal{L}\mathbb{A}$, as follows: $\mathcal{L}\mathbb{A}$ will maintain a *virtual* set of servers and their locations as well as the real set of servers in the metric space. There is a one-to-one correspondence between real servers and virtual servers. The virtual set will simulate the behavior of $\mathbb{A}$. The initial server positions of the virtual and real servers are the same. Whenever a virtual server reaches a request point, the corresponding real server is also moved to that point (unless both virtual servers reach the point simultaneously, in which case only the physically closest is moved there). Otherwise the real servers do not move.

In [8], it was observed that for any 2-server algorithm, there exists a non-crossing algorithm with the same cost on all sequences. In [15], it was observed that for an algorithm $\mathbb{A}$ and its lazy version $\mathcal{L}\mathbb{A}$, for any sequence I of requests, $\mathbb{A}(I) \geq \mathcal{L}\mathbb{A}(I)$ (we refer to this as the *laziness obervation*). Note that the laziness observation applies to the general k-server problem in metric spaces, so the results which depend on it can also be generalized beyond the baby server problem.

We define a number of algorithms by defining their behavior on the next request point, p. For all algorithms, no moves are made if a server already occupies the request point (though internal state changes are sometimes made in such a situation).

GREEDY moves the closest server to p. Note that due to the problem formulation, ties cannot occur (and the server on C is never moved).

If p is in between the two servers, Double Coverage (DC), moves both servers at the same speed in the direction of p until at least one server reaches the point. If p is on the same side of both servers, the nearest server moves to p.

We define a-DC to work in the same way as DC, except that the right-most server moves at a speed $a \leq d$ times faster than the left-most server.

We refer to the lazy version of DC as LDC and the lazy version of a-DC as a-LDC.

The balance algorithm [15], BAL, makes its decisions based on the total distance travelled by each server. For each server, s, let d_s denote the total distance travelled by s from the initiation of the algorithm up to the current point in time. On a request, BAL moves a server, aiming to obtain the smallest possible $\max_s d_s$ value *after* the move. In case of a tie, BAL moves the server which must move the furthest.

If p is in between the two servers, DUMMY moves the server that is furthest away to the request point. If p is on the same side of both servers, the nearest server moves to p. Again, due to the problem formulation, ties cannot occur (and the server on A is never moved).

2.3   Quality Measures

In analyzing algorithms for the baby server problem, we consider input sequences I of request points. An algorithm $\mathbb{A}$, which treats such a sequence has some cost, which is the total distance moved by the two servers. This cost is denoted by $\mathbb{A}(I)$. Since I is of finite length, it is clear that there exists an offline algorithm with minimal cost. By OPT, we refer to such an algorithm and OPT(I) denotes the unique minimal cost of processing I.

All of the measures described below can lead to a conclusion as to which algorithm of two is better. In contrast to the others, Bijective Analysis does not indicate how much better the one algorithm might be; it does not produce a ratio, as the others do.

Competitive Analysis: In Competitive Analysis [11,16,12], we define an algorithm $\mathbb{A}$ to be c-competitive if there exists a constant α such that for all input sequences I, $\mathbb{A}(I) \leq c \, \mathrm{OPT}(I) + \alpha$.

The Max/Max Ratio: The Max/Max Ratio [3] compares an algorithm's worst cost for any sequence of length n to OPT's worst cost for any sequence of length n. The Max/Max Ratio of an algorithm $\mathbb{A}$, $w_M(\mathbb{A})$, is $M(\mathbb{A})/M(\mathrm{OPT})$, where

$$M(\mathbb{A}) = \limsup_{t \to \infty} \max_{|I|=t} \mathbb{A}(I)/t.$$

The Random Order Ratio: Kenyon [13] defines the Random Order Ratio to be the worst ratio obtained over all sequences, comparing the expected value of an algorithm, $\mathbb{A}$, with respect to a uniform distribution of all permutations of a given sequence, to the value of OPT of the given sequence:

$$\limsup_{\mathrm{OPT}(I) \to \infty} \frac{E_\sigma\left[\mathbb{A}(\sigma(I))\right]}{\mathrm{OPT}(I)}$$

The original context for this definition is Bin Packing for which the optimal packing is the same, regardless of the order in which the items are presented. Therefore, it does not make sense to take an average over all permutations for OPT. For server problems, however, the order of requests in the sequence may very well change the cost of OPT. We choose to generalize the Random Order Ratio as shown to the left, but for the results presented here, the definition to the right would give the same:

$$\limsup_{\mathrm{OPT}(I) \to \infty} \frac{E_\sigma\left[\mathbb{A}(\sigma(I))\right]}{E_\sigma\left[\mathrm{OPT}(\sigma(I))\right]} \qquad\qquad \limsup_{\mathrm{OPT}(I) \to \infty} E_\sigma\left[\frac{\mathbb{A}(\sigma(I))}{\mathrm{OPT}(\sigma(I))}\right]$$

Bijective Analysis and Average Analysis: In [1], Bijective and Average Analysis are defined, as methods of comparing two online algorithms directly. We adapt those definitions to the notation used here. As with the Max/Max Ratio and Relative Worst Order Analysis, the two algorithms are not necessarily compared on the same sequence.

In Bijective Analysis, the sequences of a given length are mapped, using a bijection onto the same set of sequences. The performance of the first algorithm on a sequence, I, is compared to the performance of the second algorithm on the sequence I is mapped to. If I_n denotes the set of all input sequences of length n, then an online algorithm $\mathbb{A}$ is no worse than an online algorithm $\mathbb{B}$ according to Bijective Analysis if there exists an integer $n_0 \geq 1$ such that for each $n \geq n_0$, there is a bijection $f : I_n \to I_n$ satisfying $\mathbb{A}(I) \leq \mathbb{B}(f(I))$ for each $I \in I_n$.

Average Analysis can be viewed as a relaxation of Bijective Analysis. An online algorithm $\mathbb{A}$ is no worse than an online algorithm $\mathbb{B}$ according to Average Analysis if there exists an integer $n_0 \geq 1$ such that for each $n \geq n_0$, $\Sigma_{I \in I_n} \mathbb{A}(I) \leq \Sigma_{I \in I_n} \mathbb{B}(I)$.

Relative Worst Order Analysis: Relative Worst Order Analysis was introduced in [5] and extended in [6]. It compares two online algorithms directly. As with the Max/Max Ratio, it compares two algorithms on their worst sequence in the same part of a partition. The partition is based on the Random Order Ratio, so that the algorithms are compared on sequences having the same content, but possibly in different orders.

Definition 3. *Let I be any input sequence, and let n be the length of I. If σ is a permutation on n elements, then $\sigma(I)$ denotes I permuted by σ. Let $\mathbb{A}$ be any algorithm. Then, $\mathbb{A}(I)$ is the cost of running $\mathbb{A}$ on I, and*

$$\mathbb{A}_W(I) = \max_\sigma \mathbb{A}(\sigma(I)).$$

Definition 4. *For any pair of algorithms $\mathbb{A}$ and $\mathbb{B}$, we define*

$$c_l(\mathbb{A}, \mathbb{B}) = \sup\{c \mid \exists b\colon \forall I\colon \mathbb{A}_W(I) \geq c\,\mathbb{B}_W(I) - b\} \text{ and}$$
$$c_u(\mathbb{A}, \mathbb{B}) = \inf\{c \mid \exists b\colon \forall I\colon \mathbb{A}_W(I) \leq c\,\mathbb{B}_W(I) + b\} .$$

If $c_l(\mathbb{A}, \mathbb{B}) \geq 1$ or $c_u(\mathbb{A}, \mathbb{B}) \leq 1$, the algorithms are said to be comparable *and the* Relative Worst-Order Ratio *$WR_{\mathbb{A},\mathbb{B}}$ of algorithm $\mathbb{A}$ to algorithm $\mathbb{B}$ is defined. Otherwise, $WR_{\mathbb{A},\mathbb{B}}$ is undefined.*

$$\text{If } c_l(\mathbb{A}, \mathbb{B}) \geq 1, \text{ then } WR_{\mathbb{A},\mathbb{B}} = c_u(\mathbb{A}, \mathbb{B}), \text{ and}$$

$$\text{if } c_u(\mathbb{A}, \mathbb{B}) \leq 1, \text{ then } WR_{\mathbb{A},\mathbb{B}} = c_l(\mathbb{A}, \mathbb{B}) .$$

If $WR_{\mathbb{A},\mathbb{B}} < 1$, algorithms $\mathbb{A}$ and $\mathbb{B}$ are said to be comparable in $\mathbb{A}$'s favor. *Similarly, if $WR_{\mathbb{A},\mathbb{B}} > 1$, the algorithms are said to be* comparable in $\mathbb{B}$'s favor.

Definition 5. *Let c_u be defined as in Definition 4. If at least one of the ratios $c_u(\mathbb{A}, \mathbb{B})$ and $c_u(\mathbb{B}, \mathbb{A})$ is finite, the algorithms $\mathbb{A}$ and $\mathbb{B}$ are $(c_u(\mathbb{A}, \mathbb{B}), c_u(\mathbb{B}, \mathbb{A}))$-related.*

Definition 6. *Let $c_u(\mathbb{A}, \mathbb{B})$ be defined as in Definition 4. Algorithms $\mathbb{A}$ and $\mathbb{B}$ are* weakly comparable in $\mathbb{A}$'s favor, *1) if $\mathbb{A}$ and $\mathbb{B}$ are comparable in $\mathbb{A}$'s favor, 2) if $c_u(\mathbb{A}, \mathbb{B})$ is finite and $c_u(\mathbb{B}, \mathbb{A})$ is infinite, or 3) if $c_u(\mathbb{A}, \mathbb{B}) \in o(c_u(\mathbb{B}, \mathbb{A}))$.*

3   Competitive Analysis

The k-server problem has been studied using Competitive Analysis starting in [14]. In [8], it is shown that the competitive ratios of DC and LDC are k, which is optimal, and that GREEDY is not competitive.

4   The Max/Max Ratio

In [3], a concrete example is given with two servers and three non-colinear points. It is observed that the Max/Max Ratio favors the greedy algorithm over the balance algorithm, BAL.

BAL behaves similarly to LDC and identically on LDC's worst case sequences. The following theorem shows that the same conclusion is reached when the three points are on the line.

Theorem 1. GREEDY *is better than* LDC *on the baby server problem with respect to the Max/Max Ratio.*

It follows from the proof of this theorem that GREEDY is close to optimal with respect to the Max/Max Ratio, since the cost of GREEDY divided by the cost of OPT tends toward one for large d.

Since LDC and DC perform identically on their worst sequences of any given length, they also have the same Max/Max Ratio.

5 The Random Order Ratio

The Random Order Ratio correctly distinguishes between DC and LDC, indicating that the latter is the better algorithm.

Theorem 2. LDC *is better than* DC *according to the Random Order Ratio.*

Proof. For any sequence I, $E_\sigma[\text{DC}(\sigma(I))] \geq E_\sigma[\text{LDC}(\sigma(I))]$, by the laziness observation. Let $I = (ABC)^n$. Whenever the subsequence $CABC$ occurs in $\sigma(I)$, DC moves a server from C towards B and back again, while moving the other server from A to B. In contrast, LDC lets the server on C stay there, and has cost 2 less than DC. The expected number of occurrences of $CABC$ in $\sigma(I)$ is cn for some constant c. The expected costs of both OPT and LDC on $\sigma(I)$ are bounded above and below by some other constants times n. Thus, LDC's random order ratio will be less than DC's.

Theorem 3. LDC *is better than* GREEDY *on the baby server problem with regards to the Random Order Ratio.*

Proof. The Random Order Ratio is the worst ratio obtained over all sequences, comparing the expected value of an algorithm over all permutations of a given sequence to the expected value of OPT over all permutations of the given sequence.

Since the competitive ratio of LDC is two, on any given sequence, LDC's cost is bounded by two times the cost of OPT on that sequence, plus an additive constant. Thus, the Random Order Ratio is also at most two.

Consider all permutations of the sequence $(BA)^{\frac{n}{2}}$. We consider positions from 1 through n in these sequences. Refer to a maximal consecutive subsequence consisting entirely of either As or Bs as a *run*.

Given a sequence containing h As and t Bs, the expected number of runs is $1 + \frac{2ht}{h+t}$. (A problem in [10] gives that the expected number of runs of As is $\frac{h(t+1)}{h+t}$, so the expected number of runs of Bs is $\frac{t(h+1)}{h+t}$. Adding these gives the result.) Thus, with $h = t = \frac{n}{2}$, we get $\frac{n}{2} + 1$ expected number of runs.

The cost of GREEDY is equal to the number of runs if the first run is a run of Bs. Otherwise, the cost is one smaller. Thus, GREEDY's expected cost on a permutation of s is $\frac{n}{2} + \frac{1}{2}$.

The cost of OPT for any permutation of s is d, since it simply moves the server from C to B on the first request to B and has no other cost after that.

Thus, the Random Order Ratio is $\frac{n+1}{2d}$, which, as n tends to infinity, is unbounded.

6 Bijective Analysis

Bijective analysis correctly distinguishes between DC and LDC, indicating that the latter is the better algorithm. This follows from the following general theorem about lazy algorithms, and the fact that there are some sequences where one of DC's servers repeatedly moves from C towards B, but moves back to C before ever reaching B, while LDC's server stays on C.

Theorem 4. *The lazy version of any algorithm for the baby server problem is at least as good as the original algorithm according to Bijective Analysis.*

Theorem 5. GREEDY *is at least as good as any other lazy algorithm* LAZY *(including* LDC*) for the baby server problem according to Bijective Analysis.*

Proof. Since GREEDY has cost zero for the sequences consisting of only the point A or only the point C and cost one for the point B, it is easy to define a bijection f for sequences of length one, such that $\text{GREEDY}(I) \leq \text{LAZY}(f(I))$. Suppose that for all sequences of length k that we have a bijection, f, from GREEDY's sequences to LAZY's sequences, such that for each sequence I of length k, $\text{GREEDY}(I) \leq \text{LAZY}(f(I))$. To extend this to length $k+1$, consider the three sequences formed from a sequence I of length k by adding one of the three requests A, B, or C to the end of I, and the three sequences formed from $f(I)$ by adding each of these points to the end of $f(I)$. At the end of sequence I, GREEDY has its two servers on different points, so two of these new sequences have the same cost for GREEDY as on I and one has cost exactly 1 more. Similarly, LAZY has its two servers on different points at the end of $f(I)$, so two of these new sequences have the same cost for LAZY as on $f(I)$ and one has cost either 1 or d more. This immediately defines a bijection f' for sequences of length $k+1$ where $\text{GREEDY}(I) \leq \text{LAZY}(f'(I))$ for all I of length $k+1$.

If an algorithm is better than another algorithm with regards to Bijective Analysis, then it is also better with regards to Average Analysis [1].

Corollary 1. GREEDY *is the unique optimal algorithm with regards to Bijective and Average Analysis.*

Proof. Note that the proof of Theorem 5 shows that GREEDY is strictly better than any lazy algorithm which ever moves the server away from C, so it is better than any other lazy algorithm with regards to Bijective Analysis. By Theorem 4, it is better than any algorithm. By the observation above, it also holds for Average Analysis.

Theorem 6. DUMMY *is the unique worst algorithm among compliant server algorithms for the baby server problem according to Bijective Analysis.*

Lemma 1. *If $a \leq b$, then there exists a bijection $\sigma_n : \{A,B,C\}^n \to \{A,B,C\}^n$ such that $a\text{-LDC}(I) \leq b\text{-LDC}(\sigma_n(I))$ for all sequences $I \in \{A,B,C\}^n$.*

Theorem 7. *According to Bijective Analysis and Average Analysis, slower variants of* LDC *are better than faster variants for the baby server problem.*

Proof. Follows immediately from Lemma 1 and the definition of the measures.

Thus, the closer a variant of LDC is to GREEDY, the better Bijective and Average Analysis predict that it is.

7 Relative Worst Order Analysis

Similarly to the random order ratio and bijective analysis, relative worst order analysis correctly distinguishes between DC and LDC, indicating that the latter is the better algorithm. This follows from the following general theorem about lazy algorithms, and the fact that there are some sequences where one of DC's servers repeatedly moves from C towards B, but moves back to C before ever reaching B, while LDC's server stays on C. If d is just marginally larger than some integer, even on LDC's worst ordering of this sequence, it does better than DC.

Let $I_{\mathbb{A}}$ denote a worst ordering of the sequence I for the algorithm $\mathbb{A}$.

Theorem 8. *The lazy version of any algorithm for the baby server problem is at least as good as the original algorithm according to Relative Worst Order Analysis.*

Theorem 9. GREEDY *and* LDC *are* $(\infty, 2)$*-related and are thus weakly comparable in* LDC*'s favor for the baby server problem according to Relative Worst Order Analysis.*

Proof. First we show that $c_u(\text{GREEDY}, \text{LDC})$ is unbounded. Consider the sequence $(BA)^{\frac{n}{2}}$. As n tends to infinity, GREEDY's cost is unbounded, whereas LDC's cost is at most $3d$ for any permutation.

Next we turn to $c_u(\text{LDC}, \text{GREEDY})$. Since the competitive ratio of LDC is 2, for any sequence I and some constant b, $\text{LDC}(I_{\text{LDC}}) \leq 2\text{GREEDY}(I_{\text{LDC}}) + b \leq 2\text{GREEDY}(I_{\text{GREEDY}}) + b$. Thus, $c_u(\text{LDC}, \text{GREEDY}) \leq 2$.

For the lower bound of 2, consider a family of sequences $I_p = (BABA...BC)^p$, where the length of the alternating A/B-sequence before the C is $2\lfloor d \rfloor + 1$.
$\text{LDC}(I_p) = p(2\lfloor d \rfloor + 2d)$.

A worst ordering for GREEDY alternates As and Bs. Since there is no cost for the Cs and the A/B sequences start and end with Bs, $\text{GREEDY}(\sigma(I_p)) \leq p(2\lfloor d \rfloor) + 1$ for any permutation σ.

Then, $c_u(\text{LDC}, \text{GREEDY}) \geq \frac{p(2\lfloor d \rfloor + 2d)}{p(2\lfloor d \rfloor) + 1} \geq \frac{p(4d)}{p(2d) + 1}$. As p goes to infinity, this approaches 2.

Thus, GREEDY and LDC are weakly comparable in LDC's favor.

Recalling the definition of a-LDC, a request for B is served by the right-most server if it is within a virtual distance of no more than a from B. Thus, when the left-most server moves and its virtual move is over a distance of l, then the right-most server virtually moves a distance al. When the right-most server moves and its virtual move is over a distance of al, then the left-most server virtually moves a distance of l.

In the results that follow, we frequently look at the worst ordering of an arbitrary sequence.

Definition 7. *The* canonical worst ordering *of a sequence, I, for an algorithm $\mathbb{A}$ is the sequence produced by allowing the cruel adversary (the one which always lets the next request be the unique point where $\mathbb{A}$ does not currently have a server) to choose requests from the multiset defined from I. This process continues until there are no requests remaining in the multiset for the point where $\mathbb{A}$ does not have a server. The remaining points from the multiset are concatenated to the end of this new request sequence in any order.*

The canonical worst ordering of a sequence for a-LDC is as follows:

Proposition 1. *Consider an arbitrary sequence I containing n_A As, n_B Bs, and n_C Cs. A canonical worst ordering of I for a-LDC is $I_a = (BABA...BC)^{p_a}X$, where the length of the alternating A/B-sequence before the C is $2\left\lfloor\frac{d}{a}\right\rfloor+1$. Here, X is a possibly empty sequence. The first part of X is an alternating sequence of As and Bs, starting with a B, until there are not both As and Bs left. Then we continue with all remaining As or Bs, followed by all remaining Cs. Finally,*

$$p_a = \min\left\{\left\lfloor\frac{n_A}{\left\lfloor\frac{d}{a}\right\rfloor}\right\rfloor, \left\lfloor\frac{n_B}{\left\lfloor\frac{d}{a}\right\rfloor+1}\right\rfloor, n_C\right\}.$$

Theorem 10. *If $a \le b$, then a-LDC and b-LDC are $(\frac{\left\lfloor\frac{d}{a}\right\rfloor+d}{\left\lfloor\frac{d}{b}\right\rfloor+d}, \frac{(\left\lfloor\frac{d}{b}\right\rfloor+d)\left\lfloor\frac{d}{a}\right\rfloor}{(\left\lfloor\frac{d}{a}\right\rfloor+d)\left\lfloor\frac{d}{b}\right\rfloor})$-related for the baby server problem according to Relative Worst Order Analysis.*

We provide strong indication that LDC is better than b-LDC for $b \ne 1$. If $b > 1$, this is always the case, whereas if $b < 1$, it holds in many cases, including all integer values of d.

Theorem 11. *Consider the baby server problem evaluated according to Relative Worst Order Analysis. For $b > 1$, if LDC and b-LDC behave differently, then they are (r, r_b)-related, where $1 < r < r_b$. If $a < 1$, a-LDC and LDC behave differently, and d is a positive integer, then they are (r_a, r)-related, where $1 < r_a < r$.*

The algorithms a-LDC and $\frac{1}{a}$-LDC are in some sense of equal quality:

Corollary 2. *When $\frac{d}{a}$ and $\frac{d}{b}$ are integers, then a-LDC and b-LDC are (b, b)-related when $b = \frac{1}{a}$.*

We now set out to prove that LDC is an optimal algorithm in the following sense: there is no other algorithm $\mathbb{A}$ such that LDC and $\mathbb{A}$ are comparable and $\mathbb{A}$ is strictly better or such that LDC and $\mathbb{A}$ are weakly comparable in $\mathbb{A}$'s favor.

Theorem 12. *LDC is optimal for the baby server problem according to Relative Worst Order Analysis.*

Similar proofs show that a-LDC and BAL are also optimal algorithms.

In the definitions of LDC and BAL given in Sect. 2, different decisions are made as to which server to use in cases of ties. In LDC the server which is physically closer is moved in the case of a tie (equal virtual distances from the point requested). The rationale behind this is that the server which would have the least cost is moved. In BAL the server which is further away is moved to the point. The rationale behind this is that, since $d > 1$, when there is a tie, the total cost for the closer server is already significantly higher than the total cost for the other, so moving the server which is further away evens out how much total cost they have, at least temporarily. With these tie-breaking decisions, the two algorithms behave very similarly when d is an integer.

Theorem 13. LDC *and* BAL *are not comparable on the baby server problem with respect to Relative Worst Order Analysis, except when d is an integer, in which case they are equivalent.*

8   Concluding Remarks

The purpose of quality measures is to give information for use in practice, to choose the best algorithm for a particular application. What properties should such quality measures have?

First, it may be desirable that if one algorithm does at least as well as another on every sequence, then the measure decides in favor of the better algorithm. This is especially desirable if the better algorithm does significantly better on important sequences. Bijective Analysis, Relative Worst Order Analysis, and the Random Order Ratio have this property, but Competitive Analysis and the Max/Max Ratio do not. This was seen in the lazy vs. non-lazy version of Double Coverage for the baby server problem (and the more general metric k-server problem). Similar results have been presented previously for the paging problem—LRU vs. FWF and look-ahead vs. no look-ahead. See [6] for these results under Relative Worst Order Analysis and [1] for Bijective Analysis.

Secondly, it may be desirable that, if one algorithm does unboundedly worse than another on some important families of sequences, the quality measure reflects this. For the baby server problem, GREEDY is unboundedly worse than LDC on all families of sequences which consist mainly of alternating requests to the closest two points. This is reflected in Competitive Analysis, the Random Order Ratio, and Relative Worst Order Analysis, but not by the Max/Max Ratio or Bijective Analysis. Similarly, according to Bijective Analysis, LIFO and LRU are equivalent for paging, but LRU is often significantly better than LIFO, which keeps the first $k - 1$ pages it sees in cache forever. In both of these cases, Relative Worst Order Analysis says that the algorithms are weakly comparable in favor of the "better" algorithm.

Another desirable property would be ease of computation for many different problems, as with Competitive Analysis and Relative Worst Order Analysis. It is not clear that the other measures have this property.

References

1. Angelopoulos, S., Dorrigiv, R., López-Ortiz, A.: On the separation and equivalence of paging strategies. In: 18th ACM-SIAM Symposium on Discrete Algorithms, pp. 229–237 (2007)
2. Bein, W.W., Iwama, K., Kawahara, J.: Randomized competitive analysis for two-server problems. In: Halperin, D., Mehlhorn, K. (eds.) ESA 2008. LNCS, vol. 5193, pp. 161–172. Springer, Heidelberg (2008)
3. Ben-David, S., Borodin, A.: A new measure for the study of on-line algorithms. Algorithmica 11(1), 73–91 (1994)
4. Borodin, A., El-Yaniv, R.: Online Computation and Competitive Analysis. Cambridge University Press, Cambridge (1998)
5. Boyar, J., Favrholdt, L.M.: The relative worst order ratio for on-line algorithms. ACM Transactions on Algorithms 3(2), Article No. 22 (2007)
6. Boyar, J., Favrholdt, L.M., Larsen, K.S.: The relative worst order ratio applied to paging. Journal of Computer and System Sciences 73(5), 818–843 (2007)
7. Boyar, J., Irani, S., Larsen, K.S.: A comparison of performance measures for online algorithms. Technical report, arXiv:0806.0983v1 (2008)
8. Chrobak, M., Karloff, H.J., Payne, T.H., Vishwanathan, S.: New results on server problems. SIAM Journal on Discrete Mathematics 4(2), 172–181 (1991)
9. Dorrigiv, R., López-Ortiz, A.: A survey of performance measures for on-line algorithms. SIGACT News 36(3), 67–81 (2005)
10. Feller, W.: An Introduction to Probability Theory and Its Applications, 3rd edn., vol. 1. John Wiley & Sons, Inc., New York (1968); Problem 28, ch. 9, p. 240
11. Graham, R.L.: Bounds for certain multiprocessing anomalies. Bell Systems Technical Journal 45, 1563–1581 (1966)
12. Karlin, A.R., Manasse, M.S., Rudolph, L., Sleator, D.D.: Competitive snoopy caching. Algorithmica 3, 79–119 (1988)
13. Kenyon, C.: Best-fit bin-packing with random order. In: 7th Annual ACM-SIAM Symposium on Discrete Algorithms, pp. 359–364 (1996)
14. Manasse, M.S., McGeoch, L.A., Sleator, D.D.: Competitive algorithms for on-line problems. In: 20th Annual ACM Symposium on the Theory of Computing, pp. 322–333 (1988)
15. Manasse, M.S., McGeoch, L.A., Sleator, D.D.: Competitive algorithms for server problems. Journal of Algorithms 11(2), 208–230 (1990)
16. Sleator, D.D., Tarjan, R.E.: Amortized efficiency of list update and paging rules. Communications of the ACM 28(2), 202–208 (1985)

Delaunay Triangulation of Imprecise Points Simplified and Extended

Kevin Buchin[1], Maarten Löffler[2], Pat Morin[3], and Wolfgang Mulzer[4]

[1] Dep. of Mathematics and Computer Science, TU Eindhoven, The Netherlands
kbuchin@win.tue.nl
[2] Dep. of Information and Computing Sciences, Utrecht University, The Netherlands
loffler@cs.uu.nl
[3] School of Computer Science, Carleton University, Canada
morin@scs.carleton.ca
[4] Department of Computer Science, Princeton University, USA
wmulzer@cs.princeton.edu

Abstract. Suppose we want to compute the Delaunay triangulation of a set P whose points are restricted to a collection $\mathcal{R}$ of input regions known in advance. Building on recent work by Löffler and Snoeyink [21], we show how to leverage our knowledge of $\mathcal{R}$ for faster Delaunay computation. Our approach needs no fancy machinery and optimally handles a wide variety of inputs, eg, overlapping disks of different sizes and fat regions.

1   Introduction

Data imprecision is a fact of life that is often ignored in the design of geometric algorithms. The input for a typical computational geometry problem is a finite point set P in $\mathbb{R}^2$, or more generally $\mathbb{R}^d$. Traditionally, one assumes that P is known exactly, and indeed, in the 1980s and 1990s this was often justified, as much of the input data was hand-constructed for computer graphics or simulations. Nowadays, however, the input is often sensed from the real world, and thus inherently imprecise. This leads to a growing need to deal with imprecision.

An early model for imprecise geometric data, motivated by finite precision of coordinates, is ε-geometry [17]. Here, the input is a traditional point set P and a parameter ε. The true point set is unknown, but each point is guaranteed to lie in a disk of radius ε. Even though this model has proven fruitful and remains popular due to its simplicity [2, 18], it may often be too restrictive: imprecision regions could be more complicated than disks, and their shapes and sizes may even differ from point to point, eg, to model imprecision from different sources or independent imprecision in different input dimensions. The extra freedom in modeling leads to more involved algorithms, but still many results are available.

1.1   Preprocessing

The above results assume that the imprecise input is given once and simply has to be dealt with. While this holds in many applications, it is also often possible to

F. Dehne et al. (Eds.): WADS 2009, LNCS 5664, pp. 131–143, 2009.

get (more) precise estimates of the points, but they will only become available later, or they come at a higher cost. For example, in the *update complexity* model [16, 6], each data point is given imprecisely at the beginning but can always be found precisely at a certain price.

One model that has received attention lately is that of *preprocessing* an imprecise point set so that some structure can be computed faster when the exact points become available later. Here, we consider triangulations: let $\mathcal{R}$ be a collection of n planar regions, and suppose we know that the input has exactly one point from each region. The question is whether we can exploit our knowledge of $\mathcal{R}$ to quickly triangulate the exact input, once it is known. More precisely, we want to preprocess $\mathcal{R}$ into a data structure for *imprecise triangulation queries*: given a point p_i from each region $R_i \in \mathcal{R}$, compute a triangulation of $\{p_1, \ldots, p_n\}$. There are many parameters to consider; not only do we want preprocessing time, space usage, and query time to be small, but we would also like to support general classes of input regions and obtain "nice" (ie, Delaunay) triangulations. In the latter case, we speak of *imprecise Delaunay queries*.

Held and Mitchell [19] show that if $\mathcal{R}$ consists of n disjoint unit disks, it can be preprocessed in $O(n \log n)$ time into a linear-space data structure that can answer imprecise triangulation queries in linear time. This is improved by Löffler and Snoeyink [21] who can handle imprecise *Delaunay* queries with the same parameters. Both results generalize to regions with limited overlap and limited difference in shape and size—as long as these parameters are bounded, the same results hold. However, no attempt is made to optimize the dependence on the parameters.

Contrarily, van Kreveld, Löffler and Mitchell [23] study imprecise triangulation queries when $\mathcal{R}$ consists of n disjoint polygons with a total of m vertices, and they obtain an $O(m)$-space data structure with $O(m)$ query and $O(m \log m)$ preprocessing time. There is no restriction on the shapes and sizes of the individual regions (they do not even strictly have to be polygonal), only on the overlap. As these works already mention, a similar result for imprecise *Delaunay* queries is impossible. Djidjev and Lingas [15] show that if the points are sorted in any one direction, it still takes $\Omega(n \log n)$ time to compute their Delaunay triangulation. If $\mathcal{R}$ consists of vertical lines, the only information we could precompute is exactly this order (and the distances, but they can be found from the order in linear time anyway). All the algorithms above are deterministic.

1.2   Contribution

Our main concern will be imprecise Delaunay queries. First, we show that the algorithm by Löffler and Snoeyink [21] can be simplified considerably if we are happy with randomization and expected running time guarantees. In particular, we avoid the need for linear-time polygon triangulation [7], which was the main tool in the previous algorithm.

Second, though fast Delaunay queries for arbitrary regions are out of reach, we show that for realistic input we can get a better dependence on the realism parameters than in [21]. In particular, we consider k, the largest depth in the arrangement of $\mathcal{R}$, and β_f, the smallest fatness of any region in $\mathcal{R}$ (defined for a region R as the largest β such that for any disk D with center in R and intersecting ∂R, area$(R \cap D) \geq \beta \cdot$ area(D)). We can preprocess $\mathcal{R}$ in $O(n \log n)$ time into a data structure of $O(n)$ size that handles imprecise Delaunay queries in $O(n \log(k/\beta_f))$ time. We also consider β_t, the smallest thickness (defined as the fraction of the outcircle of a region occupied by it) of any region in $\mathcal{R}$, and r, the ratio between the diameters of the largest and the smallest region in $\mathcal{R}$. With the same preprocessing time and space, we can answer imprecise Delaunay queries in $O(n(\log(k/\beta_t) + \log \log r))$ time. For comparison, the previous bound is $O(nkr^2/\beta_t^2)$ [21]. Finally, we achieve similar results in various other realistic input models.

We describe two different approaches. The first, which gives the same result as [21], is extremely simple and illustrates the general idea. The second approach relies on quadtrees [12, Chapter 14] and is a bit more complicated, but generalizes easily. We extensively use a technique that has emerged just recently in the literature [9, 10, 23] and which we call *scaffolding*: in order to compute many related structures quickly, we first compute a "typical" structure—the *scaffold* Q—in a preprocessing phase. To answer a query, we insert the input points into Q and use a *hereditary* algorithm [9] to remove the scaffold efficiently. We need an algorithm for hereditary Delaunay triangulations:

Theorem 1 (Chazelle *et al* [8], see also [9]). *Let $P, Q \subseteq \mathbb{R}^2$ be two planar point sets with $|P \cup Q| = m$, and suppose that DT $(P \cup Q)$ is available. Then DT (P) can be computed in expected time $O(m)$.* □

2 Unit Disks: Simplified Algorithm

We begin with a very simple randomized algorithm for the original setting: given a sequence $\mathcal{R} = \langle R_1, \ldots, R_n \rangle$ of n disjoint unit disks, we show how to preprocess $\mathcal{R}$ in $O(n \log n)$ time into a linear-space data structure that can handle imprecise Delaunay queries in $O(n)$ *expected* time.

Let c_i denote the center of R_i and for $r > 0$ let R_i^r be the disk centered at c_i with radius r. The preprocessing algorithm creates a point set Q that for each R_i contains c_i and 7 points equally spaced on ∂R_i^2, the boundary of R_i^2. Then it computes DT (Q), the Delaunay triangulation of Q, and stores it. Since Q has $8n$ points, this takes $O(n \log n)$ time (eg, [12, Section 9]). We will need the following useful lemma about Q.

Lemma 1. *Let X be a point set with at most one point from each R_i. Any disk D of radius r contains at most $9(r + 3)^2$ points of $Q \cup X$.*

Proof. Let c be the center of D. Any point of $Q \cup X$ in D comes from some R_i with $\|c - c_i\| \leq r + 2$. The number of such R_i is at most the number of disjoint unit disks that fit into a disk of radius $r + 3$. A simple volume argument bounds this this by $(r+3)^2$. As each R_i contributes up to 9 points, the claim follows. □

Given the sequence $P = \langle p_1, \ldots, p_n \rangle$ of precise points, we construct $DT(P)$ by first inserting P into $DT(Q)$ to obtain $DT(Q \cup P)$ and then applying Theorem 1 to remove Q. To compute $DT(Q \cup P)$, we proceed as follows: for each point p_i we perform a (breadth-first or depth-first) search among the triangles of $DT(Q \cup \{p_1, \ldots, p_{i-1}\})$ that starts at some triangle incident to c_i and never leaves R_i, until we find the triangle t_i that contains p_i. Then we insert p_i into $DT(Q \cup \{p_1, \ldots, p_{i-1}\})$ by making it adjacent to the three vertices of t_i and performing *Delaunay flipping* [12, Section 9.3]. This takes time proportional to the number of triangles visited plus the degree of p_i in $DT(Q \cup \{p_1, \ldots, p_i\})$. The next lemma allows us to bound these quantities.

Lemma 2. *Let $Y = Q \cup X$ where X is any point set and consider $DT(Y)$. Then, for any point $p \in Y \cap R_i$, all neighbors of p in $DT(Y)$ lie inside R_i^3.*

Proof. Suppose there is an edge pq with $p \in R_i^1$ and $q \notin R_i^3$, see Figure 1. Then there is a disk C with p and q on its boundary and having no point of Y in its interior. The disk C contains a (generally smaller) disk C' tangent to R_i^1 and to ∂R_i^3. The intersection of C' with ∂R_i^2 is a circular arc of length $8 \arcsin(1/4) > 4\pi/7$. But this is a contradiction since one of the 7 points on ∂R_i^2 must lie in $C' \subseteq C$, so C cannot be empty. $\square$

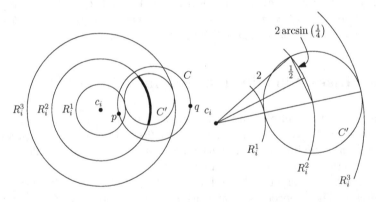

Fig. 1. C' covers a constant fraction of the boundary of R_i^2 and hence meets Q

The next two lemmas bound the number of triangles visited while inserting p_i.

Lemma 3. *Any triangle of $DT(Q \cup \{p_1, \ldots, p_i\})$ that intersects R_i has all three vertices in R_i^3.*

Proof. Let t be a triangle with a vertex q outside of R_i^3 that intersects R_i. The outcircle C of t intersects R_i, has q on its boundary, and contains no point of $Q \cup \{p_1, \ldots, p_i\}$ in its interior. As in Lemma 2, we see that C contains one of the 7 points on the boundary of R_i^2, so t cannot be Delaunay. $\square$

Lemma 4. *At most 644 triangles of* $\mathrm{DT}\,(Q \cup \{p_1, \ldots, p_i\})$ *intersect* R_i.

Proof. The triangles that intersect R_i form the of faces of a planar graph G. By Lemma 3 every vertex of G lies inside R_i^3, so by Lemma 1, there are at most $v = 9(3 + 3)^2 = 324$ vertices, and thus at most $2v - 4 = 644$ faces.   □

The final lemma bounds the degree of p_i at the time it is inserted.

Lemma 5. *The degree of* p_i *in* $\mathrm{DT}(Q \cup \{p_1, \ldots, p_i\})$ *is at most 324.*

Proof. By Lemma 2, all the neighbours of p_i are inside R_i^3 and by Lemma 1 there are at most $9(3 + 3)^2 = 324$ points of $Q \cup \{p_1, \ldots, p_i\}$ in R_i^3.   □

Thus, by Lemmas 4 and 5 each point of P can be inserted in constant time, so we require $O(n)$ time to construct $\mathrm{DT}(Q \cup P)$. A further $O(n)$ expected time is then needed to obtain $\mathrm{DT}\,(P)$ using Theorem 1. This yields the desired result.

Theorem 2. *Let* $\mathcal{R} = \langle R_1, \ldots, R_n \rangle$ *be a sequence of disjoint planar unit disks. In* $O(n \log n)$ *time and using* $O(n)$ *space we can preprocess* $\mathcal{R}$ *into a data structure that can answer imprecise Delaunay queries in* $O(n)$ *expected time.*

3   Disks of Different Size: Quadtree-Approach

We now extend Theorem 2 to differently-sized disks using a somewhat more involved approach. The main idea is the same: to answer a Delaunay query P we first construct $\mathrm{DT}\,(Q' \cup P)$ for an appropriate set Q' and split it using Theorem 1. The difference is that now we do not immediately precompute Q', but we derive a *quadtree* from $\mathcal{R}$ that can be used to find Q' and $\mathrm{DT}\,(Q' \cup P)$ efficiently later on. With the additional structure of the quadtree we can handle disks of different sizes. The following lemma follows from the empty circle property of Delaunay triangulations.

Lemma 6 (see Rajan [22]). *Let* $P \subseteq \mathbb{R}^2$ *be a planar n-point set, and let* T *be a triangulation of* P *with no obtuse angle. Then* T *is Delaunay.*   □

A *free quadtree* T is an ordered rooted tree that corresponds to a hierarchical decomposition of the plane into axis-aligned square *boxes*. Each node v of T has an associated box B_v, such that (i) if w is a descendent of v in T, then $B_w \subseteq B_v$; and (ii) if v and w are unrelated, then $B_v \cap B_w = \emptyset$. The *size* of a node v is the side length of B_v. For each node v, its *cell* C_v is the part of B_v not covered by v's children. The cells are pairwise disjoint and their union covers the root. A standard quadtree is a free quadtree with two kinds of nodes: *internal nodes* have exactly four children of half their size, and *leaf nodes* have no children. In this section we also allow *cluster nodes*, which have a single child that is smaller than its parent by at least a large constant factor 2^c, see Figure 2(a). They ensure that the complexity of T stays linear [4, Section 3.2]. Given $P \subseteq \mathbb{R}^2$, we say that T is a quadtree for P if (i) $|P \cap C_v| \leq 1$ for each leaf v of T; (ii) $P \cap C_v = \emptyset$ for all nonleaves v; (iii) P is contained in T's root box and sufficiently far away from its boundary; and (iv) T has $O(|P|)$ nodes, see Figure 2(b). The next lemma is a variant of a theorem by Bern *et al* [4] (see also [5]).

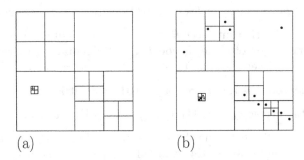

Fig. 2. (a) A quadtree. The lower left box contains a cluster node. (b) The quadtree is a valid quadtree for this set of points.

Lemma 7. *Let $P \subseteq \mathbb{R}^2$ be a planar n-point set, and let T be a quadtree for P. Then, given P and T, we can find $\mathrm{DT}(P)$ in expected time $O(n)$.*

Proof. First, we extend T into a quadtree T' that is (i) *balanced* and (ii) *separated*, ie, (i) no leaf in T' shares an edge with a leaf whose size differs by more than a factor of two and (ii) each non-empty leaf of T' is surrounded by two layers of empty boxes of the same size. This can be done by a top-down traversal of T, adding additional boxes for the balance condition and by subdividing the non-empty leaves of T to ensure separation. If after c subdivision steps a non-empty leaf B still does not satisfy separation, we place a small box around the point in B and treat it as a cluster, in which separation obviously holds.

Given T', we obtain a non-obtuse Steiner triangulation $\mathcal{T}$ for P with $O(n)$ additional vertices through a sequence of local manipulations, as described by Bern *et al* [4, Section 3]. Since all these operations involve constantly many adjacent cells, the total time for this step is linear. By Lemma 6, $\mathcal{T}$ is Delaunay, and we can use Theorem 1 to extract $\mathrm{DT}(P)$ in $O(n)$ expected time.   □

To apply Lemma 7 we need to preprocess $\mathcal{R}$ into a quadtree T such that any cell C_v in T intersects only constantly many disks. While we could consider the disks directly, we will instead use a quadtree T for a point set Q representing the disks. For each disk we include its center and top-, bottom-, left- and rightmost points in Q. Then, T can be constructed in $O(n \log n)$ time [4].

Lemma 8. *Every cell C_v of T is intersected by $O(1)$ disks in $\mathcal{R}$.*

Proof. If v is an internal node, then $C_v = \emptyset$. If v is a leaf, then $C_v = B_v$ and if a disk D intersects B_v without meeting a corner of B_v, then B_v either contains D's center or one of its four extreme points [13]. Thus, C_v intersects at most 5 disks, one for each corner and one for a point of Q it contains.

Now suppose v is a cluster node with child w. Then $C_v = B_v \setminus B_w$, and we must count the disks that intersect B_v, do not cover a corner of B_v, and have an extreme point or their center in B_w. For this, consider the at most four orthogonal neighbors of B_w in B_v (ie, copies of B_w directly to the left, to the right, above and below B_w). As we just argued, each of these neighbors meets $O(1)$ disks, and every disk D with an extreme point or center in B_w that intersects C_v also meets one of the orthogonal neighbors (if D has no extreme point or center in an orthogonal neighbor and does not cover any of its corners, it has to cover its center), which implies the claim.[1] $\square$

Theorem 3. *Let $\mathcal{R} = \langle R_1, \ldots, R_n \rangle$ be a sequence of disjoint planar disks (of possibly different size). In $O(n \log n)$ time and using $O(n)$ space we can prepro- cess $\mathcal{R}$ into a data structure that can answer imprecise Delaunay queries in $O(n)$ expected time.*

Proof. We construct Q and the quadtree T for Q as described above. For each R_i we store a list with the leaves in T that intersect it. By Lemma 8, the total size of these lists, and hence the complexity of the data structure, is linear. Now we describe how queries are handled: let $P = \langle p_1, \ldots, p_n \rangle$ be the input sequence. For each p_i, we find the node v of T such that $p_i \in C_v$ by traversing the list for R_i. This takes linear time. Since each cell of T contains at most constantly many input points, we can turn T into a quadtree for P in linear time. We now compute DT(P) via Lemma 7. $\square$

4   Overlapping Disks: Deflated Quadtrees

We extend the approach to disks with limited overlap. Now $\mathcal{R}$ contains n planar disks such that no point is covered by more than k disks. Aronov and Har- Peled [1] show that k can be approximated up to a constant factor in $O(n \log n)$ time. It is easily seen that imprecise Delaunay queries take $\Omega(n \log k)$ time in the worst case, and we show that this bound can be achieved.

The general strategy is the same as in Section 3. Let Q be the $5n$ representative points for $\mathcal{R}$, and let T be a quadtree for Q. As before, T can be found in time $O(n \log n)$ and has complexity $O(n)$. Now we use T to build a k-*deflated* quadtree T'. For an integer $\lambda > 0$, a λ-deflated quadtree T' for a point set Q has the same general structure as the quadtrees from the previous section, but it has lower complexity: each node of T' can contain up to λ points of Q in its cell and there are $O(n/\lambda)$ nodes. We have four different types of nodes: (i) *leaves* are nodes v without children, with up to λ points in C_v; (ii) *internal nodes* v have four children of half their size covering their parent, and $C_v = \emptyset$; (iii) *cluster nodes* are, as before, nodes v with a single—much smaller—child, and with no points in C_v; (iv) finally, a *deflated node* v has only one child w—possibly much smaller than its parent—and additionally C_v may contain up to λ points. Deflated nodes are a generalization of cluster nodes and ensure a more rapid progress in partitioning the point set Q.

---

[1] There is a slight subtlety concerning clusters which are close to the boundary of B_v: we require that either B_w shares an edge with B_v or that B_w's orthogonal neighbors are fully contained in B_v. This is ensured by positioning the clusters appropriately. The additional points on B_v's boundary can easily be handled, eg, by building a corresponding cluster in the adjacent box.

Algorithm DeflateTree(v)

1. If $n_v \leq \lambda$, return the tree consisting of v.
2. Let T_v be the subtree rooted in v, and let z be a node in T_v with the smallest value n_z such that $n_z > n_v - \lambda$. Note that z could be v.
3. For all children w of z, let $T'_w = $ DeflateTree(w).
4. Build a tree T'_v by picking v as the root, z as the only child of v, and linking the trees T'_w to z. If $v \neq z$, then v is a deflated node. Return T'_v as the result.

Algorithm 1. Turn a quadtree into a λ-deflated quadtree

Given a quadtree T for Q, a λ-deflated quadtree T' can be found in linear time. For every node v in T, compute $n_v = |B_v \cap Q|$. This takes $O(n)$ time. Then, T' is obtained by applying DeflateTree (Algorithm 1) to the root of T. Since DeflateTree performs a simple top-down traversal of T, it takes $O(n)$ time.

Lemma 9. *A λ-deflated quadtree T' produced by Algorithm 1 has $O(n/\lambda)$ nodes.*

Proof. Let T'' be the subtree of T' that contains all nodes v with $n_v > \lambda$, and suppose that every cluster node in T'' has been contracted with its child. We will show that T'' has $O(n/\lambda)$ nodes, which implies the claim, since no two cluster nodes are adjacent, and because all the non-cluster nodes in T' which are not in T'' must be leaves. We count the nodes in T'' as follows: (i) since the leaves of T'' correspond to disjoint subsets of Q of size at least λ, there are at most n/λ of them; (ii) the bound on the leaves also implies that T'' contains at most n/λ nodes with at least 2 children; (iii) the number of nodes in T'' with a single child that has at least 2 children is likewise bounded; (iv) when an internal node v has a single child w that also has only a single child, then by construction v and w together must contain at least λ points in their cells, otherwise they would not have been two separate nodes. Thus, we can charge $\lambda/2$ points from Q to v, and the total number of such nodes is $2n/\lambda$. $\square$

Now let T' be a k-deflated quadtree for Q. By treating deflated nodes like clusters and noting that the center and corners of each box of T' can be contained in at most k disks, the same arguments as in Lemma 8 lead to the next lemma:

Lemma 10. *Every cell C_v of T' is intersected by $O(k)$ disks of $\mathcal{R}$.* $\square$

Theorem 4. *Let $\mathcal{R} = \langle R_1, \ldots, R_n \rangle$ be a sequence of planar disks such that no point is covered by more than k disks. In $O(n \log n)$ time and using $O(n)$ space we can preprocess $\mathcal{R}$ into a data structure that can answer imprecise Delaunay queries in $O(n \log k)$ expected time.*

Proof. It remains to show how to preprocess T' to handle the imprecise Delaunay queries in time $O(n \log k)$. By Lemmas 9 and 10, the total number of disk-cell incidences in T' is $O(n)$. Thus, in $O(n)$ total time we can find for each $R \in \mathcal{R}$ the list of cells of T' it intersects. Next, we determine for each node v in T'

the portion X_v of the original quadtree T inside the cell C_v and build a point location data structure for X_v. Since X_v is a partial quadtree for at most k points, it has complexity $O(k)$, and since the X_v are disjoint, the total space requirement and construction time are linear. This finishes the preprocessing.

To handle an imprecise Delaunay query, we first locate the input points P in the cells of T' just as in Theorem 3. This takes $O(n)$ time. Then we use the point location structures for the X_v to locate P in T in total time $O(n \log k)$. Now we turn T into a quadtree for P in time $O(n \log k)$, and find the Delaunay triangulation in time $O(n)$, as before. □

5   Realistic Input Models

In this section we show that the results of the previous sections readily generalize to many realistic input models [14]. This is because a point set representing the regions—like the point set Q for the disks—exists in such models, eg, for fat regions. Thus, we directly get an algorithm for fat regions for which we provide a matching lower bound. We then demonstrate how to handle situations where a set like Q cannot be easily constructed by the example of *thick regions*.

Let β be a constant with $0 < \beta \le 1$. A planar region R is β-*fat* if for any disk D with center in R and intersecting ∂R, area$(R \cap D) \ge \beta \cdot$ area(D). A planar region R is β-*thick* if area$(R) \ge \beta \cdot$ area$(D_{\min}(R))$, where $D_{\min}$ denotes the smallest disk enclosing R. Let κ be a positive integer. A set Q of points is called a κ-*guarding set (against axis-aligned squares)* for a set of planar regions $\mathcal{R}$, if any axis-aligned square not containing a point from Q intersects at most κ regions from $\mathcal{R}$. For instance, the point set Q considered in the previous sections is a 4-guarding set for disjoint disks [13]. It is also a $4k$-guarding set for disks which do not cover any point more than k times.

The definition of a κ-guarding set Q does not explicitly state how many regions a square containing points of Q might intersect, but a square containing m points of Q can intersect only $4\kappa m$ regions [13, Theorem 2.8]. Now, assume each point in Q is assigned to a region in $\mathcal{R}$. We call Q a κ-*strong-guarding set* of $\mathcal{R}$ if any square containing m points of Q intersects at most κ regions plus the regions assigned to the m points. This definition is motivated by the following relation between fatness and guardability. A set of planar regions $\mathcal{R}$ is κ-*cluttered* if the corners of the bounding rectangles for $\mathcal{R}$ constitute a κ-guarding set. De Berg *et al* prove that a set of disjoint β-fat regions is $16/\beta$-cluttered [14, Theorem 3.1, Theorem 3.2]. Their argument actually shows strong guardability (cf [11, Lemma 2.8]) and easily extends to overlapping regions.

Lemma 11. *For a set of β-fat regions $\mathcal{R}$ that cover no point more than k times, the corners of the bounding rectangles of $\mathcal{R}$ constitute a $(16k/\beta)$-strong-guarding set for $\mathcal{R}$ (with corners assigned to the corresponding region).* □

Since the argument in the proof of Lemma 8 (and of Lemma 10) is based on axis-aligned squares, it directly generalizes to the quadtree for a guarding set.

Lemma 12. *Let $\mathcal{R}$ be a set of regions and Q a κ-strong-guarding set for $\mathcal{R}$. Let T' be a κ-deflated quadtree of Q. Then any cell of T intersects $O(\kappa)$ regions.* □

We say that $\mathcal{R}$ is *traceable* if we can find the m incidences between the n regions in $\mathcal{R}$ and the l cells of a deflated quadtree T in $O(l + m + n)$ time and the bounding rectangles of the regions in $\mathcal{R}$ in $O(n)$ time. For example, this holds for polygonal regions of total complexity $O(|\mathcal{R}|)$.

Theorem 5. *Let $\mathcal{R} = \langle R_1, \ldots, R_n \rangle$ be a sequence of traceable planar regions with linear-size κ-strong-guarding set Q, where κ is not necessarily known, but Q is. In $O(n \log n)$ time and using $O(n)$ space we can preprocess $\mathcal{R}$ into a data structure that can answer imprecise Delaunay queries in $O(n \log \kappa)$ expected time.*

Proof. Lemma 12 would directly imply the theorem if κ was known. We can find a suitable λ-deflated tree with $\lambda \in O(\kappa)$ by an exponential search on λ, i.e., for a given λ we build a λ-deflated tree and check whether any box intersects more than $c\lambda$ regions for a constant $c \geq 6$. Recall that a deflated quadtree can be computed from a quadtree in linear time. Thus, finding a suitable λ takes $O(n \log n)$ time. □

For overlapping fat regions Lemma 11 and Theorem 5 imply the following result.

Corollary 1. *Let $\mathcal{R} = \langle R_1, \ldots, R_n \rangle$ be a sequence of planar traceable β-fat regions such that no point is covered by more than k of the regions. In $O(n \log n)$ time and using $O(n)$ space we can preprocess $\mathcal{R}$ into a data structure that can answer imprecise Delaunay queries in $O(n \log(k/\beta))$ expected time.*

We next show that the $O(n \log(1/\beta))$ bound for disjoint fat regions is optimal.

Theorem 6. *For any n and $1/\beta \in [1, n]$, there exists a set $\mathcal{R}$ of $O(n)$ planar β-fat rectangles such that imprecise Delaunay queries for $\mathcal{R}$ take $\Omega(n \log(1/\beta))$ steps in the algebraic computation tree model.*

Proof. We adapt a lower bound by Djidjev and Lingas [15, Section 4]. Wlog, β^{-1} is an integer. Consider the problem β-1-CLOSENESS: we are given $k = \beta n$-sequences $\mathbf{x}_1, \ldots, \mathbf{x}_k$, each containing β^{-1} real numbers in $[0, \beta^{-2}]$, and we need to decide whether any $\mathbf{x}_i$ contains two numbers with difference at most 1. Any algebraic decision tree for β-1-CLOSENESS has cost $\Omega(n \log(1/\beta))$: let $W' \subseteq \mathbb{R}^n$ be defined as $W' = \{ (\mathbf{x}_1, \ldots, \mathbf{x}_k) \mid |x_{ij} - x_{il}| > 1 \text{ for } 1 \leq i \leq k; 1 \leq j \neq l \leq \beta^{-1} \}$, where x_{ij} is the jth coordinate of $\mathbf{x}_i$. Let $W = W' \cap [0, \beta^{-2}]^n$. Since W has at least $(\beta^{-1}!)^{\beta n}$ connected components, Ben-Or's lower bound [3, Theorem 5] implies the claim. Now, we construct $\mathcal{R}$. Let $\varepsilon = \beta/10$ and consider the $\beta^{-1} + 2$ intervals on the x-axis $B_\varepsilon[0], B_\varepsilon[\beta/3], B_\varepsilon[2\beta/3], \ldots, B_\varepsilon[1/3], B_\varepsilon[1/2]$, where $B_\varepsilon[x]$ denotes the one-dimensional closed ε-ball around x. Extend the intervals into β^3-fat rectangles with side lengths 2ε and β^{-2}. These rectangles constitute a *group*. Now, $\mathcal{R}$ consists of k congruent groups $G_1, \ldots, G_k$; sufficiently far away from each other. Let $\mathbf{x}_1, \ldots, \mathbf{x}_k$ be an instance of β-1-CLOSENESS. The input P consists of k sets $P_1, \ldots, P_k$, one for each $\mathbf{x}_i$. Each P_i contains $\beta^{-1} + 2$ points, one from every rectangle in G_i: $P_i = \langle (0,0), (1/(3\alpha), x_{i1}), \ldots, (1/3, x_{i\beta-1}), (1/2, 0) \rangle + v_i$, where v_i denotes the displacement vector for G_i. Clearly, P can be computed in

$O(n)$ time. Djidjev and Lingas [15] argue that either $\mathbf{x}_i$ contains two numbers with difference at most $1/3$, or the Voronoi cells for P_i intersect the line through the left-most rectangle in G_i according to the sorted order of $\mathbf{x}_i$. In either case, we can decide β-1-CLOSENESS in linear time from DT (P), as desired. $\square$

Finally, we consider β-thick regions. Although thickness does not give us a guarding set, we can still preprocess $\mathcal{R}$ for efficient imprecise Delaunay queries if the ratio between the largest and smallest region in $\mathcal{R}$ is bounded (we measure the size by the radius of the smallest enclosing circle).

Theorem 7. *Let $\mathcal{R}$ be a sequence of n β-thick k-overlapping regions such that the ratio of the largest and the smallest region in $\mathcal{R}$ is r. In $O(n \log n)$ time we can preprocess $\mathcal{R}$ into a linear-space data structure that can answer imprecise Delaunay queries in time $O(n(\log(k/\beta) + \log \log r))$.*

Proof. Subdivide the regions into $\log r$ groups such that in each group the radii of the minimum enclosing circles differ by at most a factor of 2. For each group $\mathcal{R}_i$, let ρ_i be the largest radius of a minimum enclosing circle for a region in $\mathcal{R}_i$. We replace every region in $\mathcal{R}_i$ by a disk of radius ρ_i that contains it. This set of disks is at most $(2k/\beta)$-overlapping, so we can build a data structure for $\mathcal{R}_i$ in $O(n_i \log n_i)$ time by Theorem 4. To answer an imprecise Delauany query, we handle each group in $O(n_i \log(k/\beta))$ time and then use Kirkpatrick's algorithm [20] to combine the triangulations in time $O(n \log \log r)$. $\square$

6   Conclusions

We give an alternative proof of the result by Löffler and Snoeyink [21] with a much simpler, albeit randomized, algorithm that avoids heavy machinery. Our approach yields optimal results for overlapping disks of different sizes and fat regions. Furthermore, it enables us to leverage known facts about guarding sets to handle many other realistic input models. We need randomization only when we apply Theorem 1 to remove the scaffold, and finding a deterministic algorithm for hereditary Delaunay triangulations remains an intriguing open problem.

Acknowledgments

The results in Section 2 were obtained at the NICTA Workshop on Computational Geometry for Imprecise Data, December 18–22, 2008 at the NICTA University of Sydney Campus. We would like to thank the other workshop participants, namely Hee-Kap Ahn, Sang Won Bae, Dan Chen, Otfried Cheong, Joachim Gudmundsson, Allan Jørgensen, Stefan Langerman, Marc Scherfenberg, Michiel Smid, Tasos Viglas and Thomas Wolle, for helpful discussions and for providing a stimulating working environment. We would also like to thank David Eppstein for answering our questions about [4] and pointing us to [5]. This research was partially supported by the Netherlands Organisation for Scientific

Research (NWO) through the project GOGO and the BRICKS/FOCUS project no. 642.065.503. W. Mulzer was supported in part by NSF grant CCF-0634958 and NSF CCF 0832797. Pat Morin was supported by NSERC, NICTA and the University of Sydney.

References

[1] Aronov, B., Har-Peled, S.: On approximating the depth and related problems. SIAM Journal on Computing 38(3), 899–921 (2008)

[2] Bandyopadhyay, D., Snoeyink, J.: Almost-Delaunay simplices: Nearest neighbor relations for imprecise points. In: SODA, pp. 403–412 (2004)

[3] Ben-Or, M.: Lower bounds for algebraic computation trees. In: STOC, pp. 80–86 (1983)

[4] Bern, M., Eppstein, D., Gilbert, J.: Provably good mesh generation. J. Comput. System Sci. 48(3), 384–409 (1994)

[5] Bern, M., Eppstein, D., Teng, S.-H.: Parallel construction of quadtrees and quality triangulations. Internat. J. Comput. Geom. Appl. 9(6), 517–532 (1999)

[6] Bruce, R., Hoffmann, M., Krizanc, D., Raman, R.: Efficient update strategies for geometric computing with uncertainty. Theory Comput. Syst. 38(4), 411–423 (2005)

[7] Chazelle, B.: Triangulating a simple polygon in linear time. Disc. and Comp. Geometry 6, 485–524 (1991)

[8] Chazelle, B., Devillers, O., Hurtado, F., Mora, M., Sacristán, V., Teillaud, M.: Splitting a Delaunay triangulation in linear time. Algorithmica 34(1), 39–46 (2002)

[9] Chazelle, B., Mulzer, W.: Computing hereditary convex structures. To appear in SoCG (2009)

[10] Clarkson, K.L., Seshadhri, C.: Self-improving algorithms for Delaunay triangulations. In: SoCG, pp. 148–155 (2008)

[11] de Berg, M.: Linear size binary space partitions for uncluttered scenes. Algorithmica 28(3), 353–366 (2000)

[12] de Berg, M., Cheong, O., van Kreveld, M., Overmars, M.: Computational geometry, 3rd edn. Springer, Berlin (2000); Algorithms and applications

[13] de Berg, M., David, H., Katz, M.J., Overmars, M.H., van der Stappen, A.F., Vleugels, J.: Guarding scenes against invasive hypercubes. Computational Geometry: Theory and Applications 26(2), 99–117 (2003)

[14] de Berg, M., van der Stappen, A.F., Vleugels, J., Katz, M.J.: Realistic input models for geometric algorithms. Algorithmica 34(1), 81–97 (2002)

[15] Djidjev, H.N., Lingas, A.: On computing Voronoi diagrams for sorted point sets. Internat. J. Comput. Geom. Appl. 5(3), 327–337 (1995)

[16] Franciosa, P.G., Gaibisso, C., Gambosi, G., Talamo, M.: A convex hull algorithm for points with approximately known positions. Internat. J. Comput. Geom. Appl. 4(2), 153–163 (1994)

[17] Guibas, L.J., Salesin, D., Stolfi, J.: Epsilon geometry: building robust algorithms from imprecise computations. In: SoCG, pp. 208–217 (1989)

[18] Guibas, L.J., Salesin, D., Stolfi, J.: Constructing strongly convex approximate hulls with inaccurate primitives. Algorithmica 9, 534–560 (1993)

[19] Held, M., Mitchell, J.S.B.: Triangulating input-constrained planar point sets. Inf. Process. Lett. 109(1), 54–56 (2008)

[20] Kirkpatrick, D.G.: Efficient computation of continuous skeletons. In: FOCS, pp. 18–27 (1979)

[21] Löffler, M., Snoeyink, J.: Delaunay triangulations of imprecise points in linear time after preprocessing. In: SoCG, pp. 298–304 (2008)

[22] Rajan, V.T.: Optimality of the Delaunay triangulation in $\mathbb{R}^d$. Disc. and Comp. Geometry 12, 189–202 (1994)

[23] van Kreveld, M.J., Löffler, M., Mitchell, J.S.B.: Preprocessing imprecise points and splitting triangulations. In: Hong, S.-H., Nagamochi, H., Fukunaga, T. (eds.) ISAAC 2008. LNCS, vol. 5369, pp. 544–555. Springer, Heidelberg (2008)

An Improved SAT Algorithm
in Terms of Formula Length*

Jianer Chen and Yang Liu

Department of Computer Science and Engineering
Texas A&M University
College Station, TX 77843, USA
{chen,yangliu}@cse.tamu.edu

Abstract. We present an improved algorithm for the general SATISFIA-
BILITY problem. We introduce a new measure, the l-value, for a Boolean
formula $\mathcal{F}$, which is defined based on weighted variable frequencies in
the formula $\mathcal{F}$. We then develop a branch-and-search algorithm for the
SATISFIABILITY problem that tries to maximize the decreasing rates in
terms of the l-value during the branch-and-search process. The complexity of the algorithm in terms of the l-value is finally converted into
the complexity in terms of the total length L of the input formula, resulting in an algorithm of running time $O(2^{0.0911L}) = O(1.0652^L)$ for
the SATISFIABILITY problem, improving the previous best upper bound
$O(2^{0.0926L}) = O(1.0663^L)$ for the problem.

1   Introduction

The SATISFIABILITY problem (briefly, SAT: given a CNF Boolean formula, decide
if the formula has a satisfying assignment) is perhaps the most famous and most
extensively studied NP-complete problem. Given the NP-completeness of the
problem [4], it has become natural to develop exponential time algorithms that
solve the problem as fast as possible.

There are three popular parameters that have been used in measuring exponential time algorithms for the SAT problem: the number n of variables, the
number m of clauses, and the *total length L* that is the sum of the clause lengths
in the input formula. Note that the parameter L is probably the most precise
parameter in terms of standard complexity theory, and that both parameters n
and m could be sublinear in instance length. Algorithms for SAT in terms of each
of these parameters have been extensively studied. See [6] for a comprehensive
review and [10] for more recent progress on the research in these directions.

In the current paper, we are focused on algorithms for SAT in terms of the parameter L. The research started 20 years ago since the first published algorithm
of time $O(1.0927^L)$ [5]. The upper bound was subsequently improved by an im-
pressive list of publications. We summarize the major progress in the following
table.

* This work was supported in part by the National Science Foundation under the
Grant CCF-0830455.

F. Dehne et al. (Eds.): WADS 2009, LNCS 5664, pp. 144–155, 2009.

Ref.	Van Gelder [5]	Kullmann *et al.* [9]	Hirsh [7]	Hirsh [8]	Wahlstom [11]
Bound	1.0927^L	1.0801^L	1.0758^L	1.074^L	1.0663^L
Year	1988	1997	1998	2000	2005

The branch-and-search method has been widely used in the development of SAT algorithms. Given a Boolean formula $\mathcal{F}$, let $\mathcal{F}[x]$ and $\mathcal{F}[\overline{x}]$ be the resulting formula after assigning TRUE and FALSE, respectively, to the variable x in the formula $\mathcal{F}$. The branch-and-search method is based on the fact that $\mathcal{F}$ is satisfiable if and only if at least one of $\mathcal{F}[x]$ and $\mathcal{F}[\overline{x}]$ is satisfiable. Most SAT algorithms are based on this method.

Unfortunately, analysis directly based on the parameter L usually does not give a good upper bound in terms of L for a branch-and-search SAT algorithm. Combinations of the parameter L and other parameters, such as the number n of variables, have been used as "measures" in the analysis of SAT algorithms. For example, the measure $L - 2n$ [12] and a more general measure that is a function $f(L, n)$ of the parameters L and n [11] have been used in the analysis of SAT algorithms whose complexity is measured in terms of the parameter L.

In the current paper, we introduce a new measure, the *l-value* of a Boolean formula $\mathcal{F}$. Roughly speaking, the measure *l*-value $l(\mathcal{F})$ is defined based on weighted variable frequencies in the input formula $\mathcal{F}$. We develop a branch-and-search algorithm that tries to maximize the decreasing rates in terms of the *l*-value during its branch-and-search process. In particular, by properly choosing the variable frequency weights so that the formula *l*-value is upper bounded by $L/2$, by adopting new reduction rules, and by applying the analysis technique of Measure and Conquer recently developed by Fomin et al. [3], we develop a new branch-and-search algorithm for the SAT problem whose running time is bounded by $O(1.1346^{l(\mathcal{F})})$ on an input formula $\mathcal{F}$. Finally, by combining this algorithm with the algorithm in [12] to deal with formulas of lower variable frequencies and by converting the measure $l(\mathcal{F})$ into the parameter L, we achieve a SAT algorithm of running time $O(1.0652^L)$, improving the previously best SAT algorithm of running time $O(1.0663^L)$ [11].

We remark that although the analysis of our algorithm is lengthy, our algorithm itself is very simple and can be easily implemented. Note that the lengthy analysis needs to be done only once to ensure the correctness of the algorithm, while the simplicity of the algorithm gives its great advantage when it is applied (many times) to determine the satisfiability of CNF Boolean formulas.

2 Preliminaries

A (Boolean) *variable* x can be assigned value either 1 (TRUE) or 0 (FALSE). The variable x has two corresponding *literals* x and $\overline{x}$. A literal z is *satisfied* if $z = 1$. A *clause* C is a disjunction of a set of literals, which can be regarded as a set of literals. Therefore, we may write $C_1 = zC_2$ to indicate that the clause C_1 consists of the literal z plus all literals in the clause C_2, and use C_1C_2 to denote the clause that consists of all literals that are in either C_1 or C_2, or both. The *length* of a clause C, denoted by $|C|$, is the number of literals in C. A clause

C is *satisfied* if any literal in C is satisfied. A (CNF Boolean) *formula* $\mathcal{F}$ is a conjunction of clauses $C_1, \ldots, C_m$, which can be regarded as a collection of the clauses. The formula $\mathcal{F}$ is *satisfied* if all clauses in $\mathcal{F}$ are satisfied. The *length* L of the formula $\mathcal{F}$ is defined as $L = |C_1| + \cdots + |C_m|$.

A literal z is an *i-literal* if z is contained in exactly i clauses, and is an *i^+-literal* if z is contained in at least i clauses. An *(i,j)-literal* z is a literal such that exactly i clauses contain z and exactly j clauses contain $\overline{z}$. A variable x is an *i-variable* (or, the *degree* of x is i) if there are exact i clauses that contain either x or $\overline{x}$. A clause C is an *k-clause* if $|C| = k$, and is an *k^+-clause* if $|C| \geq k$.

A *resolvent* on a variable x in a formula $\mathcal{F}$ is a clause of the form CD such that xC and $\overline{x}D$ are clauses in $\mathcal{F}$. The *resolution* on the variable x in the formula $\mathcal{F}$, written as $DP_x(\mathcal{F})$, is a formula that is obtained by first removing all clauses that contain either x or $\overline{x}$ from $\mathcal{F}$ and then adding all possible resolvents on the variable x into $\mathcal{F}$.

A *branching vector* is a tuple of positive real numbers. A branching vector $t = (t_1, \ldots, t_r)$ corresponds to a polynomial $1 - \sum_{i=1}^{r} x^{-t_i}$, which has a unique positive root $\tau(t)$ [1]. We say that a branching vector t' is *inferior* to a branching vector t'' if $\tau(t') \geq \tau(t'')$. In particular, if either $t_1' \leq t_1''$ and $t_2' \leq t_2''$, or $t_1' \leq t_2''$ and $t_2' \leq t_1''$, then it can be proved [1] that the branching vector $t' = (t_1', t_2')$ is inferior to the branching vector $t'' = (t_1'', t_2'')$.

The execution of a SAT algorithm based on the branch-and-search method can be represented as a search tree $\mathcal{T}$ whose root is labeled by the input formula $\mathcal{F}$. Recursively, if at a node w_0 labeled by a formula $\mathcal{F}_0$ in the search tree $\mathcal{T}$, the algorithm breaks $\mathcal{F}_0$, in polynomial time, into r smaller formulas $\mathcal{F}_1, \ldots, \mathcal{F}_r$, and recursively works on these smaller formulas, then the node w_0 in $\mathcal{T}$ has r children, labeled by $\mathcal{F}_1, \ldots, \mathcal{F}_r$, respectively. Suppose that we use a measure $\mu(\mathcal{F})$ for a formula $\mathcal{F}$, then the branching vector for this branching, with respect to the measure μ, is $t = (t_1, \ldots, t_r)$, where $t_i = \mu(\mathcal{F}_0) - \mu(\mathcal{F}_i)$ for all i. Finally, suppose that t' is a branching vector that is inferior to all branching vectors for any branching in the search tree $\mathcal{T}$, then the complexity of the SAT algorithm is bounded by $O(\tau(t')^{\mu(\mathcal{F})})$ times a polynomial of L [1].

Formally, for a given formula $\mathcal{F}$, we define the *l-value* for $\mathcal{F}$ to be $l(\mathcal{F}) = \sum_{i \geq 1} w_i n_i$, where for each i, n_i is the number of i-variables in $\mathcal{F}$, and the *frequency weight* w_i for i-variables are set by the following values:

$$w_0 = 0, \qquad w_1 = 0.32, \qquad w_2 = 0.45, \qquad (1)$$
$$w_3 = 0.997, \qquad w_4 = 1.897, \qquad w_i = i/2, \ \text{for } i \geq 5.$$

Define $\delta_i = w_i - w_{i-1}$, for $i \geq 1$. Then we can easily verify that

$$\delta_i \geq 0.5, \quad \text{for all } i \geq 3,$$
$$\delta_{\min} = \min\{\delta_i \mid i \geq 1\} = \delta_2 = 0.13, \qquad (2)$$
$$\delta_{\max} = \max\{\delta_i \mid i \geq 1\} = \delta_4 = 0.9.$$

Note that the length L of the formula $\mathcal{F}$ is equal to $\sum_{i \geq 1} i \cdot n_i$, and that $i/5 \leq w_i \leq i/2$ for all i. Therefore, we have $L/5 \leq l(\mathcal{F}) \leq L/2$.

3   The Reduction Rules

We say that two formulas $\mathcal{F}_1$ and $\mathcal{F}_2$ are *equivalent* if $\mathcal{F}_1$ is satisfiable if and only if $\mathcal{F}_2$ is satisfiable. A literal z in a formula $\mathcal{F}$ is *monotone* if the literal $\bar{z}$ does not appear in $\mathcal{F}$.

We present in this section a set of reduction rules that reduce a given formula $\mathcal{F}$ to an equivalent formula $\mathcal{F}'$ without increasing the l-value. Consider the algorithm given in Figure 1.

Algorithm Reduction($\mathcal{F}$)
INPUT: a non-empty formula $\mathcal{F}$
OUTPUT: an equivalent formula on which no further reduction is applicable

change = true;
while change **do**
    case 1. a clause C is a subset of a clause D: remove D;
    case 2. a clause C contains both x and $\bar{x}$: remove C;
    case 3. a clause C contains multiple copies of a literal z:
        remove all but one z in C;
    case 4. there is a variable x with at most one non-trivial resolvent:
        $\mathcal{F} \leftarrow DP_x(\mathcal{F})$;
    case 5. there is a 1-clause (z) or a monotone literal z: $\mathcal{F} \leftarrow \mathcal{F}[z]$;
    case 6. there exist a 2-clause $z_1 z_2$ and a clause $z_1 \bar{z}_2 C$:
        remove $\bar{z}_2$ from the clause $z_1 \bar{z}_2 C$;
    case 7. there are clauses $z_1 z_2 C_1$ and $z_1 \bar{z}_2 C_2$ and z_2 is a $(2,1)$-literal:
        remove z_1 from the clause $z_1 z_2 C_1$;
    case 8. there are clauses $z_1 z_2$ and $\bar{z}_1 \bar{z}_2 C$ such that literal $\bar{z}_1$ is a 1-literal:
        remove the clause $z_1 z_2$;
    case 9. there is a $(2,2)$-variable x with clauses $\bar{x} z_1$, $\bar{x} z_2$ and two
        3-clauses $x C_1$ and $x C_2$ such that $\bar{z}_1$ and $\bar{z}_2$ are in $C_1 \cup C_2$:
        $\mathcal{F} \leftarrow DP_x(\mathcal{F})$. Apply case 2, if possible;
    case 10. there is a 2-clause $z_1 z_2$ where z_1 is a 1-literal, or there are two
        2-clauses $z_1 z_2$ and $\bar{z}_1 \bar{z}_2$: replace z_1 with $\bar{z}_2$. Apply case 2 if possible;
    case 11. there are two clauses $C D_1$ and $C D_2$ with $|C| > 1$:
        replace $C D_1$ and $C D_2$ with $\bar{x} C$, $x D_1$, and $x D_2$, where x is a new
        variable. Apply case 2 on C if possible;
    default: change = false;

Fig. 1. The reduction algorithm

Lemma 1. *The algorithm* **Reduction**($\mathcal{F}_1$) *produces a formula equivalent to* $\mathcal{F}_1$.

Proof. It suffices to prove that in each of the listed cases, the algorithm **Reduction** on the formula $\mathcal{F}_1$ produces an equivalent formula $\mathcal{F}_2$. This can be easily verified for Cases 1, 2, 3, and 5.

The claim holds true for Cases 4 and 9 from the resolution principle [2].

In Case 6, the clause $z_1 \bar{z}_2 C$ in $\mathcal{F}_1$ is replaced with the clause $z_1 C$. If an assignment A_2 satisfies $\mathcal{F}_2$, then obviously A_2 also satisfies $\mathcal{F}_1$. On the other hand, if an assignment A_1 satisfies $\mathcal{F}_1$ but does not satisfy the clause z_1C in $\mathcal{F}_2$, then because of the clause z_1z_2 in $\mathcal{F}_1$, we must have $z_1 = 0$ and $z_2 = 1$. Since A_1 satisfies the clause $z_1\bar{z}_2C$ in $\mathcal{F}_1$, this would derive a contradiction that A_1 must satisfy z_1C. Therefore, A_1 must also satisfy the formula $\mathcal{F}_2$.

In Case 7, the clause $z_1z_2C_1$ in $\mathcal{F}_1$ is replaced with the clause z_2C_1. Again, the satisfiability of $\mathcal{F}_2$ trivially implies the satisfiability of $\mathcal{F}_1$. For the other direction, let z_2C_3 be the other clause that contains z_2 (note that z_2 is a $(2,1)$-literal). If an assignment A_1 satisfies $\mathcal{F}_1$ (thus satisfies $z_1z_2C_1$) but not $\mathcal{F}_2$ (i.e., not z_2C_1), then we must have $z_1 = 1$, $z_2 = 0$, and $C_3 = 1$ under A_1. By replacing the assignment $z_2 = 0$ with $z_2 = 1$ in A_1, we will obtain an assignment A_2' that satisfies all z_2C_1, $z_1\bar{z}_2C_2$, and z_2C_3, thus satisfies the formula $\mathcal{F}_2$.

In Case 8, the formula $\mathcal{F}_2$ is obtained from the formula $\mathcal{F}_1$ by removing the clause z_1z_2. Thus, the satisfiability of $\mathcal{F}_1$ trivially implies the satisfiability of $\mathcal{F}_2$. On the other hand, suppose that an assignment A_2 satisfying $\mathcal{F}_2$ does not satisfy $\mathcal{F}_1$ (i.e., does not satisfy the clause z_1z_2). Then A_2 must assign $z_1 = 0$ and $z_2 = 0$. We can simply replace $z_1 = 0$ with $z_1 = 1$ in A_2 and keep the assignment satisfying $\mathcal{F}_2$: this is because $\bar{z}_1\bar{z}_2C$ is the only clause in $\mathcal{F}_2$ that contains $\bar{z}_1$. Now the new assignment also satisfies z_1z_2, thus satisfies the formula $\mathcal{F}_1$.

For Case 10, it is suffice to show that we can always set $z_1 = \bar{z}_2$ in a satisfying assignment for the formula $\mathcal{F}_1$. For the subcase where z_1 is a 1-literal and z_1z_2 is a 2-clause, if a satisfying assignment A_1 for $\mathcal{F}_1$ assigns $z_2 = 1$ then we can simply let $z_1 = \bar{z}_2 = 0$ since z_1 is only contained in the clause z_1z_2. If A_1 assigns $z_2 = 0$ then because of the 2-clause z_1z_2, A_1 must assign $z_1 = \bar{z}_2 = 1$. For the other subcase, note that the existence of the 2-clauses z_1z_2 and $\bar{z}_1\bar{z}_2$ in the formula $\mathcal{F}_1$ trivially requires that every assignment satisfying $\mathcal{F}_1$ have $z_1 = \bar{z}_2$.

For Case 11, the clauses CD_1 and CD_2 in $\mathcal{F}_1$ are replaced with the clauses $\bar{x}C$, xD_1, and xD_2 in $\mathcal{F}_2$. If an assignment A_1 satisfies $\mathcal{F}_1$ (thus satisfies CD_1 and CD_2), then if $C = 0$ under A_1 we assign the new variable $x = 0$, and if $C = 1$ under A_1 we assign the new variable $x = 1$. It is easy to verify that this assignment to x plus A_1 will satisfy $\mathcal{F}_2$. For the other direction, suppose that an assignment A_2 satisfies $\mathcal{F}_2$. If A_2 assigns $x = 1$ then we have $C = 1$ under A_2 thus the assignment A_2 also satisfies CD_1 and CD_2 thus $\mathcal{F}_1$; and if A_2 assigns $x = 0$ then we have $D_1 = 1$ and $D_2 = 1$ under A_2 and again A_2 satisfies $\mathcal{F}_1$.   □

Next, we show that the algorithm **Reduction** always decreases the l-value.

Lemma 2. *Let $\mathcal{F}_1$ and $\mathcal{F}_2$ be two formulas such that $\mathcal{F}_2 = \textbf{Reduction}(\mathcal{F}_1)$, and $\mathcal{F}_1 \neq \mathcal{F}_2$. Then $l(\mathcal{F}_1) \geq l(\mathcal{F}_2) + 0.003$.*

Proof. Since $\mathcal{F}_1 \neq \mathcal{F}_2$, at least one of the cases in the algorithm **Reduction** is applicable to the formula $\mathcal{F}_1$. Therefore, it suffices to verify that each case in the algorithm **Reduction** decreases the l-value by at least 0.003.

Cases 1-8 simply remove certain literals, which decrease the degree of certain variables in the formula. Therefore, if any of these cases is applied on the formula $\mathcal{F}_1$, then the l-value of the formula is decreased by at least $\delta_{\min} = \delta_2 = 0.13$.

Consider Cases 9-11. Note that if we reach these cases then Cases 4-5 are not applicable, which implies that the formula $\mathcal{F}_1$ contains only 3^+-variables.

Case 9. If both $\overline{z}_1$ and $\overline{z}_2$ are in the same clause, say C_1, then the resolution $DP_x(\mathcal{F})$ after the next application of Case 2 in the algorithm will replace the four clauses $\overline{x}z_1$, $\overline{x}z_2$, xC_1, and xC_2 with two 3-clauses z_1C_2 and z_2C_2, which decreases the l-value by w_4 (because of removing the 4-variable x) and on the other hand increases the l-value by at most $2\delta_{max}$ (because of increasing the degree of the two variables in C_2). Therefore, in this case, the l-value is decreased by at least $w_4 - 2\delta_{max} = w_4 - 2\delta_4 = 0.097$. If $\overline{z}_1$ and $\overline{z}_2$ are not in the same clause of C_1 and C_2, say $\overline{z}_1$ is in C_1 and $\overline{z}_2$ is in C_2, then the resolution $DP_x(\mathcal{F})$ after the next application of Case 2 in the algorithm will replace the four clauses $\overline{x}z_1$, $\overline{x}z_2$, xC_1, and xC_2 with two 3-clauses z_1C_2 and z_2C_1. In this case, the l-value is decreased by exactly $w_4 = 1.897$ because of removing the 4-variable x.

Case 10. Suppose that z_1 is an i-variable and z_2 is a j-variable. Replacing z_1 by $\overline{z}_2$ removes the i-variable z_1 and makes the j-variable z_2 into an $(i+j)$-variable. However, after an application of Case 2 in the algorithm, the clause z_1z_2 in the original formula disappears, thus z_2 becomes an $(i+j-2)$-variable. Therefore, the total value decreased in the l-value is $(w_i + w_j) - w_{i+j-2}$. Because of the symmetry, we can assume without loss of generality that $i \leq j$. Note that we always have $i \geq 3$. If $i = 3$, then $w_3 + w_j = \delta_{max} + 0.097 + w_j \geq w_{j+1} + 0.097 = w_{3+j-2} + 0.097$. If $i = 4$, then $w_4 + w_j = 2\delta_{max} + w_j + 0.097 \geq w_{j+2} + 0.097 = w_{4+j-2} + 0.097$. If $i \geq 5$, then $w_i + w_j = i/2 + j/2 = (i+j-2)/2 + 1 = w_{i+j-2} + 1$. Therefore, in this case, the l-value of the formula is decreased by $(w_i + w_j) - w_{i+j-2}$, which is at least 0.097.

Case 11. Since the clauses CD_1 and CD_2 in $\mathcal{F}_1$ are replaced with $\overline{x}C$, xD_1 and xD_2, each variable in C has its degree decreased by 1. Since all variables in $\mathcal{F}_1$ are 3^+-variables and $|C| \geq 2$, the degree decrease for the variables in C makes the l-value to decrease by at least $2 \cdot \min\{\delta_i \mid i \geq 3\} = 1$. On the other hand, the introduction of the new 3-variable x and the new clauses $\overline{x}C$, xD_1 and xD_2 increases the l-value by exactly $w_3 = 0.997$. In consequence, the total l-value in this case is decreased by at least $1 - 0.997 = 0.003$. $\square$

4   The Main Algorithm

By definition, the l-value $l(\mathcal{F}_1)$ of the formula $\mathcal{F}_1$ is bounded by $L_1/2$, where L_1 is the length of the formula $\mathcal{F}_1$. By Lemma 2, each application of a case in the algorithm **Reduction** takes time polynomial in L_1 and decreases the l-value by at least 0.003. Therefore, the algorithm must stop in polynomial time and produce an equivalent formula $\mathcal{F}_2$ for which no case in the algorithm is applicable. Such a formula $\mathcal{F}_2$ will be called a *reduced formula*. Reduced formulas have a number of interesting properties, which are given below.

Lemma 3. *There are no 1-variables or 2-variables in a reduced formula.*

Lemma 3 holds true because Cases 4-5 of the algorithm **Reduction** are not applicable to a reduced formula.

Lemma 4. *Let $\mathcal{F}$ be a reduced formula and let xy be a clause in $\mathcal{F}$. Then*

(1) *No other clauses contain xy;*
(2) *No clause contains $x\overline{y}$ or $\overline{x}y$;*
(3) *There is at most one clause containing $\overline{xy}$, and that clause must be a 3^+-clause. Moreover, if y is a 3-variable or if $\overline{x}$ is a 1-literal, then no clause contains $\overline{xy}$.*

Proof. (1) Since Case 1 is not applicable, no other clause in $\mathcal{F}$ contains xy.

(2) Since Case 6 is not applicable, no clause in $\mathcal{F}$ contains either $x\overline{y}$ or $\overline{x}y$.

(3) A clause containing $\overline{xy}$ in $\mathcal{F}$ cannot be a 2-clause since Case 10 is not applicable. Thus, it must be a 3^+-clause. Moreover, since Case 11 is not applicable, there cannot be two 3^+-clauses containing $\overline{xy}$.

If y is a 3-variable and if a clause in $\mathcal{F}$ contains $\overline{xy}$, then it is not hard to verify that the resolution on y would have at most one non-trivial resolvent, and Case 4 would be applicable. Finally, if $\overline{x}$ is a 1-literal, then no clause in $\mathcal{F}$ can contain $\overline{xy}$, since Case 8 is no applicable. □

Lemma 5. *Let $\mathcal{F}$ be a reduced formula. Then for any two literals z_1 and z_2, at most one clause in $\mathcal{F}$ can contain z_1z_2. Moreover, If z_1z_2 appears in a clause in $\mathcal{F}$ and if z_2 is a literal of a 3-variable, then among $\overline{z}_1z_2$, $\overline{z}_1\overline{z}_2$, $z_1\overline{z}_2$, only $\overline{z}_1z_2$ may appear in another clause in $\mathcal{F}$.*

Proof. If z_1z_2 is a 2-clause in $\mathcal{F}$, then by Case 1, no other clause can contain z_1z_2. If z_1z_2 is contained in two 3^+-clauses, then Case 11 would be applicable. In conclusion, at most one clause in the reduced formula $\mathcal{F}$ may contain z_1z_2.

If z_1z_2 appears in a clause in $\mathcal{F}$ and if z_2 is a literal of a 3-variable, then $\overline{z}_1\overline{z}_2$ can not appear in any clause – otherwise, the resolution on z_2 would have at most one non-trivial resolvent, and Case 4 would be applicable. If $z_1\overline{z}_2$ appears in a clause in $\mathcal{F}$, then Case 7 would be applicable. □

Our main algorithm for the SAT problem is given in Figure 2. Some explanations are needed to understand the algorithm. The *degree* $d(\mathcal{F})$ of a formula $\mathcal{F}$ is defined to be the largest degree of a variable in $\mathcal{F}$. As already defined, for a literal z, $\mathcal{F}[z]$ is the formula obtained from $\mathcal{F}$ by assigning the value TRUE to the literal z. To extend this notation, for a set $\{z_1, \cdots z_h\}$ of literals, denote by $\mathcal{F}[z_1, \ldots, z_h]$ the formula obtained from $\mathcal{F}$ by assigning the value TRUE to all the literals z_1, ..., z_h. Finally, in Step 4.1 of the algorithm, where the formula $\mathcal{F}$ contains a clause $\overline{y}_2C_0$, $\mathcal{F}[C_0 = \text{true}]$ denotes the formula obtained from $\mathcal{F}$ by replacing the clause $\overline{y}_2C_0$ by the clause C_0, and $\mathcal{F}[C_0 = \text{false}]$ denotes the formula obtained from $\mathcal{F}$ by assigning the value FALSE to all literals in C_0.

Theorem 1. *The algorithm* **SATSolver**$(\mathcal{F})$ *solves the SAT problem in time $O(1.0652^L)$, where L is the length of the input formula $\mathcal{F}$.*

Proof. We first verify the correctness of the algorithm. Step 3 and Step 4.2 recursively test the satisfiability of the formulas $\mathcal{F}[x]$ and $\mathcal{F}[\overline{x}]$, which are obviously correct. To see the correctness of Step 4.1, note that the satisfiability of either $\mathcal{F}[C_0 = \text{true}]$ or $\mathcal{F}[C_0 = \text{false}]$ will trivially imply the satisfiability of $\mathcal{F}$. On the

---

Algorithm SATSolver($\mathcal{F}$)
INPUT: a CNF formula $\mathcal{F}$
OUTPUT: a report whether $\mathcal{F}$ is satisfiable

1. $\mathcal{F} = \text{Reduction}(\mathcal{F})$;
2. pick a $d(\mathcal{F})$-variable x;
3. **if** $d(\mathcal{F}) > 5$ **then**
     return SATSolver($\mathcal{F}[x]$) $\vee$ SATSolver($\mathcal{F}[\overline{x}]$);
4. **else if** $d(\mathcal{F}) > 3$ **then**
   4.1   **if** x is a $(2,2)$-variable with clauses $x\overline{y}_1 z_1$, $xz_2 z_3$, $\overline{x}y_1$, and $\overline{x}y_2$
            such that y_1 is a 4-variable and y_2 is a 3-variable **then**
            let $\overline{y}_2 C_0$ be a clause containing $\overline{y}_2$;
            return SATSolver($\mathcal{F}[C_0 = \text{true}]$) $\vee$ SATSolver($\mathcal{F}[C_0 = \text{false}]$);
   4.2   **if** both x and $\overline{x}$ are 2^+-literals **then**
            return SATSolver($\mathcal{F}[x]$) $\vee$ SATSolver($\mathcal{F}[\overline{x}]$);
   4.3   **else** (* assume the only clause containing $\overline{x}$ is $\overline{x}z_1 \cdots z_h$ *)
            return SATSolver($\mathcal{F}[x]$) $\vee$ SATSolver($\mathcal{F}[\overline{x}, \overline{z}_1, \ldots, \overline{z}_h]$);
5. **else if** $d(\mathcal{F}) = 3$ **then**
     Apply the algorithm by Wahlström [12];
6. **else** return true;

Fig. 2. The main algorithm

other hand, if an assignment satisfying $\mathcal{F}$ does not satisfy $\mathcal{F}[C_0 = \text{true}]$, then it must not satisfy the clause C_0. Therefore, it must satisfy $\mathcal{F}[C_0 = \text{false}]$.

Next, consider Step 4.3. Since the variable x passes Step 4.2, x is either a $(d, 1)$-variable or a $(1, d)$-variable, where $d \geq 3$. Therefore, without loss of generality, we can assume that x is a $(d, 1)$-variable (otherwise we simply replace x with $\overline{x}$) and that the only clause containing $\overline{x}$ is $\overline{x}z_1 \cdots z_h$. Again, the satisfiability of either $\mathcal{F}[x]$ or $\mathcal{F}[\overline{x}, \overline{z}_1, \ldots, \overline{z}_h]$ trivially implies the satisfiability of $\mathcal{F}$. On the other hand, since $\overline{x}z_1 \cdots z_h$ is the only clause in $\mathcal{F}$ that contains $\overline{x}$, $\mathcal{F}[x]$ is obtained from $\mathcal{F}$ by removing all clauses in $\mathcal{F}$ that contain x, and replacing the clause $\overline{x}z_1 \cdots z_h$ by the clause $z_1 \cdots z_h$. In particular, if an assignment π satisfying $\mathcal{F}$ does not satisfies $\mathcal{F}[x]$, then it must assign value FALSE to x and must not satisfy $z_1 \cdots z_h$. Therefore, π must assign the value TRUE to all literals $\overline{x}$, $\overline{z}_1$, $\ldots$, $\overline{z}_h$. This implies that the formula $\mathcal{F}[\overline{x}, \overline{z}_1, \ldots, \overline{z}_h]$ must be satisfiable.

Finally, if the variable x passes Steps 3-4, the degree of x is bounded by 3. In case the degree of x is 3, we apply Wahlström's algorithm [12] to solve the problem in Step 5. If the variable x also passes Step 5, then the degree of x is bounded by 2. By Lemma 3, a reduced formula has no 1-variable and 2-variable. Thus, the reduced formula obtained from Step 1 of the algorithm must be an empty formula, which is trivially satisfiable, as concluded in Step 6.

It verifies the correctness of the algorithm **SATSolver($\mathcal{F}$)**.

To see the complexity of the algorithm, note that Wahlström's algorithm on a formula $\mathcal{F}$ of degree bounded by 3 runs in time $O(1.1279^n)$, where n is the number of variables in $\mathcal{F}$ [12], which is also $O(1.1346^{l(\mathcal{F})})$ since $l(\mathcal{F}) \leq w_3 n =$

$0.997n$. The proof that the algorithm **SATSolver**$(\mathcal{F})$ runs in time $O(1.1346^{l(\mathcal{F})})$ when the degree of $\mathcal{F}$ is larger than 3 is given in the next section. The relation $O(1.1346^{l(\mathcal{F})}) = O(1.0652^L)$ is because $l(\mathcal{F}) \leq L/2$. □

5   The Analysis of the Main Algorithm

Given two formulas $\mathcal{F}_1$ and $\mathcal{F}_2$, by definition we have $l(\mathcal{F}_1) = \sum_{x \in \mathcal{F}_1} w(x)$ and $l(\mathcal{F}_2) = \sum_{x \in \mathcal{F}_2} w'(x)$, where $w(x)$ is the frequency weight of x in $\mathcal{F}_1$ and $w'(x)$ is the frequency weight of x in $\mathcal{F}_2$. The *l-value reduction* from $\mathcal{F}_1$ to $\mathcal{F}_2$ is $l(\mathcal{F}_1) - l(\mathcal{F}_2)$. The *contribution of x to the l-value reduction* from $\mathcal{F}_1$ to $\mathcal{F}_2$ is $w(x) - w'(x)$. The *contribution of a variable set S to the l-value reduction* from $\mathcal{F}_1$ to $\mathcal{F}_2$ is the sum of contributions of all variables in S.

Given a formula $\mathcal{F}$, let *reduced*$(\mathcal{F})$ be the output formula of **Reduction**$(\mathcal{F})$, and let *reduced$_p$*$(\mathcal{F})$ be the first formula during the execution of **Reduction**$(\mathcal{F})$ such that Cases 1-8 are not applicable to the formula. We first discuss the relationship among the l-values of $\mathcal{F}$, of reduced$_p$$(\mathcal{F})$, and of reduced$(\mathcal{F})$.

By Lemma 2, each application of a case in the algorithm **Reduction** decreases the l-value of the formula. This trivially gives the following lemma.

Lemma 6. $l(\mathcal{F}) \geq l(reduced_p(\mathcal{F})) \geq l(reduced(\mathcal{F}))$.

The reason that we consider reduced$_p$$(\mathcal{F})$ is that it is easier to give a bound on the l-value reduction from $\mathcal{F}$ to reduced$_p$$(\mathcal{F})$. In particular, since Cases 1-8 of the algorithm **Reduction** neither introduces new variables nor increases the degree of any variables, we have the following lemma.

Lemma 7. *The contributions of any subset of variables in a formula $\mathcal{F}$ is bounded by the l-value reduction from $\mathcal{F}$ to reduced$_p$$(\mathcal{F})$.*

From now on in this section, let $\mathcal{F}$ be the formula after step 1 in the algorithm **SATSolver**$(\mathcal{F})$. Then $\mathcal{F}$ is a reduced formula. In the algorithm **SATSolver**$(\mathcal{F})$, we break $\mathcal{F}$ into two formulas $\mathcal{F}_1$ and $\mathcal{F}_2$ of smaller l-values at step 3 or 4. To give better bound, we are interested in the branching vector from $\mathcal{F}$ to reduced$(\mathcal{F}_1)$ and reduced$(\mathcal{F}_2)$, instead of the branching vector from $\mathcal{F}$ to $\mathcal{F}_1$ and $\mathcal{F}_2$. To give feasible analysis, we focus on the branching vector from $\mathcal{F}$ to reduced$_p$$(\mathcal{F}_1)$ and reduced$_p$$(\mathcal{F}_2)$. This is valid by the following lemma, which can be easily derived using Lemma 6.

Lemma 8. *The branching vector from $\mathcal{F}$ to reduced$_p$$(\mathcal{F}_1)$ and reduced$_p$$(\mathcal{F}_2)$ is inferior to the branching vector from $\mathcal{F}$ to reduced$(\mathcal{F}_1)$ and reduced$(\mathcal{F}_2)$.*

To simplify our description, we say that a *variable x is in a clause* if either x or $\overline{x}$ is in the clause (this should be distinguished with a *literal* in a clause).

In Step 3 and Step 4.2 of the algorithm **SATSolver**$(\mathcal{F})$, we break the input formula $\mathcal{F}$ into $\mathcal{F}[x]$ and $\mathcal{F}[\overline{x}]$, where x is a variable whose degree is equal to the degree of the formula $\mathcal{F}$. Let y be a variable such that the variables x and y are in the same clause. We can bound from below the contribution of y to the l-value reduction from $\mathcal{F}$ to reduced$_p$$(\mathcal{F}[x])$.

Lemma 9. *Let y be an i-variable. The contribution of y to the l-value reduction from $\mathcal{F}$ to reduced$_p(\mathcal{F}[x])$ is at least*

(1) w_3, *if $i = 3$ and y is in a clause with the literal x;*

(2) w_i, *if y is in a 2-clause with the literal $\overline{x}$;*

(3) δ_i, *if $i > 3$ and y is only in one clause with the literal x, or $\delta_i + \delta_{i-1}$ if $i > 3$ and y is in more than one clauses with the literal x.*

Let S be the set of variables that are contained with the variable x in some clause. We do not include x in S. Let x be an h-variable in $\mathcal{F}$. We can bound the l-value reduction from $\mathcal{F}$ to reduced$_p(\mathcal{F}[x])$ with the following calculations:

Step 1: set $c_x = w_h$ and $c_y = 0$ for $y \in S$.

Step 2: for each 2-clause $\overline{x}y$ or $\overline{xy}$, where y is an i-variable in S

   (1) when $i = 3$, add w_3 to c_y,

   (2) when $i > 3$, add $w_i - \delta_i$ to c_y if there is a clause xC containing variable y, or add w_i to c_y otherwise.

Step 3: for each clause xyC, where y is an i-variable in S

   (1) when $i = 3$, add w_3 to c_y,

   (2) when $i > 3$, add δ_i to c_y.

Step 4: $c = c_x + \sum_{y \in S} c_y$.

The value c calculated above is the c-value from $\mathcal{F}$ to reduced$_p(\mathcal{F}[x])$. The c-value from $\mathcal{F}$ to reduced$_p(\mathcal{F}[\overline{x}])$ can be calculated similarly. The following lemma shows that the c-value is not larger than the l-value reduction from $\mathcal{F}$ to reduced$_p(\mathcal{F})$. The lemma can be verified directly by the definitions.

Lemma 10. *The c-value is not larger than the contribution of $S \cup \{x\}$ to the l-value reduction from $\mathcal{F}$ to reduced$_p(\mathcal{F}[x])$.*

To give a better analysis, some further notations are needed.

n_1: the number of 3^+-clauses containing literal x.

n_3: the number of 2-clauses containing 3-variables and literal x.

n_4: the number of 2-clauses containing 4-variables and literal x.

n_5: the number of 2-clauses containing 5^+-variables and literal x.

$\overline{n_1}$: the number of 3^+-clauses containing literal $\overline{x}$.

$\overline{n_3}$: the number of 2-clauses containing 3-variables and literal $\overline{x}$.

$\overline{n_4}$: the number of 2-clauses containing 4-variables and literal $\overline{x}$.

$\overline{n_5}$: the number of 2-clauses containing 5^+-variables and literal $\overline{x}$.

$m_1 = w_i + 2n_1\delta_i + (n_3 + \overline{n_3} + \overline{n_4})w_3 + n_4\delta_4 + n_5\delta_5 + \overline{n_5}w_4$.

$m_2 = w_i + 2\overline{n_1}\delta_i + (n_3 + \overline{n_3} + n_4)w_3 + \overline{n_4}\delta_4 + \overline{n_5}\delta_5 + n_5w_4$.

We have the following lemma:

Lemma 11. *The value m_1 is not larger than the l-value reduction from $\mathcal{F}$ to reduced$_p(\mathcal{F}[x])$, and the value m_2 is not larger than the l-value reduction from $\mathcal{F}$ to reduced$_p(\mathcal{F}[\overline{x}])$.*

Lemma 11 is sufficient for most cases in the following analysis. Sometimes, we may need values better than m_1. Let

$n_{1,1}$: the number of 3-clauses containing literal x.

$n_{1,2}$: the number of 4^+-clauses containing literal x.

$\overline{n_{4,1}}$: the number of 2-clauses containing literal $\overline{x}$ and variable y such that some clause containing both literal x and variable y.

$\overline{n_{4,2}}$: the number of 2-clauses containing literal $\overline{x}$ and variable y such that no clauses containing both literal x and variable y.

$m'_1 = w_i + (2n_{1,1} + 3n_{1,2})\delta_i + (n_3 + \overline{n_3} + \overline{n_{4,1}})w_3 + n_4\delta_4 + n_5\delta_5 + \overline{n_{4,2}}w_4 + \overline{n_5}w_4$.

By a proof similar to that for Lemma 11, we can prove the following lemma.

Lemma 12. *The value m'_1 is not larger than the l-value reduction from $\mathcal{F}$ to $reduced_p(\mathcal{F}[x])$, and the value m_2 is not larger than the l-value reduction from $\mathcal{F}$ to $reduced_p(\mathcal{F}[\overline{x}])$.*

Now we are ready to analyze the branching vector from $\mathcal{F}$ to $reduced_p(\mathcal{F}[x])$ and $reduced_p(\mathcal{F}(\overline{x})$.

Lemma 13. *Let $\mathcal{F}$ be a reduced formula with $d(\mathcal{F}) = i$, and let x be an i-variable in $\mathcal{F}$. Then both (m_1, m_2) and (m'_1, m_2) are inferior to the branching vector from $\mathcal{F}$ to $reduced_p(\mathcal{F}[x])$ and $reduced_p(\mathcal{F}[\overline{x}])$.*

If x is a $(i-1, 1)$-literal and if no 2-clause contains x, we can have a better branching vector.

Lemma 14. *Given a reduced formula $\mathcal{F}$ of degree i, and an $(i-1, 1)$-literal x in $\mathcal{F}$ such that no 2-clause contains x, let*

$$m'_1 = w_i + 2n_1\delta_i + n_3w_3 + n_4\delta_4 + n_5\delta_5, \qquad and \qquad m'_2 = w_i + 3w_3.$$

Then (m'_1, m'_2) is inferior to the branching vector from $\mathcal{F}$ to $reduced_p(\mathcal{F}[x])$ and $reduced_p(\mathcal{F}[\overline{x}])$. Moreover, $\overline{n_1} = 1$ and $\overline{n_3} = \overline{n_4} = \overline{n_5} = 0$.

Due to the space limit, we omit the analysis for the case where the formula degree is 4 or 5, which will be given in a complete version of the current paper. In the following, we present the analysis for formulas of degree larger than 5.

Let x be a (d_1, d_0)-literal. Then $d = d_1 + d_0 \geq 6$. Suppose there are s_1 2-clauses containing literal x, and s_0 2-clauses containing literal $\overline{x}$. Let l_1 be the l-value reduction and c_1 be the value of c from $\mathcal{F}$ to $\mathcal{F}[x]$. Let l_0 be the l-value reduction and c_0 be the c-value from $\mathcal{F}$ to $\mathcal{F}[\overline{x}]$. By lemma 10, c_1 (c_0) is not larger than l_1 (l_0). Thus $l_1 + l_2 \geq c_1 + c_2$. It can be verified that each 2-clause (containing either x or $\overline{x}$) adds at least 0.5 to both c_1 and c_0 by the calculation of c-value. Thus the $s_1 + s_2$ 2-clauses add at least $s_1 + s_0$ to $c_1 + c_2$. Moreover, the $d_1 - s_1$ 3^+-clauses containing literal x add at least $2(d_1 - s_1)\delta_i \geq d_1 - s_1$ to c_1 since $i \geq 3$ (all variables in $\mathcal{F}$ are 3-variables), and the $d_0 - s_0$ 3^+-clauses containing literal $\overline{x}$ add at least $d_0 - s_0$ to c_0. Thus the 3^+-clauses add at least $(d_1 - s_1) + (d_0 - s_0)$ to $c_1 + c_2$. Finally, x adds $w_d = 0.5d \geq 3$ to both c_1 and c_0 since x is a d-variable where $d \geq 6$. Thus x add at least $2w_d \geq 6$ to $c_1 + c_0$. Therefore, we have that $l_1 + l_0 \geq c_1 + c_0 \geq (s_1 + s_0) + (d_1 - s_1) + (d_0 - s_0) + 6 \geq 12$.

Next we prove that both of l_1 and l_0 are greater than $0.5 + d/2 = 3.5$. As shown above, the $s_1 + s_0$ 2-clauses add at least $0.5(s_1 + s_0)$ to both c_1 and c_0, the $(d_1 - s_1)$ 3^+-clauses add at least $d_1 - s_1$ to c_1, the $(d_0 - s_0)$ 3^+-clauses add at least $d_0 - s_0$ to c_0, and x add at least 3 to both c_1 and c_0. Thus $c_1 \geq 0.5(s_1 + s_0) + (d_1 - s_1) + 3$ and $c_0 \geq 0.5(s_1 + s_0) + (d_0 - s_0) + 3$. Note that $d_0 - s_1 \geq 0$ and $d_0 - s_0 \geq 0$. If $s_1 + s_0 = 1$, then both c_1 and c_0 are not less than 3.5. If $s_1 + s_0 = 0$, then $s_1 = s_0 = 0$. Since both d_1 and d_0 are not less than 1, we have that both c_1 and c_0 are not less than 4. By lemma 10, both l_1 and l_0 are not less than 3.5.

So the branching vector in this case is at least (l_1, l_2), not inferior to $(3.5, 12 - 3.5) = (3.5, 8.5)$, which leads to $O(1.1313^{l(\mathcal{F})})$.

Summarizing all the above discussions, we can verify that the worst branching vector is $t_0 = (3w_3 + 5\delta_4, 3w_3 + \delta_4) = (7.491, 3.891)$. The root of the polynomial corresponding to this branching vector is $\tau(t_0) \leq 1.1346$. In conclusion, we derive that the time complexity of the algorithm **SATSolver** is bounded by $O(1.1346^{l(\mathcal{F})})$ on an input formula $\mathcal{F}$, which completes the proof of Theorem 1.

References

1. Chen, J., Kanj, I., Jia, W.: Vertex cover: further observations and further improvements. Journal of Algorithms 41, 280–301 (2001)
2. Davis, M., Putnam, H.: A computing procedure for quantification theory. Journal of the ACM 7, 201–215 (1960)
3. Fomin, F.V., Grandoni, F., Kratsch, D.: Measure and conquer: a simple $O(2^{0.288n})$ independent set algorithm. In: Proc. 17th Annual ACM-SIAM Symposium on Discrete Algorithms, pp. 18–25 (2006)
4. Garey, M., Johnson, D.: Computers and Intractability: A Guide to the Theory of NP-Completeness. W.H.Freeman and Company, New York (1979)
5. Van Gelder, A.: A satisfiability tester for non-clausal propositional claculus. Information and Computation 79, 1–21 (1988)
6. Gu, J., Purdom, P., Wah, W.: Algorithms for the satisfiability (SAT) problem: A survey. In: Satisfiability Problem: Theory and Applications, DIMACS Series in Discrete Mathematics and Theoretical Computer Science, AMS, pp. 19–152 (1997)
7. Hirsh, E.: Two new upper bounds for SAT. In: Proc. 9th Annual ACM-SIAM Symp. on Discrete Algorithms, pp. 521–530 (1998)
8. Hirsh, E.: New Worst-Case Upper Bounds for SAT. Journal of Automated Reasoning 24, 397–420 (2000)
9. Kullmann, O., Luckhardt, H.: Deciding propositional tautologies: Algorithms and their complexity (manuscript) (1997)
10. Schöning, U.: Algorithmics in exponential time. In: Diekert, V., Durand, B. (eds.) STACS 2005. LNCS, vol. 3404, pp. 36–43. Springer, Heidelberg (2005)
11. Wahlström, M.: An algorithm for the SAT problem for formulae of linear length. In: Brodal, G.S., Leonardi, S. (eds.) ESA 2005. LNCS, vol. 3669, pp. 107–118. Springer, Heidelberg (2005)
12. Wahlstöm, M.: Faster Exact Solving of SAT Formulae with a Low Number of Occurrences per Variable. In: Bacchus, F., Walsh, T. (eds.) SAT 2005. LNCS, vol. 3569, pp. 309–323. Springer, Heidelberg (2005)

Shortest Path Problems on a Polyhedral Surface*

Atlas F. Cook IV and Carola Wenk

University of Texas at San Antonio
Department of Computer Science
One UTSA Circle, San Antonio, TX 78249-0667
acook@cs.utsa.edu, carola@cs.utsa.edu

Abstract. We develop algorithms to compute edge sequences, Voronoi diagrams, shortest path maps, the Fréchet distance, and the diameter of a polyhedral surface. Distances on the surface are measured by the length of a Euclidean shortest path. Our main result is a linear factor speedup for the computation of all shortest path edge sequences and the diameter of a convex polyhedral surface. This speedup is achieved with kinetic Voronoi diagrams. We also use the star unfolding to compute a shortest path map and the Fréchet distance of a non-convex polyhedral surface.

Keywords: Polyhedral Surface, Voronoi Diagram, Shortest Path Map, Fréchet distance, Diameter.

1 Introduction

Two questions are invariably encountered when dealing with shortest path problems. The first question is how to represent the combinatorial structure of a shortest path. In the plane with polygonal obstacles, a shortest path can only turn at obstacle vertices, so a shortest path can be combinatorially described as a sequence of obstacle vertices [16]. On a polyhedral surface, a shortest path need not turn at vertices [20], so a path is often described combinatorially by an *edge sequence* that represents the sequence of edges encountered by the path [1]. The second question is how to compute shortest paths in a problem space with M vertices. The following preprocessing schemes compute combinatorial representations of all possible shortest paths. In a simple polygon, Guibas et al. [16] give an optimal $\Theta(M)$ preprocessing scheme that permits a shortest path between two query points to be computed in $O(\log M)$ time. In the plane with polygonal obstacles, Chiang and Mitchell [9] support shortest path queries between any two points after $O(M^{11})$ preprocessing. On a *convex* polyhedral surface, Mount [21] shows that $\Theta(M^4)$ combinatorially distinct shortest path edge sequences exist, and Schevon and O'Rourke [23] show that only $\Theta(M^3)$ of

* This work has been supported by the National Science Foundation grant NSF CA-REER CCF-0643597. Previous versions of this work have appeared in [12,13].

these edge sequences are *maximal* (i.e., they cannot be extended at either end without creating a suboptimal path). Agarwal et al. [1] use these properties to compute the $\Theta(M^4)$ shortest path edge sequences in $O(M^6 2^{\alpha(M)} \log M)$ time and the *diameter* in $O(M^8 \log M)$ time, where $\alpha(M)$ is the inverse Ackermann function. The diameter is the largest shortest path distance between any two points on the surface. Our main result improves the edge sequence and diameter algorithms of [1] by a linear factor. We achieve this improvement by combining the star unfolding of [1] with the kinetic Voronoi diagram of Albers et al. [3].

A popular alternative to precomputing all combinatorial shortest paths is to precompute a shortest path map structure SPM(s) that describes all shortest paths from a fixed source s. In the plane with polygonal obstacles, Hershberger and Suri [17] use the continuous Dijkstra paradigm to support all queries from a fixed source after $\Theta(M \log M)$ preprocessing. On a (possibly non-convex) polyhedral surface, Mitchell, Mount, and Papadimitriou [20] use the continuous Dijkstra paradigm to construct SPM(s) by propagating a wavefront over a polyhedral surface in $O(M^2 \log M)$ time and $O(M^2)$ space. Chen and Han [8] solve the same polyhedral surface problem in $O(M^2)$ time and space by combining unfolding and Voronoi diagram techniques. Schreiber and Sharir [24] use the continuous Dijkstra paradigm to construct an *implicit* representation of a shortest path map for a *convex* polyhedral surface in $O(M \log M)$ time and space. In addition to the exact algorithms above, there are also various efficient algorithms to compute approximate shortest paths on weighted polyhedral surfaces, see for example Aleksandrov et al. [4].

1.1   Terminology

Throughout this paper, M is the total complexity of a problem space that contains a polyhedral surface and auxiliary objects on the surface such as points, line segments, and polygonal curves. A shortest path on a polyhedral surface between points s and t is denoted by $\pi(s,t)$, and $d(s,t)$ signifies the Euclidean length of $\pi(s,t)$. A convex polyhedral surface is denoted by $\mathcal{P}$, and a non-convex polyhedral surface is represented by $\mathcal{P}_N$. The extremely slowly growing inverse Ackermann function is signified by $\alpha(M)$. The line segment with endpoints a and b is denoted by $\overline{ab}$. The *Fréchet distance* [5] is a similarity metric for *continuous* shapes that is defined for two polygonal curves $A, B : [0,1] \to \mathbb{R}^\nu$ as $\delta_F(A, B) = \inf_{\alpha,\beta:[0,1]\to[0,1]} \sup_{t\in[0,1]} d(A(\alpha(t)), B(\beta(t)))$, where $\mathbb{R}^\nu$ is an arbitrary Euclidean vector space, α and β range over continuous non-decreasing reparameterizations, and d is a distance metric for points. For a given constant $\varepsilon \geq 0$, *free space* is $\{(s,t) \mid s \in A,\ t \in B,\ d(s,t) \leq \varepsilon\}$. A *cell* is the parameter space defined by two line segments $\overline{ab} \subset A$ and $\overline{cd} \subset B$, and the free space inside the cell consists of all points $\{(s,t) \mid s \in \overline{ab},\ t \in \overline{cd},\ d(s,t) \leq \varepsilon\}$.

1.2   Our Results

For a *convex* polyhedral surface, Agarwal et al. [1] give algorithms to compute the diameter and either the exact set or a superset of all $\Theta(M^4)$ shortest path edge sequences. All three of these algorithms are improved by a linear factor in sections 2 and 3. Section 4 contains an algorithm to compute the Fréchet distance between polygonal curves on a convex polyhedral surface, and this algorithm is a linear factor faster than the algorithm of Maheshwari and Yi [19]. In addition, section 4 contains the first algorithm to compute the Fréchet distance between polygonal curves on a *non-convex* polyhedral surface. Our motivation for studying the Fréchet distance on a polyhedral surface is that teaming up two people for safety reasons is common practice in many real-life situations, ranging from scouts in summer camp, to fire fighters and police officers, and even to astronauts exploring the moon. In all of these applications, two team members need to coordinate their movement in order to stay within "walking distance" so that fast assistance can be offered in case of an emergency. The Fréchet distance is an ideal model for this scenario. Section 5 describes shortest path maps that support queries from any point on a line segment.

2   Shortest Path Edge Sequences

This section contains superset and exact algorithms to compute the $\Theta(M^4)$ shortest path edge sequences on a convex polyhedral surface $\mathcal{P}$. Both of these algorithms improve results of Agarwal et al. [1] by a linear factor.

Let $v_1, ..., v_M$ be the vertices of $\mathcal{P}$, and let $\Pi = \{\pi(s, v_1), ..., \pi(s, v_M)\}$ be an angularly ordered set of non-crossing shortest paths from a source point $s \in \mathcal{P}$ to each vertex $v_j \in \mathcal{P}$.[1] The *star unfolding* $\mathcal{S}$ is a simple polygon [6] defined by cutting $\mathcal{P}$ along each of the shortest paths in Π and unfolding the resulting shape into the plane. Since the source point s touches all of the M cuts, $s \in \mathcal{P}$ maps to M image points $s_1, ..., s_M$ on the (two-dimensional) boundary of the unfolded simple polygon $\mathcal{S}$ (see Fig. 1).

The *equator* [8] in the star unfolding is the closed polygonal curve through the points $v_1, ..., v_M, v_1$. The region inside the equator contains no source image and is called the *core* [14].[2] The regions outside the core each contain a source image and are collectively referred to as the *anti-core* [14]. A *core edge* is the image of an edge of $\mathcal{P}$ that was not cut during the unfolding process. Each of the $O(M)$ core edges has both of its endpoints at vertices and is entirely contained in the core. An *anti-core edge* is the image of a connected portion of an edge of $\mathcal{P}$ that was defined by cuts during the unfolding process. Each of the $\Theta(M^2)$ anti-core edges either has both of its endpoints on a cut or has one endpoint on a cut and the other endpoint at a vertex. The interior of an anti-core edge is entirely contained in one anti-core region (see Fig. 1b). The dual graph of the star unfolding is a tree,[3] and this tree defines a unique edge sequence that can be used to connect any two points in the star unfolding.

---

[1] If multiple shortest paths exist from s to v_j, then any of these shortest paths can be used to represent $\pi(s, v_j)$ [8].

[2] The core has also been referred to as the *kernel* or the *antarctic* in [1,8]. Note that neither the star unfolding nor its core are necessarily star-shaped [1].

[3] In [1], this dual graph is referred to as the *pasting tree*.

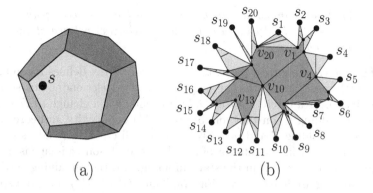

Fig. 1. (a) A convex polyhedral surface $\mathcal{P}$. (b) The star unfolding of $\mathcal{P}$ is created by cutting along shortest paths from the source point s to every vertex $v_1, ..., v_{20}$ of $\mathcal{P}$. The *core* of the star unfolding is heavily-shaded.

The star unfolding of s can be used to compute a shortest path $\pi(s,t)$ for points $s, t \in \mathcal{P}$ as follows. If an image of t lies in the anti-core region containing s_i, then the line segment in the star unfolding from s_i to the image of t is an optimal shortest path [1]. By contrast, if an image of t lies in the core, then a nearest source image can be determined with Voronoi diagram techniques, and the line segment in the star unfolding from this nearest source image to the image of t is an optimal shortest path [1,8].

Agarwal et al. [1] partition the M edges of the convex polyhedral surface $\mathcal{P}$ into $O(M^3)$ line segment *edgelets*. All source points on an edgelet can be associated with the same combinatorial star unfolding, and all source points in the interior of an edgelet have a unique shortest path to each vertex of $\mathcal{P}$ [1]. These edgelets are constructed in $O(M^3 \log M)$ time by computing a shortest path between each pair of vertices on $\mathcal{P}$ and intersecting these $O(M^2)$ shortest paths with each of the M edges of $\mathcal{P}$ [1]. Agarwal et al. [1] compute a star unfolding for each edgelet and use these structures to construct an $O(M^6)$ superset of the $\Theta(M^4)$ shortest path edge sequences for $\mathcal{P}$ in $O(M^6)$ time and space [1]. In addition, Agarwal et al. [1] show how to compute the exact set of $\Theta(M^4)$ shortest path edge sequences in $O(M^6 2^{\alpha(M)} \log M)$ time.

Although we have defined the star unfolding only for a *convex* polyhedral surface $\mathcal{P}$, the concept generalizes to a *non-convex* polyhedral surface $\mathcal{P}_N$ because the star unfolding can still be defined by an angularly ordered set of non-crossing shortest path cuts from the source to every vertex [8,20]. In addition, there are still $O(M^3)$ edgelets on $\mathcal{P}_N$ because a shortest path between each pair of vertices can intersect each edge at most once.

We show how to *maintain* a combinatorial star unfolding in $O(M^4)$ total time and space as a source point varies continuously over all $O(M^3)$ edgelets on a possibly non-convex polyhedral surface $\mathcal{P}_N$. Our approach takes advantage of small combinatorial changes between adjacent edgelets and achieves a linear factor improvement over the approach [1] of computing a separate star unfolding for each edgelet.

Theorem 1. *A star unfolding can be maintained as a source point s varies continuously over all M edges of a (possibly non-convex) polyhedral surface $\mathcal{P}_N$ in $O(M^4)$ time and space.*

Proof. The set Π of shortest paths to each corner vertex defines a combinatorial star unfolding. These paths can only change at edgelet endpoints, so Π can be maintained over a discrete set of $O(M^3)$ events. Each change to Π requires removing and adding a constant number of $O(M)$ complexity anti-core regions from the star unfolding $\mathcal{S}$ and possibly updating all $O(M)$ core edges in $\mathcal{S}$. As s varies continuously in the interior of an edgelet, each source image is parameterized along a line segment in the star unfolding, and the remaining vertices in the star unfolding are fixed [1]. See Fig. 2a. Thus, $O(M \cdot M^3)$ time and space is sufficient to maintain $\mathcal{S}$ combinatorially over all edgelets. □

The below lemma computes an implicit superset of the shortest path edge sequences on $\mathcal{P}$ in $O(M^5)$ time and space. Note that we do not attempt to compute shortest path edge sequences on a non-convex polyhedral surface $\mathcal{P}_N$ because Mount [21] has shown that there can be exponentially many shortest path edge sequences on $\mathcal{P}_N$.

Theorem 2. *An implicit superset of the $\Theta(M^4)$ shortest path edge sequences for a convex polyhedral surface $\mathcal{P}$ with M vertices can be computed in $O(M^5)$ time and space.*

Proof. Each edgelet defines a star unfolding with source images $s_1, ..., s_M$. For each s_i, use the relevant star unfolding's dual graph tree to construct an edge sequence from s_i to each of the $O(M)$ anti-core edges in the anti-core region containing s_i and to each of the $O(M)$ core edges. This yields $O(M^2)$ edge sequences per edgelet, and $O(M^5)$ edge sequences over all edgelets. The result is the desired superset because only core edges have shortest path edge sequences to multiple sites, and this approach considers all possibilities. $O(M^5)$ storage is sufficient to store a dual graph tree for each edgelet, and these trees collectively encode an implicit representation of the desired superset. □

The *exact* set of shortest path edge sequences for each combinatorial star unfolding can be determined with a kinetic Voronoi diagram that allows its defining point sites to move. In our case, the moving sites are the source images $s_1, ..., s_M$, and each source image is parameterized along a line segment as a source point varies continuously over an edgelet [1]. This behavior ensures that each *pair* of moving source images defines $O(M)$ Voronoi events [3], and Albers et al. [3] show how to maintain a kinetic Voronoi diagram in $O(\log M)$ time per event with a priority queue.

Theorem 3. *A kinetic Voronoi diagram of source images $s_1, ..., s_M$ can be maintained in $O(M^4 \log M)$ time and $O(M^4)$ space as a source point varies continuously over one edge e of a convex polyhedral surface $\mathcal{P}$.*

Proof. A kinetic Voronoi diagram for the *first* edgelet on e defines $O(M^2 \cdot M)$ events [3] due to the linear motion of $O(M^2)$ *pairs* of source images in the star unfolding. Each of the $O(M^2)$ subsequent edgelets on e can be handled by removing and adding a constant number of source image sites. All other sites continue to be parameterized along the same line segments as in the previous edgelet. Thus, each of these $O(M^2)$ edgelets contributes $M - 1$ new *pairs* of sites and $O(M \cdot M)$ new events to the priority queue. Handling each event in $O(\log M)$ time and $O(1)$ space as in [3] yields the stated runtime.   □

We construct the exact set of shortest path edge sequences as follows. For the moment, fix an edgelet α and a core vertex $v_i \in \mathcal{S}$ such that v_i touches the anti-core region that contains the source image s_i. Maintaining a kinetic Voronoi diagram for α yields a two-dimensional parameterized Voronoi cell φ_i for the source image s_i. The unique edge sequence in the star unfolding's dual graph from s_i to a core edge e represents a shortest path if and only if e intersects φ_i for some $s \in \alpha$. This follows because s_i must be a nearest source image to some point on e in order to define a shortest path to e.

Agarwal et al. [1] represent each parameterized Voronoi vertex as an algebraic curve. They triangulate the region of the core that is directly visible to core vertex v_i such that each triangle Δ has apex v_i. The dual graph D of the (fixed) core for an edgelet α is a *tree* [1] that defines candidate edge sequences. Let the portion of D inside a fixed triangle Δ be the subtree D_Δ. Agarwal et al. [1] compute each subtree D_Δ in $O(M)$ time. In the following lemma, we improve this process to $O(\log M)$ time.

Lemma 1. *A subtree D_Δ can be computed in $O(\log M)$ time.*

Proof. Assume Δ has vertices v_i, v_j, v_k. The subtree D_Δ consists of one path in D from the face containing v_i to the face containing v_j and a second path in D from the face containing v_i to the face containing v_k. Point location in the core can identify these two paths in D in $O(\log M)$ time.   □

After computing the subtree D_Δ for each triangle Δ, Agarwal et al. [1] use polar coordinates centered at core vertex v_i to compute an upper envelope μ of the algebraic curves defining the kinetic Voronoi cell φ_i. This upper envelope is then refined into a set of curve segments such that each curve segment is·contained in some triangle Δ. For each curve segment, a binary search is performed on the two paths in D_Δ. The deepest edge on each of these two paths that is intersected by a curve segment defines a *maximal* shortest path edge sequence. Repeating this technique for all core vertices defined by all edgelets yields $\Theta(M^3)$ *maximal* shortest path edge sequences. The set of all prefixes of these maximal sequences defines all $\Theta(M^4)$ shortest path edge sequences of $\mathcal{P}$ [1].

Theorem 4. *The exact set of $\Theta(M^4)$ shortest path edge sequences for a convex polyhedral surface $\mathcal{P}$ with M vertices can be explicitly constructed in $O(M^5 2^{\alpha(M)} \log M)$ time. This set can be implicitly stored in $O(M^4)$ space or explicitly stored in $O(M^5)$ space.*

Proof. Let n_i be the total number of parameterized Voronoi vertices over all edgelets, and let t_Δ be the time to process each triangle Δ. There are $O(M^5)$

possible triangles Δ because each of the $O(M^3)$ edgelets defines $O(M)$ core vertices, and each of these vertices defines $O(M)$ triangles. The technique of Agarwal et al. [1] requires $O(n_i 2^{\alpha(M)} \log M + M^5 t_\Delta)$ time. Since they assume $n_i \in O(M^6)$ and $t_\Delta \in O(M)$, this yields $O(M^6 2^{\alpha(M)} \log M)$ time.

We improve this runtime as follows. By Theorem 3, $n_i \in O(M^5)$ over all $O(M)$ edges of $\mathcal{P}$. By Lemma 1, $t_\Delta \in O(\log M)$ time. Thus, we achieve $O(M^5 2^{\alpha(M)} \log M)$ total time. The implicit space bound follows by storing the kinetic Voronoi diagram for only one edge at a time and storing each of the $\Theta(M^3)$ *maximal shortest path edge sequences* [23] in $O(M)$ space. $\square$

3 Diameter

The *diameter* of a convex polyhedral surface is the largest shortest path distance between any pair of points on the surface. O'Rourke and Schevon [22] originally gave an algorithm to compute the diameter in $O(M^{14} \log M)$ time. Subsequently, Agarwal et al. [1] showed how to compute the diameter in $O(M^8 \log M)$ time. The approach of Agarwal et al. [1] computes shortest paths between all pairs of vertices, and these shortest paths induce an arrangement of $O(M^4)$ *ridge-free regions* on the surface. Each ridge-free region can be associated with a combinatorial star unfolding that defines a set of source images in the unfolded plane. Each of these source images can be linearly parameterized according to the position of a source point s in a (two-dimensional) ridge-free region. Using these linear parameterizations, Agarwal et al. [1] represent a kinetic Voronoi diagram of the source images as a lower envelope in $\mathbb{R}^9$. The upper bound theorem for convex polyhedra ensures that this kinetic Voronoi diagram has $O(M^4)$ complexity [1].

The below approach computes the diameter a linear factor faster than [1]. Instead of representing a kinetic Voronoi diagram of parameterized source images as a high-dimensional lower envelope, we maintain a kinetic Voronoi diagram over a set of collinear and co-circular Voronoi events that are defined by a set of continuously moving sites. The idea of maintaining a kinetic Voronoi diagram for a set of continuously moving points is due to Albers et al. [3]. They show that a kinetic Voronoi diagram can be maintained in $O(\log M)$ time per event.

Theorem 5. *The diameter of a convex polyhedral surface $\mathcal{P}$ with M vertices can be computed in $O(M^7 \log M)$ time and $O(M^4)$ space.*

Proof. To compute the diameter, we maintain a kinetic Voronoi diagram of source images and return the largest distance ever attained between a source image site and any of its Voronoi vertices. Begin by picking an initial ridge-free region r and choosing linear parameterizations for the source images in the combinatorial star unfolding of r. As mentioned above, the upper bound theorem for convex polyhedra ensures that these parameterizations define $O(M^4)$ Voronoi events. Process the remaining ridge free regions in depth-first order so that the current ridge-free region r_c is always adjacent to a previously processed region r_p. Due to the definition of ridge-free regions, the star unfolding for r_c can always be obtained from the star unfolding for r_p by removing and inserting two source image sites. This implies that the kinetic Voronoi diagram for r_c involves only $O(M^3)$ Voronoi events that were not present in r_p. This follows because each of these $O(M^3)$ events must involve at least one of the two new source image sites. These bounds imply that there are a total of $O(M^7)$ Voronoi events over all $O(M^4)$ ridge-free regions, and each of these events can be handled in $O(\log M)$ time by [3]. Each parameterized Voronoi vertex v can now be associated with a function $f(v)$ that represents the distance from v to its defining source image. The diameter is the largest distance defined by any of these functions.   □

4   Fréchet Distance

Let $\delta_C(A, B)$ (resp. $\delta_N(A, B)$) denote the Fréchet distance between polygonal curves A and B on a convex (resp. non-convex) polyhedral surface. Maheshwari and Yi [19] have previously shown how to compute $\delta_C(A, B)$ in $O(M^7 \log M)$ time by enumerating all edge sequences. However, their approach relies on [18] whose key claim "has yet to be convincingly established" [1]. By contrast, we use the star unfolding from section 2 to compute $\delta_C(A, B)$ in $O(M^6 \log^2 M)$ time and $O(M^2)$ space. We build a *free space diagram* [5] to measure the distance $d(s, t)$ between all pairs of points $s \in A$ and $t \in B$. Each *cell* in our free space diagram is the parameter space defined by an edgelet $\alpha \in A$ and either a core edge or an anti-core edge in the combinatorial star unfolding for α. A cell is always interior-disjoint from all other cells.

To compute $\delta_C(A, B)$, we determine for a given constant $\varepsilon \geq 0$ all points $\{(s, t)$ | $s \in A$, $t \in B$, $d(s, t) \leq \varepsilon\}$ that define the *free space* [5]. The star unfolding $\mathcal{S}$ maps a fixed source point $s \in A$ to a set $s_1, ..., s_M$ of source image points in $\mathcal{S}$ and maps the polygonal curve B to a set $\beta_1, ..., \beta_{O(M^2)}$ of core and anti-core edges in $\mathcal{S}$. Since s maps to multiple images in $\mathcal{S}$, free space is defined by the union of a set of disks $d_1, ..., d_M$, where each disk d_i has radius ε and is centered at s_i (see Fig. 2). This follows by [6,7] because all L_2 distances in the star unfolding for a *convex* polyhedral surface are at least as large as the shortest path between those two points (even when the L_2 path does not stay inside the boundary of the star unfolding). As the source point s varies continuously over an edgelet $\alpha \in A$, the core is fixed and each s_i is parameterized along a line segment l_i in the star unfolding [1]. This is illustrated in Fig. 2a. The below $\delta_C(A, B)$ *decision problem* decides whether the Fréchet distance between polygonal curves A and B on a convex polyhedral surface is at most some constant $\varepsilon \geq 0$.

Theorem 6. *The $\delta_C(A, B)$ decision problem can be computed in $O(M^6 \log M)$ time and $O(M^2)$ space.*

Proof. Partition the polygonal curve A into $O(M^3)$ edgelets and maintain a star unfolding for these edgelets in $O(M^4)$ total time by Theorem 1. Free space for an edge β in the anti-core region containing s_i is defined by an ellipse $E_{l_i, \beta} = \{(s, t) \mid s \in l_i, \, t \in \beta, \, ||s - t|| \leq \varepsilon\}$, and free space for an edge γ in the core is defined by the union of the M ellipses $E_{l_1, \gamma}, ..., E_{l_M, \gamma}$ (see Fig. 2b). Thus, the free space defined by all $O(M^2)$ anti-core edges and $O(M)$ core edges has

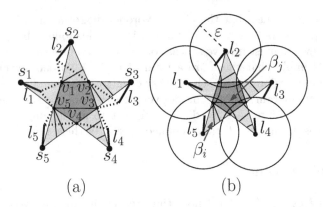

Fig. 2. The Star Unfolding of a Polyhedral Surface. Free space for an edge in the lightly-shaded anti-core is completely described by a single disk (e.g., the disk centered on l_5 is always closest to β_i). Free space for edges in the heavily-shaded core (e.g., β_j) is defined by the union of $O(M)$ disks.

$O(M^3)$ total complexity per edgelet and $O(M^6)$ complexity over all edgelets. Reachability information can be propagated through the free space diagram via plane sweep [11] in $O(M^6 \log M)$ time, and the decision problem returns true if and only if the upper right corner of the free space diagram is reachable. By storing one star unfolding, one cell, and one vertical line segment of the free space diagram at a time, $O(M^2)$ space is sufficient.   □

For a *non-convex* polyhedral surface, even the core of the star unfolding can overlap itself [12], and shortest paths can turn at vertices in the star unfolding [20]. However, the star unfolding can still be defined by a tree of angularly ordered shortest path cuts from the source to every vertex [8,20], and a core can still be defined by a polygonal equator with $O(M)$ complexity that connects adjacent leaves in the tree. An anti-core region now has an *hourglass* shape [16,10] because an anti-core region is bounded by a (possibly parameterized) vertex, a line segment, and two shortest paths in the unfolded plane. The below $\delta_N(A, B)$ *decision problem* decides whether the Fréchet distance between polygonal curves A and B on a non-convex polyhedral surface is at most some constant $\varepsilon \geq 0$.

Theorem 7. *The $\delta_N(A, B)$ decision problem can be computed in $O(M^7 \log M)$ time and $O(M^3)$ space.*

Proof. Partition the polygonal curve A into $O(M^3)$ edgelets such that all points on an edgelet can be associated with the same combinatorial star unfolding. Maintain a star unfolding for these edgelets in $O(M^4)$ total time by Theorem 1. Let C be the parameter space for an edgelet and either a core edge or an anti-core edge. Free space for an *anti-core edge* is the intersection of C with the $O(M)$ complexity free space for *one* hourglass. Free space for a *core edge* is the intersection of C with the union of the free spaces for $O(M)$ hourglasses. This union has $O(M^3)$ complexity because the free space for any pair of hourglasses has $O(M)$ complexity. Since each core edge γ is a chord of the core, the dual graph of the core is a tree. Consequently, the $O(M)$ hourglasses for γ can be defined by iteratively extending an hourglass from every vertex in the star unfolding [8] through the dual graph of the core to γ. The free space for each edgelet has $O(M^4)$ complexity because it involves $O(M^2)$ anti-core edges and $O(M)$ core edges, and this free space can be computed in $O(M^4 \log M)$ time [2]. A plane sweep [11] can be used to answer the decision problem over all $O(M^3)$ edgelets in $O(M^7 \log M)$ time. $O(M^3)$ space is sufficient to store one star unfolding, one cell, and one vertical line segment of the free space diagram at a time.   □

Theorem 8. *The Fréchet distance can be computed on a convex polyhedral surface in $O(M^6 \log^2 M)$ time and $O(M^2)$ space and on a non-convex polyhedral surface in $O(M^7 \log^2 M)$ time and $O(M^3)$ space, where M is the total complexity of the surface and the polygonal curves A, B. The free space diagram for a non-convex polyhedral surface can have $\Omega(M^4)$ complexity.*

Proof. Represent each of the $O(M^6)$ (resp. $O(M^7)$) free space vertices from Theorems 6 and 7 as an algebraic curve $\rho_i(\varepsilon)$ that has constant degree and description complexity. *Critical values* [5] are candidate values of ε that are caused by a geometric configuration change of the free space. Type (a) critical values are values of ε such that some $\rho_i(\varepsilon)$ touches a corner of the free space diagram. Type (b) critical values occur when two $\rho_i(\varepsilon)$ intersect or when free space becomes tangent to a cell boundary. Monotonicity-enforcing type (c) critical values occur when a pair of intersection points lie on a horizontal/vertical line. Parametric search [5] can be applied to the $\rho_i(\varepsilon)$ functions to compute the Fréchet optimization problem in $O(M^6 \log^2 M)$ (resp. $O(M^7 \log^2 M)$) time. The space bounds are identical to the decision problems. See [12] for our lower bound.   □

5   Shortest Path Maps

This section develops shortest path maps on convex and non-convex polyhedral surfaces. These structures support queries from any point on an arbitrary source line segment $\overline{ab}$ that lies on the surface. Throughout this section, M denotes the complexity of either a convex or non-convex polyhedral surface, and K is the complexity of any returned path.

Theorem 9. *A shortest path map $SPM(\overline{ab}, \mathcal{P})$ can be built for a convex polyhedral surface $\mathcal{P}$ in $O(M^4 \log M)$ time and $O(M^4)$ space. For all points $s \in \overline{ab} \subset \mathcal{P}$ and $t \in \mathcal{P}$, $SPM(\overline{ab}, \mathcal{P})$ can return $d(s,t)$ in $O(\log^2 M)$ time and $\pi(s,t)$ in $O(\log^2 M + K)$ time.*[4]

Proof. A kinetic Voronoi diagram can be maintained for $O(M^2)$ edgelets on $\overline{ab}$ in $O(M^4 \log M)$ total time and $O(M^4)$ space by Theorem 3. Point location in this kinetic Voronoi diagram takes $O(\log^2 M)$ time by [15].   □

---

[4] $O(\log M)$ time queries are also possible by [15] but at the cost of essentially squaring both the time and space preprocessing bounds.

Our next theorem uses the star unfolding [1] and the hourglass structure of [16] to encode all shortest paths between two line segments. Such an hourglass defines a piecewise hyperbolic free space that has $O(M)$ complexity [10].

Theorem 10. *A shortest path map $SPM(\overline{ab}, \mathcal{P}_N)$ can be built for a non-convex polyhedral surface $\mathcal{P}_N$ in $O(M^{9+\epsilon})$ time and $O(M^9)$ space for any constant $\epsilon > 0$. For all points $s \in \overline{ab} \subset \mathcal{P}_N$ and $t \in \mathcal{P}_N$, $SPM(\overline{ab}, \mathcal{P}_N)$ can return $d(s,t)$ in $O(\log M)$ time and $\pi(s,t)$ in $O(\log M + K)$ time.*

Proof. Let α be one of the $O(M^2)$ edgelets on $\overline{ab}$ (see section 2). A shortest path between a point $s \in \alpha$ and any fixed point in the *anti-core* can be resolved using one hourglass (cf. section 4). By contrast, all shortest paths between $s \in \alpha$ and a fixed point in a face of the *core* are defined by $O(M)$ hourglasses (see section 4). To support logarithmic query time for all points in a fixed face of the core, we can form $O(M^3)$ constant complexity distance functions from these hourglasses and compute their lower envelope and a vertical decomposition structure in $O(M^{6+\epsilon})$ time and $O(M^6)$ space, for any constant $\epsilon > 0$ [2]. Repeating this procedure for all $O(M)$ faces in the core yields $O(M^{7+\epsilon})$ time per edgelet and $O(M^{9+\epsilon})$ time over all $O(M^2)$ edgelets. □

6   Conclusion

We develop algorithms to compute edge sequences, Voronoi diagrams, shortest path maps, the Fréchet distance, and the diameter for a polyhedral surface. Despite efforts by Chandru et al. [7] to improve edge sequence algorithms, these runtimes had not improved since 1997. Our work speeds up the edge sequence and diameter approaches of Agarwal et al. [1] by a linear factor and introduces many new shortest path algorithms that apply to both convex and non-convex polyhedral surfaces. It would be interesting to lower the gaps between our various lower and upper bounds. In particular, future work could attempt to construct the $\Theta(M^4)$ shortest path edge sequences on a convex polyhedral surface in $o(M^5)$ time. Numerous link distance results and our overlapping core example were omitted due to space constraints and can be found in our technical report [12].

References

1. Agarwal, P.K., Aronov, B., O'Rourke, J., Schevon, C.A.: Star unfolding of a polytope with applications. SIAM Journal on Computing 26(6), 1689–1713 (1997)
2. Agarwal, P.K., Sharir, M.: Davenport–Schinzel Sequences and Their Geometric Applications. Handbook of Computational Geometry, pp. 1–47. Elsevier, Amsterdam (2000)
3. Albers, G., Mitchell, J.S.B., Guibas, L.J., Roos, T.: Voronoi diagrams of moving points. Journal of Computational Geometry & Applications 8, 365–380 (1998)
4. Aleksandrov, L., Djidjev, H., Huo, G., Maheshwari, A., Nussbaum, D., Sack, J.-R.: Approximate shortest path queries on weighted polyhedral surfaces. In: Královič, R., Urzyczyn, P. (eds.) MFCS 2006. LNCS, vol. 4162, pp. 98–109. Springer, Heidelberg (2006)

5. Alt, H., Godau, M.: Computing the Fréchet distance between two polygonal curves. Journal of Computational Geometry & Applications 5, 75–91 (1995)
6. Aronov, B., O'Rourke, J.: Nonoverlap of the star unfolding. Discrete and Computational Geometry 8(1), 219–250 (1992)
7. Chandru, V., Hariharan, R., Krishnakumar, N.M.: Short-cuts on star, source and planar unfoldings. In: Lodaya, K., Mahajan, M. (eds.) FSTTCS 2004. LNCS, vol. 3328, pp. 174–185. Springer, Heidelberg (2004)
8. Chen, J., Han, Y.: Shortest paths on a polyhedron. Journal of Computational Geometry & Applications 6(2), 127–144 (1996)
9. Chiang, Y., Mitchell, J.S.B.: Two-point Euclidean shortest path queries in the plane. In: 10th Symposium on Discrete Algorithms (SODA), pp. 215–224 (1999)
10. Cook IV, A.F., Wenk, C.: Geodesic Fréchet distance inside a simple polygon. In: 25th Symposium on Theoretical Aspects of Computer Science, STACS (2008)
11. Cook IV, A.F., Wenk, C.: Geodesic Fréchet distance with polygonal obstacles. Technical Report CS-TR-2008-010, University of Texas at San Antonio (2008)
12. Cook IV, A.F., Wenk, C.: Shortest path problems on a polyhedral surface. Technical Report CS-TR-2009-001, University of Texas at San Antonio (2009)
13. Cook IV, A.F., Wenk, C.: Shortest path problems on a polyhedral surface. In: 25th European Workshop on Computational Geometry (EuroCG) (2009)
14. Demaine, E.D., O'Rourke, J.: Geometric Folding Algorithms: Linkages, Origami, Polyhedra. Cambridge University Press, New York (2007)
15. Devillers, O., Golin, M., Kedem, K., Schirra, S.: Queries on Voronoi diagrams of moving points. Computational Geometry: Theory & Applications 6(5), 315–327 (1996)
16. Guibas, L.J., Hershberger, J., Leven, D., Sharir, M., Tarjan, R.E.: Linear-time algorithms for visibility and shortest path problems inside triangulated simple polygons. Algorithmica 2, 209–233 (1987)
17. Hershberger, J., Suri, S.: An optimal algorithm for Euclidean shortest paths in the plane. SIAM Journal on Computing 28(6), 2215–2256 (1999)
18. Hwang, Y.-H., Chang, R.-C., Tu, H.-Y.: Finding all shortest path edge sequences on a convex polyhedron. In: Dehne, F., Santoro, N., Sack, J.-R. (eds.) WADS 1989. LNCS, vol. 382. Springer, Heidelberg (1989)
19. Maheshwari, A., Yi, J.: On computing Fréchet distance of two paths on a convex polyhedron. In: 21st European Workshop on Computational Geometry (EuroCG) (2005)
20. Mitchell, J.S.B., Mount, D.M., Papadimitriou, C.H.: The discrete geodesic problem. SIAM Journal on Computing 16(4), 647–668 (1987)
21. Mount, D.M.: The number of shortest paths on the surface of a polyhedron. SIAM Journal on Computing 19(4), 593–611 (1990)
22. O'Rourke, J., Schevon, C.: Computing the geodesic diameter of a 3-polytope. In: 5th Symposium on Computational Geometry (SoCG), pp. 370–379 (1989)
23. Schevon, C., O'Rourke, J.: The number of maximal edge sequences on a convex polytope. In: 26th Allerton Conference on Communication, Control, and Computing, pp. 49–57 (1988)
24. Schreiber, Y., Sharir, M.: An optimal-time algorithm for shortest paths on a convex polytope in three dimensions. Discrete & Computational Geometry 39(1-3), 500–579 (2008)

Approximation Algorithms for
Buy-at-Bulk Geometric Network Design[*]

Artur Czumaj[1], Jurek Czyzowicz[2], Leszek Gąsieniec[3], Jesper Jansson[4],
Andrzej Lingas[5], and Pawel Zylinski[6]

[1] Centre for Discrete Mathematics and its Applications (DIMAP) and Department of Computer
Science, University of Warwick, UK
A.Czumaj@warwick.ac.uk
[2] Departement d'Informatique, Universite du Quebec en Outaouais, Gatineau,
Quebec J8X 3X7, Canada
Jurek.Czyzowicz@uqo.ca
[3] Department of Computer Science, University of Liverpool, Peach Street, L69 7ZF, UK
L.A.Gasieniec@liverpool.ac.uk
[4] Ochanomizu University, 2-1-1 Otsuka, Bunkyo-ku, Tokyo-112-8610, Japan
Jesper.Jansson@ocha.ac.jp
[5] Department of Computer Science, Lund University, 22100 Lund, Sweden
Andrzej.Lingas@cs.lth.se
[6] Institute of Computer Science, University of Gdańsk, 80-952 Gdańsk, Poland
Pawel.Zylinski@inf.univ.gda.pl

Abstract. The buy-at-bulk network design problem has been extensively studied in the general graph model. In this paper we consider the *geometric* version
of the problem, where all points in a Euclidean space are candidates for network nodes. We present the first general approach for geometric versions of basic variants of the buy-at-bulk network design problem. It enables us to obtain
quasi-polynomial-time approximation schemes for basic variants of the buy-at-
bulk geometric network design problem with polynomial total demand. Then, for
instances with few sinks and low capacity links, we design very fast polynomial-
time low-constant approximations algorithms.

1 Introduction

Consider a water heating company that plans to construct a network of pipelines to carry
warm water from a number of heating stations to a number of buildings. The company
can install several types of pipes of various diameters and prices per unit length. Typically, the prices grow with the diameter while the ratio between the pipe throughput
capacity and its unit price decreases. The natural goal of the company is to minimize
the total cost of pipes sufficient to construct a network that could carry the warm water
to the buildings, assuming a fixed water supply at each source. Similar problems can
be faced by oil companies that need to transport oil to refineries or telecommunication
companies that need to buy capacities (in bulk) from a phone company.

---

[*] Research supported in part by VR grant 621-2005-4085, the Royal Society IJP - 2006/R2, the
Centre for Discrete Mathematics and its Applications, the Special Coordination Funds for Promoting Science and Technology (Japan), and the Visby Programme Scholarship 01224/2007.

F. Dehne et al. (Eds.): WADS 2009, LNCS 5664, pp. 168–180, 2009.
© Springer-Verlag Berlin Heidelberg 2009

The common difficulty of these problems is that only a limited set of types of links (e.g., pipes) is available so the price of installing a link (or, a node respectively) to carry some volume of supply between its endpoints does not grow in linear fashion in the volume but has a discrete character. Even if only one type of link with capacity not less than the total supply is available the problem is NP-hard as it includes the minimum Steiner tree problem. Since the geometric versions of the latter problem are known to be strongly NP-complete [11], these problems cannot admit fully polynomial-time approximations schemes in the geometric setting [11].

In operations research, they are often termed as discrete cost network optimization [4,20] whereas in computer science as minimum cost network (or, link/edge) installation problems [23] or as *buy-at-bulk network design* [3]; we shall use the latter term.

In computer science, the buy-at-bulk network design problem has been introduced by Salman et al. [23], who argued that the case most relevant in practice is when the graph is defined by points in the Euclidean plane. Since then, various variants of buy-at-bulk network design have been extensively studied in the *graph model* [3,5,6,7,10,12,13,14,15,17,19] (rather than in geometric setting). Depending on whether or not the whole supply at each source is required to follow a single path to a sink they are characterized as *non-divisible* or *divisible* [23]. In terms of the warm water supply problem, the divisible graph model means that possible locations of the pipes and their splits or joints are given a priori.

In this paper, we consider the following basic *geometric* divisible variants of the buy-at-bulk network design:

▷ *Buy-at-bulk geometric network design (BGND)*: for a given set of different edge types and a given set of sources and sinks placed in a Euclidean space construct a minimum cost geometric network sufficient to carry the integral supply at sources to the sinks.

▷ *Buy-at-bulk single-sink geometric network design (BSGND)*: for a given set of different edge types, a given single-sink and given set of sources construct a minimum cost geometric network sufficient to carry the integral supply at sources to the sink.

Motivated by the practical setting in which the underlying network has to posses some basic structural properties, we distinguish also special versions of both problems where each edge of the network has to be parallel to one of the coordinate system axes, and term them as *buy-at-bulk rectilinear network design (BRND)* and *buy-at-bulk single-sink rectilinear network design (BSRND)*, respectively.

Our contributions and techniques. A classical approach for approximation algorithms for geometric optimization problems builds on the techniques developed for polynomial-time approximation schemes (PTAS) for geometric optimization problems due to Arora [1]. The main difficulty with the application of this method to the general BGND problem lies in the reduction of the number of crossings on the boundaries of the dissection squares. This is because we cannot limit the number of crossings of a boundary of a dissection square below the integral amount of supply it carries into that square. On the other hand, we can significantly limit the number of crossing locations at the expense of a slight increase in the network cost. However with this relaxed approach we cannot achieve polynomial but rather only quasi-polynomial upper bounds on the number of subproblems on the dissection squares in the dynamic programming phase but for very special cases (cf. [2]). Furthermore, the subproblems, in particular the leaf ones, become much more difficult. Nevertheless, we can solve them exactly in the case of BRND with polynomially bounded demands of the sources and nearly-optimally in the case of BGND with polynomially bounded demands of the sources and constant edge capacities, in at most quasi-polynomial time[1].

As the result, we obtain a randomized *quasi-polynomial-time approximation scheme* (QPTAS) for the *divisible buy-at-bulk rectilinear network design problem* in the Euclidean plane with polynomially bounded total supply and a randomized QPTAS for the *divisible buy-at-bulk network design problem* on the plane with polynomially bounded total supply and constant edge capacities. Both results can be derandomized and the rectilinear one can be generalized to include $O(1)$-dimensional Euclidean space. They imply that the two aforementioned variants of buy-at-bulk geometric network design are not *APX*-hard, unless $SAT \in DTIME[n^{\log^{O(1)} n}]$.

These two results are later used to prove our further results about low-constant-factor approximations for more general geometric variants. By using a method based on a novel belt decomposition for the single-sink variant, we obtain a $(2 + \varepsilon)$ approximation to the divisible buy-at-bulk rectilinear network design problem in the Euclidean plane, which is fast if there are few sinks and the capacities of links are small; e.g., it runs in $n(\log n)^{O(1)}$ time if the number of sinks and the maximum link capacity are polylogarithmic in n. Similarly, we obtain a $(2 + \varepsilon)$ approximation to the corresponding variants of the divisible buy-at-bulk network design problem in the Euclidean plane, which are fast if there are few sinks and the capacities of links are small, e.g., $n(\log n)^{O(1)}$-time if the number of sinks is polylogarithmic in n and maximum link capacity is $O(1)$. For comparison, the best known approximation factor for single-sink divisible buy-at-bulk network design in the graph model is 24.92 [13].

Related work. Salman et al. [23] initiated the algorithmic study of the single-sink buy-at-bulk network design problem. They argued that the problem is especially relevant in practice in the geometric case and they provided a polynomial-time approximation algorithm for the indivisible variant of BSGND on the input Euclidean graph (which differs from our model in that Salman et al. [23] allowed only some points on the plane to be used by the solution, whereas we allow the entire space to be used) with the approximation guarantee of $O(\log D)$, where D is total supply. Salman et al. gave also a constant factor approximation for *general graphs* in case where only one sink and one type of links is available; this approximation ratio has been improved by Hassin et al. [15]. Mansour and Peleg [18] provided an $O(\log n)$ approximation for the multi-sink buy-at-bulk network design problem when only one type of link is available. Awerbuch and Azar [3] were the first who gave a non-trivial (polylogarithmic) approximation for the general graph case for the total of n sinks and sources even in the case where different sources have to communicate with different sinks.

In the *single-sink buy-at-bulk* network design problem for general graphs, Garg et al. [12] designed an $O(K)$ approximation algorithm, where K is the number of edge types,

---

[1] Our solution method does not work in quasi-polynomial time in the case of the stronger version of BRND and BGND where specified sources must be assigned to specified sinks [3].

and later Guha et al. [14] gave the first constant-factor approximation algorithm for the (non-divisible) variant of the problem. This constant has been reduced in a sequence of papers [10,13,17,24] to reach the approximation ratio of 145.6 for the non-divisible variant and 24.92 for the divisible variant. Recently, further generalizations of the buy-at-bulk network design problem in the graph model have been studied [5,6].

2 Preliminaries

Consider a Euclidean d-dimensional space $\mathbb{E}^d$. Let $s_1, \ldots, s_{n_s}$ be a given set of n_s points in $\mathbb{E}^d$ (*sources*) and $t_1, \ldots, t_{n_t}$ be a given set of n_t points in $\mathbb{E}^d$ (*sinks*). Each source s_i supplies some integral *demand* $d(s_i)$ to the sinks. Each sink t_j is required to receive some integral *demand* $d(t_j)$. The sums $\sum_i d(s_i)$, $\sum_j d(t_j)$ are assumed to be equal and their value is termed as the *total demand* D. There are K types of edges, each type with a fixed cost and capacity. The *capacity* of an edge of type i is c_i and the *cost* of placing an edge e of ith type and length $|e|$ is $|e| \cdot \delta_i$.

The objective of the *buy-at-bulk geometric network design problem* (**BGND**) is to construct a geometric directed multigraph G in $\mathbb{E}^d$ such that:

- each copy of a multi-edge in the network is one of the K types;
- all the sources s_i and the sinks t_j belong to the set of vertices of G (the remaining vertices are called *Steiner vertices*);
- for $\ell = 1, \ldots, D$, there is a supply-demand path (*sd-path* for short) P_ℓ from a source s_i to a sink t_j such that each source s_i is a startpoint of $d(s_i)$ sd-paths, each sink t_j is an endpoint of $d(t_j)$ sd-paths, and for each directed multi-edge of the multigraph the total capacity of the copies of this edge is not less than the total number of sd-paths passing through it;
- the multigraph minimizes the total cost of the copies of its multi-edges.

If the set of sinks is a singleton then the problem is termed as the *buy-at-bulk single-sink geometric network design problem* (**BSGND** for short). If the multigraph is required to be rectilinear, i.e., only vertical and horizontal edges are allowed, then the problem is termed as the *buy-at-bulk rectilinear network design problem* (**BRND** for short) and its single-sink version is abbreviated as **BSRND**.

We assume, that the types of the edges are ordered $c_1 < \cdots < c_K$, $\delta_1 < \cdots < \delta_K$ and $\frac{\delta_1}{c_1} > \cdots > \frac{\delta_K}{c_K}$, since otherwise we can eliminate some types of the edges [23].

In this paper, we will always assume that the Euclidean space under consideration is a Euclidean plane $\mathbb{E}^2$, even though the majority of our results can be generalized to any Euclidean $O(1)$-dimensional space.

Zachariasen [25] showed that several variants and generalizations of the minimum rectilinear Steiner problem in the Euclidean plane are solvable on the *Hanan grid* of the input points, i.e., on the grid formed by the vertical and horizontal straight-lines passing through these points. The following lemma extends this to BRND.

Lemma 1. *Any optimal solution to BRND in the plane can be converted into a planar multigraph (so the sd-paths do not cross) where all the vertices lie on the Hanan grid.*

3   Approximating Geometric Buy-at-Bulk Network Design

In this section, we present our QPTAS for BRND and BGND. We begin with general-izations of several results from [1,22] about PTAS for TSP and the minimum Steiner tree in the plane. We first state a generalization of the Perturbation Lemma from [1,22].

Lemma 2. [22] *Let* $G = (V, E)$ *be a geometric graph with vertices in* $[0, 1]^2$, *and let* $U \subseteq V$. *Denote by* $E(U)$ *the set of edges incident to the vertices in* U. *One can perturb the vertices in* U *so they have coordinates of the form* $(\frac{i}{k}, \frac{j}{k})$, *where* i, j *are natural numbers not greater than a common natural denominator* k, *and the total length of* G *increases or decreases by an additive term of at most* $\sqrt{2} \cdot |E|/k$.

Consider an instance of BGND or BRND with sources $s_1 \dots s_{n_s}$ and sinks $t_1 \dots t_{n_t}$. We may assume, w.l.o.g., that the sources and the sinks are in $[0, 1)^2$.

Suppose that the total demand D is $n^{O(1)}$ where $n = n_s + n_t$. It follows that the maximum degree in a minimum cost multigraph solving the BGND or BRND is $n^{O(1)}$. Hence, the total number of copies of edges incident to the sources and sinks in the multigraph is also, w.l.o.g., $n^{O(1)} = n^{O(1)} \times n$. In the case of BRND, we infer that even the total number of copies of edges incident to all vertices, i.e., including the Steiner points, is, w.l.o.g., $n^{O(1)} = n^{O(1)} \times O(n^2)$ by Lemma 1.

Let $\delta > 0$. By using a straightforward extension of Lemma 2 to include a geometric multigraph and rescaling by $L = \frac{n^{O(1)}}{\delta}$ the coordinates of the sources and sinks, we can alter our BGND or BRND with all vertices on the Hanan grid such that:

- the sources and sinks of the BGND and BRND as well as the Steiner vertices of the BRND lie on the integer grid in $[0, L)^2$, and
- for any solution to the BGND with the original sources and sites (or, BRND with all vertices on the Hanan grid) and for any type of edge, the total length of copies of edges of this type in the solution resulting for the BGND with the sources and sinks on the integer grid (or, for BRND with all vertices on the integer grid, respectively) is at most $L(1 + \delta)$ times larger, and
- for any solution to the BGND with the sources and sinks on the integer grid (or, for BRND with all vertices on the integer grid, respectively), the total length of copies of edges of this type in the solution resulting for the BGND with the orig-inal sources and sites (or, BRND with all vertices on the Hanan grid) is at most $(1 + \delta)/L$ times larger.

Note the second and the third properties imply that we may assume further that our input instance of BGND has sources and sinks on the integer grid in $[0, L)^2$, since this assumption introduces only an additional $(1 + \delta)$ factor to the final approximation factor. We shall call this assumption the **rounding assumption**. In the case of BRND, we may assume further, w.l.o.g., not only that our input instance has sources and sinks on the integer grid but also that Steiner vertices may be located only on this grid by the second and third property, respectively. This stronger assumption in the case of BRND introduces also only an additional $(1 + \delta)$ factor to the final approximation factor by the aforementioned properties. We shall term it the **strong rounding assumption**.

Now we pick two integers a and b uniformly at random from $[0, L)$ and extend the grid by a vertical grid lines to the left and $L - a$ vertical grid lines to the right. We similarly increase the height of the grid using the random integer b, and denote the obtained grid by $L(a, b)$. Next, we define the recursive decomposition of $L(a, b)$ by dissection squares using quadtree. The dissection quadtree is a 4-ary tree whose root corresponds to the square $L(a, b)$. Each node of the tree corresponding to a dissection square of area greater than 1 is dissected into four child squares of equal side length; the four child squares are called *siblings*. The obtained quadtree decomposition is denoted by $Q(a, b)$.

We say a graph G is *r-light* if it crosses each boundary between two sibling dissection squares of $Q(a, b)$ at most r times. A multigraph H is *r-fine* if it crosses each boundary between two sibling dissection squares of $Q(a, b)$ in at most r places. For a straight-line segment ℓ and an integer r, an *r-portal* of ℓ is any endpoint of any of the r segments of equal length into which ℓ can be partitioned.

3.1   QPTAS for Buy-at-Bulk Rectilinear Network Design (BRND)

We obtain the following new theorem which can be seen as a generalization of the structure theorem from [1] to include geometric multigraphs, where the guarantee of r-lightness is replaced by the weaker guarantee of r-fineness.

Theorem 1. *For any $\varepsilon > 0$ and any BRND (or BGND, respectively) on the grid $L(a, b)$, there is a multigraph on $L(a, b)$ crossing each boundary between two sibling dissection squares of $Q(a, b)$ only at $O(\log L/\varepsilon)$-portals, being a feasible solution of BRND (BGND, respectively) and having the expected length at most $(1 + \varepsilon)$ times larger than the minimum.*

To obtain a QPTAS for an arbitrary BRND with polynomial total demand in the Euclidean plane it is sufficient to show how to find a minimum cost multigraph for BRND on $L(a, b)$ which crosses each boundary between two sibling dissection squares of $Q(a, b)$ only at r-portals efficiently, where $r = O(\log n/\varepsilon)$.

We specify a subproblem in our dynamic programming method by a dissection square occurring in some level of the quadtree $Q(a, b)$, a choice of crossing points out of the $O(r)$-portals on the sides of the dissection square, and for each of the chosen crossing points p, an integral demand $d(p)$ it should either supply to or receive from the square (instead of the pairing of the distinguished portals [1]). By the upper bound $D \leq n^{O(1)}$, we may assume, w.l.o.g., that $d(p) = n^{O(1)}$. Thus, the total number of such different subproblem specifications is easily seen to be $n^{O(r)}$. The aforementioned subproblem consists of finding a minimum cost r-fine rectilinear multigraph for the BRND within the square, where the sources are the original sources within the square and the crossing points expected to supply some demand whereas the sinks are the original sinks within the square and the crossing points expected to receive some demand.

Each leaf subproblem, where the dissection square is a cell of $L(a, b)$ and the original sources and sinks may be placed only at the corners of the dissection square, and the remaining $O(r)$ ones on the boundary of the cell, can be solved by exhaustive search and dynamic programming as follows. By Lemma 1, we may assume, w.l.o.g., that an optimal solution of the subproblem is placed on the Hanan $O(r) \times O(r)$ grid. We enumerate all directions and total capacity assignments to the edges of the grid in time

$n^{O(r)}$ by using the $n^{O(1)}$ bound on the total demand. For each such grid edge with non-zero total capacity assigned, we find (if possible) the cheapest multi-covering of this capacity with different edge types with capacity bounded by the total demand by using a pseudo-polynomial time algorithm for the integer knapsack problem [11]. Next, we compare the cost of such optimal multi-covering with the cost of using a single copy of the cheapest edge type whose capacity exceeds the total demand (if any) to choose an optimal solution. It follows that all the leaf subproblems can be solved in time $n^{O(r^2)}$.

Then, we can solve subproblems corresponding to consecutive levels of the quadtree $Q(a, b)$ in a bottom up fashion by combining optimal solutions to four compatible sub-problems corresponding to the four dissection squares which are children of the dis-section square in the subproblem to solve. The compatibility requirement is concerned with the location of the crossing points and their demand requirements. Since there are $n^{O(r)}$ subproblems, solution of a single subproblem also takes $n^{O(r)}$ time.

The bottleneck in the complexity of the dynamic programming are the leaf subprob-lems. If we could arbitrarily closely approximate their solutions in time $n^{O(r)}$ then we could compute a minimum cost r-fine multigraph for BRND on $L(a, b)$ with polyno-mially bounded total demand in time $n^{O(r)}$. The following lemma will be helpful.

Lemma 3. *For any $\varepsilon > 0$, one can produce a feasible solution to any leaf subproblem which is within $(1 + \varepsilon)$ from the minimum in time $n^{O(\log^2 r)}$.*

By halving ε both in the dynamic programming for the original problem as well as in Lemma 3 and using the method of this lemma to solve the leaf subproblems, we obtain the following lemma.

Lemma 4. *A feasible r-fine multigraph for BRND on $L(a, b)$ with polynomially bounded total demand and total cost within $1 + \varepsilon$ from the optimum is computable in time $n^{O(r)}$.*

By combining Theorem 1 with Lemma 4 for $r = O(\frac{\log n}{\varepsilon})$ and the fact that the rounding assumption introduces only an additional factor of $(1 + O(\varepsilon))$ to the approximation factor, we obtain our first result.

Theorem 2. *For any $\varepsilon > 0$, there is a randomized $n^{O(\log n/\varepsilon)}$-time algorithm for BRND in the Euclidean plane with a total of n sources and sinks and total demand polynomial in n, which yields a solution whose expected cost is within $(1 + \varepsilon)$ of the optimum.*

Theorem 2 immediately implies the following result for BGND (which will be substan-tially subsumed in Section 3.2 in the case of constant maximum edge capacity).

Corollary 1. *For any $\varepsilon > 0$, there is a randomized $n^{O(\log n/\varepsilon)}$-time algorithm for BGND in the Euclidean plane with the total of n sources and sinks and with polynomial in n total demand, which yields a solution whose expected cost is within $(\sqrt{2} + \varepsilon)$ from the optimum.*

3.2   QPTAS for the Buy-at-Bulk Geometric Network Design Problem (BGND)

We can arbitrarily closely approximate BGND analogously as BRND if it is possible to solve or very closely approximate the leaf subproblems where all the sources and sinks are placed in $O(\log n/\varepsilon)$ equidistant portals on a boundary of a dissection square, and feasible solutions are restricted to the square area. Note that such a leaf subproblem is logarithmic as for the number of sources and sinks but the total capacity of its sources or sinks might be as large as the total capacity D of all sources. We shall assume D to be polynomial in the number of sinks an sources as in the previous section.

By an h-*square BGND*, we mean BGND restricted to instances where h sources and sinks are placed on a boundary of a square. By a *logarithmic square BGND*, we mean an h-*square BGND* where the total demand of the sources is $O(\log n)$.

Lemma 5. *If there is an $n^{O(\log n)}$-time approximation scheme for a logarithmic square BGND then there is an $n^{O(\log n)}$-time approximation scheme for an $O(\log n)$-square BGND with maximum edge capacity $O(1)$.*

Proof. Let D denote the total capacity of the sources in the h-square BGND, where $h = O(\log n)$. Consider an optimal solution to the h-square BGND. It can be decomposed into D sd-paths, each transporting one unit from a source to a sink. There are $O(h^2)$ types of the sd-paths in one-to-one correspondence with the $O(h^2)$ pairs source-sink. Analogously as in the rectilinear case (see Lemma 1), we may assume, w.l.o.g., that the sd-paths do not intersect and that the minimum edge capacity is 1. Let M be the maximum edge capacity in the h-square BGND.

For a type t of sd-path, let N_t be the number of sd-paths of type t in the optimal solution. Since these sd-paths do not intersect, we can number them, say, in the cyclic ordering around their common source, with the numbers in the interval $[1, N_t]$. Note that each of these paths whose number is in the sub-interval $[M, N_t - M + 1]$ can use only edges which are solely used by sd-paths of this type in the optimal solution. Let $k = \lfloor \frac{1}{\varepsilon} \rfloor$, and let ϱ be the ratio between the cost δ_1 (per length unit) of an edge of capacity 1 and the cost δ_{max} of an edge of the maximum capacity M divided by M. Suppose that $N_t \geq M + \varrho k M + 2(M - 1)$. Let $q = \lceil (N_t - 2(M - 1))/M \rceil$.

Consider the following modification of the optimal solution. Group the consecutive bunches of M sd-paths of type t in the sub-interval $[M, qM - 1]$, and direct them through q directed edges of capacity M from the source to the sink corresponding to the type t. Remove all edges in the optimal solution used by these sd-paths in this sub-interval. Note that solely at most $M - 1$ sd-paths of the type t immediately to the left of $[M, N_t - M + 1]$ as well as at most $M - 1$ sd-paths of the type t immediately to the right of this interval can loose their connections to the sink in this way. Independently of whether such a path looses its connection or not, we direct it through a direct edge of capacity 1 from the source to the sink.

The total cost of the directed edges of capacity M from the source to the sink in the distance d is $q\delta_{max}d$. It yields the lowest possible cost per unit, sent from the source to the sink corresponding to t, equal to $\frac{\delta_{max}}{M}d$. Thus the total cost of the removed edges must be at least $q\delta_{max}d$! The additional cost of the $2(M - 1)$ direct edges of capacity 1 from the source to the sink is $\leq \varepsilon$ fraction of $q\delta_{max}d$ by our assumption on N_t.

By starting from the optimal solution and performing the aforementioned modification of the current solution for each type t of sd-path satisfying $N_t \geq M + \varrho k M + 2(M - 1)$, we obtain a solution which is at most $(1 + \varepsilon)$ times more costly than the optimal, and which is decomposed into two following parts. The first, explicitly given part includes all sd-paths of type t satisfying $N_t \geq M + \varrho k M + 2(M - 1)$ whereas the second unknown part includes all paths of types t satisfying $N_t < \varrho kM + 2(M-1)$. It follows that it is sufficient to have an $(1+\varepsilon)$-approximation of an optimal solution to the logarithmic square BGND problem solved by the second part in order to obtain an $(1+O(\varepsilon))$-approximation to the original h-square BGND.   □

Lemma 6. *For any $\varepsilon > 0$, the logarithmic square BGND problem with the total capacity of the sources D can be $(1+\varepsilon)$-approximated in time $(D/\varepsilon)^{O(D(\log D/\varepsilon))}$ if $c_{max} = O(1)$.*

By combining Lemma 5 with Lemma 6 for $D = O(\log n/\varepsilon)$ and straightforward calculations, we obtain an arbitrarily close to the optimum solutions to the $n^{O(\log n/\varepsilon)}$ leaf problems in total time $n^{O(\log n/\varepsilon^{O(1)})}$. Hence, analogously as in case of BRND, we obtain a QPTAS for BGND with polynomially bounded demand when $c_{max} = O(1)$.

Theorem 3. *BGND with polynomially bounded demand of the sources and constant maximum edge capacity admits an $n^{O(\log n)}$-time approximation scheme.*

4   Fast Low-Constant Approximation for BRND and BGND

In this section, we present another method for BGND and BRND which runs in polynomial time, gives a low-constant approximation guarantee, and does not require a polynomial bound on the total demand. The method is especially efficient if the edge capacities are small and there are few sinks.

We start with the following two simple lemmas. The first lemma is analogous to the so-called routing lower bound from [18,23] and the second follows standard arguments.

Lemma 7. *Let S be the set of sources in an instance of BGND (BRND), and for each $s \in S$, let $t(s)$ be the closest sink in this instance. The cost of an optimal solution to the BGND (BRND, respectively) is at least $\sum_{s \in S} dist(s, t(s)) \frac{\delta_K}{c_K} d(s)$, where $dist(s, t(s))$ is the Euclidean distance (the L_1 distance, respectively).*

Lemma 8. *Let S be a set of k points within a square of side length ℓ. One can find in time $O(k)$ a Steiner tree of S with length $O(\ell\sqrt{k})$.*

The following lemma describes a simple reduction procedure which yields an *almost* feasible solution to BSGND or BSRND with cost arbitrarily close to the optimum.

Lemma 9. *For any $\varepsilon > 0$, there is a reduction procedure for BSGND (or BSRND, respectively), with one sink and $n-1$ sources and the ratio between the maximum and minimum distances of a source from the sink equal to m, which returns a multigraph yielding a partial solution to the BSGND (or BSRND, respectively) satisfying the following conditions:*

- *all but $O((\frac{1}{\varepsilon})^2 c_K^2 \log m)$ sources can ship their whole demand to the sink;*
- *for each source s there are at most $c_K - 1$ units of its whole demand $d(s)$ which cannot be shipped to the sink.*

The reduction runs in time $O(\frac{c_K}{\varepsilon} \log m \log n + c_K n)$, which is $O(n/\varepsilon^2)$ if $c_K = O(1)$.

Proof. Form a rectilinear $2\lceil m \rceil \times 2\lceil m \rceil$ grid F with unit distance equal to the minimum distance between the only sink t and a source, centered around t. Let μ be a positive constant to be set later.

We divide F into the square R of size $2\lceil \mu\sqrt{c_K} \rceil$ centered in t and for $i = 0, 1, \ldots$, the *belts* B_i of squares of size 2^i within the L_∞ distance at least $2^i \lceil \mu\sqrt{c_K} \rceil$ and at most $2^{i+1}\lceil \mu\sqrt{c_K} \rceil$ from t. Note that the number of squares in the belt B_i is at most $(4\lceil \mu\sqrt{c_K} \rceil)^2 = O(\mu^2 c_K)$, hence the total number of squares in all the belts is $O(\mu^2 c_K \log m)$ by the definition of the grid.

The reduction procedure consists of two phases. In the first phase, we connect each source s by a multi-path composed of $\lfloor d(s)/c_K \rfloor$ copies of a shortest path from s to t implemented with the K-th type of edges. Observe that the average cost of such a connection per each of the $c_K \lfloor d(s)/c_K \rfloor$ demand units u shipped from s to t is $dist(s,t)\frac{\delta_K}{c_K}$ which is optimal by Lemma 7. Note that after the first phase the remaining demand for each source is at most $c_K - 1$ units.

In the second phase, for each of the squares Q in each of the belts B_i, we sum the remaining demands of the sources contained in it, and for each complete c_K-tuple of demand units in Q, we find a minimum Steiner multi-tree of their sources and connect its vertex v closest to t by a shortest path to t. The total length of the resulting multi-tree is easily seen to be $dist(v,t) + O(2^i\sqrt{c_K}) \leq (1 + O(\frac{1}{\mu}))dist(v,t)$ by the definition of the squares and Lemma 8. Hence, for each unit u in the c_K-tuple originating from its source $s(u)$, we can assign the average cost of connection to t by the multi-tree implemented with the K-th type of edges not greater than $(1 + O(\frac{1}{\mu}))dist(s(u),t)\frac{\delta_K}{c_K}$.

It follows by Lemma 7 that the total cost of the constructed network is within $(1 + O(\frac{1}{\mu}))$ from the minimum cost of a multigraph for the input BSGND. By choosing μ appropriately large, we obtain the required $1 + \varepsilon$-approximation.

Since the total number of squares different from R is $O(\mu^2 c_K \log m)$, the total number of their sources with a non-zero remaining demand (at most $c_K - 1$ units) to ship is $O(\mu^2 c_K^2 \log m)$. Furthermore, since the square R can include at most $O(\mu^2 c_K)$ sources, the number of sources with a non-zero remaining demand (at most $c_K - 1$ units) in R is only $O(\mu^2 c_K)$.

The first phase can be implemented in time linear in the number of sources. The second phase requires $O(\mu^2 c_K \log m)$ range queries for disjoint squares and $O(c_K n/c_K)$ constructions of Steiner trees on c_K vertices using the method of Lemma 8. Thus it needs $O(\mu^2 c_K \log m \log n + c_K n)$ time by [21] and Lemma 8. Since, w.l.o.g, $\mu = O(\frac{1}{\varepsilon})$, we conclude that the whole procedure takes $O(\frac{c_K}{\varepsilon^2} \log m \log n + c_K n)$ time. $\square$

Extension to BRND and BGND. We can generalize our reduction to include n_t sinks by finding the Voronoi diagram in the L_2 (or L_1 for BGND) metric on the grid, locating each source in the region of the closest sink, and then running the reduction procedure separately on each set of sources contained in a single region of the Voronoi diagram. The construction of the Voronoi diagram and the location of the sources takes time $O(n \log n)$ (see [16,21]). The n_t runs of the reduction procedure on disjoint sets of sources takes time $O((\frac{1}{\varepsilon})^2 n_t c_K \log m \log n + c_K n)$. The union of the n_t resulting multigraphs may miss to ship the whole demand only from $O((\frac{1}{\varepsilon})^2 n_t c_K^2 \log m)$ sources. This gives the following generalization of Lemma 9.

Lemma 10. *For any $\varepsilon > 0$, there is a reduction procedure for BGND (or BRND, resp.), with n_t sinks and $n - n_t$ sources and the ratio between the maximum and minimum distances of a source from the sink equal to m, which returns a multigraph yielding a partial solution to the BGND (or BRND, resp.) satisfying the following conditions:*

- *all but $O((\frac{1}{\varepsilon})^2 n_t c_K^2 \log m)$ sources can ship their whole demand to the sink;*
- *for each source s there are at most $c_K - 1$ units of its whole demand $d(s)$ which cannot be shipped to the sink.*

The reduction procedure runs in time $O((\frac{1}{\varepsilon})^2 n_t c_K \log m \log n + n(c_K + \log n))$. In particular, if $c_K = (\log n)^{O(1)}$ then the running time is $(\frac{1}{\varepsilon})^2 n \log m (\log n)^{O(1)}$.

Now, we are ready to derive our main results in this section.

Theorem 4. *For any $\varepsilon > 0$, there is a $(2 + \varepsilon)$-approximation algorithm for BRND with n_t sinks and $n - n_t$ sources in the Euclidean plane, running in time $O((\frac{1}{\varepsilon})^2 n_t c_K \log^2 n + n(\log n + c_K)) + (n_t c_K^2 \log n)^{O(\frac{\log n_t + \log c_K}{\varepsilon^2})}$, in particular in time $n(\log n)^{O(1)} + (\log n)^{O(\frac{\log \log n}{\varepsilon^2})}$ if $n_t = (\log n)^{O(1)}$ and $c_K = (\log n)^{O(1)}$.*

Proof. By the rounding assumption discussed in Section 3 we can perturb the sinks and the sources so they lie on an integer grid of polynomial size introducing only an additional $(1 + O(\varepsilon))$ factor to the final approximation factor. The perturbation can be easily done in linear time. Next, we apply the reduction procedure from Lemma 10 to obtain an almost feasible solution of total cost not exceeding $(1 + O(\varepsilon))$ of that for the optimal solution to the BSRND on the grid. Note that $m \leq n^{O(1)}$ and hence $\log m = O(\log n)$ in this application of the reduction by the polynomiality of the grid. It remains to solve the BRND subproblem for the $O((\frac{1}{\varepsilon})^2 n_t c_K \log n)$ remaining sources with total remaining demand polynomial in their number. This subproblem can be solved with the randomized $(1 + O(\varepsilon))$-approximation algorithm of Theorem 2. In fact, we can use here also its derandomized version which will run in time $(n_t c_K^2 \log n)^{O(\frac{\log n_t + \log c_K}{\varepsilon^2})}$. $\square$

As an immediate corollary from Theorem 4, we obtain a $(\sqrt{8} + \varepsilon)$-approximation algorithm for BGND with n_t sinks and $n - n_t$ sources in the Euclidean plane, running in time $O((\frac{1}{\varepsilon})^2 n_t c_K \log^2 n + n(\log n + c_K)) + (n_t c_K^2 \log n)^{O(\frac{\log n_t + \log c_K}{\varepsilon^2})}$. However, the direct method analogous to that of Theorem 4 yields a better approximation, in particular also an $(2 + \varepsilon)$-approximation if $c_K = O(1)$.

Theorem 5. *For any $\varepsilon > 0$, there is a $(1 + \sqrt{2} + \varepsilon)$-approximation algorithm for BGND with n_t sinks and $n - n_t$ sources in the Euclidean plane, running in time $O((\frac{1}{\varepsilon})^2 n_t c_K \log^2 n + n(\log n + c_K)) + (n_t c_K^2 \log n)^{O(\frac{\log n_t + \log c_K}{\varepsilon^2})}$; the running time is $n(\log n)^{O(1)} + (\log n)^{O(\frac{\log \log n}{\varepsilon^2})}$ if $n_t = (\log n)^{O(1)}$ and $c_K = (\log n)^{O(1)}$. Furthermore, if $c_K = O(1)$ then the approximation factor of the algorithm is $2 + \varepsilon$.*

5 Final Remarks

We have demonstrated that BRND and BGND in a Euclidean space admit close approximation under the assumption that the total demand is polynomially bounded. By running the first phase of the reduction procedure from Lemma 9 as a preprocessing, we could get rid of the latter assumption at the expense of worsening the approximation factors by the additive term 1.

All our approximation results for different variants of BRND in Euclidean plane derived in this paper can be generalized to include corresponding variants of BRND in a Euclidean space of fixed dimension. All our approximation schemes are randomized but they can be derandomized similarly as those in [1,8,9,22].

References

1. Arora, S.: Polynomial time approximation schemes for Euclidean traveling salesman and other geometric problems. Journal of the ACM 45(5), 753–782 (1998)
2. Arora, S., Raghavan, P., Rao, S.: Approximation schemes for Euclidean k-medians and related problems. In: Proc. 30th ACM STOC, pp. 106–113 (1998)
3. Awerbuch, B., Azar, Y.: Buy-at-bulk network design. In: Proc 38th IEEE FOCS, pp. 542–547 (1997)
4. Bienstock, D., Chopra, S., Günlük, O., Tsai, C.-Y.: Minimum cost capacity installation for multicommodity network flows. Mathematical Programming 81(2), 177–199 (1998)
5. Chekuri, C., Hajiaghayi, M.T., Kortsarz, G., Salavatipour, M.R.: Polylogarithmic approximation algorithms for non-uniform multicommodity buy-at-bulk network design. In: Proc. 47th IEEE FOCS, pp. 677–686 (2006)
6. Chekuri, C., Hajiaghayi, M.T., Kortsarz, G., Salavatipour, M.R.: Approximation algorithms for node-weighted buy-at-bulk network design. In: SODA, pp. 1265–1274 (2007)
7. Chopra, S., Gilboa, I., Sastry, S.T.: Source sink flows with capacity installation in batches. Discrete Applied Mathematics 85(3), 165–192 (1998)
8. Czumaj, A., Lingas, A.: On approximability of the minimum-cost k-connected spanning subgraph problem. In: Proc. 10th IEEE-SIAM SODA, pp. 281–290 (1999)
9. Czumaj, A., Lingas, A.: Fast approximation schemes for euclidean multi-connectivity problems. In: Welzl, E., Montanari, U., Rolim, J.D.P. (eds.) ICALP 2000. LNCS, vol. 1853, pp. 856–868. Springer, Heidelberg (2000)
10. Gupta, A., Kumar, A., Roughgarden, T.: Simpler and better approximation algorithms for network design. In: Proc. 35th ACM STOC, pp. 365–372 (2003)
11. Garey, M.R., Johnson, D.S.: Computers and Intractability. A Guide to the Theory of NP-completeness. W.H. Freeman and Company, New York (1979)
12. Garg., N., Khandekar, R., Konjevod, G., Ravi, R., Salman, F.S., Sinha, A.: On the integrality gap of a natural formulation of the single-sink buy-at-bulk network design problem. In: Proc. 8th IPCO, pp. 170–184 (2001)
13. Grandoni, F., Italiano, G.F.: Improved approximation for single-sink buy-at-bulk. In: Asano, T. (ed.) ISAAC 2006. LNCS, vol. 4288, pp. 111–120. Springer, Heidelberg (2006)
14. Guha, S., Meyerson, A., Munagala, K.: A constant factor approximation for the single sink edge installation problem. In: Proc. 33rd ACM STOC, pp. 383–399 (2001)
15. Hassin, R., Ravi, R., Salman, F.S.: Approximation algorithms for a capacitated network design problem. In: Jansen, K., Khuller, S. (eds.) APPROX 2000. LNCS, vol. 1913, pp. 167–176. Springer, Heidelberg (2000)
16. Hwang, F.K.: An O(nlogn) algorithm for rectilinear minimal spanning trees. JACM 26(2), 177–182 (1979)
17. Jothi, R., Raghavachari, B.: Improved approximation algorithms for the single-sink buy-at-bulk network design problems. In: Hagerup, T., Katajainen, J. (eds.) SWAT 2004. LNCS, vol. 3111, pp. 336–348. Springer, Heidelberg (2004)

18. Mansour, Y., Peleg, D.: An approximation algorithm for minimum-cost network design. Technical report, The Weizman Institute of Science, Revohot, Israel, CS94-22 (1994)
19. Meyerson, A., Munagala, K., Plotkin, S.: COST-DISTANCE: Two metric network design. In: Proc. 41st IEEE FOCS, pp. 624–630 (2000)
20. Minoux, M.: Discrete cost multicommodity network optimization problems and exact solution methods. Annals of Operations Research 106, 19–46 (2001)
21. Preparata, F., Shamos, M.: Computational Geometry. Springer, New York (1985)
22. Rao, S.B., Smith, W.D.: Approximating geometrical graphs via "spanners" and "banyans". In: Proc. ACM STOC, pp. 540–550 (1998); Full version appeared as TR, NEC (1998)
23. Salman, F.S., Cheriyan, J., Ravi, R., Subramanian, S.: Approximating the single-sink link-installation problem in network design. SIAM J. Optimization 11(3), 595–610 (2000)
24. Talwar, K.: Single-sink buy-at-bulk LP has constant integrality gap. In: Cook, W.J., Schulz, A.S. (eds.) IPCO 2002. LNCS, vol. 2337. Springer, Heidelberg (2002)
25. Zachariasen, M.: A catalog of Hanan grid problems. Networks 38(2), 76–83 (2001)

Rank-Sensitive Priority Queues

Brian C. Dean and Zachary H. Jones

School of Computing, Clemson University
Clemson, SC, USA
{bcdean,zjones}@cs.clemson.edu

Abstract. We introduce the rank-sensitive priority queue — a data structure that always knows the minimum element it contains, for which insertion and deletion take $O(\log(n/r))$ time, with n being the number of elements in the structure, and r being the rank of the element being inserted or deleted ($r = 1$ for the minimum, $r = n$ for the maximum). We show how several elegant implementations of rank-sensitive priority queues can be obtained by applying novel modifications to treaps and amortized balanced binary search trees, and we show that in the comparison model, the bounds above are essentially the best possible. Finally, we conclude with a case study on the use of rank-sensitive priority queues for shortest path computation.

1 Introduction

Let us say that a data structure is *min-aware* if it always knows the minimum element it contains; equivalently, the structure should support an $O(1)$ time *find-min* operation. Furthermore, we say a data structure is *dynamic* if it supports an *insert* operation for adding new elements, and a *delete* operation for removing elements (as is typical, we assume *delete* takes a pointer directly to the element being deleted, since our data structure may not support an efficient means of finding elements). The functionality of a dynamic min-aware structure captures the essence of the priority queue, the class of data structures to which the results in this paper are primarily applicable. In a priority queue, the *find-min* and *delete* operations are typically combined into an aggregate *delete-min* operation, but we will find it convenient to keep them separate and focus on the generic framework of a dynamic min-aware structure in the ensuing discussion.

In the comparison model of computation, it is obvious that either *insert* or *delete* must run in $\Omega(\log n)$ worst-case time for any dynamic min-aware structure, since otherwise we could circumvent the well-known $\Omega(n \log n)$ worst-case lower bound on comparison-based sorting by inserting n elements and repeatedly deleting the minimum. As a consequence, even the most sophisticated comparison-based priority queues typically advertise a running time bound of $O(\log n)$ for *insert* or (more commonly) *delete-min*. The sorting reduction above makes it clear that this is the best one can hope to achieve to inserting or deleting the *minimum* element in our structure, but what about other elements? Might it be possible, say, to design a comparison-based dynamic min-aware data structure with *insert*

F. Dehne et al. (Eds.): WADS 2009, LNCS 5664, pp. 181–192, 2009.

and *delete* running in only $O(1)$ time, except for the special case where we insert or delete a new minimum element, which takes $O(\log n)$ time? At first glance, these requirements no longer seem to run afoul of the sorting lower bound, but they still seem quite ambitious. This motivates the general question: if we take the *rank* of the element being inserted or deleted into consideration, how quickly can a dynamic min-aware structure support *insert* and *delete*?

In this paper, we answer the question above by providing several implementations of what we call *rank-sensitive* priority queues. These are dynamic min-aware data structures capable of performing *insert* and *delete* in $O(\log(n/r))$ time (possibly amortized or in expectation), where n denotes the number of elements in the structure and r denotes the rank of the element being inserted or deleted ($r = 1$ for the minimum, $r = n$ for the maximum[1]). Note that the structure is not explicitly told the ranks of the elements being inserted or deleted; rather, its performance simply scales in a graceful manner from $O(\log n)$ for inserting or deleting near the minimum down to $O(1)$ for, say, modifying any of the 99% of the largest elements. The resulting structure should therefore be ideally suited for the case where we want to maintain a dynamic collection of elements for which we only occasionally (say, in case of emergency) need priority queue functionality. Our various implementations of rank-sensitive priority queues will appeal to the serious data structure aficionado in that they involve elegant new twists on well-studied data structures, notably treaps, amortized balanced binary search trees, and radix heaps.

After discussing our data structures, we then give a proof that in the comparison model, $O(\log(n/r))$ is essentially the best one can hope to achieve for a rank-sensitive running time bound. There are two main challenges in doing this, the first being that it is actually not so easy to state such a lower bound theorem the right way; for example, if we are not careful, we can end up with a theorem that is vacuously true since it must hold for the special case where $r = 1$. The second main challenge is to perform a reduction that somehow manages to use our data structure to sort by removing mostly non-minimal elements.

Much of the research driving the development of fast priority queues is ultimately focused on speeding up Dijkstra's shortest path algorithm. Rank-sensitive priority queues are also worth studying in this context. Since the dominant component of the running time for Dijkstra's algorithm is typically the large number of *decrease-key* operations (in our case, implemented by *delete* followed by *insert* with a new key), we expect a rank-sensitive priority queue to perform well as long as many of our *decrease-key* operations don't move elements too close to the minimum in rank (a potentially reasonable assumption, for many types of shortest path problems). One might therefore hope that rank-sensitive priority queues might give us performance bounds in practice that match those of more complicated data structures, such as Fibonacci heaps. The last section of our

---

[1] We assume for simplicity that all elements in our structure have distinct values, so ranks are uniquely-defined. In the event of ties, we would need to define r to be the maximum rank of all tied elements, in order to ensure that our upper bounds still hold in the worst case.

paper investigates this possibility with a discussion of computational results of using rank-sensitive priority queues in shortest path computation.

Several works related to ours appear in the literature. Splay trees [12] and the unified structure [3] satisfy the static and dynamic finger theorems (see [6,5]), which (if we use the minimum element as a finger) give us amortized bounds of $O(\log r)$ for insertion or deletion of a rank-r element. This is indeed a form of rank sensitivity, but it is significantly weaker than our $O(\log(n/r))$ bound above: $O(\log r)$ evaluates to $O(\log n)$ for nearly all the elements in a structure, while $O(\log(n/r))$ evaluates to $O(1)$ for nearly all the elements. A similar bound is obtained by the *Fishspear* priority queue of Fisher and Patterson [8], where the running time for inserting and deleting element x is bounded by $O(\log m(x))$, with $m(x)$ giving the maximum rank of x over its lifetime in the structure. Iacono's *queap* data structures [11] support *delete-min(x)* operation in time $O(\log q(x))$, where $q(x)$ gives the number of elements in the structure that are older than x; this can also be considered a form of "rank sensitivity", where "rank" now has the very different meaning of "seniority" in the structure. Note that the working set property of splay trees and the unified structure leads to a symmetric bound: $O(\log y(x))$, where $y(x)$ denotes the number of elements in the structure younger than x.

2   A Randomized Approach

The first idea that comes to mind when trying to build a rank-sensitive priority queue is perhaps whether or not a standard binary heap might be sufficient. Insertion and deletion of an element x in a binary heap can be easily implemented in time proportional to the height of x, which is certainly $O(\log n)$ for the minimum element and $O(1)$ for most of the high-rank elements in the structure. However, if the maximum element in the left subtree of the root is smaller than the minimum element in the right subtree of the root, it is possible we could end up with, say, the median element being the right child of the root, for which deletion will take $O(\log n)$ time instead of the $O(1)$ time required by a rank-sensitive structure.

Randomization gives us a nice way to fix the problem above, giving a simple and elegant implementation of a rank-sensitive priority queue in which *insert* and *delete* run in $O(\log(n/r))$ expected time. Let us store our elements in a heap-ordered binary tree (not necessarily balanced), where we maintain an unordered array of pointers to the empty "NULL" spaces at the bottom of the tree, as shown in Figure 1. To insert a new element into an $(n-1)$-element tree, we place it into one of the n empty spaces at the bottom of the tree, chosen uniformly at random in $O(1)$ time, and then we sift it up (by repeatedly rotating with its parent) until the heap property is restored. To delete an element, we set its value to $+\infty$ and sift it down (by repeatedly rotating with its smallest child) until it becomes a leaf, after which it is removed.

The structure above is closely related to a *treap* [2], a hybrid between a binary search tree (BST) and heap in which every node in a binary tree stores two keys:

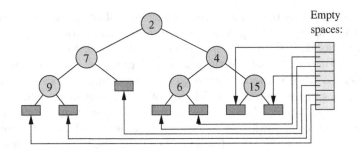

Fig. 1. The *h*-treap: a heap-ordered *n*-element binary tree augmented with an unordered array of pointers to the $n + 1$ empty "NULL" spaces at the bottom of the tree

a "BST" key, and a heap key. The structure satisfies the BST property with respect to the BST keys, and the heap property with respect to the heap keys. The primary application of treaps is to provide a simple balancing mechanism for BSTs — if we store our actual elements in the BST keys, and choose the heap keys randomly, then this forces the shape of the treap to behave probabilistically as if we had built a BST from our elements by inserting them in random order (and it is well known that such randomly-built BSTs are balanced with high probability). In our case, we are using the treap in a new "symmetric" way that does not seem to appear in the literature to date: we are storing our actual elements within the heap keys, and we are effectively assigning the BST keys randomly. However, rather than storing explicit numeric BST keys in the nodes of our tree, these random BST keys are implicit in the sequence encoded by the inorder traversal of our tree. Each time we insert a new element, we are effectively assigning it a random "BST key" since by inserting it into a randomly-chosen empty space at the bottom of the tree, we are inserting it into a randomly-chosen location within the inorder traversal sequence encoded by the tree. For lack of a better name, let us call such a structure an *h-treap* (since the actual elements are stored in the heap part of the treap), versus a standard *b-treap* in which we store our elements in the BST part. Observe that the *h*-treap behaves like a *b*-treap in that its shape is probabilistically that of a randomly-built BST, so it is balanced with high probability.

Theorem 1. *The* insert *and* delete *operations in an h-treap run in* $O(\log(n/r))$ *expected time.*

Proof. For any element x stored in an *h*-treap T, let $s(x)$ denote the number of elements present in x's subtree (including x). Since subtrees in an *h*-treap are balanced with high probability, it takes $O(\log s(x))$ time both in expectation and with high probability to delete x, since the height of x's subtree is $O(\log s(x))$ with high probability. Consider now the deletion of element x having rank r. Note that $s(x)$ is a random variable, owing to the random shape of T. If we define T_{r-1} to be the "top" part of T containing only the elements of ranks $1 \ldots r - 1$, then x will be located at one of the r empty spaces at the bottom of

T_{r-1}. Since the remaining $n - r$ elements of ranks $r + 1 \ldots n$ are equally likely to appear in each of these r spaces, we have $\mathbf{E}[s(x)] = 1 + (n - r)/r = n/r$. Due to Jensen's inequality, we now see that the expected time required to delete x is $\mathbf{E}[O(\log s(x))] = O(\log \mathbf{E}[s(x)]) = O(\log(n/r))$. The expected running time of insertion is the same due to symmetry, since the time required to insert an element of rank r is exactly the same as the time required to subsequently delete the element (the sequence of rotations performed by the deletion will be the reversal of those performed during insertion).

Since $\mathbf{E}_r[\log(n/r)] = O(1)$, we note that one can build an h-treap on n elements in $O(n)$ expected time by inserting them sequentially in random order.

3   An Amortized Approach

In [7] (problem 18-3), a simple and elegant BST balancing technique of G. Varghese is described that allows for the *insert* and *delete* operations both run in $O(\log n)$ amortized time. In this section, we build on this approach to obtain a rank-sensitive priority queue with $O(\log(n/r))$ amortized running times for *insert* and *delete*.

Let $s(x)$ denote the size (number of elements) of x's subtree in a BST. For any $\alpha \in [1/2, 1)$, we say x is α-weight-balanced if $s(left(x)) \le \alpha s(x)$ and $s(right(x)) \le \alpha s(x)$. A tree T is α-weight-balanced if all its elements are α-weight-balanced, and it is easy to show that an n-element α-weight-balanced tree has maximum height $\log_{1/\alpha} n$. The amortized rebalancing method of Varghese selects $\alpha \in (1/2, 1)$ and augments each element x in an α-weight-balanced BST with its subtree size $s(x)$. Whenever an element is inserted or deleted, we examine the elements along the path from the root down to the inserted or deleted element, and if any of these are no longer α-weight-balanced, we select the highest such node x in the tree and rebalance x's subtree in $O(s(x))$ time so it becomes $1/2$-balanced. It is easy to show that *insert* and *delete* run in only $O(\log n)$ time, because we can pay for the expensive operation of rebalancing the subtree of a non-α-weight-balanced element x by amortizing this across the $\Omega(s(x))$ intervening inserts and deletes that must have occurred within x's subtree since the last time x was $1/2$-weight-balanced.

We can build an effective rank-sensitive priority by relaxing the amortized balanced BST above so that the right subtree of every element is left "unbuilt", storing the elements of each unbuilt subtree in a circular doubly-linked list. As shown in Figure 2, only the left spine of the tree remains fully built. The only elements that effectively maintain their subtree sizes are now those on the left spine, and each one of these elements also maintains the size of its right subtree. To allow us to walk up the tree from any element, we augment every element in the circular linked list of the right subtree of element x with a pointer directly to x. In order to support *find-min* in $O(1)$ time, we maintain a pointer to the lowest element on the left spine.

Note that the mechanics of the amortized BST rebalancing mechanism above continue to function perfectly well in this relaxed setting. To insert a new

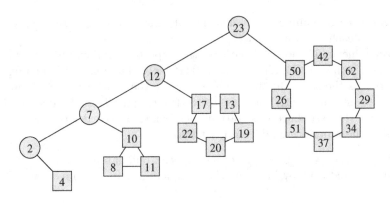

Fig. 2. An α-weight-balanced tree with all its right subtrees left unbuilt (whose elements are stored in circular doubly-linked lists)

element, we walk down the left spine until the BST property dictates into which right subtree it should be placed, unless the element is the new minimum, in which case it is placed at the bottom of the left spine. We then walk back up the tree, updating the subtree sizes and checking if rebalancing should occur (we discuss how to rebalance in a moment). Deletion is accomplished lazily, by marking elements as inactive, maintaining a count of the number of inactive elements in each subtree, and rebuilding any subtree x for which $(2\alpha - 1)s(x)$ of its elements have become inactive (if multiple subtrees satisfy this condition, we rebuild the highest one); this can be done using essentially the same mechanism we use for rebuilding in the event that elements become non-α-weight balanced.

For rebalancing, we must take some care because the standard method of rebalancing a BST in linear time exploits the fact that we can obtain a sorted ordering of its elements via an inorder traversal. Fortunately, since our structure is relaxed, we do not need to know the full sorted ordering of our elements in order to rebalance. The following procedure will make any k-element subtree $1/2$-weight-balanced in $O(k)$ time:

1. Select the median element m in the subtree in $O(k)$ time.
2. Partition the $k - 1$ remaining elements about m into two sets $S_<$ and $S_>$ such that $|S_<|$ and $|S_>|$ differ by at most one.
3. Make m the root of the rebalanced subtree, setting its right subtree to be the doubly-linked list containing elements in $S_>$.
4. Recursively build a $1/2$-weight-balanced subtree from $S_<$ and set this to be m's left subtree.

For an even simpler algorithm, we could choose m randomly instead of using a deterministic median-finding algorithm. If we do this, our running time bounds will all still hold in expectation.

Lemma 1. *In an α-weight-balanced tree with n elements, the depth of a rank-r element is at most $\log_{1/\alpha}(n/r)$.*

Proof. Let $k = \lceil \log_{1/\alpha}(n/r) \rceil$, and let $x_1, x_2, \ldots$ denote the elements down the left spine of an n-element α-weight-balanced tree, with x_1 being the root. Since $s(x_{i+1}) \leq \alpha s(x_i)$, we have $s(x_{k+1}) \leq \alpha^k s(x_1) = \alpha^k n$, so at least the $1 + (1 - \alpha^k)n \geq 1 + (1 - (r/n))n = 1 + n - r$ largest elements live at depth at most k in our tree, and this set includes the rank-r element.

Theorem 2. *The* insert *and* delete *operations in the structure above run in* $O(\log(n/r))$ *amortized time.*

Proof. To amortize the cost of rebalancing properly, let $c(x)$ denote the number of insertions and deletions in x's subtree after the last time x was made to be $1/2$-weight-balanced, and let us define a potential function $\Phi = \frac{1}{2\alpha-1} \sum_x c(x)$. The amortized cost of an operation is given by its actual ("immediate") cost plus any resulting change in potential. Assume for a moment that no rebalancing takes place. For *insert*, the preceding lemma tells us that the actual cost is $O(\log(n/r))$, and we also add up to $\frac{1}{2\alpha-1} \log_{1/\alpha}(n/r)$ units of potential. For *delete*, the actual cost is $O(1)$, and again we add up to $\frac{1}{2\alpha-1} \log_{1/\alpha}(n/r)$ new units of potential. Now consider the case where we rebalance; this can occur either if (i) some element x becomes non-α-weight-balanced, or (ii) if some element x is found to contain at least $(2\alpha - 1)s(x)$ inactive elements in its subtree. For case (i) to occur, we must have $|s(left(x)) - s(right(x))| \geq (2\alpha - 1)s(x)$, and since $c(x) \geq |s(left(x)) - s(right(x))|$, we find that in both (i) and (ii), we always have $c(x) \geq (2\alpha - 1)s(x)$ when rebalancing occurs at x, so the decrease in potential caused by setting $c(x) = 0$ is at least $s(x)$, the amount of work required to rebalance. Rebalancing is therefore essentially "free" (in the amortized sense), since we can pay for it using previously-generated credit invested in our potential function.

It is worth noting the similarity between the amortized rank-sensitive priority queue above and the well-studied *radix heap* [1]. Radix heaps are RAM data structures that store integer-valued keys in a fixed known range, but their operation is quite similar to our amortized structure above — they also leave right subtrees unbuilt, and perform periodic rebalancing when the minimum value in the heap reaches a specific threshold. In fact, one might wish to think of our structure as a natural comparison-based analog of the radix heap. Another related structure worth considering is the *scapegoat tree* [9], a more sophisticated variant of the amortized balanced BST above that manages to avoid storing any augmented information while still achieving $O(\log n)$ height at all times. Since scapegoat trees are not always α-weight-balanced (they adhere to a slightly more relaxed notion of this property), it does not appear that one can easily modify them in the manner above to obtain an analogous rank-sensitive priority queue.

If we want to build a rank-sensitive priority queue with $O(\log(n/r))$ *worst-case* performance for *insert* and *delete*, then we can do so by "de-amortizing" the structure above in a standard mechanical fashion — rebuilds are performed at an accelerated rate, several steps at a time (but still only $O(1)$ at once), in parallel with the subsequent insertions and deletions occurring after the rebuild. We omit further details until the full version of this paper. Since the resulting data structure is rather clumsy, it remains an interesting open question whether or not there is a simple and more elegant method to obtain $O(\log(n/r))$ worst-case bounds.

4   Lower Bounds

We now argue that $O(\log(n/r))$ is essentially the best bound one can hope to achieve, in the comparison model, for deletion in a rank-sensitive priority queue. It is slightly challenging to find the "right" statement of this lower bound, however. Suppose we fix a value of $\rho \in (0, 1]$ and consider an access sequence S of *insert* and *delete* operations in a dynamic min-aware data structure, all involving elements for which $r/n \leq \rho$. We would like to claim that the average cost of a *delete* operation in our access sequence must be $\Omega(\log(1/\rho))$ in the worst case, in the comparison model (henceforth, we assume we are in the comparison model). Unfortunately, this claim is trivially true since it holds for the special case where S is the access sequence arising when we use our structure to sort n elements — n *deletions* of the rank $r = 1$ element. Moreover, if we try to remedy this problem by considering only access sequences without any deletions at rank $r = 1$, then the claim above becomes false because now our data structure can now be assured that it will never need to remove the minimum element, so it can "cheat" and use less work maintaining the current minimum than it would normally need to do (e.g., it could simply maintain a pointer to the current minimum that is reset whenever a new minimum element is inserted, thereby supporting both *insert* and *delete* in $O(1)$ time).

In order to obtain a meaningful lower bound, we therefore need to consider access sequences in which deletion of a rank-1 element is possible (just to "keep the data structure on its toes" and make it do an honest amount of work in maintaining the current minimum), but where we cannot allow so many rank-1 deletions that we aggravate the comparison-based sorting lower bound and obtain a trivial $\Omega(\log n)$ worst-case lower bound per deletion that does not incorporate rank. Our solution is to consider access sequences of the following form:

Definition 1. *An access sequence S of insertions and deletions containing k deletions is ρ-graded if all k deletions S satisfy $r/n \leq \rho$, and if for every $a \geq 1$, at most k/a deletions satisfy $r/n \leq \rho/a$.*

For example, in a $\frac{1}{8}$-graded sequence, all deletions operate on elements having rank $r \leq n/8$ (i.e,. elements in the smallest $1/8$ portion of the structure), at most half the deletions can remove elements of rank $r \leq n/16$, at most a quarter can remove elements of rank $r \leq n/32$, and at most an $8/n$ fraction of these deletions can involve a rank-1 element.

A key ingredient we will need soon is a *d-limit heap* — a priority queue from which we promise to call *delete-min* at most d times. It is reasonably straightforward to implement a d-limit heap that can process x calls to *insert* and y calls to *delete-min* in $O(x) + y \log_2 d + o(y \log d)$ time; details will appear in the full version of this paper. Another ingredient we need is that for any dynamic min-aware structure M, we can instrument M so as to maintain the directed acyclic graph $G(M)$ of all its comparisons to date. The DAG $G(M)$ contains a directed edge (x, y) after M directly compares x and y and finds $x < y$. Elements inserted and removed from M are correspondingly inserted and removed from $G(M)$. If M makes c comparisons during its lifetime, then maintenance of $G(M)$ can be achieved in only $O(c)$ time, thereby leaving the asymptotic running time of M unchanged.

Lemma 2. *In any dynamic min-aware structure M, if all elements previously removed from M are smaller than those presently in M, then every non-minimal element of M must have positive in-degree in $G(M)$.*

Proof. If this were not the case, then a non-minimal element with zero in-degree would have no way of certifying its non-minimality.

The main result of this section is now the following.

Theorem 3. *For $\rho \in (0, 1/2)$, in the comparison model, any dynamic min-aware data structure (starting out empty) must spend $\Omega(|S| + n \log(1/\rho))$ worst-case time processing a ρ-graded sequence S containing n deletions.*

Proof. We show how to sort a set E containing n elements by using a small number of comparisons plus the execution of a carefully-crafted ρ-graded sequence S containing $O(n)$ deletions on a dynamic min-aware data structure M. Since the sorting problem requires at least $n \log_2 n - o(n \log n)$ comparisons in the worst case, we will be able to show that at least $\Omega(|S| + n \log(1/\rho))$ comparisons must come from M. The main challenge in our reduction is to sort with only a limited number of rank-1 deletions. To do this, we initially insert into M the contents of E followed by a set D of $d = \frac{\rho n - 1}{1 - \rho} \leq \frac{\rho}{1-\rho} n \leq n$ dummy elements that are all taken to be smaller than the elements in E. Note that the element of relative rank ρ in our structure has rank $\rho(n + d) = d + 1$, so it is the smallest of the elements in E. We maintain pointers to the d dummy elements, since we will occasionally delete and re-insert them. We now sort the contents of E by enacting n/d rounds, each of which involves these steps:

1. Initialize a new d-limit heap H.
2. For each dummy element $e \in D$, largest to smallest, delete and then re-insert e, taking the new values of the dummy elements to be larger than all the old values (but still smaller than the elements of E), so we satisfy the conditions of Lemma 2. In this step and for the rest of the phase, any element $e \in M \backslash D$ which acquires a new incoming edge in $G(M)$ from some element of D is inserted in H.
3. Repeat the following d times: call *delete-min* in H to obtain element e (the smallest remaining element in E, thanks to Lemma 2). Then delete e from M and add all e's immediate successors in $G(M)$ to H (elements e' for which (e, e') is an edge). Finally, insert a large dummy element at the end of M to keep the total element count equal to $n + d$.
4. Destroy H.

We claim that the sequence of operations we perform in M above is ρ-graded — half the deletions occur at relative rank ρ, and the other half are spread over the range $[0, \rho]$. The elements we remove from H over all n/d phases give us the contents of E in sorted order, so the procedure above must indeed require $n \log_2 n - o(n \log n)$ comparisons in the worst case. Letting c denote the total number of comparisons made by M during our procedure, we note that the total number of comparisons we make outside the operation of M is bounded by $O(c)$ except for those made by H. If we include the comparisons made by H, we find that the total number of comparisons is $O(c) + n \log_2 d + o(n \log d)$. Since this must be at least $n \log_2 n - o(n \log n)$, we have $c = \Omega(n \log n/d) = \Omega(n \log \frac{1-\rho}{\rho}) = \Omega(n \log(1/\rho))$.

5   Case Study: Shortest Path Computation

Rank-sensitive priority queues are worth considering in conjunction with Dijkstra's shortest path algorithm since they may support a fast *decrease-key* operation in practice. To decrease the key of an element, we delete it and re-insert it with a new value, and as long as this new value gives the element a new rank sufficiently far from the minimum, we expect the entire operation to run quite fast; for many shortest path instances, we expect most of the *decrease-key* invocations to run essentially in constant time, which gives the rank-sensitive priority queue the potential for matching the performance in practice of more sophisticated priority queues, such as Fibonacci heaps.

To evaluate the utility of rank-sensitive priority queues for shortest path computation, we implemented the structures outlined in Sections 2 and 3 and compared them with binary heaps and Fibonacci heaps on a variety of shortest path instances. We also tested a variant of the amortized balanced structure from Section 3 in which rebalancing was turned off, in which we only rebalanced at the lower end of the left spine in response to removal of the minimum; owing to the "well-behaved" structure of most of our inputs, this structure actually tended to perform better in practice than its counterpart with rebalancing enabled. Our implementations of the amortized balanced structures perform rebalancing of a subtree by partitioning on a randomly-chosen element, rather than by finding the median deterministically. Implementations of Dijkstra's algorithm using a binary heap and Fibonacci heap were obtained from the widely-used "splib" library [4], and inputs and random network generators were obtained from the DIMACS shortest path implementation challenge. All computational experiments were run on a 1GHz Opteron processor with 4GB memory.

Figure 3 illustrates our computational results. The graphs we tested were (a) the USA road network, with roughly 23 million nodes and 58 million edges, (b) 2d grids, (c) random graphs $G_{n,m}$ with $n = 10,000$ and m ranging from $100,000$ up to 100 million, and (d) random Euclidean graphs (defined by n nodes embedded in the 2d Euclidean plane, with edge length equal to squared Euclidean distance) with $10,000$ nodes and 1 to 100 million edges. On all of our tests, the rank-sensitive structures demonstrated performance slightly worse than a

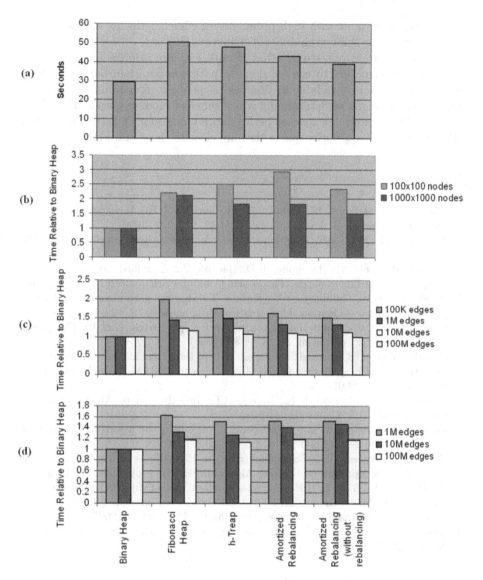

Fig. 3. Performance of rank-sensitive priority queues versus binary heaps and Fibonacci heaps on (a) the USA road network, (b) random grid graphs, (c) random $G_{n,m}$ graphs, and (d) random Euclidean graphs

binary heap and slightly better than a Fibonacci heap, with the performance gap versus the binary heap narrowing as our graphs become more dense. The random edge lengths in our grids (b) and random graphs (c) are chosen independently from a common uniform distribution, so we should not be surprised to see the standard binary heap perform so well even on dense graphs, since Goldberg and Tarjan have shown that Dijkstra's algorithm only performs $O(n \log n)$

decrease-key operations with high probability if edge lengths are independently generated from the same probability distribution [10].

We conclude from our studies that a rank-sensitive priority queue is a respectable data structure to use for shortest path computation, but most likely not the fastest choice available in practice. A possibly interesting question for future research might be determining which types of random graphs (if any) allow us to obtain provable expected performance bounds for Dijkstra's algorithm with a rank-sensitive priority queue that are close to those of a Fibonacci heap.

References

1. Ahuja, R.K., Melhorn, K., Orlin, J.B., Tarjan, R.E.: Faster algorithms for the shortest path problem. Journal of the ACM 37, 213–222 (1990)
2. Aragon, C.R., Seidel, R.: Randomized search trees. Algorithmica 16, 464–497 (1996)
3. Bădoiu, M., Cole, R., Demaine, E.D., Iacono, J.: A unified access bound on comparison-based dictionaries. Theoretical Computer Science 382(2), 86–96 (2007)
4. Cherkassky, B.V., Goldberg, A.V., Radzik, T.: Shortest path algorithms: Theory and experimental evaluation. Mathematical Programming 73, 129–174 (1996)
5. Cole, R.: On the dynamic finger conjecture for splay trees. part II: The proof. SIAM Journal on Computing 30(1), 44–85 (2000)
6. Cole, R., Mishra, B., Schmidt, J., Siegel, A.: On the dynamic finger conjecture for splay trees. part I: Splay sorting log n-block sequences. SIAM Journal on Computing 30(1), 1–43 (2000)
7. Cormen, T.H., Leiserson, C.E., Rivest, R.L.: Introduction to Algorithms. MIT Press/McGraw-Hill, Cambridge (1990)
8. Fischer, M.J., Paterson, M.S.: Fishspear: A priority queue algorithm. Journal of the ACM 41(1), 3–30 (1994)
9. Galperin, I., Rivest, R.L.: Scapegoat trees. In: Proceedings of the fourth annual ACM-SIAM Symposium on Discrete Algorithms (SODA), pp. 165–174 (1993)
10. Goldberg, A.V., Tarjan, R.E.: Expected performance of dijkstra's shortest path algorithm. Technical Report TR-96-063, NEC Research Institute (1996)
11. Iacono, J., Langerman, S.: Queaps. Algorithmica 42(1), 49–56 (2005)
12. Sleator, D.D., Tarjan, R.E.: Self-adjusting binary search trees. Journal of the ACM 32(3), 652–686 (1985)

Algorithms Meet Art, Puzzles, and Magic

Erik D. Demaine

MIT Computer Science and Artificial Intelligence Laboratory
32 Vassar Street, Cambridge, MA 02139, USA
edemaine@mit.edu

Abstract. Two years before WADS began, my father Martin Demaine and my six-year-old self designed and made puzzles as the *Erik and Dad Puzzle Company*, which distributed to toy stores across Canada. So began our journey into the interactions between algorithms and the arts. More and more, we find that our mathematical research and artistic projects converge, with the artistic side inspiring the mathematical side and vice versa. Mathematics itself is an art form, and through other media such as sculpture, puzzles, and magic, the beauty of mathematics can be brought to a wider audience. These artistic endeavors also provide us with deeper insights into the underlying mathematics, by providing physical realizations of objects under consideration, by pointing to interesting special cases and directions to explore, and by suggesting new problems to solve (such as the metapuzzle of how to solve a puzzle). This talk will give several examples in each category, from how our first font design led to a universality result in hinged dissections, to how studying curved creases in origami led to sculptures at MoMA. The audience will be expected to participate in some live magic demonstrations.

F. Dehne et al. (Eds.): WADS 2009, LNCS 5664, p. 193, 2009.

Skip-Splay: Toward Achieving the Unified Bound in the BST Model

Jonathan C. Derryberry and Daniel D. Sleator

Computer Science Department
Carnegie Mellon University

Abstract. We present skip-splay, the first binary search tree algorithm known to have a running time that nearly achieves the unified bound. Skip-splay trees require only $O(m \lg \lg n + UB(\sigma))$ time to execute a query sequence $\sigma = \sigma_1 \ldots \sigma_m$. The skip-splay algorithm is simple and similar to the splay algorithm.

1   Introduction and Related Work

Although the worst-case access cost for comparison-based dictionaries is $\Omega(\lg n)$, many sequences of operations are highly nonrandom, allowing tighter, instance-specific running time bounds to be achieved by algorithms that adapt to the input sequence. Splay trees [1] are an example of such an adaptive algorithm that operates within the framework of the binary search tree (BST) model [2], which essentially requires that all elements be stored in symmetric order in a rooted binary tree that can only be updated via rotations, and requires queried nodes to be rotated to the root. (BST algorithms that do not rotate to the root can usually be coerced into this model with just a constant factor of overhead.)

The two most general bounds proven for splay trees are the working set bound [1] and the dynamic finger bound [3], [4]. The working set bound shows that splay trees can have better than $O(\lg n)$ cost per operation when recently accessed elements are much more likely to be accessed than random elements, while the dynamic finger bound shows that splay trees have better than $O(\lg n)$ performance when each access is likely to be near the previous access.

Iacono later introduced the unified bound, which generalized both of these two bounds [5]. Roughly, a data structure that satisfies the unified bound has good performance for sequences of operations in which most accesses are likely to be near a recently accessed element. More formally, suppose the access sequence is $\sigma = \sigma_1 \ldots \sigma_m$ and each access σ_j is a query to the set $\{1, \ldots, n\}$ (we also use σ_j to refer to the actual element that is queried, as context suggests). The unified bound can be defined as follows:

$$UB(\sigma) = \sum_{j=1}^{m} \min_{j' < j} \lg(w(\sigma_{j'}, j') + |\sigma_{j'} - \sigma_j|), \qquad (1)$$

where $w(x, j)$ is, at time j, the number of distinct elements including x that have been queried since the previous query to x, or n if no such previous query exists. For a more formal definition, see the definitions that precede Lemma 1.

F. Dehne et al. (Eds.): WADS 2009, LNCS 5664, pp. 194–205, 2009.

To achieve a running time of $O(m + UB(\sigma))$, Iacono introduced a data structure called the unified structure. The unified structure did not require amortization to achieve this bound, and was later improved by Bădoiu et al. to allow insertion and deletion [6]. The unified structure was comparison-based but did not adhere to the BST model. Thus, in addition to leaving open questions regarding how powerful the BST model was, it was not clear, for example, how to achieve the unified bound while keeping track of aggregate information on subsets of elements as can be done with augmented BSTs.

These unresolved issues motivate the question of whether a BST algorithm exists that achieves the unified bound. Achieving this goal contrasts with the separate pursuit of a provably dynamically optimal BST algorithm in that it is possible for a data structure that achieves the unified bound to have the trivial competitive ratio of $\Theta(\lg n)$ to an optimal BST algorithm. Conversely, prior to this work, even if a dynamically optimal BST algorithm had been found, it would not have been clear whether it satisfied the unified bound to within any factor that was $o(\lg n)$ since dynamic optimality by itself says nothing about actual formulaic bounds, and prior to this work no competitive factor better than $O(\lg n)$ was known for the cost of the optimal BST algorithm in comparison to the unified bound. See [7], [8], and [9] for progress on dynamic optimality in the BST model.

The skip-splay algorithm presented in this paper has three important qualities. First, it conforms to the BST model and has a running time of $O(m \lg \lg n + UB(\sigma))$, just an additive term of $O(\lg \lg n)$ per query away from the unified bound. Thus, skip-splay trees nearly close the gap between what is known to be achievable in the BST model and what is achieved by the unified structure. Second, the skip-splay algorithm is very simple. The majority of the complexity of our result resides in the analysis of skip-splaying, not in the design of the algorithm itself. The unified structure, though it avoids the additional $O(\lg \lg n)$ cost per query, is significantly more complicated than skip-splay trees. Finally, skip-splaying is almost identical to splaying, which suggests that a similar analysis, in combination with new insight, might be used to prove that splay trees satisfy the unified bound, at least to within some nontrivial multiplicative factor or additive term.

2 The Skip-Splay Algorithm

We assume for simplicity that a skip-splay tree T stores all elements of $\{1, \ldots, n\}$ where $n = 2^{2^{k-1}} - 1$ for some positive integer k, and that T is initially perfectly balanced. We mark as a splay tree root every node whose height (starting at a height of 1 for the leaves) is 2^i for $i \in \{0, \ldots, k-1\}$.[1] Note that the set of all of these splay trees partitions the elements of T.

---

[1] If we allow the ratio between the initial heights of successive roots to vary, we can achieve a parameterized running time bound, but in this version of the paper we use a ratio of 2 for simplicity.

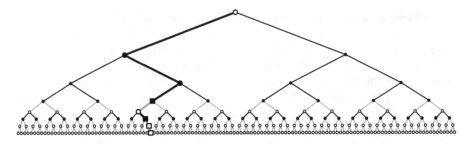

Fig. 1. An example of a four-level skip-splay tree T at the beginning of a query sequence. The white nodes are the roots of the splay trees that make up T, and the gray edges are never rotated. If the bottom element of the bold path is queried, then each of the boxed nodes is splayed to the root of its splay tree.

The following definitions will help us describe the algorithm more clearly:

1. Let T_i be the set of all keys x whose path to the root of T contains at most i root nodes, including x itself if x is marked as a root.
2. Define level i of T to be the set of keys x whose path to the root contains *exactly* i nodes. We will sometimes use the adjective "level-i" to refer to objects associated with level i in some way.
3. Let $tree(x)$ be the splay tree that contains x. Also, $tree(x)$ can represent the set of elements in $tree(x)$.

We assume that all operations are queries, and we use $\sigma = \sigma_1 \ldots \sigma_m$ to denote the sequence of queries. To query an element σ_j, we first perform binary search through T to locate σ_j. Then, we splay σ_j to the root of $tree(\sigma_j)$ and transfer the relevant root marker to σ_j. If we are at the root of T, we terminate, else we "skip" to σ_j's new parent x and repeat this process by splaying x to the root of $tree(x)$. The cost of a query is defined to be the number of nodes on the access path to σ_j.[2] Figure 1 shows an example of what a skip splay tree looks like at the beginning of an access sequence and depicts how a query is performed.

Intuitively, skip-splaying is nearly competitive to the unified bound because if the currently queried element σ_j is near to a recently queried element σ_f, then many of the elements that are splayed while querying σ_j are likely to be the same as the ones that were splayed when σ_f was queried. Therefore, by the working set bound for splay trees, these splays should be fairly cheap. The analysis in Section 3 formalizes this intuition.

3   Proving Skip-Splay Runs in Time $O(m \lg \lg n + UB(\sigma))$

Our analysis in this section consists of three lemmas that together prove that skip-splay trees run in time $O(m \lg \lg n + UB(\sigma))$. The purpose of the first lemma

---

[2] Note that this algorithm can be coerced into the BST Model defined in [2] by rotating σ_j to the root and back down, incurring only a constant factor of additional cost.

is to decompose the cost of skip-splay trees into a series of "local working set costs" with one cost term for each level in T. The second lemma is the main step of the analysis and it uses the first lemma to prove that skip-splay trees satisfy a bound that is very similar to the unified bound, plus an additive $O(\lg \lg n)$ term. The third lemma shows that this similar bound is within a constant factor of the unified bound, so our main analytical result, that skip-splay trees run in $O(m \lg \lg n + UB(\sigma))$ time, follows immediately from these three lemmas.

In the first lemma and in the rest of this paper, we will use the following custom notation for describing various parts of T:

1. Let $\rho_k = 1$ and for $i < k$ let $\rho_i = 2^{2^{k-i-1}}$ so that $\rho_i = \rho_{i+1}^2$ for $i < k - 1$. Note that if element $x \in T$ is in level i for $i < k$, then $|tree(x)| = \rho_i - 1$.
2. Let $R_i(x)$, the level-i region of $x \in T$ be defined as follows. First, define the offset $\delta_i = \delta \bmod \rho_i$, where δ is an integer that is arbitrary but fixed for all levels of T. (Our analysis will later make use of the fact that we can choose δ to be whatever we want.) Then, let $R_i(x) = R_i^*(x) \cap T$ where

$$R_i^*(x) = \left\{ \left\lfloor \tfrac{x+\delta_i}{\rho_i} \right\rfloor \rho_i - \delta_i, \ldots, \left\lfloor \tfrac{x+\delta_i}{\rho_i} \right\rfloor \rho_i - \delta_i + \rho_i - 1 \right\}.$$

Note that the level-i regions partition the elements of T and the level-$i+1$ regions are a refinement of the level-i regions. Two regions R and R' are said to be *adjacent* if they are distinct, occupy the same level, and their union covers a contiguous region of keyspace. Note that $|R_i(x)| = \rho_i$ if $R_i^*(x) \subseteq T$.
3. Let $\mathcal{R}_i(x)$, the level-i region set of x, be the set of level-i regions that are subsets of $R_{i-1}(x)$ with $\mathcal{R}_1(x)$ defined to be the set of all level-1 regions. Note that $|\mathcal{R}_i(x)| = \rho_i$ if $1 < i < k$ and $R_{i-1}^*(x) \subseteq T$.

Additionally, we give the following definitions of working set numbers and some auxiliary definitions that will also be helpful (these definitions assume we are working with a fixed query sequence σ):

1. Let $splays(j)$ be the set of elements that are splayed during query σ_j.
2. Let $p(x, j)$ represent the index of the previous access to x before time j. More formally, assuming such an access exists, let

$$p(x, j) = \max(\{1, \ldots, j - 1\} \cap \{j' \mid \sigma_{j'} = x\}).$$

We define $p(x, j) = -n$ if the argument to max is the empty set.
3. Let $p'(x, j)$ represent the index of the previous access that resulted in a splay to x before time j. More formally, assuming such an access exists, let

$$p'(x, j) = \max(\{1, \ldots, j - 1\} \cap \{j' \mid x \in splays(j')\}).$$

We define $p'(x, j) = -\rho_i$ if the argument to max is the empty set.
4. Let $p_i(x, j)$ represent the index of the previous access to region $R_i(x)$. More formally, assuming such an access exists, let

$$p_i(x, j) = \max(\{1, \ldots, j - 1\} \cap \{j' \mid R_i(\sigma_{j'}) = R_i(x)\}).$$

We define $p_i(x, j) = -\rho_i$ if the argument to max is the empty set. Also, let $p_i(R, j)$ be equivalent to $p_i(x, j)$ if $R = R_i(x)$.

5. For $x \in T$, let $w(x, j)$ represent the number of elements queried since the previous access to x. More formally, if $p(x, j) > 0$ let

$$w(x, j) = \left| \left\{ \sigma_{j'} \mid j' \in \{p(x, j), \dots, j - 1\} \right\} \right|.$$

Else, if $p(x, j) \leq 0$ then let $w(x, j) = -p(x, j)$.

6. For $x \in T$, let $w'(x, j)$ represent the working set number of x within $tree(x)$ (i.e., the number of elements splayed in $tree(x)$ since the previous query resulting in a splay to x). More formally, if $p'(x, j) > 0$ let

$$w'(x, j) = \left| tree(x) \cap \bigcup_{j' \in \{p'(x,j),\dots,j-1\}} splays(j') \right|.$$

Else, if $p'(x, j) \leq 0$ then let $w'(x, j) = -p'(x, j)$.

7. For $x \in T$, let $w_i(x, j)$ represent the number of regions in $\mathcal{R}_i(x)$ that contain a query since the previous access to a member of $R_i(x)$. More formally, if $p_i(x, j) > 0$ let

$$w_i(x, j) = \left| \left\{ R_i(\sigma_{j'}) \mid j' \in \{p_i(x, j), \dots, j - 1\} \right\} \cap \mathcal{R}_i(x) \right|.$$

Else, if $p_i(x, j) \leq 0$ then let $w_i(x, j) = -p_i(x, j)$. Also, let $w_i(R, j)$ be equivalent to $w_i(x, j)$ if $R = R_i(x)$.

8. For $x \in T$, let $w'_i(x, j)$ be the working set number of x within $tree(x)$ that is reset whenever a query is executed to a region that could cause a splay of x. More formally, let $\mathcal{R}(x)$ be the set of up to three regions R such that a query to R can cause a splay of x. If $p_i(R, j) > 0$ for some $R \in \mathcal{R}(x)$ let

$$w'_i(x, j) = \left| tree(x) \cap \bigcup_{j' \in \{\max_{R \in \mathcal{R}(x)} p_i(R,j),\dots,j-1\}} splays(j') \right|.$$

Else, if $p_i(R, j) \leq 0$ for all $R \in \mathcal{R}(x)$ then let $w'_i(x, j) = \rho_i$. Note that $w'_i(x, j) \leq 3w_i(R, j) + 1$ for $R \in (\mathcal{R}(x) \cap \mathcal{R}_i(x))$ because accesses to a region in $\mathcal{R}_i(x)$ can result in splays of at most three different elements of $tree(x)$, and at most one, the minimum element of $tree(x)$, can be splayed as the result of a query to another level-i region set. Also, note that $w'_i(x, j) \leq w'(x, j)$.

In the proof of the first lemma, we will be making use of the working set theorem in Sleator and Tarjan's original splay tree paper [1], which shows that the cost of a query sequence σ on an individual splay tree, for sufficiently large n, is bounded by $c_s(n \lg n + \sum_{j=1}^{m} \lg(w(\sigma_j, j) + 1))$, for some constant c_s. For simplicity, we assume we are starting with a minimum potential arrangement of each splay tree, so this simplifies to $\sum_{j=1}^{m} c_s \lg(w(\sigma_j, j) + 1)$. In order to make the analysis in Lemma 2 simpler, we move beyond simply associating this working set cost with each splay that is executed in T by proving the following lemma.

Lemma 1. *For query sequence σ in a skip-splay tree T with k levels, the amortized cost of query σ_j is*

$$O\left(k + \sum_{i=1}^{k} \lg w_i(\sigma_j, j)\right). \tag{2}$$

Proof. By the definition of the skip-splay algorithm and the working set theorem for splay trees, the amortized cost of query σ_j is $\sum_{x \in splays(j)} w'(x, j)$, suppressing multiplicative and additive constants. To prove Lemma 1, we will do further accounting for the cost of a query σ_j and focus on the cost associated with an arbitrary level i of T.

Note that at level i during query σ_j, one of three cases occurs with regard to which level-i node, if any, is splayed. First, if σ_j resides in a strictly shallower level than i, then no splay is performed in level i. Second, if σ_j resides within level i, then σ_j is splayed in level i. Third, if σ_j resides in a deeper level than i, then either the predecessor or the successor of σ_j in level i is splayed. (We know that at least one of these two nodes exists and is on the access path in this case.) We will use the following potential function on T to prove that the bound in Equation 2 holds regardless of which of these three cases occurs:

$$\Phi(T, j) = \phi_1(T, j) + \phi_2(T, j), \tag{3}$$

where

$$\phi_1(T, j) = \sum_{x \in T} (\lg w'(x, j+1) - \lg w_i'(x, j+1)) \tag{4}$$

and

$$\phi_2(T, j) = \sum_{(x,y) \in A} |\lg w_i'(x, j+1) - \lg w_i'(y, j+1)|, \tag{5}$$

where A is the set of pairs of level-i nodes (x, y) such that x is the maximum element in $tree(x)$, y is the minimum element in $tree(y)$, and there are no other level-i elements between x and y. For succinctness below, define $\Delta\Phi(T, j)$ to be $\Phi(T, j) - \Phi(T, j-1)$ and define $\Delta\phi_1(T, j)$ and $\Delta\phi_2(T, j)$ analogously.

First, notice that the cost of the splay, if any, that is performed on node x at level i is offset by the change in potential of $\lg w'(x, j+1) - \lg w'(x, j) = -\lg w'(x, j)$. Note that this ignores the difference $\lg w_i'(x, j) - \lg w_i'(x, j+1) = \lg w_i'(x, j)$, which will be accounted for below.

Second, define $\Delta_+\Phi(T, j)$ to be the sum of the positive terms of $\Delta\Phi(T, j)$ plus $\lg w_i'(x, j)$ in the case in which some node x is splayed during query σ_j. We will show that regardless of whether a splay is performed in level i during query σ_j, it is true that $\Delta_+\Phi(T, j)$ is at most $4\lg(3w_i(\sigma_j, j) + 1) + 2$.

To see this, let Y be the set of up to three level-i nodes that can be splayed while accessing members of the region $R_i(\sigma_j)$, and notice that if a node x is splayed at level i during query σ_j then $x \in Y$. Note that the only positive terms of $\Delta_+\Phi(T, j)$ from $\Delta\phi_1(T, j)$ are the ones that use some member of Y as an argument. This is true because $\lg w'(z, j+1) - \lg w'(z, j) \leq \lg w_i'(z, j+1) - \lg w_i'(z, j)$

for $z \in T \backslash Y$ since $w'(z,j) \geq w'_i(z,j)$ and $w'(z,j+1) - w'(z,j) \leq w'_i(z,j+1)$
$- w'_i(z,j)$. Further, note that $\Delta_+\Phi(T,j)$ contains at most two terms from
$\Delta\phi_2(T,j)$ that do not use some member of Y as an argument, and these two
terms are at most 1 each.

Now, we consider the following two cases. All additional cases are either similar
to or simpler than these two cases. First, suppose that Y contains two elements
$y_1 < y_2$ and $tree(y_1) \neq tree(y_2)$. Note that in this case we know that $\mathcal{R}_i(\sigma_j) = \mathcal{R}_i(y_1)$. Then,

$$\Delta_+\Phi(T,j) \leq \lg w'_i(y_1,j) + \lg w'_i(y_2,j) - |\lg w'_i(y_1,j) - \lg w'_i(y_2,j)| + 2$$
$$\leq 2 \lg w'_i(y_1,j) + 2$$
$$\leq 2 \lg(3 w_i(\sigma_j,j) + 1) + 2.$$

Second, suppose that Y contains three elements $y_1 < y_2 < y_3$ that all reside in
the same splay tree T', suppose y_3 is the maximum element of T', and let z be
the successor of y_3 among the level-i elements (assuming z exists in this case).
Using the fact that $|\lg w'_i(y_3,j+1) - \lg w'_i(z,j+1)| = \lg w'_i(z,j+1) = \lg w'_i(z,j)$
and the fact that $\mathcal{R}_i(y_1) = \mathcal{R}_i(y_2) = \mathcal{R}_i(y_3) = \mathcal{R}_i(\sigma_j)$, we have

$$\Delta_+\Phi(T,j) \leq \sum_{q=1}^{3} \lg w'_i(y_q,j) + \lg w'_i(z,j) - |\lg w'_i(y_3,j) - \lg w'_i(z,j)| + 2$$
$$\leq \lg w'_i(y_1,j) + \lg w'_i(y_2,j) + 2 \lg w'_i(y_3,j) + 2$$
$$\leq 4 \lg(3 w_i(\sigma_j,j) + 1) + 2. \qquad \square$$

We note that the potential function used in Lemma 1 starts at its minimum
value and the splay trees also start at their minimum potential configuration.
Therefore, the sum of the amortized costs of each query, according to Lemma 1,
is an upper bound on the cost of the sequence. Using Lemma 1, we can prove a
bound that is similar to the unified bound, plus an additive $O(\lg \lg n)$ term per
query. This bound differs from the unified bound in that the working set portion
of the cost consists not of the number of *elements* accessed since the previous
query to the relevant element, but of the number of *queries* since the previous
query to the relevant element. Before we prove this bound, we give the following
definitions, which will be useful in formally describing the bound and proving it:

1. Let f_j represent the element $\sigma_{j'}$ such that

$$j' = \underset{j'' < j}{\arg\min} \lg(w(\sigma_{j''},j) + |\sigma_j - \sigma_{j''}|).$$

   Intuitively, f_j represents the "finger" for query σ_j because it represents the
   previously-queried element that yields the smallest unified bound value for
   query σ_j.

2. For $x \in T$, let $t(x,j)$ represent the number of *queries* (rather than distinct
   elements accessed) since the previous access to x. More formally, let

$$t(x,j) = |\{p(x,j),\ldots,j-1\}| = j - p(x,j).$$

   Note that the above definition handles the case in which $p(x,j) \leq 0$.

3. For $x \in T$, let $t_i(x, j)$ represent the number of queries to all members of $\mathcal{R}_i(x)$ since the previous access to a member of $R_i(x)$. More formally, let

$$t_i(x, j) = \left| \left\{ j' \in \{\max(1, p_i(x, j)), \ldots, j - 1\} \mid R_i(\sigma_{j'}) \in \mathcal{R}_i(x) \right\} \right|,$$

with an additional $-p_i(x, j)$ added if $p_i(x, j) \leq 0$.

4. For $x \in T$, let $\hat{t}_i(x, j)$ represent the number of queries to all members of $\mathcal{R}_i(x)$ since the previous access to x. More formally, let

$$\hat{t}_i(x, j) = \left| \left\{ j' \in \{\max(1, p(x, j)), \ldots, j - 1\} \mid R_i(\sigma_{j'}) \in \mathcal{R}_i(x) \right\} \right|,$$

with an additional ρ_i^2 added if $p(x, j) \leq 0$. Note that $\hat{t}_1(x, j) \leq t(x, j) + 1$ by definition.

Next, we define $UB'(\sigma)$, a variant of the unified bound, as

$$UB'(\sigma) = \sum_{j=1}^{m} \lg(t(f_j, j) + |\sigma_j - f_j|), \tag{6}$$

and we are ready to proceed with our second lemma.

Lemma 2. *Executing the skip-splay algorithm on query sequence $\sigma = \sigma_1 \ldots \sigma_m$ costs time $O(m \lg \lg n + UB'(\sigma))$.*

Proof. In this proof, we will be making use of the bound in Lemma 1 with a randomly chosen offset δ that is selected uniformly at random from $\{0, \ldots, \rho_1 - 1\}$. We will use induction on the number of levels i from the top of the tree while analyzing the expected amortized cost of an arbitrary query σ_j. In the inductive step, we will prove a bound that is similar to the one in Lemma 2, and this similar bound will cover the cost associated with levels i and deeper. Even though we are directly proving the inductive step in expectation only, because the bound in Lemma 1 is proven for all values of δ, we know that there exists at least one value of δ such that the bound holds *without* using randomization if we amortize over the entire query sequence. Therefore, the *worst-case* bound on the *total* cost of the access sequence in Lemma 2 will follow.

Our inductive hypothesis is that the cost of skip-splaying σ_j that is associated with levels $i + 1$ and deeper according to Lemma 1 is at most

$$\alpha \lg \hat{t}_{i+1}(f_j, j) + \beta \lg \min(1 + |\sigma_j - f_j|^2, \rho_{i+1}) + \gamma(k - i), \tag{7}$$

where k, as before, represents the number of levels of splay trees in T.

We choose levels k and $k - 1$ to be our base cases. The inductive hypothesis is trivially true for these base cases as long as we choose the constants appropriately. Also, the bound for the inductive hypothesis at level 1, summed over all queries, is $O(m \lg \lg n + UB'(\sigma))$, so proving the inductive step suffices to prove the lemma.

To prove the inductive step, we assume Equation 7 holds for level $i + 1$ and use this assumption to prove the bound for level i. Thus, our goal is to prove the following bound on the cost that Lemma 1 associates with query σ_j for levels i and deeper:

$$\alpha \lg \hat{t}_i(f_j, j) + \beta \lg \min(1 + |\sigma_j - f_j|^2, \rho_i) + \gamma(k - i + 1). \tag{8}$$

As a starting point for the proof of the inductive step, Lemma 1 in addition to the inductive hypothesis allows us to prove an upper bound of

$$\lg w_i(\sigma_j, j) + \alpha \lg \hat{t}_{i+1}(f_j, j) + \beta \lg \min(1 + |\sigma_j - f_j|^2, \rho_{i+1}) + \gamma(k - i), \tag{9}$$

where we have suppressed the constant from Lemma 1 multiplying $\lg w_i(\sigma_j, j)$.

Our proof of the inductive step consists of three cases. First, if $|\sigma_j - f_j|^2 \geq \rho_i$, then substituting ρ_i for ρ_{i+1} increases the bound in Equation 9 by

$$\lg \rho_i - \lg \rho_{i+1} = \lg \left(\frac{\rho_i}{\rho_{i+1}} \right) = \lg (\rho_{i+1}) = \lg \left(\rho_i^{1/2} \right) \geq \lg \left(w_i(\sigma_j, j)^{1/2} \right), \tag{10}$$

which offsets the elimination of the cost $\lg w_i(\sigma_j, j)$ as long as $\beta \geq 2$. The other substitutions only increase the bound, so for this case we have proved the inductive step.

Second, if $|\sigma_j - f_j|^2 < \rho_i$ and $R_i(\sigma_j) \neq R_i(f_j)$, then we simply pay $\lg w_i(\sigma, j)$ which is at most $\lg \rho_i$. However, we note that the probability of this occurring for a random choice of δ is at most $\rho_i^{1/2}/\rho_i = \rho_i^{-1/2}$, so the expected cost resulting from this case is at most $\rho_i^{-1/2} \lg \rho_i$, which is at most a constant, so it can be covered by γ.

The third and most difficult case occurs when $|\sigma_j - f_j|^2 < \rho_i$ and $R_i(\sigma_j) = R_i(f_j)$, and we will spend the rest of the proof demonstrating how to prove the inductive step for this case. First, we note that $\lg t_i(f_j, j) \geq \lg w_i(f_j, j) = \lg w_i(\sigma_j, j)$, so we can replace $\lg w_i(\sigma_j, j)$ with $\lg t_i(f_j, j)$ and ρ_{i+1} with ρ_i in Equation 9 without decreasing the bound and prove a bound of

$$\lg t_i(f_j, j) + \alpha \lg \hat{t}_{i+1}(f_j, j) + \beta \lg \min(1 + |\sigma_j - f_j|^2, \rho_i) + \gamma(k - i). \tag{11}$$

It remains only to eliminate the term $\lg t_i(f_j, j)$ by substituting $\hat{t}_i(f_j, j)$ for $\hat{t}_{i+1}(f_j, j)$ while incurring an additional amortized cost of at most a constant so that it can be covered by γ.

Observe that if σ_j satisfies

$$\hat{t}_{i+1}(f_j, j) \leq \frac{\hat{t}_i(f_j, j)}{t_i(f_j, j)^{\frac{1}{2}}}, \tag{12}$$

then we have an upper bound of

$$\lg t_i(f_j, j) + \alpha(\lg \hat{t}_i(f_j, j) - \frac{\lg t_i(f_j, j)}{2}) + \beta \lg \min(1 + |\sigma_j - f_j|^2, \rho_i) + \gamma(k - i), \tag{13}$$

which would prove the inductive step if $\alpha \geq 2$. However, it is possible that $\hat{t}_{i+1}(f_j, j)$ does not satisfy the bound in Equation 12. In this latter case, we pessimistically assume that we must simply pay the additional $\lg t_i(f_j, j)$. In the rest of the proof, we show that the amortized cost of such cases is at most a constant per query in this level of the induction, so that it can be covered by the constant γ.

We first give a few definitions that will make our argument easier. A query σ_b is *R-local* if $R_i(\sigma_b) = R$. Further, if σ_b is R-local and satisfies $R_i(f_b) = R$ as well as the bound $\hat{t}_{i+1}(f_b, b) > \hat{t}_i(f_b, b)/t_i(f_b, b)^{\frac{1}{2}}$, then we define σ_b also to be *R-dense*. Note that if σ_b is R-dense then $p(f_b, b) > 0$. Finally, if σ_b additionally satisfies the inequality $\tau < t_i(f_b, b) \leq 2\tau$, then we define σ_b also to be *R-τ-bad*. Notice that all queries that have an excess cost at level i due to being in this third case and not meeting the bound in Equation 12 are R-τ-bad for some level-i region R and some value of τ (actually a range of values τ).

Our plan is to show that the ratio of R-τ-bad queries to R-local queries is low enough that the sum of the excess costs associated with the R-τ-bad queries can be spread over the R-local queries so that each R-local query is only responsible for a constant amount of these excess costs. Further, we show that if we partition the R-dense queries by successively doubling values of τ, with some constant lower cutoff, then each R-local query's share of the cost is exponentially decreasing in $\lg \tau$, so each R-local query bears only a constant amortized cost for the excess costs of all of the R-dense queries. Lastly, note that in our analysis below we are only amortizing over R-local queries for some specific but arbitrary level-i region R, so we can apply the amortization to each level-i region separately without interference.

To begin, we bound the cost associated with the R-τ-bad queries for arbitrary level-i region R and constant τ as follows. Let σ_b be the latest R-τ-bad query. First, note that the number of R-τ-bad queries σ_a where $a \in \{p(f_b, b) + 1, \ldots, b\}$ is at most $\hat{t}_i(f_b, b)/\tau$ because there are $\hat{t}_i(f_b, b)$ queries to $\mathcal{R}_i(f_b)$ in that time period, and immediately prior to each such σ_a, the previous $\tau - 1$ queries to $\mathcal{R}_i(f_b)$ are all outside of R so that $t_i(f_a, a) \geq \tau$. Second, note that because σ_b was chosen to be R-τ-bad we have

$$\hat{t}_{i+1}(f_b, b) > \frac{\hat{t}_i(f_b, b)}{t_i(f_b, b)^{1/2}} \geq \frac{\hat{t}_i(f_b, b)}{(2\tau)^{1/2}}. \tag{14}$$

Thus, the ratio of the number of R-local queries in this time period, $\hat{t}_{i+1}(f_b, b)$, to the number of R-τ-bad queries in this time period is strictly greater than

$$\frac{\hat{t}_i(f_b, b)}{(2\tau)^{1/2}} \cdot \frac{\tau}{\hat{t}_i(f_b, b)} = \left(\frac{\tau}{2}\right)^{1/2}. \tag{15}$$

The constraint that $t_i(f_a, a) \leq 2\tau$ for each of the aforementioned R-τ-bad queries σ_a implies that the excess level-i cost of each is at most $\lg(2\tau)$, so we charge each R-local query with a time index in $\{p(f_b, b) + 1, \ldots, b\}$ a cost of $\lg(2\tau)/(\frac{\tau}{2})^{1/2}$ to account for the R-τ-bad queries that occur during this time interval. Notice that we can iteratively apply this reasoning to cover the R-τ-bad queries with time indices that are at most $p(f_b, b)$ without double-charging any R-local query.

To complete the argument, we must account for all R-dense queries, not just the R-τ-bad ones for some particular value of τ. To do this, for all R-dense queries

σ_j such that $t_i(f_j, j) \leq \tau_0$, for some constant τ_0, we simply charge a cost of $\lg \tau_0$ to γ. Next, let $\tau_q = 2^q \tau_0$ for integer values $q \geq 0$. From above, we have an upper bound on the amortized cost of the R-τ_q-bad queries of $\lg(2^{q+1}\tau_0)/(2^{q-1}\tau_0)^{1/2}$, so the sum over all values of q is at most a constant and can be covered by γ. □

To complete the argument that skip-splay trees run in $O(m \lg \lg n + UB(\sigma))$ time, it suffices to show that $UB'(\sigma)$ is at most a constant factor plus a linear term in m greater than $UB(\sigma)$. Thus, the following lemma completes the proof that skip-splay trees run in time $O(m \lg \lg n + UB(\sigma))$.

Lemma 3. *For query sequence* $\sigma = \sigma_1 \ldots \sigma_m$, *the following inequality is true:*

$$\sum_{j=1}^{m} \lg(t(f_j, j) + |\sigma_j - f_j|) \leq \frac{m\pi^2 \lg e}{6} + \lg e + \sum_{j=1}^{m} 2\lg(w(f_j, j) + |\sigma_j - f_j|). \quad (16)$$

Proof. To begin, we give a new definition of a working set number that is a hybrid between $w(f_j, j)$ and $t(f_j, j)$ for arbitrary time index j. Let $h_i(f_j, j) = \max(w(f_j, j)^2, \min(t(f_j, j), j - i))$. Note that $\lg h_m(f_j, j) = 2\lg w(f_j, j)$ and $h_{-n}(f_j, j) \geq t(f_j, j)$ for all j. Also, note that if $p(f_j, j) > 0$ then $\lg h_{-n}(f_j, j) - \lg h_0(f_j, j) = 0$, else if $p(f_j, j) \leq 0$, which is true for at most n queries, then $\lg h_{-n}(f_j, j) - \lg h_0(f_j, j) \leq \lg(n^2 + n) - \lg(n^2) \leq \frac{\lg e}{n}$.

Next, note that $\lg h_i(f_j, j) - \lg h_{i+1}(f_j, j) = 0$ if $i \geq j$ or $t(f_j, j) \leq j - i - 1$ and for all j we have $\lg h_i(f_j, j) - \lg h_{i+1}(f_j, j) \leq \frac{\lg e}{w(f_j, j)^2}$. Also, we know that the number of queries for which $i < j$, $t(f_j, j) \geq j - i$, and $w(f_j, j) \leq w_0$ is at most w_0 for $w_0 \in \{1, \ldots, n\}$. This is true because each such query is to a distinct element since they all use a finger that was last queried at a time index of at most i (if two of these queries were to the same element, then the second query could use the first as a finger). If there were $w_0 + 1$ such queries, the latest such query σ_ℓ would have $w(f_\ell, j) \geq w_0 + 1$ because of the previous w_0 queries after time i to distinct elements, a contradiction. Therefore,

$$\sum_{j=1}^{m}(\lg h_i(f_j, j) - \lg h_{i+1}(f_j, j)) \leq \sum_{k=1}^{n} \frac{\lg e}{k^2} \leq \frac{\pi^2 \lg e}{6},$$

so that

$$\sum_{j=1}^{m}(\lg t(f_j, j) - 2\lg w(f_j, j)) \leq \sum_{j=1}^{m}(\lg h_{-n}(f_j, j) - \lg h_m(f_j, j)) \leq \frac{m\pi^2 \lg e}{6} + \lg e.$$

The fact that $\lg(t(f_j, j) + d) - 2\lg(w(f_j, j) + d) \leq \lg t(f_j, j) - 2\lg w(f_j, j)$ for all j and non-negative d completes the proof. □

4 Conclusions and Future Work

The ideal improvement to this result is to show that splay trees satisfy the unified bound with a running time of $O(m + UB(\sigma))$. However, achieving this ideal result could be extremely difficult since the only known proof of the dynamic finger theorem is very complicated, and the unified bound is stronger than the dynamic finger bound.

In light of this potential difficulty, one natural path for improving this result is to apply this analysis to splay trees, perhaps achieving the same competitiveness to the unified bound as skip-splay trees. Intuitively, this may work because the skip-splay algorithm is essentially identical to splaying except a few rotations are skipped to keep the elements of the tree partitioned into blocks with a particular structure that facilitates our analysis.

Additionally, it may be possible to design a different BST algorithm and show that it meets the unified bound, which would prove that we do not need to leave the BST model, and the perks such as augmentation that it provides, to achieve the unified bound. If such an algorithm is to be similar to skip-splaying, it must mix the splay trees together so that all nodes can reach constant depth.

To summarize the clearest paths for related future work, it would be significant progress to show that splay trees meet the unified bound to within any factor that is $o(\lg n)$, or to show that some BST algorithm achieves the unified bound to within better than an additive $O(\lg \lg n)$ term.

References

1. Sleator, D.D., Tarjan, R.E.: Self-adjusting binary search trees. Journal of the ACM 32, 652–686 (1985)
2. Wilber, R.: Lower bounds for accessing binary search trees with rotations. SIAM Journal on Computing 18(1), 56–67 (1989)
3. Cole, R., Mishra, B., Schmidt, J.P., Siegel, A.: On the dynamic finger conjecture for splay trees, part I: Splay sorting log n-block sequences. SIAM Journal on Computing 30(1), 1–43 (2000)
4. Cole, R.: On the dynamic finger conjecture for splay trees, part II: The proof. SIAM Journal on Computing 30(1), 44–85 (2000)
5. Iacono, J.: Alternatives to splay trees with o(log n) worst-case access times. In: Proceedings of the 12th ACM-SIAM Symposium on Discrete Algorithms, Philadelphia, PA, USA, pp. 516–522. Society for Industrial and Applied Mathematics (2001)
6. Bădoiu, M., Cole, R., Demaine, E.D., Iacono, J.: A unified access bound on comparison-based dynamic dictionaries. Theoretical Computer Science 382(2), 86–96 (2007)
7. Demaine, E.D., Harmon, D., Iacono, J., Pătraşcu, M.: Dynamic optimality—almost. SIAM Journal on Computing 37(1), 240–251 (2007)
8. Wang, C.C., Derryberry, J., Sleator, D.D.: O(log log n)-competitive dynamic binary search trees. In: Proceedings of the 17th ACM-SIAM Symposium on Discrete Algorithms, pp. 374–383. ACM, New York (2006)
9. Georgakopoulos, G.F.: Chain-splay trees, or, how to achieve and prove loglogn-competitiveness by splaying. Information Processing Letters 106(1), 37–43 (2008)

Drawing Graphs with Right Angle Crossings

(Extended Abstract)

Walter Didimo[1], Peter Eades[2], and Giuseppe Liotta[1]

[1] Dip. di Ingegneria Elettronica e dell'Informazione, Università degli Studi di Perugia
{didimo,liotta}@diei.unipg.it
[2] Department of Information Technology, University of Sydney
peter@cs.usyd.edu.au

Abstract. Cognitive experiments show that humans can read graph drawings in which all edge crossings are at right angles equally well as they can read planar drawings; they also show that the readability of a drawing is heavily affected by the number of bends along the edges. A graph visualization whose edges can only cross perpendicularly is called a *RAC (Right Angle Crossing) drawing*. This paper initiates the study of combinatorial and algorithmic questions related with the problem of computing RAC drawings with few bends per edge. Namely, we study the interplay between number of bends per edge and total number of edges in RAC drawings. We establish upper and lower bounds on these quantities by considering two classical graph drawing scenarios: The one where the algorithm can choose the combinatorial embedding of the input graph and the one where this embedding is fixed.

1 Introduction

The problem of making good drawings of relational data sets is fundamental in several application areas. To enhance human understanding the drawing must be readable, that is it must easily convey the structure of the data and of their relationships (see, for example, [4,9,10]).

A tangled rat's nest of a diagram can be confusing rather than helpful. Intuitively, one may measure the "tangledness" of a graph layout by the number of its edge crossings and by the number of its bends along the edges. This intuition has some scientific validity: experiments by Purchase *et al.* have shown that performance of humans in path tracing tasks is negatively correlated to the number of edge crossings and to the number of bends in the drawing [16,17,21].

This negative correlation has motivated intense research about how to draw a graph with few edge crossings and small *curve complexity* (i.e., maximum number of bends along an edge). As a notable example we recall the many fundamental combinatorial and algorithmic results about planar or quasi-planar straight-line drawings of graphs (see, for example, [11,12]). However, in many practical cases the relational data sets do not induce planar or quasi-planar graphs and a high number of edge crossings is basically not avoidable, especially when a particular drawing convention is adopted. How to handle these crossings in the drawing remains unanswered.

F. Dehne et al. (Eds.): WADS 2009, LNCS 5664, pp. 206–217, 2009.

Recent cognitive experiments of network visualization provide new insights in the classical correlation between edge crossings and human understanding of a network visualization. Huang *et al.* show that the edge crossings do not inhibit human task performance if the edges cross at a large angle [6,7,8]. In fact, professional graphic artists commonly use large crossing angles in network drawings. For example, crossings in hand drawn metro maps and circuit schematics are conventionally at 90° (see, for example, [20]).

This paper initiates the study of combinatorial and algorithmic questions related with the problem of computing drawings of graphs where the edges cross at 90°. Graph visualizations of this type are called *RAC (Right Angle Crossing) drawings*. We study the interplay between the curve complexity and total number of edges in RAC drawings and establish upper and lower bounds on these quantities. It is immediate to see that every graph has a RAC drawing where the edges are represented as simple Jordan curves that are "locally adjusted" around the crossings so that they are orthogonal at their intersection points. However, not every graph has a RAC drawing if small curve complexity is required.

We consider two classical graph drawing scenarios: In the *variable embedding setting* the drawing algorithm takes in input a graph G and attempts to compute a RAC drawing of G; the algorithm can choose both the circular ordering of the edges around the vertices and the sequence of crossings along each edge. In the *fixed embedding setting* the input graph G is given along with a fixed ordering of the edges around its vertices and a fixed ordering of the crossings along each edge; the algorithm must compute a RAC drawing of G that preserves these fixed orderings. An outline of our results is as follows.

- We study the combinatorial properties of straight-line RAC drawings in the variable embedding setting (Section 3). We give a tight upper bound on the number of edges of straight-line RAC drawings. Namely, we prove that straight-line RAC drawings with n vertices can have at most $4n - 10$ edges and that there exist drawings with these many edges. It might be worth recalling that straight-line RAC drawings are a subset of the quasi-planar drawings, for which the problem of finding a tight upper bound on the edge density is still open (see, for example, [1,2,14]).
- Motivated by the previous result, we study how the edge density of RAC drawable graphs varies with the curve complexity (Section 4). We show how to compute a RAC drawing whose curve complexity is three for any graph in the variable embedding setting. We also show that this bound on the curve complexity is tight by proving that curve complexity one implies $O(n^{\frac{4}{3}})$ edges and that curve complexity two implies $O(n^{\frac{7}{4}})$ edges.
- As a contrast, we show that in the fixed embedding setting the curve complexity of a RAC drawing may no longer be constant (Section 5). Namely, we establish an $\Omega(n^2)$ lower bound on the curve complexity in this scenario. We also show that if any two edges cross at most k times, it is always possible to compute a RAC drawing with $O(kn^2)$ curve complexity. This last result implies that the lower bound is tight under the assumption that the number of crossings between any two edges is bounded by a constant.

For reasons of space some proofs are sketched or omitted.

2   Preliminaries

We assume familiarity with basic definitions of graph drawing [4]. Let G be any non-planar graph. The *crossing number* of G is the minimum number of edge crossings in a plane drawing of G, and it is denoted by $cr(G)$. The following bound on $cr(G)$ for any graph G with n vertices and m edges has been proved by Pach *et al.* [13].

Lemma 1. [13] $cr(G) \geq \frac{1}{31.1} \frac{m^3}{n^2} - 1.06n$.

A *Right Angle Crossing drawing* (or *RAC drawing* for short) of G is a poly-line drawing D of G such that any two crossing segments are orthogonal. Throughout the paper we study RAC drawings such that no edge is self-intersecting and any two edges cross a finite number of times. We also assume that all graphs are simple, that is, they contain neither multiple edges nor self-loops.

The *curve complexity* of D is the maximum number of bends along an edge of D. A *straight-line RAC drawing* has curve complexity zero.

3   Straight-Line Right Angle Crossing Drawings

A *quasi-planar drawing* of a graph G is a drawing of G where no three edges are pairwise crossing [2]. If G admits a quasi-planar drawing it is called a *quasi-planar graph*. Quasi-planar graphs are sometimes called *3-quasi-planar graphs* in the literature.

Lemma 2. *Straight-line RAC drawings are a proper subset of the quasi-planar drawings.*

Proof. In a straight-line RAC drawing there cannot be any three mutually crossing edges because if two edges cross a third one, these two edges are parallel. Hence a straight-line RAC drawing is a quasi-planar drawing. The subset is proper because in a quasi-planar drawing edge crossings may not form right angles.   □

Quasi-planar drawings have been the subject of intense studies devoted to finding an upper bound on their number of edges as a function of their number of vertices (extremal problems of this type are generically called *Turán-type problems* in combinatorics and in discrete and computational geometry [12]). Agarwal *et al.* prove that quasi-planar drawings have $O(n)$ edges where n denotes the number of the vertices [2]. This result is refined by Pach, Radoicic, and Tóth, who prove that the number of edges of a quasi-planar drawing is at most $65n$ [14]. This upper bound is further refined by Ackerman and Tardos, who prove that straight-line quasi-planar drawings have at most $6.5n - 20$ edges [1]. We are not aware of any tight upper bound on the number of edges of quasi-planar drawings.

The main result of this section is a tight upper bound on the number of edges of straight-line RAC drawings with a given number of vertices.

Let G be a graph and let D be a straight-line RAC drawing of G; the *crossing graph* $G^*(D)$ of D is the intersection graph of the (open) edges of D. That is, the vertices of $G^*(D)$ are the edges of D, and two vertices of $G^*(D)$ are adjacent in $G^*(D)$ if they cross in D. The following lemma is an immediate consequence of the fact that if two edges of a straight-line RAC drawing cross a third edge, then these two edges are parallel.

Lemma 3. *The crossing graph of a straight-line RAC drawing is bipartite.*

Let E be the set of the edges of a straight-line RAC drawing D. Based on Lemma 3 we can partition E into two subsets E_1 and E_2, such that no two edges in the same set cross. We refine this bipartition by dividing E into three subsets as follows: (i) a *red* edge set E_r, whose elements have no crossings; a red edge corresponds to an isolated vertex of $G^*(D)$, (ii) a *blue* edge set $E_b = E_1 - E_r$, and (iii) a *green* edge set $E_g = E_2 - E_r$. We call this partition a *red-blue-green partition* of E. Let $D_{rb} = (V, E_r \cup E_b)$ denote the subgraph of D consisting of the red and blue edges, and let $D_{rg} = (V, E_r \cup E_g)$ denote the subgraph of D consisting of the red and green edges. Graphs D_{rb} and D_{rg} are also called the *red-blue graph* and *red-green graph* induced by D, respectively.

Since only blue and green edges can cross each other in D, it follows that both the red-blue and the red-green are planar embedded graphs. Therefore, each of them has a number of edges that is less than or equal to $3n - 6$, and so a straight-line RAC drawing has at most $6n - 12$ edges. However, to get a tight upper bound $4n - 10$ we need to count more precisely.

Let G be a graph that has a straight-line RAC drawing. We say that G is *RAC maximal* if any graph obtained from G by adding an extra edge does not admit a straight-line RAC drawing. The proof of the next lemma is omitted for reasons of space.

Lemma 4. *Let G be a RAC maximal graph, let D be any straight-line RAC drawing of G, and let D_{rb} and D_{rg} be the red-blue and red-green graphs induced by D, respectively. Every internal face of D_{rb} and every internal face of D_{rg} contains at least two red edges. Also, all edges of the external boundary of D are red edges.*

Theorem 1. *A straight-line RAC drawing with $n \geq 4$ vertices has at most $4n - 10$ edges. Also, for any $k \geq 3$ there exists a straight-line RAC drawing with $n = 3k - 5$ vertices and $4n - 10$ edges.*

Proof. Let G be a RAC maximal graph with $n \geq 4$ vertices and m edges. Let D be a straight-line RAC drawing of G. Denote by E_r, E_b, E_g the red-blue-green partition of the edges of D and let $m_r = |E_r|$, $m_b = |E_b|$, $m_g = |E_g|$. Assume (without loss of generality) that $m_g \leq m_b$. Of course $m = m_r + m_b + m_g$.

Denote by f_{rb} the number of faces in D_{rb}, and let ω be the number of edges of the external face of D_{rb}. From Lemma 4 we have that D_{rb} has $f_{rb} - 1$ faces with at least two red edges and one face (the external one) with ω red edges. Also, since every edge occurs on exactly two faces, we have

$$m_r \geq f_{rb} - 1 + \omega/2. \tag{1}$$

Graph D_{rb} is not necessarily connected, but Euler's formula guarantees that

$$m_r + m_b \leq n + f_{rb} - 2. \tag{2}$$

Substituting the inequality (1) into (2) we deduce that

$$m_b \leq n - 1 - \omega/2. \tag{3}$$

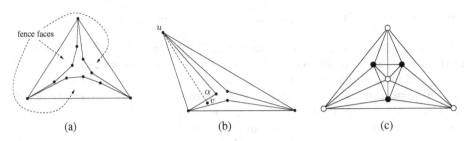

Fig. 1. (a) Fence faces. (b) Triangular fence faces; the dashed edge is a green edge. (c) A straight-line RAC drawing with $n = 7$ vertices and $m = 4n - 10$ edges.

Since D_{rg} has the same external face as D_{rb} we have

$$m_r + m_g \le 3n - 3 - \omega. \tag{4}$$

Also, from $m = m_r + m_g + m_b$, we can sum the Inequalities (3) and (4) to obtain

$$m \le 4n - 4 - 3\omega/2. \tag{5}$$

Observe that if $\omega \ge 4$ then Inequality (5) implies $m \le 4n - 10$. Thus we need only consider the case when the external face is a triangle, that is, $\omega = 3$.

Suppose that $\omega = 3$; consider the (at least one and at most three) faces that share an edge with the outside face, as in Fig. 1(a). We call these faces the *fence faces* of D. Notice that, since we are assuming that $n > 3$ then there is at least one internal vertex. Also, since the graph is RAC maximal, then it is not possible that every internal vertex is an isolated vertex. Hence every fence face has at least one internal edge.

Suppose that one of the fence faces has more than three edges. In this case, D_{rb} is a planar graph in which at least one face has at least four edges; this implies that

$$m_r + m_b \le 3n - 7. \tag{6}$$

Since we have assumed that $m_g \le m_b$, Inequality (6) implies that

$$m_r + m_g \le 3n - 7. \tag{7}$$

Summing Inequalities (3) with $\omega = 3$ and (7) yields

$$m \le 4n - 19/2; \tag{8}$$

since m is an integer the result follows.

The final case to consider is where all the fence faces are triangles. In this case there are exactly three fence faces. We show that all the edges of at least two of these faces are red. Suppose that a fence face has one blue edge. This implies that this edge must be crossed by a green edge (u, v). Note that two edges incident on a common vertex cannot be crossed by a third edge or else the perpendicularity of the crossings would be violated. From this fact and since the external face is red, it follows that (u, v) cannot cross another edge of the fence face. Therefore (u, v) must be incident to one vertex of the external face, as in Fig. 1(b).

Now (u, v) crosses at an angle of $90°$, and so the interior angle α of the triangle that it crosses is less than $90°$. However, the sum of the interior angles of the three fence faces is at least $360°$. Thus at most one of the three triangles can have an interior angle less than $90°$, and so at least two of the fence faces cannot have an edge crossing. Thus at least two of the fence faces have three red edges. Also, the outside face has three red edges, and so the drawing has at least three faces in which all three edges are red. It follows that the number of red edges is bounded from below:

$$m_r \geq f_{rb} - 3 + (3 \cdot 3)/2 = f_{rb} + 3/2. \tag{9}$$

Substituting (9) into (2), we deduce that $m_b \leq n - 7/2$, and thus $m \leq 4n - 19/2$. Since m is integer, the first part of the theorem follows.

We now prove that for each even integer $k \geq 3$, there exists a RAC maximal graph G_k with $n = 3k - 5$ vertices and $4n - 10$ edges. Graph G_k is constructed as follows (refer to Fig. 1(c) for an illustration where $k = 4$). Start from an embedded maximal planar graph with k vertices and add to this graph its dual planar graph without the face-node corresponding to the external face (in Fig. 1(c) the primal graph has white vertices and the dual graph has black vertices). Also, for each face-node u, add to G_k three edges that connect u to the three vertices of the face associated with u.

A result by Brightwell and Scheinermann about representations of planar graphs and of their duals guarantees that G_k admits a straight-line RAC drawing [3]. More precisely, Brightwell and Scheinermann show that every 3-connected planar graph G can be represented as a collection of circles, a circle for each vertex and a circle for each face. For each edge e of G, the four circles representing the two end-points of e and the two faces sharing e meet at a point, and the vertex-circles cross the face-circles at right angles. This implies that the union of G and its dual (without the face-node corresponding to the external face) has a straight-line drawing such that the primal edges cross the dual edges at right angles.

Since the number of face-nodes is $2k - 5$, then G_k has $n = 3k - 5$ vertices. The number of edges of G_k is given by $m = (3k - 6) + 3(2k - 5) + 3k - 9$, and hence $m = 12k - 30 = 4n - 10$. □

4 Poly-Line Right Angle Crossing Drawings

Motivated by Theorem 1, in the attempt of computing RAC drawings of dense graphs we relax the constraint that the edges be drawn as straight-line segments. In this section we study how the edge density of RAC drawable graphs varies with the curve complexity in the variable embedding setting.

Lemma 5. *Every graph has a RAC drawing with at most three bends per edge.*

Sketch of Proof: Papakostas and Tollis describe an algorithm to compute an orthogonal drawing H of G with at most one bend per edge and such that each vertex is represented as a box [15]. Of course, in an orthogonal drawing any two crossing segments are perpendicular. To get a RAC drawing D from H it is sufficient to replace each vertex-box with a point placed in its center and to use at most two extra bends per edge to connect the centers to the boundaries of the boxes. □

Lemma 5 naturally raises the question about whether three bends are not only sufficient but sometimes necessary. This question has a positive answer as we are going to show with the following lemmas.

Let D be a poly-line drawing of a graph G. An *end-segment* in D is an edge segment incident to a vertex. An edge segment in D that is not an end-segment is called an *internal segment*. Note that the end points of an internal segment are bends in D.

Lemma 6. *Let D be a RAC drawing of a graph G. For any two vertices u and v in G, there are at most two crossings between the end-segments incident to u and the end-segments incident to v.*

Proof. Each crossing between an end-segment incident to u and an edge segment incident to v in D occurs on the circle whose diameter is the line segment $\overline{uv}$. If there are more than two such points, then at least two crossings occur in a half circle (either from a side of $\overline{uv}$ or on the other side of $\overline{uv}$). It follows that two line segments meet at an angle of more than $90°$, and the drawing is not a RAC drawing. □

Lemma 7. *Let D be a RAC drawing of a graph G with n vertices. Then the number of crossings between all end-segments is at most $n(n-1)$.*

Proof. It follows from Lemma 6 by considering that the number of distinct pairs of vertices is $n(n-1)/2$. □

Lemma 8. *Let D be a RAC drawing and let s be any edge segment of D. The number of end-segments crossed by s is at most n.*

Proof. If s crosses more than n end-segments in D, then there are two of these segments incident to the same vertex, which is impossible in a RAC drawing. □

The previous lemmas are the ingredients to show that not all graphs admit a RAC drawing with curve complexity two.

Lemma 9. *A RAC drawing with n vertices and curve complexity two has $O(n^{\frac{7}{4}})$ edges.*

Proof. Let D be a RAC drawing with at most two bends per edge. We prove that the number m of edges of D is $m \le 36n^{\frac{7}{4}}$. Assume by contradiction that $m > 36n^{\frac{7}{4}}$. From Lemma 1, the number of crossings in D is at least $\frac{1}{31.1}\frac{m^3}{n^2} - 1.06n$. There are at most $3m$ edge segments in D because every edge has at most two bends; it follows that there is at least one edge segment s with at least $\frac{1}{93.3}\frac{m^2}{n^2} - 0.36\frac{n}{m}$ crossings. For each vertex u, at most one end-segment of an edge incident to u can cross s. Hence, there are at most n edges (u,v) that cross s in an end-segment of (u,v). This implies that the number m' of edges whose internal segments cross s is such that:

$$m' \ge \frac{1}{93.3}\frac{m^2}{n^2} - 0.36\frac{n}{m} - n \tag{10}$$

From our assumption that $m > 36n^{\frac{7}{4}}$, we can replace m on the right hand side of Equation (10) with $36n^{\frac{7}{4}}$ to obtain $m' > 13,89n^{\frac{3}{2}} - 0.01n^{-\frac{3}{4}} - n$. Since $0.01n^{-\frac{3}{4}} < 1$, it follows that $m' > 13,89n^{\frac{3}{2}} - (n+1)$. Also, since $2n^{\frac{3}{2}} \geq n+1$ (for every $n \geq 1$), it follows that:

$$m' > 11,89n^{\frac{3}{2}}. \tag{11}$$

Let D' be a sub-drawing of D consisting of m' edges that cross s with an internal segment, as well as the vertices incident to these edges. Let n' be the number of vertices in D'. Using Lemma 1 applied to D', the number of crossings in D' is at least $\frac{1}{31.1}\frac{m'^3}{n'^2} - 1.06n'$. However, the internal segments of edges in D' are all parallel (since they all cross s at an angle of $90°$). Thus, all crossings in D' involve an end-segment. From Lemmas 7 and 8, there are at most $n'(n'-1) + m'n'$ such crossings. Hence, it must be

$$n'(n'-1) + m'n' \geq \frac{1}{31.1}\frac{m'^3}{n'^2} - 1.06n'. \tag{12}$$

Since $n' < n$, and since, from Inequality (11), $m' > n-1$, we have that $2m'n \geq n'(n'-1) + m'n'$. From Inequality (12), it must also hold $2m'n \geq \frac{1}{31.1}\frac{m'^3}{n'^2} - 1.06n$, that is:

$$n \geq \frac{1}{62.2}\frac{m'^2}{n^2} - 0.53\frac{n}{m'}. \tag{13}$$

From Inequalities (11) and (13) we have $n \geq 2.27n - 0.045n^{-\frac{1}{2}}$, which is however false for any $n \geq 1$, a contradiction. □

The next lemma completes the analysis of the number of edges in poly-line RAC drawings with curve complexity smaller than three.

Lemma 10. *A RAC drawing with n vertices and curve complexity one has $O(n^{\frac{4}{3}})$ edges.*

Proof. Let D be a RAC drawing with at most one bend per edge. D contains end-segments only. Therefore, from Lemma 7, the number of crossings in D is at most $n(n-1)$. Also, from Lemma 1, the number of crossings in D must be at least $\frac{1}{31.1}\frac{m^3}{n^2} - 1.06n$. It follows that: $n(n-1) \geq \frac{1}{31.1}\frac{m^3}{n^2} - 1.06n$, which implies that $n^4 + 0.06n^3 \geq \frac{1}{1.31}m^3$, and then $m < 3.1n^{\frac{4}{3}}$, i.e., $m = O(n^{\frac{4}{3}})$. □

The following theorem summarizes the interplay between curve complexity and edge density of RAC drawings in the variable embedding setting. It is implied by Theorem 1, Lemma 5, Lemma 9, and Lemma 10.

Theorem 2. *Let G be a graph with n vertices and m edges.*

(a) *There always exists a RAC drawing of G with at most three bends per edge.*
(b) *If G admits a RAC drawing with straight-line edges then $m = O(n)$.*
(c) *If G admits a RAC drawing with at most one bend per edge then $m = O(n^{\frac{4}{3}})$.*
(d) *If G admits a RAC drawing with at most two bends per edge then $m = O(n^{\frac{7}{4}})$.*

5   Fixed Embedding Setting

A classical constraint of many algorithms that draw planar graphs is to preserve a given circular ordering of the edges around the vertices, also called a *combinatorial embedding*. In this section we consider similar constraints for RAC drawings. In contrast with Theorem 2, we show that fixed combinatorial embedding constraints may lead to RAC drawings of non-constant curve complexity, while quadratic curve-complexity is always sufficient for any graph and any fixed combinatorial embedding.

Let G be a graph and let D be a drawing of G. Since in a RAC drawing no three edges can cross each other at the same point, we shall only consider drawings whose crossings involve exactly two edges. We denote by $\overline{G}$ the planar embedded graph obtained from D by replacing each edge crossing with a vertex, and we call it a *planar enhancement* of G. A vertex of $\overline{G}$ that replaces a crossing is called a *cross vertex*. Giving a planar enhancement of G corresponds to fixing the number and the ordering of the cross vertices along each edge, the circular clockwise ordering of the edges incident to each vertex (both real and cross vertices), and the external face.

Let G be a graph along with a planar enhancement $\overline{G}$ and let D' be a drawing of G. We say that D' *preserves the planar enhancement* $\overline{G}$ if the planar enhancement of G obtained from D' coincides with $\overline{G}$.

The next theorems establish lower and upper bounds for the curve complexity of RAC drawings in the fixed embedding setting.

Theorem 3. *There are infinitely many values of n for which there exists a graph G with n vertices and a planar enhancement $\overline{G}$ such that any RAC drawing preserving $\overline{G}$ has curve complexity $\Omega(n^2)$.*

Sketch of Proof: Based on a construction of Roudneff, Felsner and Kriegel show simple arrangements of m pseudolines in the Euclidean plane forming $m(m-2)/3$ triangular faces for infinitely many values of m [5,18]. For each such values of m, let $\mathcal{A}(m)$ be the corresponding arrangement of pseudolines and let $n = 2(\lfloor \sqrt{m} \rfloor + 1)$.

We define G as a simple bipartite graph with n vertices and m edges such that every partition set of G has $\frac{n}{2}$ vertices (note that $\frac{n^2}{4} \geq m$). We also define a planar enhancement of G by constructing a drawing D where each edge uses a portion of a corresponding pseudoline of $\mathcal{A}(m)$. The planar enhancement of G obtained from D is denoted as $\overline{G}$.

We observe that the arrangement of pseudolines defined in [5,18] has the following property: There exists a circle $C(m)$ such that all crossings of $\mathcal{A}(m)$ lie inside $C(m)$ and every pseudoline of $\mathcal{A}(m)$ crosses $C(m)$ in exactly two points. Drawing D is defined as follows:

- Each vertex v of G is drawn as a distinct point $p(v)$ arbitrarily chosen outside $C(m)$.
- Let $\{\ell_1, \ldots, \ell_m\}$ be the pseudolines of $\mathcal{A}(m)$ and let $\{e_1, \ldots, e_m\}$ be the edges of G. Let p_i^1 and p_i^2 be the points of intersection between $C(m)$ and ℓ_i and let $e_i = (v_i^1, v_i^2)$ $(1 \leq i \leq m)$. Edge e_i is drawn as the union of: (i) the portion of ℓ_i inside $C(m)$ that connects p_i^1 with p_i^2; (ii) a simple curve that connects p_i^1

with $p(v_i^1)$ and that does not cross the interior of $C(m)$; (iii) a simple curve that connects p_i^2 with $p(v_i^2)$ and that does not cross the interior of $C(m)$.

Since drawing D maintains all triangular faces of $\mathcal{A}(m)$ and $m = \Theta(n^2)$, it follows that D (and hence $\overline{G}$) has $\Theta(n^4)$ triangular faces inside $C(m)$. Also, the vertices of each triangular face inside $C(m)$ are cross vertices in $\overline{G}$. Therefore, any RAC drawing of G that preserves $\overline{G}$ has at least one bend for each triangular face inside $C(m)$. Hence any RAC drawing of G preserving $\overline{G}$ has $\Omega(n^4)$ bends and curve complexity $\Omega(n^2)$. □

The next theorem proves that the lower bound of Theorem 3 is tight for those graphs that can be drawn in the plane such that the number of crossings between any two edges is bounded by a given constant.

Theorem 4. *Let G be a graph with n vertices and let $\overline{G}$ be a planar enhancement of G obtained from a drawing where any two edges cross at most k times, for some $k \geq 1$. There exists a RAC drawing of G that preserves $\overline{G}$ and that has $O(kn^2)$ curve complexity.*

Sketch of Proof: Let m be the number of edges of G and let $\overline{n}$ and $\overline{m}$ be the number of vertices and edges of $\overline{G}$, respectively. From the hypothesis that two distinct edges cross at most k times and that an edge cannot cross itself, we have that $\overline{n} \leq n + k(m-1)m$. Namely, every edge of G is subdivided in $\overline{G}$ by at most $k(m-1)$ cross vertices, i.e., it is formed by at most $k(m-1) + 1 = km - k + 1$ edges of $\overline{G}$.

  Assume first that G has vertex degree at most four (which of course implies that also $\overline{G}$ has vertices of degree at most four). In this case one can compute a planar orthogonal drawing $\overline{D}$ of $\overline{G}$ with the technique described by Tamassia and Tollis [19]. This technique first computes a visibility representation of the graph, i.e., a planar drawing in which each vertex is drawn as a horizontal segment and each edge is drawn as a vertical segment between its end-vertices. Then it replaces each horizontal segment of a vertex v with a point p_v, and connects p_v to the vertical segments representing the incident edges of v, by a local transformation that uses at most two bends per edge around p_v (see, e.g., Fig. 2(a)). Hence an edge can get at most four bends (two for each local transformation around an end-vertex). Therefore, this technique guarantees at most 4 bends per edge. Also, observe that since it is always possible to compute a visibility representation of an embedded planar graph that preserves its planar embedding and since the local transformations do not change this embedding, the technique described above can be applied so that the embedding of $\overline{G}$ is preserved.

  When cross vertices are replaced by cross points, we get from $\overline{D}$ an orthogonal drawing D of G that preserves $\overline{G}$ and that has at most $4(km - k + 1)$ bends per edge. Since $m < \frac{n^2}{2}$ and since D is a RAC drawing, the theorem follows in this case.

  If $\overline{G}$ has vertices of degree greater than four then we can apply a variant of the algorithm of Tamassia and Tollis. Namely, after the computation of a visibility representation of $\overline{G}$ we apply the same transformations as before around the vertices of degree at most four. For a vertex v of degree greater than four we replace the horizontal segment of v with a point p_v, and then locally modify the edges incident to v as shown in Fig. 2(b), by using at most one bend per edge. The drawing $\overline{D}$ obtained in this way is not an orthogonal drawing but it still has at most four bends per edge. By replacing

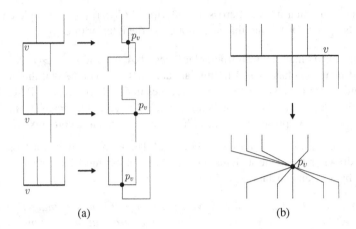

Fig. 2. Local transformations from a visibility representation to a RAC drawing: (a) for vertices of degree at most four; (b) for vertices of degree greater than four

cross vertices with cross points, we still get from $\overline{D}$ a drawing D of G that preserves $\overline{G}$ and that has at most $4(km-k+1)$ bends per edge. Also, since a cross vertex has degree four in $\overline{D}$, we are guaranteed that D is a RAC drawing, because for these vertices we have applied an orthogonal drawing transformation.   □

6   Conclusion and Open Problems

This paper has studied RAC drawings of graphs, i.e. drawings where edges can cross only at right angles. In fact, many crossings are unavoidable when drawing large graphs and recent perceptual studies have shown that right angle crossings do not have impact on the readability of a diagram. We have focused on the interplay between edge density and curve complexity of RAC drawings and have proved lower and upper bounds for these quantities. There are several open questions that we consider of interest about RAC drawings. Among them we mention the following.

1. By Theorem 1, a graph that has a straight-line RAC drawing has at most $4n - 10$ edges. How difficult is it to recognize whether a graph with $m \leq 4n - 10$ edges has a straight-line RAC drawing?
2. Find tight upper bounds to the number of edges of RAC drawings with curve complexity one and two. See also Theorem 2.
3. Study the area requirement of RAC drawings. For example, can all planar graphs be drawn in $o(n^2)$ area if edges are allowed to cross at right angles?

Acknowledgements

We thank the anonymous referees for their valuable comments.

References

1. Ackerman, E., Tardos, G.: On the maximum number of edges in quasi-planar graphs. J. Comb. Theory, Ser. A 114(3), 563–571 (2007)
2. Agarwal, P.K., Aronov, B., Pach, J., Pollack, R., Sharir, M.: Quasi-planar graphs have a linear number of edges. Combinatorica 17(1), 1–9 (1997)
3. Brightwell, G., Scheinerman, E.R.: Representations of planar graphs. SIAM J. Discrete Math. 6(2), 214–229 (1993)
4. Di Battista, G., Eades, P., Tamassia, R., Tollis, I.G.: Graph Drawing. Prentice Hall, Upper Saddle River (1999)
5. Felsner, S.: Triangles in euclidean arrangements. Discrete & Computational Geometry 22(3), 429–438 (1999)
6. Huang, W.: Using eye tracking to investigate graph layout effects. In: APVIS, pp. 97–100 (2007)
7. Huang, W.: An eye tracking study into the effects of graph layout. CoRR, abs/0810.4431 (2008)
8. Huang, W., Hong, S.-H., Eades, P.: Effects of crossing angles. In: PacificVis, pp. 41–46 (2008)
9. Jünger, M., Mutzel, P. (eds.): Graph Drawing Software. Springer, Heidelberg (2003)
10. Kaufmann, M., Wagner, D. (eds.): Drawing Graphs. Springer, Heidelberg (2001)
11. Nishizeki, T., Rahman, M.S.: Planar Graph Drawing. World Scientific, Singapore (2004)
12. Pach, J.: Geometric graph theory. In: Handbook of Discrete and Computational Geometry, pp. 219–238. CRC Press, Boca Raton (2004)
13. Pach, J., Radoicic, R., Tardos, G., Tóth, G.: Improving the crossing lemma by finding more crossings in sparse graphs. Discrete & Computational Geometry 36(4), 527–552 (2006)
14. Pach, J., Radoicic, R., Tóth, G.: Relaxing planarity for topological graphs. In: Akiyama, J., Kano, M. (eds.) JCDCG 2002. LNCS, vol. 2866, pp. 221–232. Springer, Heidelberg (2003)
15. Papakostas, A., Tollis, I.G.: Efficient orthogonal drawings of high degree graphs. Algorithmica 26(1), 100–125 (2000)
16. Purchase, H.C.: Effective information visualisation: a study of graph drawing aesthetics and algorithms. Interacting with Computers 13(2), 147–162 (2000)
17. Purchase, H.C., Carrington, D.A., Allder, J.-A.: Empirical evaluation of aesthetics-based graph layout. Empirical Software Engineering 7(3), 233–255 (2002)
18. Roudneff, J.-P.: The maximum number of triangles in arrangements of pseudolines. J. Comb. Theory, Ser. B 66(1), 44–74 (1996)
19. Tamassia, R., Tollis, I.G.: Planar grid embedding in linear time. IEEE Trans. Circuit Syst. CAS-36(9), 1230–1234 (1989)
20. Vignelli, M.: New york subway map,
http://www.mensvogue.com/design/articles/2008/05/vignelli
21. Ware, C., Purchase, H.C., Colpoys, L., McGill, M.: Cognitive measurements of graph aesthetics. Information Visualization 1(2), 103–110 (2002)

Finding a Hausdorff Core of a Polygon: On Convex Polygon Containment with Bounded Hausdorff Distance*

Reza Dorrigiv[1], Stephane Durocher[1,2], Arash Farzan[1], Robert Fraser[1],
Alejandro López-Ortiz[1,**], J. Ian Munro[1], Alejandro Salinger[1],
and Matthew Skala[1,3]

[1] Cheriton School of Computer Science, University of Waterloo, Waterloo, Canada
{rdorrigiv,sdurocher,afarzan,r3fraser,alopez-o,
imunro,ajsalinger,mskala}@cs.uwaterloo.ca
[2] Department of Computer Science, University of Manitoba, Winnipeg, Canada
durocher@cs.umanitoba.ca
[3] Department of Computer Science, University of Toronto, Toronto, Canada
mskala@ansuz.sooke.bc.ca

Abstract. Given a simple polygon P, we consider the problem of finding a convex polygon Q contained in P that minimizes $H(P, Q)$, where H denotes the Hausdorff distance. We call such a polygon Q a *Hausdorff core* of P. We describe polynomial-time approximations for both the minimization and decision versions of the Hausdorff core problem, and we provide an argument supporting the hardness of the problem.

1 Introduction

Traditional hierarchical representations allow for efficient storage, search and representation of spatial data. These representations typically divide the search space into areas for which membership can be tested efficiently. If the query region does not intersect a given area, the query can proceed without further consideration of that area. When a space or object has certain structural properties, the data structure built upon it can benefit from those properties. For example, the data structure of Kirkpatrick [12] is designed to index planar subdivisions answering point queries in time $O(\log n)$ and space $O(n)$, with preprocessing time $O(n \log n)$.

Our study is motivated by the problem of path planning in the context of navigation at sea. In this application, a plotted course must be tested against bathymetric soundings to ensure that the ship will not run aground. We suppose the soundings have been interpolated into contour lines [1] and the plotted course is given as a polygonal line. There is no requirement of monotonicity or even

---

* Funding for this research was made possible by the NSERC strategic grant on Optimal Data Structures for Organization and Retrieval of Spatial Data.

** Part of this work took place while the fifth author was on sabbatical at the Max-Planck-Institut für Informatik in Saarbrücken, Germany.

F. Dehne et al. (Eds.): WADS 2009, LNCS 5664, pp. 218–229, 2009.

continuity between contour lines in the map. A given line might be a maximum, minimum or a falling slope. Similarly, we observe that in general there are several disconnected contour lines with the same integer label (depth).

Although contour lines can be arbitrarily complicated, typical shipping routes run far from potential obstacles for the majority of their trajectories, and only short segments require more careful route planning. As a result, most intersection checks should be easy: we should be able to subdivide the map into areas such that most of our intersection tests are against conveniently-shaped areas, reserving more expensive tests for the rare cases where the path comes close to intersecting the terrain.

The search for easily-testable areas motivates the study of the simplification of a contour line into a simpler object which is either entirely contained within the contour line or fully contains it. In this paper we consider the case in which the simplified polygon must be convex and contained.

1.1 Definitions

A *polygon* P is a closed region in the plane bounded by a finite sequence of line segments or *edges*. We restrict our attention to *simple polygons*, in which the intersection of any two edges is either empty or an endpoint of each edge and the intersection of any three edges is empty. Finally, recall that a region P is *convex* if for all points p and q in P, the line segment $\overline{pq}$ is contained in P.

Given a simple polygon P and a metric d (defined on polygons), a *d-core* of P is a convex polygon Q contained in P that minimizes $d(P, Q)$. Examples of metrics d of interest include the area of the region $P \setminus Q$, the Hausdorff distance between P and Q, and the link distance (which is a discrete distance metric). A common measure of distance between two sets P and Q is given by

$$d(P, Q) = \max \left\{ \max_{p \in P} \min_{q \in Q} \operatorname{dist}(p, q), \max_{q \in Q} \min_{p \in P} \operatorname{dist}(p, q) \right\}.$$

When P and Q are polygons in the plane and $\operatorname{dist}(p, q)$ denotes the Euclidean (ℓ_2) distance between points p and q, $d(P, Q)$ corresponds to the *Hausdorff distance* between sets P and Q, which we denote by $H(P, Q)$. We define the corresponding *d-core* as the *Hausdorff core*. We consider both the minimization and decision versions of problem of finding a Hausdorff core for a given simple polygon P:

Input. A simple polygon P.

Question. Find a Hausdorff core of P.

Input. A simple polygon P and a non-negative integer k.

Question. Does there exist a convex polygon Q contained in P such that $H(P, Q) \leq k$?

The *1-centre* of a polygon P (also known as Euclidean centre) is the point c that minimizes the maximum distance from c to any point in P. In this work we are only interested in the 1-centre inside P, also known as constrained Euclidean centre. Although the unconstrained 1-centre is unique, this is not necessarily true for the constrained version [6]. A constrained 1-centre of a polygon P of n vertices can be computed in time $O(n \log n + k)$, where k is the number of intersections between P and the furthest point Voronoi diagram of the vertices of P [6]. For simple polygons $k \in O(n^2)$. Note that the constrained 1-centre of P is a point $c \in P$ that minimizes $H(P, c)$. Throughout the rest of the paper, when we refer to a 1-centre, we specifically mean a constrained 1-centre.

1.2   Related Work

We can divide the problem of approximating polygons into two broad classes: inclusion problems seek an approximation contained in the original polygon, while enclosure problems determine approximation that contains the original polygon. Formally, let $\mathcal{P}$ and $\mathcal{Q}$ be classes of polygons and let μ be a function on polygons such that for polygons P and Q, $P \subseteq Q \Rightarrow \mu(P) \leq \mu(Q)$. Chang and Yap [7] define the inclusion and enclosure problems as:

- $Inc(\mathcal{P}, \mathcal{Q}, \mu)$: Given $P \in \mathcal{P}$, find $Q \in \mathcal{Q}$ included in P, maximizing $\mu(Q)$.
- $Enc(\mathcal{P}, \mathcal{Q}, \mu)$: Given $P \in \mathcal{P}$, find $Q \in \mathcal{Q}$ enclosing P, minimizing $\mu(Q)$.

The best known enclosure problem is the convex hull, which we may state formally as $Enc(\mathcal{P}_{\mathrm{simple}}, \mathcal{P}_{\mathrm{con}}, \mathrm{area})$, where $\mathcal{P}_{\mathrm{simple}}$ is the family of simple polygons and $\mathcal{P}_{\mathrm{con}}$ is the family of convex polygons. Given a convex polygon P, many problems are tractable in linear time: $Enc(\mathcal{P}_{\mathrm{con}}, \mathcal{P}_3, \mathrm{area})$ [16], $Enc(\mathcal{P}_{\mathrm{con}}, \mathcal{P}_3, \mathrm{perimeter})$ [5], and $Enc(\mathcal{P}_{\mathrm{con}}, \mathcal{P}_{\mathrm{par}}, \mathrm{area})$ [17], where $\mathcal{P}_{\mathrm{par}}$ is the family of parallelograms. For general k-gons, $Enc(\mathcal{P}_{\mathrm{con}}, \mathcal{P}_k, \mathrm{area})$ can be solved in $O(kn + n \log n)$ time [3].

Perhaps the best known inclusion problem is the potato-peeling problem of Chang and Yap [7], defined as $Inc(\mathcal{P}_{\mathrm{simple}}, \mathcal{P}_{\mathrm{con}}, \mathrm{area})$. There is an $O(n^7)$ time algorithm for this problem, and an $O(n^6)$ time algorithm when the measure is the perimeter, $Inc(\mathcal{P}_{\mathrm{simple}}, \mathcal{P}_{\mathrm{con}}, \mathrm{perimeter})$, where n is the number of vertices of P [7]. The problem of finding the triangle of maximal area included in a convex polygon, $Inc(\mathcal{P}_{con}, \mathcal{P}_3, \mathrm{area})$, can be solved in linear time [9]. The generalization of this problem to any k-gon can be solved in time $O(kn + n \log n)$ [2]. If the input polygon is not restricted to be convex, $Inc(\mathcal{P}_{con}, \mathcal{P}_3, \mathrm{area})$ can be found in time $O(n^4)$ [15].

The inclusion and enclosure problems can also be formulated as minimizing or maximizing a measure $d(P, Q)$. Note that in the case when $\mu(Q)$ is the area, maximizing or minimizing $\mu(Q)$ for the inclusion and enclosure problems, respectively, is equivalent to minimizing the difference in areas $(d(P, Q) = |\mu(P) - \mu(Q)|)$. Both the inclusion and enclosure problems using the Hausdorff distance as a measure were studied by Lopez and Reisner [14], who present polynomial-time algorithms to approximate a convex polygon minimizing the Hausdorff distance to within an arbitrary factor of the optimal. Since the input polygon is convex, the approximating solution is restricted to a maximum number of vertices. In the

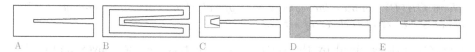

Fig. 1. A. The input polygon P. **B.** "Shrinking" the polygon. **C.** Shrink until the convex hull is contained in P. **D.** The solution returned by the Chassery and Coeurjolly algorithm. **E.** An actual solution.

same work, the authors also studied the *min-#* version of the problem, where the goal is to minimize the number of vertices of the approximating polygon, given a maximum allowed error. For this setting, they show that the inclusion and enclosure problems can be approximated to within one vertex of the optimal in $O(n \log n)$ time and $O(n)$ time, respectively.

The inclusion problem that minimizes the Hausdorff distance where the input is a simple (not necessarily convex) polygon was addressed in [8]. They present an algorithm that returns a Hausdorff core for the case when the point 1-centre is contained in the input polygon P. The algorithm shrinks the input polygon P until its convex hull is contained in the original P. If the shrunken polygon P' is not convex, the region in which the convex hull P' intersects P is removed from P'. The procedure is repeated starting with P' until a convex polygon is obtained. In general, the algorithm does not return a Hausdorff core if the point 1-centre is not contained in P. A counterexample is illustrated in Figure 1. To the best of our knowledge, no algorithm for finding a Hausdorff core of an arbitrary simple polygon, $Inc(\mathcal{P}_{\text{simple}}, \mathcal{P}_{\text{con}}, \text{Hausdorff})$, has appeared in the literature.

2   Preliminary Observations

In this section we make several observations about properties of polygons, convex polygons, and the Hausdorff distance in the context of the discussed problem. These observations will be useful in later sections in establishing our main results. Due to lack of space, we omit the proofs.

Given a polygon P and a convex polygon Q inside P, it suffices to optimize the maximum distance from points $p \in P$ to polygon Q to obtain a Q with a minimum Hausdorff distance:

Observation 1. *Given any simple polygon P and any convex polygon Q contained in P, $\max_{p \in P} \min_{q \in Q} d(p, q) \geq \max_{q \in Q} \min_{p \in P} d(p, q)$. Therefore,*

$$H(P, Q) = \max_{p \in P} \min_{q \in Q} d(p, q).$$

Among the points of P and Q, the Hausdorff distance is realized at the vertices of P. Furthermore, it occurs between Q and vertices that lie on the convex hull of P:

Lemma 1. *Given a simple polygon P and a convex polygon Q contained in P,*

$$H(P, Q) = H(CH(P)_V, Q),$$

where $CH(P)$ denotes the convex hull of set P and for any polygon A, A_V denotes the set of vertices of set A.

$H(P, Q)$ is determined by the vertices of P that lie on the convex hull of P, however all vertices and edges of P must be considered to determine whether Q is contained in P. The decision version of the Hausdorff core problem with parameter k is defined as follows; we consider circles of radius k centered at vertices $CH(P)_V$ and ask whether there exists a convex polygon Q such that it intersects all such circles:

Observation 2. *Let $C_k(p)$ denote a circle of radius k centred at p. Given a simple polygon P and a convex polygon Q contained in P,*

$$H(P, Q) \leq k \Leftrightarrow \forall p \in CH(P), \ C_k(p) \cap Q \neq \varnothing.$$

Finally, we wish to know some point contained in Q. If the 1-centre of P is not in Q, then Q intersects some vertex of P:

Lemma 2. *Given a simple polygon P and a convex polygon Q contained in P, let P_{1c} be the constrained 1-centre of P. At least one point in the set $\{P_{1c}, P_V\}$ is contained in Q if Q is a Hausdorff core of P. Let a point chosen arbitrarily from this set be Q_p.*

3   Hausdorff Core Minimization Problem

In this section we outline an algorithm to solve the Hausdorff core problem which operates by shrinking circles centred on selected vertices of P (which vertices have circles is discussed shortly). Invariant 1 must hold for a solution to exist:

Invariant 1. There exists a set of points $\{p_1, p_2, \ldots, p_k\}$, where k is the current number of circles, such that $\forall i \ p_i \in C_i$ and $\forall i, j, i \neq j \ \overline{p_i p_j}$ does not cross outside the original simple polygon.

Invariant 1 implies that a solution Q with $H(P, Q) = r$ exists, where r is the radius of the circles. We sketch the solution in Algorithm 1, and we illustrate an example of the operation of the algorithm in Figure 2. We find P_{1c} using the technique of [6]; there may be multiple such vertices, but we can choose one arbitrarily. A solution is not unique in general, but we find a polygon Q which minimizes $H(P, Q)$.

3.1   Proof of Correctness

The solution Q is a convex polygon that intersects every circle. If each circle C_i touches the solution convex polygon Q, we know that the distance from each vertex with a circle to Q is at most r, the radius of C_i. If a vertex $v \in CH(P)_V$ does not have a circle, then $\text{dist}(v, Q_p) \leq r$. Therefore, given a simple polygon P, this algorithm finds a convex polygon Q contained in P such that $\forall p \in CH(P)_V, \exists q \in Q$ s.t. $d(p, q) \leq r$. By Lemma 1, we know that Q is a solution

Algorithm 1. Hausdorff Core Minimization Algorithm

HCORE(P)

$Q = \emptyset, r_{\min} = \infty$

for each $Q_p \in \{P_{1c}, P_V\}$ **do**

Begin with circles of radius r_0 centred on the vertices $v \in CH(P)_V$, where $r_0 = \text{dist}(v_f, Q_p)$ and $v_f = \arg\max_{p \in P} \text{dist}(p, Q_p)$.

Any circle centred at a vertex v where $\text{dist}(Q_p, v) < r$ contains Q_p; such circles are ignored for now.

Reduce the radius such that at time $t_i \in [0, 1]$, each circle has radius $r(t_i) = r_0 \times (1 - t_i)$. Let $Q(t_i)$ be a solution at time t_i, if it exists. The radius is reduced until one of three events occurs:

(1) $r(t_i) = \text{dist}(Q_p, v_n)$, where v_n is the farthest vertex from Q_p that is not the centre of a circle. Add a circle centred at v_n with radius $r(t_i)$.

(2) $Q(t_i)$ cannot cover Q_p. In this case, we break and if $r(t_i) < r_{\min}$, then set $Q = Q(t_i)$ and $r_{\min} = r(t_i)$.

(3) A further reduction of r will prevent visibility in P between two circles. Again, we break and if $r(t_i) < r_{\min}$, then set $Q = Q(t_i)$ and $r_{\min} = r(t_i)$.

end for

return Q

---

where $H(P, Q) = r$. It remains to be shown that there does not exist a convex polygon Q' such that $\text{dist}(p, q') \leq r'$, where $r' < r$. This cannot be the case, for if the circles were shrunk any further, no convex polygon could intersect some pair of the circles by Invariant 1. Therefore, the polygon would necessarily be of distance $\text{dist}(p, q') > r'$ for some vertex p.

Finally, the optimality of the algorithm is guaranteed since we search different possibilities for the point Q_p which is contained in the solution Q. By Lemma 2, we know that at least one such point Q_p is contained in the optimal solution. By trying all possibilities, we ensure that the globally optimal solution is obtained.

4   Algorithmic Complexity of the Problem

The decision version of the exact problem consists of determining whether we can draw a polygon with one vertex in or on each circle and each successive pair of vertices is able to see each other around the obstructions formed by vertices of the input. For any fixed choice of the obstructing vertices, this consists of a system of quadratic constraints of the form "variable point in circle" and "two variable points collinear with one constant point." For the optimization version we need only make the circle radius a variable and minimize that. This is a simple mathematical programming problem, potentially tractable with a general solver.

Solving systems that include quadratic constraints is in general NP-hard; we can easily reduce from 0-1 programming by means of constraints of the form $x(x - 1) = 0$. Nonetheless, some kinds of quadratic constraints can be addressed by known efficient algorithms. Lobo et al. [13] describe many applications for second-order cone programming, a special case of semidefinite programming. The

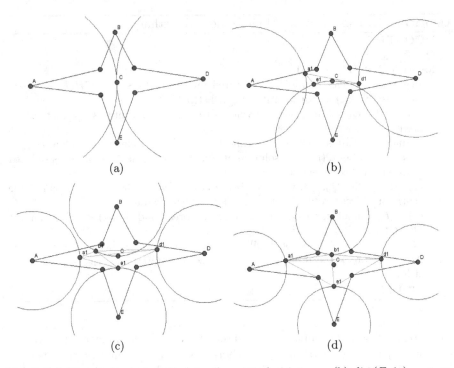

(a)   (b)

(c)   (d)

Fig. 2. (a) Two circles are centred on the critical points v_f. (b) dist$(E, 1c) = r$, so we add a new circle centred on E of radius r. The orange (light) lines indicate lines of visibility between the circles. (c) Another circle is added centred at point B. (d) We cannot shrink the circles any further, otherwise Invariant 1 would be violated. Therefore, a solution can be composed from the orange line segments.

"point in circle" constraints of our problem can be easily expressed as second-order cone constraints, so we might hope that our problem could be expressed as a second-order cone program and solved by their efficient interior point method.

However, the "two variable points collinear with one constant point" constraints are not so easy to handle. With (x_1, y_1) and (x_2, y_2) the variable points and (x_C, y_C) the constant point, we have the following:

$$\frac{y_1 - y_C}{x_1 - x_C} = \frac{y_2 - y_C}{x_2 - x_C} \tag{1}$$

$$x_2 y_1 - x_2 y_C - x_C y_1 = x_1 y_2 - x_1 y_C - x_C y_2 \tag{2}$$

This constraint is hyperbolic because of its cross-product terms. The techniques of Lobo et al. [13] can be applied to some hyperbolic constraints, subject to limitations whose basic purpose is to keep the optimization region convex.

As shown in Figure 3, it is possible for our problem to have two disconnected sets of solutions, even with as few as four circles. For a point A on the first circle, we can trace the polygon through the constant point B to that edge's intersection with the second circle at C, then through the constant point D and

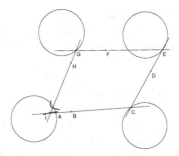

Fig. 3. Two disconnected solution intervals

so on around to H. The lines AB and GH intersect at I, which is our choice
for one vertex of the polygon, the others being C, E, and G. If I is inside the
circle, we have a feasible solution. But the heavy curves show the locus of I for
different choices of A, and the portion of it inside the circle is in two disjoint
pieces. The set of solutions to the problem as shown is disjoint, corresponding to
a slice (for a constant value of the circle-radius variable) through a non-convex
optimization region. As a result, neither second-order cone programming nor any
other convex optimization technique is immediately applicable.

5   An Approximation Algorithm Hausdorff Core

5.1   The Decision Problem

First we discuss the decision version of the approximation algorithm, where were
are given a distance r and wish to know whether there is an approximate Hausdorff core solution with $H(P, Q') \leq r + 2\varepsilon'$. This approximation scheme seeks to
grow circles by an additive factor ε', and determine whether there exists a solution for these expanded circles. We still require that the approximate solution Q'
must not cross outside P, and that Invariant 1 holds. Given ε as input, where ε is
a small positive constant, we calculate $\varepsilon' = d_{vf} \cdot \varepsilon$ as the approximation factor of
$H(P, Q)$. Recall that d_{vf} is the distance from the constrained 1-centre P_{1c} to the
most distant vertex $v_f \in P$. Notice that this method of approximation maintains
a scale invariant approximation factor, and the size of the of the approximation
factor for a given P is constant, regardless of Q and the magnitude of r.

The strategy behind this approximation scheme is that by growing the circles
by ε', they may be discretized. Consequently, it is possible to check for strong visibility between discrete intervals, which avoids some of the problems faced by the
exact formulation of the problem. One of the challenges of this approach is the selection of the length of the intervals on the new circles of radius $r + \varepsilon'$. We require
that the intervals be small enough so that we will find a solution for the approximation if one existed for the original circle radius. In other words, given an exact
solution Q for the original radius r such that $H(P, Q) \leq r$, we are guaranteed that
at least one interval on each of the expanded circles will be contained inside Q.

First we determine whether the polygon can be approximated by a single line segment. We construct an arc segment of radius $2d_{vf}$ (the maximum diameter of P) and arc length ε'. The interior angle of the circular segment C_φ formed by this arc is $\varphi = \varepsilon'/2d_{vf} = \varepsilon/2$. If an interior angle of Q' is less than or equal to φ, then Q' may be fully covered by C_φ since Q' is convex. In this case, there exists a line segment Q_ℓ which approximates Q' such that $H(Q', Q_\ell) < \varepsilon'$.

To determine whether Q can be approximated by a line segment, we grow all existing circles by a further factor of ε', so that they have radius $r^* = r + 2\varepsilon'$. Since Q is convex, this operation means that a line segment which approximates Q will now intersect at least one arc from each circle. By Lemma 2, we know that $P_c \in \{P_{1c}, P_V\}$ is contained in Q. Therefore, we attempt to find a line intersecting a point P_c and a segment of each circle of radius r^* for each P_c. For a selected P_c, we build an interval graph in the range $[0...\pi]$. For each circle C_i, if a line at angle θ mod π from an arbitrary reference line intersects a segment of C_i contained in P before intersecting P itself, then C_i covers θ in the interval graph. If there is a non-zero intersection between all circles in the interval graph, then the solution is a line segment Q_ℓ at angle θ to the reference line, intersecting P_c with endpoints at the last circles that Q_ℓ intersects. Therefore, if there exists a solution $H(P, Q) \leq r$ where Q can be approximated by a line segment Q_ℓ with $H(Q, Q_\ell) < 2\varepsilon'$, then we will find Q_ℓ.

If we have not found a solution Q_ℓ, we know that all interior angles of Q are greater than φ, and so we wish to determine an approximating polygon Q'. If we divide the expanded circle of radius $r + \varepsilon'$ into $12\pi/(\varepsilon^2 d_{vf})$ equal intervals, at least one would be fully contained in Q regardless of where the intervals are placed on the circle. Now finding Q' is simply a matter of finding a set of intervals such that there exists one interval on each circle which has strong visibility with an interval on all the other circles, and then selecting one point from each interval. A solution has the form $Q' = \{q_1 \ldots q_k\}$, where q_i is a point on C_i in the interval contained in the solution.

We use a dynamic programming algorithm to find a solution given a set of circles in the input polygon. We use a table $A[i, j]$ that stores, for a pair of intervals i and j in different circles, a range of possible solutions that include those intervals (See Figure 4). We find the convex polygon that includes intervals i and j by combining two convex polygons, one that includes i and an interval k^* and another that includes j and k^*. In order to compute $A[i, j]$ we lookup the entries for $A[i, k_1] \ldots A[i, k_m]$ and $A[k_1, j] \ldots A[k_m, j]$, where $k_1, \ldots, k_m$ are the intervals of a circle k, to determine if there is such k^* for which there are solutions $A[i, k^*]$ and $A[k^*, j]$ that can be combined into one convex polygon. There are many solutions that include a certain pair of intervals, but we store only $O(n)$ solutions for each pair. For example, for the entry $A[i, j]$ we would store the edge coming out of j that minimizes the angle Θ for each choice of an edge coming out of interval i, as shown in Figure 4. This would be done recursively at each level, which would make partial solutions easier to combine with other solutions while keeping convexity. Note that a particular choice of pairs of circles to form the solution Q' corresponds to a triangulation of Q', and since there are $O(n)$ pairs

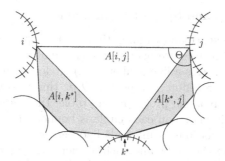

Fig. 4. The convex polygon that includes intervals i and j is built by combining a polygon that includes i and k^* and one that includes j and k^* (painted in grey)

of vertices joined in the triangulation, we need to store entries for the intervals of $O(n)$ pairs of circles. Given the clique of strongly visible intervals, we may now freely choose a point from each interval to obtain the solution polygon Q'. We run the dynamic programming algorithm iteratively for each $P_c \in \{P_{1c}, P_V\}$, using only circles centred on vertices $v \in P_V$ where $\text{dist}(v, P_c) < r$. If no solution Q' is found for any P_c, then there is no solution where $H(P, Q) = r$.

We present the following observations pertaining to Q and Q':

- $\exists Q \Rightarrow \exists Q'$, $\neg \exists Q' \Rightarrow \neg \exists Q$. The intervals are defined such that at least one interval from each circle will be contained in Q'.
- $\exists Q' \nRightarrow \exists Q$. The existence of Q' does not imply the existence of Q because the optimal solution may have circles of radius $r + \nu d_{vf}$, where $\nu < \varepsilon$.

5.2   The Minimization Problem

Given an optimal solution polygon Q where $H(P, Q) = r_{OPT}$, our algorithm finds an approximate solution Q' such that $H(P, Q') < r_{OPT} + 3\varepsilon'$. To determine a value of r' such that $r' \leq r_{OPT} + 3\varepsilon'$, it suffices to perform a binary search over possible values for r' in the range of $[0 \dots v_f]$ executing the decision approximation algorithm at each iteration. At the ith iteration of the algorithm, let the current radius be r_i. If the algorithm finds a solution Q_i such that $H(P, Q_i) = r_i$, we shrink the circles and use $r_{i+1} = r_i - dvf/2^i$. If the algorithm fails to find a solution, we use $r_{i+1} = r_i + dvf/2^i$. Initially, $r_0 = d_{vf}$, and the stopping condition is met when we find an approximate solution for radius r, and the approximate decision algorithm fails for radius $r - \varepsilon'$. Thus, the minimization version of the approximation algorithm requires $O(\log(\varepsilon^{-1}))$ iterations of the decision algorithm to find a solution. In the decision version, we showed that $H(Q, Q') < 2\varepsilon'$, if Q exists. In the minimization version, the best solution for a value of r may approach ε' less than the optimal value located on one of the radius intervals. Therefore, the minimization algorithm returns a solution Q' where $H(P, Q') < r_{OPT} + 3\varepsilon'$.

5.3   Running Time and Space Requirements

First we estimate the space and running time of the approximate decision algorithm. We compute the 1-centre using the technique in [6], which takes $O(n^2)$ time. The line solution tests a line against $O(n)$ circles, each of which may have $O(n)$ segments. This procedure is repeated $O(n)$ times, so this requires $O(n^3)$ time in total. In the dynamic programming table, there are $O(n)$ pairs of circles. The number of intervals on each circle is bounded by $O(\varepsilon^{-2})$, so we have $O(\varepsilon^{-4})$ possible combinations of intervals between two circles. Therefore there are $O(n\varepsilon^{-4})$ entries in the table, and each of them stores a description of $O(n)$ solutions. Hence the table needs roughly $O(n^2\varepsilon^{-4})$ space. If the number of entries in the table is $O(n\varepsilon^{-4})$, the dynamic programming algorithm should run in time $O(n\varepsilon^{-6})$, since in order to calculate each entry we need to check all the $O(\varepsilon^{-2})$ intervals of one circle. The algorithm may require $O(n)$ iterations to test each value of P_c, so the approximate decision algorithm requires $O(n^3 + n^2\varepsilon^{-6})$ time. Finally, the minimization version of the algorithm performs $O(\log(\varepsilon^{-1}))$ iterations of the approximate decision algorithm, so the complete algorithm requires $O((n^3 + n^2\varepsilon^{-6})\log(\varepsilon^{-1}))$ time to find an approximate solution.

6   Discussion and Directions for Future Research

The d-core problem is defined for any metric on polygons; we chose the Hausdorff metric, but many others exist. A natural extension of the Hausdorff metric might consider the *average* distance between two polygons instead of the *maximum*. This metric could be defined as follows:

$$H'(P, Q) = \max \left\{ \int_{p \in P} \min_{q \in Q} \operatorname{dist}(p, q) \; dp, \int_{q \in Q} \min_{p \in P} \operatorname{dist}(p, q) \; dq \right\},$$

where $\operatorname{dist}(p, q)$ denotes the Euclidean (ℓ_2) distance between points p and q. If Q is a point, then finding a point Q that minimizes $H'(P, Q)$ for a given polygon P corresponds to the continuous Weber problem, also known as the continuous 1-median problem. In the discrete setting, no algorithm is known for finding the exact position of the Weber point [4]. Furthermore, the problem is not known to be NP-hard nor polynomial-time solvable [11]. That suggests our problem may be equally poorly-behaved. Fekete et al. [10] considered the continuous Weber problem under the ℓ_1 distance metric.

In our original application, we hoped to create a hierarchy of simplified polygons, from full-resolution contour lines down to the simplest possible approximations. Then we could test paths against progressively more accurate, and more expensive, approximations until we got a definitive answer. We would hope to usually terminate in one of the cheaper levels. But our definition of d-core requires the core to be convex. Convexity has many useful consequences and so is of theoretical interest, but it represents a compromise to the original goal because it only provides one non-adjustable level of approximation. It would be interesting to consider other related problems that might provide more control over the approximation level.

Therefore, a direction for further work would be to define some other constraint to require of the simplified polygon. For instance, we could require that it be star-shaped, i.e. there is some point $p \in P$ such that every $q \in P$ can see p. A similar but even more general concept might be defined in terms of link distance.

Acknowledgements. The authors would like to thank Diego Arroyuelo and Barbara Macdonald for their participation in early discussions of the problem, and the anonymous reviewers for their useful comments and suggestions.

References

1. Agarwal, P.K., Arge, L., Murali, T.M., Varadarajan, K.R., Vitter, J.S.: I/O-efficient algorithms for contour-line extraction and planar graph blocking. In: Proc. SODA, pp. 117–126. SIAM, Philadelphia (1998)
2. Aggarwal, A., Klawe, M.M., Moran, S., Shor, P., Wilber, R.: Geometric applications of a matrix-searching algorithm. Algorithmica 2(1), 195–208 (1987)
3. Aggarwal, A., Park, J.: Notes on searching in multidimensional monotone arrays. In: Proc. SFCS, pp. 497–512. IEEE Computer Society Press, Los Alamitos (1988)
4. Bajaj, C.: The algebraic degree of geometric optimization problems. Disc. & Comp. Geom. 3, 177–191 (1988)
5. Bhattacharya, B.K., Mukhopadhyay, A.: On the minimum perimeter triangle enclosing a convex polygon. In: Akiyama, J., Kano, M. (eds.) JCDCG 2002. LNCS, vol. 2866, pp. 84–96. Springer, Heidelberg (2003)
6. Bose, P., Toussaint, G.: Computing the constrained Euclidean geodesic and link center of a simple polygon with applications. In: Proc. CGI, p. 102. IEEE, Los Alamitos (1996)
7. Chang, J.S., Yap, C.K.: A polynomial solution for the potato-peeling problem. Disc. & Comp. Geom. 1(1), 155–182 (1986)
8. Chassery, J.-M., Coeurjolly, D.: Optimal shape and inclusion. In: Mathematical Morphology: 40 Years On, vol. 30, pp. 229–248. Springer, Heidelberg (2005)
9. Dobkin, D.P., Snyder, L.: On a general method for maximizing and minimizing among certain geometric problems. In: Proc. SFCS, pp. 9–17 (1979)
10. Fekete, S.P., Mitchell, J.S.B., Weinbrecht, K.: On the continuous Fermat-Weber problem. Oper. Res. 53, 61–76 (2005)
11. Hakimi, S.L.: Location theory. In: Rosen, Michaels, Gross, Grossman, Shier (eds.) Handbook Disc. & Comb. Math. CRC Press, Boca Raton (2000)
12. Kirkpatrick, D.: Optimal search in planar subdivisions. SIAM J. Comp. 12(1), 28–35 (1983)
13. Lobo, M.S., Vandenberghe, L., Boyd, S., Lebret, H.: Applications of second-order cone programming. Lin. Alg. & App. 284(1–3), 193–228 (1998)
14. Lopez, M.A., Reisner, S.: Hausdorff approximation of convex polygons. Comp. Geom. Theory & App. 32(2), 139–158 (2005)
15. Melissaratos, E.A., Souvaine, D.L.: On solving geometric optimization problems using shortest paths. In: Proc. SoCG, pp. 350–359. ACM Press, New York (1990)
16. O'Rourke, J., Aggarwal, A., Maddila, S., Baldwin, M.: An optimal algorithm for finding minimal enclosing triangles. J. Alg. 7, 258–269 (1986)
17. Schwarz, C., Teich, J., Vainshtein, A., Welzl, E., Evans, B.L.: Minimal enclosing parallelogram with application. In: Proc. SoCG, pp. 434–435. ACM Press, New York (1995)

Efficient Construction of Near-Optimal Binary and Multiway Search Trees

Prosenjit Bose and Karim Douïeb*

School of Computer Science, Carleton University, Herzberg Building
1125 Colonel By Drive, Ottawa, Ontario, K1S 5B6 Canada
{jit,karim}@cg.scs.carleton.ca
http://cg.scs.carleton.ca

Abstract. We present a new linear-time algorithm for constructing multiway search trees with near-optimal search cost whose running time is independent of the size of the node in the tree. With the analysis of our construction method, we provide a new upper bound on the average search cost for multiway search trees that nearly matches the lower bound. In fact, it is tight for infinitely many probability distributions. This problem is well-studied in the literature for the case of binary search trees. Using our new construction method, we are able to provide the tightest upper bound on the average search cost for an optimal binary search tree.

1 Introduction

Search trees are fundamental data structures widely used to store and retrieve elements from a set of totally ordered keys. The problem of building static search trees optimizing various criteria is well-studied in the literature.

Consider a set $x_1, x_2, \ldots, x_n$ of ordered keys. We are given $2n + 1$ weights $q_0, p_1, q_1, p_2, q_2, \ldots, p_n, q_n$ such that $\sum_{i=1}^{n} p_i + \sum_{i=0}^{n} q_i = 1$. Each p_i is the probability that we query the key x_i (a successful search) and q_i is the probability that we query a key lying between x_i and x_{i+1} (an unsuccessful search). Note that q_0 is the probability that we query a key that is less than all keys in the set. Similarly, q_n is the probability we query a key that is greater than all keys in the set.

A static k-*ary tree* (or a multiway search tree) is a generalization of most static search tree structures. Each internal node of a k-ary tree contains at least one key and at most $k - 1$ keys, so it has between 2 and k children. A leaf in a k-ary tree does not contain any key. A successful search ends up at the internal node of the k-ary tree containing the requested key, whereas an unsuccessful search ends up in one of the $n + 1$ leaves of the k-ary tree. A standard measure of the average number of nodes traversed in a k-ary tree T is the *average path-length* (or weighted path-length) defined as follows:

---

* Research partially supported by NSERC and MRI.

F. Dehne et al. (Eds.): WADS 2009, LNCS 5664, pp. 230–241, 2009.
© Springer-Verlag Berlin Heidelberg 2009

$$\sum_{i=1}^{n} p_i(d_T(x_i) + 1) + \sum_{i=0}^{n} q_i d_T(x_{i-1}, x_i), \qquad (1)$$

where $d_T(x_i)$ is the depth of the internal node containing the key x_i, which is the number of edges on the path from the internal node to the root. The depth of the leaf reached at the end of an unsuccessful search for a key lying between x_{i-1} and x_i is denoted by $d_T(x_{i-1}, x_i)$. In the context of binary search trees in the comparison-based model, the path length corresponds to the average number of comparisons performed during a search. In the external memory model, the path length corresponds to the average number of I/Os performed during a search.

In this paper, we provide a tight upper bound on the average path-length of a static optimal k-ary tree and then find an efficient algorithm to build a k-ary tree given the access probability distribution whose average path-length matches this bound. The construction algorithm is also shown to be near-optimal. For the case of binary search trees (where $k=2$), we show that our construction algorithm improves on the current best upper bound on the average path-length.

1.1   Related Work

Knuth [11] has shown that an optimal binary search tree can be built in $O(n^2)$ time using $O(n^2)$ space. Mehlhorn [14] gave an $O(n)$ time algorithm to build a binary search tree that is near-optimal. His analysis of the construction algorithm provided the first upper bound on the average path-length of an optimal binary search tree. Currently, the tightest upper bound on the average path-length of an optimal binary search tree is due to De Prisco and De Santis [15].

With respect to k-ary trees, Vaishnavi et al. [18] and Gotlieb [8] independently showed that an optimal k-ary tree can be built in $O(kn^3)$ time. Becker [2] gave an $O(kn^\alpha)$ time algorithm, where $\alpha = 2 + \log_k 2$, to build an optimal k-ary tree in the model proposed by Bayer and McCreight [1] (B-tree) where every leaf in the k-ary tree has the same depth and every internal node contains between $(k-1)/2$ and $k-1$ keys except possibly the root. In the remainder of the paper, we only consider the general k-ary tree model and not the B-tree model. Becker [3] presented a method to build a k-ary tree in $O(Dkn)$ time where D is the height of the resulting tree. He provided empirical evidence to suggest that the k-ary tree produced by his method is near-optimal, however no theoretical bound was given on the average path-length of the resulting tree nor was any approximation ratio proven. In the restricted model of (a,b) trees where a node has between a and b children with $2 \le a \le \lceil b/2 \rceil$ and considering only successful searches, Feigenbaum and Tarjan [6] showed how to achieve a good upper bound on the optimal path length of an (a,b) tree (this improved [4,5]).

The problem of building an optimal search tree when only unsuccessful searches occur, i.e., when $\sum_{i=1}^{n} p_i = 0$, is called the optimal *alphabetic search tree* problem. An alternate view of this problem is to consider that the keys only occur in the leaves and the internal nodes simply serve as indicators to help guide the search to the appropriate leaf. Hu and Tucker [9] developed an $O(n^2)$ time and $O(n)$ space algorithm for constructing an optimal alphabetic binary search tree. This was subsequently improved by two algorithms; the first was developed by Knuth [10] and the second by Garsia and Wachs [7]. Both algorithms require $O(n \lg n)$ [1] time and $O(n)$ space. Yeung [19] provided an $O(n)$ time algorithm to build a near-optimal alphabetic binary search tree. Yeung's upper bound was then improved by De Prisco and De Santis [15].

1.2 Our Results

In Section 2, we describe our linear-time algorithm to build a near-optimal k-ary tree whose running time is independent of the size of the node in the tree. Let P_{opt} represent the average path-length of the optimal k-ary tree and P_T represent the average path-length of the tree built using our algorithm. We prove that

$$\frac{H}{\lg(2k-1)} \leq P_{opt} \leq P_T \leq \frac{H}{\lg k} + 1 + \sum_{i=0}^{n} q_i - q_0 - q_n - \sum_{i=0}^{m} q_{rank[i]},$$

where $H = -\sum_{i=1}^{n} p_i \lg p_i - \sum_{i=0}^{n} q_i \lg q_i$ is the entropy of the access probability distribution. The value $m = \max\{n - 3P, P\} - 1 \geq \frac{n}{4} - 1$ where P is the number of increasing or decreasing sequences in the access probability distribution on the ordered leaves. The value $q_{rank[i]}$ is the ith smallest access probability among the leaves except for the extremal ones (i.e. we exclude q_0 and q_n from consideration). The upper and lower bounds are explained in subsection 2.4 and 2.5 respectively.

We provide a better upper bound on the average path-length of an optimal k-ary tree. For $k = 2$, i.e. binary search trees, our new construction provides a better upper bound than the algorithm by Mehlhorn [14]. Moreover, when restricted to alphabetic trees, our method provides a better upper bound than the method of Yeung [19]. The current best upper bound on the average path-length of binary search trees is derived through a construction algorithm by De Prisco and De Santis [15]. Since at the core of their algorithm they use Yeung's method, we are able to provide a tighter upper bound on the average path-length by incorporating our construction algorithm instead. We show precisely how to do this in Section 4.

2   New Method to Construct Near-Optimal k-Ary Trees

We introduce a technique to build near-optimal multiway search trees inspired by Mehlhorn's technique [14] to build near-optimal binary search trees. To the best of our knowledge, our technique is the first to provide an upper bound on the average path-length of an optimal k-ary tree. The construction algorithm by Becker [3] was only shown empirically to be near-optimal. No theoretical proof is given. In the case of binary search trees (k=2), our technique improves the upper bound on the path-length given by Mehlhorn [14] and by Yeung [19] (for alphabetic tree).

---

[1] $\lg x$ in this paper is defined as $\log_2 x$.

We begin with an overview of the main steps of our technique: given the access probabilities $p_1, p_2, \ldots, p_n$ (for the internal nodes) and $q_0, q_1, \ldots, q_n$ (for the leaves), we first build a new sequence of probabilities, $p'_1, \ldots, p'_n$, by transferring the weights of the leaves to the internal nodes. We then build an initial tree using these new weights on the internal nodes. Finally, we attach the leaves to this tree, resulting in the final tree. The key to ensuring that the final tree is near-optimal is in the weight transfer. When a leaf with access probability q_i is attached, the weight transfer guarantees that it does not fall much deeper than $\log_k \frac{1}{q_i}$.

2.1 Transfer of Leaf Weights

Consider the access probability distribution on the leaves $q_0, q_1, \ldots, q_n$. This distribution is decomposed into maximal increasing or decreasing sequences of probabilities. The sequences are defined from left to right such that the probability of each leaf in an increasing sequence is less than or equal to the probability of its right adjacent leaf, i.e. $q_i \leq q_{i+1}$. For decreasing sequences, $q_i \geq q_{i+1}$. We define a *peak* (resp. a *valley*) to be the last leaf of an increasing (resp. decreasing) sequence. Let P be the number of peaks in a given distribution.

The leaf weights are transfered to the internal nodes in the following way: The weight of an extremal leaf, i.e., $(-\infty, x_1)$ or (x_n, ∞), is transferred entirely to its unique adjacent internal node. Thus, q_0 is added to p_1 and q_n is added to p_n. The weight of a peak or a valley leaf is split equally between its two adjacent internal nodes, i.e. $q_i/2$ is added to p_i and to p_{i+1}. Finally, the weight of a leaf in an increasing (resp. decreasing) sequence is transfered entirely to its adjacent left (resp. right) internal node. Let p'_i be the weight of the internal node x_i after the weights of the leaves have been transferred. Note that by construction, we have that $p'_i \geq p_i$ for all node x_i and $\sum_{i=1}^{n} p'_i = 1$. See Fig. 1 for an example of weight transfer.

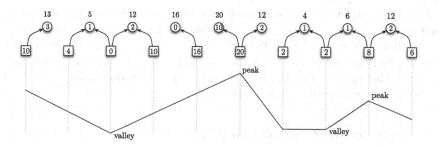

Fig. 1. Example of transferring the weights of the leaves (squares) into the internal nodes (disks). The number on top of an internal node is its weight after the transfer. The weights are represented as percentages.

Lemma 1. *After transferring the weights, the following is satisfied:*

$$q_i \leq 2\min\{p_i', p_{i+1}'\} \quad \text{for} \quad i = 1, \ldots, n-1, \tag{2}$$

$$q_0 \leq p_1', \tag{3}$$

$$q_n \leq p_n', \tag{4}$$

$$q_i \leq \min\{p_i', p_{i+1}'\} \quad \text{for at least } \max\{n - 3P, P\} - 1 \text{ leaves.} \tag{5}$$

Proof. First consider the extremal leaves $(-\infty, x_1)$ and (x_n, ∞). Their weights, q_0 and q_n are entirely transferred to the internal nodes x_1 and x_n, respectively. Hence $q_0 \leq p_1'$ and $q_n \leq p_n'$, thereby proving (3) and (4). For the remainder of this proof when referring to a leaf, we exclude the extremal leaves from consideration.

The weight q_i of a valley leaf (x_i, x_{i+1}) is divided equally between its two adjacent internal nodes x_i and x_{i+1}. Since (x_i, x_{i+1}) is a valley leaf, by definition, its two adjacent leaves (x_{i-1}, x_i) and (x_{i+1}, x_{i+2}) have a weight that is greater than or equal to q_i. Moreover, leaf (x_{i-1}, x_i) is either a peak or on a decreasing sequence, which means that it transfers at least half of its weight to x_i. Similarly, (x_{i+1}, x_{i+2}) is either a peak or on an increasing sequence and transfers at least half of its weight to x_{i+1}. Thus, p_i' and p_{i+1}' are each greater than or equal to q_i. Note that the number of valley leaves is at least $P - 1$. This partially proves (5), namely that at least $P - 1$ leaves satisfy $q_i \leq \min\{p_i', p_{i+1}'\}$.

A peak leaf (x_i, x_{i+1}) divides its weight equally between the internal nodes x_i and x_{i+1}. Thus, we have $\frac{q_i}{2} \leq \min\{p_i', p_{i+1}'\}$.

Call a leaf adjacent to a peak, a *subpeak* if it is not already a valley. For a subpeak (x_i, x_{i+1}), note that either $\frac{q_i}{2} \leq p_i'$ or $\frac{q_i}{2} \leq p_{i+1}'$ depending on whether or not the subpeak is on the left or right side of the peak. Without loss of generality, assume it is on the left. Since the peak is to its right, we have that $\frac{q_i}{2} \leq p_{i+1}'$. As the subpeak is part of an increasing sequence, all of the weight of q_i is transferred to x_i, which implies that $q_i \leq p_i'$.

Finally, a leaf (x_i, x_{i+1}) that is neither a peak, a subpeak nor a valley transfers its entire weight either to x_i or x_{i+1} depending on whether it is on an increasing or decreasing sequence. Without loss of generality, assume it is on an increasing sequence. The entire weight of (x_i, x_{i+1}) is transferred to the internal node x_i, thus $q_i \leq p_i'$. Since it is not a subpeak, the weight of (x_{i+1}, x_{i+2}) is greater than or equal to q_i and entirely transferred to the internal node x_{i+1}, thus $q_i \leq p_{i+1}'$. So (5) is guaranteed for every leaf except for the peaks and subpeaks, i.e., for $n - 3P - 1$ leaves (the extremal leaves are not counted). This proves (2 and 5).   □

2.2   Construction of the Internal Tree

After transferring the leaf weights, we have an ordered set of keys $x_1, \ldots, x_n$ where each key x_i has a weight p_i' such that $\sum_{i=1}^{n} p_i' = 1$. We now begin the construction of the k-ary tree. The main idea is to ensure that after each step of search the weight is divided by k. Our construction is recursive. The keys $x_{k_1}, \ldots, x_{k_\ell}$ (with $\ell < k$) are selected to form the root node of the internal tree provided that they adhere to the following properties:

$$\sum_{i=1}^{k_1-1} p_i' \le \frac{1}{k}, \quad \sum_{i=k_j+1}^{k_{j+1}-1} p_i' \le \frac{1}{k} \quad \text{for} \quad j = 2, \ldots, \ell-1, \quad \text{and} \quad \sum_{i=k_\ell+1}^{n} p_i' \le \frac{1}{k}.$$

We outline precisely how to efficiently select a set of keys having these properties in Section 3. The procedure is then repeated recursively in each subtree of the root. The probabilities are normalized at each iteration.

Lemma 2. *A tree T built according to our construction guarantees*

$$d_T(x_i) \le \left\lfloor \log_k \frac{1}{p_i'} \right\rfloor \quad \text{for} \quad i = 1, \ldots, n.$$

Proof. By construction, the sum of the weights p' of the nodes contained in a subtree at depth j is at most $1/k^j$. This implies $d_T(x_i) \le \lfloor \log_k \frac{1}{p_i'} \rfloor$.   □

2.3   Attach the Leaves

Consider $x_{k_1}, x_{k_2}, \ldots, x_{k_\ell}$ the set of keys contained in a node y of the k-ary tree T constructed so far. We define the left or the right child of a key x_{k_i} as the child of y that defines a subtree containing keys in the ranges $[x_{k_{i-1}}, x_{k_i}]$ or $[x_{k_i}, x_{k_{i+1}}]$ respectively. The left child of x_{k_1} and the right child of x_{k_ℓ} corresponds to the child of y that defines a subtree containing keys smaller than x_{k_1} or respectively greater than x_{k_ℓ}. Note that two adjacent keys inside an internal node share their right and left child.

Once the internal tree T is built, the leaf nodes are attached. This is done by performing an inorder traversal of the internal tree adding the leaves sequentially to the tree. If within an internal node, a key x_i has no left or right child, then the leaf (x_{i-1}, x_i) or (x_i, x_{i+1}), respectively, is attached to it. Since a leaf (x_{i-1}, x_i) is necessarily a child of an internal node containing either the key x_{i-1}, x_i or both, we note the following:

$$d_T(x_{i-1}, x_i) \le \max\{d_T(x_{i-1}), d_T(x_i)\} + 1. \tag{6}$$

Building the internal tree and attaching the leaves to it are two steps that can easily be done at the same time.

2.4   Upper Bound

Theorem 1. *The average path-length of a k-ary tree built using the method presented above is at most*

$$\frac{H}{\lg k} + 1 + \sum_{i=0}^{n} q_i - q_0 - q_n - \sum_{i=1}^{m} q_{rank[i]},$$

where $H = -\sum_{i=1}^{n} p_i \lg p_i - \sum_{i=0}^{n} q_i \lg q_i$ is the entropy of the access probability distribution. The value $m = \max\{n - 3P, P\} - 1 \ge \frac{n}{4} - 1$ and $q_{rank[i]}$ is the ith smallest access probability among leaves except the extremal leaves.

Proof. According to Lemma 2 the depth of an internal node x_i is

$$d_T(x_i) \leq \lfloor \log_k \frac{1}{p'_i} \rfloor \leq \lfloor \log_k \frac{1}{p_i} \rfloor.$$

Equation (6) says that the depth of a leaf (x_{i-1}, x_i) is

$$d_T(x_{i-1}, x_i) \leq \max\{d_T(x_{i-1}), d_T(x_i)\} + 1 \leq \lfloor \log_k \max\{\frac{1}{p'_{i-1}}, \frac{1}{p'_i}\} \rfloor + 1.$$

By Lemma 1, $d_T(x_{i-1}, x_i) \leq \lfloor \log_k \frac{2}{q_i} \rfloor + 1$ for every leaf (x_{i-1}, x_i) and for at least m of them $d_T(x_{i-1}, x_i) \leq \lfloor \log_k \frac{1}{q_i} \rfloor + 1$ (possibly the ones with the smallest weights). This is also the case for the extremal leaves. Putting this in equation (1) proves the theorem. □

Note that for $k = 2$, i.e., for binary search trees, this new construction improves the performance of the original method of Mehlhorn [14]. That one guarantees $H + \sum_{i=1}^{n} p_i + \sum_{i=0}^{n} 2q_i$, thus we decrease this upper bound by $(q_0 + q_n) + \sum_{i=0}^{m} q_{rank[i]}$. When restricted to alphabetic trees our method improves over the method of Yeung [19], we decrease his bound by $\sum_{i=0}^{m} q_{rank[i]}$. Since the best method to build near-optimal BST developed by De Prisco and De Santis [15] uses Yeung method, we consequently improve it. We will see more precisely how in Section 4.

The construction of an (a,b) tree developed by Feigenbaum and Tarjan [6] can easily be adapted to handle unsuccessful searches so that the path length is upper bounded by $\frac{H}{\lg a} + 3 + \sum_{i=0}^{n} q_i$. This gives in the general case of k-ary tree model a path length of at most $\frac{H}{\lg(k)+1} + 3 + \sum_{i=0}^{n} q_i$ since b is the maximum number of children in the tree and $a \leq b/2$.

2.5   Lower Bound

Theorem 2. *The path length of an optimal k-ary tree is at least $\frac{H}{\lg(2k-1)}$, where $H = -\sum_{i=1}^{n} p_i \lg p_i - \sum_{i=0}^{n} q_i \lg q_i$ is the entropy of the access probability distribution.*

Proof. This corresponds in fact to an information theoretic result due to Shannon [17]. A k-ary tree corresponds to a code tree of degree $(2k - 1)$ derived by moving the weight of each key inside a node to a leaf extending from this node. Then the variable length coding theorem for codes using an alphabet of size $(2k - 1)$ gives the lower bound, i.e., $\frac{H}{\lg(2k-1)}$. □

Now we show that our lower bound on the path-length of any k-ary tree is sharp for precise access probability distributions. The argument used is a generalization of [13]. A k-ary tree is said to be *full* if each of its internal nodes contains exactly $k - 1$ keys. Consider any full k-ary tree T and define an access probability distribution on the keys in T as follows:

$$p_i = 1/(2k-1)^{d(x_i)+1},$$
$$q_i = 1/(2k-1)^{d(x_{i-1},x_i)},$$

where $d(x_i)$ is the depth of the key x_i in the tree T and $d(x_{i-1}, x_i)$ is the depth of the leaf (x_{i-1}, x_i) in T. It is easy to show by induction that $\sum_{i=1}^{n} p_i + \sum_{i=0}^{n} q_i = 1$, i.e., the weights p_i and q_i are indeed probabilities. The path-length of T is

$$\frac{H}{\lg(2k-1)} \leq P_{opt} \leq P_T = \sum_{i=1}^{n} p_i(d(x_i)+1) + \sum_{i=0}^{n} q_i d(x_{i-1}, x_i)$$

$$= \sum_{i=1}^{n} p_i \log_{(2k-1)} p_i + \sum_{i=0}^{n} q_i \log_{(2k-1)} q_i$$

$$= \frac{H}{\lg(2k-1)}.$$

This shows that our lower bound on the path-length of a k-ary tree is sharp for infinitely many distributions. Note that for those distributions our algorithm constructs the optimal tree. For the particular case of binary search trees (k=2) tighter lower bounds on the path length are known [12,13,14,16] as well as for the alphabetic case [15,19].

3   Efficient Implementation

We now discuss the complexity of the construction algorithm. The set S of keys $x_{k_1}, \ldots, x_{k_\ell}$ (with $\ell < k$) associated with the root node of the tree is chosen using exponential and binary search techniques combined together similar to what Mehlhorn [14] did to construct near-optimal binary search trees. Here we generalize this approach to construct near-optimal k-ary trees. We first build in $O(n)$ time a table of size n where the ith element of the table, associated with the key x_i, is set to $s_i = \sum_{j=1}^{i} p'_j$.

Consider the smallest element x that satisfies $s_x \geq \frac{1}{k}$ and the greatest element y that satisfies $s_y \leq \frac{k-1}{k}$ (x could possibly be equal to y). In order to find one of these elements, we perform an exponential search simultaneously from both ends of the table. That is, we search for the smallest value i, with $i = 0, 1, 2, \ldots$, where either $s_{2^i} \geq \frac{1}{k}$ or $s_{n+1-2^i} \leq \frac{k-1}{k}$. Suppose j is the minimum value of i. Either $2^{j-1} \leq x \leq 2^j$ or $n+1-2^j \leq y \leq n+1-2^{j-1}$, in the first case we perform a binary search for x between the positions 2^{j-1} and 2^j, in the second case we perform a binary search for y between the positions $n+1-2^j$ and $n+1-2^{j-1}$. If we are in the first case, we include the key in position x into the set S of keys associated with the root of the tree, otherwise we include the key in position y. We then iterate the process on the table minus the part from its beginning to the element x or from the element y to the end of the table. The values s_i are virtually updated when the left part of the table is removed, i.e., s_i is set to $s_i - s_x$ (we only have to remember an offset value). The number of element has to be virtually updated as well.

By virtually updated, we mean that the algorithm retains two parameters, the first one is the weight of elements removed from consideration and the second is the number of elements remaining in consideration. We simply update values on the fly using these two parameters.

We stop the process when the remaining elements have a total weight of at most $1/k$. The process can be repeated at most $k-1$ times since at least $1/k$ of the weight is removed at each iteration. Let $I_1, I_2, \ldots, I_\ell$ be the sizes of successive parts removed from the array during each iteration in the selection of keys associated with the root described above. That is, in the first iteration, I_1 is the number of elements removed from the array (i.e. either all elements whose value is less than x or greater than y depending on which position is found first in the double-ended search). Note that ℓ is the number of keys in the root node. The complexity of the selection procedure is given by

$$\sum_{i=1}^{\ell} c \lg I_i,$$

where c is a constant. By construction, we have $I_i \leq \frac{n-\sum_{j=1}^{i-1} I_j+1}{2}$, this implies $n - \sum_{j=1}^{\ell} I_j \geq \frac{n-2^\ell+1}{2^\ell}$ (proof by induction on ℓ). Once the keys associated with the root are found the construction of the k-ary tree continues recursively in each part of size I_i. The time $T(n)$ required to build a k-ary tree of n keys using our construction technique is given by the following recurrence:

$$T(0) = 0,$$

$$T(n) = T(I_1 - 1) + \ldots + T(I_\ell - 1) + T(n - \sum_{i=1}^{\ell} I_i) + c \sum_{i=1}^{\ell} \lg I_i.$$

Theorem 3. *Our construction technique guarantees* $T(n) \leq c(n - \lg(n+1))$ *which is independent of k.*

Proof. We prove this theorem by induction on the number n of keys contained in the k-ary tree. For $n = 0$, we have $T(0) \leq 0$ which is true. For $n > 0$ we have

$$T(n) = T(I_1 - 1) + \ldots + T(I_\ell - 1) + T(n - \sum_{i=1}^{\ell} I_i) + c \sum_{i=1}^{\ell} \lg I_i.$$

We apply the induction hypothesis and obtain

$$T(n) \leq c(n - \ell) - c \sum_{i=1}^{\ell} \lg I_i - c \lg(n - \sum_{i=1}^{\ell} I_i + 1) + c \sum_{i=1}^{\ell} \lg I_i$$

$$= cn - c(\ell + \lg(n - \sum_{i=1}^{\ell} I_i + 1))$$

$$\leq cn - c(\ell + \lg(\frac{n - 2^\ell + 1}{2^\ell} + 1))$$

$$= c(n - \lg(n+1)). \qquad \square$$

Our method builds a near-optimal k-ary tree in $O(n)$ time. Note that the running time is independent of k. This improves over Becker's construction both in running time since his method has $O(kDn)$ construction time, where D is the height of the output tree, and more importantly, Becker's construction is only a heuristic (with empirical evidence to support its near-optimal average path length) whereas our bound is proven theoretically. Remark that Feigenbaum and Tarjan [6] did not specifically describe a tree construction algorithm. However it is easy to use their data structure to build a multiways tree since their structure is dynamic. Simply insert successively elements into an initially empty structure. The construction time of this algorithm is $O(n \lg_k n)$.

4   Improving Upper Bound on Optimal BST

In this section we show how to improve the method developed by De Prisco and De Santis [15] to build near-optimal binary search trees. We briefly describe their method. It consists of three phases:

Phase 1: Modify the instance of the problem by introducing $2n$ *auxiliary* keys $k_1, k_2, \ldots, k_{2n}$ with a zero access probability and by considering each original key as a leaf. The key k_i is adjacent to the $\lceil \frac{i}{2} \rceil$th original key and $\lfloor \frac{i}{2} \rfloor$th original leaf. Given this new instance of the problem, an alphabetic tree is built using the technique of Yeung [19].

Phase 2: Each original key x_i is moved up in the tree to replace an auxiliary key corresponding to the lowest common ancestor of the original leaves (x_{i-1}, x_i) and (x_i, x_{i+1}).

Phase 3: The last phase removes the remaining auxiliary keys left after the second phase. A node containing an auxiliary key and its two children are contracted into one node containing the original key present in one of the children.

Every phase takes $O(n)$ time. Basically, phase 1 builds a tree T on a given instance. Then phases 2 and 3 modify T to decrease by at least 2 the depth of every original key and by at least 1 the depth of every original leaf except for one. An example originally given by De Prisco and De Santis illustrating the different phases is shown in Fig. 2.

In Section 2 our method has been shown to achieve a tighter upper bound than that given by Yeung [19]. Therefore by using our method instead of Yeung's method in phase 1, we improve the construction and the bound given by De Prisco and De Santis [15] by $\sum_{i=0}^{m'} pq_{rank[i]}$.

Theorem 4. *The path length of an optimal binary search tree is at most*

$$H + 1 - q_0 - q_n + q_{max} - \sum_{i=0}^{m'} pq_{rank[i]},$$

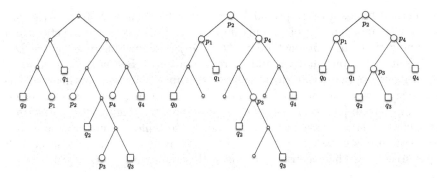

Fig. 2. Situation after each phase of the construction method

where $H = -\sum_{i=1}^{n} p_i \lg p_i - \sum_{i=0}^{n} q_i \lg q_i$ is the entropy of the access probability distribution. The value $m' = \max\{2n - 3P, P\} - 1 \geq \frac{n}{2} - 1$, $pq_{rank[i]}$ is the ith smallest access probability among every key and every leaf (except the extremal leaves) and q_{max} is the greatest leaf probability including external leaves.

Proof. Given the new instance constructed in phase 1, i.e., $2n$ auxiliary keys and $2n + 1$ leaves, our method guarantees that the depth of every leaf with an access probability α is smaller than $\lfloor \lg \frac{1}{\alpha} \rfloor$ plus 1 for at least $\max\{2n - 3P, P\} - 1$ leaves (for the extremal leaves as well) and plus 2 for the remaining leaves. Here P is defined as the number of peaks in the access probability distribution of the original keys and leaves combined, i.e., $q_0, p_1, q_1, \ldots, p_n, q_n$. Thus after applying the last two phases of the method of De Prisco and De Santis, each original key decrease its depth by at least 2 and each original leaf decrease it depth by at least 1 except for one (possibly the one with the greatest probability among the leaves). Therefore this binary search tree construction guarantees an upper bound on the path length of $H + 1 - q_0 - q_n + q_{max} - \sum_{i=0}^{m} pq_{rank[i]}$, $\square$

Remark that this result only improves the upper bound on the cost of optimal binary search trees. For the restricted case of alphabetic trees the upper bound matches the previous best one. This is because in the alphabetic case every $p_i = 0$ and the probability distribution on the leaves of the modified instance is as follows $q_0, 0, q_1, 0, \ldots, 0, q_n$. Thus the number of peaks in this distribution is exactly n minus the number of $q_i = 0$ which implies $\sum_{i=0}^{m'} pq_{rank[i]} = 0$. Thus our method only guarantees the same upper bound.

Corollary 1. *If all access probabilities are of the form 2^{-i} with $i \in \mathbb{N}$ then the upper bounds on the path length can be improved to $H - q_0 - q_n + q_{max} - \sum_{i=0}^{m} pq_{rank[i]} + pq_{min}$, where pq_{min} is the smallest non-null probability among all access probabilities. This is an improvement of $1 - pq_{min}$.*

Proof. Due to lack of space we omit the proof.

References

1. Bayer, R., McCreight, E.M.: Organization and maintenance of large ordered indexes. Acta Informatica 1(3), 173–189 (1972)
2. Becker, P.: A new algorithm for the construction of optimal B-trees. Nordic Journal of Computing 1, 389–401 (1994)
3. Becker, P.: Construction of nearly optimal multiway trees. In: Jiang, T., Lee, D.T. (eds.) COCOON 1997. LNCS, vol. 1276, pp. 294–303. Springer, Heidelberg (1997)
4. Bent, S., Sleator, D., Tarjan, R.: Biased 2-3 trees. In: Annual Symposium on Foundations of Computer Science, pp. 248–254 (1980)
5. Bent, S., Sleator, D., Tarjan, R.: Biased search trees. SIAM J. Comput. 14(3), 545–568 (1985)
6. Feigenbaum, J., Tarjan, R.: Two new kinds of biased search trees. Bell Syst. Tech. J. 62, 3139–3158 (1983)
7. Garsia, A.M., Wachs, M.L.: A new algorithm for minimum cost binary trees. SIAM Journal on Computing 6, 622–642 (1977)
8. Gotlieb, L.: Optimal multi-way search trees. SIAM Journal of Computing 10(3), 422–433 (1981)
9. Hu, T.C., Tucker, A.C.: Optimal computer search trees and variable-length alphabetical codes. SIAM Journal on Applied Mathematics 21(4), 514–532 (1971)
10. Knuth, D.: The Art of Computer Programming: Sorting and Searching, vol. 3. Addison-Wesley, Reading (1973)
11. Knuth, D.: Optimum binary search trees. Acta Informatica 1, 79–110 (1971)
12. Mehlhorn, K.: Data structures and algorithms: Sorting und Searching. EATCS Monographs on Theoretical Computer Science, vol. 1. Springer, Berlin (1984)
13. Mehlhorn, K.: Nearly optimal binary search trees. Acta Inf. 5, 287–295 (1975)
14. Mehlhorn, K.: A best possible bound for the weighted path length of binary search trees. SIAM Journal on Computing 6, 235–239 (1977)
15. Prisco, R.D., Santis, A.D.: On binary search trees. Information Processing Letters 45(5), 249–253 (1993)
16. Prisco, R.D., Santis, A.D.: New lower bounds on the cost of binary search trees. Theor. Comput. Sci. 156(1&2), 315–325 (1996)
17. Shannon, C.: A mathematical theory of communication. Bell System Technical Journal 27, 379–423, 623–656 (1948)
18. Vaishnavi, V.K., Kriegel, H.P., Wood, D.: Optimum multiway search trees. Acta Informatica 14(2), 119–133 (1980)
19. Yeung, R.: Alphabetic codes revisited. IEEE Transactions on Information Theory 37(3), 564–572 (1991)

On the Approximability of Geometric and Geographic Generalization and the Min-Max Bin Covering Problem

Wenliang Du[1], David Eppstein[2], Michael T. Goodrich[2], and George S. Lueker[2]

[1] Department of Electrical Engineering and Computer Science, Syracuse University, 3-114 Sci-Tech Building. Syracuse, NY 13244.
wedu@syr.edu
[2] Dept. of Computer Science, Univ. of California, Irvine, CA 92697-3435.
{eppstein,goodrich,lueker}@ics.uci.edu.

Abstract. We study the problem of abstracting a table of data about individuals so that no selection query can identify fewer than k individuals. We show that it is impossible to achieve arbitrarily good polynomial-time approximations for a number of natural variations of the generalization technique, unless $P = NP$, even when the table has only a single quasi-identifying attribute that represents a geographic or unordered attribute:

- *Zip-codes*: nodes of a planar graph generalized into connected subgraphs
- *GPS coordinates*: points in $\mathbf{R}^2$ generalized into non-overlapping rectangles
- *Unordered data*: text labels that can be grouped arbitrarily.

These hard single-attribute instances of generalization problems contrast with the previously known NP-hard instances, which require the number of attributes to be proportional to the number of individual records (the rows of the table). In addition to impossibility results, we provide approximation algorithms for these difficult single-attribute generalization problems, which, of course, apply to multiple-attribute instances with one that is quasi-identifying. Incidentally, the generalization problem for unordered data can be viewed as a novel type of bin packing problem—*min-max bin covering*—which may be of independent interest.

1 Introduction

Data mining is an effective means for extracting useful information from various data repositories, to highlight, for example, health risks, political trends, consumer spending, or social networking. In addition, some public institutions, such as the U.S. Census Bureau, have a mandate to publish data about U.S. communities, so as to benefit socially-useful data mining. Thus, there is a public interest in having data repositories available for public study through data mining.

Unfortunately, fulfilling this public interest is complicated by the fact that many databases contain confidential or personal information about individuals. The publication of such information is therefore constrained by laws and policies governing privacy protection. For example, the U.S. Census Bureau must limit its data releases to those that reveal no information about any individual. Thus, to allow the public to benefit from the knowledge that can be gained through data mining, a privacy-protecting transformation should be performed on a database before its publication.

F. Dehne et al. (Eds.): WADS 2009, LNCS 5664, pp. 242–253, 2009.
© Springer-Verlag Berlin Heidelberg 2009

One of the greatest threats to privacy faced by database publication is a *linking attack* [14, 13]. In this type of attack, an adversary who already knows partial identifying information about an individual (e.g., a name and address or zip-code) is able to identify a record in another database that belongs to this person. A linking attack occurs, then, if an adversary can "link" his prior identifying knowledge about an individual through a non-identifying attribute in another database. Non-identifying attributes that can be subject to such linking attacks are known as *quasi-identifying* attributes.

To combat linking attacks, several researchers [9, 13, 14, 11, 1, 3, 18, 2] have proposed *generalization* as a way of specifying a quantifiable privacy requirement for published databases. The generalization approach is to group attribute values into equivalence classes, and replace each individual attribute value with its class name. In this paper we focus on generalization methods that try to minimize the maximum size of any equivalence class, subject to lower bounds on the size of any equivalence class.

Related Prior Results. The concept of k-anonymity [14, 13], although not a complete solution to linking attacks, is often an important component of such solutions. In this application of generalization, the equivalence classes are chosen to ensure that each combination of replacement attributes that occurs in the generalized database occurs in at least k of the records. Several researchers have explored heuristics, extensions, and adaptations for k-anonymization (e.g., see [9, 1, 3, 18, 2, 16]).

The use of heuristics, rather than exact algorithms, for performing generalization is motivated by claims that k-anonymization-based generalization is NP-hard. Meyerson and Williams [11] assume that an input dataset has been processed into a database or table in which identical records from the original dataset have been aggregated into a single row of the table, with a count representing its frequency. They then show that if the number of aggregated rows is n and the number of attributes (table columns) is at least $3n$, then generalization for k-anonymization is NP-hard. Unfortunately, their proof does not show that generalization is NP-hard in the strong sense: the difficult instances generated by their reduction have frequency counts that are large binary numbers, rather than being representable in unary. Therefore, their result doesn't actually apply to the original k-anonymization problem. Aggarwal *et al.* [1] address this deficiency, showing that k-anonymization is NP-hard in the strong sense for datasets with at least $n/3$ quasi-identifying attributes. Their proof uses cell suppression instead of generalization, but Byun *et al.* [3] show that the proof can be extended to generalization. As in the other two NP-hardness proofs, Byun *et al.* require that the number of quasi-identifying attributes be proportional to the number of records, which is typically not the case. Park and Shim [12] present an NP-hardness proof for a version of k-anonymization involving cell suppression in place of generalization, and Wong *et al.* [17] show an anonymity problem they call (α, k)-anonymity to be NP-hard.

Khanna *et al.* [8] study a problem, RTILE, which is closely related to generalization of geographic data. RTILE involves tiling an $n \times n$ integer grid with at most p rectangles so as to minimize the maximum weight of any rectangle. They show that no polynomial-time approximation algorithm can achieve an approximation ratio for RTILE of better than 1.25 unless P=NP. Unlike k-anonymization, however, this problem does not constrain the minimum weight of a selected rectangle.

Our Results. In this paper, we study instances of k-anonymization-based generalization in which there is only a single quasi-identifying attribute, containing geographic or unordered data. In particular, we focus on the following attribute types:

- *Zip-codes:* nodes of a planar graph generalized into connected subgraphs
- *GPS coordinates:* points in $\mathbf{R}^2$ generalized into non-overlapping rectangles
- *Unordered data*: text labels that can be grouped arbitrarily (e.g., disease names).

We show that even in these simple instances, k-anonymization-based generalization is NP-complete in the strong sense. Moreover, it is impossible to approximate these problems to within $(1 + \epsilon)$ of optimal, where $\epsilon > 0$ is an arbitrary fixed constant, unless $P = NP$. These results hold *a fortiori* for instances with multiple quasi-identifying attributes of these types, and they greatly strengthen previous NP-hardness results which require unrealistically large numbers of attributes. Nevertheless, we provide a number of efficient approximation algorithms and show that they achieve good approximation ratios. Our approximation bounds for the zip-codes problem require that the graph has sufficiently strong connectivity to guarantee a sufficiently low-degree spanning tree.

The intent of this paper is not to argue that single-attribute generalization is a typical application of privacy protection. Indeed, most real-world anonymization applications will have dozens of attributes whose privacy concerns vary from hypersensitive to benign. Moreover, the very notion of k-anonymization has been shown to be insufficient to protect against all types of linking attack, and has been extended recently in various ways to address some of those concerns (e.g., see [5, 10, 17]); some work also argues against any approach similar to k-anonymization [?]. We do not attempt to address this issue here. Rather, our results should be viewed as showing that even the simplest forms of k-anonymization-based generalization are difficult but can be approximated. We anticipate that similar results may hold for its generalizations and extensions as well.

In addition, from an algorithmic perspective, our study of k-anonymization-based generalization has uncovered a new kind of bin-packing problem (e.g., see [4]), which we call *Min-Max Bin Covering*. In this variation, we are given a collection of items and a nominal bin capacity, k, and we wish to distribute the items to bins so that each bin has total weight at least k while minimizing the maximum weight of any bin. This problem may be of independent interest in the algorithms research community.

2   Zip-Code Data

The first type of quasi-identifying information we consider is that of zip-codes, or analogous numeric codes for other geographic regions. Suppose we are given a database consisting of n records, each of which contains a single quasi-identifying attribute that is itself a zip-code. A common approach in previous papers using generalization for zip-code data (e.g., see [3, 18]) is to generalize consecutive zip-codes. That is, these papers view zip-codes as character strings or integers and generalize based on this data type. Unfortunately, when zip-codes are viewed as numbers or strings, geographic adjacency information can be lost or misleading: consecutive zip codes may be far apart geographically, and geographically close zip codes may be numerically far, leading to generalizations that have poor quality for data mining applications.

We desire a generalization algorithm for zip-codes that preserves geographic adjacency. Formally, we assume each zip-code is the name of a node in a planar graph, G. The most natural generalization in this case is to group nodes of G into equivalence classes that are connected subgraphs. This is motivated, in the zip-code case, by a desire to group adjacent regions in a country, which would naturally have more likelihood to be correlated according to factors desired as outcomes from data mining, such as health or buying trends. So the optimization problem we investigate in this section is one in which we are given a planar graph, G, with non-negative integer weights on its nodes (representing the number of records for each node), and we wish to partition G into connected subgraphs so that the maximum weight of any subgraph is minimized subject to the constraint that each has weight at least k.

Generalization for Zip-codes is Hard. Converting this to a decision problem, we can add a parameter K and ask if there exists a partition into connected subgraphs such that the weight of each subgraph in G is at least k and at most K. In this section, we show that this problem is NP-complete even if the weights are all equal to 1 and $k = 3$. Our proof is based on a simple reduction that sets $K = 3$, so as to provide a reduction from the following problem:

3-Regular Planar Partition into Paths of Length 2 (3PPPL2): Given a 3-regular planar graph G, can G be partitioned into paths of length 2? That is, is there a spanning forest for G such that each connected component is a path of length 2?

This problem is a special case of the problem, "Partition into Paths of Length-2 (PPL2)", whose NP-completeness is included as an exercise in Garey and Johnson [7]. Like PPL2, 3PPPL2 is easily shown to be in NP. To show that 3PPPL2 is NP-hard, we provide a reduction from the 3-dimensional matching (3DM) problem:

3-Dimensional Matching (3DM): Given three sets X, Y, and Z, each of size n, and a set of triples $\{(x_1, y_1, z_1), \ldots, (x_m, y_m, z_m)\}$, is there a subset S of n triples such that each element in X, Y, and Z is contained in exactly one of the triples?

Suppose we are given an instance of 3DM. We create a vertex for each element in X, Y, and Z. For each tuple, (x_i, y_i, z_i), we create a tuple subgraph gadget as shown in Figure 1a, with nodes $t_{i,x}$, $t_{i,y}$, and $t_{i,z}$, which correspond to the representatives x_i, y_i, and z_i, respectively, in the tuple. We then connect each $t_{i,x}$, $t_{i,y}$ and $t_{i,z}$ vertex to the corresponding element vertex from X, Y, and Z, respectively, using the connector gadget in Figure 1b.

This construction is, in fact, a version of the well-known folklore reduction from 3DM to PPL2, which solves an exercise in Garey and Johnson [7]. Note, for example, that the vertices in the triangle in the tuple gadget must all three be completely included in a single group or must all be in separate groups. If they are all included, then grouping the degree-1 vertices requires that the corresponding x, y, and z elements must all be included in a group with the degree-1 vertex on the connector. If they are all not included, then the corresponding x, y, and z elements must be excluded from a group in this set of gadgets.

Continuing the reduction to an instance of 3PPPL2, we make a series of transformations. The first is to embed the graph in the plane in such a way that the only crossings occur in connector gadgets. We then take each crossing of a connector, as shown in Figure 2a, and replace it with the cross-over gadget shown in Figure 2b.

There are four symmetric ways this gadget can be partitioned into paths of length 2. Note that the four ways correspond to the four possible ways that connector "parity" can be transmitted and that they correctly perform a cross-over of these two parities. In particular, note that it is impossible for opposite connectors to have the same parity in any partition into paths of length 2. Thus, replacing each crossing with a cross-over gadget completes a reduction of 3DM to planar partition in paths of length 2.

Next, note that all vertices of the planar graph are degree-3 or less except for the "choice" vertices at the center of cross-over gadgets and possibly some nodes corresponding to elements of X, Y, and Z. For each of these, we note that all the edges incident on such nodes are connectors. We therefore replace each vertex of degree-4 or higher with three connector gadgets that connect the original vertex to three binary trees whose respective edges are all connector gadgets. This allows us to "fan out" the choice semantics of the original vertex while exclusively using degree-3 vertices. To complete the reduction, we perform additional simple transformations to the planar graph to make it 3-regular. This completes the reduction of 3DM to 3PPPL2.

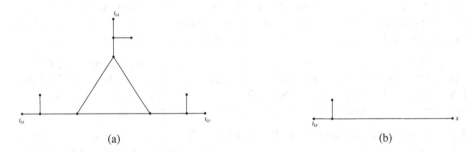

(a)   (b)

Fig. 1. Gadgets for reducing 3DM to PPL2. (a) the tuple gadget; (b) the connector.

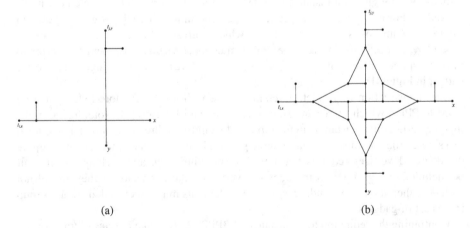

(a)   (b)

Fig. 2. Dealing with edge crossings. (a) a connector crossing; (b) the cross-over gadget.

In the preprint version of this paper [6], we complete the proof of the following:

Theorem 1. *There is a polynomial-time approximation algorithm for k-anonymization on planar graphs that guarantees an approximation ratio of 4 for 3-connected planar graphs and 3 for 4-connected planar graphs. It is not possible for a polynomial-time algorithm to achieve an approximation ratio better than 1.33, even for 3-regular planar graphs, unless P=NP.*

3   GPS-Coordinate Data

Next we treat geographic data that is given as geographic coordinates rather than having already been generalized to zip-codes. Suppose we are given a table consisting of n records, each of which contains a single quasi-identifying attribute that is itself a GPS coordinate, that is, a point (x, y) in the plane. Suppose further that we wish to generalize such sets of points using axis-aligned rectangles.

Generalizing GPS-Coordinates is Hard.   Converting this to a decision problem, we can add a parameter K and ask whether there exists a partition of the plane into rectangles such that the weight of the input points within each rectangle is at least k and at most K. We show that this problem is NP-complete even when we set k and K equal to three. Our proof, which is given in the preprint version of this paper [6], is based on a simple reduction from 3-dimensional matching (3DM).

We also provide in the preprint version of this paper [6] the following:

Theorem 2. *There is a polynomial-time approximation algorithm for rectangular generalization, with respect to k-anonymization in the plane, that achieves an approximation ratio of 5 in the worst case. It is not possible for a polynomial-time algorithm to achieve an approximation ratio better than 1.33 unless P=NP.*

4   The Min-Max Bin Covering Problem

In this section, we examine single-attribute generalization, with respect to the problem of k-anonymization for unordered data, where quasi-identifying attribute values are arbitrary labels that come from an unordered universe. (Note that if the labels were instead drawn from an ordered universe, and we required the generalization groups to be intervals, the resulting one-dimensional k-anonymization problem could be solved optimally in polynomial time by a simple dynamic programming algorithm.) Our optimization problem, then, is to generalize the input labels into equivalence classes so as to minimize the maximum size of any equivalence class, subject to the k-anonymization constraint.

It is convenient in this context to use the terminology of bin packing; henceforth in this section we refer to the input labels as *items*, the equivalence classes as *bins*, and the entire generalization as a *packing*. The *size* of an item corresponds in this way to the number of records having a given label as their attribute value. Thus the problem becomes the following, which we call the *Min-Max Bin Covering Problem*:

Input: Positive integers $x_1, x_2, \ldots, x_n$ and an integer nominal bin capacity $k > 0$.
Output: a partition of $\{1, 2, \ldots, n\}$ into subsets S_j, satisfying the constraint that,
for each j,

$$\sum_{i \in S_j} x_i \geq k, \tag{1}$$

and minimizing the objective function

$$\max_j \sum_{i \in S_j} x_i. \tag{2}$$

We will say that a partition satisfying (1) for all j is *feasible*, and the function shown in
(2) is the *cost* of this partition. Note that any feasible solution has cost at least k.

Hardness Results. In this subsection, we show that Min-Max Bin Covering is NP-hard
in the strong sense. We begin by converting the problem to a decision problem by adding
a parameter K, which is intended as an upper bound on the size of any bin: rather than
minimizing the maximum size of an bin, we ask whether there exists a solution in which
all bins have size at most K. This problem is clearly in NP.

We show that Min-Max Bin Covering is NP-hard by a reduction from the following
problem, which is NP-complete in the strong sense [7].

- **3-Partition.** Given a value B, and a set S of $3m$ weights $w_1, w_2, \ldots, w_{3m}$ each
  lying in $(B/4, B/2)$, such that $\sum_{i=1}^{3m} w_i = mB$, can we partition $\{1, 2, \ldots, 3m\}$
  into sets S_j such that for each j, $\sum_{i \in S_j} w_i = B$? (Note that any such family of sets
  S_j would have to have exactly m members.)

For the reduction we simply let $x_i = w_i$ and $k = K = B$. If the 3-Partition problem
has answer yes, then we can partition the items into m sets each of total size $k = K = B$ so the Min-Max Bin Covering problem has answer yes. If, on the other hand, the
3-Partition problem has answer no, no such partition is possible, so we have

Theorem 3. *Min-Max Bin Covering is NP-complete in the strong sense.*

In the preprint version of this paper [6], we show that there are limits on how well we
can approximate the optimum solution (unless P = NP):

Theorem 4. *Assuming $P \neq NP$, there does not exist a polynomial-time algorithm for
Min-Max Bin-Covering that guarantees an approximation ratio better than 2 (when
inputs are expressed in binary), or better than 4/3 (when inputs are expressed in unary).*

Achievable Approximation Ratios. While the previous section shows that sufficiently
small approximation ratios are hard to achieve, in this section we show that we can establish larger approximation bounds with polynomial time algorithms. The algorithms
in this section can handle inputs that are expressed either in unary or binary, so they
are governed by the stronger lower bound of 2 on the approximation ratio given in Theorem 4. If A is some algorithm for Min-Max Bin Covering Problem, and I is some
instance, let $A(I)$ denote the cost of the solution obtained by A. Let **Opt**(I) denote the
optimum cost for this instance.

Note that if $\sum_{i=1}^{n} x_i < k$, there is no feasible solution; we will therefore restrict our attention to instances for which

$$\sum_{i=1}^{n} x_i \geq k. \tag{3}$$

An approximation ratio of three is fairly easy to achieve.

Theorem 5. *Assuming* (3) *there is a linear-time algorithm A guaranteeing that*

$$A(I) \leq \max(k - 1 + \max_{i=1}^{n} x_i, 3k - 3).$$

Proof. Put all items of size k or greater into their own bins, and then, with new bins, use the Next Fit heuristic for bin covering (see [?]) for the remaining items, i.e., add the items one at a time, moving to a new bin once the current bin is filled to a level of at least k. Then all but the last bin in this packing have level at most $2k - 2$, as they each have level at most $k - 1$ before the last item value is added and this last item has size less than k. There may be one leftover bin with level less than k which must be merged with some other bin, leading to the claimed bound.   □

With a bit more effort we can improve the approximation ratio. For convenience, in the remainder of this section we scale the problem by dividing the item sizes by k. Thus each bin must have level at least 1, and the item sizes are multiples of $1/k$.

Lemma 1. *Suppose we are given a list of numbers $x_1, x_2, \ldots, x_n$, with each $x_i \leq 1/2$ and $\sum_{i=1}^{n} x_i = 1$. Then we can partition the list into three parts each having a sum of at most $1/2$.*

Proof. Omitted.

Theorem 6. *There is a polynomial algorithm to solve Min-Max Bin Packing with an approximation factor of $5/2$.*

Proof. We will assume without loss of generality that $\mathbf{Opt}(I) \geq 6/5$, since otherwise the algorithm of Theorem 5 could give a $5/2$-approximation.

Assume the items are numbered in order of decreasing size. Pack them greedily in this order into successive bins, moving to a new bin when the current bin has level at least 1. Note that then all of the bins will have levels less than 2, and all of the bins except the last will have level at least 1. If the last bin also has level at least 1, this packing is feasible and has cost less than 2, so it is within a factor of 2 of the optimum.

Next suppose that the last bin has level less than 1. We omit the details for the case in which we have formed at most 3 bins, and subsequently we assume we have formed at least 4 bins.

Now let f be size of the largest item in the final bin, and let r be the total size of the other items in the last bin. Call an item *oversize* if its size is at least 1, *large* if its size is in $(1/2, 1)$, and *small* if its size is at most $1/2$. Consider two cases.

Case 1. $f \leq 1/2$. Then all items in the last bin are small, so by Lemma 1 we can partition them into three sets, each of total size at most $1/2$. Add each of these sets to one of the first three bins, so no bin is filled to more than $5/2$, unless it was one of the bins containing an oversize item. (We no longer use the last bin.) Thus we have achieved an approximation ratio of 5/2.

Case 2. $f > 1/2$. Note that in this case there must be an odd number of large items, since each bin except the last has either zero or exactly two large items. Note also that r in this case is the total size of the small items, and $r < 1/2$. Let x_1 be the first large item packed. If x_1 lies in the last bin, we must have packed at least one oversize item. Then moving all of the items from the last bin (which will no longer be used) into the bin with this oversize item guarantees a 2-approximation. Thus assume x_1 is not in the last bin.

Case 2.1. $x_1 + r \geq 1$. Then swap items x_1 and f, so the last bin will be filled to a level $x_1 + r \in [1, 2]$. Also, the bin now containing f will contain two items of size in the range $[1/2, 1]$ and thus have a level in the range $[1,2]$. Thus we have a solution that meets the constraints and has cost at most 2.

Case 2.2. $x_1 + r < 1$. Since r is the total size of the small items, if any bin had only one large item it could not have level at least 1 (as required for feasibility) and at most 6/5 (as required since $\mathbf{Opt}(I) \leq 6/5$). Thus the optimum solution has no bin containing only one large item. Since there are an odd number of large items, this means that the optimum solution has at least one bin with 3 or more large items, so the cost of the optimum solution is at least 3/2. But then since the simple algorithm of Theorem 5 gives a solution of cost less than 3, it provides a solution that is at most twice the optimum. □

A Polynomial Time Approximation Guaranteeing a Ratio Approaching 2. With more effort we can come arbitrarily close to the lower bound of 2 on the approximation factor given in Theorem 4 for the binary case, with a polynomial algorithm.

Theorem 7. *For each fixed $\epsilon > 0$, there is a polynomial time algorithm A_ϵ that, given some instance I of Min-Max Bin Covering, finds a solution satisfying*

$$A_\epsilon(I) \leq (1+\epsilon)(\mathbf{Opt}(I) + 1). \tag{4}$$

(The degree of the polynomial becomes quite large as ϵ becomes small.)

Proof. The idea of the proof is similar to many approximation algorithms for bin packing (see in particular [15, Chapter 9]); for the current problem, we have to be especially careful to ensure that the solution constructed is feasible.

We can assume that the optimum cost is at most 3, by the following reasoning. Say an item is *nominal* if its size is less than 1, and *oversize* if its size is greater than or equal to 1. First suppose the total size of the nominal items is at least 1 and some oversize item has size at least 3. Then the greedy algorithm of Theorem 5 achieves an optimum solution, so we are done. Next suppose the sum of the nominal items is at least 1 and no oversize item has size at least 3. Then the greedy algorithm of Theorem 5 achieves a solution of cost at most 3, so the optimum is at most 3. Finally suppose that the total size of the nominal items is less than 1. Then there must be an optimum solution in which every bin contains exactly one oversize item (and possibly some nominal items). Let t_0 (resp. t_1) be the size of the smallest (resp. largest) oversize item. If $t_1 - t_0 \geq 1$, then we can form an optimum solution by putting all nominal items in a bin with t_0. If on the other hand $t_1 - t_0 < 1$, we can reduce the size of all oversize items by $t_0 - 1$ without changing the structure of the problem, after which all oversize items will have size at most 2, and the optimum will be at most 3.

Now call those items that have size greater than or equal to ϵ *large*, and the others *small*. Let $b = \sum_{i=1}^{n} x_i$; note that $b \leq 3n$, and any feasible partition will have at most b bins. Let N be the largest integer for which $\epsilon(1+\epsilon)^N$ is less than three; note that N is a constant depending only on ϵ. Let

$$\hat{S} = \{\epsilon(1+\epsilon)^\ell : \ell \in \{0, 1, 2, \ldots, N\}\}.$$

For any item size x, define $round(x)$ to be the largest value in $\hat{S}$ that is less than or equal to x. Let the *type* of a packing P, written $type(P)$, be the result of discarding all small items in P, and replacing each large x_i by $round(x_i)$. Note that any type can be viewed as a partial packing in which the bins contain only items with sizes in $\hat{S}$.

Since, for fixed ϵ, there are only a constant number of item sizes in $\hat{S}$, and each of these is at least ϵ, there are only finitely many ways of packing a bin to a level of at most 3 using the rounded values; call each of these ways a *configuration* of a bin. Since the ordering of the bins does not matter, we can represent the type of a packing by the number of times it uses each configuration. It is not hard to show that for fixed ϵ, as in the proof of [15, Lemma 9.4], there are only polynomially many types having at most b bins. (Of course, for small ϵ, this will be a polynomial of very high degree.) We will allow types that leave some of the bins empty, allowing them to be filled later.

The algorithm proceeds as follows. Enumerate all possible types T that can be formed using the rounded large item sizes. For each such type T carry out the following steps:

1. Let T' be the result of replacing each item x in T, which resulted from rounding some original input item x_i, by any one of the original items x_j such that $x = round(x_j)$, in such a way that the set of items in T' is the same as the set of large items in the original input. Note that there is no guarantee that $x_i = x_j$, since the rounding process does not maintain the distinct identities of different items that round to the same value in $\hat{S}$. However, we do know that $round(x_i) = round(x_j)$, so we can conclude that $x_j/x_i \in \left((1+\epsilon)^{-1}, 1+\epsilon\right)$.
2. Pack the small items into T' by processing them in an arbitrary order, placing each into the bin with the lowest current level. Call this the *greedy completion* of T.
3. Finally, while any bin has a level less than 1, merge two bins with the lowest current levels. Note that this will lead to a feasible packing because of (3). Call the resulting packing $\mathcal{F}(T)$, and let $cost(\mathcal{F}(T))$ be the maximum level to which any bin is filled.

Return the packing $\mathcal{F}(T)$ that minimizes $cost(\mathcal{F}(T))$ over all T.

We now show that (4) holds. Let P^* be a feasible packing achieving $\mathbf{Opt}(I)$, and let P^*_{large} be the result of discarding the small items in P^* (retaining any bins that become empty). Consider the type T obtained by rounding all large items in P^*_{large} down to a size in $\hat{S}$. Note that this must be one of the types, say T, considered in the algorithm. When we perform step 1 on T, we obtain a packing T' such that $cost(T') \leq (1+\epsilon)cost(P^*)$.

If any level in the greedy completion is greater than $(1 + \epsilon)\mathbf{Opt}(I) + \epsilon$, then during the greedy completion all bins must have reached a level greater than $(1 + \epsilon)\mathbf{Opt}(I)$, so their total size would be greater than $(1 + \epsilon)\sum_{i=1}^{n} x_i$, contradicting the fact that the greedy completion uses each of the original items exactly once. Thus all bins in the greedy completion have level at most $(1 + \epsilon)\mathbf{Opt}(I) + \epsilon$. Also, it cannot be that all bins in the greedy completion have level less than 1, since then the total size of the items would be less than the number of bins, contradicting the fact that the optimum solution covers all the bins.

During step 3, as long as at least two bins have levels below 1, two of them will be merged to form a bin with a level at most 2. If then only one bin remains with a level below 1, it will be merged with a bin with level in $[1, (1 + \epsilon)\mathbf{Opt}(I) + \epsilon)$ to form a feasible packing with no bin filled to a level beyond $(1 + \epsilon)\mathbf{Opt}(I) + 1 + \epsilon$, as desired.

$\qquad\square$

Note that the bound of Theorem 7 implies $A_\epsilon(I) \leq 2(1 + \epsilon)\mathbf{Opt}(I)$.

We also note that if one is willing to relax both the feasibility constraints and the cost of the solution obtained, a polynomial-time $(1 + \epsilon)$ approximation scheme of sorts is possible. (Of course, this would not guarantee k-anonymity.)

Theorem 8. *Assume that all item sizes x_i in the input are expressed in binary, and let $\epsilon > 0$ be a fixed constant. There is a polynomial time algorithm that, given some instance I of Min-Max Bin Covering, finds a partition of the items into disjoint bins S_j such that*

$$\forall j \sum_{i \in S_j} x_i \geq 1 - \epsilon, \quad \text{and} \quad \max_j \sum_{i \in S_j} x_i \leq (1 + \epsilon)\mathbf{Opt}(I). \tag{5}$$

Proof (sketch). Roughly, one can use an algorithm similar to that of the previous theorem but omitting the last phase in which we merge bins to eliminate infeasibility. We omit the details. $\qquad\square$

Acknowledgments. We would like to thank Padhraic Smyth and Michael Nelson for several helpful discussions related to the topics of this paper. This research was supported in part by the NSF under grants 0830403, 0847968, and 0713046.

References

1. Aggarwal, G., Feder, T., Kenthapadi, K., Motwani, R., Panigrahy, R., Thomas, D., Zhu, A.: Anonymizing tables. In: Eiter, T., Libkin, L. (eds.) ICDT 2005. LNCS, vol. 3363, pp. 246–258. Springer, Heidelberg (2005)
2. Bayardo, R.J., Agrawal, R.: Data privacy through optimal k-anonymization. In: Proc. of 21st Int. Conf. on Data Engineering (ICDE), pp. 217–228. IEEE Computer Society Press, Los Alamitos (2005)
3. Byun, J.-W., Kamra, A., Bertino, E., Li, N.: Efficient k-anonymization using clustering techniques. In: Kotagiri, R., Radha Krishna, P., Mohania, M., Nantajeewarawat, E. (eds.) DASFAA 2007. LNCS, vol. 4443, pp. 188–200. Springer, Heidelberg (2007)
4. Coffman Jr., E.G., Garey, M.R., Johnson, D.S.: Approximation algorithms for bin packing: a survey. In: Approximation algorithms for NP-hard problems, pp. 46–93. PWS Publishing Co., Boston (1997)

5. Domingo-Ferrer, J., Torra, V.: A critique of k-anonymity and some of its enhancements. In: ARES 2008: Proceedings of the, Third International Conference on Availability, Reliability and Security, Washington, DC, USA, pp. 990–993. IEEE Computer Society Press, Los Alamitos (2008)

6. Du, W., Eppstein, D., Goodrich, M.T., Lueker, G.S.: On the approximability of geometric and geographic generalization and the min-max bin covering problem. Electronic preprint arxiv:0904.3756 (2009)

7. Garey, M.R., Johnson, D.S.: Computers and Intractability: A Guide to the Theory of NP-Completeness. W. H. Freeman, New York (1979)

8. Khanna, S., Muthukrishnan, S., Paterson, M.: On approximating rectangle tiling and packing. In: Proceedings of the 9th Annual ACM-SIAM Symposium on Discrete Algorithms SODA 1998, pp. 384–393. ACM Press, New York (1998)

9. LeFevre, K., Dewitt, D.J., Ramakrishnan, R.: Incognito:efficient full-domain k-anonymity. In: Proceedings of the 2005 ACM SIGMOD, June 12-16 (2005)

10. Machanavajjhala, A., Kifer, D., Gehrke, J., Venkitasubramaniam, M.: L-diversity: Privacy beyond k-anonymity. ACM Trans. Knowl. Discov. Data 1(1), 3 (2007)

11. Meyerson, A., Williams, R.: On the complexity of optimal k-anonymity. In: PODS 2004: Proceedings of the Twenty-Third ACM SIGMOD-SIGACT-SIGART Symposium on Principles of Database Systems, pp. 223–228. ACM Press, New York (2004)

12. Park, H., Shim, K.: Approximate algorithms for K-anonymity. In: SIGMOD 2007: Proceedings of the, ACM SIGMOD International Conference on Management of Data, pp. 67–78. ACM Press, New York (2007)

13. Samarati, P.: Protecting respondents' identities in microdata release. IEEE Transactions on Knowledge and Data Engineering 13(6) (2001)

14. Samarati, P., Sweeney, L.: Protecting privacy when disclosing information: k-anonymity and its enforcement through generalization and suppression. Technical report, SRI (1998)

15. Vazirani, V.V.: Approximation Algorithms. Springer, Berlin (2003)

16. Wang, K., Fung, B.C.M.: Anonymizing sequential releases. In: KDD 2006: Proceedings of the 12th ACM SIGKDD International Conference on Knowledge Discovery and Data Mining, pp. 414–423. ACM Press, New York (2006)

17. Wong, R.C.-W., Li, J., Fu, A.W.-C., Wang, K.: (α, k)-anonymity: an enhanced k-anonymity model for privacy preserving data publishing. In: KDD 2006: Proceedings of the 12th ACM SIGKDD International Conference on Knowledge Discovery and Data Mining, pp. 754–759. ACM Press, New York (2006)

18. Zhong, S., Yang, Z., Wright, R.N.: Privacy-enhancing k-anonymization of customer data. In: PODS 2005: Proceedings of the Twenty-Fourth ACM SIGMOD-SIGACT-SIGART Symposium on Principles of Database Systems, pp. 139–147. ACM Press, New York (2005)

On Reconfiguration of Disks in the Plane and Related Problems

Adrian Dumitrescu[1,*] and Minghui Jiang[2,**]

[1] Department of Computer Science, University of Wisconsin-Milwaukee
WI 53201-0784, USA
ad@cs.uwm.edu
[2] Department of Computer Science, Utah State University
Logan, UT 84322-4205, USA
mjiang@cc.usu.edu

Abstract. We revisit two natural reconfiguration models for systems of disjoint objects in the plane: translation and sliding. Consider a set of n pairwise interior-disjoint objects in the plane that need to be brought from a given start (initial) configuration S into a desired goal (target) configuration T, without causing collisions. In the translation model, in one move an object is translated along a fixed direction to another position in the plane. In the sliding model, one move is sliding an object to another location in the plane by means of an arbitrarily complex continuous motion (that could involve rotations). We obtain several combinatorial and computational results for these two models:

(I) For systems of n congruent disks in the translation model, Abellanas et al. [1] showed that $2n - 1$ moves always suffice and $\lfloor 8n/5 \rfloor$ moves are sometimes necessary for transforming the start configuration into the target configuration. Here we further improve the lower bound to $\lfloor 5n/3 \rfloor - 1$, and thereby give a partial answer to one of their open problems.

(II) We show that the reconfiguration problem with congruent disks in the translation model is NP-hard, in both the labeled and unlabeled variants. This answers another open problem of Abellanas et al. [1].

(III) We also show that the reconfiguration problem with congruent disks in the sliding model is NP-hard, in both the labeled and unlabeled variants.

(IV) For the reconfiguration with translations of n arbitrary convex bodies in the plane, $2n$ moves are always sufficient and sometimes necessary.

1 Introduction

A *body* (or *object*) in the plane is a compact connected set in $\mathbb{R}^2$ with nonempty interior. Two initially disjoint bodies *collide* if they share an interior point at some time during their motion. Consider a set of n pairwise interior-disjoint

* Supported in part by NSF CAREER grant CCF-0444188.
** Supported in part by NSF grant DBI-0743670.

F. Dehne et al. (Eds.): WADS 2009, LNCS 5664, pp. 254–265, 2009.

objects in the plane that need to be brought from a given start (initial) configuration S into a desired target (goal) configuration T, without causing collisions. The *reconfiguration* problem for such a system is that of computing a sequence of object motions (a schedule, or motion plan) that achieves this task. Depending on the existence of such a sequence of motions, we call that instance of the problem *feasible* and respectively, *infeasible*.

Our reconfiguration problem is a simplified version of the multi-robot motion planning problem, in which a system of robots are operating together in a shared workplace and once in a while need to move from their initial positions to a set of target positions. The workspace is often assumed to extend throughout the entire plane, and has no obstacles other than the robots themselves. In many applications, the robots are indistinguishable (unlabeled), so each of them can occupy any of the specified target positions. Beside multi-robot motion planning, another application which permits the same abstraction is moving around large sets of heavy objects in a warehouse [10]. Typically, one is interested in minimizing the number of moves and designing efficient algorithms for carrying out the motion plan. It turns out that moving a set of objects from one place to another is related to certain *separability* problems [4,6,7,9]; see also [11]. There are several types of moves that make sense to study, as dictated by specific applications. Here we focus on systems of convex bodies in the plane.

Next we formulate these models for systems of disks, since they are simpler and most of our results are for disks. These rules can be extended (not necessarily uniquely) for arbitrary convex bodies in the plane. The decision problems we refer to below, pertaining to various reconfiguration problems we discuss here, are in standard form, and concern systems of (arbitrary or congruent) disks. For instance, the *Reconfiguration Problem* U-SLIDE-RP for congruent disks is: Given a start configuration and a target configuration, each with n unlabeled congruent disks in the plane, and a positive integer k, is there a reconfiguration motion plan with at most k sliding moves? It is worth clarifying that for the unlabeled variant, if the start and target configuration contain subsets of congruent disks, there is freedom is choosing which disks will occupy target positions. However in the labeled variant, this assignment is uniquely determined by the labeling; of course a valid labeling must respect the size of the disks.

1. *Sliding model*: one move is sliding a disk to another location in the plane without colliding with any other disk, where the disk center moves along an arbitrary continuous curve. This model was introduced in [3]. The labeled and unlabeled variants are L-SLIDE-RP and U-SLIDE-RP, respectively.
2. *Translation model*: one move is translating a disk to another location in the plane along a fixed direction without colliding with any other disk. This is a restriction imposed to the sliding model above for making each move as simple as possible. This model was introduced in [1]. The labeled and unlabeled variants are L-TRANS-RP and U-TRANS-RP, respectively.
3. *Lifting model*: one move is lifting a disk and placing it back in the plane anywhere in the free space. This model was introduced in [2]. The labeled and unlabeled variants are L-LIFT-RP and U-LIFT-RP, respectively. (We have only included this model for completeness.)

Our results are:

(1) For any n, there exist pairs of start and target configurations each with n disks, that require $\lfloor 5n/3 \rfloor - 1$ translation moves for reconfiguration (Theorem 1 in Section 2). This improves the previous bound of $\lfloor 8n/5 \rfloor$ due to Abellanas et al. and thereby gives a partial answer to their first open problem regarding the translation model [1].

(2) The reconfiguration problem with congruent disks in the translation model, in both the labeled and unlabeled variants, is NP-hard. That is, L-TRANS-RP and U-TRANS-RP are NP-hard (Theorem 2 and Theorem 3 in Section 3). This answers the second open problem of Abellanas et al. regarding the translation model [1].

(3) The reconfiguration problem with congruent disks in the sliding model, in both the labeled and unlabeled variants, is NP-hard. That is, L-SLIDE-RP and U-SLIDE-RP are NP-hard (Theorem 4 and Theorem 5 in Section 4).

(4) For the reconfiguration with translations of n arbitrary convex bodies in the plane, $2n$ moves are always sufficient and sometimes necessary (Theorem 6 in Section 5).

2   A New Lower Bound for Translating Unlabeled Congruent Disks

In this section we consider the problem of moving n disks of unit radius, here also referred to as coins, to n target positions using translation moves. Abellanas et al. [1] have shown that $\lfloor 8n/5 \rfloor$ moves are sometimes necessary. Their lower bound construction is shown in Fig. 1. Here we further improve this bound to $\lfloor 5n/3 \rfloor - 1$.

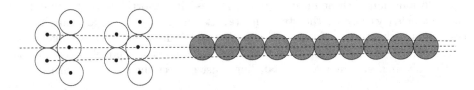

Fig. 1. Two groups of five disks with their targets: part of the old $\lfloor 8n/5 \rfloor$ lower bound construction for translating disks. The disks are white and their targets are shaded.

Theorem 1. *For every $m \geq 1$, there exist pairs of start and target configurations each with $n = 3m + 2$ disks, that require $5m + 3$ translation moves for reconfiguration. Consequently, for any n, we have pairs of configurations that require $\lfloor 5n/3 \rfloor - 1$ translation moves.*

Proof. A move is a *target move* if it moves a disk to a final target position. Otherwise, it is a *non-target* move. We also say that a move is a *direct target move* if it moves a disk from its start position directly to its target position.

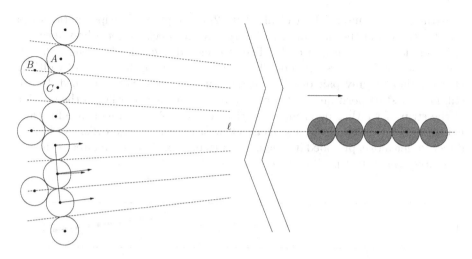

Fig. 2. Illustration of the lower bound construction for translating congruent unlabeled disks, for $m = 3$, $n = 11$. The disks are white and their targets are shaded. Two consecutive partially overlapping parallel strips of width 2 are shown.

Let $n = 3m + 2$. The start and target configurations, each with n disks, are shown in Fig. 2. The n target positions are all on a horizontal line ℓ, with the disks at these positions forming a horizontal chain, $T_1, \ldots, T_n$, consecutive disks being tangent to each other. Let o denote the center of the median disk, $T_{\lfloor n/2 \rfloor}$. Let $r > 0$ be very large. The start disks are placed on two very slightly convex chains (two concentric circular arcs):

- $2m + 2$ disks in the first layer (chain). Their centers are $2m + 2$ equidistant points on a circular arc of radius r centered at o.
- m disks in the second layer. Their centers are m equidistant points on a concentric circular arc of radius $r \cos \alpha + \sqrt{3}$. Each pair of consecutive points on the circle of radius r subtends an angle of 2α from the center of the circle (α is very small).

The parameters of the construction are chosen to satisfy: $\sin \alpha = 1/r$ and $2n \sin n\alpha \leq 2$. Set for instance $\alpha = 1/n^2$, which results in $r = \Theta(n^2)$.

Alternatively, the configuration can be viewed as consisting of m groups of three disks each, plus two disks, one at the higher and one at the lower end of the chain along the circle of radius r. Denote the three pairwise tangent start disks in a group by A, B and C, with their centers making an equilateral triangle, and the common tangent of A and C passing through o. Disks A and C are on the first layer, and the "blocking" disk B on the second layer. The groups are numbered from the top. We therefore refer to the three start disks of group i by A_i, B_i and C_i, where $i = 1, \ldots, m$.

For each pair of tangent disks on the first chain, consider the open strip of width 2 of parallel lines orthogonal to the line segment connecting their centers. By selecting r large enough we ensure the following crucial property of the construction: the intersection of all these $2m + 1$ parallel strips contains the set of n centers of the targets in its interior. More precisely, let a be the center of T_1 and let b be the center of T_n. Then the closed segment ab of length $2n - 2$ is contained in the intersection of all the $2m + 1$ open parallel strips of width 2. Observe that for any pair of adjacent disks in the first layer, if both disks are still in their start position, neither can move so that its center lies in the interior of the strip of width 2 determined by their centers. As a consequence for each pair of tangent disks on the first chain at most one of the two disks can have a direct target move, provided its neighbor tangent disks have been already moved away from their start positions. See also Fig. 3.

Fig. 3. The n disks at the target positions as viewed from a parallel strip of a pair of start positions below the horizontal line ℓ in Fig. 2. The targets are shown denser than they are: the chain of targets is in fact longer.

Recall that there are $2m + 2$ disks in the first layer and m disks in the second layer. We show that the configuration requires at least $2m + 1$ non-target moves, and consequently, at least $3m + 2 + 2m + 1 = 5m + 3$ moves are required to complete the reconfiguration. Throughout the process let:

- k be the number of disks in the first layer that are in their start positions,
- c be the number of connected components in the intersection graph of *these* disks, i.e., disks in the first layer that are still in their start positions,
- x be the number of non-target moves executed so far.

Let t denote the number of moves executed. Consider the value $\Phi = k - c$ after each move. Initially, $k = 2m + 2$ and $c = 1$, so the initial value of $k - c$ is $\Phi_0 = 2m + 1$. In the end, $k = c = 0$, hence the final value of $k - c$ is $\Phi_t = 0$, and $x = x_t$ represents the total number of non-target moves executed for reconfiguration. Consider any reconfiguration schedule. It is enough to show that after any move that reduces the value of Φ by some amount, the value of x increases by at least the same amount. Since the reduction of Φ equals $2m + 1$, it implies that $x \geq 2m + 1$, as desired.

Observe first that a move of a coin in the second layer does not affect the values of k and c, and therefore leaves Φ unchanged. Consider now any move of a coin in the first layer, and examine the way Φ and x are modified as a result of this move and possibly some preceding moves. The argument is essentially a charging scheme that converts the reduction in the value of Φ into non-target moves.

Case 0. If the moved coin is the only member of its component, then k and c decrease both by 1, so the value of Φ is unchanged.

Assume now that the moved coin is from a component of size at least two (in the first layer). We distinguish two cases:

Case 1. The coin is an *end* coin, i.e., one of the two coins at the upper or the lower end of the component (chain) in the current step. Then k decreases by 1 and c is unchanged, thus Φ decreases by 1. By the property of the construction, the current move is a non-target move, thus x increases by 1.

Case 2. The coin is a *middle* coin, i.e., any other coin in the first layer in its start position that is not an end coin in the current step. By the property of the construction, this is necessarily a non-target move (from a component of size at least 3). As a result, k decreases by 1 and c increases by 1, thus Φ decreases by 2. Before the middle coin (A_i or C_i) can be moved by the non-target move, its blocking coin B_i in the second layer must have been moved by a previous non-target move. Observe that this previous non-target move is uniquely assigned to the current non-target move of the middle coin, because the middle coins of different moves cannot be adjacent! Indeed, as soon a middle coin is moved, it breaks up the connect component, and its two adjacent coins cannot become middle coins in subsequent moves. We have therefore found two non-target moves uniquely assigned to this middle coin move, which contribute an increase by 2 to the value of x.

This exhausts the possible cases, and thereby completes the analysis. The lower bounds for values of n other than $3m + 2$ are immediately obtainable from the above: for $n = 3m$, at least $5m - 1$ moves are needed, while for $n = 3m + 1$, at least $5m + 1$ moves are needed. This completes the proof of the theorem. $\square$

3 Hardness Results for Translating Congruent Disks

Theorem 2. *The unlabeled version of the disk reconfiguration problem with translations U-TRANS-RP is NP-hard even for congruent disks.*

Proof. Here we adapt for our purpose the reduction in [5] showing that the reconfiguration problem with unlabeled chips in graphs is NP-complete. We reduce 3-SET-COVER to U-TRANS-RP. The problem 3-SET-COVER is a restricted variant of SET-COVER. An instance of SET-COVER consists of a family $\mathcal{F}$ of subsets of a finite set U, and a positive integer k. The problem is to decide whether there is a set cover of size k for $\mathcal{F}$, i.e., a subset $\mathcal{F}' \subseteq \mathcal{F}$, with $|\mathcal{F}'| \le k$, such that every element in U belongs to at least one member of $\mathcal{F}'$. In the variant 3-SET-COVER the size of each set in $\mathcal{F}$ is bounded from above by 3. Both the standard and the restricted variants are known to be NP-hard [8].

Consider an instance of 3-SET-COVER represented by a bipartite graph $(B \cup C, E)$, where $B = \mathcal{F}$, $C = U$, and E describes the membership relation. First construct a "broom" graph G with vertex set $A \cup B \cup C$, where $|A| = |C|$, as shown in Fig. 4. Place a start (unlabeled) chip at each element of $A \cup B$, and let each element of $B \cup C$ be a target position. A move in the graph is defined as shifting a chip from v_1 to v_2 ($v_1, v_2 \in V(G)$) along a "free" path in G, so that no intermediate vertices are occupied; see also [5]. Positions in B are called

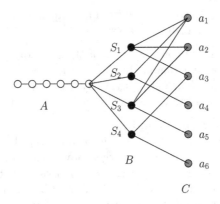

Fig. 4. The "broom" graph G corresponding to a 3-SET-COVER instance with $m = |B| = |\mathcal{F}| = 4$ and $n = |A| = |C| = |U| = 6$. Chips-only: white; obstacles: black; target-only: gray.

obstacles, since any obstacle position that becomes free during reconfiguration must be finally filled by one of the chips. Write $m = |\mathcal{F}|$, and $n = |U|$. Then $|B| = m$ and $|A| = |C| = n$.

Now construct a start and target configuration, each with $O((m+n)^8)$ disks, that represents G in a suitable way. The start positions are (correspond to) $S = A \cup B$ and the target positions are $T = B \cup C$. The positions in B are also called *obstacles*, since any obstacle position that becomes free during reconfiguration must be finally filled by one of the disks. Let z be a parameter used in the construction, to be set later large enough. Consider an axis-parallel rectangle R of width $2z \cdot \max\{m+1, n\}$ and height $z \cdot \max\{m+1, n\}$. Fig. 5 shows a scaled-down version for a smaller value of z ($z = 10$). Initially place an obstacle disk centered at each grid point in R. The obstacle chips in B from the graph G are represented by m obstacles disks, denoted by $S_1, \ldots, S_m$ (the m sets), whose centers are on on the top side of R at distances $2z, 4z, 6z, \ldots$ from the left side of R. Next we (i) delete some of the obstacle disks in R, (ii) add a set of target-only disks in n connected areas (legs) below R, and (iii) change the positions of some of the obstacle disks in B, as described next:

(i) Consider the obstacles whose centers are on on the bottom side of R at distances $z, 3z, 5z, \ldots$ from the left side of R. Let these be denoted by $a_1, \ldots, a_n$ in Fig. 5 and Fig. 6. For each edge $S_i a_j$ in the bipartite graph $(B \cup C, E)$, consider the convex hull H_{ij} of the two disks S_i and a_j; see Fig. 6(middle). We refer to these H_{ij}s as *roads*. Delete now from R any obstacle disk D, except the disks S_i, that intersects some H_{ij} in its interior (the disks $a_1, \ldots, a_n$ are also deleted).

(ii) The target-only chips in C from the graph G are represented by n^2 target-only disks located in n connected areas (legs) extending rectangle R from below. Each leg is a thin rectangle of unit width. These legs extend vertically below the bottom side of R at distances $z, 3z, 5z, \ldots$ from the left side of R,

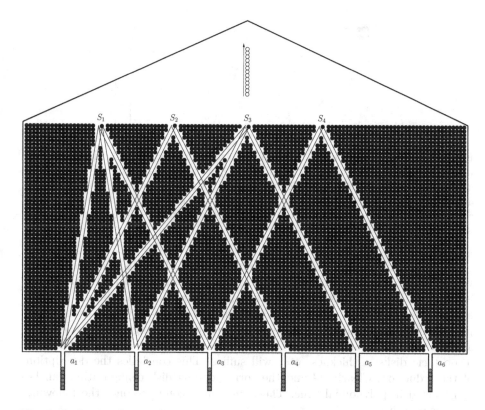

Fig. 5. Reduction from 3-SET-COVER to U-TRANS-RP implementing a disk realization of the "broom" graph G in Fig. 4 ($z = 10$). $|A| = |C| = n = 6$, and $|B| = m = 4$. The start-only disks are white, the obstacle disks are black, and the target-only disks are gray. Obstacle disks on the side of the roads are pushed tangent to the roads (not shown here). The pentagon with 6 legs that encloses the main part of the construction is further enclosed by another thick layer of obstacles. Only 13 out of the 36 start-only disks are shown. Notice the displaced final obstacle positions for the disks S_i; see also Fig. 6(left). An optimal reconfiguration takes $3 \times 36 + 3 = 111$ translation moves (S_2, S_3, and S_4 form an optimal set cover).

       exactly below the obstacles $a_1, \ldots, a_n$ that were previously deleted. Let these legs be denoted by $L_1, \ldots, L_n$ in Fig. 6. In each of the legs we place n target-only disk positions. Since each leg is vertical, its vertical axis is not parallel to any of the road directions H_{ij}.

(iii) For each road H_{ij}, push the disk obstacles next to the road sides closer to the roads, so that they become tangent to the road sides. Displace now each of the obstacle disks S_i to a nearby position that partially blocks any of the (at most 3) roads H_{ij} incident to S_i. See Fig. 6. The new obstacle positions prohibit any S_i to reach any a_j position in one translation move. This special position of these obstacles is important, as the reduction wouldn't work otherwise (that is, if the obstacle would be placed at the intersection of the outgoing roads), at least not in this way.

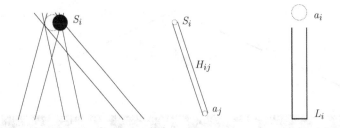

Fig. 6. Left: the actual position of the disk S_i (black) partially blocks the three outgoing roads H_{ij} from S_i. The white dotted disk (not part of the construction) is contained in the intersection of the three incident roads H_{ij}. Middle: a road H_{ij}. Right: a white dotted disk (not part of the construction) representing a_i. This element is used in road construction, for the placement of the incident incoming roads. The corresponding leg is L_i.

The start-only disks in A form a vertical chain of n^2 disks placed on the vertical line ℓ, which is a vertical symmetry axis of R. The position of start disks is such that no start disk is on any of the road directions H_{ij}. Finally enclose all the above start-only disks, obstacle disks, and target-only disks by a closed pentagonal shaped chain with n legs of tangent (or near tangent) obstacle disks, as shown in Fig. 5. Surround all the above construction by another thick layer of obstacle disks; a thickness of z will suffice. This concludes the description of the reduction. Clearly, G and the corresponding disk configuration can be constructed in polynomial time. The reduction is complete once the following claim is established.

Claim. There is a set cover consisting of at most q sets if and only if the disk reconfiguration can be done using at most $3n^2 + q$ translations.   □

A similar reduction can be made for the labeled version by adapting the idea used in [5] to show that the labeled version for reconfiguration of chips in graphs is NP-hard.

Theorem 3. *The labeled version of the disk reconfiguration problem with translations L-TRANS-RP is NP-hard even for congruent disks.*

4   Hardness Results for Sliding Congruent Disks

We start with the unlabeled variant, and adapt for our purpose the reduction in [5] showing that the reconfiguration problem with unlabeled chips in an infinite grid is NP-complete. We reduce the *Rectilinear Steiner Tree* problem R-STEINER to U-SLIDE-RP. An instance of R-STEINER consists of a set S of n points in the plane, and a positive integer bound q. The problem is to decide whether there is a rectilinear Steiner tree (RST), that is, a tree with only horizontal and vertical edges that includes all the points in S, along with possibly some extra *Steiner points*, of total length at most q. For convenience the points can be chosen with integer coordinates. R-STEINER is known to be Strongly NP-complete [8], so we can assume that all point coordinates are given in unary.

Theorem 4. *The unlabeled version of the sliding disks reconfiguration problem in the plane U-SLIDE-RP is NP-hard even for congruent disks.*

Proof. Assume the disks have unit diameter. In our construction each (start and target) disk will be centered at an integer grid point. A disk position (i.e., the center of a disk) that is both a start and a target position is called an *obstacle*. We have four types of grid points: free positions, start positions, target positions, and obstacles.

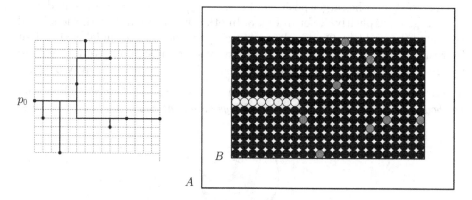

Fig. 7. Left: an instance of R-STEINER with $n = 9$ points, and a rectilinear Steiner tree for it. Right: the configuration of start positions (white), target positions (gray), and obstacle positions (black).

Consider an instance $P = \{p_0, p_1, \ldots, p_{n-1}\}$ of R-STEINER with n points. Assume that $p_0 = (0, 0)$ is a leftmost point in P, see Fig. 7(left). The instance of U-SLIDE-RP is illustrated in Fig. 7(right). Choose $n - 1$ start positions with zero y-coordinate and x-coordinates $0, -1, -2, \ldots, -(n-2)$, i.e., in a straight horizontal chain extending from p_0 to the left. Choose $n - 1$ target positions at the remaining $n - 1$ points $\{p_1, \ldots, p_{n-1}\}$ of the R-STEINER instance. Let B be a smallest axis-parallel rectangle containing the $2n - 2$ disks at the start and target positions, and Δ be the length of the longer side of B. Consider a sufficiently large axis-parallel rectangle A enclosing B: the boundary of A is at distance $2n\Delta$ from the boundary of B. Place obstacle disks centered at each of the remaining grid points in the rectangle A. The number of disks in the start and target configurations is $O(n^2\Delta^2)$. This construction is done in time polynomial in Δ, which is polynomial in the size of the R-STEINER instance since the coordinates are given in unary. The reduction is complete once the following claim is established.

Claim. There is a rectilinear Steiner tree of length at most q for P if and only if the disk reconfiguration can be done using at most q sliding moves.   □

A similar reduction can be made for the labeled version by adapting the idea used in [5] to show that the labeled version for reconfiguration of chips in grids is NP-hard.

Theorem 5. *The labeled version of the sliding disks reconfiguration problem in the plane L-SLIDE-RP is NP-hard even for congruent disks.*

5   Translating Convex Bodies

In this section we consider the general problem of reconfiguration of convex bodies with translations. When the convex bodies have different shapes, sizes, and orientations, assume that the correspondence between the start positions $\{S_1, \ldots, S_n\}$ and the target positions $\{T_1, \ldots, T_n\}$, where T_i is a translated copy of S_i, is given explicitly. Refer to Fig. 8. In other words, we deal with the labeled variant of the problem. Our result can be easily extended to the unlabeled variant by first computing a valid correspondence by shape matching. We first extend the $2n$ upper bound for translating arbitrary disks to arbitrary convex bodies:

Theorem 6. *For the reconfiguration with translations of n labeled disjoint convex bodies in the plane, $2n$ moves are always sufficient and sometimes necessary.*

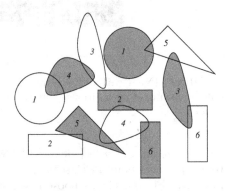

Fig. 8. Reconfiguration of convex bodies with translations. The start positions are unshaded; the target positions are shaded.

5.1   Translating Unlabeled Axis-Parallel Unit Squares

Throughout a translation move, the moving square remains axis-parallel, however the move can be in any direction. We have the following bounds:

Theorem 7. *For the reconfiguration with translations of n unlabeled axis-parallel unit squares in the plane, $2n - 1$ moves are always sufficient, and $\lfloor 3n/2 \rfloor$ moves are sometimes necessary.*

Theorem 8. *For the reconfiguration with translations of n unlabeled axis-parallel unit squares in the plane from the third quadrant to the first quadrant, n moves are always sufficient and sometimes necessary.*

References

1. Abellanas, M., Bereg, S., Hurtado, F., Olaverri, A.G., Rappaport, D., Tejel, J.: Moving coins. Comput. Geometry: Theory and Applications 34(1), 35–48 (2006)
2. Bereg, S., Dumitrescu, A.: The lifting model for reconfiguration. Discrete & Computational Geometry 35(4), 653–669 (2006)
3. Bereg, S., Dumitrescu, A., Pach, J.: Sliding disks in the plane. International Journal of Computational Geometry & Applications 15(8), 373–387 (2008)
4. de Bruijn, N.G.: Aufgaben 17 and 18 (in Dutch), Nieuw Archief voor Wiskunde 2, 67 (1954)
5. Călinescu, G., Dumitrescu, A., Pach, J.: Reconfigurations in graphs and grids. SIAM Journal on Discrete Mathematics 22(1), 124–138 (2008)
6. Chazelle, B., Ottmann, T.A., Soisalon-Soininen, E., Wood, D.: The complexity and decidability of separation. In: Paredaens, J. (ed.) ICALP 1984. LNCS, vol. 172, pp. 119–127. Springer, Heidelberg (1984)
7. Fejes Tóth, L., Heppes, A.: Über stabile Körpersysteme (in German). Compositio Mathematica 15, 119–126 (1963)
8. Garey, M., Johnson, D.: Computers and Intractability: A Guide to the Theory of NP-Completeness. W. H. Freeman and Company, New York (1979)
9. Guibas, L., Yao, F.F.: On translating a set of rectangles. In: Preparata, F. (ed.) Computational Geometry. Advances in Computing Research, vol. 1, pp. 61–67. JAI Press, London (1983)
10. Hopcroft, J., Schwartz, J., Sharir, M.: On the complexity of motion planning for multiple independent objects; PSPACE- Hardness of the Warehouseman's Problem. The International Journal of Robotics Research 3(4), 76–88 (1984)
11. O'Rourke, J.: Computational Geometry in C, 2nd edn. Cambridge University Press, Cambridge (1998)

Orientation-Constrained Rectangular Layouts

David Eppstein[1] and Elena Mumford[2]

[1] Department of Computer Science, University of California, Irvine, USA
[2] Department of Mathematics and Computer Science, TU Eindhoven, The Netherlands

Abstract. We construct partitions of rectangles into smaller rectangles from an input consisting of a planar dual graph of the layout together with restrictions on the orientations of edges and junctions of the layout. Such an orientation-constrained layout, if it exists, may be constructed in polynomial time, and all orientation-constrained layouts may be listed in polynomial time per layout.

1 Introduction

Consider a partition of a rectangle into smaller rectangles, at most three of which meet at any point. We call such a partition a *rectangular layout*. Rectangular layouts are an important tool in many application areas. In VLSI design rectangular layouts represent floorplans of integrated circuits [8], while in architectural design they represent floorplans of buildings [2, 12]. In cartography they are used to visualize numeric data about geographic regions, by stylizing the shapes of a set of regions to become rectangles, with areas chosen to represent statistical data about the regions; such visualizations are called *rectangular cartograms*, and were first introduced in 1934 by Raisz [11].

The *dual graph* or *adjacency graph* of a layout is a plane graph $G(\mathcal{L})$ that has a vertex for every region of $\mathcal{L}$ and an edge for every two adjacent regions. In both VLSI design and in cartogram construction, the adjacency graph G is typically given as input, and one has to construct its *rectangular dual*, a rectangular layout for which G is the dual graph. Necessary and sufficient conditions for a graph to have a rectangular dual are known [7], but graphs that admit a rectangular dual often admit more than one. This fact allows us to impose additional requirements on the rectangular duals that we select, but it also leads to difficult algorithmic questions concerning problems of finding layouts with desired properties. For example, Eppstein et al [3] have considered the search for *area-universal* layouts, layouts that can be turned into rectangular cartograms for any assignment of positive weights to their regions.

In this paper, we consider another kind of constrained layouts. Given a graph G we would like to know whether G has a rectangular dual that satisfies certain constraints on the orientations of the adjacencies of its regions; such constraints may be particularly relevant for cartographic applications of these layouts. For example, in a cartogram of the U.S., we might require that a rectangle representing Nevada be right of or above a rectangle representing California, as geographically Nevada is east and north of California. We show that layouts with orientation constraints of this type may be constructed in polynomial time. Further, we can list all layouts obeying the constraints in polynomial time per layout. Our algorithms can handle constraints (such as the one above) limiting the allowed orientations of a shared edge between a pair of adjacent regions,

F. Dehne et al. (Eds.): WADS 2009, LNCS 5664, pp. 266–277, 2009.

as well as more general kind of constraints restricting the possible orientations of the three rectangles meeting at any junction of the layout. We also discuss the problem of finding area-universal layouts in the presence of constraints of these types. A version of the orientation-restricted layout problem was previously considered by van Kreveld and Speckmann [14] but they required a more restrictive set of constraints and searched exhaustively through all layouts rather than developing polynomial time algorithms.

Following [3], we use Birkhoff's representation theorem for finite distributive lattices to associate the layouts dual to G with partitions of a related partial order into a lower set and an upper set. The main idea of our new algorithms is to translate the orientation constraints of our problem into an equivalence relation on this partial order. We form a quasiorder by combining this relation with the original partial order, partition the quasiorder into lower and upper sets, and construct layouts from these partitions. However, the theory as outlined above only works directly on dual graphs with no nontrivial separating 4-cycles. To handle the general case we must do more work to partition G by its 4-cycles into subgraphs and to piece together the solutions from each subgraph.

2 Preliminaries

Kozminski and Kinnen [7] demonstrated that a plane triangulated graph G has a rectangular dual if it can be augmented with four external vertices $\{l,t,r,b\}$ to obtain an *extended graph* $E(G)$ in which every inner face is a triangle, the outer face is a quadrilateral, and $E(G)$ does not contain any separating 3-cycles (a *separating k-cycle* is a k-cycle that has vertices both inside and outside of it). A graph G that can be extended in this way is said to be *proper*—see Fig. 1 for an example. The extended graph $E(G)$ is sometimes referred to as a *corner assignment* of G, since it defines which vertices of G become corner rectangles of the corresponding dual layout. In the rest of the paper we assume that we are given a proper graph with a corner assignment. For proper graphs without a corner assignment, one can always test all possible corner assignments, as their number is polynomial in the number of external vertices of the graph.

A rectangular dual L induces a labeling for the edges of its graph $G(L)$: we color each edge blue if the corresponding pair of rectangles share a vertical line, or red if the corresponding pair of rectangles share a horizontal border; we direct the blue edges from left to right and red edges from bottom to top. For each inner vertex v of $G(L)$ the incident edges with the same label form continuous blocks around v: all incoming blue edges are followed (in clockwise order) by all outgoing red, all outgoing blue and finally all incoming red edges. All edges adjacent to one of the four external vertices l,t,r,b have a single label. A labeling that satisfies these properties is called a *regular edge*

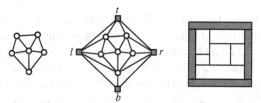

Fig. 1. A proper graph G, extended graph $E(G)$, and rectangular dual L of $E(G)$, from [3]

labeling [6]. Each regular edge labeling of a proper graph corresponds to an equivalence class of rectangular duals of G, considering two duals to be *equivalent* whenever every pair of adjacent regions have the same type of adjacency in both duals.

2.1 The Distributive Lattice of Regular Edge Labelings

A *distributive lattice* is a partially ordered set in which every pair (a, b) of elements has a unique supremum $a \wedge b$ (called the *meet* of a and b) and a unique infimum $a \vee b$ (called the *join* of a and b) and where the join and meet operations are distributive over each other. Two comparable elements that are closest neighbours in the order are said to be a *covering pair*, the larger one is said to *cover* the smaller one.

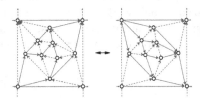

Fig. 2. Recoloring the interior of an alternatingly-colored four-cycle in a regular edge labeling

All regular edge labelings of a proper graph form a distributive lattice [4, 5, 13] in which the covering pairs of layouts are the pairs in which the layouts can be transformed from one to the other by means of a *move*—changing the labeling of the edges inside an *alternating four-cycle* (a 4-cycle C whose edge colors alternate along the cycle). Each red edge within the cycle becomes blue and vice versa; the orientations of the edges are adjusted in a unique way such that the cyclic order of the edges around each vertex is as defined above—see Fig. 2 for an example. In terms of layouts, the move means rotating the sublayout formed by the inner vertices of C by 90 degrees. A move is called *clockwise* if the borders between the red and blue labels of each of the four vertices of the cycle move clockwise by the move, and called *counterclockwise* otherwise. A counterclockwise move transforms a layout into another layout higher up the lattice.

We can represent the lattice by a graph in which each vertex represents a single layout and each edge represents a move between a covering pair of layouts, directed from the lower layout to the higher one. We define a *monotone path* to be a path in this graph corresponding to a sequence of counterclockwise moves.

2.2 The Birkhoff Representation of the Lattice of Layouts

For any finite distributive lattice D, let $\mathcal{P}$ be the partial order of join-irreducible elements (elements that cover only one other element of D), and let $J(\mathcal{P})$ be the lattice of partitions of $\mathcal{P}$ into sets L and U, where L is downward closed and U is upward closed and where meets and joins in $J(\mathcal{P})$ are defined as intersections and unions of these sets. Birkhoff's representation theorem [1] states that D is isomorphic to $J(\mathcal{P})$.

Eppstein et al. [3] show that when $E(G)$ has no nontrivial separating four-cycles (four-cycles with more than one vertex on the inside) the partial order of join-irreducible

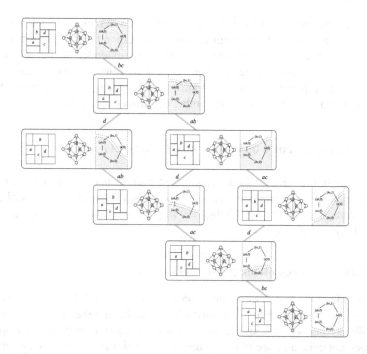

Fig. 3. The rectangular layouts dual to a given extended graph $E(G)$ and the corresponding regular edge labelings and partial order partitions. Two layouts are shown connected to each other by an edge if they differ by reversing the color within a single alternatingly-colored four-cycle; these moves are labeled by the edge or vertex within the four-cycle. From [3].

elements of the lattice of rectangular duals of $E(G)$ is order-isomorphic to the partial order $\mathcal{P}$ on pairs (x, i), where x is a *flippable item* of $E(G)$, and i is a *flipping number* of x. A *flippable item* x is either a degree-four vertex of G or an edge of G that is not adjacent to a degree-four vertex, such that there exist two regular edge labelings of $E(G)$ in which x has different labels (when x is a vertex we refer to the labels of its four adjacent edges). Given a layout L and a flippable item x, the number $f_x(L)$ is the number of times that x has been flipped on a monotone path in the distributive lattice from its bottom element to L; this number, which we call the *flipping number* of x in L, is well defined, since it is independent of the path by which L has been reached. For every flippable item x, $\mathcal{P}$ contains pairs (x, i) for all i such that there exist a layout L where $f_x(L) = i - 1$. A pair (x, i) is associated with the transition of x from state i to $i+1$. A pair (x, i) is less than a pair (y, j) in the partial order if is not possible to flip y for the jth time before flipping x for the ith time. If (x, i) and (y, j) form a covering pair in $\mathcal{P}$, the flippable items x and y belong to the same triangular face of $E(G)$.

As Eppstein et al. show, the layouts dual to $E(G)$ correspond one-for-one with partitions of the partial order $\mathcal{P}$ into a lower set L and an upper set U. The labeling of the layout corresponding to a given partition of $\mathcal{P}$ can be recovered by starting from the minimal layout and flipping each flippable item x represented in the lower set L of the partition $n_x + 1$ times, where (x, n_x) is the highest pair involving x in L. The downward moves that can be performed from L correspond to the maximal elements of L, and the upward moves that can be performed from L correspond to the minimal elements of U. Fig. 3 depicts the lattice of layouts of a 12-vertex extended dual graph, showing for each layout the corresponding partition of the partial order into two sets L and U. The partial order of flippable items has at most $O(n^2)$ elements and can be constructed in time polynomial in n, where n is the number of vertices in G [3].

3   The Lattice Theory of Constrained Layouts

As we describe in this section, in the case where every separating 4-cycle in $E(G)$ is trivial, the orientation-constrained layouts of $E(G)$ may themselves be described as a distributive lattice, a sublattice (although not in general a connected subgraph) of the lattice of all layouts of $E(G)$.

3.1   Sublattices from Quotient Quasiorders

We first consider a more general order-theoretic problem. Let $\mathcal{P}$ be a partial order and let C be a (disconnected) undirected *constraint graph* having the elements of $\mathcal{P}$ as its vertices. We say that a partition of $\mathcal{P}$ into a lower set L and an upper set U *respects C* if there does not exist an edge of C that has one endpoint in L and the other endpoint in U. As we now show, the partitions that respect C may be described as a sublattice of the distributive lattice $J(\mathcal{P})$ defined via Birkhoff's representation theorem from $\mathcal{P}$.

We define a quasiorder (that is, reflexive and transitive binary relation) Q on the same elements as $\mathcal{P}$, by adding pairs to the relation that cause certain elements of $\mathcal{P}$ to become equivalent to each other. More precisely, form a directed graph that has the elements of $\mathcal{P}$ as its vertices , and that has a directed edge from x to y whenever either $x \leq y$ in $\mathcal{P}$ or xy is an edge in C, and define Q to be the transitive closure of this directed graph: that is, (x, y) is a relation in Q whenever there is a path from x to y in the directed graph. A subset S of Q is downward closed (respectively, upward closed) if there is no pair (x, y) related in Q for which $S \cap \{x, y\} = \{y\}$ (respectively, $S \cap \{x, y\} = \{x\}$).

Denote by $J(Q)$ the set of partitions of Q into a downward closed and an upward closed set. Each strongly connected component of the directed graph derived from $\mathcal{P}$ and C corresponds to a set of elements of Q that are all related bidirectionally to each other, and Q induces a partial order on these strongly connected components. Therefore, by Birkhoff's representation theorem, $J(Q)$ forms a distributive lattice under set unions and intersections.

Lemma 1. *The family of partitions in $J(Q)$ is the family of partitions of $\mathcal{P}$ into lower and upper sets that respect C.*

Proof. We show the lemma by demonstrating that every partition in $J(Q)$ corresponds to a partition of $J(\mathcal{P})$ that respects C and the other way round.

In one direction, let (L, U) be a partition in $J(Q)$. Then, since $Q \supset \mathcal{P}$, it follows that (L, U) is also a partition of $\mathcal{P}$ into a downward-closed and an upward-closed subset. Additionally, (L, U) respects C, for if there were an edge xy of C with one endpoint in

L and the other endpoint in U then one of the two pairs (x,y) or (y,x) would contradict the definition of being downward closed for L.

In the other direction, let (L',U') be a partition of $\mathcal{P}$ into upper and lower sets that respects C, let (x,y) be any pair in Q, and suppose for a contradiction that $x \in U'$ and $y \in L'$. Then there exists a directed path from x to y in which each edge consists either of an ordered pair in $\mathcal{P}$ or an edge in C. Since $x \in U'$ and $y \in L'$, this path must have an edge in which the first endpoint is in U' and the second endpoint is in L'. But if this edge comes from an ordered pair in $\mathcal{P}$, then (L',U') is not a partition of $\mathcal{P}$ into upper and lower sets, while if this edge comes from C then (L',U') does not respect C. This contradiction establishes that there can be no such pair (x,y), so (L',U') is a partition of Q into upper and lower sets as we needed to establish.

If $\mathcal{P}$ and C are given as input, we may construct Q in polynomial time: by finding strongly connected components of Q we may reduce it to a partial order, after which it is straightforward to list the partitions in $J(Q)$ in polynomial time per partition.

3.2   Edge Orientation Constraints

Consider a proper graph G with corner assignment $E(G)$ and assume that each edge e is given with a set of *forbidden labels*, where a labels is a color-orientation combination for an edge, and let $\mathcal{P}$ be the partial order whose associated distributive lattice $J(\mathcal{P})$ has its elements in one-to-one correspondence with the layouts of $E(G)$. Let x be the flippable item corresponding to e—that is either the edge itself of the degree-four vertex e is adjacent to. Then in any layout L, corresponding to a partition $(L,U) \in J(\mathcal{P})$, the orientation of e in L may be determined from i mod 4, where i is the largest value such that $(x,i) \in L$. Thus if we would like to exclude a certain color-orientation combination for x, we have find the corresponding value $k \in \mathbb{Z}_4$ and exclude the layouts L such that $f_x(L) = k$ mod 4 from consideration. Thus the set of flipping values for x can be partitioned into *forbidden* and *legal* values for x; instead of considering color-orientation combinations of the flippable items we may consider their flipping values. We formalize this reasoning in the following lemma.

Lemma 2. *Let $E(G)$ be a corner assignment of a proper graph G. Let x be a flippable item in $E(G)$, let L be an element of the lattice of regular edge labelings of $E(G)$, and let (L,U) be the corresponding partition of $\mathcal{P}$.*

Then L satisfies the constraints described by the forbidden labels if and only if for every flippable item x one of the following is true:

- *The highest pair involving x in L is (x,i), where $i+1$ is not a forbidden value for x, or*
- *$(x,0)$ is in the upper set and 0 is not a forbidden value for x.*

Lemma 1 may be used to show that the set of all constrained layout is a distributive lattice, and that all constrained layouts may be listed in polynomial time per layout. For technical reasons we augment $\mathcal{P}$ to a new partial order $A(\mathcal{P}) = \mathcal{P} \cup \{-\infty, +\infty\}$, where the new element $-\infty$ lies below all other elements and the new element $+\infty$ lies above all other elements. Each layout of $E(G)$ corresponds to a partition of $\mathcal{P}$ into lower and

Fig. 4. The family of rectangular layouts dual to a given extended graph $E(G)$ satisfying the constraints that the edge between rectangles a and b must be vertical (cannot be colored red) and that the edge between rectangles b and c must be horizontal (cannot be colored blue). The green regions depict strongly connected components of the associated quasiorder Q. The four central shaded elements of the lattice correspond to layouts satisfying the constraints.

upper sets, which can be mapped into a partition of $A(\mathcal{P})$ by adding $-\infty$ to the lower set and $+\infty$ to the upper set. The distributive lattice $J(A(\mathcal{P}))$ thus has two additional elements that do not correspond to layouts of $E(G)$: one in which the lower set is empty and one in which the upper set is empty. We define a constraint graph C having as its vertices the elements of $A(\mathcal{P})$, with edges defined as follows:

– If (x, i) and $(x, i+1)$ are both elements of $A(\mathcal{P})$ and $i+1$ is a forbidden value for x, we add an edge from (x, i) to $(x, i+1)$ in C.
– If (x, i) is an element of $A(\mathcal{P})$ but $(x, i+1)$ is not, and $i+1$ is a forbidden value for x, we add an edge from (x, i) to $+\infty$ in C.
– If 0 is a forbidden value for x, we add an edge from $-\infty$ to $(x, 0)$ in C.

All together, this brings us to the following result:

Lemma 3. *Let $E(G)$ be an extended graph without nontrivial separating 4-cycles and with a given set of forbidden orientations, and let Q be the quasiorder formed from the transitive closure of $A(\mathcal{P}) \cup C$ as described in Lemma 1. Then the elements of $J(Q)$*

corresponding to partitions of Q into two nonempty subsets correspond to exactly the layouts that satisfy the forbidden orientation constraints.

Proof. By Lemma 2 and the definition of C, a partition in $J(\mathcal{P})$ corresponds to a constrained layout if and only if it respects each of the edges in C. By Lemma 1, the elements of $J(Q)$ correspond to partitions of $A(\mathcal{P})$ that respect C. And a partition of $A(\mathcal{P})$ corresponds to an element of $J(\mathcal{P})$ if and only if its lower set does not contain $+\infty$ and its upper set does not contain $-\infty$.

Corollary 1. *Let $E(G)$ be an extended graph without nontrivial separating 4-cycles and with a given set of forbidden orientations. There exists a constrained layout for $E(G)$ if and only if there exists more than one strongly connected component in Q.*

Corollary 2. *The existence of a constrained layout for a given extended graph $E(G)$ without nontrivial separating 4-cycles can be proved or disproved in polynomial time.*

Corollary 3. *All constrained layouts for a given extended graph $E(G)$ without nontrivial separating 4-cycles can be listed in polynomial time per layout.*

Figure 4 depicts the sublattice resulting from these constructions for the example from Figure 3, with constraints on the orientations of two of the layout edges.

3.3 Junction Orientation Constraints

So far we have only considered forbidding certain edge labels. However the method above can easily be extended to different types of constraints. For example, consider two elements of $\mathcal{P}$ (x,i) and (y,j) that are a covering pair in $\mathcal{P}$; this implies that x and y are two of the three flippable items surrounding unique a T-junction of the layouts dual to $E(G)$. Forcing (x,i) and (y,j) to be equivalent by adding an edge from (x,i) to (y,j) in the constraint graph C can be used for more general constraints: rather than disallowing one or more of the four orientations for any single flippable item, we can disallow one or more of the twelve orientations of any T-junction. For instance, by adding equivalences of this type we could force one of the three rectangles at the T-junction to be the one with the 180-degree angle.

Any internal T-junction of a layout for $E(G)$ (dual to a triangle of G) has 12 potential orientations: each of its three rectangles can be the one with the 180-degree angle, and with that choice fixed there remain four choices for the orientation of the junction. In terms of the regular edge labeling, any triangle of G may be colored and oriented in any of 12 different ways. For a given covering pair (x,i) and (y,j), let $C_{x,y}^{i,j}$ denote the set of edges between pairs $(x,i+4k)$ and $(y,j+4k)$ for all possible integer values of k, together with an edge from $-\infty$ to $(y,0)$ if $j \bmod 4 = 0$ and an edge from $(x,i+4k)$ to $+\infty$ if $i+4k$ is the largest value of i' such that (x,i') belongs to $\mathcal{P}$. Any T-junction is associated with 12 of these edge sets, as there are three ways of choosing a pair of adjacent flippable items and four ways of choosing values of i and j (mod 4) that lead to covering pairs. Including any one of these edge sets in the constraint graph C corresponds to forbidding one of the 12 potential orientations of the T-junction.

Thus, Lemma 3.2 and its corollaries may be applied without change to dual graphs $E(G)$ with junction orientation constraints as well as edge orientation constraints, as long as $E(G)$ has no nontrivial separating 4-cycles.

4   Constrained Layouts for Unconstrained Dual Graphs

Proper graphs with nontrivial separating 4-cycles still have finite distributive lattices of layouts, but it is no longer possible to translate orientation constraints into equivalences between members of an underlying partial order. The reason is that, for a graph without trivial separating 4-cycles, the orientation of a feature of the layout changes only for a flip involving that feature, so that the orientation may be determined from the flip count modulo four. For more general graphs the orientation of a feature is changed not only for flips directly associated with that feature, but also for flips associated with larger 4-cycles that contain the feature, so the flip count of the feature no longer determines its orientation. For this reason, as in [3], we treat general proper graphs by decomposing them into *minimal separation components* with respect to separating 4-cycles and piecing together solutions found separately within each of these components.

For each separating four-cycle C in a proper graph G with a corner assignment $E(G)$ consider two minors of G defined as follows. The *inner separation component* of C is a graph G_C and its extended graph $E(G_C)$, where G_C is the subgraph of G induced by the vertices inside C and $E(G_C)$ adds the four vertices of the cycle as corners of the extended graph. The *outer separation component* of C is a graph formed by contracting the interior of C into a single supervertex. A *minimal separation component* of G is a minor of G formed by repeatedly splitting larger graphs into separation components until no nontrivial separating four-cycles remain. A partition tree of $E(G)$ into minimal separation components may be found in linear time [3].

We use the representation of a graph as a tree of minimal separation components in our search for constrained layouts for G. We first consider each such minimal component separately for every possible mapping of vertices of C to $\{l, t, r, b\}$ (we call these mappings the *orientation* of $E(G)$). Different orientations imply different flipping values of forbidden labels for the given constraint function, since the flipping numbers are defined with respect to the orientation of $E(G)$. Having that in mind we are going to test the graph $E(G)$ for existence of a constrained layout in the following way:

For each piece in a bottom-up traversal of the decomposition tree and for each orientation of the corners of the piece:

1. Find the partial order $\mathcal{P}$ describing the layouts of the piece
2. Translate the orientation constraints within the piece into a constraint graph on the augmented partial order $A(\mathcal{P})$.

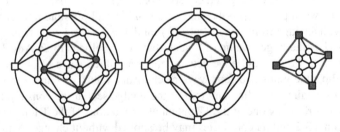

Fig. 5. An extended graph with a nontrivial separating four-cycle (left), its outer separation component (center), and its inner separation component (right). From [3].

3. Compute the strongly connected components of the union of $A(\mathcal{P})$ with the constraint graph, and form a binary relation that is a subset of Q and that includes all covering relations in Q by finding the components containing each pair of elements in each covering relation in $\mathcal{P}$.

4. Translate the existence or nonexistence of a layout into a constraint on the label of the corresponding degree-4 vertex in the parent piece of the decomposition. That is, if the constrained layout for a given orientation of $E(\mathcal{G}')$ does not exist, forbid (in the parent piece of the decomposition) the label of the degree-four vertex corresponding to that orientation.

If the algorithm above confirms the existence of a constrained layout, we may list all layouts satisfying the constraints as follows. For each piece in the decomposition tree, in top-down order:

1. List all lower sets of the corresponding quasiorder Q.
2. Translate each lower set into a layout for that piece.
3. For each layout, and each child of the piece in the decomposition tree, recursively list the layouts in which the child's corner orientation matches the labeling of the corresponding degree-four vertex of the outer layout.
4. Glue the inner and outer layouts together.

Theorem 1. *The existence of a constrained layout for a proper graph $\mathcal{G}$ can be found in polynomial time in $|\mathcal{G}|$. The set of all constrained layouts for graph can be found in polynomial time per layout.*

As described in [3], the partial order $\mathcal{P}$ describing the layouts of each piece has a number of elements and covering pairs that is quadratic in the number of vertices in the dual graph of the piece, and a description of this partial order in terms of its covering pairs may be found in quadratic time. The strongly connected component calculation within the algorithm takes time linear in the size of $\mathcal{P}$, and therefore the overall algorithm for testing the existence of a constrained layout takes time $O(n^2)$, where n is the number of vertices in the given dual graph.

5   Finding Area-Universal Constrained Layouts

Our previous work [3] included an algorithm for finding area-universal layouts that is fixed-parameter tractable, with the maximum number of separating four-cycles in any piece of the separation component decomposition as its parameter. It is not known whether this problem may be solved in polynomial time for arbitrary graphs. But as we outline in this section, the same fixed parameter tractability result holds for a combination of the constraints from that paper and from this one: the problem of searching for an area-universal layout with constrained orientations.

These layouts correspond to partitions of $\mathcal{P}$ such that all flippable items that are minimal elements of the upper set and all flippable items that are maximal items of the lower set are all degree-four vertices. A brute force algorithm can find these partitions by looking at all sets of degree-four vertices as candidates for extreme sets for partitions of $\mathcal{P}$. Instead, in our previous work on this problem we observed that a flippable edge is not *free* in a layout (i.e. cannot be flipped in the layout), if an only if it is *fixed* by a so-called *stretched pair*. A stretched pair is a pair of two degree-four vertices (v, w), such that on any monotone path up from L w is flipped before w, and on any monotone path down from L v is flipped before w. If $f_v(L)$ is the maximal flipping value of v, then we declare $(v, \emptyset)$ to be stretched (where $\emptyset$ is a special symbol) and declare $(\emptyset, w)$ to be a stretched pair if $f_w(L) = 0$. An edge is *fixed* by a stretched pair (v, w) in L if x if every monotone path up from L moves w before moving x, and every monotone path down from L moves v before moving x. So instead of looking for extreme sets, we could check every set of degree-four vertices for existence of a layout in which every pair in the set is stretched, and check which edges each such a set fixes. Each set H of pairs can be checked for stretchability by starting at the bottom of the lattice and flipping the corresponding items of $\mathcal{P}$ up the lattice until every pair in H is stretched or the maximal elements of the lattice is reached. If there are k degree-four vertices (or equivalently separating 4-cycles) in a piece, there are $2^{O(k^2)}$ sets of stretched pairs we need consider, each of which takes polynomial time to test, so the overall algorithm for searching for unconstrained area-universal layouts takes time $2^{O(k^2)} n^{O(1)}$.

For constrained layouts, a similar approach works. Within each piece of the separation decomposition, we consider $2^{O(k^2)}$ sets of stretched pairs in $\mathcal{P}$, as before. However, to test one of these sets, we perform a monotonic sequence of flips in $J(Q)$, at each point either flipping an element of Q that contains the upper element of a pair that should be stretched, or performing a flip that is a necessary prerequisite to flipping such an upper element. Eventually, this process will either reach an area-universal layout for the piece or the top element of the lattice; in the latter case, no area-universal layout having that pattern of stretched pairs exists. By testing all sets of stretched pairs, we may find whether an area-universal layout matching the constraints exists for any corner coloring of any piece in the separation decomposition. These constrained layouts for individual pieces can then be combined by the same tree traversal of the separation decomposition tree that we used in the previous section, due to the observation from [3] that a layout is area-universal if and only if the derived layout within each of its separation components is area-universal. The running time for this fixed-parameter tractable algorithm is the same as in [3].

6   Conclusions and Open Problems

We have provided efficient algorithms for finding rectangular layouts with orientation constraints on the features of the constraints, and we have outlined how to combine our approach with the previous algorithms for finding area-universal layouts so that we can find orientation-constrained area-universal layouts as efficiently as we can solve the unconstrained problem.

An important problem in the generation of rectangular layouts with special properties, that has resisted our lattice-theoretic approach, is the generation of sliceable layouts. If we are given a graph G, can we determine whether it is the graph of a sliceable layout in polynomial time? Additionally, although our algorithms are polynomial time, there seems no reason intrinsic to the problem for them to take as much time as they do:

can we achieve subquadratic time bounds for finding orientation-constrained layouts, perhaps by using an algorithm based more on the special features of the problem and less on general ideas from lattice theory?

Moving beyond layouts, there are several other important combinatorial constructions that may be represented using finite distributive lattices, notably the set of matchings and the set of spanning trees of a planar graph, and certain sets of orientations of arbitrary graphs [9]. It would be of interest to investigate whether our approach of combining the underlying partial order of a lattice with a constraint graph produces useful versions of constrained matching and constrained spanning tree problems, and whether other algorithms that have been developed in the more general context of distributive finite lattices [10] might fruitfully be applied to lattices of rectangular layouts.

Acknowledgements

Work of D. Eppstein was supported in part by NSF grant 0830403 and by the Office of Naval Research under grant N00014-08-1-1015.

References

1. Birkhoff, G.: Rings of sets. Duke Mathematical Journal 3(3), 443–454 (1937)
2. Earl, C.F., March, L.J.: Architectural applications of graph theory. In: Wilson, R., Beineke, L. (eds.) Applications of Graph Theory, pp. 327–355. Academic Press, London (1979)
3. Eppstein, D., Mumford, E., Speckmann, B., Verbeek, K.: Area-universal rectangular layouts. In: Proc. 25th ACM Symp. Computational Geometry (to appear, 2009)
4. Fusy, É.: Transversal structures on triangulations, with application to straight-line drawing. In: Healy, P., Nikolov, N.S. (eds.) GD 2005. LNCS, vol. 3843, pp. 177–188. Springer, Heidelberg (2006)
5. Fusy, É.: Transversal structures on triangulations: A combinatorial study and straight-line drawings. Discrete Mathematics (to appear, 2009)
6. Kant, G., He, X.: Regular edge labeling of 4-connected plane graphs and its applications in graph drawing problems. Theoretical Computer Science 172(1-2), 175–193 (1997)
7. Koźmiński, K., Kinnen, E.: Rectangular duals of planar graphs. Networks 5(2), 145–157 (1985)
8. Liao, C.-C., Lu, H.-I., Yen, H.-C.: Compact floor-planning via orderly spanning trees. Journal of Algorithms 48, 441–451 (2003)
9. Propp, J.: Lattice structure for orientations of graphs. Electronic preprint arxiv:math/0209005 (1993)
10. Propp, J.: Generating random elements of finite distributive lattices. Electronic J. Combinatorics 4(2), R15 (1997)
11. Raisz, E.: The rectangular statistical cartogram. Geographical Review 24(2), 292–296 (1934)
12. Rinsma, I.: Rectangular and orthogonal floorplans with required rooms areas and tree adjacency. Environment and Planning B: Planning and Design 15, 111–118 (1988)
13. Tang, H., Chen, W.-K.: Generation of rectangular duals of a planar triangulated graph by elementary transformations. In: IEEE Int. Symp. Circuits and Systems, vol. 4, pp. 2857–2860 (1990)
14. van Kreveld, M., Speckmann, B.: On rectangular cartograms. Computational Geometry: Theory and Applications 37(3), 175–187 (2007)

The h-Index of a Graph and Its
Application to Dynamic Subgraph Statistics

David Eppstein[1] and Emma S. Spiro[2]

[1] Computer Science Department, University of California, Irvine
[2] Department of Sociology, University of California, Irvine

Abstract. We describe a data structure that maintains the number of triangles in a dynamic undirected graph, subject to insertions and deletions of edges and of degree-zero vertices. More generally it can be used to maintain the number of copies of each possible three-vertex subgraph in time $O(h)$ per update, where h is the *h-index* of the graph, the maximum number such that the graph contains h vertices of degree at least h. We also show how to maintain the h-index itself, and a collection of h high-degree vertices in the graph, in constant time per update. Our data structure has applications in social network analysis using the exponential random graph model (ERGM); its bound of $O(h)$ time per edge is never worse than the $\Theta(\sqrt{m})$ time per edge necessary to list all triangles in a static graph, and is strictly better for graphs obeying a power law degree distribution. In order to better understand the behavior of the h-index statistic and its implications for the performance of our algorithms, we also study the behavior of the h-index on a set of 136 real-world networks.

1 Introduction

The *exponential random graph model* (ERGM, or p^* model) [18, 35, 30] is a general technique for assigning probabilities to graphs that can be used both to generate simulated data for social network analysis and to perform probabilistic reasoning on real-world data. In this model, one fixes the vertex set of a graph, identifies certain *features* f_i in graphs on that vertex set, determines a *weight* w_i for each feature, and sets the probability of each graph G to be proportional to an exponential function of the sum of its features' weights, divided by a normalizing constant Z:

$$\Pr(G) = \frac{\exp \sum_{f_i \in G} w_i}{Z}.$$

Z is found by summing over all graphs on that vertex set:

$$Z = \sum_G \exp \sum_{f_i \in G} w_i.$$

For instance, if each potential edge is considered to be a feature and all edges have weight $\ln \frac{p}{1-p}$, the normalizing constant Z will be $(1-p)^{-n(n-1)/2}$, and the probability of any particular m-edge graph will be $p^m(1-p)^{n(n-1)/2-m}$, giving rise to the familiar Erdős-Rényi $G(n,p)$ model. However, the ERG model is much more general than

F. Dehne et al. (Eds.): WADS 2009, LNCS 5664, pp. 278–289, 2009.

the Erdős-Rényi model: for instance, an ERGM in which the features are whole graphs can represent arbitrary probabilities. The generality of this model, and its ability to define probability spaces lacking the independence properties of the simpler Erdős-Rényi model, make it difficult to analyze analytically. Instead, in order to generate graphs in an ERG model or to perform other forms of probabilistic reasoning with the model, one typically uses a Markov Chain Monte Carlo method [31] in which one performs a large sequence of small changes to sample graphs, updates after each change the counts of the number of features of each type and the sum of the weights of each feature, and uses the updated values to determine whether to accept or reject each change. Because this method must evaluate large numbers of graphs, it is important to develop very efficient algorithms for identifying the features that are present in each graph.

Typical features used in these models take the form of small subgraphs: *stars* of several edges with a common vertex (used to represent constraints on the degree distribution of the resulting graphs), *triangles* (used in the triad model [19], an important predecessor of ERG models, to represent the likelihood that friends-of-friends are friends of each other), and more complicated subgraphs used to control the tendencies of simpler models to generate unrealistically extremal graphs [32]. Using highly local features of this type is important for reasons of computational efficiency, matches well the type of data that can be obtained for real-world social networks, and is well motivated by the local processes believed to underly many types of social network. Thus, ERGM simulation leads naturally to problems of *subgraph isomorphism*, listing or counting all copies of a given small subgraph in a larger graph.

There has been much past algorithmic work on subgraph isomorphism problems. It is known, for instance, that an n-vertex graph with m edges may have $\Theta(m^{3/2})$ triangles and four-cycles, and all triangles and four-cycles can be found in time $O(m^{3/2})$ [22, 6]. All cycles of length up to seven can be counted rather than listed in time of $O(n^{\omega})$ [3] where $\omega \approx 2.376$ is the exponent from the asymptotically fastest known matrix multiplication algorithms [7]; this improves on the previous $O(m^{3/2})$ bounds for dense graphs. Fast matrix multiplication has also been used for more general problems of finding and counting small cliques in graphs and hypergraphs [10, 24, 26, 34, 36]. In planar graphs, or more generally graphs of bounded local treewidth, the number of copies of any fixed subgraph may be found in linear time [13, 14], even though this number may be a large polynomial of the graph size [11]. Approximation algorithms for subgraph isomorphism counting problems based on random sampling have also been studied, with motivating applications in bioinformatics [9, 23, 29]. However, much of this subgraph isomorphism research makes overly restrictive assumptions about the graphs that are allowed as input, runs too slowly for the ERGM application, depends on impractically complicated matrix multiplication algorithms, or does not capture the precise subgraph counts needed to accurately perform Markov Chain Monte Carlo simulations.

Markov Chain Monte Carlo methods for ERGM-based reasoning process a sequence of graphs each differing by a small change from a previous graph, so it is natural to seek additional efficiency by applying *dynamic graph algorithms* [15, 17, 33], data structures to efficiently maintain properties of a graph subject to vertex and edge insertions and deletions. However, past research on dynamic graph algorithms has focused on problems of connectivity, planarity, and shortest paths, and not on finding the features needed in ERGM calculations. In this paper, we apply dynamic graph algorithms to subgraph isomorphism problems important in ERGM feature identification. To our knowledge, this is the first work on dynamic algorithms for subgraph isomorphism.

A key ingredient in our algorithms is the h-index, a number introduced by Hirsch [21] as a way of balancing prolixity and impact in measuring the academic achievements of individual researchers. Although problematic in this application [1], the h-index can be defined and studied mathematically, in graph-theoretic terms, and provides a convenient measure of the uniformity of distribution of edges in a graph. Specifically, for a researcher, one may define a bipartite graph in which the vertices on one side of the bipartition represent the researcher's papers, the vertices on the other side represent others' papers, and edges correspond to citations by others of the researcher's papers. The h-index of the researcher is the maximum number h such that at least h vertices on the researcher's side of the bipartition each have degree at least h. We generalize this to arbitrary graphs, and define the h-index of any graph to be the maximum h such that the graph contains h vertices of degree at least h. Intuitively, an algorithm whose running time is bounded by a function of h is capable of tolerating arbitrarily many low-degree vertices without slowdown, and is only mildly affected by the presence of a small number of very high degree vertices; its running time depends primarily on the numbers of intermediate-degree vertices. As we describe in more detail in Section 7, the h-index of any graph with m edges and n vertices is sandwiched between m/n and $\sqrt{2m}$, so it is sublinear whenever the graph is not dense, and the worst-case graphs for these bounds have an unusual degree distribution that is unlikely to arise in practice.

Our main result is that we may maintain a dynamic graph, subject to edge insertions, edge deletions, and insertions or deletions of isolated vertices, and maintain the number of triangles in the graph, in time $O(h)$ per update where h is the h-index of the graph at the time of the update. This compares favorably with the time bound of $\Theta(m^{3/2})$ necessary to list all triangles in a static graph. In the same $O(h)$ time bound per update we may more generally maintain the numbers of three-vertex induced subgraphs of each possible type, and in constant time per update we may maintain the h-index itself. Our algorithms are randomized, and our analysis of them uses amortized analysis to bound their expected times on worst-case input sequences. Our use of randomization is limited, however, to the use of hash tables to store and retrieve data associated with keys in $O(1)$ expected time per access. By using either direct addressing or deterministic integer searching data structures instead of hash tables we may avoid the use of randomness at an expense of either increased space complexity or an additional factor of $O(\log \log n)$ in time complexity; we omit the details.

We also study the behavior of the h-index, both on scale-free graph models and on a set of real-world graphs used in social network analysis. We show that for scale-free graphs, the h-index scales as a power of n, less than its square root, while in the real-world graphs we studied the scaling exponent appears to have a bimodal distribution.

2   Dynamic h-Indexes of Integer Functions

We begin by describing a data structure for the following problem, which generalizes that of maintaining h-indexes of dynamic graphs. We are given a set S, and a function

f from S to the non-negative integers, both of which may vary discretely through a sequence of updates: we may insert or delete elements of S (with arbitrary function values for the inserted elements), and we may make arbitrary changes to the function value of any element of S. As we do so, we wish to maintain a set H such that, for every $x \in H$, $f(x) \geq |H|$, with H as large as possible with this property. We call $|H|$ the *h-index* of S and f, and we call the partition of S into the two subsets $(H, S \setminus H)$ an *h-partition* of S and f.

To do so, we maintain the following data structures:

- A dictionary F mapping each $x \in S$ to its value under f: $F[x] = f(x)$.
- The set H (stored as a dictionary mapping members of H to an arbitrary value).
- The set $B = \{x \in H \mid f(x) = |H|\}$.
- A dictionary C mapping each non-negative integer i to the set $\{x \in S \setminus B \mid f(x) = i\}$. We only store these sets when they are non-empty, so the situation that there is no x with $f(x) = i$ can be detected by the absense of i among the keys of C.

To insert an element x into our structure, we first set $F[x] = f(x)$, and add x to $C[f(x)]$ (or add a new set $\{x\}$ at $C[f(x)]$ if there is no existing entry for $f(x)$ in C). Then, we test whether $f(x) > |H|$. If not, the *h*-index does not change, and the insertion operation is complete. But if $f(x) > |H|$, we must include x into H. If B is nonempty, we choose an arbitrary $y \in B$, remove y from B and from H, and add y to $C[|H|]$ (or create a new set $\{y\}$ if there is no entry for $|H|$ in C). Finally, if $f(x) > |H|$ and B is empty, the insertion causes the *h*-index ($|H|$) to increase by one. In this case, we test whether there is an entry for the new value of $|H|$ in C. If so, we set B to equal the identity of the set in $C[|H|]$ and delete the entry for $|H|$ in C; otherwise, we set B to the empty set.

To remove x from our structure, we remove its entry from F and we remove it from B (if it belongs there) or from the appropriate set in $C[f(x)]$ otherwise. If x did not belong to H, the *h*-index does not change, and the deletion operation is complete. Otherwise, let h be the value of $|H|$ before removing x. We remove x from H, and attempt to restore the lost item from H by moving an element from $C[h]$ to B (deleting $C[h]$ if this operation causes it to become empty). But if C has no entry for h, the *h*-index decreases; in this case we store the identity of set B into $C[h]$, and set B to be the empty set.

Changing the value of $f(x)$ may be accomplished by deleting x and then reinserting it, with some care so that we do not update H if x was already in H and both the old and new values of $f(x)$ are at least equal to $|H|$.

Theorem 1. *The data structure described above maintains the h-index of S and f, and an h-partition of S and f, in constant time plus a constant number of dictionary operations per update.*

We defer the proof to the full version of the paper [16].

3 Gradual Approximate *h*-Partitions

Although the vector *h*-index data structure of the previous section allows us to maintain the *h*-index of a dynamic graph very efficiently, it has a property that would be undesirable were we to use it directly as part of our later dynamic graph data structures:

the h-partition $(H, S \setminus H)$ changes too frequently. Changes to the set H will turn out to be such an expensive operation that we only wish them to happen, on average, $O(1/h)$ times per update. In order to achieve such a small amount of change to H, we need to restrict the set of updates that are allowed: now, rather than arbitrary changes to f, we only allow it to be incremented or decremented by a single unit, and we only allow an element x to be inserted or deleted when $f(x) = 0$. We now describe a modification of the H-partition data structure that has this property of changing more gradually for this restricted class of updates.

Specifically, along with all of the structures of the H-partition, we maintain a set $P \subset H$ describing a partition $(P, S \setminus P)$. When an element of x is removed from H, we remove it from P as well, to maintain the invariant that $P \subset H$. However, we only add an element x to P when an update (an increment of $f(x)$ or decrement of $f(y)$ for some other element y) causes $f(x)$ to become greater than or equal to $2|H|$. The elements to be added to P on each update may be found by maintaining a dictionary, parallel to C, that maps each integer i to the set $\{x \in H \setminus P \mid f(x) = i\}$.

Theorem 2. *Let σ denote a sequence of operations to the data structure described above, starting from an empty data structure. Let h_t denote the value of h after t operations, and let $q = \sum_i 1/h_i$. Then the data structure undergoes $O(q)$ additions and removals of an element to or from P.*

We defer the proof to the full version of the paper [16]. For our later application of this technique as a subroutine in our triangle-finding data structure, we will need a more local analysis. We may divide a sequence of updates into *epochs*, as follows: each epoch begins when the h-index reaches a value that differs from the value at the beginning of the previous epoch by a factor of two or more. Then, as we show in the full version, an epoch with h as its initial h-index lasts for at least $\Omega(h^2)$ steps. Due to this length, we may assign a full unit of credit to each member of P at the start of each epoch, without changing the asymptotic behavior of the total number of credits assigned over the course of the algorithm. With this modification, it follows from the same analysis as above that, within an epoch of s steps, with an h-index of h at the start of the epoch, there are $O(s/h)$ changes to P.

4   Counting Triangles

We are now ready to describe our data structure for maintaining the number of triangles in a dynamic graph. It consists of the following information:

- A count of the number of triangles in the current graph
- A set E of the edges in the graph, indexed by the pair of endpoints of the edge, allowing constant-time tests for whether a given pair of endpoints are linked by an edge.
- A partition of the graph vertices into two sets H and $V \setminus H$ as maintained by the data structure from Section 3.
- A dictionary P mapping each pair of vertices u, v to a number $P[u, v]$, the number of two-edge paths from u to v via a vertex of $V \setminus H$. We only maintain nonzero values for this number in P; if there is no entry in P for the pair u, v then there exist no two-edge paths via $V \setminus H$ that connect u to v.

Theorem 3. *The data structure described above requires space $O(mh)$ and may be maintained in $O(h)$ randomized amortized time per operation, where h is the h-index of the graph at the time of the operation.*

Proof. Insertion and deletion of vertices with no incident edges requires no change to most of these data structures, so we concentrate our description on the edge insertion and deletion operations.

To update the count of triangles, we need to know the number of triangles uvw involving the edge uv that is being deleted or inserted. Triangles in which the third vertex w belongs to H may be found in time $O(h)$ by testing all members of H, using the data structure for E to test in constant time per member whether it forms a triangle. Triangles in which the third vertex w does not belong to H may be counted in time $O(1)$ by a single lookup in P.

The data structure for E may be updated in constant time per operation, and the partition into H and $V \setminus H$ may be maintained as described in the previous sections in constant time per operation. Thus, it remains to describe how to update P. If we are inserting an edge uv, and u does not belong to H, it has at most $2h$ neighbors; we examine all other neighbors w of u and for each such neighbor increment the counter in $P[v, w]$ (or create a new entry in $P[v, w]$ with a count of 1 if no such entry already exists). Similarly if v does not belong to H we examine all other neighbors w of v and for each such neighbor increment $P[u, w]$. If we are deleting an edge, we similarly decrement the counters or remove the entry for a counter if decrementing it would leave a zero value. Each update involves incrementing or decrementing $O(h)$ counters and therefore may be implemented in $O(h)$ time.

Finally, a change to the graph may lead to a change in H, which must be reflected in P. If a vertex v is moved from H to $V \setminus H$, we examine all pairs u, w of neighbors of v and increment the corresponding counts in $P[u, w]$, and if a vertex v is moved from $V \setminus H$ to H we examine all pairs u, w of neighbors of v and decrement the corresponding counts in $P[u, w]$. This step takes time $O(h^2)$, because v has $O(h)$ neighbors when it is moved in either direction, but as per the analysis in Section 3 it is performed an average of $O(1/h)$ times per operation, so the amortized time for updates of this type, per change to the input graph, is $O(h)$.

The space for the data structure is $O(m)$ for E, $O(n)$ for the data structure that maintains H, and $O(mh)$ for P because each edge of the graph belongs to $O(h)$ two-edge paths through low-degree vertices. □

5   Subgraph Multiplicity

Although the data structure of Theorem 3 only counts the number of triangles in a graph, it is possible to use it to count the number of three-vertex subgraphs of all types, or the number of induced three-vertex subgraphs of all types. In what follows we let $p_i = p_i(G)$ denote the number of paths of length i in G, and we let $c_i = c_i(G)$ denote the number of cycles of length i in G.

The set of all edges in a graph G among a subset of three vertices $\{u, v, w\}$ determine one of four possible induced subgraphs: an independent set with no edges, a graph with a single edge, a two-star consisting of two edges, or a triangle. Let g_0, g_1, g_2, and g_3

denote the numbers of three-vertex subgraphs of each of these types, where g_i counts the three-vertex induced subgraphs that have i edges.

Observe that it is trivial to maintain for a dynamic graph, in constant time per operation, the three quantities n, m, and p_2, where n denotes the number of vertices of the graph, m denotes the number of edges, and p_2 denotes the number of two-edge paths that can be formed from the edges of the graph. Each change to the graph increments or decrements n or m. Additionally, adding an edge uv to a graph where u and v already have d_u and d_v incident edges respectively increases p_2 by $d_u + d_v$, while removing an edge uv decreases p_2 by $d_u + d_v - 2$. Letting c_3 denote the number of triangles in the graph as maintained by Theorem 3, the quantities described above satisfy the matrix equation

$$\begin{bmatrix} 1 & 1 & 1 & 1 \\ 0 & 1 & 2 & 3 \\ 0 & 0 & 1 & 3 \\ 0 & 0 & 0 & 1 \end{bmatrix} \begin{bmatrix} g_0 \\ g_1 \\ g_2 \\ g_3 \end{bmatrix} = \begin{bmatrix} n(n-1)(n-2)/6 \\ m(n-2) \\ p_2 \\ c_3 \end{bmatrix}.$$

Each row of the matrix corresponds to a single linear equation in the g_i values. The equation from the first row, $g_0 + g_1 + g_2 + g_3 = \binom{n}{3}$, can be interpreted as stating that all triples of vertices form one graph of one of these types. The equation from the second row, $g_1 + 2g_2 + 3g_3 = m(n-2)$, is a form of double counting where the number of edges in all three-vertex subgraphs is added up on the left hand side by subgraph type and on the right hand side by counting the number of edges (m) and the number of triples each edge participates in ($n - 2$). The third row's equation, $g_2 + 3g_3 = p_2$, similarly counts incidences between two-edge paths and triples in two ways, and the fourth equation $g_3 = c_3$ follows since each three vertices that are connected in a triangle cannot form any other induced subgraph than a triangle itself.

By inverting the matrix we may reconstruct the g values:

$$g_3 = c_3$$
$$g_2 = p_2 - 3g_3$$
$$g_1 = m(n-2) - (2g_2 + 3g_3)$$
$$g_0 = \binom{n}{3} - (g_1 + g_2 + g_3).$$

Thus, we may maintain each number of induced subgraphs g_i in the same asymptotic time per update as we maintain the number of triangles in our dynamic graph. The numbers of subgraphs of different types that are not necessarily induced are even easier to recover: the number of three-vertex subgraphs with i edges is given by the ith entry of the vector on the right hand side of the matrix equation.

As we detail in the full version of the paper [16], it is also possible to maintain efficiently the numbers of star subgraphs of a dynamic graph, and the number of four-vertex paths in a dynamic graph.

6   Weighted Edges and Colored Vertices

It is possible to generalize our triangle counting method to problems of weighted triangle counting: we assign each edge uv of the graph a weight w_{uv}, define the weight of a triangle to be the product of the weights of its edges, and maintain the total weight of all triangles. For instance, if $0 \leq w_{uv} \leq 1$ and each edge is present in a subgraph with probability w_{uv}, then the total weight gives the expected number of triangles in that subgraph.

Theorem 4. *The total weight of all triangles in a weighted dynamic graph, as described above, may be maintained in time $O(h)$ per update.*

Proof. We modify the structure $P[u, v]$ maintained by our triangle-finding data structure, so that it stores the weight of all two-edge paths from u to v. Each update of an edge uv in our structure involves a set of individual triangles uvx involving vertices $x \in H$ (whose weight is easily calculated) together with the triangles formed by paths counted in $P[u, v]$ (whose total weight is $P[u, v]w_{uv}$). The same time analysis from Theorem 3 holds for this modified data structure.   □

For social networking ERGM applications, an alternative generalization may be appropriate. Suppose that the vertices of the given dynamic graph are colored; we wish to maintain the number of triangles with each possible combination of colors. For instance, in graphs representing sexual contacts [25], edges between individuals of the same sex may be less frequent than edges between individuals of opposite sexes; one may model this in an ERGM by assigning the vertices two different colors according to whether they represent male or female individuals and using feature weights that depend on the colors of the vertices in the features. As we now show, problems of counting colored triangles scale well with the number of different groups into which the vertices of the graph are classified.

Theorem 5. *Let G be a dynamic graph in which each vertex is assigned one of k different colors. Then we may maintain the numbers of triangles in G with each possible combination of colors, in time $O(h + k)$ per update.*

Proof. We modify the structure $P[u, v]$ stored by our triangle-finding data structure, to store a vector of k numbers: the ith entry in this vector records the number of two-edge paths from u to v through a low-degree vertex with color i. Each update of an edge uv in our structure involves a set of individual triangles uvx involving vertices $x \in H$ (whose colors are easily observed) together with the triangles formed by paths counted in $P[u, v]$ (with k different possible colorings, recorded by the entries in the vector $P[u, v]$). Thus, the part of the update operation in which we compute the numbers of triangles for which the third vertex has low degree, by looking up u and v in P, takes time $O(k)$ instead of $O(1)$. The same time analysis from Theorem 3 holds for all other aspects of this modified data structure.   □

Both the weighting and coloring generalizations may be combined with each other without loss of efficiency.

7   How Small Is the *h*-Index of Typical Graphs?

It is straightforward to identify the graphs with extremal values of the *h*-index. A split graph in which an *h*-vertex clique is augmented by adding $n - h$ vertices, each connected only to the vertices in the clique, has n vertices and $m = h(n-1)$ edges, achieving an h-index of $m/(n-1)$. This is the minimum possible among any graph with n vertices and m edges: any other graph may be transformed into a split graph of this type, while increasing its number of edges and not decreasing h, by finding an h-partition $(H, V \setminus H)$ and repeatedly replacing edges that do not have an endpoint in H by edges that do have such an endpoint. The graph with the largest h-index is a clique with m edges together with enough isolated vertices to fill out the total to n; its h-index is $\sqrt{2m}(1 + o(1))$. Thus, for sparse graphs in which the numbers of edges and vertices are proportional to each other, the h-index may be as small as $O(1)$ or as large as $\Omega(\sqrt{n})$. At which end of this spectrum can we expect to find the graphs arising in social network analysis?

One answer can be provided by fitting mathematical models of the *degree distribution*, the relation between the number of incident edges at a vertex and the number of vertices with that many edges, to social networks. For many large real-world graphs, observers have reported *power laws* in which the number of vertices with degree d is proportional to $nd^{-\gamma}$ for some constant $\gamma > 1$; a network with this property is called *scale-free* [2,25,27,28]. Typically, γ lies in or near the interval $2 \le \gamma \le 3$ although more extreme values are possible. The h-index of these graphs may be found by solving for the h such that $h = nh^{-\gamma}$; that is, $h = \Theta(n^{1/(1+\gamma)})$. For any $\gamma > 1$ this is an asymptotic improvement on the worst-case $O(\sqrt{n})$ bound for graphs without power-law degree distributions. For instance, for $\gamma = 2$ this would give a bound of $h = O(n^{1/3})$ while for $\gamma = 3$ it would give $h = O(n^{1/4})$. That is, by depending on the h-index as it does, our algorithm is capable of taking advantage of the extra structure inherent in scale-free graphs to run more quickly for them than it does in the general case.

To further explore h-index behavior in real-world networks, we computed the h-index for a collection of 136 network data sets typical of those used in social network analysis. These data sets were drawn from a variety of sources traditionally viewed as common repositories for such data. The majority of our data sets were from the well known Pajek datasets [4]. Pajek is a program used for the analysis and visualization of large networks. The collection of data available with the Pajek software includes citation networks, food-webs, friendship network, etc. In addition to the Pajek data sets, we included network data sets from UCINET [5]. Another software package developed for network analysis, UCINET includes a corpus of data sets that are more traditional in the social sciences. Many of these data sets represent friendship or communication relations; UCINET also includes various social networks for non-human animals. We also used network data included as part of the statnet software suite [20], statistical modeling software in R. statnet includes ERGM functionality, making it a good example for data used specifically in the context of ERG models. Finally, we included data available on the UCI Network Data Repository [8], including some larger networks such as the WWW, blog networks, and other online social networks. By using this data we hope to understand how the h-index scales in real-world networks.

Details of the statistics for these networks are presented in the full version of the paper [16]; a summary of the statistics for network size and h-index are in Table 1, below. For this sample of 136 real-world networks, the h-index ranges from 2 to 116. The row of summary statistics for $\log h / \log n$ suggests that, for many networks, h scales as a sublinear power of n. The one case with an h-index of 116 represents the ties among

Table 1. Summary statistics for real-world network data

	min.	median	mean	max.
network size (n)	10	67	535.3	10616
h-index (h)	2	12	19.08	116
$\log n$	2.303	4.204	4.589	9.270
$\log h$	0.6931	2.4849	2.6150	4.7536
$\log h/\log n$	0.2014	0.6166	0.6006	1.0000

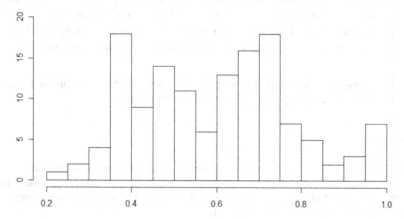

Fig. 1. A frequency histogram for $\log h/\log n$

Slovenian magazines and journals between 1999 and 2000. The vertices of this network represent journals, and undirected edges between journals have an edge weight that represents the number of shared readers of both journals; this network also includes self-loops describing the number of all readers that read this journal. Thus, this is a dense graph, more appropriately handled using statistics involving the edge weights than with combinatorial techniques involving the existence or nonexistence of triangles. However, this is the only network from our dataset with an *h*-index in the hundreds. Even with significantly larger networks, the *h*-index appears to scale sublinearly in most cases.

A histogram of the *h*-index data in Figure 1 clearly shows a bimodal distribution. Additionally, as the second peak of the bimodal distribution corresponds to a scaling exponent greater than 0.5, the graphs corresponding to that peak do not match the predictions of the scale-free model. However we were unable to discern a pattern to the types of networks with smaller or larger *h*-indices, and do not speculate on the reasons for this bimodality. We look more deeply at the scaling of the *h*-index using standard regression techniques in the full version of the paper [16].

8   Discussion

We have defined an interesting new graph invariant, the *h*-index, presented efficient dynamic graph algorithms for maintaining the *h*-index and, based on them, for maintaining the set of triangles in a graph, and studied the scaling behavior of the h-index both on theoretical scale-free graph models and on real-world network data.

There are many directions for future work. For sparse graphs, the h-index may be larger than the *arboricity*, a graph invariant used in static subgraph isomorphism [6,12]; can we speed up our dynamic algorithms to run more quickly on graphs of bounded arboricity? We handle undirected graphs but the directed case is also of interest. We would like to find efficient data structures to count larger subgraphs such as 4-cycles, 4-cliques, and claws; dynamic algorithms for these problems are likely to be slower than our triangle-finding algorithms but may still provide speedups over static algorithms. Another network statistic related to triangle counting is the clustering coefficient of a graph; can we maintain it efficiently? Additionally, there is an opportunity for additional work in implementing our data structures and testing their efficiency in practice.

Acknowledgements. This work was supported in part by NSF grant 0830403 and by the Office of Naval Research under grant N00014-08-1-1015.

References

1. Adler, R., Ewing, J., Taylor, P.: Citation Statistics: A report from the International Mathematical Union (IMU) in cooperation with the International Council of Industrial and Applied Mathematics (ICIAM) and the Institute of Mathematical Statistics. In: Joint Committee on Quantitative Assessment of Research (2008)
2. Albert, R., Jeong, H., Barabasi, A.-L.: The diameter of the world wide web. Nature 401, 130–131 (1999)
3. Alon, N., Yuster, R., Zwick, U.: Finding and counting given length cycles. Algorithmica 17(3), 209–223 (1997)
4. Batagelj, V., Mrvar, A.: Pajek datasets (2006),
   http://vlado.fmf.uni-lj.si/pub/networks/data/
5. Borgatti, S.P., Everett, M.G., Freeman, L.C.: UCINet 6 for Windows: Software for social network analysis. Analytic Technologies, Harvard, MA (2002)
6. Chiba, N., Nishizeki, T.: Arboricity and subgraph listing algorithms. SIAM J. Comput. 14(1), 210–223 (1985)
7. Coppersmith, D., Winograd, S.: Matrix multiplication via arithmetic progressions. Journal of Symbolic Computation 9(3), 251–280 (1990)
8. DuBois, C.L., Smyth, P.: UCI Network Data Repository (2008),
   http://networkdata.ics.uci.edu
9. Duke, R.A., Lefmann, H., Rödl, V.: A fast approximation algorithm for computing the frequencies of subgraphs in a given graph. SIAM J. Comput. 24(3), 598–620 (1995)
10. Eisenbrand, F., Grandoni, F.: On the complexity of fixed parameter clique and dominating set. Theoretical Computer Science 326(1–3), 57–67 (2004)
11. Eppstein, D.: Connectivity, graph minors, and subgraph multiplicity. Journal of Graph Theory 17, 409–416 (1993)
12. Eppstein, D.: Arboricity and bipartite subgraph listing algorithms. Information Processing Letters 51(4), 207–211 (1994)
13. Eppstein, D.: Subgraph isomorphism in planar graphs and related problems. Journal of Graph Algorithms & Applications 3(3), 1–27 (1999)
14. Eppstein, D.: Diameter and treewidth in minor-closed graph families. Algorithmica 27, 275–291 (2000)

15. Eppstein, D., Galil, Z., Italiano, G.F.: Dynamic graph algorithms. In: Atallah, M.J. (ed.) Algorithms and Theory of Computation Handbook, ch. 8, CRC Press, Boca Raton (1999)
16. Eppstein, D., Spiro, E.S.: The h-index of a graph and its application to dynamic subgraph statistics. Electronic preprint arxiv:0904.3741 (2009)
17. Feigenbaum, J., Kannan, S.: Dynamic graph algorithms. In: Rosen, K. (ed.) Handbook of Discrete and Combinatorial Mathematics. CRC Press, Boca Raton (2000)
18. Frank, O.: Statistical analysis of change in networks. Statistica Neerlandica 45, 283–293 (1999)
19. Frank, O., Strauss, D.: Markov graphs. J. Amer. Statistical Assoc. 81, 832–842 (1986)
20. Handcock, M.S., Hunter, D., Butts, C.T., Goodreau, S.M., Morris, M.: statnet: An R package for the Statistical Modeling of Social Networks (2003), http://www.csde.washington.edu/statnet
21. Hirsch, J.E.: An index to quantify an individual's scientific research output. Proc. National Academy of Sciences 102(46), 16569–16572 (2005)
22. Itai, A., Rodeh, M.: Finding a minimum circuit in a graph. SIAM J. Comput. 7(4), 413–423 (1978)
23. Kashtan, N., Itzkovitz, S., Milo, R., Alon, U.: Efficient sampling algorithm for estimating subgraph concentrations and detecting network motifs. Bioinformatics 20(11), 1746–1758 (2004)
24. Kloks, T., Kratsch, D., Müller, H.: Finding and counting small induced subgraphs efficiently. Information Processing Letters 74(3–4), 115–121 (2000)
25. Liljeros, F., Edling, C.R., Amaral, L.A.N., Stanley, H.E., Åberg, Y.: The web of human sexual contacts. Nature 411, 907–908 (2001)
26. Nešetřil, J., Poljak, S.: On the complexity of the subgraph problem. Commentationes Mathematicae Universitatis Carolinae 26(2), 415–419 (1985)
27. Newman, M.E.J.: The structure and function of complex networks. SIAM Review 45, 167–256 (2003)
28. desolla Price, D.J.: Networks of scientific papers. Science 149(3683), 510–515 (1965)
29. Pržulj, N., Corneil, D.G., Jurisica, I.: Efficient estimation of graphlet frequency distributions in protein–protein interaction networks. Bioinformatics 22(8), 974–980 (2006)
30. Robins, G., Morris, M.: Advances in exponential random graph (p*) models. Social Networks 29(2), 169–172 (2007); Special issue of journal with four additional articles
31. Snijders, T.A.B.: Markov chain Monte Carlo estimation of exponential random graph models. Journal of Social Structure 3(2), 1–40 (2002)
32. Snijders, T.A.B., Pattison, P.E., Robins, G., Handcock, M.S.: New specifications for exponential random graph models. Sociological Methodology 36(1), 99–153 (2006)
33. Thorup, M., Karger, D.R.: Dynamic graph algorithms with applications. In: Halldórsson, M.M. (ed.) SWAT 2000. LNCS, vol. 1851, pp. 667–673. Springer, Heidelberg (2000)
34. Vassilevska, V., Williams, R.: Finding, minimizing and counting weighted subgraphs. In: Proc. 41st ACM Symposium on Theory of Computing (2009)
35. Wasserman, S., Pattison, P.E.: Logit models and logistic regression for social networks, I: an introduction to Markov graphs and p*. Psychometrika 61, 401–425 (1996)
36. Yuster, R.: Finding and counting cliques and independent sets in r-uniform hypergraphs. Information Processing Letters 99(4), 130–134 (2006)

Optimal Embedding into Star Metrics

David Eppstein and Kevin A. Wortman

Department of Computer Science
Universitiy of California, Irvine
{eppstein,kwortman}@ics.uci.edu

Abstract. We present an $O(n^3 \log^2 n)$-time algorithm for the following problem: given a finite metric space X, create a star-topology network with the points of X as its leaves, such that the distances in the star are at least as large as in X, with minimum dilation. As part of our algorithm, we solve in the same time bound the *parametric negative cycle detection problem*: given a directed graph with edge weights that are increasing linear functions of a parameter λ, find the smallest value of λ such that the graph contains no negative-weight cycles.

1   Introduction

A *metric space* is a set of sites separated by symmetric positive distances that obey the triangle inequality. If X and Y are metric spaces and $f : X \mapsto Y$ does not decrease the distance between any two points, the *dilation* or *stretch factor* of f is

$$\sup_{x_1, x_2 \in X} \frac{d(f(x_1), f(x_2))}{d(x_1, x_2)}.$$

We define a *star metric* to be a metric space in which there exists a *hub* h such that, for all x and y, $d(x, y) = d(x, h) + d(h, y)$. Given the distance matrix of an n-point metric space X, we would like to construct a function f that maps X into a star metric Y, that does not decrease distances, and that has as small a dilation as possible. In this paper we describe an algorithm that finds the optimal f in time $O(n^3 \log^2 n)$. Our problem may be seen as lying at the confluence of three major areas of algorithmic research:

Spanner construction. A *spanner* for a metric space X is a graph G with the points of X as its vertices and weights (lengths) on its edges, such that path lengths in G equal or exceed those in X; the dilation of G is measured as above as the maximum ratio between path length and distance in X. The construction of sparse spanners with low dilation has been extensively studied [9] but most papers in this area limit themselves to bounding the dilation of the spanners they construct rather than constructing spanners of optimal dilation. Very few optimal spanner construction problems are known to be solvable in polynomial time; indeed, some are known to be NP-complete [15] and others NP-hard [3, 8]. Our problem can be viewed as constructing a spanner in the form of a star (a tree with one non-leaf node) that has optimal dilation.

Metric embedding. There has been a large amount of work within the algorithms community on *metric embedding* problems, in which an input metric space is to be embedded into a simpler target space with minimal distortion [16]; typical target spaces

F. Dehne et al. (Eds.): WADS 2009, LNCS 5664, pp. 290–301, 2009.

for results of this type include spaces with L_p norms and convex combinations of tree metrics. As with spanners, there are few results of this type in which the minimum dilation embedding can be found efficiently; instead, research has concentrated on proving bounds for the achievable dilation. Our result provides an example of a simple class of metrics, the star metrics, for which optimal embeddings may be found efficiently. As with embeddings into low-dimensional L_p spaces, our technique allows an input metric with a quadratic number of distance relationships to be represented approximately using only a linear amount of information.

Facility location. In many applications one is given a collection of *demand points* in some space and must select one or more *supply points* that maximize some objective function. For instance, the 1-median (minimize the sum of all distances from demand points to a single supply point) and 1-center (minimize the greatest distance between any destination point and a single supply point) can be applied to operational challenges such as deciding where to build a radio transmitter or railroad hub so as to maximize its utility [7]. In a similar vein the problem discussed in this paper may be seen as selecting a single supply point to serve as the hub of a star-topology network. In this context dilation corresponds to the worst multiplicative cost penalty imposed on travel between any pair of input points due to the requirement that all travel is routed through the hub (center) point. Superficially, our problem differs somewhat from typical facility location problems in that the star we construct has a hub that is not given as part of the input. However, it is possible to show that the hub we find belongs to the *tight span* of the input metric space [6], a larger metric space that has properties similar to those of L_∞ spaces. Viewing our problem as one of selecting the optimal hub point from the tight span gives it the format of a facility location problem.

Previously [10] we considered similar minimum dilation star problems in which the input and output were both confined to low-dimensional Euclidean spaces. As we showed, the minimum-dilation star with unrestricted hub location may be found in $O(n \log n)$ expected time in any bounded dimension, and for $d = 2$ the optimal hub among the input points may be selected in expected time $O(n 2^{\alpha(n)} \log^2 n)$, where $\alpha(n)$ is the inverse Ackermann function. For the general metric spaces considered here, the difficulty of the problems is reversed: it is trivial to select an input point as hub in time $O(n^3)$, while our results show that an arbitrary hub may be found in time $O(n^3 \log^2 n)$.

As we discuss in Section 2, the minimum dilation star problem can be represented as a linear program; however solving this program directly would give a running time that is a relatively high order polynomial in n and in the number of bits of precision of the input matrix. In this paper we seek a faster, purely combinatorial algorithm whose running time is strongly polynomial in n. Our approach is to first calculate the dilation λ^* of the optimal star. We do this by forming a λ-*graph* $G(\lambda)$: a directed graph with weights in the form $w(e) = \lambda \cdot m_e + b_e$ for parameters $m_e \geq 0$ and b_e determined from the input metric. $G(\lambda)$ has the property that it contains no negative weight cycles if and only if there exists a star with dilation λ. Next we calculate λ^*, the smallest value such that $G(\lambda^*)$ contains no negative-weight cycles, which is also the dilation of the star we will eventually create. Finally we use $G(\lambda)$ and λ^* to compute the lengths of the edges from the star's center to each site, and output the resulting star.

Fig. 1. Example of a metric space and its optimal star, which has dilation $\lambda^* = 8/5$

Our algorithm for computing λ^*, the smallest parameter value admitting no negative cycles in a parametrically weighted graph, warrants independent discussion. To our knowledge no known strongly polynomial algorithm solves this problem in full generality. Karp and Orlin [14] gave an $O(mn)$ time algorithm for a problem in which the edge weights have the same form $w(e) = \lambda \cdot m_e + b_e$ as ours, but where each m_e is restricted to the set $\{0,1\}$. If all $m_e = 1$, the problem is equivalent to finding the minimum mean cycle in a directed graph [13], for which several algorithms run in $O(mn)$ time [4]. In our problem, each m_e may be any nonnegative real number; it is not apparent how to adapt the algorithm of Karp and Orlin to our problem. Gusfield provided an upper bound [12] on the number of breakpoints of the function describing the shortest path length between two nodes in a λ-graph, and Carstensen provided a lower bound [2] for the same quantity; both bounds have the form $n^{\Theta(\log n)}$. Hence any algorithm that constructs a piecewise linear function that fully describes path lengths for the entire range of λ values takes at least $n^{\Theta(\log n)}$ time. In Section 4 we describe our algorithm, which is based on a dynamic programming solution to the all pairs shortest paths problem. Our algorithm maintains a compact piecewise linear function representing the shortest path length for each pair of vertices over a limited range of λ values, and iteratively contracts the range until a unique value λ^* can be calculated. Thus it avoids Carstensen's lower bound by finding only the optimal λ^*, and not the other breakpoints of the path length function, allowing it to run in $O(n^3 \log^2 n)$ time.

2   Linear Programming Formulation

In this section we formally define the overall minimum dilation star problem and describe how to solve it directly using linear programming. Our eventual algorithm never solves nor even constructs this linear program directly; however stating the underlying linear program and its related terminology will aid our later exposition.

The input to our algorithm is a finite *metric space*. Formally, a metric space X is a tuple $X = (X, d_X)$, where X is a set of sites and the function d_X maps any pair of sites to the nonnegative, real distance between them. The following *metric conditions* also hold for any $x, y, z \in X$:

1. $d_X(x,y) = 0$ if and only if $x = y$ (positivity);
2. $d_X(x,y) = d_X(y,x)$ (symmetry); and
3. $d_X(x,y) + d_X(y,z) \geq d_X(x,z)$ (the triangle inequality).

The input to our algorithm is a finite metric space $S = (S, d_S)$; we assume that the distance $d_S(x,y)$ between any $x,y \in S$ may be reported in constant time, for instance by a lookup matrix.

A *star* is a connected graph with one *center* vertex. A star contains an edge between the center and every other vertex, but no other edges. Hence any star is a tree of depth 1, and every vertex except the center is a leaf. Our algorithm must output a weighted star H whose leaves are the elements S from the input. The edge weights in H must be at least as large as the distances in S, and must obey reflexivity and the triangle inequality. In other words, if $d_H(x,y)$ is the length of a shortest path from x to y in H, then $d_H(x,y) \geq d_S(x,y)$, $d_H(x,y) = d_H(y,x)$, and $d_H(x,y) + d_H(y,z) \geq d_H(x,z)$ for any vertices x,y,z in H.

We also ensure that the *dilation* of H is minimized. For any two vertices u,v in some weighted graph G whose vertices are points in a metric space, the dilation between u and v is

$$\delta_G(u,v) = \frac{d_G(u,v)}{d_S(u,v)}.$$

The dilation of the entire graph G is the largest dilation between any two vertices, i.e.

$$\Delta_G = \max_{u,v \in G} \delta_G(u,v).$$

Our output graph H is a star; hence every path between two leaves has two edges, so if we apply the definition of dilation to H, we obtain

$$\delta_H(u,v) = \frac{d_H(u,c) + d_H(c,v)}{d_S(u,v)} = \frac{w_{u,c} + w_{c,v}}{d_S(u,v)}$$

where $w_{x,y}$ is the weight of the edge connecting x and y in H. Hence the dilation of H may be computed by

$$\Delta_H = \max_{u,v \in H} \frac{w_{u,c} + w_{c,v}}{d_S(u,c)}.$$

This equation lays the foundation for our formulation of the minimum dilation star problem as a linear program.

Definition 1. *Let $\mathcal{L}$ be the following linear program, defined over the variables λ and c_v for every $v \in S$:*

$$Minimize\ \lambda$$

such that for any $v \in S$,

$$c_v \geq 0, \tag{1}$$

and for any $v,w \in S$,

$$c_v + c_w \geq d_S(v,w) \tag{2}$$
$$c_v + c_w \leq \lambda \cdot d_S(v,w). \tag{3}$$

Let λ^* be the value assigned to λ in the optimal solution to $\mathcal{L}$. In other words, λ^* is the smallest dilation admitted by any set of distances satisfying all the constraints of $\mathcal{L}$.

$\mathcal{L}$ is clearly feasible. For example, if $D = \max_{x,y \in S} d_S(x,y)$, then the solution $\forall v\ c_v = D$ and $\lambda = 2D/\min_{x,y \in S} d_S(x,y)$ is a feasible, though poor, solution.

Lemma 1. *For any optimal solution of L, the value of λ gives the minimum dilation of any star network spanning S, and the c_v values give the edge lengths of an optimal star network spanning S.*

Proof. Each variable c_v corresponds to the weight $w_{v,c}$ of the edge between c and v in H. Inequality 1 ensures that the distances are nonnegative, Inequality 2 ensures that they obey the triangle inequality, and Inequality 3 dictates that λ is a largest dilation among any pair of sites from S. The value of λ is optimal since L is defined to minimize λ.

Unfortunately L contains $O(n)$ variables and $O(n^2)$ constraints. Such a program could be solved using general purpose techniques in a number of steps that is a high-order polynomial in n and the number of bits of precision used, but our objective is to obtain a fast algorithm whose running time is strongly polynomial in n. Megiddo showed [19] that linear programs with at most two variables per inequality may be solved in strongly polynomial time; however our type (3) inequalities have three variables, so those results cannot be applied to our problem.

3   Reduction to Parameteric Negative Weight Cycle Detection

In this section we describe a subroutine that maps the set of sites S to a directed, parametrically-weighted λ-graph $G(\lambda)$. Every edge of $G(\lambda)$ is weighted according to a nondecreasing linear function of a single graph-global variable λ. An important property of $G(\lambda)$ is that the set of values of λ that cause $G(\lambda)$ to contain a negative weight cycle is identical to the set of values of λ that cause the linear program L to be infeasible. Thus any assignment of λ for which $G(\lambda)$ contains no negative weight cycles may be used in a feasible solution to L.

Definition 2. *A λ-graph is a connected, weighted, directed graph, where the weight $w(e)$ of any edge e is defined by a linear function in the form*

$$w(e) = \lambda \cdot m_e + b_e,$$

where m_e and b_e are real numbers and $m_e \geq 0$.

Definition 3. *Let $G(\lambda)$ be the λ-graph corresponding to a particular set of input sites S. $G(\lambda)$ has vertices $\bar{s}$ and $\underline{s}$ for each $s \in S$. For $s,t \in S$, $G(\lambda)$ has an edge of length $-d_S(s,t)$ from $\underline{s}$ to $\bar{t}$, and for $s \neq t$, $G(\lambda)$ has an edge of length $\lambda \cdot d_S(s,t)$ from $\bar{s}$ to $\underline{t}$.*

Note that an edge from $\underline{s}$ to $\bar{t}$ has weight $-d_S(s,s) = 0$ when $s = t$. An example λ-graph $G(\lambda)$ for $n = 3$ is shown in Figure 2.

Lemma 2. *$G(\lambda)$ may be constructed in $O(n^2)$ time.*

Proof. $G(\lambda)$ has $2n$ vertices and $O(n^2)$ edges, each of which may be initialized in constant time.

Lemma 3. *If $\lambda \geq 1$ is assigned such that L has a feasible solution, then $G(\lambda)$ contains no negative weight cycle.*

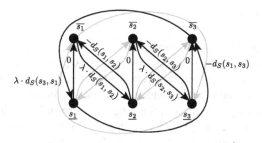

Fig. 2. The graph $G(\lambda)$ for $n = 3$. The weights of grayed edges are omitted.

Proof. Since $G(\lambda)$ is bipartite, any sequence of edges M traversed by a cycle in $G(\lambda)$ has even length. Depending on which partition M begins with, the sequence either takes the form

$$M = \langle(\overline{s_{i_1}}, s_{i_2}), (s_{i_2}, \overline{s_{i_3}}), (\overline{s_{i_3}}, s_{i_4}), \ldots, (s_{i_k}, \overline{s_{i_1}})\rangle$$

or

$$M = \langle(s_{i_1}, \overline{s_{i_2}}), (\overline{s_{i_2}}, s_{i_3}), (s_{i_3}, \overline{s_{i_4}}), \ldots, (\overline{s_{i_k}}, s_{i_1})\rangle,$$

where $s_{i_1}, s_{i_2}, \ldots, s_{i_k}$ are vertices from $G(\lambda)$. In either case, the cycle has weight

$$w(M) = \lambda \cdot d_S(s_{i_1}, s_{i_2}) - d_S(s_{i_2}, s_{i_3}) + \lambda \cdot d_S(s_{i_3}, s_{i_4}) - \ldots - d_S(s_{i_k}, s_{i_1}) \qquad (4)$$

by the commutativity of addition. Since $\mathcal{L}$ is feasible, there exists some set of distances C satisfying the constraints of $\mathcal{L}$, i.e.

$$c_x + c_y \leq \lambda \cdot d_S(x, y) \Rightarrow (c_x + c_y)/\lambda \leq d_S(x, y) \qquad (5)$$

and

$$c_x + c_y \geq d_S(x, y) \Rightarrow -(c_x + c_y) \leq -d_S(x, y). \qquad (6)$$

Substituting (5) and (6) into (4), we obtain

$$\begin{aligned}
w(M) &\geq \lambda((c_{i_1} + c_{i_2})/\lambda) - (c_{i_2} + c_{i_3}) + \lambda((c_{i_3} + c_{i_4})) - \ldots - (c_{i_k} + c_{i_1}) \\
&\geq (c_{i_1} + c_{i_2}) - (c_{i_2} + c_{i_3}) + (c_{i_3} + c_{i_4}) - \ldots - (c_{i_k} + c_{i_1}) \\
&\geq c_{i_1} - c_{i_1} + c_{i_2} - c_{i_2} + \ldots + c_{i_k} - c_{i_k} \\
&\geq 0.
\end{aligned}$$

Theorem 1. *Any set S of n sites from a metric space may be mapped to a λ-graph $G(\lambda)$ with $O(n)$ vertices, such that for any $\lambda \geq 1$, $G(\lambda)$ contains a negative weight cycle if and only if $\mathcal{L}$ is infeasible for that value of λ. The mapping may be accomplished in $O(n^2)$ time.*

Proof. By Lemma 2, $G(\lambda)$ may be created in $O(n^2)$ time, and by Lemma 3, feasibility of $\mathcal{L}$ implies an absence of negative cycles in $G(\lambda)$. Section 5 describes an algorithm that, given a value λ for which $G(\lambda)$ has no negative cycle, generates an edge length c_v for every $v \in S$ that obeys the constraints of $\mathcal{L}$. Thus, by the correctness of that algorithm, an absence of negative cycles in $G(\lambda)$ implies feasibility of $\mathcal{L}$.

4   Searching for λ^*

We now turn to the problem of computing the quantity λ^*. This problem is an example of *parametric negative weight cycle detection*: given a λ-graph $G(\lambda)$, find λ^*, the smallest value such that $G(\lambda^*)$ contains no cycles of negative weight. Our algorithm functions by maintaining a range $[\lambda_1, \lambda_2]$ which is known to contain λ^*. Initially the range is $[-\infty, +\infty]$; over $O(\log n)$ iterations, the range is narrowed until it is small enough that λ^* may be calculated easily. This approach is similar in spirit to Megiddo's general parametric search framework [17, 18], which, in loose terms, searches for the solution to an optimization problem by simulating the execution of a parallel algorithm for the corresponding decision problem.

Our algorithm is presented in Listing 1. It is an adaptation of a parallel all pairs shortest paths algorithm based on matrix squaring [20]. The original algorithm uses a matrix $D_i(u, v)$, which stores the weight of the shortest path from u to v among paths with at most 2^i edges. Each $D_i(u, v)$ may be defined as the smallest sum of two cells of D_{i-1}, and $D_{\lceil \log_2 n \rceil}$ defines the shortest paths in the graph. In the context of that original algorithm, edges and paths had real-number lengths, so it was sufficient to store real numbers in D_i. In the context of this paper, an edge's weight is a linear function of a variable λ; hence the weight of a path is a linear function of λ. Unfortunately the minimum-cost path between u and v may be different for varying values of λ, so the weight of the shortest path from u to v is defined by the minima of one or more linear functions of λ. Such a lower envelope of linear functions may be represented by a piecewise linear function; hence each element of D_i must store a piecewise linear function. Without further attention the number of breakpoints in these piecewise linear functions would grow at every iteration, and eventually operating on them would dominate our algorithm's running time. To address this, at every iteration we choose a new interval $[\lambda_1, \lambda_2]$ that contains no breakpoints, so that every D_i may be compacted down to a single linear function.

Lemma 4. *For any $\lambda \in [\lambda_1, \lambda_2]$, the function $D_i(u, v)$ as computed in the listing evaluates to the weight of the shortest path from u to v among paths with at most 2^i edges, or $+\infty$ if no such path exists.*

Proof. We argue by induction on i. In the base case $i = 0$, $D_i(u, v)$ must represent the weight of shortest path from u to v that includes up to $2^0 = 1$ edges. The only such paths are trivial paths, for which $u = v$ and $D_i(u, v) = 0$, and single edge paths, for which the path length equals the edge length.

For $i \geq 1$, each $D_i(u, v)$ is first defined as the lower envelope of two entries of D_{i-1} in line 10, then redefined as a strictly linear function over the new smaller range $[\lambda_1, \lambda_2]$ in line 16, so we argue that the lemma holds after each assignment. In the first assignment, $D_i(u, v)$ is defined to be the lower envelope of $[D_{i-1}(u, w) + D_{i-1}(w, v)]$ for all $w \in V$; in other words, every $w \in V$ is considered as a potential "layover" vertex, and $D_i(u, v)$ is defined as a piecewise linear function that may be defined by differing layover vertices throughout the range $[\lambda_1, \lambda_2]$. By the inductive hypothesis, the D_{i-1} values represent weights of minimum cost paths with at most 2^{i-1} edges; hence the resulting D_i values represent weights of minimum cost paths with at most $2^{i-1} + 2^{i-1} = 2^i$ edges.

Listing 1. Computing the quantity λ^*.

1: **INPUT:** A λ-graph $G(\lambda)$ with n vertices V.
2: **OUTPUT:** λ^*, the smallest value of λ such that $G(\lambda)$ has no negative-weight cycles.
3: Let $\lambda_1 = -\infty$ and $\lambda_2 = +\infty$.
4: **INVARIANT:** $\lambda_1 \le \lambda^* \le \lambda_2$
5: **INVARIANT:** $D_i(u,v)$ contains a linear function that represents the length of the shortest path from u to v among the subset of paths that use at most 2^i edges, as a function of λ, for any $\lambda \in [\lambda_1, \lambda_2]$
6: Let D_0 be an $n \times n$ matrix of piecewise linear functions.
7: Initialize $D_0(u,v) \equiv \begin{cases} 0 & \text{if } u = v \\ \lambda \cdot m_e + b_e & \text{if } G(\lambda) \text{ contains an edge } e \text{ from } u \text{ to } v \\ +\infty & \text{otherwise} \end{cases}$
8: **for** $i = 1, 2, \ldots, \lceil \log_2 n \rceil$ **do**
9:     **for** $u, v \in V$ **do**
10:         $D_i(u,v) \equiv \min_{w \in V}[D_{i-1}(u,w) + D_{i-1}(w,v)]$
11:     **end for**
12:     Let B be the set of breakpoints of the piecewise linear functions stored in the entries of D_i.
13:     Perform a binary search among the values in B, seeking an interval bounded by two consecutive breakpoints that contains λ^*. At each step, the test value of the binary search is less than λ^* if and only if setting λ equal to the test value causes $G(\lambda)$ to contain a negative cycle; use the Bellman–Ford shortest paths algorithm to determine whether this is the case.
14:     Set λ_1 and λ_2 to the endpoints of the interval found in the previous step.
15:     **for** $u, v \in V$ **do**
16:         Replace the piecewise linear function $D_i(u,v)$ with the equivalent linear function over the range $[\lambda_1, \lambda_2]$.
17:     **end for**
18: **end for**
19: Compute λ^*, the smallest value in the range $[\lambda_1, \lambda_2]$, such that $D_k(v,v) \ge 0$ for every $v \in V$.
20: **Return** λ^*.

When $D_i(u,v)$ is reassigned in line 16, the range endpoints λ_1 and λ_2 have been contracted such that no entry of D_i contains breakpoints in the range $[\lambda_1, \lambda_2]$. Hence any individual $D_i(u,v)$ has no breakpoints in that range, and is replaced by a simple linear function. This transformation preserves the condition that $D_i(u,v)$ represents the weight of the shortest path from u to v for any $\lambda \in [\lambda_1, \lambda_2]$.

Lemma 5. *Given two values λ_1 and λ_2 such that $\lambda_1 < \lambda_2$, it is possible to decide whether $\lambda^* < \lambda_1$, $\lambda^* > \lambda_2$, or $\lambda^* \in [\lambda_1, \lambda_2]$, in $O(n^3)$ time.*

Proof. By Lemma 3, for any value λ', if $G(\lambda')$ contains a negative cycle when $\lambda = \lambda'$, then $\lambda' < \lambda^*$. So we can determine the ordering of λ_1, λ_2, and λ^* using the Bellman–Ford shortest paths algorithm [1, 11] to detect negative cycles, as follows. First run Bellman–Ford, substituting $\lambda = \lambda_2$ to evaluate edge weights. If we find a negative cycle, then report that $\lambda^* > \lambda_2$. Otherwise run Bellman–Ford for $\lambda = \lambda_1$; if we find a negative cycle, then λ^* must be in the range $[\lambda_1, \lambda_2]$. If not, then $\lambda^* < \lambda_1$. This decision process invokes the Bellman–Ford algorithm once or twice, and hence takes $O(n^3)$ time.

Lemma 6. *The algorithm presented in Listing 1 runs in* $O(n^3 \log^2 n)$ *time.*

Proof. Each $D_{i-1}(u, v)$ is a linear function, so each $[D_{i-1}(u, w) + D_{i-1}(w, v)]$ is a linear function as well. $D_i(u, v)$ is defined as the lower envelope of n such linear functions, which may be computed in $O(n \log n)$ time [5]. So each $D_i(u, v)$ may be computed is $O(n \log n)$ time, and all $O(n^2)$ iterations of the first inner for loop take $O(n^3 \log n)$ total time. Each $D_i(u, v)$ represents the lower envelope of $O(n)$ lines, and hence has $O(n)$ breakpoints. So the entries of D_i contain a total of $O(n^3)$ breakpoints, and they may all be collected and sorted into B in $O(n^3 \log n)$ time. Once sorted, any duplicate elements may be removed from B in $O(|B|) = O(n^3)$ time.

Next our algorithm searches for a new, smaller $[\lambda_1, \lambda_2]$ range that contains λ^*. Recall that λ^* is the value of λ for which $G(\lambda^*)$ contains no negative weight cycle, and every entry of D_i is a piecewise linear function comprised of non-decreasing linear segments; so it is sufficient to search for the segment that intersects the $\lambda = 0$ line. We find this segment using a binary search in B. At every step in the search, we decide which direction to seek using the decision process described in Lemma 5. Each decision takes $O(n^3)$ time, and a binary search through the $O(n^2)$ elements of B makes $O(\log n)$ decisions, so the entire binary search takes $O(n^3 \log n)$ time.

Replacing an entry of D_i with a (non-piecewise) linear function may be done naively in $O(n)$ time by scanning the envelope for the piece that defines the function in the range $[\lambda_1, \lambda_2]$. So the second inner for loop takes $O(n^3)$ total time, and the outer for loop takes a total of $O(n^3 \log^2 n)$ time.

The initialization before the outer for loop takes $O(n^2)$ time. The last step of the algorithm is to compute λ^*, the smallest value in the range $[\lambda_1, \lambda_2]$ such that $D_k(v, v) \geq 0$ for every $v \in V$. At this point each $D_i(u, v)$ is a non-piecewise increasing linear function, so this may be done by examining each of the n linear functions $D_k(v, v)$, solving for its λ-intercept, and setting λ^* to be the largest intercept. This entire process takes $O(n^2)$ time, so the entire algorithm takes $O(n^3 \log^2 n)$ time.

Theorem 2. *The algorithm presented in Listing 1 calculates* λ^* *in* $O(n^3 \log^2 n)$ *time.*

5   Extracting the Edge Weights

Once λ^* has been calculated, all that remains is to calculate the weight of every edge in the output star. Our approach is to create a new graph G', which is a copy of $G(\lambda)$ with the addition of a new source node s with an outgoing weight 0 edge to every $\bar{v}$ (see Figure 3). We then compute the single source shortest paths of G' starting at s, and define each c_v to be a function of the shortest path lengths to $\bar{v}$ and $\underline{v}$. This process is a straightforward application of the Bellman–Ford algorithm, and hence takes $O(n^3)$ time. The remainder of this section is dedicated to proving the correctness of this approach.

Definition 4. *Let G' be a copy of the graph $G(\lambda)$ described in Definition 3, with all edge weights evaluated to real numbers for $\lambda = \lambda^*$, and the addition of a source vertex s with an outgoing 0-weight edge to every $\bar{v} \in G'$. Let $P(v)$ be a shortest path from s to v for any vertex $v \in G'$, and let $l(v)$ be the total weight of any such $P(v)$. The operation $P(v) \cup w$ yields the path formed by appending the edge (v, w) to $P(v)$.*

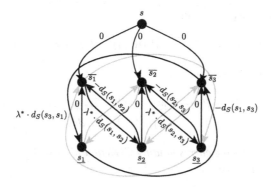

Fig. 3. The graph G' for $n = 3$. The weights of grayed edges are omitted.

Definition 5. *Define* $c_v = \frac{l(\underline{v}) - l(\overline{v})}{2}$.

We now show that our choice of c_v satisfies all three metric space properties.

Lemma 7. *Every* c_v *satisfies* $c_v \geq 0$.

Proof. For each vertex $v \in G'$ there exists an edge from $\underline{v}$ to $\overline{v}$ with weight 0.

Lemma 8. *Every distinct* c_v *and* c_w *satisfy* $c_v + c_w \geq d_S(v, w)$.

Proof. By the definition of shortest paths, we have

$$l(\overline{w}) \leq l(\underline{v}) - d_S(v, w)$$
$$d_S(v, w) \leq l(\underline{v}) - l(\overline{w}).$$

and by symmetric arguments,

$$d_S(w, v) \leq l(\underline{w}) - l(\overline{v}).$$

Adding these inequalities, we obtain

$$d_S(v, w) + d_S(w, v) \leq l(\underline{v}) - l(\overline{w}) + l(\underline{w}) - l(\overline{v})$$
$$d_S(v, w) \leq \frac{l(\underline{v}) - l(\overline{v})}{2} + \frac{l(\underline{w}) - l(\overline{w})}{2}$$
$$d_S(v, w) \leq (c_v) + (c_w).$$

Lemma 9. *Every distinct* c_v *and* c_w *satisfy* $c_v + c_w \leq \lambda \cdot d_S(v, w)$.

Proof. Observe that the path $P(\overline{w}) \cup \underline{v}$ is a path to $\underline{v}$ with weight $l(\overline{w}) + \lambda \cdot d_S(w, v)$, and that the path $P(\overline{v}) \cup \underline{w}$ is a path to $\underline{w}$ with weight $l(\overline{v}) + \lambda \cdot d_S(v, w)$. By definition $P(\underline{v})$ is a shortest path to $\underline{v}$, and similarly $P(\underline{w})$ is a shortest path to $\underline{w}$, so we have

$$l(\underline{v}) \leq l(\overline{w}) + \lambda \cdot d_S(v, w)$$

and

$$l(\underline{w}) \leq l(\overline{v}) + \lambda \cdot d_S(v, w).$$

Adding these inequalities, we obtain

$$l(\underline{v}) + l(\underline{w}) \leq (l(\overline{w}) + \lambda \cdot d_S(w,v)) + (l(\overline{v}) + \lambda \cdot d_S(v,w)).$$

By assumption $d_S(w,v) = d_S(v,w)$, so

$$l(\underline{v}) - l(\overline{v}) + l(\underline{w}) - l(\overline{w}) \leq 2\lambda \cdot d_S(v,w)$$
$$(c_v) + (c_w) \leq \lambda \cdot d_S(v,w).$$

Theorem 3. *Given S and the corresponding $G(\lambda)$ and λ^*, a set C of edge lengths c_v for each $v \in S$, such that for every $v \in S$*

$$c_v \geq 0$$

and for every distinct $v, w \in S$

$$c_v + c_w \geq d_S(v,w)$$

$$c_v + c_w \leq \lambda \cdot d_S(v,w)$$

may be computed in $O(n^3)$ time.

Theorem 3 establishes that for any λ^* there exists a set C of valid edge lengths. This completes the proof of Theorem 1.

6   Conclusion

Finally we codify the main result of the paper as a theorem.

Theorem 4. *Given a set $S \subseteq X$ of n sites from a metric space $X = (X,d)$, it is possible to generate a weighted star H such that the distances between vertices of H obey the triangle inequality, and such that H has the smallest possible dilation among any such star, in $O(n^3 \log^2 n)$ time.*

Acknowledgements. This work was supported in part by NSF grant 0830403 and by the Office of Naval Research under grant N00014-08-1-1015.

References

1. Bellman, R.: On a routing problem. Quarterly in Applied Mathematics 16(1), 87–90 (1958)
2. Carstensen, P.J.: Parametric cost shortest chain problem, Bellcore (manuscript) (1984)
3. Cheong, O., Haverkort, H., Lee, M.: Computing a minimum-dilation spanning tree is NP-hard. In: Proc. 13th Australasian Symp. Theory of Computing, pp. 15–24 (2007)
4. Dasdan, A., Irani, S., Gupta, R.: An experimental study of minimum mean cycle algorithms. Technical report, University of California, Irvine (1998)
5. de Berg, M., van Kreveld, M., Overmars, M., Schwarzkopf, O.: Computational Geometry: Algorithms and Applications, 2nd edn. Springer, Heidelberg (1998)

6. Dress, A.W.M., Huber, K.T., Moulton, V.: Metric spaces in pure and applied mathematics. In: Proc. Quadratic Forms LSU, Documenta Mathematica, pp. 121–139 (2001)
7. Drezner, Z., Hamacher, H.: Facility Location: Applications and Theory. Springer, Heidelberg (2002)
8. Edmonds, J.: Embedding into L_∞^2 is Easy, Embedding into L_∞^3 is NP-Complete. Discrete and Computational Geometry 39(4), 747–765 (2008)
9. Eppstein, D.: Spanning trees and spanners. In: Sack, J.-R., Urrutia, J. (eds.) Handbook of Computational Geometry, ch. 9, pp. 425–461. Elsevier, Amsterdam (2000)
10. Eppstein, D., Wortman, K.A.: Minimum dilation stars. Computational Geometry 37(1), 27–37 (2007); Special Issue on the Twenty-First Annual Symposium on Computational Geometry - SoCG 2005
11. Ford, L.R., Fulkerson, D.: Flows in Networks. Princeton University Press, Princeton (1962)
12. Gusfield, D.: Sensitivity analysis for combinatorial optimization. Technical Report UCB/ERL M80/22, Electronics Research Laboratory, Berkeley (1980)
13. Karp, R.M.: A characterization of the minimum cycle mean in a digraph. Discrete Mathematics 23(3), 309–311 (1978)
14. Karp, R.M., Orlin, J.B.: Parametric Shortest Path Algorithms with an Application to Cyclic Staffing. Technical Report OR 103-80, MIT Operations Research Center (1980)
15. Klein, R., Kutz, M.: Computing geometric minimum-dilation graphs Is NP-hard. In: Kaufmann, M., Wagner, D. (eds.) GD 2006. LNCS, vol. 4372, pp. 196–207. Springer, Heidelberg (2007)
16. Linial, N.: Finite metric spaces–combinatorics, geometry and algorithms. In: Proc. International Congress of Mathematicians, Beijing, vol. 3, pp. 573–586 (2002)
17. Megiddo, N.: Combinatorial optimization with rational objective functions. In: Proc. 10th ACM Symp. Theory of computing, pp. 1–12. ACM Press, New York (1978)
18. Megiddo, N.: Applying Parallel Computation Algorithms in the Design of Serial Algorithms. J. ACM 30(4), 852–865 (1983)
19. Megiddo, N.: Towards a Genuinely Polynomial Algorithm for Linear Programming. SIAM Journal on Computing 12(2), 347–353 (1983)
20. Savage, C.: Parallel Algorithms for Graph Theoretic Problems. PhD thesis, University of Illinois, Urbana-Champaign (1977)

Online Square Packing

Sándor P. Fekete, Tom Kamphans*, and Nils Schweer

Braunschweig University of Technology
Department of Computer Science, Algorithms Group
Mühlenpfordtstrasse 23, 38106 Braunschweig, Germany

Abstract. We analyze the problem of packing squares in an online fashion: Given a semi-infinite strip of width 1 and an unknown sequence of squares of side length in [0, 1] that arrive from above, one at a time. The objective is to pack these items as they arrive, minimizing the resulting height. Just like in the classical game of Tetris, each square must be moved along a collision-free path to its final destination. In addition, we account for gravity in both motion and position. We apply a geometric analysis to establish a competitive factor of 3.5 for the bottom-left heuristic and present a $\frac{34}{13} \approx 2.6154$-competitive algorithm.

1 Introduction

In this paper, we consider online *strip packing* of squares. Squares arrive from above in an online fashion, one at a time, and have to be moved to their final positions in a semi-infinite, vertical strip of unit width. On its path, a square may move only through unoccupied space; in allusion to the well-known computer game, this is called the *Tetris constraint*. In addition, an item is not allowed to move upwards and has to be supported from below when reaching its final position (i.e., the bottom side of the square touches either another square or the bottom side of the strip). These conditions are called *gravity constraints*. Note that the gravity constraints make the problem harder, because we are not allowed to "hang squares in the air". The objective is to minimize the total height of the packing. Applications of this problem arise whenever physical access to the packed items is required. For example, objects stored in a warehouse need to be packed in a way such that the final positions can be accessed. Moreover, gravity is—obviously—a natural constraint in real-world packing applications.

Related Work. The strip packing problem was first considered by Baker et al. [1]. They showed that for online packing of *rectangles*, the bottom left heuristic does not necessarily guarantee a constant competitive ratio. For the offline case they proved an upper bound of 3 for a sequence of rectangles, and of 2 for squares. Kenyon and Rémila designed a FPTAS [2] for packing rectangles. For the online case, Csirik and Woeginger [3] gave a lower bound of 1.69103 on rectangle packings and an algorithm whose asymptotic worst-case ratio comes arbitrarily close to this value.

---

* Supported by DFG grant FE 407/8-3, project "ReCoNodes".

F. Dehne et al. (Eds.): WADS 2009, LNCS 5664, pp. 302–314, 2009.

For the one-dimensional online version of bin packing (pack a set of items into a minimum number of unit-capacity bins), the current best algorithm is $1,58889$-competitive [4]. The best known lower bound is $1,54014$ [5]. Coppersmith and Raghavan considered this problem in higher dimensions $d \geq 2$ [6]. They introduced a 2.6875-competitive algorithm and a lower bound of $4/3$ for $d = 2$. Epstein and van Stee improved both bounds by designing an optimal, bounded-space, 2.3722-competitive algorithm [7].

Life gets considerably harder if items cannot be packed at arbitrary positions, but must be placed from above avoiding previously packed objects as obstacles—just like in the classical game of Tetris. In this setting, no item can ever move upward, no collisions between objects must occur, an item will come to a stop if and only if it is supported from below, and each placement has to be fixed before the next item arrives. Tetris is PSPACE-hard, even for the original game with a limited set of different objects; see Breukelaar et al. [8].

Azar and Epstein [9] considered tetris-constraint online packing of rectangles into a strip. For the case without rotation, they showed that no constant competitive ratio is possible, unless there is a fixed-size lower bound of ε on the side length of the objects, in which case there is an upper bound of $O(\log \frac{1}{\varepsilon})$. For the case with rotation, they showed a 4-competitive strategy, based on shelf-packing methods; until now, this is also the best deterministic upper bound for squares. Observe that their strategy does not take the gravity constraints into account, as items are allowed to be placed at appropriate levels, even if they are unsupported. Coffmann et al. [10] considered probabilistic aspects of online rectangle packing with Tetris constraint, without allowing rotations. If rectangle side lengths are chosen uniformly at random from the interval $[0, 1]$, they showed that there is a lower bound of $(0.3138...)n$ on the expected height of the strip. Using another strategy, which arises from the bin-packing–inspired *Next Fit Level*, they established an upper bound of $(0.3697...)n$ on the expected height.

Our Results. In this paper, we demonstrate that it pays off to take a closer look at the geometry of packings. We analyze a natural and simple heuristic called *BottomLeft*, similar to the one introduced by Baker et al. [1]. We show that it is possible to give a better competitive guarantee than 4 (as achieved by Azar and Epstein), even in the presence of gravity. We obtain an asymptotic competitive ratio of 3.5 for *BottomLeft*, implying an asymptotic density of at least $0.2857...$ Improving this ratio even further, we introduce the strategy *SlotAlgorithm* and establish a competitive ratio $34/13 = 2.6154...$

2   Preliminaries

We are given a vertical strip, S, of width 1 and a sequence, $\mathcal{A} = (A_1, \ldots, A_n)$, of squares with side lengths $a_i \leq 1$. Our goal is to find a non-overlapping, axis-parallel placement of squares in the strip that keeps the height of the strip as low as possible. A packing has to fulfill the Tetris and the gravity constraints. Moreover, we consider the online problem.

We denote the bottom (left, right) side of the strip by B_S (R_S, L_S; respectively), and the sides of a square, A_i, by B_{A_i}, T_{A_i}, R_{A_i}, L_{A_i} (bottom, top, right, left; respectively). The x-coordinates of the left and right side of A_i in a packing are l_{A_i} and r_{A_i}; the y-coordinates of the top and bottom side are t_{A_i} and b_{A_i}, respectively. Let the *left neighborhood*, $N_L(A_i)$, be the set of squares that touch the left side of A_i. In the same way we define the bottom, top, and right neighborhoods, denoted by $N_B(A_i)$, $N_T(A_i)$, and $N_R(A_i)$, respectively.

A packing may leave areas of the strip empty. We call a maximal connected component of the strip's empty area a *hole*, denoted by H_h, $h \in \mathbb{N}$. A point, P, is called *unsupported*, if there is a vertical line segment from P downwards whose interior lies inside a hole. Otherwise, P is *supported*. A section of a line segment is supported, if every point in this section is supported. For an object ξ we refer to the boundary by $\partial\xi$, to the interior by ξ°, and to its area by $|\xi|$.

3   The Strategy *BottomLeft*

In this section, we analyze the packing generated by the strategy *BottomLeft*, which works as follows: We place the current square as close as possible to the bottom of the strip (provided that there is a collision-free path from the top of the strip to the desired position that never moves in positive y-direction). We break ties by choosing the leftmost among all possible bottommost positions.[1]

For a simplified analysis, we finish the packing with an additional square, A_{n+1}, of side length 1. This implies that all holes have a closed boundary. Let $H_1, \ldots, H_s$ be the holes in the packing. Then the height of the packing produced by *BottomLeft* is $BL = \sum_{i=1}^{n} a_i^2 + \sum_{h=1}^{s} |H_h|$. In the following sections, we prove $\sum_{h=1}^{s} |H_h| \leq 2.5 \cdot \sum_{i=1}^{n+1} a_i^2$. Because any strategy needs at least a height of $\sum_{i=1}^{n} a_i^2$, our bound implies that asymptotically $BL \leq 3.5 \cdot OPT$.

We proceed as follows. First, we state some properties of the generated packing (Section 3.1). In Section 3.2 we simplify the shape of the holes by partitioning a hole into several disjoint new parts.[2] In the packing, these new holes are open at their top side, so we introduce *virtual lids* that close these holes. Afterwards, we estimate the area of a hole in terms of the squares that enclose the hole (Section 3.3). Summing up the charges to a single square (Table 1) we get

Theorem 1. BottomLeft *is (asymptotically) 3.5-competitive.*

3.1   Basic Properties of the Generated Packing

In this section, we analyze structural properties of the boundary of a hole. We say that a square, A_i, *contributes* to the boundary of a hole, H_h, iff ∂A_i and ∂H_h intersect in more than one point. Let $\tilde{A}_1, \ldots, \tilde{A}_k$ denote the squares on the

---

[1] To implement the strategy, we can use robot-motion-planning techniques. For k placed squares, this can be done in time $O(k \log^2 k)$; see de Berg et al. [11].

[2] Let the new parts replace the original hole, so that we do not have to distinguish between 'holes' and 'parts of a hole'.

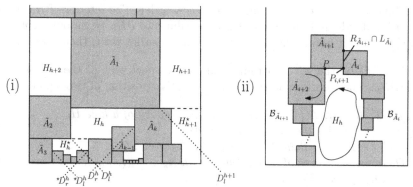

Fig. 1. (i) A packing produced by *BottomLeft*. The squares $\tilde{A}_1, \ldots, \tilde{A}_k$ contribute to the boundary of the hole H_h, which is split into a number of subholes. In the shown example one new subhole $H_h^\star$ is created. Note that the square $\tilde{A}_1$ also contributes to the holes H_{h+1} and H_{h+2} and serves as a virtual lid for $H_{h+1}^\star$. (ii) The hole H_h with the two squares $\tilde{A}_i$ and $\tilde{A}_{i+1}$ and their bottom sequences. Here, $\tilde{A}_{i+1}$ is $\tilde{A}_1$. If ∂H_h is traversed in ccw order, then $\partial H_h \cap \partial \tilde{A}_{i+2}$ is traversed in cw order w.r.t. to $\partial \tilde{A}_{i+2}$.

boundary of H_h in counterclockwise order starting with the upper left square.[3] We call $\tilde{A}_1$ the *lid* of H_h and define $\tilde{A}_{k+1} = \tilde{A}_1$, $\tilde{A}_{k+2} = \tilde{A}_2$ and so on. By $P_{i,i+1}$ we denote the point where ∂H_h leaves $\partial \tilde{A}_i$ and enters $\partial \tilde{A}_{i+1}$.

Let A_i be a packed square. We define the *left (bottom) sequence*, $\mathcal{L}_{A_i}$, $(\mathcal{B}_{A_i})$, of A_i, as follows: The first element of $\mathcal{L}_{A_i}$ $(\mathcal{B}_{A_i})$ is A_i. The next element is chosen as an arbitrary left (bottom) neighbor of the previous element. The sequence ends if no such neighbor exists. We call the polygonal chain from the upper right corner of the first element of $\mathcal{L}_{A_i}$ to the upper left corner of the last element while traversing the boundary of the sequence in counterclockwise order the *skyline*, $\mathcal{S}_{A_i}$, of A_i. Obviously, $\mathcal{S}_{A_i}$ has an endpoint on L_S. Further, $\mathcal{S}_{A_i}^\circ \cap H_h^\circ = \emptyset$.

Lemma 1. *Let $\tilde{A}_i$ be a square that contributes to ∂H_h. Then,*
(i) $\partial H_h \cap \partial \tilde{A}_i$ is a single curve, and
(ii) if ∂H_h is traversed in counterclockwise (clockwise) order, then $\partial H_h \cap \partial \tilde{A}_i$ is traversed in clockwise (counterclockwise) order w.r.t. $\partial \tilde{A}_i$; see Fig. 1(ii).

Proof. (i) Assume that $\partial H_h \cap \partial \tilde{A}_i$ consists of (at least) two curves, c_1 and c_2. Consider a simple curve, C, that lies inside H_h and has one endpoint in c_1 and the other one in c_2. We add the straight line between the endpoints to C and obtain a simple closed curve C'. As c_1 and c_2 are not connected, there is a square $\tilde{A}_j$ inside C' that is a neighbor of $\tilde{A}_i$. If $\tilde{A}_j$ is a left, right or bottom neighbor of $\tilde{A}_i$ this contradicts the existence of $\mathcal{B}_{\tilde{A}_j}$; if it is a top neighbor this contradicts the existence of $\mathcal{L}_{\tilde{A}_j}$. Hence, $\partial H_h \cap \partial \tilde{A}_i$ is a single curve.

---

[3] It is always clear from the context which hole defines this sequence of squares. Thus, we chose not to introduce an additional superscript referring to the hole.

(ii) Imagine that we walk along ∂H_h in ccw order: The interior of H_h lies on our left-hand side and all squares that contribute to ∂H_h lie on our right-hand side. Hence, their boundaries are traversed in cw order w.r.t. their interior. □

Let P and Q be the left and right endpoint, respectively, of the line segment $\partial \tilde{A}_1 \cap \partial H_h$. The next lemma restricts the relative position of two squares:

Lemma 2. *Let $\tilde{A}_i$, $\tilde{A}_{i+1}$ contribute to the boundary of a hole H_h.*
(i) If $\tilde{A}_{i+1} \in N_L(\tilde{A}_i)$ then either $\tilde{A}_{i+1} = \tilde{A}_1$ or $\tilde{A}_i = \tilde{A}_1$.
(ii) If $\tilde{A}_{i+1} \in N_T(\tilde{A}_i)$ then $\tilde{A}_{i+1} = \tilde{A}_1$ or $\tilde{A}_{i+2} = \tilde{A}_1$.
(iii) There are two types of holes: Type I with $\tilde{A}_k \in N_R(\tilde{A}_{k-1})$, and Type II with $\tilde{A}_k \in N_T(\tilde{A}_{k-1})$; see Fig. 3.

Proof. (i) Let $\tilde{A}_{i+1} \in N_L(\tilde{A}_i)$. Consider the endpoints of the vertical line $R_{\tilde{A}_{i+1}} \cap L_{\tilde{A}_i}$; see Fig. 1(ii). We traverse ∂H_h in counterclockwise order starting in P. By Lemma 1, we traverse $\partial \tilde{A}_i$ in clockwise order and, therefore, $P_{i,i+1}$ is the lower endpoint of $R_{\tilde{A}_{i+1}} \cap L_{\tilde{A}_i}$. Now, $\mathcal{B}_{\tilde{A}_i}$, $\mathcal{B}_{\tilde{A}_{i+1}}$, and the segment of B_S completely enclose an area that completely contains the hole, H_h. If the sequences share a square, A_j, we consider the area enclosed up to the first intersection. Therefore, if $b_{\tilde{A}_{i+1}} \geq b_{\tilde{A}_i}$ then $\tilde{A}_{i+1} = \tilde{A}_1$ else $\tilde{A}_i = \tilde{A}_1$ by the definition of $\overline{PQ}$.

The proof of (ii) follows almost directly from (i). Let $\tilde{A}_{i+1} \in N_T(\tilde{A}_i)$. We know that $\partial \tilde{A}_{i+1}$ is traversed in clockwise order and we know that $\tilde{A}_{i+1}$ has to be supported to the left. Therefore, $\tilde{A}_{i+2} \in N_L(\tilde{A}_{i+1}) \cup N_B(\tilde{A}_{i+1})$ and the result follows from (i). For (iii) we traverse ∂H_h from P in clockwise order. From the definition of $\overline{PQ}$ and Lemma 1 we know that $P_{k,1}$ is a point on $L_{\tilde{A}_k}$. If $P_{k-1,k} \in L_{\tilde{A}_k}$, then $\tilde{A}_k \in N_R(\tilde{A}_{k-1})$; if $P_{k-1,k} \in B_{\tilde{A}_k}$, then $\tilde{A}_k \in N_T(\tilde{A}_{k-1})$. In any other case $\tilde{A}_k$ does not have a lower neighbor. □

3.2 Splitting Holes

Let H_h be a hole whose boundary does not touch the boundary of the strip. We define two lines: The *left diagonal*, D_l^h, is defined as the straight line with slope -1 starting in $P_{2,3}$ if $P_{2,3} \in R_{\tilde{A}_2}$ or, otherwise, in the lower right corner of $\tilde{A}_2$; see Fig. 3. We denote the point in which D_l^h starts by P'. The *right diagonal*, D_r^h, is defined as the line with slope 1 starting in $P_{k-1,k}$ if $\tilde{A}_k \in N_R(\tilde{A}_{k-1})$ (Type I) or in $P_{k-2,k-1}$, otherwise (Type II). Note that $P_{k-2,k-1}$ lies on $L_{\tilde{A}_{k-1}}$, otherwise there would not be a left neighbor of $\tilde{A}_{k-1}$. We denote the point in which D_r^h starts by Q'. If h is clear or does not matter, we omit the superscript.

Lemma 3. *Let H_h be a hole, D_r its right diagonal. Then $D_r \cap H_h^\circ = \emptyset$ holds.*

Proof. Consider the left sequence, $\mathcal{L}_{\tilde{A}_k} = (\tilde{A}_k = \alpha_1, \alpha_2, \dots)$ or $\mathcal{L}_{\tilde{A}_{k-1}} = (\tilde{A}_{k-1} = \alpha_1, \alpha_2, \dots)$ for H_h being of Type I or II, respectively. By induction, the upper left corners of the α_i's lie above D_r: If D_r intersects $\partial \alpha_i$ at all, the first intersection is on R_{α_i}, the second on B_{α_i}. Thus, at least the skyline separates D_r and H_h. □

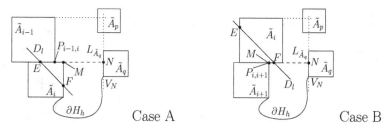

Fig. 2. D_l can intersect $\tilde{A}_i$ (for the second time) in two different ways: on the right side or on the bottom side. In Case A, the square $\tilde{A}_{i-1}$ is on top of $\tilde{A}_i$; in Case B, $\tilde{A}_i$ is on top of $\tilde{A}_{i+1}$.

It is a simple observation that if D_l intersects a square $\tilde{A}_i$ in a nontrivial way[4] then either $F \in R_{\tilde{A}_i}$ and $E \in T_{\tilde{A}_i}$ or $F \in B_{\tilde{A}_i}$ and $E \in L_{\tilde{A}_i}$. To break ties, we define that an intersection in the lower right corner of $\tilde{A}_i$ belongs to $B_{\tilde{A}_i}$ (Fig. 1(i) and 2). Unfortunately, Lemma 3 does not hold for D_l. Therefore, we split our hole, H_h, into two new holes, $H_h^{(1)}$ and $H_h^\star$, as follows: Let F be the first nontrivial intersection point of ∂H_h and D_l while traversing ∂H_h in counterclockwise fashion, starting in P. We consider two cases, $F \in R_{\tilde{A}_i} \setminus B_{\tilde{A}_i}$ (Case A) and $F \in B_{\tilde{A}_i}$ (Case B); see Fig. 2.

Let E be the other intersection point of D_l and $\partial \tilde{A}_i$. In Case A, let $\tilde{A}_{\text{up}} := \tilde{A}_{i-1}$ and $\tilde{A}_{\text{low}} := \tilde{A}_i$, in Case B $\tilde{A}_{\text{up}} := \tilde{A}_i$ and $\tilde{A}_{\text{low}} := \tilde{A}_{i+1}$. The horizontal ray that emanates from the upper right corner of $\tilde{A}_{\text{low}}$ to the right is subdivided into supported and unsupported sections. Let $U = \overline{MN}$ be the leftmost unsupported section. Now we split H_h into two parts, $H_h^\star$ below $\overline{MN}$ and $H_h^{(1)} := H_h \setminus H_h^\star$.

We split $H_h^{(1)}$ into $H_h^{(2)}$ and $H_h^{\star\star}$ etc., until there is no further intersection between the boundary of $H_h^{(z)}$ and D_l^h. Every split is caused by a pair of squares. It can be shown that $\overline{MN} < \tilde{a}_{up}$ and, therefore, a copy of $\tilde{A}_{\text{up}}$, denoted by $\tilde{A}'_{\text{up}}$, placed on $\overline{MN}$ can serve as a *virtual lid* for the hole below. Moreover, a square serves as a virtual lid for at most one hole. Regarding the holes, they are either of Type I or Type II and, thus, can be analyzed in the same way as original holes. See the full version of this paper for a rigorous proof.

3.3   Computing the Area of a Hole

We eliminated all intersections of D_l^h with the boundary of the hole $H_h^{(z)}$ by splitting the hole. Thus, we have a set of holes $\hat{H}_h$, $h = 1, \ldots, s^*$, that fulfill $\partial \hat{H}_h \cap D_l^h = \emptyset$ and have either a non-virtual or a virtual lid.

Our aim is to bound $|\hat{H}_h|$ by the areas of the squares that contribute to $\partial \hat{H}_h$. A square A_i may contribute to more than one hole. It is too expensive to use

---

[4] An intersection, $p \in \Gamma \cap \Delta$, of a Jordan curve Δ and a line, ray, or line segment Γ is called *nontrivial*, iff there is a line segment ℓ of length $\varepsilon > 0$ on the line through Γ such that p is in the interior of ℓ and the endpoints of ℓ lie on different sides of Δ.

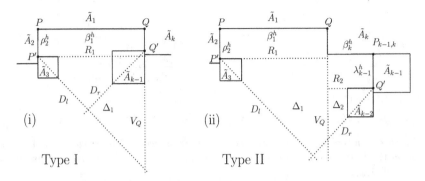

Fig. 3. Holes of Type I and Type II with their left and right diagonals

its total area a_i^2 in the bound for a single hole. Instead, we charge only fractions of a_i^2 per hole. Moreover, we charge every edge of A_i separately. By Lemma 1, $\partial \hat{H}_h \cap \partial A_i$ is connected. In particular, every side of A_i contributes at most one line segment to $\partial \hat{H}_h$. For the left (bottom, right) side of a square A_i, we denote the length of the line segment contributed to $\partial \hat{H}_h$ by λ_i^h (β_i^h, ρ_i^h; respectively).[5]

Let $c_{h,i}^{\{\lambda,\beta,\rho\}}$ be appropriate coefficients, such that the area of a hole can be charged against the area of the adjacent squares; i.e., $|\hat{H}_h| \leq \sum_{i=1}^{n+1} c_{h,i}^\lambda (\lambda_i^h)^2 + c_{h,i}^\beta (\beta_i^h)^2 + c_{h,i}^\rho (\rho_i^h)^2$. As each point on ∂A_i is on the boundary of at most one hole, the line segments are pairwise disjoint. Thus, for the left side of A_i, the two squares inside A_i induced by the line segments λ_i^h and λ_i^g of two different holes, $\hat{H}_h$ and $\hat{H}_g$, do not overlap. Therefore, we obtain $\sum_{h=1}^{s^*} c_{h,i}^\lambda \cdot (\lambda_i^h)^2 \leq c_i^\lambda \cdot a_i^2$, where $c_i^\lambda := \max_h c_{h,i}^\lambda$. We call c_i^λ the *charge of* L_{A_i} and define c_i^β and c_i^ρ analogously.

We use virtual copies of some squares as lids. However, for every square, A_i, there is at most one copy, A_i'. We denote the line segments and charges corresponding to A_i' by $\lambda_{i'}^h$, $c_{h,i'}^\lambda$ and so on. The *total charge of* A_i is given by $c_i = c_i^\lambda + c_i^\beta + c_i^\rho + c_{i'}^\lambda + c_{i'}^\beta + c_{i'}^\rho$. Altogether, we bound $\sum_{h=1}^{s^*} |\hat{H}_h| \leq \sum_{i=1}^{n+1} c_i \cdot a_i^2 \leq \sum_{i=1}^{n+1} c \cdot a_i^2$, with $c = \max_i c_i$. Next, we want to find an upper bound on c.

Holes with a Non-Virtual Lid. We removed all intersections of $\hat{H}_h$ with its diagonal D_l^h. Therefore, $\hat{H}_h$ lies completely inside the polygon formed by D_l^h, D_r^h and the part of $\partial \hat{H}_h$ that is clockwise between P' and Q'; see Fig. 3. If $\hat{H}_h$ is of Type I, we consider the rectangle, R_1, of area $\rho_2^h \cdot \beta_1^h$ induced by P, P' and Q. Let Δ_1 be the triangle below R_1 formed by the bottom side of R_1, D_l^h, and the vertical line V_Q passing through Q; see Fig. 3(i). Obviously, $|\hat{H}_h| \leq |R_1| + |\Delta_1|$. As D_l^h has slope -1, we get $|\Delta_1| = \frac{1}{2}(\beta_1^h)^2$. We have $|R_1| = \rho_2^h \cdot \beta_1^h \leq \frac{1}{2}(\rho_2^h)^2 + \frac{1}{2}(\beta_1^h)^2$. Thus, for a Type I-hole we get $|\hat{H}_h| \leq (\beta_1^h)^2 + \frac{1}{2}(\rho_2^h)^2$, i.e., we charge the bottom side of $\tilde{A}_1$ with 1 and the right side of $\tilde{A}_2$ with $\frac{1}{2}$. In this case, we get $c_{h,1}^\beta = 1$

---

[5] If a side of a square does not contribute to a hole, the corresponding length of the line segment is defined to be zero.

and $c_{h,2}^\rho = \frac{1}{2}$. For a Type II hole, we additionally get a rectangle R_2 and a triangle, Δ_2, as in Fig. 3(ii). Using similar arguments as above we get charges $c_{h,1}^\beta = c_{h,k}^\beta = 1$ and $c_{h,2}^\rho = \frac{1}{2} = c_{h,k-1}^\lambda = \frac{1}{2}$.

Holes with a Virtual Lid. Let $\hat{H}_h$ be a hole with a virtual lid, $\hat{H}_g$ be immediately above $\hat{H}_h$, $\tilde{A}_{up}$ be the square whose copy, $\tilde{A}'_{up}$, becomes a new lid, and $\tilde{A}_{low}$ the bottom neighbor of $\tilde{A}_{up}$. We show that $\tilde{A}'_{up}$ increases the charge of $\tilde{A}_{up}$ by at most $\frac{1}{2}$: If $\tilde{A}_{up}$ does not exceed $\tilde{A}_{low}$ to the left, it cannot serve as a lid for any other hole (Fig. 4). Hence, the charge of the bottom side of $\tilde{A}_{up}$ is 0; like in the preceding section, we obtain a charge ≤ 1 to the bottom of $\tilde{A}'_{up}$. If it exceeds $\tilde{A}_{low}$ to the left, we know that the part $B_{\tilde{A}_{up}} \cap T_{\tilde{A}_{low}}$ of $B_{\tilde{A}_{up}}$ is not charged by another hole, because it does not belong to a hole and the lid is defined uniquely.

We define points P and P' for $\hat{H}_h$ in the same way as in the preceding section. Independent of $\hat{H}_h$'s type, $\tilde{A}'_{up}$ gets charged only for the rectangle R_1 induced by P, P' and N, as well as for the triangle below R_1 (Fig. 3). Now we show that we do not have to charge $\tilde{A}'_{up}$ for R_1, since the part of R_1 above D_l^g is already included in the bound for $\hat{H}_g$, and the remaining part can be charged to $B_{\tilde{A}_{up}}$ and $R_{\tilde{A}_{low}}$. $\tilde{A}'_{up}$ gets charged $\frac{1}{2}$ for the triangle.

D_l^g splits R_1 into a part that is above this line, and a part that is below this line. The latter part of R_1 is not included in the bound for $\hat{H}_g$. Let F be the intersection of $\partial \hat{H}_g$ and D_l^g that caused the creation of $\hat{H}_h$. If $F \in R_{\tilde{A}_{low}}$, this part is at most $\frac{1}{2}(\rho_{low}^h)^2$, where ρ_{low}^h is the length of $\overline{P'F}$. We charge $\frac{1}{2}$ to $R_{\tilde{A}_{low}}$.

If $F \in B_{\tilde{A}_{up}}$, the part of R_1 below D_l^g can be split into a rectangular part of area $\rho_{low}^h \cdot \beta_{up}^h$, and a triangular part of area $\frac{1}{2}(\rho_{low}^h)^2$. Here β_{up}^h is the length of $\overline{PF}$. The cost of the triangle is charged to $R_{\tilde{A}_{low}}$. Note that the part of $B_{\tilde{A}_{up}}$ that exceeds $\tilde{A}_{low}$ to the right is not charged and ρ_{low}^h is not larger than $B_{\tilde{A}_{up}} \cap T_{\tilde{A}_{low}}$ (i.e., the part of $B_{\tilde{A}_{up}}$ that was not charged before). Thus, we can charge the rectangular part completely to $B_{\tilde{A}_{up}}$. Hence, $\tilde{A}'_{up}$ is charged $\frac{1}{2}$ in total.

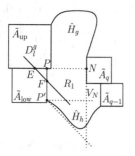 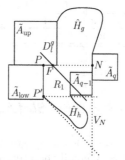

Fig. 4. The holes $\hat{H}_g$ and $\hat{H}_h$ and the rectangle R_1 which is divided into two parts by D_l^g. The upper part is already included in the bound for $\hat{H}_g$. The lower part is charged completely to $R_{\tilde{A}_{low}}$ and $B_{\tilde{A}'_{up}}$. Here, P and P' are defined w.r.t. $\hat{H}_h$.

Holes Containing Parts of ∂S. We show in this section that holes that touch ∂S are just special cases of the ones discussed in the preceding sections.

Because the top side of a square never gets charged for a hole, it does not matter whether a part of B_S belongs to the boundary. Moreover, for any hole $\hat{H}_h$ either L_S or R_S can be a part of $\partial \hat{H}_h$, because otherwise there exits a curve with one endpoint on L_S and the other endpoint on R_S, with the property that this curve lies completely inside of $\hat{H}_h$. This contradicts the existence of the bottom sequence of a square lying above the curve.

For a hole $\hat{H}_h$ touching L_S, $L_S \cap \partial \hat{H}_h$ is a single line segment (similar to Lemma 1). Let P be the topmost point of this line segment and $\tilde{A}_1$ be the square containing P. The existence of $\mathcal{B}_{\tilde{A}_1}$ implies that $\tilde{A}_1$ is the lid of $\hat{H}_h$. As $\tilde{A}_1$ must have a bottom neighbor, $\tilde{A}_k$, and $\tilde{A}_k$ must have a right neighbor, $\tilde{A}_{k-1}$, we get $P_{k,1} \in B_{\tilde{A}_1}$ and $P_{k-1,k} \in L_{\tilde{A}_k}$, respectively. We define the right diagonal D_r and the point Q' as above and conclude that $\hat{H}_h$ lies completely inside the polygon formed by $L_S \cap \partial \hat{H}_h$, D_r and the part of $\partial \hat{H}_h$ that is between P and Q' (in clockwise order). We split this polygon into a rectangle and a triangle in order to obtain charges of 1 to $B_{\tilde{A}_1}$ and $\frac{1}{2}$ to $L_{\tilde{A}_k}$.

Now consider a hole where a part of R_S belongs to $\partial \hat{H}_h$. We denote the topmost point on $R_S \cap \partial \hat{H}_h$ by Q, and the square containing Q by $\tilde{A}_1$. $\tilde{A}_1$ is the lid of this hole. As above, we eliminate the intersections of D_l and $\partial \hat{H}_h$ by creating new holes. After this, the modified hole $\hat{H}_h^{(z)}$ can be viewed as a hole of Type II, for which the part on the right side of V_Q has been cut off. We obtain charges of 1 to $B_{\tilde{A}_1}$, $\frac{1}{2}$ to $R_{\tilde{A}_2}$, and $\frac{1}{2}$ to the bottom of a virtual lid.

Table 1. Charges to different sides of a single square. The charges depend on the type of the adjacent hole (Type I, II, touching or not touching the strip's boundary), but the maximal charge dominates the other one. Moreover, the square may also serve as a virtual lid. These charges sum up to a total charge of 2.5 per square.

	Non-virtual Lid					Virtual Lid				Total
	Type I	Type II	L_S	R_S	Max.	Type I	Type II	R_S	Max.	
Left side	0	0.5	0.5	0	0.5	0	0	0	0	0.5
Bottom side	1	1	1	1	1	0.5	0.5	0.5	0.5	1.5
Right Side	0.5	0.5	0	0.5	0.5	0	0	0	0	0.5
Total					2				0.5	2.5

4   The Strategy *SlotAlgorithm*

Consider two vertical lines going upward from the bottom side of S and parallel to sides of S. We call the area between these lines a *slot*, the lines the slot's *left* and *right boundary*, and the distance between the lines the *width* of the slot.

Our strategy *SlotAlgorithm* works as follows: We divide the strip S of width 1 into one slot of width 1, two slots of width 1/2, four slots of width 1/4 etc.

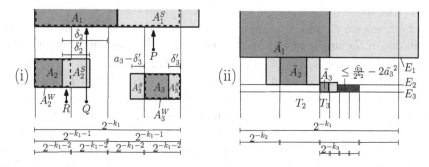

Fig. 5. (i) Squares A_i with *shadows* A_i^S and *widenings* A_i^W. $\delta_2' = a_2$ and $\delta_3' = \delta_3$. P and Q are charged to A_1. R is charged to A_2. (ii) The first three squares of the sequence. Here, $\tilde{A}_2$ is the smallest square that bounds $\tilde{A}_1$ from below. $\tilde{A}_3$ is the smallest one that intersects E_2 in an active slot (w.r.t. E_2) of width $1/2^{k_2}$. T_2 is nonactive (w.r.t. E_2) and also w.r.t. all E_j, $j \geq 3$. The part of $F_{\tilde{A}_1}$ (darkest gray) between E_2 and E_3 in an active slot of width 2^{-k_2} is $\leq \tilde{a}_3/2^{k_2} - 2\tilde{a}_3^2$ as points in $\tilde{A}_3^W$ are not charged to $\tilde{A}_1$.

(i.e., creating 2^j of width 2^{-j}). Note that a slot of width 2^{-j} contains 2 slots of width 2^{-j-1}; see Fig. 5(i). For every square A_i we round the side length a_i to the smallest number $1/2^{k_i}$ that is larger than or equal to a_i. We place A_i in the slot of width 2^{-k_i} that allows A_i to be placed as near to the bottom of S as possible by moving A_i down along the left boundary of the chosen slot until another square is reached. *SlotAlgorithm* satisfies the Tetris and the Gravity constraints.

Theorem 2. SlotAlgorithm *is (asymptotically) 2.6154-competitive.*

Proof. Let A_i be a square placed by *SlotAlgorithm* in a slot T_i of width 2^{-k_i}. Let δ_i be the distance between the right side of A_i and the right boundary of the slot of width 2^{-k_i+1} that contains A_i and $\delta_i' := \min\{a_i, \delta_i\}$. We call the area obtained by enlarging A_i by δ_i' to the right and by $a_i - \delta_i'$ to the left the *shadow* of A_i and denote it by A_i^S. Thus, A_i^S is an area of the same size as A_i and lies completely inside a slot of twice the width of A_i's slot. Moreover, we define the *widening* of A_i as $A_i^W = (A_i \cup A_i^S) \cap T_i$; see Fig. 5(i).

Now, consider a point P in T_i that is not inside an A_j^W for any square A_j. We charge P to the square A_i if A_i^W is the first widening that intersects the vertical line going upwards from P. Let F_{A_i} be the set of all points charged to A_i. For the analysis, we place a closing square, A_{n+1}, of side length 1 on top of the packing. Therefore, every point in the packing that does not lie inside an A_j^W is charged to a square. Because A_i and A_i^S have the same area, we can bound the height of the packing by $2\sum_{i=1}^{n} a_i^2 + \sum_{i=1}^{n+1} |F_{A_i}|$.

The height of an optimal packing is at least $\sum_{i=1}^{n} a_i^2$ and, therefore, it suffices to show $|F_{A_i}| \leq 0.6154 \cdot a_i^2$. We construct for every A_i a sequence of squares $\tilde{A}_1^i, \tilde{A}_2^i, \ldots, \tilde{A}_m^i$ with $\tilde{A}_1^i = A_i$. (To ease notation, we omit the superscript i in the following.) We denote by E_j the extension of the bottom side of $\tilde{A}_j$ to the left and to the right; see Fig. 5(ii). We will show that by an appropriate choice of the sequence we can bound the area of the part of $F_{\tilde{A}_1}$ that lies between a consecutive pair of extensions, E_j and E_{j+1}, in terms of $\tilde{A}_{j+1}$ and the slot widths. From this we will derive the upper bound on the area of $F_{\tilde{A}_1}$. We assume throughout the proof that the square $\tilde{A}_j$, $j \geq 1$, is placed in a slot, T_j, of width 2^{-k_j}. Note that $F_{\tilde{A}_1}$ is completely contained in T_1. A slot is called *active (w.r.t. E_j and $\tilde{A}_1$)* if there is a point in the slot that lies below E_j and that is charged to $\tilde{A}_1$ and *nonactive* otherwise. If it is clear from the context we leave out the $\tilde{A}_1$.

The sequence of squares is chosen as follows: $\tilde{A}_1$ is the first square and $\tilde{A}_{j+1}$, $j = 1, \ldots, m - 1$ is chosen as the smallest one that intersects or touches E_j in an active slot (w.r.t. E_j and $\tilde{A}_1$) of width 2^{-k_j} and that is not equal to $\tilde{A}_j$. The sequence ends if all slots are nonactive w.r.t. to an extension E_m. We claim:

(i) $\tilde{A}_{j+1}$ exists for $j + 1 \leq m$ and $\tilde{a}_{j+1} \leq 2^{-k_j-1}$ for $j + 1 \leq m - 1$.
(ii) The number of active slots (w.r.t. E_j) of width 2^{-k_j} is at most 1 for $j = 1$ and $\prod_{i=2}^{j} \left(\frac{1}{2^{k_i-1}} 2^{k_i} - 1 \right)$ for $j \geq 2$.
(iii) The area of the part of $F_{\tilde{A}_1}$ that lies in an active slot of width 2^{-k_j} between E_j and E_{j+1} is at most $2^{-k_j} \tilde{a}_{j+1} - 2\tilde{a}_{j+1}^2$.

We prove the claims by induction. Assume that the $(j+1)$st element does not exist for $j + 1 \leq m$. Let T' be an active slot in T_1 (w.r.t. E_j) of width 2^{-k_j} for which E_j is not intersected by a square in T'. If there is a rectangle of height ε below $T' \cap E_j$ for which every point is charged to $\tilde{A}_1$, SlotAlgorithm would have chosen this slot for $\tilde{A}_j$. Hence, at least

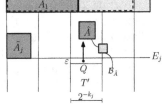

one point, Q, below E_j is not charged to $\tilde{A}_1$. Consider the bottom sequence (see Section 3.1) of the square, $\hat{A}$, to which Q is charged. This sequence has to intersect E_j outside of T' (by choice of T'). But then one of its elements has to intersect the left or the right boundary of T' and we can conclude that this square has at least the width of T', because (by the algorithm) a square with rounded side length $2^{-\ell}$ cannot cross a slot's boundary of width larger than $2^{-\ell}$. Hence, a setting as shown in the figure is not possible. In turn, a square larger than T' completely covers T' and T' cannot be active w.r.t. to E_j and $\tilde{A}_1$. Thus, all points in T' below E_j are charged to this square; a contradiction. This proves the existence of $\tilde{A}_{j+1}$. Because we chose $\tilde{A}_{j+1}$ to be of minimal side length, $\tilde{a}_{j+1} \geq 2^{-k_j}$ would imply that all slots inside T are nonactive (w.r.t. E_j). Therefore, if $\tilde{A}_{j+1}$ is not the last element of the sequence, $\tilde{a}_{j+1} \leq 2^{-k_j-1}$ holds.

By the induction hypothesis there are at most $(2^{-k_1} 2^{k_2} - 1) \cdot (2^{-k_2} 2^{k_3} - 1) \cdot \ldots \cdot (2^{-k_{j-2}} 2^{k_{j-1}} - 1)$ active slots of width $2^{-k_{j-1}}$ (w.r.t. E_{j-1}). Each of these slots contains $2^{k_j-k_{j-1}}$ slots of width 2^{-k_j} and in every active slot of width $2^{-k_{j-1}}$ at least one slot of width 2^{-k_j} is nonactive because we chose $\tilde{A}_j$ to be of minimum side length. Hence, the number of active slots (w.r.t. E_j) is a factor of $\left(\frac{1}{2^{k_{j-1}}} 2^{k_j} - 1 \right)$ larger than the number of active slots (w.r.t. E_{j-1}).

By the choice of $\tilde{A}_{j+1}$ and the fact that in every active slot of width 2^{-k_j} there is at least one square that intersects E_j (points below its widening are not charged to $\tilde{A}_1$) we conclude that the area of $F_{\tilde{A}_1}$ between E_j and E_{j+1} is at most $2^{-k_j}\tilde{a}_{j+1} - 2\tilde{a}_{j+1}^2$ in every active slot of width 2^{-k_j} (Fig. 5). We get $|F_{\tilde{A}_1}| \le \frac{\tilde{a}_2}{2^{k_1}} - 2\tilde{a}_2^2 + \sum_{j=2}^{m}\left[\left(\frac{\tilde{a}_{j+1}}{2^{k_j}} - 2\tilde{a}_{j+1}^2\right)\prod_{i=1}^{j-1}\left(\frac{2^{k_{i+1}}}{2^{k_i}} - 1\right)\right]$. This is maximized for $\tilde{a}_{i+1} = 1/2^{k_i+2}$, $1 \le i \le m$ implying $k_i = k_1 + 2(i-1)$. We get $|F_{\tilde{A}_1}| \le \sum_{i=0}^{\infty}\frac{3^i}{2^{2k_1+4i+3}}$. $|F_{\tilde{A}_1}|/\tilde{a}_1^2$ is maximized for $\tilde{a}_1$ as small as possible; i.e., $\tilde{A}_1 = 2^{-(k_1+1)} + \varepsilon$. We get: $\frac{|F_{\tilde{A}_1}|}{\tilde{a}_1^2} \le \sum_{i=0}^{\infty}\frac{2^{2k_1+2}\cdot 3^i}{2^{2k_1+4i+3}} = \frac{1}{2}\sum_{i=0}^{\infty}\left(\frac{3}{16}\right)^i = \frac{8}{13} = 0.6154...$ □

5 Conclusion

We have demonstrated that geometric analysis improves the best competitive guarantee for online square packing. We believe that this is not the end of the line: It should be possible to combine this type of analysis with more sophisticated, shelf-based algorithms. Our best lower bound for *BottomLeft* is a competitive factor of 5/4: Consider a sequence of small items of total width 1/3, followed by two items slightly larger than 1/3. Asymptotically, this yields a lower bound of 5/4 by taking turns with unit squares.

The bottleneck in our analysis are squares that have large holes at their right, left, and bottom side and also serve as a virtual lid; see Fig. 1(i). This worst case can happen to only a few squares, but never to all of them, so it may be possible to transfer charges between squares. It may also be possible to apply better lower bounds than the total area, e.g., similar to [12].

We also presented an algorithm that is 2.6154-competitive. We believe that our algorithm can be improved (as the best known lower bound is only 1.2). Moreover, we believe that our approach can be extended to higher dimensions. Rectangles may require a slightly different analysis. These topics will be the subject of future research. It is an open question whether our analysis is tight or can be improved. The best lower bound for *SlotAlgorithm* known to us is 2.

References

1. Baker, B.S., Coffman Jr., E.G., Rivest, R.L.: Orthogonal packings in two dimensions. SIAM Journal on Computing 9(4), 846–855 (1980)
2. Kenyon, C., Rémila, E.: Approximate strip packing. In: Proc. 37th Annu. IEEE Sympos. Found. Comput. Sci., pp. 31–36 (1996)
3. Csirik, J., Woeginger, G.J.: Shelf algorithms for on-line strip packing. Information Processing Letters 63, 171–175 (1997)
4. Seiden: On the online bin packing problem. JACM: Journal of the ACM 49 (2002)
5. van Vliet, A.: An improved lower bound for online bin packing algorithms. Information Processing Letters 43, 277–284 (1992)
6. Coppersmith, D., Raghavan, P.: Multidimensional online bin packing: Algorithms and worst case analysis. orl 8, 17–20 (1989)
7. Epstein, L., van Stee, R.: Optimal online bounded space multidimensional packing. In: Proc. 15th Symp. on Discrete Algorithms (SODA), pp. 214–223. ACM/SIAM (2004)

8. Breukelaar, R., Demaine, E.D., Hohenberger, S., Hoogeboom, H.J., Kosters, W.A., Liben-Nowell, D.: Tetris is hard, even to approximate. Internat. J. Comput. Geom. Appl. 14, 41–68 (2004)
9. Azar, Y., Epstein, L.: On two dimensional packing. J. Algor. 25, 290–310 (1997)
10. Coffman Jr., E., Downey, P.J., Winkler, P.: Packing rectangles in a strip. Acta Inform. 38, 673–693 (2002)
11. de Berg, M., van Kreveld, M., Overmars, M., Schwarzkopf, O.: Computational Geometry: Algorithms and Applications. Springer, Berlin (2000)
12. Fekete, S.P., Schepers, J.: New classes of lower bounds for the bin packing problem. Math. Progr. 91, 11–31 (2001)

Worst-Case Optimal Adaptive Prefix Coding

Travis Gagie[1],* and Yakov Nekrich[2]

[1] Research Group for Combinatorial Algorithms in Bioinformatics
University of Bielefeld, Germany
travis.gagie@gmail.com
[2] Department of Computer Science
University of Bonn, Germany
yasha@cs.uni-bonn.de

Abstract. A common complaint about adaptive prefix coding is that it is much slower than static prefix coding. Karpinski and Nekrich recently took an important step towards resolving this: they gave an adaptive Shannon coding algorithm that encodes each character in $O(1)$ amortized time and decodes it in $O(\log H + 1)$ amortized time, where H is the empirical entropy of the input string s. For comparison, Gagie's adaptive Shannon coder and both Knuth's and Vitter's adaptive Huffman coders all use $\Theta(H + 1)$ amortized time for each character. In this paper we give an adaptive Shannon coder that both encodes and decodes each character in $O(1)$ worst-case time. As with both previous adaptive Shannon coders, we store s in at most $(H + 1)|s| + o(|s|)$ bits. We also show that this encoding length is worst-case optimal up to the lower order term. In short, we present the first algorithm for adaptive prefix coding that encodes and decodes each character in optimal worst-case time while producing an encoding whose length is also worst-case optimal.

1 Introduction

Adaptive prefix coding is a well studied problem whose well known and widely used solution, adaptive Huffman coding, is nevertheless not worst-case optimal. Suppose we are given a string s of length m over an alphabet of size n. For static prefix coding, we are allowed to make two passes over s but, after the first pass, we must build a single prefix code, such as a Shannon code [19] or Huffman code [11], and use it to encode every character. Since a Huffman code minimizes the expected codeword length, static Huffman coding is optimal (ignoring the asymptotically negligible $O(n \log n)$ bits needed to write the code). For adaptive prefix coding, we are allowed only one pass over s and must encode each character with a prefix code before reading the next one, but we can change the code

---

* The first author was funded by the Italy-Israel FIRB Project "Pattern Discovery Algorithms in Discrete Structures, with Applications to Bioinformatics" while at the University of Eastern Piedmont and by the Sofja Kovalevskaja Award from the Alexander von Humboldt Foundation and the German Federal Ministry of Education and Research while at Bielefeld University.

F. Dehne et al. (Eds.): WADS 2009, LNCS 5664, pp. 315–326, 2009.

Table 1. Bounds for adaptive prefix coding: the times to encode and decode each character and the total length of the encoding. Bounds in the first row and last column are worst-case; the others are amortized.

	Encoding	Decoding	Length
this paper	$O(1)$	$O(1)$	$(H+1)m + o(m)$
Karpinski and Nekrich [12]	$O(1)$	$O(\log H + 1)$	$(H+1)m + o(m)$
Gagie [6]	$O(H+1)$	$O(H+1)$	$(H+1)m + o(m)$
Vitter [21]	$O(H+1)$	$O(H+1)$	$(H+1+h)m + o(m)$
Knuth [13] Gallager [9] Faller [3]	$O(H+1)$	$O(H+1)$	$(H+2+h)m + o(m)$

after each character. Assuming we compute each code deterministically from the prefix of s already encoded, we can later decode s symmetrically. Table 1 presents a summary of bounds for adaptive prefix coding. The most intuitive solution is to encode each character using a Huffman code for the prefix already encoded; Knuth [13] showed how to do this in time proportional to the length of the encoding produced, taking advantage of a property of Huffman codes discovered by Faller [3] and Gallager [9]. Shortly thereafter, Vitter [21] gave another adaptive Huffman coder that also uses time proportional to the encoding's length; he proved his coder stores s in fewer than m more bits than static Huffman coding, and that this is optimal for any adaptive Huffman coder. With a similar analysis, Milidiú, Laber and Pessoa [15] later proved Knuth's coder uses fewer than $2m$ more bits than static Huffman coding. In other words, Knuth's and Vitter's coders store s in at most $(H+2+h)m + o(m)$ and $(H+1+h)m + o(m)$ bits, respectively, where $H = \sum_a (\text{occ}(a,s)/m) \log(m/\text{occ}(a,s))$ is the empirical entropy of s (i.e., the entropy of the normalized distribution of characters in s), $\text{occ}(a,s)$ is the number of occurrences of the character a in s, and $h \in [0,1)$ is the redundancy of a Huffman code for s; therefore, both adaptive Huffman coders use $\Theta(H+1)$ amortized time to encode and decode each character of s. Turpin and Moffat [20] gave an adaptive prefix coder that uses canonical codes, and showed it achieves nearly the same compression as adaptive Huffman coding but runs much faster in practice. Their upper bound was still $O(H+1)$ amortized time for each character but their work raised the question of asymptotically faster adaptive prefix coding. In all of the above algorithms the encoding and decoding times are proportional to the *bit length* of the encoding. This implies that we need $O(H+1)$ time to encode/decode each symbol; since the entropy H depends on the size of the alphabet, the running times grow with the alphabet size.

The above results for adaptive prefix coding are in contrast to the algorithms for the prefix coding in the static scenario. The simplest static Huffman coders use $\Theta(H+1)$ amortized time to encode and decode each character but, with a lookup table storing the codewords, it is not hard to speed up encoding to take $O(1)$ worst-case time for each character. We can also decode an arbitrary prefix code in $O(1)$ time using a look-up table, but the space usage and initialization time for such a table can be prohibitively high, up to $O(m)$. Moffat and Turpin [16] described a practical algorithm for decoding prefix codes in $O(1)$ time; their algorithm works for a special class of prefix codes, the *canonical codes* introduced by Schwartz and Kallick [18].

While all adaptive coding methods described above maintain the optimal Huffman code, Gagie [6] described an adaptive prefix coder that is based on sub-optimal Shannon coding; his method also needs $O(H + 1)$ amortized time per character for both encoding and decoding. Although the algorithm of [6] maintains a Shannon code that is known to be worse than the Huffman code in the static scenario, it achieves $(H + 1)m + o(m)$ upper bound on the encoding length that is better than the best known upper bounds for adaptive Huffman algorithms. Karpinski and Nekrich [12] recently reduced the gap between static and adaptive prefix coding by using quantized canonical coding to speed up an adaptive Shannon coder of Gagie [6]: their coder uses $O(1)$ amortized time to encode each character and $O(\log H)$ amortized time to decode it; the encoding length is also at most $(H + 1)m + o(m)$ bits. Nekrich [17] implemented a version of their algorithm and showed it performs fairly well on the Calgary and Canterbury corpora.

In this paper we describe an algorithm that both encodes and decodes each character in $O(1)$ worst-case time, while still using at most $(H+1)m+o(m)$ bits. It can be shown that the encoding length of any adaptive prefix coding algorithm is at least $(H + 1)m - o(m)$ bits in the worst case. Thus our algorithm works in optimal worst-case time independently of the alphabet size and achieves optimal encoding length (up to the lower-order term). As is common, we assume $n \ll m$; in fact, our main result is valid if $n = o(m/\log^{5/2} m)$; for two results in section 6 we need a somewhat stronger assumption that $n = o(\sqrt{m}/\log m)$. For simplicity, we also assume s contains at least two distinct characters and m is given in advance; Karpinski and Nekrich [12, Section 6.1] described how to deal with the case when m is not given. Our model is a unit-cost word RAM with $\Omega(\log m)$-bit words on which it takes $O(1)$ time to input or output a word. Our encoding algorithm uses only addition and bit operations; our decoding algorithm also uses multiplication and finding the most significant bit in $O(1)$ time. We can also implement the decoding algorithm, so that it uses AC^0 operations only. Encoding needs $O(n)$ words of space, and decoding needs $O(n \log m)$ words of space. The decoding algorithm can be implemented with bit operations only at a cost of higher space usage and additional pre-processing time. For an arbitrary constant $\alpha > 0$, we can construct in $O(m^\alpha)$ time a look-up table that uses $O(m^\alpha)$ space; this look-up table enables us to implement multiplications with $O(1)$ table look-ups and bit operations.

While the algorithm of [12] uses quantized coding, i.e., coding based on the quantized symbol probabilities, our algorithm is based on delayed probabilities: encoding of a symbol $s[i]$ uses a Shannon code for the prefix $s[1..i - d]$ for an appropriately chosen parameter d; henceforth $s[i]$ denotes the i-th symbol in the string s and $s[i..j]$ denotes the substring of s that consists of symbols $s[i]s[i+1]\ldots s[j]$. In Section 2 we describe canonical Shannon codes and explain how they can be used to speed up Shannon coding. In Section 3 we describe two useful data structures that allow us to maintain the Shannon code efficiently. We present our algorithm and analyze the number of bits it needs to encode a string in Section 4. In Section 5 we prove a matching lower bound by extending Vitter's lower bound from adaptive Huffman coders to all adaptive prefix coders.

2   Canonical Shannon Coding

Shannon [19] defined the entropy $H(P)$ of a probability distribution $P = p_1, \ldots, p_n$ to be $\sum_{i=1}^{n} p_i \log_2(1/p_i)$.[1] He then proved that, if P is over an alphabet, then we can assign each character with probability $p_i > 0$ a prefix-free binary codeword of length $\lceil \log_2(1/p_i) \rceil$, so the expected codeword length is less than $H(P) + 1$; we cannot, however, assign them codewords with expected length less than $H(P)$.[2] Shannon's proof of his upper bound is simple: without loss of generality, assume $p_1 \geq \cdots \geq p_n > 0$; for $1 \leq i \leq n$, let $b_i = \sum_{j=1}^{i-1} p_j$; since $|b_i - b_{i'}| \geq p_i$ for $i' \neq i$, the first $\lceil \log(1/p_i) \rceil$ bits of b_i's binary representation uniquely identify it; let these bits be the codeword for the ith character. The codeword lengths do not change if, before applying Shannon's construction, we replace each p_i by $1/2^{\lceil \log(1/p_i) \rceil}$. The code then produced is canonical [18]: i.e., if a codeword is the cth of length r, then it is the first r bits of the binary representation of $\sum_{\ell=1}^{r-1} W(\ell)/2^\ell + (c-1)/2^r$, where $W(\ell)$ is the number of codewords of length ℓ. For example,

1) 000	7) 1000
2) 001	8) 1001
3) 0100	9) 10100
4) 0101	10) 10101
5) 0110	$\vdots$
6) 0111	16) 11011

are the codewords of a canonical code. Notice the codewords are always in lexicographic order.

Static prefix coding with a Shannon code stores s in $(H + 1)m + o(m)$ bits. An advantage to using a canonical Shannon code is that we can easily encode each character in $O(1)$ worst-case time (apart from first pass) and decode it symmetrically in $O(\log \log m)$ worst-case time (see [16]). To encode a symbol $s[i]$ from s, it suffices to know the pair $\langle r, c \rangle$ such that the codeword for $s[i]$ is the c-th codeword of length r, and the first codeword l_r of length r. Then the codeword for $s[i]$ can be computed in $O(1)$ time as $l_r + c$. We store the pair $\langle r, c \rangle$ for the k-th symbol in the alphabet in the k-th entry of the array C. The array $L[1..\lceil \log m \rceil]$ contains first codewords of length l for each $1 \leq l \leq \lceil \log m \rceil$. Thus, if we maintain arrays L and C we can encode a character from s in $O(1)$ time.

For decoding, we also need a data structure D of size $\log m$ and a matrix M of size $n \times \lceil \log m \rceil$. For each l such that there is at least one codeword of

---

[1] We assume $0 \log(1/0) = 0$.

[2] In fact, these bounds hold for any size of code alphabet; we assume throughout that codewords are binary, and by log we always mean $\log_2$.

length l, the data structure D contains the first codeword of length l padded with $\lceil \log m \rceil - l$ 0's. For an integer q, D can find the *predecessor* of q in D, $\mathrm{pred}(q, D) = \max\{x \in D | x \le q\}$. The entry $M[r, c]$ of the matrix M contains the symbol s, such that the codeword for s is the c-th codeword of length r. The decoding algorithm reads the next $\lceil \log m \rceil$ bits into a variable w and finds the predecessor of w in D. When $\mathrm{pred}(w, D)$ is known, we can determine the length r of the next codeword, and compute its index c as $(w - L[r]) \gg (\lceil \log m \rceil - r)$ where $\gg$ denotes the right bit shift operation.

The straightforward binary tree solution allows us to find predecessors in $O(\log \log m)$ time. We will see in the next section that predecessor queries on a set of $\log m$ elements can be answered in $O(1)$ time. Hence, both encoding and decoding can be performed in $O(1)$ time in the static scenario. In the adaptive scenario, we must find a way to maintain the arrays C and L efficiently and, in the case of decoding, the data structure D.

3 Data Structures

It is not hard to speed up the method for decoding we described in Section 2. For example, if the augmented binary search tree we use as D is optimal instead of balanced then, by Jensen's Inequality, we decode each character in $O(\log H)$ amortized time. Even better, if we use a data structure by Fredman and Willard [4], then we can decode each character in $O(1)$ worst-case time; due to space constraints, we postpone the proof to the full version of this paper.

Lemma 1 (Fredman and Willard, 1993). *Given $O(\log^{1/6} m)$ keys, in $O(\log^{2/3} m)$ worst-case time we can build a data structure that stores those keys and supports predecessor queries in $O(1)$ worst-case time.*

Corollary 1. *Given $O(\log m)$ keys, in $O(\log^{3/2} m)$ worst-case time we can build a data structure that stores those keys and supports predecessor queries in $O(1)$ worst-case time.*

In Lemma 1 and Corollary 1 we assume that multiplication and finding the most significant bit of an integer can be performed in constant time. As shown in [1], we can implement the data structure of Lemma 1 using AC^0 operations only. We can restrict the set of elementary operations to bit operations and table look-ups by increasing the space usage and preprocessing time to $O(m^\varepsilon)$. In our case all keys in the data structure D are bounded by m; hence, we can construct in $O(m^\varepsilon)$ time a look-up table that uses $O(m^\varepsilon)$ space and allows us to multiply two integers or find the most significant bit of an integer, in constant time.

Corollary 1 is useful to us because the data structure it describes not only supports predecessor queries in $O(1)$ worst-case time but can also be built in time polylogarithmic in m; the latter property will let our adaptive Shannon coder keep its data structures nearly current by regularly rebuilding them. The array C and matrix M cannot be built in $o(n)$ time, however, so we combine them in a data structure that can be updated incrementally. Arrays $C[]$ and $L[]$, and the matrix M defined in the previous section, can be rebuilt as described in the next lemma, which will be proved in the full version of this paper.

Lemma 2. *If codeword lengths of $f \geq \log m$ symbols are changed, we can rebuild arrays $C[]$ and $L[]$, and update the matrix M in $O(f)$ time.*

4   Algorithm

The main idea of the algorithm of [12], that achieves $O(1)$ amortized encoding cost per symbol, is *quantization* of probabilities. The Shannon code is maintained for the probabilities $\tilde{p}_j = \frac{\lceil i/q \rceil}{\lceil occ(a_i, s[1..i])/q \rceil}$ where $occ(a_j, s[1..i])$ denotes the number of occurrences of the symbol a_j in the string $s[1..i]$ and the parameter $q = \Theta(\log m)$. The symbol a_i must occur q times before the denominator of the fraction $\tilde{p}_i$ is incremented by 1. Roughly speaking, the value of $\tilde{p}_i$, and hence the codeword length of a_i, changes at most once after $\log m$ occurrences of a_i. As shown in Lemma 2, we can rebuild the arrays $C[]$ and $L[]$ in $O(\log m)$ time. Therefore encoding can be implemented in $O(1)$ amortized time. However, it is not clear how to use this approach to obtain constant worst-case time per symbol.

In this paper a different approach is used. Symbols $s[i+1], s[i+2], \ldots, s[i+d]$ are encoded with a Shannon code for the prefix $s[1]s[2] \ldots s[i-d]$ of the input string. Recall that in a traditional adaptive code the symbol $s[i+1]$ is encoded with a code for $s[1] \ldots s[i]$. While symbols $s[i+1] \ldots s[i+d]$ are encoded, we build an optimal code for $s[1] \ldots s[i]$. The next group of symbols, i.e., $s[i+d+1] \ldots s[i+2d]$ will be encoded with a Shannon code for $s[1] \ldots s[i]$, and the code for $s[1] \ldots s[i+d]$ will be simultaneously rebuilt in the background. Thus every symbol $s[j]$ is encoded with a Shannon code for the prefix $s[1] \ldots s[j-t]$, $d \leq t < 2d$, of the input string. That is, when a symbol $s[i]$ is encoded, its codeword length equals

$$\left\lceil \log \frac{i + n - t}{\max\left(occ(s[i], s[1..i-t]), 1\right)} \right\rceil .$$

We increased the enumerator of the fraction by n and the denominator is always at least 1 because we assume that every character is assigned a codeword of length $\lceil \log n - d \rceil$ before encoding starts. We make this assumption only to simplify the description of our algorithm. There are other methods of dealing with characters that occur for the first time in the input string that are more practically efficient, see e.g. [13]. The method of [13] can also be used in our algorithm, but it would not change the total encoding length.

Later we will show that the delay of at most $2d$ increases the length of encoding only by a lower order term. Now we turn to the description of the procedure that updates the code, i.e., we will show how the code for $s[1] \ldots s[i]$ can be obtained from the code for $s[1] \ldots s[i-d]$.

Let $\mathcal{C}$ be a code for $s[1] \ldots s[i-d]$ and $\mathcal{C}'$ be a code for $s[1] \ldots s[i]$. As shown in section 2, updating the code is equivalent to updating the arrays $C[]$ and $L[]$, the matrix M, and the data structure D. Since a group of d symbols contains at most d different symbols, we must change codeword lengths of at most d codewords. The list of symbols $a_1, \ldots, a_k$ such that the codeword length of a_k must be changed can be constructed in $O(d)$ time. We can construct an array $L[]$ for the code C' in $O(\max(d, \log m))$ time by Lemma 2. The matrix M and the array $C[]$ can be updated in $O(d)$ time because only $O(d)$ cells of M are modified. However, we cannot build new versions of M and $C[]$ because they contain $\Theta(n \log m)$ and $\Theta(n)$ cells respectively. Since we must obtain the new version of M while the old version is still used, we modify M so that each cell of M is allowed to contain two different values, an old one and a new one. For each cell (r, c) of M we store two values $M[r, c].old$ and $M[r, c].new$ and the separating value $M[r, c].b$: when the symbol $s[t]$, $t < M[r, c].b$, is decoded, we use $M[r, c].old$; when the symbol $s[t]$, $t \geq M[r, c].b$, is decoded, we use $M[r, c].new$. The procedure for updating M works as follows: we visit all cells of M that were modified when the code C was constructed. For every such cell we set $M[r, c].old = M[r, c].new$ and $M[r, c].b = +\infty$. Then, we add the new values for those cells of M that must be modified. For every cell that must be modified, the new value is stored in $M[r, c].new$ and $M[r, c].b$ is set to $i + d$. The array $C[]$ can be updated in the same way. When the array $L[]$ is constructed, we can construct the data structure D in $O(\log^{3/2} m)$ time.

The algorithm described above updates the code if the codeword lengths of some of the symbols $s[i - d + 1] \ldots s[i]$ are changed. But if some symbol a does not occur in the substring $s[i - d + 1] \ldots s[i]$, its codeword length might still change in the case when $\log(i) > \log(i_a)$ where $i_a = \max\{j < i | s[j] = a\}$. We can, however, maintain the following invariant on the codeword length l_a:

$$\left\lceil \log \frac{i + n}{\max(occ(a, s[1..i - 2d]), 1)} \right\rceil \leq l_a \leq \left\lceil \log \frac{i + 2n}{\max(occ(a, s[1..i - 2d]), 1)} \right\rceil .$$
(1)

When the codeword length of a symbol a must be modified, we set its length to $\lceil \log \frac{i+2n}{\max(occ(a, s[1..i]), 1)} \rceil$. All symbols a are also stored in the queue Q. When the code C' is constructed, we extract the first d symbols from Q, check whether their codeword lengths must be changed, and append those symbols at the end of Q. Thus the codeword length of each symbol is checked at least once when an arbitrary substring $s[u] \ldots s[u + n]$ of the input string s is processed. Clearly, the invariant 1 is maintained.

Thus the procedure for obtaining the code C' from the code C consists of the following steps:

1. check symbols $s[i - d + 1] \ldots s[i]$ and the first d symbols in the queue Q; construct a list of codewords whose lengths must be changed; remove the first d symbols from Q and append them at the end of Q
2. traverse the list $\mathcal{N}$ of modified cells in the matrix M and the array $C[]$, and remove the old values from those cells; empty the list $\mathcal{N}$
3. update the array $C[]$ for code C'; simultaneously, update the matrix M and construct the list $\mathcal{N}$ of modified cells in M and $C[]$
4. construct the array L and the data structure D for the new code C'

Each of the steps described above, except the last one, can be performed in $O(d)$ time; the last step can be executed in $O(\max(d, \log^{3/2} m))$ time. For $d = \lfloor \log^{3/2} m \rfloor / 2$, code $\mathcal{C}$ can be constructed in $O(d)$ time. If the cost of constructing $\mathcal{C}'$ is evenly distributed among symbols $s[i-d], \ldots, s[i]$, then we spend $O(1)$ extra time when each symbol $s[j]$ is processed. Since $\mathrm{occ}(a, s[1..i - \lfloor \log^{3/2} m \rfloor]) \geq \max(\mathrm{occ}(a, s[1..i]) - \lfloor \log^{3/2} m \rfloor, 1)$, we need at most

$$\left\lceil \log \frac{i + 2n}{\max\left(\mathrm{occ}(s[i], s[1..i]) - \lfloor \log^{3/2} m \rfloor, 1\right)} \right\rceil$$

bits to encode $s[i]$.

Lemma 3. *We can keep an adaptive Shannon code such that, for $1 \leq i \leq m$, the codeword for $s[i]$ has length at most*

$$\left\lceil \log \frac{i + 2n}{\max\left(\mathrm{occ}(s[i], s[1..i]) - \lfloor \log^{3/2} m \rfloor, 1\right)} \right\rceil$$

and we use $O(1)$ worst-case time to encode and decode each character.

We can combine several inequalities by Gagie [6] and Karpinski and Nekrich [12] into the following lemma, which will be proved in the full version of this paper:

Lemma 4

$$\sum_{i=1}^{m} \left\lceil \log \frac{i + 2n}{\max\left(\mathrm{occ}(s[i], s[1..i]) - \lfloor \log^{3/2} m \rfloor, 1\right)} \right\rceil \leq (H+1)m + O(n \log^{5/2} m).$$

Together with Lemma 3, this immediately yields our result.

Theorem 1. *We can encode s in at most $(H+1)m + o(m)$ bits with an adaptive prefix coding algorithm that encodes and decodes each character in $O(1)$ worst-case time.*

5   Lower Bound

It is not difficult to show that any prefix coder uses at least $(H+1)m - o(m)$ bits in the worst case (e.g., when s consists of $m-1$ copies of one character and 1 copy of another, so $Hm < \log m + \log e$). However, this does not rule out the possibility of an algorithm that always uses, say, at most $m/2$ more bits than static Huffman coding. Vitter [21] proved such a bound is unachievable with an adaptive Huffman coder, and we now extend his result to all adaptive prefix coders. This implies that for an adaptive prefix coder to have a stronger worst-case upper bound than ours (except for lower-order terms), that bound can be in terms of neither the empirical entropy nor the number of bits used by static Huffman coding.[3]

Theorem 2. *Any adaptive prefix coder stores s in at least $m - o(m)$ more bits in the worst case than static Huffman coding.*

Proof. Suppose $n = m^{1/2} = 2^\ell + 1$ and the first n characters of s are an enumeration of the alphabet. For $n < i \leq m$, when the adaptive prefix coder reaches $s[i]$, there are at least two characters assigned codewords of length at least $\ell + 1$. To see why, consider the prefix code used to encode $s[i]$ as a binary tree — a code-tree — whose leaves are labelled with the characters in the alphabet and whose branches correspond to the codewords; since any binary tree on $2^\ell + 1$ leaves has at least two leaves at depth at least $\ell + 1$, there must be two codewords of length at least $\ell + 1$. It follows that, in the worst case, the coder uses at least $(\ell + 1)m - o(m)$ bits to store s. On the other hand, a static prefix coder can assign codewords of length ℓ to the $n - 2$ most frequent characters and codewords of length $\ell + 1$ to the two least frequent ones, and thus use at most $\ell m + o(m)$ bits. Therefore, since a Huffman code minimizes the expected codeword length, any adaptive prefix coder uses at least $m - o(m)$ more bits in the worst case than static Huffman coding.   □

6   Other Coding Problems

Several variants of the prefix coding problem have been considered and extensively studied. In the alphabetic coding problem [10], codewords must be sorted lexicographically, i.e., $i < j \Rightarrow c(a_i) < c(a_j)$, where $c(a_k)$ denotes the codeword of a_k. In the length-limited coding problem, the maximum codeword length is limited by a parameter $F > \log n$. In the coding with unequal letter costs problem, one symbol in the code alphabet costs more than another and we want to minimize the average cost of a codeword. All of the above problems were studied in the static scenario. Adaptive prefix coding algorithms for those problems were considered in [5]. In this section we show that the good upper bounds on the length of the encoding can be achieved by algorithms that encode in $O(1)$ worst-case time. The main idea of our improvements is that we encode a symbol $s[i]$ in the input string with a code that was constructed for the prefix $s[1..i - d]$ of the input string, where the parameter d is chosen in such a way that a corresponding (almost) optimal code can be constructed in $O(d)$ time. Using the same arguments as in the proof of Lemma 4 we can show that encoding with delays increases the length of encoding by an additive term of $O(d \cdot n \log m)$ (the analysis is more complicated in the case of coding with unequal letter costs). We will provide proofs in the full version of this paper.

---

[3] Notice we do not exclude the possibility of natural probabilistic settings in which our algorithm is suboptimal — e.g., if s is drawn from a memoryless source for which a Huffman code has smaller redundancy than a Shannon code, then adaptive Huffman coding almost certainly achieves better asymptotic compression than adaptive Shannon coding — but in this paper we are interested only in worst-case bounds.

Alphabetic Coding. Gagie [6] noted that, by replacing Shannon's construction by a modified construction due to Gilbert and Moore [10], his coder can be made alphabetic — i.e., so that the lexicographic order of the possible encodings is the same as that of the inputs. This modification increases the worst-case length of the encoding to $(H + 2)m + o(m)$ bits, and increases the time to encode and decode each character to $O(\log n)$. We can use our results to speed up Gagie's adaptive alphabetic prefix coder when $n = o(\sqrt{m}/\log m)$.

Whereas Shannon's construction assigns a prefix-free binary codeword of length $\lceil \log(1/p_i) \rceil$ to each character with probability $p_i > 0$, Gilbert and Moore's construction assigns a codeword of length $\lceil \log(1/p_i) \rceil + 1$. Building a code according to Gilbert and Moore's algorithm takes $O(n)$ time because, unlike Shannon's construction, we do not need to sort the characters by probability. Hence, although we don't know how to update the alphabetic code efficiently, we can construct it from scratch in $O(n)$ time. We can apply the same approach as in previous sections, and use rebuilding with delays: while we use the alphabetic code for $s[1..i - n/2]$ to encode symbols $s[i], s[i + 1], \ldots, s[i + n/2]$, we construct the code for $s[1..i]$ in the background. If we use $O(1)$ time per symbol to construct the next code, it will be completed when $s[i + n/2]$ is encoded. Hence, we encode each character using at most $\left\lceil \log \frac{i+n}{\max(\mathrm{occ}(s[i], s[1..i]) - n, 1)} \right\rceil + 1$ bits. Unfortunately we cannot use the encoding and decoding methods of Section 2 because the alphabetic coding is not canonical. When an alphabetic code for the following group of $O(n)$ symbols is constructed, we also create in $O(n)$ time a table that stores the codeword of each symbol a_i. Such a table can be created from the alphabetic code in $O(n)$ time; hence, the complexity of the encoding algorithm is not increased. We can decode the next codeword by searching in the data structure that contains all codewords. Using a data structure due to Andersson and Thorup [2] we can decode in $O(\min(\sqrt{\log n}, \log \log m)$ time per symbol. A detailed description of the algorithm will be given in the full version of this paper.

Theorem 3. *There is an algorithm for adaptive alphabetic prefix coding that encodes and decodes each symbol of a string s in $O(1)$ and $O(\min(\sqrt{\log n}, \log \log m))$ time respectively. If $n = o(\sqrt{m}/\log m)$, the encoding length is $((H + 2)m + o(m)$.*

Coding with unequal letter costs. Krause [14] showed how to modify Shannon's construction for the case in which code letters have different costs, e.g., the different durations of dots and dashes in Morse code. Consider a binary channel and suppose $\mathrm{cost}(0)$ and $\mathrm{cost}(1)$ are constants with $0 < \mathrm{cost}(0) \leq \mathrm{cost}(1)$. Krause's construction gives a code such that, if a symbol has probability p, then its codeword has cost less than $\ln(p)/C + \mathrm{cost}(1)$, where the channel capacity C is the largest real root of $e^{-\mathrm{cost}(0) \cdot x} + e^{-\mathrm{cost}(1) \cdot x} = 1$ and e is the base of the natural logarithm. It follows that the expected codeword cost in the resulting code is $H \ln 2/C + \mathrm{cost}(1)$, compared to Shannon's bound of $H \ln 2/C$. Based on Krause's construction, Gagie gave an algorithm that produces an encoding of s with total cost at most $\left(\frac{H \ln 2}{C} + \mathrm{cost}(1) \right) m + o(m)$ in $O(m \log n)$ time. Since the code of Krause [14] can be constructed in $O(n)$ time, we can use the encoding with delay n and achieve $O(1)$ worst-case time. Since the costs are constant and the minimum probability is $\Omega(1/m)$, the maximum codeword length is $O(\log m)$. Therefore, we can decode using the data structure described above.

Theorem 4. *There is an algorithm for adaptive prefix coding with unequal letter costs that encodes and decodes each symbol of a string s in $O(1)$ and $O(\min(\sqrt{\log n}, \log \log m))$ time respectively. If $n = o(\sqrt{m}/\log m)$, the encoding length is $\left(\frac{H \ln 2}{C} + \text{cost}(1)\right) m + o(m)$.*

Length-limited coding. Finally, we can design an algorithm for adaptive length-limited prefix coding by modifying the algorithm of section 4. Using the same method as in [5] — i.e., smoothing the distribution by replacing each probability with a weighted average of itself and $1/n$ — we set the codeword length of symbol $s[i]$ to $\lceil \log \frac{2^f}{(2^f-1)x+1/n} \rceil$ instead of $\lceil \log \frac{1}{x} \rceil$, where $x = \frac{\max\left(\text{occ}(s[i],s[1..i])-\lfloor \log^{3/2} m \rfloor, 1\right)}{i+2n}$ and $f = F - \log n$. We observe that the codeword lengths l_i of this modified code satisfy the Kraft-McMillan inequality:

$$\sum_i 2^{-l_i} \leq \sum_x \frac{(2^f-1)x+1/n}{2^f} = \sum_x \frac{2^f x}{2^f} - \sum_x \frac{x}{2^f} + n\frac{1/n}{2^f} < 1 .$$

Therefore we can construct and maintain a canonical prefix code with codeword lengths l_i. Since $\frac{2^f}{(2^f-1)x+1/n} \leq \min(\frac{2^f}{(2^f-1)x}, \frac{2^f}{1/n})$, $\lceil \log \frac{2^f}{(2^f-1)x+1/n} \rceil \leq \min(\lceil \log \frac{2^f}{(2^f-1)x} \rceil, \log n + f)$. Thus the codeword length is always smaller than F. We can estimate the encoding length by bounding the first part of the above expression: $\lceil \log \frac{2^f}{(2^f-1)x} \rceil < x+1+\log \frac{2^f+1}{2^f}$ and $\log \frac{2^f+1}{2^f} = (1/2^f) \log(1+\frac{1}{2^f})^{2^f} \leq \frac{1}{2^f \ln 2}$. Summing up by all symbols $s[i]$, the total encoding length does not exceed

$$\sum_{i=1}^{m} \log \frac{i + 2n}{\max\left(\text{occ}(s[i], s[1..i]) - \lfloor \log^{3/2} m \rfloor, 1\right)} + m + \frac{m}{2^f \ln 2} .$$

We can estimate the first term in the same way as in Lemma 4; hence, the length of the encoding is $(H+1+\frac{1}{2^f \ln 2})m+O(n \log^{5/2} m)$. We thus obtain the following theorem:

Theorem 5. *There is an algorithm for adaptive length-limited prefix coding that encodes and decodes each symbol of a string s in $O(1)$ time. The encoding length is $(H+1+\frac{1}{2^f \ln 2})m+O(n \log^{5/2} m)$, where $f = F - \log n$ and F is the maximum codeword length.*

7   Open Problems

In a recent paper [7] we gave an adaptive prefix coding algorithm that lets us trade off the quality of the compression against the amount of memory used, while using $O(\log \log n)$ worst-case time to encode and decode each character; combining that algorithm with the one in this paper, to improve the tradeoff and reduce the time bound to $O(1)$ is an interesting open question. Our algorithms for alphabetic and unequal letter cost coding assume that $n = o(\sqrt{m}/\log m)$. Design of efficient adaptive codes for the case $n = \Omega(\sqrt{m}/\log m)$ remains an open problem. In [8] we showed how a similar approach can be used to obtain tight bounds for the online stable sorting problem in the case when $n = o(\sqrt{m}/\log m)$. Thus this question is also relevant for the online stable sorting problem.

References

1. Andersson, A., Bro Miltersen, P., Thorup, M.: Fusion trees can be implemented with AC^0 instructions only. Theoretical Computer Science 215(1–2), 337–344 (1999)
2. Andersson, A., Thorup, M.: Dynamic ordered sets with exponential search trees. Journal of the ACM 54(3) (2007)
3. Faller, N.: An adaptive system for data compression. In: Record of the 7th Asilomar Conference on Circuits, Systems and Computers, pp. 593–597 (1973)
4. Fredman, M.L., Willard, D.E.: Surpassing the information theoretic bound with fusion trees. Journal of Computer and System Sciences 47(3), 424–436 (1993)
5. Gagie, T.: Dynamic Shannon coding. In: Proceedings of the 12th European Symposium on Algorithms, pp. 359–370 (2004)
6. Gagie, T.: Dynamic Shannon coding. Information Processing Letters 102(2–3), 113–117 (2007)
7. Gagie, T., Karpinski, M., Nekrich, Y.: Low-memory adaptive prefix coding. In: Proceedings of the Data Compression Conference, pp. 13–22 (2009)
8. Gagie, T., Nekrich, Y.: Tight bounds for online stable sorting and comparison-based guessing games (submitted)
9. Gallager, R.G.: Variations on a theme by Huffman. IEEE Transactions on Information Theory 24(6), 668–674 (1978)
10. Gilbert, E.N., Moore, E.F.: Variable-length binary encodings. Bell System Technical Journal 38, 933–967 (1959)
11. Huffman, D.A.: A method for the construction of minimum-redundancy codes. Proceedings of the IRE 40(9), 1098–1101 (1952)
12. Karpinski, M., Nekrich, Y.: A fast algorithm for adaptive prefix coding. Algorithmica (to appear)
13. Knuth, D.E.: Dynamic Huffman coding. Journal of Algorithms 6(2), 163–180 (1985)
14. Krause, R.M.: Channels which transmit letters of unequal durations. Information and Control 5(1), 13–24 (1962)
15. Milidiú, R.L., Laber, E.S., Pessoa, A.A.: Bounding the compression loss of the FGK algorithm. Journal of Algorithms 32(2), 195–211 (1999)
16. Moffat, A., Turpin, A.: On the implementation of minimum redundancy prefix codes. IEEE Transactions on Communications 45(10), 1200–1207 (1997)
17. Nekrich, Y.: An efficient implementation of adaptive prefix coding. In: Proceedings of the Data Compression Conference, p. 396 (2007)
18. Schwartz, E.S., Kallick, B.: Generating a canonical prefix encoding. Communications of the ACM 7(3), 166–169 (1964)
19. Shannon, C.E.: A mathematical theory of communication. Bell System Technical Journal 27, 379–423, 623–656 (1948)
20. Turpin, A., Moffat, A.: On-line adaptive canonical prefix coding with bounded compression loss. IEEE Transactions on Information Theory 47(1), 88–98 (2001)
21. Vitter, J.S.: Design and analysis of dynamic Huffman codes. Journal of the ACM 1987(4), 825–845 (1987)

New Results on Visibility in Simple Polygons

Alexander Gilbers and Rolf Klein

Institute of Computer Science
University of Bonn
53117 Bonn, Germany
{gilbers,rolf.klein}@cs.uni-bonn.de

Abstract. We show that (A), 14 points on the boundary of a Jordan curve, and (B), 16 points in convex position encircled by a Jordan curve, cannot be shattered by interior visibility domains. This means that there always exists a subset of the given points, for which no point of the curve's interior domain sees all points of the subset and no point of its complement. As a consequence, we obtain a new result on guarding art galleries. If each point of the art gallery sees at least an r-th part of the gallery's boundary, then the art gallery can be covered by $13 \cdot C \cdot r \log r$ guards placed on the boundary. Here, C is the constant from the ϵ-net theorem.

Keywords: Computational geometry, art galleries, guards, VC-dimension, visibility.

1 Introduction

Visibility of objects and, in particular, the guarding of art galleries, belong to the most famous topics in computational geometry. A wealth of results can be found, e. g., in O'Rourke [6] and Urrutia [7].

A classic result states that a simple polygon on n vertices can be guarded by at most $\lfloor \frac{n}{3} \rfloor$ stationary guards with 360° view. The well-known example depicted in Figure 1 shows that this number of guards may also be necessary.

This art gallery has the significant property that it contains many points from which only a small portion of the total area is visible. It is quite natural to ask if galleries that do not have this property can be guarded by fewer guards. More precisely, let us assume that R is a simple closed Jordan curve in the plane[1] and that, for each point p in R the condition

$$\text{area}(\text{vis}(p)) \geq \frac{1}{r} \cdot \text{area}(R)$$

holds, for some parameter $r > 1$. Here

$$\text{vis}(p) = \{\ x \in R \mid \overline{xp} \subset R\ \}$$

denotes the visibility domain of p in R.

---

[1] For simplicity, we use R to also denote the interior domain plus the curve itself.

F. Dehne et al. (Eds.): WADS 2009, LNCS 5664, pp. 327–338, 2009.

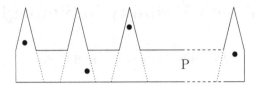

Fig. 1. An art gallery over $3m + 2$ vertices that requires m guards

One might be tempted to think that, if each guard see an r-th part of R, then r guards should be sufficient to cover all of R. Although this argument ignores that visibility regions can overlap, so that their areas cannot simply be added up, it is, surprisingly, not too far off. Indeed, Kalai and Matoušek [2] have shown that the Vapnik-Chervonenkis dimension

$$\dim_{\mathrm{VC}}(R, F)$$

of a simple closed curve R and of the set F of all visibility domains $\mathrm{vis}(w)$, where $w \in R$, is bounded from above by some constant[2].

By definition,

$$d_{\max} := \max_R \dim_{\mathrm{VC}}(R, F)$$

is the smallest number for which the following holds. If one draws an arbitrary simple closed curve R and places an arbitrary set S of $d_{\max} + 1$ points inside R, then S cannot be "shattered" by F, in the following sense. There always exists a subset T of S, such that no point w in R can see each point of T and no point of $S \setminus T$. In other words, subset T cannot be visually distinguished.

By the theory of ϵ-nets, there exists an $1/r$-net N for (R, F) of size

$$C \cdot d_{\max} \cdot r \log r \qquad (1)$$

This means that each visibility domain $\mathrm{vis}(w)$ of area at least $1/r$ times the area of R contains a point p of N. If we assume that *each* visibility domain $\mathrm{vis}(w)$ fulfills this size condition, we can conclude that each $w \in R$ is visible from some point p of N, that is, the $C \cdot d_{\max} \cdot r \log r$ many points of N are a system of guards for R. One should observe that only parameter r depends on the actual curve R, while C and $d_{\max}$ are independent constants. (Thus, the false argument from above is wrong by only a $\log r$ factor.) Since ϵ-nets can be found by random sampling, such results have strong algorithmic consequences.

Complete proofs of these facts can be found in Matoušek [5], Chapter 10. While they result in an upper bound of 13.08 for C, the resulting upper bound for $d_{\max}$ is $\approx 10^{12}$, due to costly dualization. A better bound on $d_{\max}$ was presented by Valtr [8]. He shows that

$$6 \leq d_{\max} \leq 23$$

---

[2] In general, for a set F of subsets of some set X, $\dim_{\mathrm{VC}}(X, F)$ is defined as the maximum size of a subset $Y \subseteq X$ such that each $Y' \subseteq Y$ can be obtained by intersecting Y with a set in F.

holds, leaving a challenging gap that has been open now for a decade. While one strongly feels that the true value of $d_{\max}$ is closer to 6 than to 23, it seems not easy to improve on Valtr's arguments. Only for the case of external visibility have some new results been obtained. Isler, Kannan, Daniilidis, and Valtr [1] proved, among other results, that no set of 6 boundary points of R can be shattered by visibility domains whose centers are situated outside the convex hull of R. They gave an example of a set of 5 boundary points that can be shattered in this way. King [3] considered 1.5-dimensional terrains and showed that here at most 4 points can be shattered by visibility.

With this paper we start a new attack on the original, apparently more difficult problem of internal visibility. We prove the following results.

(A) No set of 14 boundary points of R can be shattered, and
(B) no set of 16 arbitrary points of R in convex position can be shattered

by visibility domains $\mathrm{vis}(w)$, where $w \in R$. Result (A) gives also rise to a new bound on guard numbers.

(C) If each point $w \in R$ sees at least an r-th part of the boundary of R, then all of R can be guarded by $C \cdot 13 \cdot r \log r$ guards placed on the boundary of R.

Here, C is the constant for the size of ϵ-nets from (1) above. Result (C) is similar in spirit to the following theorem of Kirkpatrick's [4]. If each vertex of a simple polygon sees at least an r-th part of its boundary, as few as $64r \log \log r$ vertex guards are sufficient to cover the *boundary* of R. One should, however, observe that guards covering the boundary are not necessarily able to see each interior point.

In our proof of (A) we are using new techniques that are different from the cell decompositions employed by Matoušek [5] and Valtr [8]. We start with proving a simple, yet powerful combinatorial fact that may be interesting in its own right; see Lemma 1. This fact is applied to separate, by lines, points from potential viewpoints; see Section 2 for details. Then we show that, for certain points p, those viewpoints w' that can see p form a contiguous radial subsequence of the set of all viewpoints w. This allows us to apply a bound on the dual VC-dimension of wedges by Isler et al. [1].

2   Preliminaries

Let R be the closed domain encircled by a Jordan curve, and let F denote the set of all visibility domains $\mathrm{vis}(w)$, where $w \in R$.

A finite point set $S \subset R$ is said to be "shattered" by F if, for each subset T of S, there exists a viewpoint w in R such that $T = \mathrm{vis}(w) \cap S$. In this case we say that viewpoint w *sees* subset T (and no other point of S); cf. Figure 2 for an example. We are interested in establishing that point sets S of at least a certain size cannot be shattered.

The following fact will be useful in Section 3. It holds for arbitrary sets E and their power sets, $P(E)$. We are not aware that this fact has been established before.

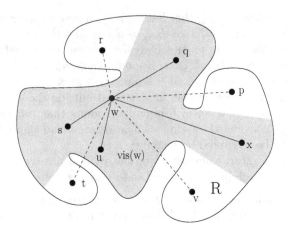

Fig. 2. Viewpoint w sees $\{q, s, u, x\}$. Solid segments indicate visibility, while dashed segments connect points that are not mutually visible.

Lemma 1. *Let $E = E' \uplus E''$ and $P(E) = P' \uplus P''$ be disjoint decompositions of a set E and of its power set, $P(E)$, respectively. Then P' shatters E', or P'' shatters E''.*

Proof. Assume that P' does not shatter E'. Then there exists a subset $S_0' \subset E'$ such that for all $A' \in P'$ the inequality $S_0' \neq A' \cap E'$ holds. Since each set $A' := S_0' \uplus S''$, for arbitrary $S'' \in P''$, does fulfill $S_0' = A' \cap E'$, such A' cannot be in P'. We conclude

$$\text{for each } S'' \in P'' : \quad S_0' \uplus S'' \in P''. \tag{2}$$

Moreover, let us assume that P'' does not shatter E'' either. By a symmetric argument, we obtain that there exists a subset $S_0'' \subset E''$ such that

$$\text{for each } S' \in P' : \quad S' \uplus S_0'' \in P'. \tag{3}$$

Now we specialize $S'' := S_0''$ in (2) and $S' := S_0'$ in (3), and obtain the contradiction

$$S_0' \uplus S_0'' \in P' \cap P'' = \emptyset.$$

Now we explain the main idea of the proof of Result (A), which states that 14 boundary points cannot be shattered. We shall illustrate how Lemma 1 will be put to work in this proof.

Kalai and Matoušek [2] obtained a finite upper bound on the VC-dimension of interior visibility domains by proving that, for shattered point sets beyond a certain size, the impossible configuration depicted in Figure 3 (i) would occur[3].

---

[3] Again, solid segments indicate visibility, while dashed segments are crossed by the boundary of P, so that their endpoints cannot see each other.

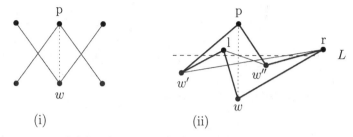

$$\text{(i)} \qquad\qquad\qquad\qquad\qquad \text{(ii)}$$

Fig. 3. In (i), segment $\overline{pw}$ cannot be accessed by the boundary of R which, by assumption, encircles these six points. Hence, this configuration is impossible. For the same reason, the dotted segment in (ii) connecting p to w must be a solid visibility segment.

We shall use similar configurations, but in a different way. Let us consider the set V of all viewpoints from which two points, l and r, are visible, like w', w, w'' in Figure 3 (ii). If w' and w'' can also see point p, then w must be able to see p, too, as this figure illustrates. Consequently, those viewpoints of V that are able to see p form a contiguous subsequence, in radial order around p. Our proof of Result (A) draws on this "wedge" property, that will be formally defined in Definition 1 below.

But how can we guarantee this property? To ensure that segment $\overline{pw}$ is indeed encircled by visibility segments, as shown in Figure 3 (ii), two conditions should be fulfilled. First, to all viewpoints should l, p, r appear in the same clockwise order. This condition is met because l, p, r are situated on the boundary of R. Second, l, p, r and the view points should see each other (in the absence of R) at angles $< 180°$. This condition is ensured by proving, with the help of Lemma 1, that a line L exists that separates points from view points.

Now we formally define the wedge property.

Definition 1. *Let p be a point in S, G a subset of $S \setminus \{p\}$, and V a set of view points all of which can see at least the points of G. Moreover, let L denote a line. We say that (p, G, V, L) has the* wedge property *iff the following conditions hold for V and for the subset*

$$V_p := \{w \in V \mid p \text{ is visible from } w\}$$

1. *Point p and view point set V are separated by line L.*
2. *For all $w', w'' \in V_p$ and for all $w \in V$: if w lies inside the wedge with apex p defined by w', w'' then $w \in V_p$ holds.*

In Figure 3 (ii) we have $G = \{l, r\}$ and $V = \{w', w, w''\}$. In general, the sets G, V and the line L may depend on p. If we are in a situation where (p, G, V, L) has the wedge property, we call p a *wedge point* and the points in G *wing points*. As the name suggests, the wedge property allows us to define a geometric wedge in the following way. Let w_l, w_r be those view points of V_p that maximize $\angle(w_l, p, w_r)$; this angle is less than $180°$ because p and w_l, w_r are separated

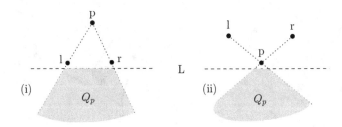

Fig. 4. Two configurations addressed in Lemma 2

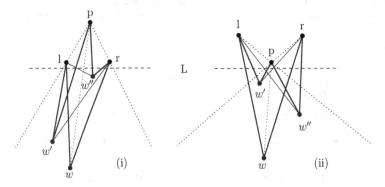

Fig. 5. In either case, segment $\overline{pw}$ is encirled by visibility edges. Thus, w sees p.

by a line. Now, let W_p denote the wedge with apex p whose angles are defined by w_l and w_R. Then the wedge property ensures that W_p contains exactly those points w of V that are able to see p. That is, $V_p = V \cap W_p$ holds.

Now we show that two configurations, which will be used repeatedly in our proofs, enjoy the wedge property.

Lemma 2. *Let l, p, r, L, and Q_p be situated as shown in Figure 4 (i) or (ii). Let $G := \{l, r\}$, let V be a set of view points below L each of which sees G, and assume that $V_p \subset Q_p$ holds. Then (p, G, V, L) has the wedge property.*

Proof. Suppose that $w', w, w'' \subset Q_p$ appear in counterclockwise order around p, as shown in Figure 5. Moreover, suppose that $w', w'' \in V_p$ holds, that is, w' and w'' can see l, p, r. If $w \in V$ then segment $\overline{pw}$ is contained in an interior domain of the cycle

$$\overline{wl} \cdot \overline{lw'} \cdot \overline{w'p} \cdot \overline{pw''} \cdot \overline{w''r} \cdot \overline{rw}.$$

We observe that this fact is independent of the position of w' with respect to the line through l, w. Similarly, it does not matter if w'' lies to the left or to the right of the line through w and r. Hence, w sees p, that is, $w \in V_p$. This proves the wedge property.

If, in Figure 4 (i), line L passes through l and r, or, in (ii), if L passes through p, we can draw the same conclusions if $V \cap L = \emptyset$. But without a separating line, the wedge property would not always be granted, as the example depicted in Figure 6 shows.

3 Boundary Points

In this section we first prove Result (A).

Theorem 1. *No set of 14 points on a simple closed curve can be shattered by interior visibility regions.*

Proof. Let R denote the closed interior domain of a Jordan curve, and let S be a set of 14 points on the boundary of R

We choose two points $l, r \in S$ such that the two boundary segments, τ and β, of P between l and r contain 6 points of S each. Let $T := \tau \cap S$ and $B := \beta \cap S$. We may assume that the line L through l and r is horizontal; see Figure 7 for a sketch.

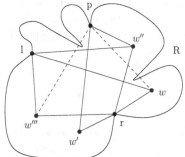
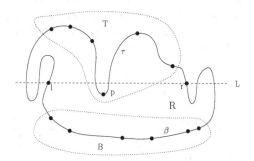

Fig. 6. Here the viewpoints seeing l, p, r do not appear consecutively around p

Fig. 7. Some notations used

Now let us assume that S can be shattered by the visibility regions vis(w), where $w \in R$. This implies, in particular, the following. For each $U \subseteq T \cup B$ there is a viewpoint $w_U \in R$ such that vis(w_U) = $\{l, r\} \cup U$.

Let $E := T \cup B$. Let P' denote the set of all subsets $U \subseteq E$ where w_U is situated *below* line L, and let P'' be the set of those $U \subseteq E$ for which w_U lies *above* L. Clearly, $P' \cup P''$ equals the power set $P(E)$ of E.

An application of Lemma 1 yields, up to symmetry, the following fact. For each $U' \subseteq T$ there exists a viewpoint $w_U \in R$ below line L such that $U' = T \cap U$.

That is, for each subset U' of T there is a point $v_{U'} := w_U$ in R that (i), lies *below* line L, (ii), sees l, r, and, (iii), sees of T exacly the subset U'^4. We define

$$V := \{v_{U'} | U' \subseteq T\} \quad \text{and, for } p \in T, \quad V_p := \{v_{U'} \in V | p \in U'\}.$$

---

[4] We need not care which points of B are visible from $v_{U'}$.

Thus, from each viewpoint in V_p at least l, r and p are visible. Now, our main task is in proving the following fact.

Lemma 3. *For each $p \in T$ there exists a wedge W_p with apex p such that $V_p = W_p \cap V$ holds.*

Proof. We distinguish two cases.

Case 1: p lies above line L. We define the wedge W_p by p and the two half-lines from p through those viewpoints $v_l, v_r \in V_p$ that maximize $\angle(v_l, p, v_r)$. Clearly, direction "$\subseteq$" of our lemma holds by definitions of W_p and V_p.

Since p and v_l, v_r are separated by line L, both visibility segments $\overline{pv_l}$ and $\overline{pv_r}$ must cross L. We claim that both crossing points, c_l, c_r must be situated on L between l and r. In fact, all other orderings would lead to a contradiction. Namely, if r were lying in between c_l and c_r, then r could not be contained in the boundary of R, which cannot intersect visibility segments. In Figure 8 the combinatorially different configurations are depicted. If both c_l and c_r were to the right of r, then v_l and v_r could not be contained in R; see Figure 9. This settles the claim: c_l, c_r must indeed be between l and r, see Figure 10. We are in the situation depicted in Figure 4 (i), and conclude from Lemma 2 that $(p, \{l, r\}, V, \mathrm{line}(l, r))$ has the wedge property. Thus, $V_p \supseteq W_p \cap V$, and the proof of Case 1 is complete.

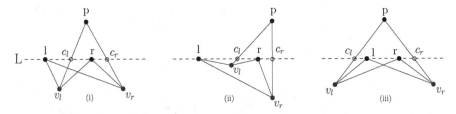

Fig. 8. If r were between c_l and c_r it could not be reached by the boundary of R

Case 2: p lies below line L. Let Q_p denote the wedge with apex p defined by the halflines from l and r through p; see Figure 11 (i). We claim that $V_p \subset Q_p$ holds. Indeed, if some point $v' \in V_p$ were situated outside Q_p then R could not contain v', because the boundary of R visits l, r, p in counterclockwise order; see Figure 11 (ii).

Now let v_l, v_r be those viewpoints in V_p that maximize $\angle(v_l, p, v_r)$. As in Case 1, let W_p be the wedge with apex p defined by the halflines from p trough v_l, v_r, respectively; see Figure 11 (iii). We have $W_p \subseteq Q_p$ and are in the situation depicted in Figure 4 (ii). As in Case 1, Lemma 2 implies $V_p \supseteq W_p \cap V$. This completes the proof of Lemma 3.

Now let $U' \subseteq T$ and $p \in T$. Thanks to Lemma 3 we obtain the following equivalence for the viewpoint $v_{U'} \in V$ of U'.

$$v_{U'} \in W_p \iff v_{U'} \in V_p \iff p \in U'.$$

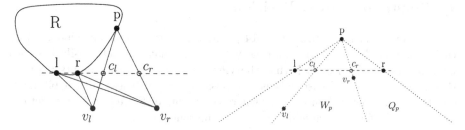

Fig. 9. If c_l, c_r were to the right of r then the boundary of R, which visits l, r, p in counterclockwise order, could not encircle v_l, v_r

Fig. 10. Viewpoint set V_p is contained in wedge $W_p \subset Q_p$

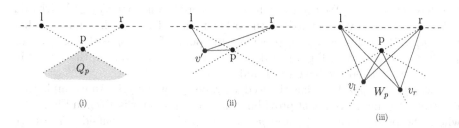

Fig. 11. (i) Wedge Q_p is defined by p and the halflines from l, r through p. (ii) As the boundary of R visits l, r, p in counterclockwise order, it could not encircle v'. (iii) Wedge W_p is defined by the viewpoints v_l, v_r of V_p that make a maximum angle at p.

That is, for each subset U' of T there exists a point (namely, $v_{U'}$) that is contained in exactly those wedges W_p where $p \in U'$. But this is impossible because of $|T| = 6$. Namely, Isler et al. [1] have shown that for at most 5 wedges can each subset be stabbed by one point. This concludes the proof of Theorem 1.

As in the Introduction, let C denote the constant from the ϵ-net theorem.

Theorem 2. *If each point encircled by a simple closed curve sees at least an rth part of the curve, then its interior domain can be covered by $13 \cdot C \cdot r \log r$ many guards on the curve.*

Proof. Let J denote the curve, let R be its interior domain plus J, and $F := \{\text{vis}(w) \cap J | w \in R\}$. From Theorem 1 we infer that $\dim_{\text{VC}}(J, F) \leq 13$ holds. By ϵ-net theory, there exists an $1/r$-net $N \subset J$ of size at most $13\, C\, r \log r$. By assumption, each set in F contains a point p of N. That is, each $w \in R$ is visible from some point $p \in N$ on the curve.

We think that this statement is quite interesting because a set of curve points covering the whole curve is not necessarily able to guard each point of the interior.

4   Points in Convex Position

One key property for showing that 14 points on a curve cannot be shattered by their visibility domains was, informally speaking, that we could easily find two *wing points* l, r from S such that the viewpoints see all other points between them. For the general case this is not possible. Nevertheless with a bit more effort we can show that in the case where the points of S are in convex position, we can find different such wing points for different points.

Theorem 3. *No set of 16 points in convex position inside a simple closed curve can be shattered by interior visibility regions.*

We give only an outline of the proof. If $P = \{p_1, \ldots, p_{16}\}$ is ordered along its convex hull boundary, we consider the viewpoint v_E, that sees exactly the points that have even index. We can then show that there is some point $p_i \in P$ with odd index that sees only about half of the other points of P. We then find another point $p_j \in P$, so that p_i and p_j serve as wing points for all p_k with $i < k < j$ or $j < k < i$, respectively. Figure 12 depicts a typical configuration, that arises in this situation. There we have $i = 1$ and $j = 7$.

Then $(p_k, \{p_i, p_j\}, V, L)$ has the wedge property for every k. An example that illustrates the main argument providing the wedge property is given in Figure 13, where the points that see p_1 and p_7 but not p_3 have to lie outside the wedge $w'p_3w''$.

In general the wedge points obtained by this first step will be less than 6. But by repeating the procedure and forming the union of the sets of wedge points and wing points, respectively, we always get at least 6 wedge points in the end.

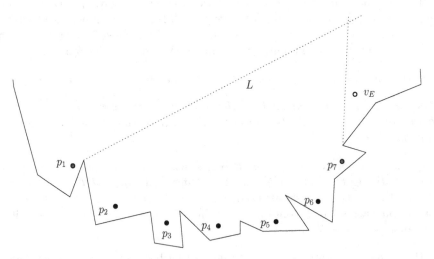

Fig. 12. p_1 and p_7 serve as wing points for the points between them

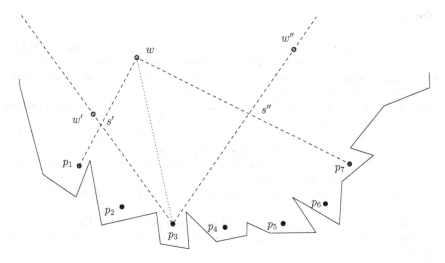

Fig. 13. Visibility edge p_3w cannot be broken because it is inside the quadrilateral $p_3s'ws''$

Therefore the full proof amounts to careful bookkeeping, to make sure that 16 points are always sufficient to obtain a shattered set of 6 wedge points, which leads to a contradiction because of [1]. We omit it due to lack of space.

5   Conclusions

Understanding visibility inside simple polygons is a fundamental challenge to computational geometry. In this paper we have shown that no set of 14 boundary points, and no set of 16 points in convex position, can be shattered by visibility domains. The former fact implies a new result on the number of boundary guards needed to guard a polygon. Our proofs are based on two new techniques. First, we use Lemma 1 to find lines that separate points from viewpoints. Second, we single out some wing points, like l, r in Lemma 2, that must be seen by all view points considered; these visibility edges serve to protect other point-viewpoint edges from boundary intrusion. Costwise, the first technique increases the number of points by a factor of two, while the second technique adds a constant. It will be interesting to see to what extent these techniques can be generalized or strenghtened.

Acknowledgement

The authors would like to thank Adrian Dumitrescu for many helpful discussions.

References

1. Isler, V., Kannan, S., Daniilidis, K., Valtr, P.: VC-dimension of exterior visibility. IEEE Trans. Pattern Analysis and Machine Intelligence 26(5), 667–671 (2004)
2. Kalai, G., Matoušek, J.: Guarding Galleries Where Every Point Sees a Large Area. Israel Journal of Mathematics 101, 125–139 (1997)
3. King, J.: VC-Dimension of Visibility on Terrains. In: Proc. 20th Canadian Conf. on Computational Geometry (CCCG 2008), pp. 27–30 (2008)
4. Kirkpatrick, D.: Guarding Galleries with No Nooks. In: Proc. 12th Canadian Conf. on Computational Geometry (CCCG 2000), pp. 43–46 (2000)
5. Matoušek, J.: Lectures on Discrete Geometry. Graduate Texts in Mathematics, vol. 212. Springer, Heidelberg (2002)
6. O'Rourke, J.: Art Gallery Theorems and Algorithms. Oxford University Press, Oxford (1987)
7. Urrutia, J.: Art Gallery and Illumination Problems. In: Sack, J.-R., Urrutia, J. (eds.) Handbook of Computational Geometry, pp. 973–1023. North-Holland, Amsterdam (2000)
8. Valtr, P.: Guarding Galleries Where No Point Sees a Small Area. Israel Journal of Mathematics 104, 1–16 (1998)

Dynamic Graph Clustering
Using Minimum-Cut Trees*

Robert Görke, Tanja Hartmann, and Dorothea Wagner

Faculty of Informatics, Universität Karlsruhe (TH)
Karlsruhe Institute of Technology (KIT)
{rgoerke,hartmn,wagner}@informatik.uni-karlsruhe.de

Abstract. Algorithms or target functions for graph clustering rarely admit quality guarantees or optimal results in general. Based on properties of minimum-cut trees, a clustering algorithm by Flake et al. does however yield such a provable guarantee. We show that the structure of minimum-s-t-cuts in a graph allows for an efficient dynamic update of minimum-cut trees, and present a dynamic graph clustering algorithm that maintains a clustering fulfilling this quality guarantee, and that effectively avoids changing the clustering. Experiments on real-world dynamic graphs complement our theoretical results.

1 Introduction

Graph clustering has become a central tool for the analysis of networks in general, with applications ranging from the field of social sciences to biology and to the growing field of complex systems. The general aim of graph clustering is to identify dense subgraphs in networks. Countless formalizations thereof exist, however, the overwhelming majority of algorithms for graph clustering relies on heuristics, e.g., for some NP-hard optimization problem, and do not allow for any structural guarantee on their output. For an overview and recent results on graph clustering see, e.g., [2,1] and references therein. Inspired by the work of Kannan et al. [8], Flake et al. [3] recently presented a clustering algorithm which does guarantee a very reasonable bottleneck-property. Their elegant approach employs minimum-cut trees, pioneered by Gomory and Hu [4], and is capable of finding a hierarchy of clusterings by virtue of an input parameter. There has been an attempt to dynamize this algorithm, by Saha and Mitra [9], however, we found it to be erroneous beyond straightforward correction. We are not aware of any other dynamic graph-clustering algorithms in the literature.

In this work we develop the first correct algorithm that efficiently and dynamically maintains a clustering for a changing graph as found by the method of Flake et al. [3], allowing arbitrary atomic changes in the graph, and keeping consecutive clusterings similar (a notion we call *temporal smoothness*). Our algorithms build upon partially updating an intermediate minimum-cut tree of a graph in the spirit of Gusfield's [6] simplification of the Gomory-Hu algorithm [4]. We show that, with only slight modifications, our techniques can update entire min-cut trees. We corroborate our theoretical

---

* This work was partially supported by the DFG under grant WA 654/15-1.

F. Dehne et al. (Eds.): WADS 2009, LNCS 5664, pp. 339–350, 2009.

results on clustering by experimentally evaluating the performance of our procedures compared to the static algorithm on a real-world dynamic graph.

We briefly give our notational conventions and one fundamental lemma in Sec. 1. Then, in Sec. 2, we revisit some results from [4,6,3], convey them to a dynamic scenario, and derive our central results. In Section 3 we give actual update algorithms, which we analyse in Sec. 4, concluding in Sec. 5. We do not include any proof in this extended abstract, without further notice these can all be found in the full version [5].

Preliminaries and Notation. Throughout this work we consider an undirected, weighted graph $G = (V, E, c)$ with vertices V, edges E and a non-negative edge weight function c, writing $c(u, v)$ as a shorthand for $c(\{u, v\})$ with $u \sim v$, i.e., $\{u, v\} \in E$. We reserve the term *node* (or *super-node*) for compound vertices of abstracted graphs, which may contain several basic vertices; however, we identify singleton nodes with the contained vertex without further notice. Dynamic modifications of G will solely concern edges; the reason for this is, that vertex insertions and deletions are trivial for a disconnected vertex. Thus, a modification of G always involves edge $\{b, d\}$, with $c(b, d) = \Delta$, yielding $G_\oplus$ if $\{b, d\}$ is newly inserted into G, and $G_\ominus$ if it is deleted from G. For simplicity we will not handle changes to edge weights, since this can be done almost exactly as deletions and additions. Bridges in G require special treatment when deleted or inserted. However, since they are both simple to detect and to deal with, we ignore them by assuming the dynamic graph to stay connected at all times.

The *minimum-cut tree* $T(G) = (V, E_T, c_T)$ of G is a tree on V and represents for any node pair $\{u, v\} \in \binom{V}{2}$ a minimum-u-v-cut $\theta_{u,v}$ in G by the cheapest edge on the u-v-path in $T(G)$. For $b, d \in V$ we always call this path γ (as a set of edges). An edge $e_T = \{u, v\}$ of T induces the cut $\theta_{u,v}$ in G, sometimes denoted θ_v if the context identifies u. We sometimes identify e_T with the cut it induces in G.

A *contraction* of G by $N \subseteq V$ means replacing set N by a single super-node η, and leaving η adjacent to all former adjacencies u of vertices of N, with edge weight equal to the sum of all former edges between N and u. Analogously we can contract by a set $M \subseteq E$. A *clustering* $\mathscr{C}(G)$ of G is a partition of V into *clusters* C_i, usually conforming to the paradigm of *intra-cluster density and inter-cluster sparsity*. We start by giving some fundamental insights, which we will rely on in the following.

Lemma 1. *Let $e = \{u, v\} \in E_T$ be an edge in $T(G)$.*
Consider $G_\oplus$: If $e \notin \gamma$ then e is still a min-u-v-cut with weight $c(\theta_e)$. If $e \in \gamma$ then its cut-weight is $c(\theta_e) + \Delta$, it stays a min-u-v-cut iff $\forall u$-v-cuts θ' in G that do not separate b, d: $c(\theta') \geq c(\theta_e) + \Delta$.
Consider $G_\ominus$: If $e \in \gamma$ then e remains a min-u-v-cut, with weight $c(\theta_e) - \Delta$. If $e \notin \gamma$ then it retains weight $c(\theta_e)$, it stays a min-u-v-cut iff $\forall u$-v-cuts θ' in G that separate b, d: $c(\theta') \geq c(\theta_e) + \Delta$.

2   Theory

The Static Algorithm. Finding communities in the world wide web or in citation networks are but example applications of graph clustering techniques. In [3] Flake et al. propose and evaluate an algorithm which clusters such instances in a way that yields a certain guarantee on the quality of the clusters. The authors base their quality measure on the expansion of a cut $(S, \bar{S})$ due to Kannan et al. [8]:

$$\Psi = \frac{\sum_{u \in S, v \in \bar{S}} w(u, v)}{\min\{|S|, |\bar{S}|\}} \qquad (expansion \text{ of cut } (S, \bar{S})) \qquad (1)$$

Inspired by a bicriterial approach for good clusterings by Kannan et al. [8], which bases on the related measure conductance[1], Flake et al. [3] design a graph clustering algorithm that, given parameter α, asserts:[2]

$$\underbrace{\frac{c(C, V \setminus C)}{|V \setminus C|}}_{\text{intercluster cuts}} \leq \alpha \leq \underbrace{\frac{c(P, Q)}{\min\{|P|, |Q|\}}}_{\text{intracluster cuts}} \qquad \forall C \in \mathscr{C} \quad \forall P, Q \neq \emptyset \quad P \dot\cup Q = C \qquad (2)$$

These quality guarantees—simply called quality in the following—are due to special properties of min-cut trees, which are used by the clustering algorithm, as given in Alg. 1 (comp. [3], we omit a textual description). In the following, we will call the fact that a clustering can be computed by this procedure the invariant. For the proof that CUT-CLUSTERING yields a clustering that obeys Eq. (2) and for a

---

Algorithm 1. CUT-CLUSTERING

Input: Graph $G = (V, E, c)$, α
1 $V_\alpha := V \cup \{t\}$
2 $E_\alpha := E \cup \{\{t, v\} \mid v \in V\}$
3 $c_\alpha|_E := c, c_\alpha|_{E_\alpha \setminus E} := \alpha$
4 $G_\alpha := (V_\alpha, E_\alpha, c_\alpha)$
5 $T(G_\alpha) :=$ min-cut tree of G_α
6 $T(G_\alpha) \leftarrow T(G_\alpha) - t$
7 $\mathscr{C}(G) \leftarrow$ components of $T(G_\alpha)$

---

number of other interesting properties, we refer the reader to [3]. In the following we will use the definition of $G_\alpha = (V_\alpha, E_\alpha, c_\alpha)$, denoting by $G_\alpha^\ominus$ and $G_\alpha^\oplus$ the corresponding augmented *and* modified graphs. For now, however, general $G_{\oplus(\ominus)}$ are considered.

A Dynamic Attempt. Saha and Mitra [9] published an algorithm that aims at the same goal as our work. Unfortunately, we discovered a methodical error in this work. Roughly speaking, the authors implicitly (and erroneously) assume an equivalence between quality and the invariant, e.g., in their *CASE2* of the procedure for dynamic inter-addition: their proof of correctness requires the invariant but guarantees only quality; there is no immediate remedy for this error. We scrutinize these issues alongside counter-examples and correct parts in the full versions [5,7]. A full description is beyond the scope of this extended abstract, but we sketch out a counter-example in Fig. 1.

Minimum-Cut Trees and the Gomory-Hu Algorithm. Although we heavily build upon the construction of a min-cut tree as proposed by Gomory and Hu [4] we cannot accomodate a detailed description of their algorithm and refer the reader to their work. The algorithm builds the min-cut tree of a graph by iteratively finding min-u-v-cuts for vertices that have not yet been separated by a previous min-cut. An *intermediate* min-cut tree $T_*(G) = (V_*, E_*, c_*)$ (or simply T_* if the context is clear) is initialized as an isolated, edgeless super-node containing all original nodes. Then, until no node S of

---

[1] conductance is similar to expansion but normalizes cuts by total incident edge weight.

[2] The disjoint union $A \cup B$ with $A \cap B = \emptyset$ is denoted by $A \dot\cup B$.

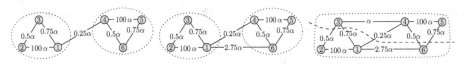

Fig. 1. Example showing error in [9]. Left: initial instance, clustered via static algorithm; middle: updated clustering after one edge-addition, preserving quality but *not* the invariant; right: update after second edge-addition, quality is violated, dashed cut weighs $11/4\alpha < \alpha \min\{|P|, |V \setminus |P|\}$.

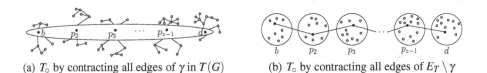

(a) $T_\circ$ by contracting all edges of γ in $T(G)$ (b) $T_\circ$ by contracting all edges of $E_T \setminus \gamma$

Fig. 2. Sketches of intermediate min-cut trees $T_\circ$; for $G_\oplus$ (a) we contract γ to a node, and for $G_\ominus$ (b) we contract each connected component induced by $E_T \setminus \gamma$

T_* contains more than one vertex, a node S is *split*. To this end, nodes $S_i \neq S$ are dealt with by contracting in G whole subtrees N_j of S in T_*, connected to S via edges $\{S, S_j\}$, to single nodes η_j before cutting, which yields G_S—a notation we will continue using in the following. The split of S into (S_u, S_v) is then defined by a min-u-v-cut in G_S. Afterwards, N_j is reconnected, again by S_j, to either S_u or S_v depending on which side of the cut η_j, containing S_j, ended up. Note that this cut in G_S can be proven to induce a min-u-v-cut in G. An *execution* GH $= (G, F, K)$ of GOMORY-HU is characterized by graph G, sequence F of $n - 1$ *step pairs* of nodes and sequence K of *split cuts*. Pair $\{u, v\} \subseteq V$ is a *cut pair* of edge e of cut-tree T if θ_e is a min-u-v-cut in G.

Theorem 1. *Consider a set $M \subseteq E_T$ and let $T_\circ(G) = (V_\circ, M, c_\circ)$ be $T(G)$ with $E_T \setminus M$ contracted. Let f and f' be sequences of the elements of M and $E_T \setminus M$, respectively, and k and k' the corresponding sequences of edge-induced cuts of G. GH $= (G, f' \cdot f, k' \cdot k)^3$ has $T_\circ(G)$ as intermediate min-cut tree (namely after f).*

In the following we will denote by $T_\circ$ an intermediate min-cut tree which serves as a starting point, and by T_* a working version. This theorem states that if for some reason we can only be sure about a subset of the edges of a min-cut tree, we can contract all other edges to super-nodes and consider the resulting tree $T_\circ$ as the correct intermediate result of some GH, which can then be continued. One such reason could be a dynamic change in G, such as the insertion of an edge, which by Lem. 1 maintains a subset of the old min-cuts. This already hints at an idea for an effort-saving update of min-cut trees.

Using Arbitrary Minimum Cuts in G. Gusfield [6] presented an algorithm for finding min-cut trees which avoids complicated contraction operations. In essence he provided rules for adjusting iteratively found min-u-v-cuts in G (instead of in G_S) that potentially cross, such that they are consistent with the Gomory-Hu procedure and thus noncrossing, but still minimal. We need to review and generalize some of these ideas as to

---

[3] The term $b \cdot a$ denotes the concatenation of sequences b and a, i.e., a happens first.

fit our setting. The following lemma essentially tells us, that at any time in GOMORY-HU, for any edge e of $T_\circ$ there exists a cut pair of e in the two nodes incident to e.

Lemma 2 (Gus. [6], Lem. 4[4]). *Let S be cut into S_x and S_y, with $\{x,y\}$ being a cut pair (not necessarily the step pair). Let now $\{u,v\} \subseteq S_x$ split S_x into S_{xu} and S_{xv}, wlog. with $S_y \sim S_{xu}$ in T_*. Then, $\{x,y\}$ remains a cut pair of edge $\{S_y, S_{xu}\}$ (we say edge $\{S_x, S_y\}$ gets reconnected). If $x \in S_{xv}$, i.e., the min-u-v-cut separates x and y, then $\{u,y\}$ is also a cut pair of $\{S_{xu}, S_y\}$.*

In the latter case of Lem. 2, we say that pair $\{x,y\}$ gets *hidden*, and, in the view of vertex y, its former counterpart x gets *shadowed* by u (or by S_u). It is not hard to see that during GOMORY-HU, step pairs remain cut pairs, but cut pairs need not stem from step pairs. However, each edge in T has at least one cut pair in the incident nodes. We define the *nearest cut pair* of an edge in T_* as follows: As long as a step pair $\{x,y\}$ is in adjacent nodes S_x, S_y, it is the nearest cut pair of edge $\{S_x, S_y\}$; if a nearest cut pair gets hidden in T_* by a step of GOMORY-HU, as described in Lem. 2 if $x \in S_{xv}$, the nearest cut pair of the reconnected edge $\{S_y, S_{xu}\}$ becomes $\{u,y\}$ (which are in the adjacent nodes S_y, S_{xu}). The following theorem basically states how to iteratively find min-cuts as GOMORY-HU, without the necessity to operate on a contracted graph.

Theorem 2 (Gus. [6], Theo. 2[5]). *Let $\{u,v\}$ denote the current step pair in node S during some GH. If $(U, V \setminus U)$, $(u \in U)$ is a min-u-v-cut in G, then there exists a min-u-v-cut $(U_S, V_S \setminus U_S)$ of equal weight in G_S such that $S \cap U = S \cap U_S$ and $S \cap (V \setminus U) = S \cap (V_S \setminus U_S)$, $(u \in U_S)$.*

Being an ingredient to the original proof of Theo. 2, the following Lem. 3 gives a constructive assertion, that tells us how to arrive at a cut described in the theorem by inductively adjusting a given min-u-v-cut in G. Thus, it is the key to avoiding contraction and using cuts in G by rendering min-u-v-cuts non-crossing with other given cuts.

Lemma 3 (Gus. [6], Lem. 1[5]). *Let $(Y, V \setminus Y)$ be a min-x-y-cut in G $(y \in Y)$. Let $(H, V \setminus H)$ be a min-u-v-cut, with $u, v \in V \setminus Y$ and $y \in H$. Then the cut $(Y \cup H, (V \setminus Y) \cap (V \setminus H))$ is also a min-u-v-cut.*

Given a cut as by Theo. 2, Gomory and Hu state a simple mechanism which reconnects a former neighboring subtree N_j of a node S to either of its two split parts; when avoiding contraction, this criterion is not available. For this purpose, Gusfield iteratively defines *representatives* $r(S_i) \in V$ of nodes S_i of T_*, and states his Theorem 3: For $u, v \in S$ let *any* min-u-v-cut $(U, V \setminus U)$, $u \in U$, in G split node S into $S_u \ni u$ and $S_v \ni v$ and let $(U_S, V \setminus U_S)$ be this cut adjusted via Lem. 3 and Theo. 2; then a neighboring subtree N_j of S, formerly connected by edge $\{S, S_j\}$, lies in U_S iff $r(S_j) \in U$. We do not have such representatives and thus need to adapt this, namely using nearest cut pairs as representatives:

Theorem 3 (comp. Gus. [6], Theo. 3[5]). *In any T_* of a GH, suppose $\{u,v\} \subseteq S$ is the next step pair, with subtrees N_j of S connected by $\{S, S_j\}$ and nearest cut pairs $\{x_j, y_j\}$, $y_j \in S_j$. Let $(U, V \setminus U)$ be a min-u-v-cut in G, and $(U_S, V \setminus U_S)$ its adjustion. Then $\eta_j \in U_S$ iff $y_j \in U$.*

---

[4] This lemma is also proven in [6] and [4], we thus omit a proof.

[5] This Lemma alongside Lemma 3, Theo. 2 and a simpler version of our Theo. 3 have been discussed in [6] and the lemma also in [4].

Finding and Shaping Minimum Cuts in the Dynamic Scenario. In this section we let graph G change, i.e., we consider the addition of an edge $\{b,d\}$ or its deletion, yielding $G_{\oplus}$ or $G_{\ominus}$. First of all we define valid representatives of the nodes on $T_{\circ}$(omitting proofs). By Lem. 1 and Theo. 1, given an edge addition, $T_{\circ}$ consists of a single super-node and many singletons, and given edge deletion, $T_{\circ}$ consists of a path of super-nodes; for examples see Fig. 2.

Definition 1 (Representatives in $T_{\circ}$)
Edge addition: *Set singletons to be representatives of themselves; for the only super-node S choose an arbitrary $r(S) \in S$.*
Edge deletion: *For each node S_i, set $r(S_i)$ to be the unique vertex in S_i which lies on γ.*
New nodes during algorithm, and the choice of step pairs: *On a split of node S require the step pair to be $\{r(S),v\}$ with an arbitrary $v \in S, v \neq r(S)$. Let the split be $S = S_{r(S)} \dot\cup S_v, v \in S_v$, then define $r(S_{r(S)}) := r(S)$ and $r(S_v) := v$.*

Following Theo. 1, we define the set M of "good" edges of the old tree $T(G)$, i.e., edges that stay valid due to Lem. 1, as $M := E_T \setminus \gamma$ for the insertion of $\{b,d\}$ and to $M := \gamma$ for the deletion. Let $T_{\circ}(G_{\oplus(\ominus)})$ be $T(G)$ contracted by M. As above, let f be any sequence of the edges in M and k the corresponding cuts in G. We now state a specific variant of the setting in Theo. 1 which is the basis of our updating algorithms, founded on $T_{\circ}$s as in Fig. 2, using arbitrary cuts in $G_{\oplus(\ominus)}$ instead of actual contractions.

Lemma 4. *Given an edge addition (deletion) in G. The Gomory-Hu execution $\mathrm{GH}_{\oplus(\ominus)} = (G_{\oplus(\ominus)}, f_{\oplus(\ominus)} \cdot f, k_{\oplus(\ominus)} \cdot k)$ is feasible for $G_{\oplus(\ominus)}$ yielding $T_{\circ}(G)$ as the intermediate min-cut tree after sequence f, if $f_{\oplus(\ominus)}$ and $k_{\oplus(\ominus)}$ are feasible sequences of step pairs and cuts on $T_{\circ}(G_{\oplus(\ominus)})$.*

Cuts That Can Stay. The non-crossing nature of min-u-v-cuts allow for more effort-saving and temporal smoothness. There are several circumstances which imply that a previous cut is still valid after a graph modification, making its recomputation unnecessary. The following lemma gives one such assertion (we omit a few others here), based on the definition of a *treetop* and of *wood* (comp. Fig. 3): Consider edge $e = \{u,v\}$ off γ, and cut $\theta = (U, V \setminus U)$ in G induced by e in $T(G)$ with γ contained in U. In $G_{\ominus}(S)$, $S \cap (V \setminus U)$ is called the *treetop* $\Uparrow_e$, and $S \cap U$ the *wood* $\#_e$ of e. The subtrees of S are N_b and N_d, containing b and d, respectively.

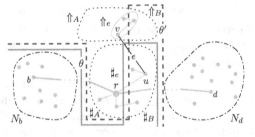

Fig. 3. Special parts of $G_{\ominus}$: γ (fat) connects b and d, with r on it; wood $\#_e$ and treetop $\Uparrow_e$ (dotted) of edge e, both cut by θ' (dashed), adjusted to θ (solid) by Lem. 6. Both $\#_e$ and $\Uparrow_e$ are part of some node S, with representative r, outside subtrees of r are N_b and N_d (dash-dotted). Compare to Fig. 2(b).

Lemma 5. *In $G_{\ominus}$, let $(U, V \setminus U)$ be a min-u-v-cut not separating $\{b,d\}$, with γ in $V \setminus U$. Then, a cut induced by edge $\{g,h\}$ of the old $T(G)$, with $g,h \in U$, remains a min separating cut for all its previous cut pairs within U in $G_{\ominus}$, and a min g-h-cut in particular.*

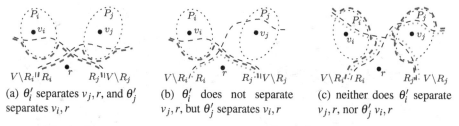

(a) θ_i' separates v_j, r, and θ_j' separates v_i, r

(b) θ_i' does not separate v_j, r, but θ_j' separates v_i, r

(c) neither does θ_i' separate v_j, r, nor θ_j' v_i, r

Fig. 4. Three cases concerning the positions of θ_i' and θ_j', and their adjustments

As a corollary from Lem. 5 we get that in $T(G_\ominus)$ the entire treetops of reconfirmed edges of $T(G)$ are also reconfirmed. This saves effort and encourages smoothness; however new cuts can also be urged to behave well, as follows.

The Shape of New Cuts. In contrast to the above lemmas, during a Gomory-Hu execution for $G_\ominus$, we might find an edge $\{u,v\}$ of the old $T(G)$ that is *not* reconfirmed by a computation in $G_\ominus$, but a new, cheaper min-u-v-cut $\theta' = (U, V(S) \setminus U)$ is found. For such a new cut we can still make some guarantees on its shape to resemble its "predecessor": Lemma 6 tells us, that for any such min-u-v-cut θ', there is a min-u-v-cut $\theta = (U \setminus \Uparrow_e, (V(S) \setminus U) \cup \Uparrow_e)$ in $G_\ominus$ that (a) does not split $\Uparrow_e$, (b) but splits $V \setminus \Uparrow_e$ exactly as θ' does. Figure 3 illustrates such cuts θ (solid) and θ' (dashed).

Lemma 6. *Given $e = \{u,v\}$ within S (off γ) in $G_\ominus(S)$. Let $(\Uparrow_A, \Uparrow_B)$ be a cut of $\Uparrow_e$ with $v \in \Uparrow_A$. Then $c_\ominus(N_b \cup \Uparrow_e, N_d \cup \#_e) \leq c_\ominus(N_b \cup \Uparrow_A, N_d \cup \#_e \cup \Uparrow_B)$. Exchanging N_b and N_d is analogous. This result can be generalized in that both considered cuts are also allowed to cut the wood $\#_e$ in some arbitrary but fixed way.*

While this lemma can be applied in order to retain treetops, even if new cuts are found, we now take a look at how new, cheap cuts can affect the treetops of *other* edges. In fact a similar treetop-conserving result can be stated. Let G' denote an undirected, weighted graph and $\{r, v_1, \ldots, v_z\}$ a set of designated vertices in G'. Let $\Pi := \{P_1, \ldots, P_z\}$ be a partition of $V \setminus r$ such that $v_j \in P_j$. We now assume the following partition-property to hold: For each v_j it holds that for any v_j-r-cut $\theta_j' := (R_j, V \setminus R_j)$ (with $r \in R_j$), the cut $\theta_j := (R_j \setminus P_j, (V \setminus R_j) \cup P_j)$ is of at most the same weight. The crucial observation is, that Lem. 6 implies this partition-property for $r(S)$ and its neighbors in $T(G)$ that lie inside S of $T_\circ$ in $G_\ominus$. Treetops thus are the sets P_j. However, we keep things general for now: Consider a min-v_i-r-cut $\theta_i' := (R_i, V \setminus R_i)$, with $r \in R_i$, that does not split P_i and an analog min-v_j-r-cut θ_j' (by the partition-property they exist). We distinguish three cases, given in Fig. 4, which yield the following possibilities of reshaping min-cuts:

Case (a): As cut θ_i' separates v_j and r, and as v_j satisfies the partition-property, the cut $\theta_i := (R_i \setminus P_j, (V \setminus R_i) \cup P_j)$ (red dashed) has weight $c(\theta_i) \leq c(\theta_i')$ and is thus a min-v_i-r-cut, which does not split $P_i \cup P_j$. For θ_j' an analogy holds.

Case (b): For θ_j' Case (a) applies. Furthermore, by Lem. 3, the cut $\theta_{\text{new}(j)} := (R_i \cap R_j, (V \setminus R_i) \cup (V \setminus R_j))$ (green dotted) is a min-v_j-r-cut, which does not split $P_i \cup P_j$. By Lem. 2 the previous split cut θ_i' is also a min-v_i-v_j-cut, as $\theta_{\text{new}(j)}$ separates v_i, r.

Case (c): As in case (b), by Lem. 3 the cut $\theta_{\text{new}(i)} := ((V \setminus R_j) \cup R_i, (V \setminus R_i) \cap R_j)$ (green dotted) is a min-v_i-r-cut, and the cut $\theta_{\text{new}(j)} := ((V \setminus R_i) \cup R_j, (V \setminus R_j) \cap R_i)$

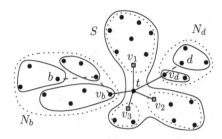
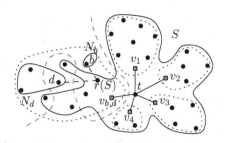

Fig. 5. $T_\circ(G_{\bar{\alpha}}^{\ominus})$ for an inter-del.; t's neighbors off γ need inspection. Cuts of v_b and v_d are correct, but might get shadowed.

Fig. 6. $T_\circ(G_{\bar{\alpha}}^{\ominus})$ for an intra-del.; edge $\{v_{b,d},t\}$ defines a treetop (t's side). The dashed cut could be added to Θ.

(green dotted) is a min-v_j-r-cut. These cuts do not cross. So as v_i and v_j both satisfy the partition-property, cut $\theta_i := (((V \setminus R_j) \cup R_i) \setminus P_i, ((V \setminus R_i) \cap R_j) \cup P_i)$ and $\theta_j := (((V \setminus R_i) \cup R_j) \setminus P_j, ((V \setminus R_j) \cap R_i) \cup P_j)$ (both red dashed) are non-crossing min separating cuts, which neither split P_i nor P_j.

To summarize the cases discussed above, we make the following observation.

Observation 1. *During a* GH *starting from* $T_\circ$ *for* $G_\ominus$, *whenever we discover a new, cheaper min-v_i-$r(S)$-cut θ' ($v_i \sim r(S)$ in node S) we can iteratively reshape θ' into a min-v_i-$r(S)$-cut θ which neither cuts $\Uparrow_i$ nor any other treetop $\Uparrow_j$ ($v_i \sim r(S)$ in S).*

3   Update Algorithms for Dynamic Clusterings

In this section we put the results of the previous sections to good use and give algorithms for updating a min-cut tree clustering, such that the invariant is maintained and thus also the quality. By concept, we merely need to know all vertices of $T(G)$ adjacent to t; we call this set $W = \{v_1, \ldots, v_z\} \cup \{v_b, v_d\}$, with $\{v_b, v_d\}$ being the particular vertex/vertices on the path from t to b and d, respectively. We call the corresponding set of non-crossing min-v_i-t-cuts that isolate t, Θ. We will thus focus on dynamically maintaining only this information, and sketch out how to unfold the rest of the min-cut tree. From Lem. 4, for a given edge insertion or deletion, we know $T_\circ$, and we know in which node of $T_\circ$ to find t, this is the node we need to examine. We now give algorithms for the deletion and the insertion of an edge running inside or between clusters.

Algorithm 2. INTER-CLUSTER EDGE DELETION

    Input: $W(G), \Theta(G)$ $G_{\bar{\alpha}}^{\ominus} = (V_\alpha, E_\alpha \setminus \{\{b,d\}\}, c_{\bar{\alpha}}^{\ominus})$, edge $\{b,d\}$ with weight Δ
    Output: $W(G_\ominus), \Theta(G_\ominus)$
1 $L(t) \leftarrow \emptyset, l(t) \leftarrow \emptyset$
2 **for** $i = 1, \ldots, z$ **do** Add v_i to $L(t)$, $D(v_i) \leftarrow \emptyset$ // old cut-vertices, shadows
3 $\Theta(G_\ominus) \leftarrow \{\theta_b, \theta_d\}$, $W(G_\ominus) \leftarrow \{v_b, v_d\}$
4 **return** CHECK CUT-V. $(W(G), \Theta(G), W(G_\ominus), \Theta(G_\ominus), G_{\bar{\alpha}}^{\ominus}, \{b,d\}, D, L(t))$

Algorithm 3. CHECK CUT-VERTICES

Input: $W(G), \Theta(G), W(G_\ominus), \Theta(G_\ominus), G_\alpha^\ominus, \{b,d\}, D, L(t)$
Output: $W(G_\ominus), \Theta(G_\ominus)$

1 **while** $L(t)$ *has next element* v_i **do**
2      $\theta_i \leftarrow$ first min-v_i-t-cut given by FLOWALGO(v_i, t)   // small side for v_i
3   **if** $c_\alpha^\ominus(\theta_i) = c_\alpha(\theta_i^{old})$ **then** Add θ_i^{old} to $l(t)$ // retain old cut?
4   **else**
5   Add θ_i to $l(t)$ // pointed at by v_i
6   **while** $L(t)$ *has next element* $v_j \neq v_i$ **do**   // test vs. other new cuts
7   **if** θ_i *separates* v_j *and* t **then**   // v_j shadowed by Lem. 3
8   Move v_j from $L(t)$ to $D(v_i)$
9   **if** $l(t) \ni \theta_j$, *pointed at by* v_j **then** Delete θ_j from $l(t)$

10 **while** $L(t)$ *has next element* v_i **do**   // make new cuts cluster-preserving
11   set $(R, V_\alpha \setminus R) := \theta_i$ with $t \in R$ for $\theta_i \in l(t)$ pointed at by v_i
12      $\theta_i \leftarrow (R \setminus C_i, (V_\alpha \setminus R) \cup C_i)$ // by partition-property (Lem. 6)
13   **forall** $v_j \in D(v_i)$ **do** $\theta_i \leftarrow (R \setminus C_j, (V_\alpha \setminus R) \cup C_j)$ // Cases (a) and (b)
14   **forall** $v_j \neq v_i$ in $L(t)$ **do** $\theta_i \leftarrow (R \cup C_j, (V_\alpha \setminus R) \setminus C_j)$ // Case (c)

15 Add all vertices in $L(t)$ to $W(G_\ominus)$, and their cuts from $l(t)$ to $\Theta(G_\ominus)$

Edge Deletion. Our first algorithm handles inter-cluster deletion (Alg. 2). Just like its three counterparts, it takes as an input the old graph G and its sets $W(G)$ and $\Theta(G)$ (not the entire min-cut tree $T(G_\alpha)$), furthermore it takes the changed graph, augmented by t, $G_\alpha^\ominus$, the deleted edge $\{b,d\}$ and its weight Δ. Recall that an inter-cluster deletion yields t on γ, and thus, $T_\circ(G_\alpha)$ contains edges $\{v_b, t\}$ and $\{v_d, t\}$ cutting off the subtrees N_b and N_d of t by cuts θ_b, θ_d, as shown in Fig. 5. All clusters contained in node $S \ni t$ need to be changed or reconfirmed. To this end Algorithm 2 lists all cut vertices in S, $v_1, \ldots, v_z$, into $L(t)$, and initializes their shadows $D(v_i) = \emptyset$. The known cuts θ_b, θ_d are already added to the final list, as are v_b, v_d (line 3). Then the core algorithm, CHECK CUT-VERTICES is called, which—roughly speaking—performs those GH-steps that are necessary to isolate t, using (most of) the above lemmas derived.

First of all, note that if $|\mathscr{C}| = 2$ ($\mathscr{C} = \{N_b, N_d\}$ and $S = \{t\}$) then $L(t) = \emptyset$ and Alg. 2 lets CHECK CUT-VERTICES (Alg. 3) simply return the input cuts and terminates. Otherwise, it iterates the set of former cut-vertices $L(t)$ once, thereby possibly shortening it. We start by computing a new min-v_i-t-cut for v_i. We do this with a max-v_i-t-flow computation, which is known to yield *all* min-v_i-t-cuts, taking the *first* cut found by a breadth-first search from v_i (lines 2). This way we find a cut which minimally interferes with other treetops, thus encouraging temporal smoothness. If the new cut is non-cheaper, we use the old one instead, and add it to the tentative list of cuts $l(t)$ (lines 3-3). Otherwise we store the new, cheaper cut θ_i, and examine it for later adjustment. For any candidate v_j still in $L(t)$ that is separated from t by θ_i, Case (a) or (b) applies (line 7). Thus, v_j will be in the shadow of v_i, and not a cut-vertex (line 8). In case v_j has already been processed, its cut is removed from $l(t)$. Once all cut-vertex candidates are processed, each one either induces the same cut as before, is new and shadows other former cut-vertices or is itself shadowed by another cut-vertex. Now that we have

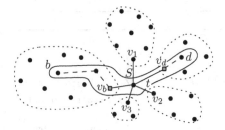 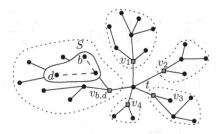

Fig. 7. $T_\circ(G_\alpha^\oplus)$ for an inter-cluster addition. At least v_b and v_d need inspection.

Fig. 8. $T_\circ(G_\alpha^\oplus)$ for an intra-cluster addition. All relevant min-v-t-cuts persist.

collected these relations, we actually apply Cases (a,b,c) and Lem. 6 in lines 10-14. Note that for retained, old cuts, no adjustment is actually performed here. Finally, all non-shadowed cut-vertices alongside their adjusted cuts are added to the final lists.

Unfortunately we must completely omit *intra*-cluster edge deletion here, please find it detailed in the full versions [5,7]. In a way roughly analogue to the former case we again call CHECK CUT-VERTICES and after that merely have to clean up a potentially "wild" area of leftover vertices from cluster $C_{b,d}$.

Edge Addition. The good news for handling $G_\oplus$ is, that an algorithm INTRA-CLUSTER EDGE ADDITION only needs to return the old clustering: By Lem. 1 and Theo. 1, in $T_\circ$, only path γ is contracted. But since γ lies within a cluster, the cuts in G_α, defining the old clustering, all remain valid in $G_\alpha^\oplus$, as depicted in Fig. 8 with dotted clusters and affected node S. By contrast, adding an edge between clusters is more demanding. Again, γ is contracted, see region S in Fig. 7; however, t lies on γ in this case. A sketch of what needs to be done resembles the above algorithms: We compute new min-v_b-t- and min-v_d-t-cuts (or possibly only one, if it immediately shadows the other), and keep the old v_i-t-cuts. Then—proceeding as usual—we note which cuts shadow which others and reconnect nodes by Theo. 3.

Updating Entire Min-Cut Trees. An interesting topic on its own right and more fundamental than clustering, is the dynamic maintenance of min-cut trees. In fact the above clustering algorithms are surprisingly close to methods that update min-cut trees. Since all the results from Sec. 2 still apply, we only need to unfold treetops or subtrees of t—which we gladly accept as super-nodes for the purpose of clustering—and take care to correctly reconnect subtrees. This includes, that merely examining the neighbors of t does not suffice, we must iterate through all nodes S_i of $T_\circ$. For the sake of brevity we must omit further details on such algorithms and refer the reader to the full version [5].

4   Performance of the Algorithm

Temporal Smoothness. Our secondary criterion—which we left unformalized—to preserve as much of the previous clustering as possible, in parts synergizes with effort-saving, an observation foremost reflected in the usage of $T_\circ$. Lemmas 5 and 6, using *first cuts* and Observation 1 nicely enforce temporal smoothness. However, in some

Table 1. Bounds on the number of max-flow calculations

	worst case	old clustering still valid		
		lower bound	upper bound	guaran. smooth
Inter-Del	$|\mathscr{C}(G)| - 2$	$|\mathscr{C}(G)| - 2$	$|\mathscr{C}(G)| - 2$	Yes
Intra-Del	$|\mathscr{C}(G)| + |C_{b,d}| - 1$	1	$|\mathscr{C}(G)| + |C_{b,d}| - 1$	No (1)
Inter-Add	$|C_b| + |C_d|$	1	$|C_b| + |C_d|$	No (2)
Intra-Add	0	0	0	Yes

cases we must cut back on this issue, e.g., when we examine which other cut-vertex candidates are shadowed by another one, as in line 7 of Alg. 3. Here it entails many more cut-computations and a combinatorially non-trivial problem to find an ordering of $L(t)$ to optimally preserve old clusters. Still we can state the following lemma:

Lemma 7. *Let $\mathscr{C}(G)$ fulfill the invariant for $G_\ominus$, i.e., let the old clustering be valid for $G_\ominus$. In the case of an inter-cluster deletion, Alg 2 returns $\mathscr{C}(G)$. For an intra-cluster deletion we return a clustering $\mathscr{C}(G_\ominus) \supseteq \mathscr{C}(G) \setminus C_{b,d}$, i.e., only $C_{b,d}$ might become fragmented. Intra-cluster addition retains a valid old clusterings.*

Running Times. We express running times of our algorithms in terms of the number of max-flow computations, leaving open how these are done. A summary of tight bounds is given in Tab. 1 (for an in-depth discussion thereof see the full version). The columns *lower bound/upper bound* denote bounds for the—possibly rather common—case that the old clustering is still valid after some graph update. As discussed in the last subsection, the last column (*guaran. smooth*) states whether our algorithms *always* return the previous clustering, in case its valid; the numbers in brackets denotes a tight lower bound on the running time, in case our algorithms do find that previous clustering. Note that a computation from scratch entails a tight upper bound of $|V|$ max-flow computations for all four cases, in the worst case.

Further Speed-Up. For the sake of brevity we leave a few ideas and lemmas for effort-saving to the full version. One heuristic is to decreasingly order vertices in the list $L(t)$, e.g., in line 2 of Alg. 2; for their static algorithm Flake et al. [3] found that this effectively reduces the number of cuts necessary to compute before t is isolated. Since individual min-u-v-cuts are constantly required, another dimension of effort-saving lies in dynamically maintaining max-u-v-flows. We detail two approaches based on dynamic residual graphs in the full version.

Experiments. In this brief section, we very roughly describe some experiments we made with an implementation of the update algorithms described above, just for a first proof of concept. The instance we use is a network of e-mail communications within the Fakultät für Informatik at Universität Karlsruhe. Vertices represent members and edges correspond to e-mail contacts, weighted by the number of e-mails sent between two individuals during the last 72 hours. We process a queue of 12560 elementary modifications, 9000 of which are actual edge modifications, on the initial graph G ($|V| = 310, |E| = 450$). This queue represents about one week, starting on Saturday (21.10.06); a spam-attack lets the graph slightly grow/densify over the course.

We delete zero-weight edges and isolated nodes. Following the recommendations of Flake et al. [3] we choose $\alpha = 0.15$ for the initial graph, yielding 45 clusters. For the $9K$ proper steps, static computation needed $\approx 2M$ max-flows, and our dynamic update needed $\approx 2K$, saving more than 90% max-flows, such that in 96% of all modifications, the dynamic algorithm was quicker. Surprisingly, inter-additions had the greatest impact on effort-saving, followed by the trivial intra-additions. Out of the $9K$ operations, 49 of the inter-, and 222 of the intra-cluster deletions were the only ones, where the static algorithm was quicker. See the full versions [5,7] for details on these results.

5   Conclusion

We have proven a number of results on the nature of min-u-v-cuts in changing graphs, which allow for feasible update algorithms of a minimum-cut tree. In particular we have presented algorithms which efficiently update specific parts of such a tree and thus fully dynamically maintain a graph clustering based on minimum-cut trees, as defined by Flake et al. [3] for the static case, under arbitrary atomic changes. The striking feature of graph clusterings computed by this method is that they are guaranteed to yield a certain expansion—a bottleneck measure—within and between clusters, tunable by an input parameter α. As a secondary criterion for our updates we encourage temporal smoothness, i.e., changes to the clusterings are kept at a minimum, whenever possible. Furthermore, we disprove an earlier attempt to dynamize such clusterings [9]. Our experiments on real-world dynamic graphs affirm our theoretical results and show a significant practical speedup over the static algorithm of Flake et al. [3]. Future work on dynamic minimum-cut tree clusterings will include a systematic comparison to other dynamic clustering techniques and a method to dynamically adapt the parameter α.

References

1. Brandes, U., Delling, D., Gaertler, M., Görke, R., Höfer, M., Nikoloski, Z., Wagner, D.: On Modularity Clustering. IEEE TKDE 20(2), 172–188 (2008)
2. Brandes, U., Erlebach, T. (eds.): Network Analysis. LNCS, vol. 3418. Springer, Heidelberg (2005)
3. Flake, G.W., Tarjan, R.E., Tsioutsiouliklis, K.: Graph Clustering and Minimum Cut Trees. Internet Mathematics 1(4), 385–408 (2004)
4. Gomory, R.E., Hu, T.: Multi-terminal network flows. Journal of the Society for Industrial and Applied Mathematics 9(4), 551–570 (1961)
5. Görke, R., Hartmann, T., Wagner, D.: Dynamic Graph Clustering Using Minimum-Cut Trees. Technical report, Informatics, Universität Karlsruhe (2009)
6. Gusfield, D.: Very simple methods for all pairs network flow analysis. SIAM Journal on Computing 19(1), 143–155 (1990)
7. Hartmann, T.: Clustering Dynamic Graphs with Guaranteed Quality. Master's thesis, Universität Karlsruhe (TH), Fakultät für Informatik (October 2008)
8. Kannan, R., Vempala, S., Vetta, A.: On Clusterings - Good, Bad and Spectral. In: Proc. of FOCS 2000, pp. 367–378 (2000)
9. Saha, B., Mitra, P.: Dynamic Algorithm for Graph Clustering Using Minimum Cut Tree. In: Proc. of the, SIAM Int. Conf. on Data Mining, pp. 581–586 (2007)

Rank-Balanced Trees

Bernhard Haeupler[2], Siddhartha Sen[1,*], and Robert E. Tarjan[1,3,*]

[1] Princeton University, {sssix,ret}@cs.princeton.edu
[2] Massachusetts Institute of Technology, haeupler@mit.edu
[3] HP Laboratories, Palo Alto CA 94304

Abstract. Since the invention of AVL trees in 1962, a wide variety of ways to balance binary search trees have been proposed. Notable are red-black trees, in which bottom-up rebalancing after an insertion or deletion takes $O(1)$ amortized time and $O(1)$ rotations worst-case. But the design space of balanced trees has not been fully explored. We introduce the *rank-balanced tree*, a relaxation of AVL trees. Rank-balanced trees can be rebalanced bottom-up after an insertion or deletion in $O(1)$ amortized time and at most two rotations worst-case, in contrast to red-black trees, which need up to three rotations per deletion. Rebalancing can also be done top-down with fixed lookahead in $O(1)$ amortized time. Using a novel analysis that relies on an exponential potential function, we show that both bottom-up and top-down rebalancing modify nodes exponentially infrequently in their heights.

1 Introduction

Balanced search trees are fundamental and ubiquitous in computer science. Since the invention of AVL trees [1] in 1962, many alternatives [2,3,4,5,7,10,9,11] have been proposed, with the goal of simpler implementation or better performance or both. Simpler implementations of balanced trees include Andersson's implementation [2] of Bayer's binary B-trees [3] and Sedgewick's related left-leaning red-black trees [4,11]. These data structures are asymmetric, which simplifies rebalancing by eliminating roughly half the cases. Andersson further simplified the implementation by factoring rebalancing into two procedures, *skew* and *split*, and by adding a few other clever ideas. Standard red-black trees [7], on the other hand, have update algorithms with guaranteed efficiency: rebalancing after an insertion or deletion takes $O(1)$ rotations worst-case and $O(1)$ time amortized [13,15]. As a result of these developments, one author [12, p. 177] has said, "AVL... trees are now passé."

Yet the design and analysis of balanced trees is a rich area, not yet fully explored. We continue the exploration. Our work yields both a new design and new analyses, and suggests that AVL trees are anything but passé. Our new design is the *rank-balanced tree*, a relaxation of AVL trees that has properties similar to those of red-black trees but

* Research at Princeton University partially supported by NSF grants CCF-0830676 and CCF-0832797 and US-Israel Binational Science Foundation grant 2006204. The information contained herein does not necessarily reflect the opinion or policy of the federal government and no official endorsement should be inferred.

F. Dehne et al. (Eds.): WADS 2009, LNCS 5664, pp. 351–362, 2009.
© Springer-Verlag Berlin Heidelberg 2009

better in several ways. If no deletions occur, a rank-balanced tree is exactly an AVL tree; with deletions, its height is at most that of an AVL tree with the same number of insertions but no deletions. Rank-balanced trees are a proper subset of red-black trees, with a different balance rule and different rebalancing algorithms. Insertion and deletion take at most two rotations in the worst case and $O(1)$ amortized time; red-black trees need three rotations in the worst case for a deletion. Insertion and deletion can be done top-down with fixed look-ahead in $O(1)$ amortized rebalancing time per update.

Our new analyses use an exponential potential function to measure the amortized efficiency of operations on a balanced tree as a function of the heights of its nodes. We use this method to show that rebalancing in rank-balanced trees affects nodes exponentially infrequently in their heights. This is true of both bottom-up and top-down rebalancing.

This paper contains five sections in addition to this introduction. Section 2 gives our tree terminology. Section 3 introduces rank-balanced trees and presents and analyzes bottom-up rebalancing methods for insertion and deletion. Section 4 presents and analyzes top-down rebalancing methods. Section 5 develops our method of using an exponential potential function for amortized analysis, and with it shows that rebalancing affects nodes with a frequency that is exponentially small in their heights. The concluding Section 6 compares rank-balanced trees with red-black trees.

2   Tree Terminology

A *binary tree* is an ordered tree in which each node x has a *left child* $left(x)$ and a *right child* $right(x)$, either or both of which may be missing. Missing nodes are also called *external*; non-missing nodes are *internal*. Each node is the *parent* of its children. We denote the parent of a node x by $p(x)$. The *root* is the unique node with no parent. A *leaf* is a node with both children missing. The *ancestor*, respectively *descendant* relationship is the reflexive, transitive closure of the parent, respectively child relationship. If node x is an ancestor of node y and $y \neq x$, x is a *proper ancestor* of y and y is a *proper descendant* of x. If x is a node, its *left*, respectively *right* subtree is the binary tree containing all descendants of $left(x)$, respectively $right(x)$. The *height* $h(x)$ of a node x is defined recursively by $h(x) = 0$ if x is a leaf, $h(x) = \max\{h(left(x)), h(right(x))\} + 1$ otherwise. The height h of a tree is the height of its root.

We are most interested in binary trees as search trees. A binary search tree stores a set of *items*, each of which has a *key* selected from a totally ordered universe. We shall assume that each item has a distinct key; if not, we break ties by item identifier. In an *internal binary search tree*, each node is an item and the items are arranged in *symmetric order*: the key of a node x is greater, respectively less than those of all items in its left, respectively right subtree. Given such a tree and a key, we can search for the item having that key by comparing the key with that of the root. If they are equal, we have found the desired item. If the search key is less, respectively greater than that of the root, we search recursively in the left, respectively right subtree of the root. Each key comparison is a *step* of the search; the *current node* is the one whose key is compared with the search key. Eventually the search either locates the desired item or reaches a missing node, the left or right child of the last node reached by the search in the tree.

To insert a new item into such a tree, we first do a search on its key. When the search reaches a missing node, we replace this node with the new item. Deletion is a little harder. First we find the item to be deleted by doing a search on its key. If neither child of the item is missing, we find either the next item or the previous item, by walking down through left, respectively right children of the right, respectively left child of the item until reaching a node with a missing left, respectively right child. We swap the item with the item found. Now the item to be deleted is either a leaf or has one missing child. In the former case, we replace it by a missing node; in the latter case, we replace it by its non-missing child. An access, insertion, or deletion takes $O(h + 1)$ time in the worst case, if h is the tree height.

An alternative kind of search tree is an *external binary search tree*: the external nodes are the items, the internal nodes contain keys but no items, and all the keys are in symmetric order. Henceforth by a binary tree we mean an internal binary search tree, with each node having pointers to its children. Our results extend to external binary search trees and to other binary tree data structures. We denote by n the number of nodes and by m and d, respectively, the number of insertions and the number of deletions in a sequence of intermixed searches, insertions, and deletions that starts with an empty tree. These numbers are related: $d = m - n$.

3 Rank-Balanced Trees

To make search, insertion, and deletion efficient, we limit the tree height by imposing a *rank rule* on the tree. A *ranked binary tree* is a binary tree each of whose nodes x has an integer *rank* $r(x)$. We adopt the convention that missing nodes have rank -1. The *rank* of a ranked binary tree is the rank of its root. If x is a node with parent $p(x)$, the *rank difference* of x is $r(p(x)) - r(x)$. We call a node an *i-child* if its rank difference is i, and an *i, j-node* if its children have rank differences i and j; the latter definition does not distinguish between left and right children and allows children to be missing.

Our initial rank rule is that every node is a 1,1-node or a 1,2-node. This rule gives exactly the AVL trees: each leaf is a 1,1-node of rank zero, the rank of each node is its height, and the left and right subtrees of a node have heights that differ by at most one. To encode ranks we store with each non-root node a bit indicating whether it is a 1- or 2-child. This is Brown's representation [5] of an AVL tree; in the original representation [1], each node stores one of three states, indicating whether its left or right subtree is higher or they have equal heights. The rank rule guarantees a logarithmic height bound. Specifically, the minimum number of nodes n_k in an AVL tree of rank k satisfies the recurrence $n_0 = 1, n_1 = 2, n_k = 1 + n_{k-1} + n_{k-2}$ for $k > 1$. This recurrence gives $n_k = F_{k+3} - 1$, where F_k is the kth Fibonacci number. Since $F_{k+2} \geq \phi^k$ [8], where $\phi = (1 + \sqrt{5})/2$ is the golden ratio, $k \leq \log_\phi n \leq 1.4404 \lg n$[1].

AVL trees support search in $O(\log n)$ time, but an insertion or deletion may cause a violation of the rank rule. To restore the rule, we change the ranks of certain nodes and do rotations to rebalance the tree. A *promotion*, respectively *demotion* of a node x increases, respectively decreases its rank by one. A *rotation* at a left child x with parent

---

[1] We denote by lg the base-two logarithm.

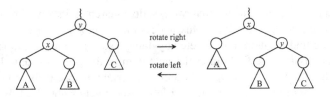

Fig. 1. Right rotation at node x. Triangles denote subtrees. The inverse operation is a left rotation at y.

y makes y the right child of x while preserving symmetric order; a rotation at a right child is symmetric. (See Figure 1.) A rotation takes $O(1)$ time.

In the case of an insertion, if the parent of the newly inserted node was previously a leaf, the new node will have rank difference zero and hence violate the rank rule. Let q be the newly added node, and let p be its parent if it exists, null if not. After adding q, we rebalance the tree by repeating the following step until a case other than promotion occurs (see Figure 2):

Insertion Rebalancing Step at p:

Stop: Node p is null or q is not a 0-child. Stop.

In the remaining cases q is a 0-child. Let s be the sibling of q, which may be missing.

Promotion: Node s is a 1-child. Promote p. This repairs the violation at q but may create a new violation at p. Node p now has exactly one child of rank difference one, namely q. Replace q by p. Let p be the parent of q if it exists, null if not.

In the remaining cases s is a 2-child. Assume q is the left child of p; the other possibility is symmetric. Let t be the right child of q, which may be missing.

Rotation: Node t is a 2-child. Rotate at q and demote p. This repairs the violation without creating a new one. Stop.

Double Rotation: Node t is a 1-child. Rotate at t twice, making q its left child and p its right child. Promote t and demote p and q. This repairs the violation without creating a new one. Stop.

During rebalancing there is at most one violation of the rank rule: node q may be a 0-child. Rebalancing walks up the path from the newly inserted node to the root, doing zero or more promotion steps followed by one non-promotion step. The first step is either a stop or a promotion. After one promotion step, node q is always a 1,2-node. The *rank* of the insertion is the rank of p in the last step, just before the step occurs; if p is null, the rank of the insertion is the rank of q in the last step (a stop).

One can do a deletion in an AVL tree similarly [6] [8, pp. 465-468], but the rebalancing may require a logarithmic number of rotations, rather than the one or two needed for an insertion. To reduce this number, we relax the rank rule to allow non-leaf 2,2-nodes as well as 1,1- and 1,2-nodes; leaves must still be 1,1-nodes. We call the resulting

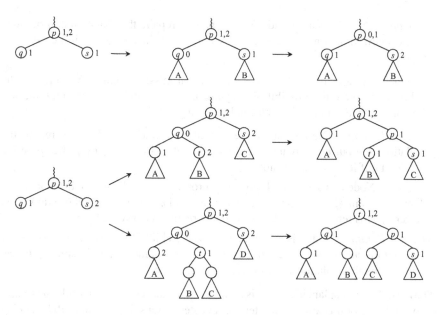

Fig. 2. Rebalancing after an insertion. Numbers are rank differences. The first case is non-terminating.

trees *rank-balanced trees* or *rb*-trees (not to be confused with red-black trees, which are equivalent to ranked binary trees with a different rank rule). One bit per non-root node still suffices to encode the rank differences. An AVL tree is just an rb-tree with no 2,2-nodes. The rank of an rb-tree is at least its height and at most twice its height.

Theorem 1. *The rank and hence the height of an rb-tree is at most* $2 \lg n$.

Proof. The minimum number of nodes n_k in an rb-tree of rank k satisfies the recurrence $n_0 = 1, n_1 = 2, n_k = 1 + 2n_{k-2}$ for $k \geq 2$. By induction $n_k \geq 2^{\lceil k/2 \rceil}$.   □

Insertion is the same in rb-trees as in AVL trees: insertion rebalancing steps do not create 2,2-nodes (but can destroy them). A deletion in an rb-tree can violate the rank rule by creating a node of rank one with two missing children or a node of rank two with a (missing) 3-child. Let q be the node that replaces the deleted node (q can be a missing node), and let p be its parent if it exists, null if not. We repair the violation by walking up the path to the root, repeating the following step until a case other than demotion or double demotion occurs (see Figure 3):

Deletion Rebalancing Step at p:

Stop: Node p is null, or q is not a 3-child and p is not a 2,2-node of rank 1. Stop.

In the remaining cases node q is a 2- or 3-child. Let s be the sibling of q, which may be missing.

Demotion: Node s is a 2-child. Demote p. This repairs the violation at q but may create a new violation at p. Replace q by p. Let p be the parent of q if it exists, null if not.

In the remaining cases q is a 3-child and s is a non-missing 1-child. Assume q is the right child of p; the other possibility is symmetric. Let t and u be the right and left children of s, either or both of which can be missing.

Double Demotion: Nodes t and u are 2-children: Demote p and s. This repairs the violation at q but may create a new violation at p. Replace q by p. Let p be the parent of q if it exists, null if not.

Rotation: Node u is a 1-child. Rotate at s, promote s, and demote p. If t is missing, demote p again. (In this case q is also missing, and p is now a leaf, whose rank must be zero.) This repairs the violation without creating a new one. Stop.

Double Rotation: Node t is a 1-child and u is a 2-child. Rotate at t twice, making s its left child and p its right child. Promote t twice, demote s, and demote p twice. This repairs the violation without creating a new one. Stop.

During deletion rebalancing, there is at most one violation of the rank rule: p is a 2,2-node of rank one or q is a 3-child; after the first step, q must be a 3-child. Rebalancing walks up the path from the node that replaces the deleted node toward the root, doing zero or more demotion and double demotion steps followed by a stop, a rotation, or a double rotation. The *rank* of the deletion is the rank of p in the last step, just before the step occurs; if p is null, the rank of the deletion is the rank of p in the next-to-last step, just before the step occurs, or the rank of the deleted node if there is no next-to-last step (the root is deleted).

Deletion in rb-trees is only slightly more complicated than insertion, with two non-terminal cases instead of one. Deletion takes at most two rotations, the same as insertion.

The rebalancing process needs access to the affected nodes on the search path. To facilitate this, we can either add parent pointers to the tree or store the search path, either in a separate stack or by reversing pointers along the path. A third method is to maintain a *trailing node* during the search. This node is the topmost node that will be affected by rebalancing. In the case of an insertion, it is either the root or the parent of the nearest ancestor of the last node reached by the search that is a 2-child or a 1,2-node. In the case of a deletion, it is either the root or the parent of the nearest ancestor of the current node that is a 1-child or a 1,2-node whose 1-child is not a 2,2-node. In both cases, we initialize the trailing node to be the root and update it as the search proceeds. Once the search reaches the bottom of the tree, we do rebalancing steps (appropriately modified) top-down, starting from the trailing node. This method needs only $O(1)$ extra space, but it incurs additional overhead during the search and during the rebalancing, to maintain the trailing node and to determine the next node on the search path, respectively. Its big advantage is that it extends to top-down rebalancing with finite look-ahead, as we discuss in the next section.

With any of these methods, a search, insertion, or deletion takes $O(\log n)$ time worst-case. The number of rebalancing steps in an insertion or deletion is $\Theta(\log n)$ worst-case but $O(1)$ amortized. To obtain this bound, we use a standard method of amortized

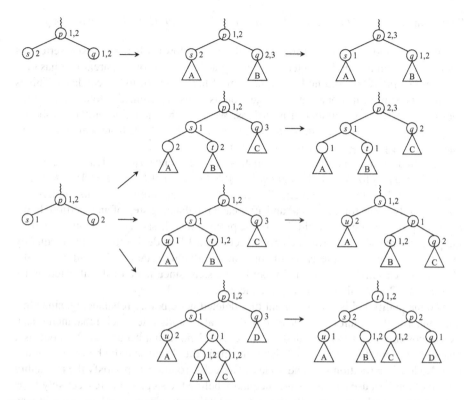

Fig. 3. Rebalancing after a deletion. Numbers are rank differences. The first two cases are non-terminating. If q is a 2-child, the first case only applies if p is a leaf. The third case assumes t is not a missing node; if it is, p is a leaf and is demoted.

analysis [14]. We assign to each state of the data structure a non-negative *potential* that is zero for an empty (initial) structure. We define the *amortized cost* of an operation to be its actual cost plus the net increase in potential it causes. Then the sum of the amortized costs is an upper bound on the sum of the actual costs for any sequence of operations that begins with an empty structure.

To analyze rb-tree rebalancing, we define the potential of a tree to be the number of non-leaf 1,1-nodes plus twice the number of 2,2-nodes. Each non-terminal insertion rebalancing step decreases the potential by one by converting a 1,1-node into a 1,2-node. Each non-terminal deletion rebalancing step except possibly the first decreases the potential by at least one, by converting a 2,2-node into a 1,2- or 1,1-node. The first deletion rebalancing step can increase the potential by one, by converting a 1,2-node into a 1,1-node. A terminal insertion or deletion rebalancing step increases the potential by at most two or three, respectively. This gives the following theorem:

Theorem 2. *The total number of rebalancing steps is at most $3m + 6d$.*

We conclude this section by deriving a bound on the height of rb-trees that is close to that of AVL trees unless there are almost as many deletions as insertions.

Theorem 3. *With bottom-up rebalancing, the height of an rb-tree is at most* $\log_\phi m$.

Proof. We define a *count* $\kappa(x)$ for each node x, as follows: when x is first inserted, its count is 1; when a node is deleted, its count is added to that of its parent if it has one. The *total count* $K(x)$ of a node x is the sum of the counts of its descendants. This is equal to the sum of its count and the total counts of its children. The total count of the root is at most m, the number of insertions. We prove by induction on the number of rebalancing steps that if a node x has rank k, $K(x) \geq F_{k+3} - 1$, from which it follows that $m \geq F_{k+3} - 1 \geq \phi^k$, giving the theorem.

We noted earlier that $F_{k+3} - 1$ satisfies the recurrence $x_0 = 1, x_1 = 2, x_k = 1 + x_{k-1} + x_{k-2}$ for $k > 1$. This gives $K(x) \geq F_{k+3} - 1$ if $k = 0$; $k = 1$; or $k > 1$, x is a 1,1- or 1,2-node, and the inequality holds for both children of x. This implies that the inequality holds for a new leaf and after each rebalancing step of an insertion. In the case of a promotion step, the children of the promoted node satisfy the inequality before the promotion; since the promoted node becomes a 1,2-node, it satisfies the inequality after the promotion. In the cases of rotation and double rotation, the children of the affected nodes satisfy the inequality before the step; since none of the affected nodes becomes a 2,2-node, they all satisfy the inequality after the step.

The inequality holds for the parent of a deleted node before rebalancing, since this node inherits the count of the deleted node. It also holds after each rebalancing step except possibly at a newly created 2,2-node. A 1,2-node that becomes a 2,2-node as a result of the demotion of a child satisfies the inequality because it did before the demotion. Node s in a rotation step and node t in a double rotation step satisfy the inequality after the step because p satisfies the inequality before the step, and s, respectively t has the same rank and count after the step as p did before it. The only other case of a new 2,2-node is node p in a rotation step if p is not a leaf. For p to become a 2,2-node, q cannot be missing. Either q was demoted by the previous rebalancing step, or q is a leaf whose parent was deleted. In the former case, q satisfies the inequality at rank $k - 1$ before its demotion, where k is the new rank of p. Since t, the other child of p, satisfies the inequality at rank $k - 2$, p satisfies the inequality as well. In the latter case p has new rank two and has total count at least four, so it satisfies the inequality. $\qquad\square$

4   Top-Down Rebalancing

The method of rebalancing using a trailing node described in Section 3 does the rebalancing top-down rather than bottom-up. We can modify this method to use fixed look-ahead. If the look-ahead is sufficiently large, the amortized number of rebalancing steps per update remains O(1). The idea is to force a reset of the trailing node after sufficiently many search steps. In an insertion, if the current node of the search is a 1,1-node whose parent is a 1,1-node, we can force the next search step to do a reset by promoting the current node and rebalancing top-down from the trailing node. In a deletion, if the current node is a 2,2 node or a 1,2-node whose 1-child is a 2,2-node, we can force the next step to do a reset by demoting the current node in the former case or the current node and its 1-child in the latter case, and rebalancing top-down from the trailing node.

Forcing a reset as often as possible minimizes the look-ahead, but if we force a reset less often we can guarantee $O(1)$ amortized rebalancing steps per update. To demonstrate this, we use the same potential function as in Section 3. In an insertion, if a search step does not do a reset, every node along the search path from the grandchild of the trailing node to the parent of the current node is a 1,1-node. Thus if we force a reset after five search steps that do not do a reset (by promoting the fifth 1,1-node in a row), each rebalancing to force a reset decreases the potential: the potential of the current node increases by one, each of the four non-terminal rebalancing steps decreases the potential by one, and the last rebalancing step increases it by at most two. A forced reset takes $O(1)$ time including rebalancing. If we scale this time to be at most one, the amortized time of a forced reset is non-positive. In a deletion, if a search step does not do a reset, every node along the search path from the grandchild of the trailing node to the parent of the current node is a 2,2-node or a 1,2-node whose 1-child is a 2,2-node. If we force a reset after five search steps that do not do a reset (by doing a demotion or a double demotion at the fifth node in a row that is a 2,2-node or a 1,2-node whose 1-child is a 2,2-node), each rebalancing to force a reset decreases the potential: decreasing the rank of the current node and possibly that of its child does not increase the potential, each of the four non-terminal rebalancing steps decreases the potential by one, and the last step increases it by at most three. In either an insertion or deletion, any rebalancing at the bottom of the search path takes $O(1)$ amortized time. This gives the following theorem:

Theorem 4. *Top-down rebalancing with sufficiently large fixed look-ahead does $O(1)$ amortized rebalancing steps per insertion or deletion.*

Theorem 4 remains true as long as every forced reset reduces the potential. One disadvantage of top-down rebalancing is that the proof of Theorem 3 breaks down: the induction does not apply to the 2,2-nodes created by forced resets during insertions.

5   Rank-Based Analysis

The amortized analysis of bottom-up rebalancing in Section 3 implies that most rebalancing steps are low in the tree: in a sequence of m insertions and d deletions, there are $O((m+d)/k)$ insertions and deletions of rank k or greater. Something much stronger is true, however: for some $b > 1$, there are only $O((m+d)/b^k))$ insertions and deletions of rank k or greater. That is, the frequency of rebalancing steps decreases exponentially with height. This is true (and easy to prove) for weight-balanced trees if one ignores the need to update size information, but to our knowledge ours is the first such result for trees that use some form of height balance, and it covers rank changes as well as rotations. The result also holds for top-down rebalancing with sufficiently large fixed look-ahead, for a value of b that depends on the look-ahead.

It is convenient to assign potential to 1,2-nodes as well as to 1,1- and 2,2-nodes. We assign to a node of rank k a potential of Φ_k if it is a 1,1- or 2,2-node, or Φ_{k-2} if it is a 1,2-node, where Φ is a non-decreasing function such that $\Phi_0 = \Phi_{-1} = 0$, to be chosen later. The potential of a tree is the sum of the potentials of its nodes.

With this choice of potential, the potential change of a sequence of non-terminal rebalancing steps telescopes. Specifically, a non-terminal insertion rebalancing step at a node of rank k decreases the potential by $\Phi_k - \Phi_{k-1}$ and promotes the node to rank $k + 1$. Consecutive insertion rebalancing steps are at nodes that differ in rank by one. Thus a sequence of non-terminal insertion rebalancing steps starting at a node of rank 0 and ending at a node of rank k decreases the potential by $\Phi_k - \Phi_{-1} = \Phi_k$, since $\Phi_{-1} = 0$. A non-terminal deletion rebalancing step at a node of rank k decreases the potential by $\Phi_k - \Phi_{k-3}$ if it is a demotion of a non-2,2-node and by $\Phi_{k-1} + \Phi_{k-2} - \Phi_{k-2} - \Phi_{k-3} = \Phi_{k-1} - \Phi_{k-3}$ if it is a double demotion. Since Φ is non-decreasing, the potential decrease is at least $\Phi_{k-1} - \Phi_{k-3}$ in either case. Consecutive deletion rebalancing steps are at nodes that differ in rank by two. If the first rebalancing step is a demotion of a 2,2-node of rank one, the step does not change the potential, because the demoted node was a 1,2-node of rank one before the deletion. Thus a sequence of non-terminal deletion rebalancing steps starting at a node of rank 1 or 2 and ending at a node of rank k decreases the potential by at least $\Phi_{k-1} - \Phi_0 = \Phi_{k-1}$.

We can compute the total potential change caused by an insertion or deletion by combining the effect of the sequence of non-terminal rebalancing steps with that of the initialization and the terminal step. In an insertion, the initialization consists of adding a new leaf, which has potential $\Phi_0 = 0$. Let k be the rank of the insertion. Consider the last rebalancing step. (See Figure 2.) Suppose this step is a stop. If p is null, then either the insertion promotes the root and decreases the potential by at least Φ_k, or $k = 0$, the insertion is into an empty tree, and it does not change the potential. If p is not null but becomes a 1,2-node, the insertion decreases the potential by at least Φ_k; if p is not null but becomes a 1,1-node, the insertion increases the potential by at most $\Phi_k - 2\Phi_{k-2}$. Suppose the last step is a rotation. Then the insertion increases the potential by at most $\Phi_k - 2\Phi_{k-2}$. Finally, suppose the last step is a double rotation. Then the insertion increases the potential by at most $\Phi_k - 2\Phi_{k-2}$: node t is a 1,1- or 1,2-node before the last step. In all cases the potential increase is at most $\max\{-\Phi_k, \Phi_k - 2\Phi_{k-2}\}$.

In a deletion, the initialization consists of deleting a leaf, or deleting a node with one child and replacing it by its child. The deleted node has potential zero before it is deleted. Let k be the rank of the deletion. Consider the last rebalancing step. (See Figure 3.) Suppose this step is a stop. If p is null, then either the deletion demotes the root and decreases the potential by at least Φ_{k-1}, or $k \leq 1$ and the deletion deletes the root and does not change the potential. If p is not null but becomes a 1,2-node, the deletion decreases the potential by at least Φ_k; if p is not null but becomes a 2,2-node, the deletion increases the potential by at most $\Phi_k - 2\Phi_{k-2}$. Suppose the last step is a rotation. Then the deletion increases the potential by at most $\Phi_{k-3} - \Phi_{k-1} - \Phi_{k-3} \leq 0$ if node s is a 1,1-node, by at most $\Phi_{k-1} - 2\Phi_{k-3}$ if s is a 1,2-node of rank at least two, and by Φ_2 if s is a 1,2-node of rank one; in the third case, node p is a leaf after the rotation and is demoted. Finally, suppose the last step is a double rotation. Then the deletion increases the potential by at most $\Phi_k - 2\Phi_{k-3}$ if node u is a 1,1- or 1,2-node before the last step or by at most $\Phi_k + 2\Phi_{k-4} - 2\Phi_{k-2} - 2\Phi_{k-3} \leq \Phi_k - 2\Phi_{k-2}$ if node u is a 2,2-child before the rotation. In all cases the potential increase is at most $\max\{-\Phi_{k-1}, \Phi_k - 2\Phi_{k-3}\}$.

For $i \geq 1$, let $\Phi_i = b^i$, where $b = 2^{1/3}$. An insertion or deletion of rank at most 3 increases the potential by $O(1)$. Since $b^2 - 2 < 0$ and $b^3 - 2 = 0$, an insertion or deletion of rank 4 or more does not increase the potential. We prove that insertions and deletions of a given rank occur exponentially infrequently by stopping the growth of the potential at a corresponding rank. Specifically, for a fixed rank $k \geq 4$ and arbitrary $i \geq 1$, let $\Phi_i = b^{\min\{i,k-3\}}$. Then an insertion or deletion of rank at most 3 increases the potential by $O(1)$, and an insertion or deletion of rank greater than 3 and less than k does not increase the potential, but an insertion or deletion of rank k or greater decreases the potential by at least b^{k-3}. This gives the following theorem:

Theorem 5. *In a sequence of m insertions and d deletions with bottom-up rebalancing in an initially empty rank-balanced tree, there are $O((m + d)/2^{k/3})$ insertions and deletions of rank k or more, for any k.*

The base of the exponent in Theorem 5 can be increased to $1.32+$ by separately analyzing insertions and deletions (proof omitted). A result similar to Theorem 5 holds for top-down rebalancing (proof omitted):

Theorem 6. *A sequence of m insertions and d deletions with top-down rebalancing in an initially empty tree does $O((m + d)/b^k)$ rebalancing steps at nodes of rank k if forced resets occur after six search steps in an insertion, four in a deletion, where $b = 1.13+$. The base b can be increased arbitrarily close to $2^{1/3}$ by increasing the fixed lookahead.*

It is possible to improve the base in both Theorem 5 and Theorem 6 at the cost of making deletion rebalancing a little more complicated, specifically by changing the double rotation step of deletion rebalancing to promote p if it is a 1,1-node after the step, or promote s if it but not p is a 1,1-node after the step. With this change Theorem 5 holds for a base of $2^{1/2}$, and Theorem 6 holds for a base of $b = 1.17+$ even if deletion does a forced reset after only three search steps. By increasing the fixed lookahead, the base in Theorem 6 can be increased arbitrarily close to $2^{1/2}$. Unfortunately this change in deletion rebalancing invalidates the proof of Theorem 3.

6   Rank-Balanced Trees versus Red-Black Trees

Rank-balanced trees have properties similar to those of red-black trees but better in several respects. Rank-balanced trees are a proper subset of red-black trees (proof omitted):

Theorem 7. *The nodes of an rb-tree can be assigned colors to make it a red-black tree. The nodes of a red-black tree can be assigned ranks to make it an rb-tree if and only if it does not contain a node x such that there is a path of all black nodes from x to a leaf and another path of nodes alternating in color from x to a red leaf.*

The height bound for AVL trees holds for rb-trees as long as there are no deletions, and holds in weakened form even with deletions (Theorem 3) if rebalancing is bottom-up. On the other hand, the height of a red-black tree can be $2 \lg n - O(1)$ even without deletions. Red-black trees need up to three rotations per deletion, rb-trees only two.

We conclude that the differences between rb-trees and red-black trees favor rb-trees, especially the height bound of Theorem 3. Guibas and Sedgewick, in their classic paper on red-black trees [7], considered in passing the alternative of allowing rank differences

1 and 2 instead of 0 and 1, but said, "We have chosen to use zero weight links because the algorithms appear somewhat simpler." Our results demonstrate the advantages of the alternative. We think that rank-balanced trees will prove efficient in practice, and we intend to do experiments to investigate this hypothesis.

References

1. Adel'son-Vel'skii, G.M., Landis, E.M.: An algorithm for the organization of information. Sov. Math. Dokl. 3, 1259–1262 (1962)
2. Andersson, A.: Balanced search trees made simple. In: Dehne, F., Sack, J.-R., Santoro, N. (eds.) WADS 1993. LNCS, vol. 709, pp. 60–71. Springer, Heidelberg (1993)
3. Bayer, R.: Binary B-trees for virtual memory. In: SIGFIDET, pp. 219–235 (1971)
4. Bayer, R.: Symmetric binary B-trees: Data structure and maintenance algorithms. Acta Inf. 1, 290–306 (1972)
5. Brown, M.R.: A storage scheme for height-balanced trees. Inf. Proc. Lett., 231–232 (1978)
6. Foster, C.C.: A study of avl trees. Technical Report GER-12158, Goodyear Aerospace Corp. (1965)
7. Guibas, L.J., Sedgewick, R.: A dichromatic framework for balanced trees. In: FOCS, pp. 8–21 (1978)
8. Knuth, D.E.: The Art of Computer Programming, Sorting and Searching, vol. 3. Addison-Wesley, Reading (1973)
9. Nievergelt, J., Reingold, E.M.: Binary search trees of bounded balance. SIAM J. on Comput., 33–43 (1973)
10. Olivié, H.J.: A new class of balanced search trees: Half balanced binary search trees. ITA 16(1), 51–71 (1982)
11. Sedgewick, R.: Left-leaning red-black trees,
    http://www.cs.princeton.edu/rs/talks/LLRB/LLRB.pdf
12. Skiena, S.S.: The Algorithm Design Manual. Springer, Heidelberg (1998)
13. Tarjan, R.E.: Updating a balanced search tree in $O(1)$ rotations. Inf. Proc. Lett. 16(5), 253–257 (1983)
14. Tarjan, R.E.: Amortized computational complexity. SIAM J. Algebraic and Disc. Methods 6, 306–318 (1985)
15. Tarjan, R.E.: Efficient top-down updating of red-black trees. Technical Report TR-006-85, Department of Computer Science, Princeton University (1985)

Approximation Algorithms for Finding a Minimum Perimeter Polygon Intersecting a Set of Line Segments

Farzad Hassanzadeh and David Rappaport*

Queen's University, Kingston, Ontario, Canada
{farzad,daver}@cs.queensu.ca

Abstract. Let S denote a set of line segments in the plane. We say that a polygon P intersects S if every segment in S has a non-empty intersection with the interior or boundary of P. Currently, the best known algorithm finding a minimum perimeter polygon intersecting a set of line segments has a worst case exponential running time. It is also still unknown whether this problem is NP-hard. In this note we explore several approximation algorithms. We present efficient approximation algorithms that yield good empirical results, but can perform very poorly on pathological examples. We also present an $O(n \log n)$ algorithm with a guaranteed worst case performance bound that is at most $\pi/2$ times that of the optimum.

Keywords: Computational Geometry, Line Segment, Intersecting Polygon, Approximation Algorithm.

1 Introduction

The convex hull of a set of points in the plane is often defined as the smallest convex subset of the plane that intersects all of the points. The qualifying term smallest can be thought of as the area of this convex subset, or alternatively, smallest can refer to the perimeter of the set. Using the latter point of view we can say the convex hull of a set of points is the polygon with the smallest perimeter that intersects the points. We examine an analogous situation where the input is a set of line segments. The convex hull of a set of line segments contains the segments. In this paper the goal is to find a polygon, with minimum perimeter whose interior and boundary have a non-empty intersection with a set of segments. We use the term *minimum perimeter intersecting polygon* abbreviated as MPIP. The convex hull of a set of segments is unique. However, an MPIP may not be unique. It is well known that the convex hull of a set of segments can be determined in $O(n \log n)$ time by computing the convex hull of the endpoints of the segments, see for example [4,7,9]. However, to our knowledge there is no known polynomial time algorithm to determine a minimum perimeter polygon

---

* Supported by an NSERC of Canada Discovery Grant.

F. Dehne et al. (Eds.): WADS 2009, LNCS 5664, pp. 363–374, 2009.

that intersects a set of line segments, and furthermore the problem is not known to be NP-hard [10,15].

When the input line segments are in a fixed number of directions then a smallest perimeter polygon intersecting the segments can be found in $O(n \log n)$ time [15]. However, this algorithm incurs a cost that is exponential in the number of directions. A recent paper by Löffler and van Kreveld [10] has recently examined this very same problem in the context of computing with imprecise points. Among a variety of other algorithms (to be described in the next paragraph), they present an $O(n \log n)$ algorithm to determine a smallest perimeter polygon intersecting a set of parallel (uni-directional) line segments. This result mirrors the result obtained by Meijer and Rappaport in 1990 [13].

Löffler and van Kreveld [12] look at different computational geometry problems for imprecise points. For a set of regions $\mathcal{L}$ and a problem that takes a set of points as input and returns a real number R, they are interested in finding a way to place one point in each region of $\mathcal{L}$ such that the resulting set of points minimizes or maximizes R. The imprecision for a point can be modelled in different ways. A disc or a square can represent a point such that the point can be anywhere inside or on its boundary. The imprecision can also be modelled as a line segment, in which the point can be anywhere on the line segment. Löffler and van Kreveld [12] study the problem of minimizing and maximizing the diameter, smallest bounding box, smallest enclosing circle, etc, for regions modelled as squares and discs. They propose polynomial time algorithms for some of the problems and prove NP-hardness for the others. Recently, Mukhopadhyay et al., [14] have published an algorithm that is similar to that of Löffler and van Kreveld for determining a minimum area polygon that intersects a set of parallel segments.

In a similar vein the problem of determining whether the boundary of a convex polygon intersects with a set of line segments was proposed by Tamir [16]. A special case of the problem was solved by Goodrich and Snoeyink [6] who present an $O(n \log n)$ algorithm to determine whether the boundary of a convex polygon intersects with a set of parallel line segments, and an $O(n^2)$ algorithm to obtain such a polygon with minimum area, or perimeter.

The remainder of the paper is organized as follows. First we set down some requisite definitions and notation. We then review the algorithm in [15] as it forms a basis for heuristic approximation methods. These heuristics have been implemented and we report on empirical results that were obtained. Finally we present an approximation algorithm with guaranteed performance bounds.

2   Preliminaries and Definitions

We begin by introducing some notation and definitions. For the most part we adopt the notation used in [15].

Let S denote a set of line segments, where each segment $s \in S$ is the convex combination of the endpoints of s $a(s)$ and $b(s)$. It will be useful to allow degenerate segments, that is, segments that consist of a single unique endpoint.

We wish to determine a simple polygon P with minimum perimeter, such that the intersection of the interior and boundary of P has a non-empty intersection with every segment in S. If S admits a line transversal, that is the segments are intersected by a single line, then we can apply the algorithm of Battacharya and Toussaint [2] to find a shortest stabbing segment of S. Therefore, we assume throughout that the input we receive has passed through a preliminary step and we only proceed if S does not admit a line transversal.

Let $H^l(L)$ be the closed left half-plane bounded by a directed line L. We say that $H^l(L)$ is a *stabbing half-plane* of the set of segments S, if $H^l(L)$ has a non empty intersection with every $s \in S$. We say that L is an *extreme* line if $H^l(L)$ is a stabbing half-plane and no proper subset of $H^l(L)$ is. An endpoint $a(s)$ of a line segment s is a *critical* point if $a(s)$ lies on the extreme line and no point of s intersects the interior of $H^l(L)$. A line segment $s \in S$ with one or more critical points is a *critical segment*. An extreme line is a *critical extreme line*, if it passes through at least two critical points of two different line segments.

We partition the segments in S into two classes. The *normal segments* are segments with at most one point contained by the same extreme line, and *rim segments*, are non-degenerate segments collinear with an extreme line. We further partition the rim segments. If there is a directed line L passing through a rim segment s such that all critical points are contained in $H^l(L)$ then we say that s is a *nice* rim segment. If a rim segment is not nice then we call it a *pesky rim segment*. See Fig. 1.

This classification of line segments plays a crucial role in determining an MPIP of a set of segments. In [15] an $O(n \log n)$ algorithm is given to compute an MPIP for a set of line segments S with no pesky rim segments. On the other hand if the set of segments does contain some positive number, say k, pesky rim segments, then the computational complexity explodes to being exponential in k. The crux of the matter is that there is a local decision that can be made regarding the interaction of an MPIP and a normal or nice rim segment. For the pesky segment there are four possible outcomes and there is no known method to decide locally which outcome to adopt. A brute force method of testing all

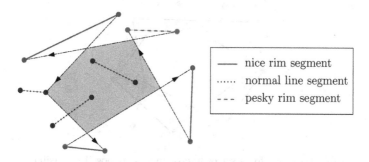

Fig. 1. Directed segments denote critical extreme line. The nice rim segments are shown as solid line segments. The normal line segments are dotted and the pesky segments are dashed.

possible outcomes leads to the exponential behaviour of the algorithm. We will propose three distinct methods for obtaining approximations for an MPIP of a set of segments. The first two are straight forward modifications of the algorithm proposed in [15]. We begin by sketching the algorithm.

It is well known that a ray of light that hits a reflective surface will have equal angles of incidence and reflection. In [15] it is proved that every MPIP of S must intersect a rim segment at a single point. This point may be an endpoint, or it may be interior to the segment. For any nice rim segment $s \in S$, if the MPIP intersects s at an interior point then the two exterior angles that the MPIP makes with s are equal, thus this vertex is called a *reflection vertex*. After sorting the critical extreme lines, starting at any nice rim segment, we can choose a point (reflection vertex) on that nice rim segment, move to the next nice rim segment (using the sorted critical extreme lines [15]), and meanwhile, cross (or include) all the normal line segments in between. In this traversal, two nice rim segments are called *neighbours* if there is no nice rim segment between them.

Consider two neighbour nice rim segments s and t. In [15] it was shown that the part of the boundary between s and t of every MPIP is constrained to lie in a region formed by taking the intersection of the left half planes of all the critical extreme lines between s and t and subtracting it from the convex hull of s, t and the critical points between them. We denote this region as a *feasible polygon*. This situation repeats for every pair of neighbouring nice rim segments, yielding a union of feasible polygons we call a *feasible cycle*. See Fig. 2. The task at hand is to find a minimum perimeter polygon lying within the feasible cycle.

One way to do this is to *flatten* the feasible cycle and determine a number of shortest paths. Consider a sequence of three consecutive nice rim segments r, s and t. This will yield two feasible polygons, defined respectively by the pairs r and s and s and t, with a common edge s. The flattening process flips about the common edge s so that the intersection point of any straight line passing through s will result in a reflection vertex in the pre-flipped feasible polygons. A repetition of r in the sequence results in three flattened feasible polygons with two copies of the edge r, say r' and r''. The example in Fig. 3 shows the

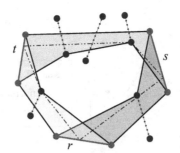

Fig. 2. A feasible polygons for a set of line segments. A possible feasible cycle is also shown as a dash-dotted cycle. Since this cycle is located inside the feasible polygons, it is guaranteed that the cycle includes at least one point of each line segment inside or on its boundary.

Fig. 3. Here we see the flattened feasible polygons, illustrating how there are two copies, r' and r'' of the segment r

situation with exactly 3 nice rim segments. Every point on r' has its twin on r''. In [15] it is shown that every shortest path from a point on r' to its twin on r'' lying within the flattened polygon is a polygon transversal of the segments. Furthermore a point and its twin that realizes an MPIP can be found using the flattened polygon in $O(n \log n)$ time.

The occurrence of pesky rim segments makes the previous approach to finding an MPIP unwieldy. For nice rim segment s an MPIP either passes through an endpoint of s or has a reflection vertex incident to s. Both these occurrences are handled by the method using the flattened polygon. A pesky rim segment s may interact with an MPIP P in one of four ways. If the input consists of k pesky rim segments the algorithm presented in [15] considers all possible combinations of the choices resulting in an algorithm that is exponential in k, and k may be in $O(n)$.

It will be most convenient to describe how a pesky rim segment p^* is handled in terms of its interaction with the an MPIP P, and in terms of constructing a flattened polygon.

Case 1. MPIP P has no vertices incident to p^*, there is no critical polygon associated to p^*, see Fig. 4(A).

Case 2. MPIP P has a reflection vertex incident to p^*, we treat p^* as a nice rim segment and construct the appropriate critical polygons, see Fig. 4(B).

Case 3, 4. MPIP P is incident to an endpoint of p^*, we treat each endpoint of p^* independently as if it were the endpoint of a normal segment, see Fig. 5.

3   The Incremental Approach

The time complexity of the brute force algorithm is exponential as a result of testing all possible outcomes. The worst case example occurs when all the line segments are pesky rim segments. In this section we introduce an approximation method, the incremental approach, which is a direct modification of the brute force algorithm explained in the previous section [15]. The idea behind the incremental approach is to locally resolve the pesky rim segments in order to decrease the time complexity of the algorithm.

The incremental approach consists of two steps. In the first step, the *preparation* step, we remove all the unnecessary line segments from the set of line segments S. A line segment is unnecessary if and only if none of its endpoints

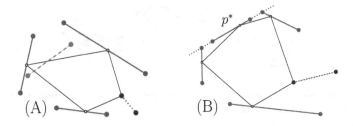

Fig. 4. (A) Case 1: Ignoring a pesky rim segment completely may have no effect on the MPIP, as shown in this example. (B) Case 2: If we eliminate all critical points from S that are on the wrong side of the line through the pesky segment p^* then we can consider the segment as if it is a nice rim segment. In this case we convert a pesky rim segment into a nice rim segment by clipping its neighbour rim segments.

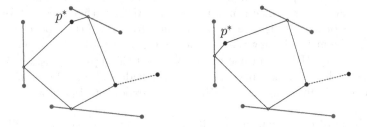

Fig. 5. Cases 3 and 4: Replacing p^* with one of its endpoints, yields results that differ from including the entire segment as in case 2

are critical. Then we replace each normal line segment l with an endpoint of l that is critical. In the second step, the *resolving* step, we remove the pesky rim segments from S one by one until S includes no pesky rim segments. The removed pesky rim segments are stored in Q. Then for each pesky rim segments p_i in Q, we add p_i back to S, we investigate the four different cases explained in Section 2 in order to resolve p_i and we replace p_i with the best case. After resolving every $p_i \in Q$ we end up with an approximated MPIP.

The preparation step of the incremental approach is trivial and can be done in $O(n)$ time. In the resolving step, we first remove all the pesky rim segments in $O(n)$. Then for each pesky rim segment p_i, we resolve p_i in $O(n \log n)$ time [15]. In the worst case, where all the line segments are pesky rim segments, the time complexity of the incremental approach is $O(n^2 \log n)$.

4   The Trimming Approach

In the incremental approach we locally check the four different cases for each pesky rim segment in order to approximate an MPIP of a set of line segments S. In the trimming approach, we only check the second case for each pesky rim segment. More precisely, for a set of line segments S with k pesky rim segments, only one combination out of the 4^k possible combinations of the pesky rim segments is investigated, and that is when all the pesky rim segments are treated as nice rim segments. In order to do so, we choose a pesky rim segments p_i and convert it to a nice rim segment by trimming off the parts of all the other line segments that do not intersect $H^l(p_i)$. We repeat this step for each remaining pesky rim segment. Note that after each iteration (trimming), one or more pesky rim segments may become non-pesky as a result of trimming the other segments (consult Fig. 6). The iteration stops when there is no pesky rim segment left. Then an MPIP of the remaining set of line segments is computed using the algorithm discussed in Section 2.

In the first phase of the trimming algorithm we may have to perform a trim operation $O(n^2)$ times. After trimming we simply apply a special case of the algorithm described in Section 2 and since there are no more pesky segments the cost of this phase is in $O(n \log n)$.

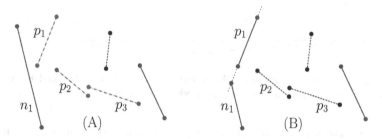

Fig. 6. (A) a set of line segments. (B) Converting p_1 into a nice rim segment by trimming n_1. Notice how p_2 and p_3 become non-pesky as a result of trimming n_1.

5   Experimental Results

For a specific set of line segments S, we feed S to the brute force algorithm and also to our two heuristics, and denote the results by $\mathcal{R}$ and $\mathcal{R}'$ respectively. Then we define the percentage error as $\frac{\mathcal{R}'-\mathcal{R}}{\mathcal{R}} \times 100$.

Table 1 shows some empirical results indicating the percentage error for different numbers of pesky rim segments. Each error value in the table is the average of the error values for 20 random sets of line segments with a specific number of pesky rim segments. Although on average these heuristics yield close-to-optimum results, the errors are not bounded in the worst case.

There are examples in which the incremental approach results in arbitrarily bad approximations. Consider the example in Fig. 7. The MPIP of the line

Table 1. The percentage error of the incremental approach and the trimming approach

segments	1	2	3	4	5	6	7	8	9	10
incremental	0	0.028	0.092	0.154	0.120	0.183	0.169	0.181	0.164	0.131
trimming	1.176	0.172	0.515	0.286	0.762	0.866	0.726	1.005	0.926	0.848

Fig. 7. On the left a set of segments with optimal MPIP. In the centre and right we see two successive steps of the incremental approach that lead to a poor approximation of the optimal MPIP.

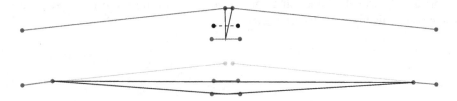

Fig. 8. Above we see the optimal MPIP, and below a solution obtained by applying the trimming approach that is a poor approximation of the optimum

segments is shown on the left. In the centre we the results after inserting one of the pesky segments and on the right we see the results after both pesky segments have been inserted. In general it is easy to construct examples where the incremental approach may yield an approximation that is arbitrarily bad.

The trimming approach is also prone to unbounded error on some pathological examples. Consider the example shown in Fig. 8. By making the upper two nice rim segments more horizontal the approximation error can be increased without bound.

Our experiments suggest that these heuristics perform quite well on some small randomly generated examples, however, in some cases these heuristics may produce very large errors. In the next section we present an entirely different approximation algorithm that is guaranteed to produce an approximate solution that is not too far from the optimum.

6   Minimum Spanning Circle Approach

In this section, we propose a completely different approach to approximate the MPIP of a set of line segments S. The algorithm relies on finding a minimum spanning circle (MSC) of the segments, that is, given a set of line segments S we find a smallest perimeter circle that is a transversal of S. The MSC of a set of (possibly overlapping) line segments can be found in $O(n \log n)$ time [1]. We show that this approach results in an intersecting polygon with a perimeter that is at most $\frac{\pi}{2}$ times greater than the optimum answer.

6.1   Approximation Bounds

We begin by establishing the relation between the perimeter of an MPIP of S and the perimeter of an MSC of S. Throughout we use $w(\cdot)$ to denote the perimeter of the boundary of a planar region.

The following lemmas provide useful direction towards the main result:

Lemma 1. *For any convex polygon P with diameter D, $2D < w(P)$.*

Proof. Let P be a polygon with diameter D realized by a pair of vertices of P, a and b. Let c be any other vertex of P such that a, b, c, is a triangle, $\triangle abc$. Clearly the perimeter of P, $w(P)$, is at least the perimeter of $\triangle abc$. Furthermore, by the triangle inequality $2D$ is less than the perimeter of $\triangle abc$. Therefore, $2D < w(P)$. □

Observation 1. *Let P be a polygon and C an MSC of P. The boundary of C either passes through exactly two vertices of P, a, and b, such that a and b realize the diameter of P, or C passes through three vertices of P such that $\triangle abc$, is an acute triangle containing the centre of C in its interior. [8]*

Theorem 2. *Let P^* be an MPIP of a set of line segments S and circle C^* with diameter D be a Minimum Spanning Circle of S. Then $2D < w(P^*) < \pi D$.*

Proof. Observe that the perimeter of C^* is at least the perimeter of P^*, thus $w(P^*) < \pi D$.

Using elementary calculus one can show that an acute angled triangle inscribed in a circle has perimeter at least $2D$. Thus we can conclude that $2D < w(P^*)$. □

6.2   Approximation Result

In the previous section we proved that the perimeter of an MPIP of a set of line segments S is within a constant times the diameter of an MSC of S. What remains to be done is to find an MPIP of S such that its perimeter is bounded. We will use the MSC of S in order to find such a polygon transversal.

Observation 3. *For any two convex shapes in the plane $\mathcal{H}_1$ and $\mathcal{H}_2$, $\mathcal{H}_1 \subset \mathcal{H}_2$, the perimeter of $\mathcal{H}_1$ is less than the perimeter of $\mathcal{H}_2$.*

According to the definition of a Minimum Spanning Circle C^* of a set of line segments S, we know that there is at least one point of each line segment $s \in S$ that is inside or on the boundary of C^*. In our approach, for each $s \in S$, we simply choose the point on s that is closest to the centre of C^* and we keep the chosen points in S'. The minimum distance from any $s \in S$ to the centre of C^* is no more than the radius of C^*, that is $D/2$. Then we compute the convex hull of S' in $O(n \log n)$ [7]. Let's call this convex hull CH′. Since all the points in S' are included in C^*, CH′ is also included inside C^*, so according to Observation 3

$$w(\text{CH}') < w(C^*) \Rightarrow w(\text{CH}') < \pi D \tag{1}$$

Suppose that P^* is an MPIP of S. Then the perimeter of CH$'$ cannot be smaller than the perimeter of P^*.

$$w(P^*) \leq w(\text{CH}') \tag{2}$$

According to Theorem. 2, Equation 1 and Equation 2 we can conclude that

$$2\mathcal{D} < w(P^*) \leq w(\text{CH}') < \pi\mathcal{D} \tag{3}$$

We summarize the result of this section with the following theorem:

Theorem 4. *Given a set of line segments S with MPIP P^*, we can obtain a convex polygon P in $O(n \log n)$ time that spans all the line segments in S and satisfies the inequality*

$$w(P) \leq \frac{\pi}{2} \times w(P^*) \tag{4}$$

6.3 The Worst Cases

According to Theorem 2, the approximated result is at most $\frac{\pi}{2}$ times greater than the optimal result. Fig. 9 shows an example in which the result of the approximation is $\frac{\pi+2}{4}$ times greater than the optimal result. In this figure, all the line segments are pesky rim segments, except for the top and the bottom line segments. The MPIP is almost vertical, crossing all the line segments. The perimeter of the MPIP is approximately $2d$. Note that according to Theorem 2, the perimeter of the MPIP in this example cannot be less than $2d$. The Approximated convex hull is almost a half circle and its perimeter is approximately $\frac{\pi d}{2} + d$. This yields a $\frac{\pi+2}{4}$ ratio.

In Fig. 10 the ratio between the optimal answer and the approximated answer is also $\frac{\pi+2}{4}$, but in this case the MPIP is located *inside* the MSC. Fig. 10(A) shows the set of line segments. Note that these line segments are non-intersecting and except for the top and the bottom line segments, all the other ones are pesky rim segments. The MPIP of this set of line segments is almost a vertical line connecting the top and the bottom line segments and crossing the rest of the line segments. The perimeter of the MPIP is approximately $2d$. In Fig. 10(B) The

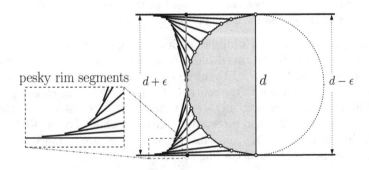

Fig. 9. Approximated MPIP and the optimal MPIP

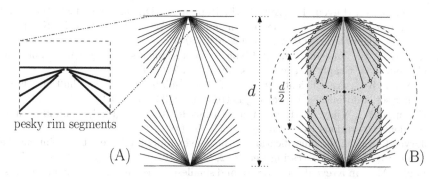

Fig. 10. Approximated MPIP and the optimal MPIP

white point on the line segment is the closest point on that line segment to the centre of the MSC (the big dashed circle). For the top half of the line segments, the closest points are located on a circle with diameter $\frac{d}{2}$ which is centred on the diameter of the MSC. For the bottom half of the line segments, there exists a circle similar to the top half. These two circles are shown as dotted circles inside the MSC in Fig. 10(B). The convex hull of the white points is shown as a grey polygon in Fig. 10(B). The perimeter of this polygon is approximately $\frac{\pi d}{2}+d$.

Although there are no proofs that the examples in Fig. 9 and Fig. 10 are the worst case examples that can be found, they motivate us to investigate whether the upper bound can be reduced from $\pi\mathcal{D}$ to $\frac{\pi\mathcal{D}}{2}+\mathcal{D}$ or even less. It is worth mentioning that for these worst case examples, the incremental approach may produce better results, compared to the MSC approach.

7   Future Work

Observe that our worst case approximations generalize, that is, for any set of objects O, we can apply the MSC approximation in order to approximate the MPIP of O. So if we can find an MSC of O, then $\frac{\text{approximate}}{\text{optimal}} \leq \frac{\pi}{2}$.

Whether the problem of finding a minimum perimeter polygon intersecting a set of line segments is NP-hard, remains unknown.

References

1. Aurenhammer, F., Drysdale, R.L.S., Krasser, H.: Farthest line segment Voronoi diagrams. Information Processing Letters 100(6), 220–225 (2006)
2. Bhattacharya, B., Toussaint, G.: Computing shortest transversals. In: Proc. 18th Internat. Colloq. Automat. Lang. Program., pp. 649–660. Springer, Heidelberg (1991)
3. Burkard, R.E., Rote, G., Yao, E.Y., Yu, Z.L.: Shortest polygonal paths in space. Computing 45, 5–68 (1990)
4. Chan, T.M.: Optimal output-sensitive convex hull algorithms in two and three dimensions. Discrete and Computational Geometry 16, 36–368 (1996)

5. Czyzowicz, J., Egyed, P., Everett, H., Rappaport, D., Shermer, T.C., Souvaine, D.L., Toussaint, G.T., Urrutia, J.: The aquarium keeper's problem. In: Symp. on Discrete Algorithms (SODA), pp. 459–464 (1999)
6. Goodrich, M., Snoeyink, J.: Stabbing parallel segments with a convex polygon. Comput. Vis. Graph. Image Process. 49, 152–170 (1990)
7. Graham, R.L.: An efficient algorithm for determining the convex hull of a finite planar set. Information Processing Letters 1(4), 132–133 (1972)
8. Hearn, D.W., Vijay, J.: Efficient algorithms for the (weighted) minimum circle problem. Operations Research 30(4), 777–795 (1982)
9. Jarvis, R.A.: On the identification of the convex hull of a finite set of points in the plane. Information Processing Letters 2, 18–21 (1973)
10. Löffler, M., van Kreveld, M.: Largest and smallest convex hulls for imprecise points. Algorithmica (in press)
11. Löffler, M., van Kreveld, M.: Approximating largest convex hulls for imprecise points. In: Kaklamanis, C., Skutella, M. (eds.) WAOA 2007. LNCS, vol. 4927, pp. 89–102. Springer, Heidelberg (2008)
12. Löffler, M., van Kreveld, M.: Largest bounding box, smallest diameter, and related problems on imprecise points. In: Dehne, F., Sack, J.-R., Zeh, N. (eds.) WADS 2007. LNCS, vol. 4619, pp. 446–457. Springer, Heidelberg (2007)
13. Meijer, H., Rappaport, D.: Minimum polygon covers of parallel line segments. In: Proc. 2nd Canadian Conference on Comp. Geom., pp. 324–327 (1990)
14. Mukhopadhyay, A., Kumar, C., Greene, E., Bhattacharya, B.: On intersecting a set of parallel line segments with a convex polygon of minimum area. Information Processing Letters 105, 58–64 (2008)
15. Rappaport, D.: Minimum polygon transversals of line segments. Int. J. Comput. Geometry Appl. 5(3), 243–256 (1995)
16. Tamir, A.: Improved complexity bounds for center location problems on networks by using dynamic data structures. SIAM J. Discrete Math. 1(3), 377–396 (1988)

Reconfiguration of
List Edge-Colorings in a Graph

Takehiro Ito[1], Marcin Kamiński[2], and Erik D. Demaine[3]

[1] Graduate School of Information Sciences, Tohoku University,
Aoba-yama 6-6-05, Sendai, 980-8579, Japan
`takehiro@ecei.tohoku.ac.jp`
[2] Department of Computer Science, Université Libre de Bruxelles,
CP 212, Bvd. du Triomphe, 1050 Bruxelles, Belgium
`marcin.kaminski@ulb.ac.be`
[3] MIT Computer Science and Artificial Intelligence Laboratory,
32 Vassar St., Cambridge, MA 02139, USA
`edemaine@mit.edu`

Abstract. We study the problem of reconfiguring one list edge-coloring of a graph into another list edge-coloring by changing one edge color at a time, while at all times maintaining a list edge-coloring, given a list of allowed colors for each edge. First we show that this problem is PSPACE-complete, even for planar graphs of maximum degree 3 and just six colors. Then we consider the problem restricted to trees. We show that any list edge-coloring can be transformed into any other under the sufficient condition that the number of allowed colors for each edge is strictly larger than the degrees of both its endpoints. This sufficient condition is best possible in some sense. Our proof yields a polynomial-time algorithm that finds a transformation between two given list edge-colorings of a tree with n vertices using $O(n^2)$ recolor steps. This worst-case bound is tight: we give an infinite family of instances on paths that satisfy our sufficient condition and whose reconfiguration requires $\Omega(n^2)$ recolor steps.

1 Introduction

Reconfiguration problems arise when we wish to find a step-by-step transformation between two feasible solutions of a problem such that all intermediate results are also feasible. Recently, Ito *et al.* [8] proposed a framework of reconfiguration problems, and gave complexity and approximability results for reconfiguration problems derived from several well-known problems, such as INDEPENDENT SET, CLIQUE, MATCHING, etc. In this paper, we study a reconfiguration problem for list edge-colorings of a graph.

An (ordinary) *edge-coloring* of a graph G is an assignment of colors from a color set C to each edge of G so that every two adjacent edges receive different colors. In *list edge-coloring*, each edge e of G has a set $L(e)$ of colors, called the *list* of e. Then, an edge-coloring f of G is called an *L-edge-coloring* of G if

F. Dehne et al. (Eds.): WADS 2009, LNCS 5664, pp. 375–386, 2009.
© Springer-Verlag Berlin Heidelberg 2009

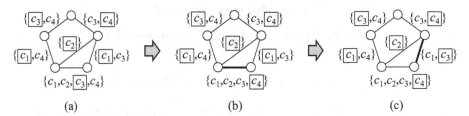

Fig. 1. A sequence of L-edge-colorings of a graph

$f(e) \in L(e)$ for each edge e, where $f(e)$ denotes the color assigned to e by f. Fig.1 illustrates three L-edge-colorings of the same graph with the same list L; the color assigned to each edge is surrounded by a box in the list. Clearly, an edge-coloring is an L-edge-coloring for which $L(e)$ is the same color set C for every edge e of G, and hence list edge-coloring is a generalization of edge-coloring.

Suppose now that we are given *two* L-edge-colorings of a graph G (e.g., the leftmost and rightmost ones in Fig.1), and we are asked whether we can transform one into the other via L-edge-colorings of G such that each differs from the previous one in only one edge color assignment. We call this problem the LIST EDGE-COLORING RECONFIGURATION problem. For the particular instance of Fig.1, the answer is "yes," as illustrated in Fig.1, where the edge whose color assignment was changed from the previous one is depicted by a thick line. One can imagine a variety of practical scenarios where an edge-coloring (e.g., representing a feasible schedule) needs to be changed (to use a newly found better solution or to satisfy new side constraints) by individual color changes (preventing the need for any coordination) while maintaining feasibility (so that nothing breaks during the transformation). Reconfiguration problems are also interesting in general because they provide a new perspective and deeper understanding of the solution space and of heuristics that navigate that space.

Reconfiguration problems have been studied extensively in recent literature [1,3,4,6,7,8], in particular for (ordinary) vertex-colorings. For a positive integer k, a *k-vertex-coloring* of a graph is an assignment of colors from $\{c_1, c_2, \ldots, c_k\}$ to each vertex so that every two adjacent vertices receive different colors. Then, the k-VERTEX-COLORING RECONFIGURATION problem is defined analogously. Bonsma and Cereceda [1] proved that k-VERTEX-COLORING RECONFIGURATION is PSPACE-complete for $k \geq 4$, while Cereceda *et al.* [4] proved that k-VERTEX-COLORING RECONFIGURATION is solvable in polynomial time for $1 \leq k \leq 3$. Edge-coloring in a graph G can be reduced to vertex-coloring in the "line graph" of G. However, by this reduction, we can solve only a few instances of LIST EDGE-COLORING RECONFIGURATION; all edges e of G must have the same list $L(e) = C$ of size $|C| \leq 3$ although any edge-coloring of G requires at least $\Delta(G)$ colors, where $\Delta(G)$ is the maximum degree of G. Furthermore, the reduction does not work the other way, so we do not obtain any complexity results.

In this paper, we give three results for LIST EDGE-COLORING RECONFIGURATION. The first is to show that the problem is PSPACE-complete, even for planar graphs of maximum degree 3 and just six colors. The second is to give a sufficient condition for which there exists a transformation between any two L-edge-colorings of a tree. Specifically, for a tree T, we prove that any two L-edge-colorings of T can be transformed into each other if $|L(e)| \geq \max\{d(v), d(w)\} + 1$ for each edge $e = vw$ of T, where $d(v)$ and $d(w)$ are the degrees of the endpoints v and w of e, respectively. Our proof for the sufficient condition yields a polynomial-time algorithm that finds a transformation between given two L-edge-colorings of T via $O(n^2)$ intermediate L-edge-colorings, where n is the number of vertices in T. On the other hand, as the third result, we give an infinite family of instances on paths that satisfy our sufficient condition and whose transformation requires $\Omega(n^2)$ intermediate L-edge-colorings.

Our sufficient condition for trees was motivated by several results on the well-known "list coloring conjecture" [9]: it is conjectured that any graph G has an L-edge-coloring if $|L(e)| \geq \chi'(G)$ for each edge e, where $\chi'(G)$ is the chromatic index of G, that is, the minimum number of colors required for an ordinary edge-coloring of G. This conjecture has not been proved yet, but some results are known for restricted classes of graphs [2,5,9]. In particular, Borodin *et al.* [2] proved that any bipartite graph G has an L-edge-coloring if $|L(e)| \geq \max\{d(v), d(w)\}$ for each edge $e = vw$. Because any tree is a bipartite graph, one might think that it would be straightforward to extend their result [2] to our sufficient condition. However, this is not the case, because the focus of reconfiguration problems is not the *existence* (as in the previous work) but the *reachability* between two feasible solutions; there must exist a transformation between any two L-edge-colorings if our sufficient condition holds.

Finally we remark that our sufficient condition is best possible in some sense. Consider a star $K_{1,n-1}$ in which each edge e has the same list $L(e) = C$ of size $|C| = n - 1$. Then, $|L(e)| = \max\{d(v), d(w)\}$ for all edges $e = vw$, and it is easy to see that there is no transformation between any two L-edge-colorings of the star.

2   PSPACE-Completeness

Before proving PSPACE-completeness, we introduce some terms and define the problem more formally. In Section 1, we have defined an L-edge-coloring of a graph $G = (V, E)$ with a list L. We say that two L-edge-colorings f and f' of G are *adjacent* if $|\{e \in E : f(e) \neq f'(e)\}| = 1$, that is, f' can be obtained from f by changing the color assignment of a single edge e; the edge e is said to be *recolored* between f and f'. A *reconfiguration sequence* between two L-edge-colorings f_0 and f_t of G is a sequence of L-edge-colorings $f_0, f_1, \ldots, f_t$ of G such that f_{i-1} and f_i are adjacent for $i = 1, 2, \ldots, t$. We also say that two L-edge-colorings f and f' are *connected* if there exists a reconfiguration sequence between f and f'. Clearly, any two adjacent L-edge-colorings are connected. Then, the LIST EDGE-COLORING RECONFIGURATION problem is to determine whether given two L-edge-colorings of a graph G are connected. The *length* of a reconfiguration sequence is the number of L-edge-colorings in the sequence, and hence the length of the reconfiguration sequence in Fig.1 is 3.

The main result of this section is the following theorem.

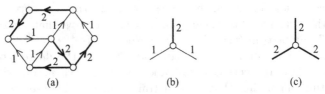

Fig. 2. (a) A configuration of an NCL machine, (b) NCL AND vertex, and (c) NCL OR vertex

Theorem 1. LIST EDGE-COLORING RECONFIGURATION *is PSPACE-complete for planar graphs of maximum degree* 3 *whose lists are chosen from six colors.*

In order to prove Theorem 1, we give a reduction from Nondeterministic Constraint Logic (NCL) [7]. An NCL "machine" is specified by a *constraint graph*: an undirected graph together with an assignment of weights from $\{1, 2\}$ to each edge of the graph. A *configuration* of this machine is an orientation (direction) of the edges such that the sum of weights of incoming edges at each vertex is at least 2. Fig.2(a) illustrates a configuration of an NCL machine, where each weight-2 edge is depicted by a thick line and each weight-1 edge by a thin line. A *move* from one configuration is simply the reversal of a single edge which results in another (feasible) configuration. Given an NCL machine and its two configurations, it is PSPACE-complete to determine whether there exists a sequence of moves which transforms one configuration into the other [7].

In fact, the problem remains PSPACE-complete even for AND/OR *constraint graphs*, which consist only of two types of vertices, called "NCL AND vertices" and "NCL OR vertices." A vertex of degree 3 is called an *NCL AND vertex* if its three incident edges have weights 1, 1 and 2. (See Fig.2(b).) An NCL AND vertex behaves as a logical AND, in the following sense: the weight-2 edge can be directed outward if and only if both weight-1 edges are directed inward. Note that, however, the weight-2 edge is not necessarily directed outward even when both weight-1 edges are directed inward. A vertex of degree 3 is called an *NCL OR vertex* if its three incident edges have weights 2, 2 and 2. (See Fig.2(c).) An NCL OR vertex behaves as a logical OR: one of the three edges can be directed outward if and only if at least one of the other two edges is directed inward. It should be noted that, although it is natural to think of NCL AND and OR vertices as having inputs and outputs, there is nothing enforcing this interpretation; especially for NCL OR vertices, the choice of input and output is entirely arbitrary since NCL OR vertices are symmetric. From now on, we call an AND/OR constraint graph simply an NCL machine.

Proof of Theorem 1

It is easy to see that LIST EDGE-COLORING RECONFIGURATION can be solved in (most conveniently, nondeterministic [10]) polynomial space. Therefore, in the remainder of this section, we show that the problem is PSPACE-hard by giving a reduction from NCL. This reduction involves constructing two types of gadgets which correspond to NCL AND and OR vertices. We call an edge

(a)   (b)

Fig. 3. (a) an NCL edge uv and (b) its corresponding edges ux and xv of a graph with lists $L(ux) = \{c_1, c_2\}$ and $L(xv) = \{c_1, c_3\}$

of an NCL machine an *NCL edge*, and say simply an *edge* of a graph for LIST EDGE-COLORING RECONFIGURATION.

Assume in our reduction that the color c_1 corresponds to "directed inward," and that both colors c_2 and c_3 correspond to "directed outward." Consider an NCL edge uv directed from u to v. (See Fig.3(a).) Then, the NCL edge is directed outward for u, but is directed inward for v. However, in list edge-coloring, each edge can receive only one color, of course. Therefore, we need to split one NCL edge uv into two edges ux and xv of a graph with lists $L(ux) = \{c_1, c_2\}$ and $L(xv) = \{c_1, c_3\}$, as illustrated in Fig.3(b). It is easy to see that one of ux and xv can be colored with c_1 if and only if the other edge is colored with either c_2 or c_3. This property represents that an NCL half-edge can be directed inward if and only if the other half is directed outward.

Fig.4 illustrates three kinds of "AND gadgets," each of which corresponds to an NCL AND vertex; two edges $u_x x$ and $u_y y$ correspond to two weight-1 NCL half-edges, and the edge $u_z z$ corresponds to a weight-2 NCL half-edge. Since NCL AND and OR vertices are connected together into an arbitrary NCL machine, there should be eight kinds of AND gadgets according to the choice of lists $\{c_1, c_2\}$ and $\{c_1, c_3\}$ for three edges $u_x x$, $u_y y$ and $u_z z$. However, since the two weight-1 NCL edges are symmetric, it suffices to consider these three kinds: all the three edges have the same list, as in Fig.4(a); $u_z z$ has a different list from the other two edges, as in Fig.4(b); and one of $u_x x$ and $u_y y$ has a different list from the other two edges, as in Fig.4(c).

We denote by $\mathcal{F}(A; c_x, c_y, c_z)$ the set of all L-edge-colorings f of an AND gadget A such that $f(u_x x) = c_x$, $f(u_y y) = c_y$ and $f(u_z z) = c_z$. Since a triple (c_x, c_y, c_z) defines the direction of the three corresponding NCL half-edges, all the L-edge-colorings in $\mathcal{F}(A; c_x, c_y, c_z)$ correspond to the same configuration

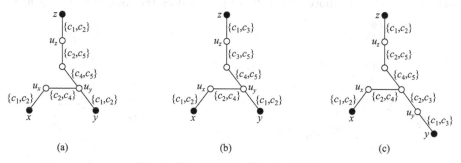

(a)   (b)   (c)

Fig. 4. Three kinds of AND gadgets

of the NCL AND vertex. We now check that the three kinds of AND gadgets satisfy the same constraints as an NCL AND vertex; we check this property by enumerating all possible L-edge-colorings of the AND gadgets. For example, in the AND gadget A of Fig.4(a), $u_z z$ can be colored with c_2 (directed outward) if and only if both $u_x x$ and $u_y y$ are colored with c_1 (directed inward); in other words, $|\mathcal{F}(A; c_x, c_y, c_2)| \geq 1$ if and only if $c_x = c_y = c_1$. In addition, every AND gadget A satisfies the following two properties:

(i) For a triple (c_x, c_y, c_z), if $|\mathcal{F}(A; c_x, c_y, c_z)| \geq 2$, then any two L-edge-colorings f and f' in $\mathcal{F}(A; c_x, c_y, c_z)$ are "internally connected," that is, there exists a reconfiguration sequence between f and f' such that all L-edge-colorings in the sequence belong to $\mathcal{F}(A; c_x, c_y, c_z)$; and

(ii) For every two triples (c_x, c_y, c_z) and (c'_x, c'_y, c'_z) which differ in a single coordinate, if $|\mathcal{F}(A; c_x, c_y, c_z)| \geq 1$ and $|\mathcal{F}(A; c'_x, c'_y, c'_z)| \geq 1$, then there exist two L-edge-colorings f and f' such that f and f' are adjacent, $f \in \mathcal{F}(A; c_x, c_y, c_z)$ and $f' \in \mathcal{F}(A; c'_x, c'_y, c'_z)$.

Then, it is easy to see that the reversal of a single NCL half-edge in an NCL AND vertex can be simulated by a reconfiguration sequence between two L-edge-colorings each of which is chosen arbitrarily from the set $\mathcal{F}(A; c_x, c_y, c_z)$, where (c_x, c_y, c_z) corresponds to the direction of the three NCL half-edges.

Fig.5 illustrates two kinds of "OR gadgets," each of which corresponds to an NCL OR vertex; three edges $u_x x$, $u_y y$ and $u_z z$ correspond to three weight-2 NCL half-edges. Since an NCL OR vertex is entirely symmetric, it suffices to consider these two kinds: all the three edges have the same list, as in Fig.5(a); and one edge has a different list from the other two edges, as in Fig.5(b). Then, similarly as AND gadgets, it is easy to see that both kinds of OR gadgets satisfy the same constraints as an NCL OR vertex, and that the reversal of a single NCL half-edge in an NCL OR vertex can be simulated by a reconfiguration sequence between corresponding two L-edge-colorings.

Given NCL machine, we construct a corresponding graph G with a list L by connecting the vertices x, y and z of AND or OR gadgets. Then, an L-edge-coloring of G corresponds to a configuration of the NCL machine. On the other hand, every configuration of the NCL machine can be mapped to at least one (in general, to exponentially many) L-edge-colorings of G. We can choose an arbitrary one for each of given two configurations, because each AND gadget satisfies Property (i) above and each OR gadget does the counterpart. It is now

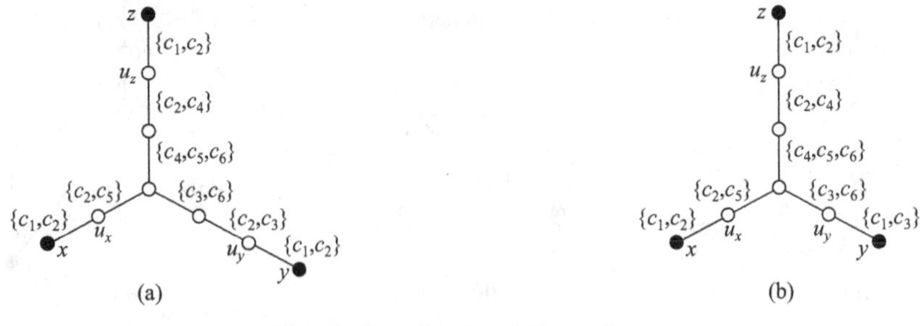

Fig. 5. Two kinds of OR gadgets

easy to see that there is a sequence of moves which transforms one configuration into the other if and only if there is a reconfiguration sequence between the two L-edge-colorings of G. Since NCL remains PSPACE-complete even if an NCL machine is planar [7], G is a planar graph of maximum degree 3. Furthermore, each list $L(e)$ is a subset of $\{c_1, c_2, \ldots, c_6\}$.   □

3   Trees

Since LIST EDGE-COLORING RECONFIGURATION is PSPACE-complete, it is rather unlikely that the problem can be solved in polynomial time for general graphs. However, in Section 3.1, we give a sufficient condition for which any two L-edge-colorings of a tree T are connected; our sufficient condition can be checked in polynomial time. Moreover, our proof yields a polynomial-time algorithm that finds a reconfiguration sequence of length $O(n^2)$ between given two L-edge-colorings, where n is the number of vertices in T. In Section 3.2, we then give an infinite family of instances on paths that satisfy our sufficient condition and whose reconfiguration sequence requires length $\Omega(n^2)$.

3.1   Sufficient Condition and Algorithm

The main result of this subsection is the following theorem, whose sufficient condition is best possible in some sense as we mentioned in Section 1.

Theorem 2. *For a tree T with n vertices, any two L-edge-colorings f and f' of T are connected if $|L(e)| \geq \max\{d(v), d(w)\} + 1$ for each edge $e = vw$ of T. Moreover, there is a reconfiguration sequence of length $O(n^2)$ between f and f'.*

Since $\Delta(T) \geq \max\{d(v), d(w)\}$ for all edges vw of a tree T, Theorem 2 immediately implies the following sufficient condition for which any two (ordinary) edge-colorings of T are connected. Note that, for a positive integer k, a k-edge-coloring of a tree T is an L-edge-coloring of T for which all edges e have the same list $L(e) = \{c_1, c_2, \ldots, c_k\}$.

Corollary 1. *For a tree T with n vertices, any two k-edge-colorings f and f' of T are connected if $k \geq \Delta(T) + 1$. Moreover, there is a reconfiguration sequence of length $O(n^2)$ between f and f'.*

It is obvious that the sufficient condition of Corollary 1 is also best possible in some sense; consider a star $K_{1,n-1}$ in Section 1.

In the remainder of this subsection, as a proof of Theorem 2, we give a polynomial-time algorithm that finds a reconfiguration sequence of length $O(n^2)$ between given two L-edge-colorings f_0 and f_t of a tree T if our condition holds.

We first give an outline of our algorithm. By the breadth-first search starting from an arbitrary vertex r of degree 1, we order all edges $e_1, e_2, \ldots, e_{n-1}$ of a tree T. At the ith step, $1 \leq i \leq n-1$, the algorithm recolors e_i from the current color to its target color $f_t(e_i)$, as follows. From the current L-edge-coloring f, we first obtain an L-edge-coloring f' of T such that

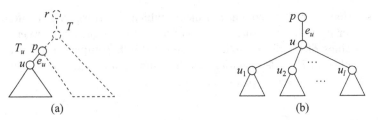

Fig. 6. (a) Subtree T_u in the whole tree T and (b) inside of T_u

(i) there is no edge which is adjacent with e_i and is colored with $f_t(e_i)$; and
(ii) there exists a reconfiguration sequence between f and f' in which any of the edges $e_1, e_2, \ldots, e_{i-1}$ is not recolored.

Then, we recolor e_i to $f_t(e_i)$. Therefore, e_i is never recolored after the ith step, while e_i may be recolored before the ith step even if e_i is colored with $f_t(e_i)$. We will show later that every edge of T can be recolored in such a way, and hence we eventually obtain the target L-edge-coloring f_t. We will also show later that the algorithm recolors each edge e_j with $j \geq i$ at most once in the ith step, and hence we can recolor e_i by recoloring at most $n - i$ edges. Our algorithm thus finds a reconfiguration sequence of total length $\sum_{i=1}^{n-1}(n - i) = O(n^2)$.

Suppose that we are given a tree T with a list L such that

$$|L(e)| \geq \max\{d(v), d(w)\} + 1 \tag{1}$$

for each edge $e = vw$ in $E(T)$. We choose an arbitrary vertex r of degree 1 as the root of T, and regard T as a rooted tree. For a vertex u in $V(T) \setminus \{r\}$, let p be the parent of u in T. We denote by T_u the subtree of T which is rooted at p and is induced by p, u and all descendants of u in T. (See Fig.6(a).) It should be noted that T_u includes the edge $e_u = pu$, but does not include the other edges incident to p. Therefore, T_u consists of a single edge if u is a leaf of T. We always denote by e_u the edge which joins u and its parent p. For an internal vertex u of T, let $u_1, u_2, \cdots, u_l$ be the children of u ordered arbitrarily, as illustrated in Fig.6(b). Then, the subtree T_u consists of e_u and the subtrees T_{u_i}, $1 \leq i \leq l$.

For a vertex u of T, we denote by $L_u = L|T_u$ the *restriction* of the list L of T to the subtree T_u, that is, $L_u(e) = L(e)$ for each edge $e \in E(T_u)$. Clearly, $d(v, T) \geq d(v, T_u)$ for each vertex $v \in V(T_u)$, where $d(v, T)$ and $d(v, T_u)$ denote the degrees of v in T and T_u, respectively. Therefore, for each edge $e = vw$ in $E(T_u)$, by Eq. (1) we have

$$\begin{aligned}
|L_u(e)| &= |L(e)| \\
&\geq \max\{d(v, T), d(w, T)\} + 1 \\
&\geq \max\{d(v, T_u), d(w, T_u)\} + 1.
\end{aligned}$$

The list L_u of T_u thus satisfies Eq. (1). For an L-edge-coloring f of T, we denote by $g = f|T_u$ the *restriction* of f to T_u, that is, g is an L_u-edge-coloring of T_u such that $g(e) = f(e)$ for each edge e in $E(T_u)$. For an L_u-edge-coloring g of T_u, an edge vw of T_u and its endpoint v, we define a subset $C_{av}(g, vw, v)$ of $L_u(vw)$, as follows:

$$C_{av}(g, vw, v) = L_u(vw) \setminus \{g(vx) : vx \in E(T_u)\}. \tag{2}$$

Then, $C_{av}(g, vw, v)$ is the set of all colors in $L_u(vw)$ available on v for vw. Therefore, $C_{av}(g, vw, v) \cap C_{av}(g, vw, w)$ is the set of all colors in $L_u(vw)$ available for vw when we wish to recolor vw from g.

Algorithm

We are now ready to describe our algorithm. Assume that all edges $e_1, e_2, \ldots,$ e_{n-1} of a tree T are ordered by the breadth-first search starting from the root r of T. At the ith step, $1 \le i \le n - 1$, the algorithm recolors e_i to its target color $f_t(e_i)$. Consider the ith step of the algorithm, and let f be the current L-edge-coloring of T obtained by $i - 1$ steps of the algorithm; let $f = f_0$ for the first step, that is, $e_i = e_1$. Then, we wish to recolor $e_i = pp'$ from $f(e_i)$ to $f_t(e_i)$. There are the following two cases to consider.

Case (a): $f_t(e_i) \in C_{av}(f, e_i, p) \cap C_{av}(f, e_i, p')$

In this case, $f_t(e_i)$ is available for e_i, that is, there is no edge which is adjacent with e_i and is colored with $f_t(e_i)$. We thus simply recolor e_i from $f(e_i)$ to $f_t(e_i)$, and obtain an L-edge-coloring f' of T: for each edge e in $E(T)$,

$$f'(e) = \begin{cases} f(e) & \text{if } e \in E(T) \setminus \{e_i\}; \\ f_t(e_i) & \text{if } e = e_i. \end{cases}$$

Case (b): $f_t(e_i) \notin C_{av}(f, e_i, p) \cap C_{av}(f, e_i, p')$

In this case, there are at most two edges pu and $p'u'$ which are colored with $f_t(e_i)$ and are sharing the endpoints p and p' with e_i, respectively.

If there is an edge pu which is colored with $f_t(e_i)$ and is sharing p with e_i, then we recolor $e_u = pu$ to a different available color c, as follows. By Eqs. (1) and (2) we have

$$\begin{aligned} |C_{av}(f, e_u, p)| &\ge |L(e_u)| - |\{f(px) : px \in E(T)\}| \\ &\ge \max\{d(p), d(u)\} + 1 - d(p) \\ &\ge 1. \end{aligned}$$

Therefore, there exists at least one color c in $L(e_u)$ which is available on p for e_u. Clearly, $c \ne f(e_i)$ and $c \ne f_t(e_i)$ since both colors are in $\{f(px) : px \in E(T)\}$. It should be noted that $c \in C_{av}(f, e_u, u)$ does not necessarily hold: c is not always available for e_u. Let $g = f|T_u$ be the restriction of f to T_u. Then, by using the procedure RECOLOR (which is described in the next page), we recolor e_u from $g(e_u) (= f(e_u))$ to c without recoloring any edge in $E(T) \setminus E(T_u)$. More precisely, we have the following lemma, whose proof is omitted due to the page limitation.

Lemma 1. *Let T be a tree with a list L satisfying Eq. (1), and let f be an L-edge-coloring of T. For a vertex u of T, let $L_u = L|T_u$ and $g = f|T_u$ be the restrictions of L and f to the subtree T_u, respectively. Then, for an arbitrary* color c in $L_u(e_u) \setminus \{g(e_u)\}$, RECOLOR$(T_u, g, c)$ returns a sequence $\mathcal{RS}$ of L_u-edge-colorings $g_1, g_2, \ldots, g_q$ of T_u which satisfies the following three properties:

(i) g and g_1 are adjacent;

(ii) g_{k-1} and g_k are adjacent for each k, $2 \le k \le q$; and

(iii) $g_k(e_u) = g(e_u)$ for each k, $1 \le k \le q-1$, and $g_q(e_u) = c$.

Procedure 1. RECOLOR(T_u, g, c)

1: $\mathcal{RS} \Leftarrow \emptyset$ {$\mathcal{RS}$ does not contain g}
2: **if** $c \in C_{av}(g, e_u, u)$ **then** {See also Fig.6(b)}
3:    {The color c is not assigned to any of the edges $uu_1, uu_2, \ldots, uu_l$}
4:    Recolor e_u from $g(e_u)$ to c, and obtain an L_u-edge-coloring g' of T_u
5:    **return** $\{g'\}$
6: **else** {The color c is assigned to one of the edges $uu_1, uu_2, \ldots, uu_l$}
7:    Let $e_j = uu_j$ be the edge such that $g(e_j) = c$
8:    Choose an arbitrary color $c' \in C_{av}(g, e_j, u)$
9:    {Recolor e_j to c' via L_j-edge-colorings of T_{u_j}, where $L_j = L_u|T_{u_j}$}
10:    $\mathcal{RS}' \Leftarrow$ RECOLOR$(T_{u_j}, g|T_{u_j}, c')$
11:    **for** each L_j-edge-coloring h_k in $\mathcal{RS}'$ (in the same order) **do**
12:      {Extend an L_j-edge-coloring h_k of T_{u_j} to an L_u-edge-coloring g_k of T_u}
13:      Let $g_k(e) = \begin{cases} g(e) & \text{if } e \in E(T_u) \setminus E(T_{u_j}); \\ h_k(e) & \text{if } e \in E(T_{u_j}) \end{cases}$
14:      $\mathcal{RS} \Leftarrow \mathcal{RS} \cup \{g_k\}$
15:    **end for**
16:    {e_j is now colored with c', and hence c is available for e_u}
17:    Recolor e_u from $g(e_u)$ to c, and obtain an L_u-edge-coloring g' of T_u
18:    **return** $\mathcal{RS} \Leftarrow \mathcal{RS} \cup \{g'\}$
19: **end if**

Since c has been chosen from $C_{av}(f, e_u, p)$ and g is the restriction of f to T_u, by Property (iii) of Lemma 1 we can easily extend each L_u-edge-coloring g_k of T_u in $\mathcal{RS}$ to an L-edge-coloring f_k of T, as follows: for each edge e in $E(T)$,

$$f_k(e) = \begin{cases} f(e) & \text{if } e \in E(T) \setminus E(T_u); \\ g_k(e) & \text{if } e \in E(T_u). \end{cases}$$

Clearly, the sequence $f, f_1, f_2, \ldots, f_q$ of L-edge-colorings of T is a reconfiguration sequence which recolors e_u from $f(e_u)$ ($= f_t(e_i)$) to c. Moreover, in the reconfiguration sequence, any of the edges in $E(T) \setminus E(T_u)$ is not recolored.

Similarly, if there is an edge $p'u'$ which is colored with $f_t(e_i)$ and is sharing the other endpoint p' with e_i, then we recolor $p'u'$ to a different color which is available on p' for $p'u'$ without recoloring any edge in $E(T) \setminus E(T_{u'})$.

Then, in the current L-edge-coloring of T, $f_t(e_i)$ is available for e_i. Therefore, we can finally recolor e_i from $f(e_i)$ to $f_t(e_i)$.

Proof of Theorem 2

Remember that all edges $e_1, e_2, \ldots, e_{n-1}$ of a tree T are ordered by the breadth-first search starting from the root r of T, and that the algorithm recolors e_i to its target color $f_t(e_i)$ at the ith step, $1 \le i \le n-1$.

We first show that e_i is never recolored after the ith step of the algorithm, as in the following lemma. (The proof is omitted due to the page limitation.)

Lemma 2. *The algorithm does not recolor any edge e_j with $j < i$ in the ith step.*

Using Lemma 1 we have shown that the algorithm can recolor e_i to $f_t(e_i)$ at the ith step, and hence Lemma 2 implies that the algorithm terminates with the target L-edge-coloring f_t. Therefore, the algorithm finds a reconfiguration sequence between given two L-edge-colorings f_0 and f_t of T if L satisfies Eq. (1).

We now estimate the length of a reconfiguration sequence found by our algorithm. Clearly, the algorithm recolors an edge at most once in each step. Therefore, by Lemma 2, at most $n - i$ edges are recolored in the ith step. The total length of the reconfiguration sequence is thus $\sum_{i=1}^{n-1}(n - i) = O(n^2)$.   □

3.2   Length of Reconfiguration Sequence

We showed in Section 3.1 that any two L-edge-colorings of a tree T are connected via a reconfiguration sequence of length $O(n^2)$ if our sufficient condition holds. In this subsection, we show that this worst-case bound on the length is tight: we give an infinite family of instances on paths that satisfy our sufficient condition and whose reconfiguration sequence requires length $\Omega(n^2)$.

Consider a path $P = \{v_0, v_1, \ldots, v_{3m+1}\}$ of $3m+1$ edges in which every edge e has the same list $L(e) = \{c_1, c_2, c_3\}$. Clearly, the list L satisfies Eq. (1), and hence any two L-edge-colorings of P are connected. We construct two L-edge-colorings f_0 and f_t of P, as follows:

$$f_0(v_iv_{i+1}) = \begin{cases} c_3 & \text{if } i \equiv 0 \mod 3; \\ c_2 & \text{if } i \equiv 1 \mod 3; \\ c_1 & \text{if } i \equiv 2 \mod 3 \end{cases} \tag{3}$$

for each edge v_iv_{i+1}, $0 \le i \le 3m$, and

$$f_t(v_iv_{i+1}) = \begin{cases} c_3 & \text{if } i \equiv 0 \mod 3; \\ c_1 & \text{if } i \equiv 1 \mod 3; \\ c_2 & \text{if } i \equiv 2 \mod 3 \end{cases} \tag{4}$$

for each edge v_iv_{i+1}, $0 \le i \le 3m$. Then, we have the following theorem, whose proof is omitted from this extended abstract.

Theorem 3. *For a path P and its two L-edge-colorings f_0 and f_t defined above, every reconfiguration sequence between f_0 and f_t requires length $\Omega(n^2)$, where n is the number of vertices in P.*

4   Concluding Remarks

A reconfiguration sequence can be represented by a sequence of "recolor steps" (e, c), where a pair (e, c) denotes one recolor step which recolors an edge e to some color $c \in L(e)$. Then, the algorithm in Section 3.1 can be easily implemented so that it runs in time $O(n^2)$: we store and compute a sequence of recolor steps (e, c) together with only the current L-edge-coloring of a tree T. On the other hand, Theorem 3 suggests that it is difficult to improve the time-complexity $O(n^2)$ of the algorithm if we wish to find an actual reconfiguration sequence explicitly.

One may expect that our sufficient condition for trees holds also for some larger classes of graphs, such as bipartite graphs, bounded treewidth graphs, etc. However, consider the following even-length cycle, which is bipartite and whose treewidth is 2. For an even integer m, let C be the cycle of $3m$ edges obtained by identifying the edge $v_0 v_1$ with the edge $v_{3m} v_{3m+1}$ of P in Section 3.2, and let f_0 and f_t be L-edge-colorings of C defined similarly as in Eqs. (3) and (4), respectively. Then, we cannot recolor any edge in the cycle, and hence there is no reconfiguration sequence between f_0 and f_t even though $|L(e)| = \max\{d(v), d(w)\} + 1$ holds for each edge $e = vw$.

Acknowledgments

We thank the Algorithms Research Group of Université Libre de Bruxelles, especially Jean Cardinal, Martin Demaine and Raphaël Jungers, for fruitful discussions.

References

1. Bonsma, P.S., Cereceda, L.: Finding paths between graph colourings: PSPACE-completeness and superpolynomial distances. In: Kučera, L., Kučera, A. (eds.) MFCS 2007. LNCS, vol. 4708, pp. 738–749. Springer, Heidelberg (2007)
2. Borodin, O.V., Kostochka, A.V., Woodall, D.R.: List edge and list total colourings of multigraphs. J. Combinatorial Theory, Series B 71, 184–204 (1997)
3. Călinescu, G., Dumitrescu, A., Pach, J.: Reconfigurations in graphs and grids. SIAM J. Discrete Mathematics 22, 124–138 (2008)
4. Cereceda, L., van den Heuvel, J., Johnson, M.: Finding paths between 3-colourings. In: Proc. of IWOCA 2008, pp. 182–196 (2008)
5. Fujino, T., Zhou, X., Nishizeki, T.: List edge-colorings of series-parallel graphs. IEICE Trans. Fundamentals E86-A, 1034–1045 (2003)
6. Gopalan, P., Kolaitis, P.G., Maneva, E.N., Papadimitriou, C.H.: The connectivity of Boolean satisfiability: computational and structural dichotomies. In: Bugliesi, M., Preneel, B., Sassone, V., Wegener, I. (eds.) ICALP 2006. LNCS, vol. 4051, pp. 346–357. Springer, Heidelberg (2006)
7. Hearn, R.A., Demaine, E.D.: PSPACE-completeness of sliding-block puzzles and other problems through the nondeterministic constraint logic model of computation. Theoretical Computer Science 343, 72–96 (2005)
8. Ito, T., Demaine, E.D., Harvey, N.J.A., Papadimitriou, C.H., Sideri, M., Uehara, R., Uno, Y.: On the complexity of reconfiguration problems. In: Hong, S.-H., Nagamochi, H., Fukunaga, T. (eds.) ISAAC 2008. LNCS, vol. 5369, pp. 28–39. Springer, Heidelberg (2008)
9. Jensen, T.R., Toft, B.: Graph Coloring Problems. Wiley Interscience, New York (1995)
10. Savitch, W.J.: Relationships between nondeterministic and deterministic tape complexities. J. Computer and System Sciences 4, 177–192 (1970)

The Simultaneous Representation Problem for Chordal, Comparability and Permutation Graphs

Krishnam Raju Jampani and Anna Lubiw

David R. Cheriton School of Computer Science,
University of Waterloo, Ontario, Canada
{krjampani,alubiw}@uwaterloo.ca

Abstract. We introduce the *simultaneous representation problem*, defined for any graph class C characterized in terms of representations, e.g.
any class of intersection graphs. Two graphs G_1 and G_2, sharing some
vertices X (and the corresponding induced edges), are said to have a *simultaneous representation* with respect to a graph class C, if there exist
representations R_1 and R_2 of G_1 and G_2 that are "consistent" on X.
Equivalently (for the classes C that we consider) there exist edges E'
between $G_1 - X$ and $G_2 - X$ such that $G_1 \cup G_2 \cup E'$ belongs to class C.

Simultaneous representation problems are related to graph sandwich
problems, probe graph recognition problems and simultaneous planar
embeddings and have applications in any situation where it is desirable
to consistently represent two related graphs.

In this paper we give efficient algorithms for the simultaneous representation problem on chordal, comparability and permutation graphs.
These results complement the recent poly-time algorithms for recognizing probe graphs for the above classes and imply that the graph sandwich
problem for these classes is solvable for an interesting special case: when
the set of optional edges induce a complete bipartite graph. Moreover for
comparability and permutation graphs, our results can be extended to
solve a generalized version of the simultaneous representation problem
when there are k graphs any two of which share a common vertex set
X. This generalized version is equivalent to the graph sandwich problem
when the set of optional edges induce a k-partite graph.

Keywords: Simultaneous graphs, Sandwich graphs, Chordal graphs,
Comparability graphs, Permutation graphs.

1   Introduction

We explore the idea of finding a simultaneous representation for two graphs
with respect to a graph class, when the graphs share some vertices and edges.
We define this precisely for intersection graph classes, but the concept is rich
enough to apply more broadly.

Let C be any intersection graph class (such as interval graphs, chordal graphs,
permutation graphs, etc) and let G_1 and G_2 be two graphs in C, sharing some

F. Dehne et al. (Eds.): WADS 2009, LNCS 5664, pp. 387–398, 2009.
© Springer-Verlag Berlin Heidelberg 2009

vertices X and the edges induced by X. G_1 and G_2 are said to be *simultaneous C-representable graphs* or *simultaneous C graphs* if there exist intersection representations R_1 and R_2 of G_1 and G_2 such that any vertex of X is represented by the same object in both R_1 and R_2. The *simultaneous representation problem* for class C asks whether G_1 and G_2 are simultaneous C graphs.

Comparability graphs do not have an intersection representation, but the simultaneous representation problem can be defined in the obvious way: Two comparability graphs G_1 and G_2 sharing some vertices X and the edges induced by X are said to be simultaneous comparability graphs if there exist transitive orientations T_1 and T_2 of G_1 and G_2 (respectively) such that any edge $e \in E(X)$ is oriented in the same way in both T_1 and T_2. For example, Figure 1(left) shows a pair of simultaneous comparability graphs, with the property that their union is not a comparability graph. Figure 1(right) shows a pair of graphs that are not simultaneous comparability graphs, though each one is a comparability graph on its own.

The main results in this paper are polynomial time algorithms for the simultaneous representation problem on chordal graphs, permutation graphs, and comparability graphs. These classes of graphs are of enduring interest because of their many applications [15], [8]. Simultaneous representation problems arise

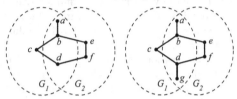

Fig. 1. The graphs on the left are simultaneous comparability graphs while the graphs on the right are not

in any situation where two related graphs should be represented consistently. A main instance is for temporal relationships, where an old graph and a new graph share some common parts. Pairs of related graphs also arise in many other situations, for example: two social networks that share some members; overlap graphs of DNA fragments of two similar organisms, etc. Simultaneous chordal graphs have an application in computational biology as a special case of reconstructing phylogenies (tree structures that model genetic mutations) when part of the information is missing. (see [1]).

The simultaneous representation problem has previously been studied for straight-line planar graph drawings: two graphs that share some vertices and edges (not necessarily induced) have a *simultaneous geometric embedding* [3] if they have planar straight-line drawings in which the common vertices are represented by common points. Thus edges may cross, but only if they are in different graphs. Deciding if two graphs have a simultaneous geometric embedding is NP-hard [6].

Simultaneous representation problems are also closely related to some graph sandwich problems. For comparability graphs and for any intersection graph class we show that the simultaneous representation problem is equivalent to a graph augmentation problem: given two graphs G_1 and G_2, sharing vertices X and the corresponding induced edges, do there exist edges E' between $G_1 - X$ and

$G_2 - X$ so that the augmented graph $G_1 \cup G_2 \cup E'$ belongs to class $\mathcal{C}$. Intuitively, the simultaneous representation problem does not specify relationships between $G_1 - X$ and $G_2 - X$, so these are the edges that can freely be added to produce a graph in class $\mathcal{C}$.

The *graph sandwich problem* [9] is a more general augmentation problem defined for any graph class $\mathcal{C}$: given graphs $G_1 = (V, E_1)$ and $G_2 = (V, E_2)$, is there a set of edges E with $E_1 \subseteq E \subseteq E_2$ so that the graph $G = (V, E)$ belongs to class $\mathcal{C}$. This problem has a wealth of applications but is NP-complete for interval, chordal, comparability and permutation graphs [9].

The simultaneous representation problem is the special case where $E_2 - E_1$ forms a complete bipartite subgraph. A related special case where $E_2 - E_1$ forms a clique is the problem of recognizing *probe graphs*: a graph G with a specified independent set N is a *probe graph* for class $\mathcal{C}$ if there exist edges $E' \subseteq N \times N$ so that the augmented graph $G \cup E'$ belongs to class $\mathcal{C}$. Probe graphs have many applications [13,10] and have received much attention recently. There are polynomial time algorithms to recognize probe interval graphs [12], probe chordal graphs [2], and probe comparability and permutation graphs [5].

For comparability and permutation graphs, our results can be directly extended to solve a generalized version of the simultaneous representation problem when there are k graphs any of two of which share a common vertex set X. This implies that the graph sandwich problem for these graph classes is solvable for the special case when the optional edges induce a k-partite graph.

Our paper is organized as follows: Section 1.1 gives notation and preliminaries. We prove the equivalence of the two problem formulations here. In sections 2, 3 and 4 we study the simultaneous representation problem for chordal, comparability and permutation graphs respectively.

1.1 Notation and Preliminaries

An *intersection graph* is one that has an intersection representation consisting of an object for each vertex such that there is an edge between two vertices if and only if the corresponding objects intersect. An intersection graph class restricts the possible objects, for example, interval graphs are intersection graphs of line segments on a line.

For a graph G, we use $V(G)$ and $E(G)$ to denote its vertex set and edge set respectively. Given a vertex v and a set of edges A, we use $N_A(v)$ to denote the *neighbors of v* w.r.t A i.e. the vertex set $\{u : (u, v) \in A\}$. If v is a vertex of G then we use $E_G(v)$ to denote the edges incident to v i.e. the edge set $\{(u, v) : u \in V(G), (u, v) \in E(G)\}$ and $G - v$ to denote the graph obtained by removing v and its incident edges from G.

Let $G_1 = (V_1, E_1)$ and $G_2 = (V_2, E_2)$ be two graphs sharing some vertices X and the edges induced by X. To be precise, $V_1 \cap V_2 = X$ and the subgraphs of G_1 and G_2 induced by vertex set X are the same. Let $A \subseteq (V_1 - X) \times (V_2 - X)$ be a set of edges. We use the notation (G_1, G_2, A) to denote the graph whose vertex set is $V_1 \cup V_2$ and whose edge set is $E_1 \cup E_2 \cup A$. Let $G = (G_1, G_2, A)$. An edge $e \in (V_1 - X) \times (V_2 - X)$ is said to be an *augmenting* edge of G.

Theorem 1. *Let $G_1 = (V_1, E_1)$ and $G_2 = (V_2, E_2)$ be two graphs belonging to intersection class $\mathcal{C}$ and sharing some vertices X and the edges induced by X. G_1 and G_2 are simultaneous $\mathcal{C}$ graphs if and only if there exists a set $A \subseteq (V_1 - X) \times (V_2 - X)$ of edges such that the graph $G = (G_1, G_2, A)$ belongs to class $\mathcal{C}$.*

The proof is straightforward and can be found in the full version of our paper [11]. Theorem 1 implies that the simultaneous representation problem for intersection classes is a special case of the graph sandwich problem in which the set of optional edges induce a complete bipartite graph. In section 3, we show that this alternative formulation holds for comparability graphs also.

2   Simultaneous Chordal Graphs

A graph is said to be *chordal* if it does not contain any induced cycles of length greater than 3. We use the following well-known results about chordal graphs [8]. Chordal graphs satisfy the hereditary property: Any induced subgraph of a chordal graph is chordal. A chordal graph always has a *simplicial* vertex: a vertex x such that $N(x)$ induces a clique. A *perfect elimination ordering* is an ordering $v_1, \ldots, v_n$ of the vertices such that each v_i is simplicial in the subgraph induced by $\{v_i, \cdots, v_n\}$. Chordal graphs are characterized by the existence of a perfect elimination ordering. Any chordal graph is the intersection graph of a family of subtrees of a tree.

Let $G_1 = (V_1, E_1)$ and $G_2 = (V_2, E_2)$ be two chordal graphs sharing some common vertices $X = V_1 \cap V_2$ and the edges induced by X. Then (by Theorem 1), the simultaneous chordal graph problem asks whether there exists a set A of augmenting edges such that the graph (G_1, G_2, A) is chordal. We solve the following generalized problem: Given G_1, G_2 and X (as above), and a set F of *forced* augmenting edges, does there exist a set A of augmenting edges such that the graph $(G_1, G_2, F \cup A)$ is chordal.

We need the following additional notation. For a vertex v in $G = (G_1, G_2, F)$, we use $N_1(v)$ and $N_2(v)$ to denote the sets $N_{E(G)}(v) \cap V(G_1)$ and $N_{E(G)}(v) \cap V(G_2)$ respectively. Note that if $v \in V_1 - X$ then $N_2(v)$ may be non-empty because of F. Finally, we use $C(v)$ to denote the edge set $\{(x, y) : x \in N_1(v) - X, y \in N_2(v) - X\}$. A vertex $v \in G = (G_1, G_2, F)$ is said to be an *S-elimination vertex* of G if $N_1(v)$ and $N_2(v)$ induce cliques in G_1 and G_2 respectively.

Lemma 1. *If $G = (G_1, G_2, F)$ is augmentable to a chordal graph then there exists an S-elimination vertex v of G.*

Proof. Let A be a set of augmenting edges such that the graph $G' = (G_1, G_2, F \cup A)$ is chordal. Because G' is chordal it has a simplicial vertex, i.e. a vertex v such that $N_{E(G')}(v)$ induces a clique in G'. This in turn implies that $N_1(v)$ and $N_2(v)$ induce cliques. □

Theorem 2. *Let $G_1 = (V_1, E_1)$ and $G_2 = (V_2, E_2)$ be two graphs sharing some vertices X and the edges induced by X. Let $G = (G_1, G_2, F)$ and let v be any*

S-elimination vertex of G. G is augmentable to a chordal graph if and only if the graph $G_v = (G_1, G_2, F \cup C(v)) - v$ is augmentable to a chordal graph.

Proof. If G_v is augmentable to a chordal graph, then there exists a set A of augmenting edges such that $G'_v = (G_1, G_2, F \cup C(v) \cup A) - v$ is chordal. We claim that $G' = (G_1, G_2, F \cup C(v) \cup A)$ is chordal. Note that $N_{E(G')}(v) = N_1(v) \cup N_2(v)$, which forms a clique in G'. Thus v is simplicial in G'. Furthermore, $G' - v = G'_v$ is chordal. This proves the claim. Thus G can be augmented to a chordal graph by adding the edges $C(v) \cup A$.

To prove the other direction, assume without loss of generality that $v \in V_1$. Let A be a set of augmenting edges of G such that the graph $G' = (G_1, G_2, F \cup A)$ is chordal. Consider a subtree representation of G'. In this representation, each node $x \in V_1 \cup V_2$ is associated with a subtree T_x and two nodes are adjacent in G' if and only if the corresponding subtrees intersect. We now alter the subtrees as follows.

For each node $x \in N_1(v) - X$ we replace T_x with $T'_x = T_x \cup T_v$. Note that T'_x is a (connected) tree since T_x and T_v intersect. Consider the chordal graph G'' defined by the (intersections of) resulting subtrees. Our goal is to show that the chordal graph $G'' - v$ is an augmentation of G_v, which will complete our proof. First note that $E(G'') \supseteq C(v)$ because for every $x \in N_1(v) - X$, subtree T'_x intersects every subtree T_y for $y \in N_{E(G')}(v)$. The only remaining thing is to show that the edges that are in G'' but not in G' are augmenting edges, i.e. edges from $V_1 - X$ to $V_2 - X$. By construction, any edge added to G'' goes from some $x \in N_1(v) - X$ to some $y \in N_{E(G')}(v)$. Thus $x \in V_1 - X$, and we only need to show that $y \in V_2 - X$. Note that (y, v) is an edge of $E(G')$. Now if $y \in V_1$ then $(y, v) \in E_1$ and thus x, y are both in the clique $N_1(v)$ and are already joined by an edge in G (and hence G'). Therefore $y \in V_2 - X$ and we are done. $\square$

Theorem 2 leads to the following algorithm for recognizing simultaneous chordal graphs

Algorithm 1
1. Let G_1 and G_2 be the input graphs and let $F = \{\}$.
2. **While** there exists an S-elimination vertex v of $G = (G_1, G_2, F)$ **Do**
3. $F \leftarrow F \cup C(v)$
4.   Remove v and its incident edges from G_1, G_2, F.
5. **End**
6. **If** G is empty return YES **else** return NO

Note that the above algorithm also computes the augmented graph for the YES instances. We show in the full version [11] that Algorithm 1 can be implemented to run in time $O(n^3)$.

3   Simultaneous Comparability Graphs

Recall that a *comparability graph* is defined as a graph whose edges can be transitively oriented. Golumbic [8] gave an $O(nm)$ time algorithm for recognizing comparability graphs and constructing the transitive orientation if it exists. In this section we extend Golumbic's [8] results to simultaneous comparability graphs and show that the simultaneous comparability problem can also be solved in $O(nm)$ time. We begin by proving the equivalence of the original definition of the simultaneous comparability graph problem and the augmenting edges version of the problem. This is the analogue of Theorem 1 which only applied to intersection classes.

We use the following additional notation. A directed edge from u to v is denoted by $\overrightarrow{uv}$. If A is a set of directed edges, then we use A^{-1} to denote the set of edges obtained by reversing the direction of each edge in A. We use $\hat{A}$ to denote the union of A and A^{-1}. A is said to be *transitive* if for any three vertices a, b, c, we have $\overrightarrow{ab} \in A$ and $\overrightarrow{bc} \in A \Rightarrow \overrightarrow{ac} \in A$. Our edge sets never include loops, so note the implication that if A is transitive then it cannot contain a directed cycle and must satisfy $A \cap A^{-1} = \emptyset$ (because if it contained both $\overrightarrow{ab}$ and $\overrightarrow{ba}$ it would contain $\overrightarrow{aa}$). By definition a *transitive orientation* assigns a direction to each edge in such a way that the resulting set of directed edges is transitive. We use $G - A$ to denote the graph obtained by undirecting A and removing it from graph G.

Let $G_1 = (V_1, E_1)$ and $G_2 = (V_2, E_2)$ be two comparability graphs sharing some vertices X and the edges induced by X. If G_1 and G_2 are simultaneous comparability graphs, then there exist transitive orientations T_1 and T_2 of G_1 and G_2 (respectively) that are consistent on $E(X)$. We call $T = T_1 \cup T_2$ a *pseudo-transitive orientation* of $G_1 \cup G_2$. Note that the orientation induced by V_1 (and V_2) in T is transitive. If $W \subseteq \hat{E}_1 \cup \hat{E}_2$, then W is said to be *pseudo-transitive* if $W \cap \hat{E}_1$ and $W \cap \hat{E}_2$ are both transitive. We can show that any pseudo-transitive orientation of $G_1 \cup G_2$ can be augmented to a transitive orientation, which is the main ingredient in the proof (see [11]) of the following equivalence theorem.

Theorem 3. *Let $G_1 = (V_1, E_1)$ and $G_2 = (V_2, E_2)$ be two comparability graphs sharing some vertices X and the edges induced by X. G_1 and G_2 are simultaneous comparability graphs if and only if there exists a set $A \subseteq (V_1 - X) \times (V_2 - X)$ of edges such that the graph $G = (V_1 \cup V_2, E_1 \cup E_2 \cup A)$ is a comparability graph.*

We now sketch a high level overview of Golumbic's algorithm for recognizing comparability graphs and compare it with our approach. Golumbic's recognition algorithm is conceptually quite simple: orient one edge (call it a "seed" edge), and follow implications to orient further edges. If this process results in an edge being oriented both forwards and backwards, the input graph is rejected. Otherwise, when there are no further implications, the set of oriented edges (called an "implication class") is removed and the process repeats with the remaining graph. The correctness proof is not so simple, requiring an analysis of implication classes, and of how deleting one implication class changes other implication classes. Golumbic proves the following theorem.

Theorem 4. *(Golumbic [8]) Let $G = (V, E)$ be an undirected graph and let $\hat{E}(G) = \hat{B}_1 + \hat{B}_2 + \cdots \hat{B}_j$ be any "G-decomposition" where for each $k \in \{1, \cdots, j\}$,*

B_k is an implication class of $G - \cup_{1 \leq l < k} \hat{B}_l$. The following statements are equivalent:

1. G is a comparability graph.
2. $I \cap I^{-1} = \emptyset$ for all implication classes I of G.
3. $B_k \cap B_k^{-1} = \emptyset$ for $k = 1, \cdots, j$.

We follow a similar strategy except that the "seed" edges must be chosen carefully for our proof to work. We define the concept of a "composite class" which is analogous to an implication class. We further classify a composite class as a "base class" or a "super class" depending on whether it is disjoint from $E(G_1) \cap E(G_2)$ or not. Our algorithm works as follows: As long as there is a base class remove it and recursively orient the remaining graph. Otherwise (when there are no base classes left) as long as there is a super class remove it and recursively orient the remaining graph.

We prove the following theorem.

Theorem 5. Let $G_1 = (V_1, E_1)$ and $G_2 = (V_2, E_2)$ be two comparability graphs sharing some vertices X and the edges induced by X. Let $\hat{E}_1 \cup \hat{E}_2 = \hat{B}_1 + \hat{B}_2 + \cdots + \hat{B}_i + \hat{S}_{i+1} + \hat{S}_{i+2} + \cdots + \hat{S}_j$ be a "S-decomposition" of $G_1 \cup G_2$ where for each $k \in \{1, \cdots, i\}$, B_k is a base class of $G - \cup_{1 \leq l < k} \hat{B}_l$ and for each $k \in \{i+1, \cdots, j\}$, S_k is a super class of $G - \cup_{1 \leq l \leq i} \hat{B}_l - \cup_{i+1 \leq l < k} \hat{S}_l$

The following statements are equivalent.

1. G_1 and G_2 are simultaneous comparability graphs
2. $C \cap C^{-1} = \emptyset$ for all composite classes C of $G = G_1 \cup G_2$.
3. $B_k \cap B_k^{-1} = \emptyset$ for $k = 1, \cdots, i$ and $S_k \cap S_k^{-1} = \emptyset$ for $k = i+1, \cdots, j$.

We now formalize and justify the above defined notions. Given an undirected graph H, we can replace each undirected edge (u, v) by two directed edges $\overrightarrow{uv}$ and $\overrightarrow{vu}$ and define a relation Γ on the directed edges as follows. $\overrightarrow{ij} \Gamma \overrightarrow{i'j'}$ if ($i = i'$ and $(j, j') \notin E(H)$) or ($j = j'$ and $(i, i') \notin E(H)$). Γ can be viewed as a constraint that directs the (i, j) edge from i to j if and only if the edge (i', j') is directed from i' to j'. It is easy to see that the transitive closure of Γ, denoted by Γ_t, is an equivalence relation. We refer to the partitions of Γ_t as *implication classes*. The following Lemmas capture some of the fundamental properties of implication classes.

Lemma 2. ([8]) Let A be an implication class of a graph H. If H has a transitive orientation F, then either $F \cap \hat{A} = A$ or $F \cap \hat{A} = A^{-1}$ and in either case, $A \cap A^{-1} = \emptyset$.

Lemma 3. ([8]) Let the vertices a, b, c induce a triangle in H and let $\overrightarrow{bc}$, $\overrightarrow{ca}$ and $\overrightarrow{ba}$ belong to implication classes A, B and C respectively. If $A \neq C$ and $A \neq B^{-1}$, then

1. If $\overrightarrow{b'c'} \in A$ then $\overrightarrow{b'a} \in C$ and $\overrightarrow{c'a} \in B$
2. No edge of A is incident with a.

Lemma 4. *([8]) Let A be an implication class of a graph H. If $A \cap A^{-1} = \emptyset$, then A is transitive.*

Note that in Lemma 2, if the directions of one or more edges of triangle abc are reversed, then the Lemma can still be applied by inversing the corresponding implication classes. For e.g when $\overrightarrow{ab} \in C$, $\overrightarrow{ac} \in B$ and $\overrightarrow{bc} \in A$, if $A \neq C^{-1}$ and $A \neq B$, then condition (1) becomes: If $\overrightarrow{b'c'} \in A$ then $\overrightarrow{ab'} \in C$ and $\overrightarrow{ac'} \in B$.

Let $G_1 = (V_1, E_1)$ and $G_2 = (V_2, E_2)$ be two comparability graphs sharing some vertices X and the edges induced by X. Let $G = G_1 \cup G_2$. We define a relation Γ' on the (directed) edges of G as follows: $\overrightarrow{e} \Gamma' \overrightarrow{f}$ if $\overrightarrow{e} \Gamma \overrightarrow{f}$ and $(\{\overrightarrow{e}, \overrightarrow{f}\} \subseteq \hat{E}_1$ or $\{\overrightarrow{e}, \overrightarrow{f}\} \subseteq \hat{E}_2)$. It is easy to see that the transitive closure of Γ' denoted by Γ_t' is an equivalence relation. We refer to the partitions of Γ_t' as "composite classes". A composite class C is said to be *pseudo-transitive* if $(\overrightarrow{ab} \in C$ and $\overrightarrow{bc} \in C) \Rightarrow \overrightarrow{ac} \in C$ whenever $\{a, b, c\} \in V_1$ or $\{a, b, c\} \in V_2$.

From the definition it follows that each composite class is a union of one or more of the implication classes of G_1 and the implication classes of G_2. If a composite class C of G has an edge that belongs to $E(X)$, then we use the term "super class" to refer to C. Otherwise C is said to be a "base class". Thus any base class is a single implication class of G_1 or G_2 and is contained in $\hat{E}_1 - \hat{E}(X)$ or $\hat{E}_2 - \hat{E}(X)$.

Observation: *Note that every implication class of a super class contains an edge $\overrightarrow{e} \in \hat{E}(X)$.*

The following Lemmas for composite classes are analogous to Lemmas 2, 3 and 4. We provide the proofs in the full version [11].

Lemma 5. *Let A be a composite class of $G = G_1 \cup G_2$. If F is a pseudo-transitive orientation of G then either $F \cap \hat{A} = A$ or $F \cap \hat{A} = A^{-1}$ and in either case, $A \cap A^{-1} = \emptyset$.*

Lemma 6. *Let the vertices $a \in X$, b and c induce a triangle in $G = G_1 \cup G_2$, such that $\overrightarrow{bc}$, $\overrightarrow{ca}$ and $\overrightarrow{ba}$ belong to composite classes A, B and C respectively. If $A \neq C$ and $A \neq B^{-1}$, then*

1. *If $\overrightarrow{b'c'} \in A$ then $\overrightarrow{b'a} \in C$ and $\overrightarrow{c'a} \in B$.*
2. *No edge of A is incident with a.*

Lemma 7. *Let the vertices a, b, c form a triangle in G and let the edges $\overrightarrow{bc}, \overrightarrow{ca}$ and $\overrightarrow{ba}$ belong to composite classes A, B and C respectively with $A \neq C$, $A \neq B^{-1}$ and $B \neq C$. If B and C are both super classes then there exists a triangle a', b', c' in X with $\overrightarrow{b'c'} \in A$, $\overrightarrow{c'a'} \in B$ and $\overrightarrow{b'a'} \in C$ and hence A is a super class.*

Lemma 8. *Let A be a composite class of a graph $G = G_1 \cup G_2$. If $A \cap A^{-1} = \emptyset$, then $A \cap \hat{E}_1$ and $A \cap \hat{E}_2$ are both transitive and hence A is pseudo-transitive.*

Corollory 1. *Let A be a composite class of a graph $G = G_1 \cup G_2$. Then A is pseudo-transitive iff $A \cap A^{-1} = \emptyset$.*

Recall that our approach involves deleting a composite class A from G. Any composite class of $G - A$ is a union of composite classes of G formed by successive "merging". Two composite classes B and C of G are said to be *merged* by the deletion of class A, if deleting A creates a (Γ') relation between a B-edge and a C-edge. Note that for this to happen there must exist a triangle a, b, c in G with $(b, c) \in \hat{A}$ and either $\overrightarrow{ba} \in C$ and $\overrightarrow{ca} \in B$ or $\overrightarrow{ab} \in C$ and $\overrightarrow{ac} \in B$. We first iteratively delete all the base classes followed by the (remaining) super classes. The following Lemmas (proved in the full version [11]) examine what happens when a base or super class gets deleted by the algorithm.

Lemma 9. *If the composite classes of $G = G_1 \cup G_2$ are all pseudo-transitive and A is a base class of G then the composite classes of $G - A$ are also pseudo-transitive.*

Lemma 10. *Let each of the composite classes of $G = G_1 \cup G_2$ be super and pseudo-transitive. If A is any super class of G then each of the composite classes of $G - A$ is pseudo-transitive.*

A partition of the edge set $\hat{E}_1 \cup \hat{E}_2 = \hat{B}_1 + \hat{B}_2 + \cdots + \hat{B}_i + \hat{S}_{i+1} + \hat{S}_{i+2} + \cdots + \hat{S}_j$ is said to be a *S-decomposition* of $G = G_1 \cup G_2$, if for each $k \in \{1, \cdots, i\}$, B_k is a base class of $G - \cup_{1 \leq l < k} \hat{B}_l$ and for each $k \in \{i+1, \cdots, j\}$, S_k is a super class of $G - \cup_{1 \leq l \leq i} \hat{B}_l - \cup_{i+1 \leq l < k} \hat{S}_l$
    We are now ready to prove the main theorem.

Theorem 5. *Let $G_1 = (V_1, E_1)$ and $G_2 = (V_2, E_2)$ be two comparability graphs sharing some vertices X and the edges induced by X. Let $\hat{E}_1 \cup \hat{E}_2 = \hat{B}_1 + \hat{B}_2 + \cdots + \hat{B}_i + \hat{S}_{i+1} + \hat{S}_{i+2} + \cdots + \hat{S}_j$ be a S-decomposition of $G_1 \cup G_2$. The following statements are equivalent.*

1. *G_1 and G_2 are simultaneous comparability graphs*
2. *Every composite class of $G = G_1 \cup G_2$ is pseudo-transitive, i.e. $C \cap C^{-1} = \emptyset$ for all composite classes C of $G = G_1 \cup G_2$.*
3. *Every partition of the S-decomposition is pseudo-transitive, i.e. $B_k \cap B_k^{-1} = \emptyset$ for $k = 1, \cdots, i$ and $S_k \cap S_k^{-1} = \emptyset$ for $k = i+1, \cdots, j$.*

Proof. $(1) \Rightarrow (2)$ follows from Lemmas 5 and 8.
    $(2) \Rightarrow (3)$ is a direct consequence of Lemmas 9 and 10.
    $(3) \Rightarrow (1)$
Let $T = B_1 + B_2 + \cdots + B_i + \cdots S_{i+1} + S_{i+2} + \cdots S_j$. We now claim that T is pseudo-transitive. For $k = 1, \cdots, j$, define C_k as $C_k = B_k$ if $k \leq i$ and $C_k = S_k$ otherwise. Thus $T = C_1 + \cdots C_j$.
    For $k = 1 \cdots j$, let $T_k = C_k + \cdots C_j$. (Thus $T_1 = T$) and $H_k = \hat{C}_k + \cdots \hat{C}_j$. Thus C_k is a composite class of H_k. Now it is enough to show that T_k is pseudo-transitive for any k. Assume inductively that $T_{k+1} = T_k - C_k$ is pseudo-transitive.

Note that $\hat{T}_{k+1} \cap \hat{C}_k = \emptyset$. Now we claim that $T_k = T_{k+1} \cup C_k$ is also pseudo-transitive.

Suppose not. Then there exist vertices a, b, c all in G_1 or all in G_2 such that $\overrightarrow{ab} \in T_k$, $\overrightarrow{bc} \in T_k$ and $\overrightarrow{ac} \notin T_k$. Since T_{k+1} and C_k are pseudo-transitive we only have to consider the case when $\overrightarrow{ab} \in T_{k+1}$ and $\overrightarrow{bc} \in C_k$ (the other case $\overrightarrow{ab} \in C_k$ and $\overrightarrow{bc} \in T_{k+1}$ is symmetric).

Now if the edge (a, c) is not present in H_k then $\overrightarrow{bc}\Gamma'\overrightarrow{ba}$ and thus $\overrightarrow{ba} \in C_k$, contradicting that $\hat{T}_{k+1} \cap \hat{C}_k = \emptyset$. So either $\overrightarrow{ca} \in T_{k+1}$ or $\overrightarrow{ca} \in C_k$. This implies (by the pseudo-transitivity of T_{k+1} and C_k) that $\overrightarrow{cb} \in T_{k+1}$ or $\overrightarrow{ba} \in C_k$. In both cases we get a contradiction to $\hat{T}_{k+1} \cap \hat{C}_k = \emptyset$.

Thus all four cases lead to a contradiction and we conclude that T_k is pseudo-transitive.   □

Theorem 5 gives rise to the following $O(nm)$ algorithm for recognizing simultaneous comparability graphs: Given graphs G_1 and G_2 check whether all composite classes of $G = G_1 \cup G_2$ are pseudo-transitive. If so return YES otherwise return NO. Further, if G_1 and G_2 are simultaneous comparability graphs then the following algorithm computes an S-decomposition of $G_1 \cup G_2$. As shown in the proof of Theorem 5, this immediately gives a pseudo-transitive orientation.

Algorithm 2
1. Compute all the base classes of $G = G_1 \cup G_2$.
2. **While** there is a base class B remaining in G **Do**
3.   Add B to the solution, delete it from G and update the remaining base classes
4. **While** there is a super class S remaining in G **Do**
5.   Add S to the solution and delete it from G.

Algorithm 2 can be implemented to run in $O(nm)$ time. We present the detailed algorithm along with its analysis in the full version [11].

Remark: Note that if T is a pseudo-transitive orientation of $G_1 \cup G_2$, then T can be augmented to a transitive orientation by computing $T' = T^2$ (as shown in the proof of Theorem 3). The complexity of this step is same as the complexity of matrix multiplication: $O(n^{2.376})$. Hence computing an augmented comparability graph takes $O(nm + n^{2.376})$ steps.

4   Simultaneous Permutation Graphs

A graph $G = (V, E)$ on vertices $V = \{1, \cdots, n\}$ is said to be a *permutation graph* if there exists a permutation π of the numbers $1, 2, \cdots, n$ such that for all $1 \le i < j \le n$, $(i, j) \in E$ if and only if $\pi(i) > \pi(j)$. Equivalently, $G = (V, E)$ is a permutation graph if and only if there are two orderings L and P of V such that $(u, v) \in E$ iff u and v appear in the opposite order in L and in P. We call $\langle L, P \rangle$ an *order-pair* for G. The intersection representation for permutation graphs follows immediately: $G = (V, E)$ is a permutation graph iff there are two parallel lines l and p and a set of line segments each connecting a distinct point on l with a distinct point on p such that G is the intersection graph of the line segments. Observe that L and P correspond to the ordering of the endpoints of the line segments on l and p respectively. Since permutation graphs are a class of intersection graphs the equivalence theorem 1 is applicable for this class.

Let $G_1 = (V_1, E_1)$ and $G_2 = (V_2, E_2)$ be two permutation graphs sharing some common vertices X and the edges induced by X. We begin with a "relaxed" characterization of simultaneous permutation graphs in terms of order-pairs.

Lemma 11. *G_1 and G_2 are simultaneous permutation graphs iff there exist order-pairs $\langle L_1, P_1 \rangle$ and $\langle L_2, P_2 \rangle$ for G_1 and G_2 (respectively) such that every pair of vertices $u, v \in X$ appear in the same order in L_1 as in L_2 AND appear in the same order in P_1 as in P_2.*

Proof. The forward direction is clear. For the reverse direction, we create a total order L on $V_1 \cup V_2$ consistent with both L_1 and L_2. This is possible because L_1 and L_2 are consistent on X. We do the same for P. The orderings L and P provide the endpoints of line segments for the simultaneous intersection representations of G_1 and G_2. □

It is well-known that a graph G is a permutation graph if and only if G and $\bar{G}$ are both comparability graphs [7]. Using this we can prove the following analogous result for simultaneous permutation graphs. The proof is available in the full version [11].

Theorem 6. *Let $G_1 = (V_1, E_1)$ and $G_2 = (V_2, E_2)$ be two undirected graphs sharing some vertices X, and the edges induced by X. G_1 and G_2 are simultaneous permutation graphs if and only if G_1 and G_2 are simultaneous comparability graphs and simultaneous co-comparability graphs.*

Since simultaneous comparability graphs can be recognized in $O(nm)$ time, Theorem 6 implies that simultaneous permutation graphs can be recognized in $O(n^3)$ time. We also note that a similar approach was used in [5] to recognize probe permutation graphs.

5   Discussion

A main contribution of this paper is the introduction of the simultaneous representation problem, which is closely related to the probe graph recognition and graph sandwich problems. We gave poly-time algorithms for solving the problem for chordal, comparability and permutation graphs. The running time of our algorithm for comparability graphs is $O(nm)$. For chordal and permutation graphs both of our algorithms run in $O(n^3)$. Our techniques for simultaneous comparability and permutation graphs can be extended to solve the graph sandwich problem for these classes when the set of optional edges induce a k-partite graph.

We believe that the simultaneous representation problem for interval graphs is also solvable in polynomial time, but it seems substantially more difficult than for chordal and comparability graphs. We have a solution in progress [14].

References

1. Berry, A., Golumbic, M.C., Lipshteyn, M.: Two tricks to triangulate chordal probe graphs in polynomial time. In: SODA 2004: Proceedings of the fifteenth annual ACM-SIAM symposium on Discrete algorithms, pp. 992–969 (2004)
2. Berry, A., Golumbic, M.C., Lipshteyn, M.: Recognizing chordal probe graphs and cycle-bicolorable graphs. SIAM J. Discret. Math. 21(3), 573–591 (2007)
3. Brass, P., Cenek, E., Duncan, C.A., Efrat, A., Erten, C., Ismailescu, D.P., Kobourov, S.G., Lubiw, A., Mitchell, J.S.B.: On simultaneous planar graph embeddings. Comput. Geom. Theory Appl. 36(2), 117–130 (2007)
4. Chandler, D.B., Chang, M.-S., Kloks, T., Liu, J., Peng, S.-L.: On probe permutation graphs (extended abstract). In: Cai, J.-Y., Cooper, S.B., Li, A. (eds.) TAMC 2006. LNCS, vol. 3959, pp. 494–504. Springer, Heidelberg (2006)
5. Chandler, D.B., Chang, M.-S., Kloks, T., Liu, J., Peng, S.-L.: Partitioned probe comparability graphs. Theor. Comput. Sci. 396(1-3), 212–222 (2008)
6. Estrella-Balderrama, A., Gassner, E., Jünger, M., Percan, M., Schaefer, M., Schulz, M.: Simultaneous geometric graph embeddings. In: Hong, S.-H., Nishizeki, T., Quan, W. (eds.) GD 2007. LNCS, vol. 4875, pp. 280–290. Springer, Heidelberg (2008)
7. Even, S., Pnueli, A., Lempel, A.: Permutation graphs and transitive graphs. J. Assoc. Comput. Mach. 19, 400–410 (1972)
8. Golumbic, M.C.: Algorithmic Graph Theory And Perfect Graphs. Academic Press, New York (1980)
9. Golumbic, M.C., Kaplan, H., Shamir, R.: Graph sandwich problems. J. Algorithms 19(3), 449–473 (1995)
10. Golumbic, M.C., Lipshteyn, M.: Chordal probe graphs. Discrete Appl. Math. 143(1-3), 221–237 (2004)
11. Jampani, K.R., Lubiw, A.: Simultaneous membership problem for chordal, comparability and permutation graphs (2009)
12. McConnell, R.M., Spinrad, J.P.: Construction of probe interval models. In: SODA 2002: Proceedings of the thirteenth annual ACM-SIAM symposium on Discrete algorithms, Philadelphia, PA, USA, pp. 866–875. Society for Industrial and Applied Mathematics (2002)
13. McMorris, F.R., Wang, C., Zhang, P.: On probe interval graphs. Discrete Appl. Math. 88(1-3), 315–324 (1998)
14. Jampani, K.R.: Simultaneous membership and related problems, Ph.D thesis, University of Waterloo (in progress)
15. Spinrad, J.: Efficient Graph Representations. Fields Institute Monographs. American Mathematical Society (2003)

Two for One: Tight Approximation of 2D Bin Packing*

Klaus Jansen, Lars Prädel, and Ulrich M. Schwarz

Institut für Informatik
Christian-Albrechts-Universität zu Kiel
Olshausenstr. 40
24098 Kiel, Germany
{kj,lap,ums}@informatik.uni-kiel.de

Abstract. In this paper, we study the two-dimensional geometrical bin packing problem (2DBP): given a list of rectangles, provide a packing of all these into the smallest possible number of 1×1 bins without rotating the rectangles.

We present a 2-approximate algorithm, which improves over the previous best known ratio of 3, matches the best results for the rotational case and also matches the known lower bound of approximability. Our approach makes strong use of a recently-discovered PTAS for a related knapsack problem and a new algorithm that can pack instances into $OPT + 2$ bins for any constant OPT.

Keywords: bin packing, approximation, rectangle packing.

1 Introduction

In recent years, there has been increasing interest in extensions of packing problems such as strip packing [1–5], knapsack [6–8] and bin packing [9–13], to multiple criteria (vector packing) or multiple dimensions (geometric packing).

Two-dimensional bin packing, both with and without rotations, is one of the very classical problems in combinatorial optimization and its study has begun several decades ago. This is not only due to its theoretical appeal, but also to a large number of applications, ranging from print and web layout [14] (putting all ads and articles onto the minimum number of pages) to office planning (putting a fixed number of office cubicles into a small number of floors), to transportation problems (packing goods into the minimum number of standard-sized containers) and VLSI design [15].

It is easy to see that two-dimensional bin packing without rotation (2DBP) is strongly NP-hard as a generalization of its one-dimensional counterpart, hence the main focus is on algorithms with provable approximation quality.

* Work supported by EU project "AEOLUS: Algorithmic Principles for Building Efficient Overlay Computers", EU contract number 015964, and DFG project JA612/12-1, "Design and analysis of approximation algorithms for two- and threedimensional packing problems".

F. Dehne et al. (Eds.): WADS 2009, LNCS 5664, pp. 399–410, 2009.
© Springer-Verlag Berlin Heidelberg 2009

Consider an algorithm A for 2DBP, and denote for each instance I with $A(I)$ the number of bins A produces and with $\mathrm{OPT}(I)$ the smallest number of bins into which I can be packed. A is an α-*approximation* for 2DBP if we have $\sup_I \{A(I)/\mathrm{OPT}(I)\} \leq \alpha$ over all instances I, and an *asymptotical α-approximation* if we have $\limsup_{\mathrm{OPT}(I) \to \infty} A(I)/\mathrm{OPT}(I) \leq \alpha$. A polynomial-time approximation scheme (PTAS) is a family $\{A_\epsilon : \epsilon > 0\}$ of $(1 + \epsilon)$-approximation algorithms.

The best previously known result for the non-rotational case was a cubic-time 3-approximation by Zhang [11]; for the rotational case, Harren and van Stee have recently given a 2-approximation in [13], the same ratio can be achieved using techniques by Jansen and Solis-Oba [5].

As to asymptotical approximation ratios, Bansal and Sviridenko showed in [10] that 2DBP does not admit an asymptotical PTAS. Caprara gave an algorithm with ratio of $1.69\ldots$ in [9], breaking the important barrier of 2. More recently, Bansal, Caprara and Sviridenko improved the rate to $1.52\ldots$ in [12] for both the rotational and non-rotational case.

A closely related problem is two-dimensional knapsack: here, every rectangle also has a profit and the objective is to pack a subset of high profit into a constant number (usually one) of target bins. The best currently known results here are a $(2 + \epsilon)$-approximation by Jansen and Zhang [16] for the general case, and a PTAS by Jansen and Solis-Oba [17] if all items are squares. For our purposes, the special case that the profit equals the item's area is important. We have recently shown in [8] that this problem admits a PTAS, and this algorithm is one of the corner stones of the algorithm presented here.

Our Contribution. We study the non-rotational geometric two-dimensional bin packing problem, i.e. we are given a list of rectangular items $\{r_i = (w_i, h_i) : i = 1, \ldots, n\}$ with all w_i, h_i taken from the interval $]0, 1]$, and the objective is to find a non-rotational non-overlapping packing of all items into the minimum number of containers of size 1×1. The main result of this paper is the following theorem:

Theorem 1. *There is a polynomial-time 2-approximation for two-dimensional geometric bin packing.*

Observing that the problem to decide whether a given instance fits in one bin or needs two is NP-complete as a generalization of 3PARTITION, this settles the matter of absolute approximation ratio.

The result is achieved using an asymptotic approximation algorithm for large optimal values; smaller (i.e. constant) values are solved by a recent breakthrough in the approximability of two-dimensional *knapsack* problems in [8, 18, 19]: there exists a PTAS for maximizing the area covered by rectangles within a 1×1 bin. This can be combined with other packing algorithms if the optimum is constant, but at least 2, to generate a packing into $\mathrm{OPT} + 2$ bins. If the optimal packing uses only one bin, we conduct a case study, again starting from a packing that covers $(1 - \epsilon)$ of the bin and generate a packing into $\mathrm{OPT} + 1 = 2$ bins.

Hence, the other corner stone of the new result will be the following lemma, proven in Section 3.

Lemma 1. *There is a polynomial-time algorithm that finds a packing into two bins, provided that a packing into one bin exists.*

The case of larger optimum values is handled in the next section.

2 Large Optimal Value

As noted above, Bansal, Caprara and Sviridenko [12] have given a polynomial-time algorithm for 2DBP which has an asymptotic approximation ratio of $1.525 < 2$. From this, we immediately obtain a (non-asymptotic) approximation ratio of 2 for instances with a large optimum:

Corollary 1. *There is a constant K so that for every instance I with $\mathrm{OPT}(I) \geq K$, the algorithm of Bansal et al. yields a packing into at most $2\mathrm{OPT}(I)$ bins.*

If, for a given instance, this is not the case, we can by enumeration try the constant number of possible optimum values and find a packing into $\mathrm{OPT} + 2$ bins using the following two key statements:

Theorem 2. *There is an algorithm that, given a set of rectangles $I = \{r_i = (w_i, h_i) : i = 1, \ldots, n\}$, a constant ϵ and a constant k such that there exists a packing of I into k bins, produces in polynomial time a packing of a subset $I' \subseteq I$ into k bins such that the total unpacked area $\sum\{w_i h_i : r_i \in I \setminus I'\}$ is bounded by $\epsilon \cdot \sum\{w_i h_i : r_i \in I\}$. Furthermore, all unpacked items are bounded by ϵ in one direction.*

The proof, which is a straightforward extension of [8] using $k - 1$ extra items and a rescaling argument, can be found in the full version. The other ingredient is a routine for packing all of the remaining items into at most 2 bins.

Lemma 2. *There is an algorithm that packs a set of rectangles with total area at most $1/2$ into 2 bins.*

The proof is an easy application of Steinberg's classical result:

Theorem 3 (Steinberg [3]). *We can pack a set of items $\{r_i = (w_i, h_i), i = 1, \ldots, n\}$ into a target area of size $u \times v$ if the following conditions hold:*

$$\max\{w_i : i = 1, \ldots, n\} \leq u, \quad \max\{h_i : i = 1, \ldots, n\} \leq v,$$

$$2\sum_{i=1}^{n} w_i h_i \leq uv - (2 \max_{i=1,\ldots,n} w_i - u)^+ (2 \max_{i=1,\ldots,n} h_i - v)^+,$$

where $(\cdot)^+$ denotes $\max\{\cdot, 0\}$.

Combining Theorem 2 and Lemma 2, we obtain:

Theorem 4. *For every constant k, we can check in polynomial time whether there exists a packing into $k + 2$ bins.*

An algorithm to solve the overall problem can now first run the algorithm of Bansal et al., and it can then run for each guessed OPT $= k < K$, of which there is only a constant number, a polynomial-time algorithm that tries to find a packing into $k + 2$ bins. For the case of OPT $= k = 1$, we generate a packing into two bins. Finally, the algorithm returns the best solution amongst these. The details of the algorithms for OPT $= 1$ are given in the next sections.

3   Solving for OPT $= 1$

In this section, we will show how an instance that admits a packing into one bin can be packed into two bins in polynomial time. For simplicity, we assume the total area of items is exactly 1. In the following, all statements still hold when interchanging width and height, unless specifically noted otherwise.

We will study several cases separately and solve each in polynomial time. The algorithm will check which case applies and use the corresponding subroutine. We will mean "we can pack" to imply a step admits polynomial-time algorithms and "can/cannot be packed" to imply general feasibility or non-feasibility.

3.1   Mostly Tall and Wide Items

Let us start by making a simple observation on the arrangement of tall items in the optimal solution:

Remark 1. Consider the set of 'tall' items of height more than $1/2$. We can always pack these items into a bin along with one arbitrary extra item r_i.

Proof. Note that no two tall items can intersect the same vertical line $x = x_0$. In particular, the total width of these items is at most 1. We sort the items by decreasing height, starting at the bottom-left corner, cf. Fig. 1a, and place the extra item in the top right corner. Assume that r_i intersects the tall items. In particular, this means that items with height larger than $1 - h_i$ have total width of more than $1 - w_i$, which means that no feasible packing of these items along with r_i exists.   □

Noting that Steinberg's algorithm will pack any set of items of total area at most $1/2$ into a bin if all items are bounded by $1/2$ in the same direction, we obtain:

Corollary 2. *Consider the set of 'tall' items of height more than $1/2$. If their total area is at least $1/2 - \beta$ for some $-1/2 \le \beta \le 1/2$, then either every other item has area less than β or we can pack all items into two bins.*

As a slight generalization, we can show:

Lemma 3. *For $0 < \gamma \le 1/2$ arbitrary, set $W := \sum\{w_j : j \in \{1, \ldots, n\}, h_j > 1 - \gamma\}$. Then,*

$$\sum\{w_j h_j : j \in \{1, \ldots, n\}, w_j > 1 - W, h_j \le 1 - \gamma\} \le 2\gamma. \tag{1}$$

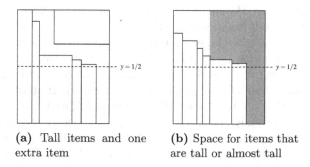

(a) Tall items and one extra item

(b) Space for items that are tall or almost tall

Fig. 1. Arrangements of large items

Proof. Consider a horizontal scanline $y = y_0$ for $y_0 \in [\gamma, 1 - \gamma]$ in any packing. Such a scanline intersects all items of height at least γ, hence, it will not admit an item of width larger than $1 - W$. Hence, all such items must be in the outermost regions of height γ at the top and bottom of the bin.   □

Lemma 4. *If the total width of items taller than $1/2$ is larger than $1 - \delta$ for $\delta = 3/4 - \sqrt{1/2} \approx 0.042$, we can pack all items into two bins.*

Proof. We pack the tall items into the first bin, sorted by non-increasing height. They must fit next to each other, since no two of them can be atop one another, and the total area covered by these items is at least $1/2 - \delta/2$. Note that by Cor. 2, all other items have individual area bounded by $\delta/2$, or we are done. In particular, every other item is bounded by $\sqrt{\delta/2}$ in at least one direction.

We define $\delta' := -1/4 + \sqrt{1/16 + \delta/2}$. It is easy to verify the following statements:

1. $\sqrt{\delta/2} \leq (1/2 - \delta')/2$,
2. $(1/2 - \delta')^2 \geq 2\delta$,
3. we can pack a "virtual item" of size $(1/2 - \delta') \times (1/2 - \delta')$ into the upper right corner without intersecting the tall items,
4. we can pack a "virtual item" of size $1/2 \times (2\sqrt{\delta})$ into the upper right corner without intersecting the tall items.

Using the first three and Steinberg's algorithm, we are done if there are items of width at most $\sqrt{\delta/2}$ and height at most $1/2 - \delta'$ with total area at least $\delta/2$, and we are also done if there are items of height at most $\sqrt{\delta/2}$ and width at most $1/2 - \delta'$ with total area at least $\delta/2$.

Hence assume neither is the case. The total area of items not yet considered is at least $1 - 1/2 - \delta/2 - \delta/2 = 1/2 - \delta$. Let us turn to items whose height is in $]1/2 - \delta', 1/2]$, i.e. they are "almost tall". Keeping in mind they cannot be packed atop an item of height larger than $1/2 + \delta'$ in any packing, the total width of areas that can accomodate them in our packing, shaded in Fig. 1b, is large enough for us to pack all but one of them by arguments similar to Graham's

LPT analysis [20]. Since the single item has area at most $\delta/2$, we are done if the total area of almost tall items is at least δ.

If this is not the case, we know that all remaining items, i.e. those of height at most $\sqrt{\delta/2}$ and width larger than $1/2 - \delta'$, cover a total area of at least $1/2 - 2\delta$. If the subset of these of width at most $1/2$ is at least $\delta/2$, we can use property 4 to pack some of them in the top right corner. If all of this does not happen, we know that all items of width larger than $1/2$ cover an area of at least $1/2 - 2.5\delta$. We now finally claim that we can pack a selection of these items of area at least $\delta/2$ along with the tall items. Namely, greedily select wide items of minimal width until their area is at least $\delta/2$ (and at most δ), and pack them vertically, starting in the top-right corner, in non-increasing order of width. The total height of this stack is at most 2δ. Assume this stack overlaps the tall items at some coordinate (x, y). In particular, this means that items taller than y have total width at least x. By Lemma 3, this means that there is only a total area of at most $2 - 2y \le 4\delta$ wider than $1 - x$. At the same time, all unselected wide items are at least this wide and have total area at least $1/2 - 3.5\delta$, which contradicts the fact that $\delta < 1/15$. □

3.2 One Big Item

In this and all subsequent cases, we will first apply the algorithm in [8], which will pack items with total area at least $1 - \epsilon$ into one bin. The remaining items have the additional property that each item is bounded in at least one direction by ϵ.

While we know these items can be packed into the one more bin we are allowed, it is in general NP-hard to find such a packing, since the items could encode two instances of PARTITION even in total area ϵ.

Our angle of attack must hence be different: we will identify a suitable strip of height at least 2ϵ in the packed bin and move all items that intersect this strip into the second bin. This creates empty space that can be used to pack all unpacked items of height no more than ϵ in the first bin using Steinberg's algorithm. The remaining unpacked items have height at least ϵ, width at most ϵ and total area at most ϵ, and we will add them in the second bin.

In this section, we consider instances which contain one *big* item r_1 with $w_1, h_1 \ge 1/2$, located at (x_1, y_1). To help rearranging items which are very limited in one dimension, the following observation will be useful:

Lemma 5. *Given a set $\{a_1, \ldots, a_n\}$ of numbers and a minimal target value T such that $S := \sum_{i=1}^{n} a_i \ge 2T + \max_{i=1\ldots,n} a_i$, we can identify a subset $I \subseteq \{1, \ldots, n\}$ such that $\sum_{i \in I} a_i \ge T$ and $\sum_{i \notin I} a_i \ge T$.*

Proof. Consider a fractional optimal solution to the PARTITION problem on the a_i's, for example obtained by greedy packing. Note the solution contains only at most one fractional item a_j. Since $a_j \le \max\{a_i : i = 1 \ldots n\}$, both parts have size at least T even if we assign a_j to the other. □

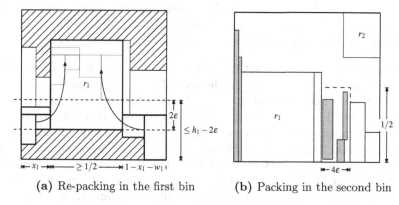

(a) Re-packing in the first bin   (b) Packing in the second bin

Fig. 2. Packing with one big item

By symmetry, we assume that $y_1 \leq 1 - h_1 - y_1$, i.e. r_1 is somewhat in the lower half of the bin. We consider the horizontal strip from y_1 to $y_1 + 2\epsilon$ and all the items that intersect it for movement to the second bin, see Fig. 2, and note that all items that do not intersect $y = y_1 + 2\epsilon$ are of bounded height at most $y_1 + 2\epsilon \leq h_1 - 2\epsilon$. In particular, we can pack these items into the 'hole' left in the first bin by moving r_1 without obstructing the horizontal strip at its bottom.

By previous discussion, we can assume that the total width of items higher than $1/2$ that intersect $y = y_1 + 2\epsilon$ is at most $1 - \delta$, in particular, we assume that there is a (continuous by reordering) interval of length at least δ which contains only items of height at most $1/2$. If we can apply Lemma 5 with target value 2ϵ, we have cleared a vertical strip of width 2ϵ.

Otherwise, we know that there is an item r_2 of width at least $\delta - 4\epsilon > 2\epsilon$ and height at most $1/2$ on this scanline. We can now construct a packing as follows: we pack all tall items, including r_1 and those that were not yet packed at all, along with r_2 as extra item, by Lemma 1. Note that there might be horizontal overlap of r_2 and the tall items, but it is bounded by 2ϵ, since it is only caused by tall unpacked items. We can pack the non-tall items of the scanline below r_2 in width 2ϵ; the remaining unpacked non-tall items, we pack into a container sized $4\epsilon \times 1/2$, which we can position in the lower-right corner, since $8\epsilon \leq w_j$ for $\epsilon \leq \delta/10$.

3.3 One Medium-Sized Item

For this case, assume that there exists an item $r_i = (w_i, h_i)$ such that $w_i, h_i \geq 12\epsilon$ and $h_i \leq 1/2$ at some position (x_i, y_i). Without loss of generality, suppose that $y_i \leq 1/2 - 1/2h_i$. We will now consider three consecutive strips I, II and III that intersect the item, cf. Fig. 3a. Since $h_i \geq 12\epsilon$, we may assume that the bottom strip is still 2ϵ away from the bottom of the bin. It is also easy to verify that in any case, the top of the third strip has a distance of at least h_i to the top of the bin.

Remark 2. If one of Strips I, II, III contains items other than r_i of height at most $1 - h_i$ and total width 2ϵ that totally bisect the strip, we can pack the instance into two bins.

Proof. Move the strip in question to the second bin. Assume by reordering that all bisecting items are adjacent to r_i. It is then possible to shift r_i to one side by at least 2ϵ, which frees a vertical strip of width 2ϵ.   □

In the following, we assume that Remark 2 cannot be applied. We move Strip II and all items that intersect it to the second bin. By reordering, we assume that $x_i + w_i = 1$, and we re-set $y_i := 1 - h_i$. All other items in the second bin can be partitioned into three groups: B, the set of items that completely bisect the strip, A, the set of items that intersect the upper boundary of the strip, but not the lower (i.e. they are 'above the strip'), and U, the set of remaining items, which are entirely in or partially under the strip. By reordering, we assume all elements of B are at the left side of the bin, ordered by non-increasing height.

Note that since Strip II is in the lower half of the bin, all items in U are bounded in height by $1/2$, and in particular by $1 - h_i$. Since Remark 2 did not apply to Strip I, those items in U that bisected Strip I have total width at most

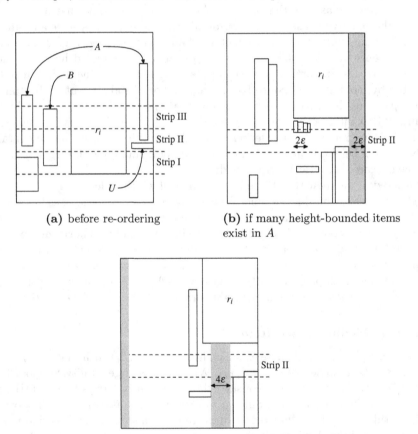

(a) before re-ordering

(b) if many height-bounded items exist in A

(c) if few height-bounded items exist in A

Fig. 3. Packing with one medium item

2ϵ and can be packed below r_i. All others are bounded in height by 4ϵ. Since Strip I still had at least 2ϵ space beneath it, we can drop these items by 2ϵ. As a consequence, none of them intersects Strip II anymore, so we re-sort the items in A by non-increasing height. We now consider those items in A and B that bisected Strip III entirely and have height at most $1 - h_i$. Again by Remark 2, their total width is bounded by 2ϵ and we pack them below r_i as well.

At this point, all items in Strip II that are not re-packed below r_i have height larger than $1 - h_1$ or belong to A and have height at most 4ϵ. If the total width of the latter is at least 2ϵ, we can shift r_i and all items packed below it to the left by 2ϵ, generating a vertical strip of width 2ϵ at the right of the bin, cf. Fig. 3b. Otherwise, we can pack them below r_i. The total free width remaining below r_i is now at least 6ϵ, which we use to shift the entire packing of U to the right by 2ϵ and finally to pack all unpacked items of height at most $1 - h_i$, width at most ϵ and total area at most ϵ into a target area sized $4\epsilon \times (1 - h_i)$. Now, all items that are not packed below r_i, including the remaining unpacked items, have height larger than $1 - h_1 > 1/2$, and hence, by Lemma 1, they fit next to r_1 as sketched in Fig. 3c.

3.4 All Small or Elongated Items

If the previous discussion does not apply, we know that all items in the instance are bounded by 12ϵ in one direction. We show that there are few items of 'medium' sidelength, and items that are small in both directions can be packed efficiently using NFDH:

Lemma 6. *The total area of packed items of height at least 12ϵ and at most $1/2$ and width at most 12ϵ is bounded by 54ϵ, or else we can pack all items into two bins.*

Proof. Partition the bins into horizontal strips of height 2ϵ, and note that each item of height at least 12ϵ will bisect at least four of these strips and intersect up to two more. In particular, at least $2/3$ of its area is used to completely bisect strips. If there is a total width of 36ϵ of such bisecting items in one strip, we can use Lemma 5 to re-pack them, freeing a vertical strip of width 2ϵ, and we obtain a packing. Otherwise, the total area of the items is at most $36\epsilon \cdot 3/2 = 54\epsilon$.   □

From this, we conclude that in the only case left to consider, the majority of items are either tall or wide or very small in both dimensions. We will use this knowledge to construct an entire packing from scratch. More precisely, denote with A_{wide} the total area of items of width at least $1/2$ and with H their total height, with A_{tall} the total area of items of height at least $1/2$ and with W their total width, and with A_{small} the total area of items which are bounded in both directions by 12ϵ. By the previous lemma, we obtain that

$$A_{tall} + A_{wide} + A_{small} \geq 1 - \epsilon - 108\epsilon. \tag{2}$$

For convenience, we assume $A_{wide} \leq A_{tall}$ and construct the packing illustrated in Fig. 4 for the first bin: all tall items are packed at the left side in

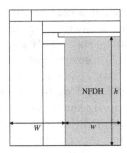

Fig. 4. Packing elongated items

non-increasing order. In the top right corner, we pack a subset of wide items of height almost $H/2$ (but for one fractional item): by arguments similar to the proof of Lemmas 4 and 3, they do not intersect the tall items. Their area is at least $A_{wide}/3 - 12\epsilon$, since wider items might all have width 1. In the bottom right corner, we reserve a target area of size $w \times h$, at least $\delta \times \delta$, which touches the tall and wide items. We fill this area with small items using NFDH. If we run out of small items doing this, the area remaining for the second bin is bounded by $108\epsilon + 2/3 A_{wide} + 12\epsilon \leq 120\epsilon + 1/3$, which is less than $1/2$ for $\epsilon \leq 1/720$, and hence, we can pack the second bin with Steinberg's algorithm.

If we do not run out of small items, we have covered at least an area of $1/2$ in the first bin: the left part of the bin is filled at least half by tall items, the top part is filled at least half by wide items. The remaining part is filled with NFDH. Note that each shelf will be packed to at least $w - 12\epsilon$. We might lose 12ϵ at the top of the area and not account for one shelf by standard shifting arguments, but still, the covered area is at least

$$(w - 12\epsilon)(h - 24\epsilon) > wh - 36\epsilon \geq wh/2 \tag{3}$$

if we set $\epsilon < \delta^2/72 \approx 2.4 \cdot 10^{-5}$. The unpacked items can hence be packed into the second bin using Steinberg's algorithm.

Summing up, the overall algorithm works as outlined in Fig. 5.

1. Run the algorithm of Bansal et al. [12].
2. For each $k = 2, \ldots, K$, try to generate a packing into $k + 2$ bins.
3. If the area of all items is at most 1:
   (a) Apply Lemma 4, if possible.
   (b) Else, generate a packing of $(1 - \epsilon)$ area in the first bin using [8].
   (c) If this packing contains a big item, apply the algorithm in Sect. 3.2.
   (d) Else, if this packing contains an item of at least 12ϵ in both directions, apply the algorithm in Sect. 3.3.
   (e) Else, apply the algorithm in Sect. 3.4.
4. Return the best solution found.

Fig. 5. The overall algorithm

4   Conclusion

We have presented an algorithm that generates 2-approximate solutions for two-dimensional geometric bin packing, which matches the rate known for the rotational problem. Since both the rotational and non-rotational problem are not approximable to any $2 - \epsilon$ unless $P = NP$, this concludes the study of absolute approximability of these problems. For practical applications, it would be interesting to find faster algorithms: our algorithm relies heavily on the knapsack PTAS in [8] and techniques in [5] with a doubly-exponential dependency on ϵ, in particular when compared to the running time of Zhang's 3-approximation in [11]. Still, our result is an important step in the study of two-dimensional packing problems.

Another important open problem is the gap in asymptotic behaviour between the non-existence of an APTAS and the best known algorithm with asymptotic quality of 1.525....

References

1. Baker, B.S., Coffman Jr., E.G., Rivest, R.L.: Orthogonal packings in two dimensions. SIAM J. Comput. 9(4), 846–855 (1980)
2. Baker, B.S., Brown, D.J., Katseff, H.P.: A 5/4 algorithm for two-dimensional packing. J. Algorithms 2(4), 348–368 (1981)
3. Steinberg, A.: A strip-packing algorithm with absolute performance bound 2. SIAM J. Comput. 26(2), 401–409 (1997)
4. Kenyon, C., Rémila, E.: A near-optimal solution to a two-dimensional cutting stock problem. Math. Oper. Res. 25(4), 645–656 (2000)
5. Jansen, K., Solis-Oba, R.: New approximability results for 2-dimensional packing problems. In: Kučera, L., Kučera, A. (eds.) MFCS 2007. LNCS, vol. 4708, pp. 103–114. Springer, Heidelberg (2007)
6. Jansen, K., Zhang, G.: On rectangle packing: maximizing benefits. In: Munro, J.I. (ed.) SODA 2004, pp. 204–213. SIAM, Philadelphia (2004)
7. Leung, J.Y.T., Tam, T.W., Wong, C.S., Young, G.H., Chin, F.Y.L.: Packing squares into a square. J. Parallel Distrib. Comput. 10(3), 271–275 (1990)
8. Jansen, K., Prädel, L.: How to maximize the total area of rectangles packed into a rectangle? Technical Report 0908, Christian-Albrechts-Universität zu Kiel (2009)
9. Caprara, A.: Packing 2-dimensional bins in harmony. In: IEEE (ed.) FOCS 2002, pp. 490–499. IEEE Computer Society Press, Los Alamitos (2002)
10. Bansal, N., Sviridenko, M.: New approximability and inapproximability results for 2-dimensional bin packing. In: Munro, J.I. (ed.) SODA 2004, pp. 196–203. SIAM, Philadelphia (2004)
11. Zhang, G.: A 3-approximation algorithm for two-dimensional bin packing. Oper. Res. Lett. 33(2), 121–126 (2005)
12. Bansal, N., Caprara, A., Sviridenko, M.: Improved approximation algorithms for multidimensional bin packing problems. In: IEEE (ed.) FOCS 2006, pp. 697–708. IEEE Computer Society, Los Alamitos (2006)
13. Harren, R., van Stee, R.: Packing rectangles into 2opt bins using rotations. In: Gudmundsson, J. (ed.) SWAT 2008. LNCS, vol. 5124, pp. 306–318. Springer, Heidelberg (2008)

14. Freund, A., Naor, J.: Approximating the advertisement placement problem. J. Scheduling 7(5), 365–374 (2004)
15. Hochbaum, D.S., Maass, W.: Approximation schemes for covering and packing problems in robotics and VLSI. In: Fontet, M., Mehlhorn, K. (eds.) STACS 1984. LNCS, vol. 166, pp. 55–62. Springer, Heidelberg (1984)
16. Jansen, K., Zhang, G.: Maximizing the total profit of rectangles packed into a rectangle. Algorithmica 47(3), 323–342 (2007)
17. Jansen, K., Solis-Oba, R.: A polynomial time approximation scheme for the square packing problem. In: Lodi, A., Panconesi, A., Rinaldi, G. (eds.) IPCO 2008. LNCS, vol. 5035, pp. 184–198. Springer, Heidelberg (2008)
18. Bansal, N., Caprara, A., Sviridenko, M.: A structural lemma in 2-dimensional packing, and its implications on approximability. Technical Report RC24468, IBM (2008)
19. Bansal, N., Caprara, A., Jansen, K., Prädel, L., Sviridenko, M.: A structural lemma in 2-dimensional packing, and its implications on approximability (unpublished manuscript) (2009)
20. Graham, R.L.: Bounds on multiprocessing timing anomalies. SIAM Journal of Applied Mathematics 17(2), 416–429 (1969)

Fault Tolerant External Memory Algorithms

Gerth Stølting Brodal, Allan Grønlund Jørgensen*, and Thomas Mølhave*

BRICS, MADALGO**, Department of Computer Science, Aarhus University,
Denmark
{gerth,jallan,thomasm}@madalgo.au.dk

Abstract. Algorithms dealing with massive data sets are usually designed for I/O-efficiency, often captured by the I/O model by Aggarwal and Vitter. Another aspect of dealing with massive data is how to deal with memory faults, *e.g.* captured by the adversary based faulty memory RAM by Finocchi and Italiano. However, current fault tolerant algorithms do not scale beyond the internal memory. In this paper we investigate for the first time the connection between I/O-efficiency in the I/O model and fault tolerance in the faulty memory RAM, and we assume that both memory and disk are unreliable. We show a lower bound on the number of I/Os required for any deterministic dictionary that is resilient to memory faults. We design a static and a dynamic deterministic dictionary with optimal query performance as well as an optimal sorting algorithm and an optimal priority queue. Finally, we consider scenarios where only cells in memory or only cells on disk are corruptible and separate randomized and deterministic dictionaries in the latter.

1 Introduction

In this paper we conduct the first study of algorithms and data structures for external memory in the presence of an unreliable internal and external memory.

Contemporary memory devices such as SRAM and DRAM [1,2] can be unreliable due to a number of factors, such as power failures, radiation, and cosmic rays. The content of a cell in unreliable memory can be silently altered and in standard memory circuits there is no direct way of detecting these types of corruptions.

Corrupted content in memory cells can greatly affect many standard algorithms. For instance, in a typical binary search in a sorted array, a single corruption encountered in the early stages of the search can cause the search path to end $\Omega(N)$ locations away from its correct position. Replication of data can help in dealing with corruptions, but is not always feasible, since the time and space overheads of storing and fetching replicated values can be significant. Memory corruptions have been addressed in various ways, both at the hardware and software level. At the software level, soft memory errors are dealt with using several

---

* Supported in part by an Ole Roemer Scholarship from the Danish National Science Research Council.
** Center for Massive Data Algorithmics, a Center of the Danish National Research Foundation.

different low-level techniques [3,4,5,6]. However, most of these handle instruction corruptions rather than data corruptions. Corruptions can also often be discovered by existing hardware techniques, but even these techniques can fail and let some corrupted data take part of computations.

Finocchi and Italiano [7] introduced the *faulty-memory random access machine*, based on the traditional RAM model. In this model, memory corruptions can occur at any time and at any place in memory during the execution of an algorithm, and corrupted memory cells cannot be distinguished from uncorrupted cells. In the faulty-memory RAM, it is assumed that there is an adaptive adversary, that chooses how, where, and when corruptions occur. The model is parametrized by an upper bound, δ, on the number of corruptions the adversary can perform during the lifetime of an algorithm, and $\alpha \leq \delta$ denotes the actual number of corruptions that takes place. Motivated by the fact that registers in the processor are considered incorruptible, $O(1)$ safe memory locations are provided. Moreover, it is assumed that reading a word from memory is an atomic operation. In randomized computation, as defined in [7], the adversary does not see the random bits used by an algorithm. An algorithm is *resilient* if it works correctly on the set of uncorrupted cells in the input. For instance, a resilient sorting algorithm outputs all uncorrupted elements in sorted order while corrupted elements can appear at arbitrary positions in the output. A resilient searching algorithm must return yes if there is an uncorrupted element matching the search key.

Memory corruptions are of particular concern for applications dealing with massive amounts of data since such applications typically run for a very long time, and are thus more likely to encounter memory cells containing corrupted data. However, algorithms designed in the RAM model assume that an infinite amount of memory cells are available. This is not true for typical computers where internal memory is limited and elements are transferred between the memory and a much larger, but dramatically slower, hard drive in large consecutive blocks. This means that it is important to design algorithms with a high degree of locality in their memory access pattern, that is, algorithms where data accessed close in time is also stored close in memory. This situation is modeled in the *I/O model* of computation [8]. In this model a disk of unlimited size and a memory of size M are available. Elements are transferred between disk and memory in blocks of size B and computation is performed on elements in memory only. The complexity measure is the number of block transfers (I/Os) performed.

Previous Work: Several problems have been addressed in the faulty-memory RAM, see a very recent survey [9] for more information. In [10,7], matching upper and lower bounds for resilient sorting and randomized searching were given. Sorting N elements requires $\Theta(N \log N + \alpha\delta)$ time [7]. Searching in a sorted array requires $\Omega(\log N + \delta)$ time, and an optimal deterministic algorithm matching that bound is described in [11]. It has been empirically shown that resilient algorithms are of practical interest [12]. Recently, in [11,13,14] resilient data structures were proposed, in particular a resilient dynamic dictionary supporting searches in optimal $\Theta(\log N + \delta)$ time with an amortized update cost of $O(\log N + \delta)$ was

Table 1. The first column shows the I/O upper bounds presented in our paper with the assumptions shown in the second column. The third and fourth column shows how many corruptions the algorithms can tolerate while still matching the optimal algorithms in the I/O and comparison model respectively. Note that the restriction imposed by the time bounds are orders of magnitude stronger than the ones imposed by the I/O bounds for realistic values of M, B and N.

	I/O Complexity	Assumptions	I/O Tolerance (max δ)	Time Tolerance (max δ)
Det. Dict.	$O\left(\frac{1}{\varepsilon}\log_B N + \frac{\delta}{B^{1-\varepsilon}}\right)$	$\frac{1}{\log B} < \varepsilon < 1$	$O(B^{1-\varepsilon}\log_B N)$	$O(\log N)$
Ran. Dict.	$O(\log_B N + \frac{\delta}{B})$	Memory Safe	$O(B\log_B N)$	$O(\log N)$
P. Queue	$O(\frac{1}{1-\varepsilon}\frac{1}{B}\log_{M/B}(N/M))$	$\delta \leq M^\varepsilon, \varepsilon < 1$	$O(M^\varepsilon)$	$O(\log N)$
Sorting	$O(\frac{1}{1-\varepsilon}\mathrm{Sort}(N))$	$\delta \leq M^\varepsilon, \varepsilon < 1$	$O(M^\varepsilon)$	$O(\sqrt{N}\log N)$

presented in [11], and a priority queue supporting operations in $O(\log N + \delta)$ time was presented in [14].

For the I/O model, a comprehensive list of results have been achieved. It is shown in [8] that sorting N elements requires $\mathrm{Sort}(N) = \Theta(N/B \log_{M/B}(N/B))$ I/Os. See recent surveys [15,16] for an overview of other results. In the I/O model, a comparison based dictionary with optimal queries can be achieved with a B-tree [17], which supports queries and updates in $O(\log_B N)$ I/Os.

Current resilient algorithms do not scale past the internal memory of a computer and thus, it is currently not possible to work with large sets of data I/O-efficiently while maintaining resiliency to memory corruptions. Since both models become increasingly interesting as the amount of data increases, it is natural to consider whether it is possible to achieve resilient algorithms that use the disk optimally. Very recently, this was also proposed as an interesting direction of research by Finocchi *et al.* [9,10].

Our Contribution: The work in this paper combines the faulty memory RAM and the external memory model in the natural way. The model has three levels of memory: a disk, an internal memory of size M, and $O(1)$ CPU registers. All computation takes place on elements placed in the registers. The content of any cell on disk or in internal memory can be corrupted at any time, but at most δ corruptions can occur. Moving elements between memory and registers takes constant time and transferring a chunk of B consecutive elements between disk and memory costs one I/O. Transfers between the different levels are atomic, no data can be corrupted while it is being copied. Correctness of an algorithm is proved with the assumption that an adaptive adversary may perform corruptions during execution. For randomized algorithms we assume that the random bits are hidden from the adversary. In two natural variants of our model it is assumed that corruptions take place only on disk, or only in memory.

In this paper, we present I/O-efficient solutions to all problems that, to the best of our knowledge, have previously been considered in the faulty memory RAM. It is not clear that resilient algorithms can be optimal both in time and in I/O-complexity. Most techniques for designing I/O-efficient algorithms naturally try to arrange data on disk such that few blocks need to be read in order to extract the information needed, whereas resilient algorithms try to put little emphasis on individual, potentially corrupted, memory cells.

It is known that any resilient comparison based search algorithm must examine $\Omega(\log N + \delta)$ memory cells [10]. Combining this with the well-known $\Omega(\log_B N)$ I/O lower bound on external memory comparison based searching, we get a simple lower bound of $\Omega\left(\log_B N + \frac{\delta}{B}\right)$ I/Os, and $\Omega(\log N + \delta)$ time. In Section 2 we prove a stronger lower bound of $\Omega\left(\frac{1}{\varepsilon}\log_B N + \frac{\delta}{B^{1-\varepsilon}}\right)$ I/Os for a search, for all $\log_B N \leq \delta \leq B \log N$ and ε given by the equation $\delta = \frac{B^{1-\varepsilon}}{\varepsilon}\log_B N$. In the case where $\delta = \Theta(\frac{B}{\log B}\log_{\log B} N)$, setting $\varepsilon = \frac{\log \log B}{\log B}$ gives a lower bound of $\Omega(\log_{\log B} N + \frac{\delta}{B}\log B)$ which is $\omega(\log_B N + \frac{\delta}{B})$. We come to the interesting conclusion that no deterministic resilient dictionary can obtain an I/O bound of $O(\log_B N + \frac{\delta}{B})$ without some assumptions on δ. The lower bound is valid for randomized algorithms as long as the internal memory is unreliable. For deterministic algorithms, the lower bound also holds if the internal memory is reliable and corruptions only occur on disk.

In Section 3 we construct a resilient dictionary supporting searches using expected $O\left(\log_B N + \frac{\delta}{B}\right)$ I/Os and $O(\log N + \delta)$ time for any δ if corruptions occur exclusively on disk. Thus, we have an interesting separation between the I/O complexity of resilient randomized and resilient deterministic searching algorithms. This also proves that it is important whether it is the disk or the internal memory that is unreliable.

In Section 4 we present an optimal resilient static dictionary supporting queries in $O\left(\frac{1}{c}\log_B N + \frac{\alpha}{B^{1-c}} + \frac{\delta}{B}\right)$ I/Os and $O(\log N + \delta)$ time when $\log N \leq \delta \leq B \log N$ and $\frac{1}{\log B} \leq c \leq 1$. Queries use $O(\log_B N + \frac{\delta}{B})$ I/Os and $O(\log n + \delta)$ time for $\delta \leq \log N$ and $\delta > B \log N$. Additionally, we construct randomized and deterministic dynamic dictionaries with optimal query bounds using our static dictionaries.

Finally, in Section 5 we describe a resilient multi-way merging algorithm. We use this algorithm to design an optimal resilient sorting algorithm using $O(\frac{1}{1-\varepsilon}\text{Sort}(N))$ I/Os and $O(N \log N + \alpha\delta)$ time under the assumption that $\delta \leq M^\varepsilon$, for $0 \leq \varepsilon < 1$. The multi-way merging algorithm is also used to design a resilient priority queue for the case $\delta \leq M^\varepsilon$, where $0 \leq \varepsilon < 1$. Our priority queue supports *insert* and *delete-min* in optimal $O(\frac{1}{1-\varepsilon}(1/B)\log_{M/B}(N/M))$ I/Os amortized, matching the bounds for non-resilient external memory priority queues. The amortized time bound for both operations is $O(\log N + \delta)$ matching the time bounds of the optimal resilient priority queue of [14].

Table 1 shows an overview of the upper bounds in this paper. The two last columns in the table shows how many corruptions our algorithms can tolerate while still achieving optimal bounds in the I/O model and comparison model respectively. Note that the bounds on δ required to get optimal time are orders of magnitude smaller than the bounds required to get optimal I/O performance for realistic values of N, M and B. We conclude that it is possible, under realistic assumptions, to get resilient algorithms that are optimal in both the I/O-model and the comparison model without restricting δ more than what was required to obtain optimal time bounds in the faulty memory RAM.

Preliminaries: Throughout the paper we use the notion of a *reliable value*, which is a value stored in unreliable memory that can be retrieved reliably in spite of possible corruptions. This is achieved by replicating the given value in $2\delta + 1$ consecutive cells. Since at most δ of the copies can be corrupted, the majority of the $2\delta + 1$ elements are uncorrupted. The value can be retrieved using $O(\frac{\delta}{B})$ I/Os and $O(\delta)$ time with the majority algorithm in [18], which scans the $2\delta+1$ values keeping a single majority candidate and a counter in reliable memory. A sequence is *faithfully ordered* if the uncorrupted elements form a sorted subsequence.

2   Lower Bound for Dictionaries

Any resilient searching algorithm must examine $\Omega(\log N +)$ memory cells in the comparison model [10]. The $\Omega(\log N)$ term follows from the comparison model lower bound for searching. It is well-known that comparison based searching in the I/O model requires expected $\Omega(\log_B N)$ I/Os. Since any resilient searching algorithm must read at least $\Omega(\delta)$ elements to ensure at least some non-corrupted information is the basis for the output, we get the following trivial lower bound.

Lemma 1. *For any comparison based randomized resilient dictionary the average-case expected search cost is* $\Omega\left(\log_B N + \frac{\delta}{B}\right)$ *I/Os.*

In this section we prove a stronger lower bound on the worst-case number of I/Os required for any deterministic resilient static dictionary in the comparison model. We do not make any assumptions on the data structure used by the dictionary, nor on the space it uses. Additionally, we do not bound the amount of computation time used in a query and we assume that the total order of all elements stored in the dictionary are known by the algorithm initially. During the search for an element e, an algorithm gains information by performing block I/Os, each I/O reading B elements from disk. Before a block of B elements is read into memory the adversary can corrupt the elements in the block. The adversary is allowed to corrupt up to δ elements during the query operation, but does not have to reveal when it chooses to do so. Also, the adversary adaptively decides what the rank of the search element has among the N dictionary elements. Of course, the rank must be consistent with the previous uncorrupted elements read by the algorithm.

Theorem 1. *Given N and δ, any deterministic resilient static dictionary requires worst-case* $\Omega\left(\frac{1}{\varepsilon}\log_B N\right)$ *I/Os for a search, for all ε where* $\frac{1}{\log B} \le \varepsilon \le 1$ *and* $\delta \ge \frac{1}{\varepsilon}B^{1-\varepsilon}\log_B N$.

Proof. We design an adversary that uses corruptions to control how much information any correct query algorithm gains from each block transfer.

Let ε be a constant such that $\frac{1}{\log B} \le \varepsilon \le 1$. The strategy of the adversary is as follows. For each I/O, the adversary narrows the *candidate interval* where

e can be contained in by a factor B^ε. Initially, the candidate interval consists of all N elements. For each I/O, the adversary implicitly divides the sorted set of elements in the candidate interval into B^ε slabs of equal size. Since the search algorithm only reads B elements in an I/O, there must be at least one slab containing at most $B^{1-\varepsilon}$ of these elements. The adversary corrupts these elements, such that they do not reveal any information, and decides that the search element resides in this slab. The remaining elements transferred are not corrupted and are automatically consistent with the interval chosen for e. The game is then played recursively on the elements of the selected slab, until all elements in the final candidate interval have been examined.

For each I/O, the candidate interval decreases by a factor B^ε. The algorithm has no information regarding elements in the slab except for the corrupted elements from the I/Os performed so far. After k I/Os the candidate interval has size $\frac{N}{(B^\varepsilon)^k}$ and the adversary has introduced at most $kB^{1-\varepsilon}$ corruptions. The game continues as long as there is at least one uncorrupted element among the elements remaining in the candidate interval, which the adversary can choose as the search element. All corrupted elements may reside in the current candidate interval, and the game ends when the size of the candidate interval, $\frac{N}{(B^\varepsilon)^k}$, becomes smaller than or equal to the total number of introduced corruptions, $kB^{1-\varepsilon}$. It follows that at least $\Omega\left(\log_{B^\varepsilon} \frac{N}{B^{1-\varepsilon}}\right) = \Omega\left(\frac{1}{\varepsilon}\log_B N\right)$ I/Os are required. The adversary introduces at most $B^{1-\varepsilon}$ corruptions in each step. If ε satisfies $\frac{1}{\varepsilon}B^{1-\varepsilon}\log_B N \leq \delta$, then the adversary can play the game for at least $\frac{1}{\varepsilon}\log_B N$ rounds and the theorem follows. $\square$

For deterministic algorithms it does not matter whether elements can be corrupted on disk or in internal memory. Since the adversary is adaptive it knows which block of elements an algorithm will read into internal memory next, and may choose to corrupt the elements on disk just before they are loaded into memory, or corrupt the elements in internal memory just after they have been written there. In randomized algorithms where the adversary does not know the algorithm's random choices it cannot determine which block of elements will be fetched from disk before the transfer has started. Therefore, the adversary can follow the strategy above only if it can corrupt elements in internal memory.

By setting $\delta = \frac{1}{\varepsilon}B^{1-\varepsilon}\log_B N$ in Theorem 1, we get the following corollary.

Corollary 1. *Any deterministic resilient static dictionary requires worst-case $\Omega(\frac{1}{\varepsilon}\log_B N) = \Omega(\frac{\delta}{B^{1-\varepsilon}})$ I/Os for a search, where $\delta \in [\log_B N, B\log N]$, and ε given by $\delta = \frac{1}{\varepsilon}B^{1-\varepsilon}\log_B N$.*

The trivial I/O lower bound for a resilient searching algorithm is $\Omega\left(\log_B N + \frac{\delta}{B}\right)$. Setting $\varepsilon = \frac{\log\log B}{\log B}$ in Theorem 1 shows that this is not optimal.

Corollary 2. *For $\delta = \frac{B}{\log B}\log_{\log B} N$ any deterministic resilient static dictionary requires worst-case $\Omega(\frac{\log B}{\log\log B}(\log_B N + \frac{\delta}{B}))$ I/Os for a search.*

3   Randomized Static Dictionary

In this section we describe a simple I/O-efficient randomized static dictionary, that is resilient to corruptions on the disk. Corruptions in memory are not allowed, thus the adversarial lower bound in Theorem 1 does not apply. The dictionary supports queries using expected $O\left(\log_B N + \frac{\delta}{B}\right)$ I/Os and $O(\log N + \delta)$ time. The algorithm is similar to the randomized binary search algorithm in [10]. Remember that, if only elements on disk can be corrupted, the lower bound from Theorem 1 also holds for deterministic algorithms. This means that deterministic and randomized algorithms are separated by the result in this section.

The idea is to store the N elements in the dictionary in sorted order in an array S and to build 2δ B-trees [17], denoted $T_1, \ldots, T_{2\delta}$, of size $\lfloor \frac{N}{2\delta} \rfloor$. The i'th B-tree T_i stores the $2\delta j + i$'th element in S for $j = 0, \ldots, \lfloor \frac{N}{2\delta} \rfloor - 1$. Each node in each tree is represented by a faithfully ordered array of $\Theta(B)$ search keys. The nodes of the B-tree are laid out in left to right breadth first order, to avoid storing pointers, i.e. the c'th child of the node at index k has index $Bk + c - (B - 1)$.

The search for an element e proceeds as follows. A random number $r_1 \in \{1, \ldots, 2\delta\}$ is generated, and the root block of T_{r_1} is fetched into the internal memory. In this block, a binary search is performed among the search keys resulting in an index, i, of the child where the search should continue. A new random number $r_2 \in \{1, \ldots, 2\delta\}$ is generated, and the i'th child of the root in tree T_{r_2} is fetched and the algorithm proceeds iteratively as above. The search terminates when a leaf is reached and two keys $S[2\delta j + i]$ and $S[2\delta(j + 1) + i]$ have been identified such that $S[2\delta j + i] \leq e < S[2\delta(j + 1) + i]$. If the binary search was not mislead by corruptions of elements, then e is located in the subarray $S[2\delta j + i, \ldots, 2\delta(j+1)+i]$. To check whether the search was mislead, the following *verification procedure* is performed. Consider the neighborhoods $L = S[2\delta(j-1)+i-1, \ldots, 2\delta j + i - 1]$ and $R = S[2\delta(j+1)+i+1, \ldots, 2\delta(j+2)+i+1]$, containing the $2\delta+1$ elements in S situated to the left of $S[2\delta j + i]$ and to the right of $S[2\delta(j+1)+i]$ respectively. The number $s_L = |\{z \in L \mid z \leq e\}|$ of elements in L that are smaller than e is computed by scanning L. Similarly, the number s_R of elements in R that are larger than e is computed. If $s_L \geq \delta+1$ and $s_R \geq \delta+1$, and the search key is not encountered in L or R, we decide whether it is contained in the dictionary or not by scanning the subarray $S[2\delta j, \ldots, 2\delta(j+1)]$. If s_L or s_R is smaller than $\delta + 1$, at least one corruption has misguided the search. In this case, the search algorithm is restarted.

Theorem 2. *The data structure described is a linear space randomized dictionary supporting searches in expected $O\left(\log_B N + \frac{\delta}{B}\right)$ I/Os and $O(\log N + \delta)$ time assuming that memory cells are incorruptible and block transfers are atomic.*

Details will appear in the full paper. If memory cells were corruptible the atomic transfer assumption would be of little use. The adversary could simply corrupt the elements in the internal memory after the block transfer completes, decreasing the benefit of the randomization.

4   Optimal Deterministic Static Dictionary

In this section we present a linear space deterministic resilient static dictionary. Let c be a constant such that $\frac{1}{\log B} \le c \le 1$. The dictionary supports queries in $O\left(\frac{1}{c}\log_B N + \frac{\alpha}{B^{1-c}} + \frac{\delta}{B}\right)$ I/Os and $O(\log N + \delta)$ time. In Section 2 we proved a lower bound on the I/O complexity of resilient dictionaries, and by choosing c in the above bound to minimize the expression for $\alpha = \delta$, this bound matches the lower bound. Thus, this dictionary is optimal.

Our data structure is based on the B-tree and the resilient binary search algorithm from [11]. In a standard B-tree search one corrupted element can misguide the algorithm, forcing at least one I/O in the *wrong* part of the tree. To circumvent this problem, each guiding element in each internal node is determined by taking majority of B^{1-c} copies. This gives a trade-off between the number of corruptions required to misguide a search, and the fan-out of the tree, which becomes B^c. Additionally, each node stores $2\delta + 1$ copies of the minimum and maximum element contained in the subtree, such that the search algorithm can reliably check whether it is on the correct path in each step. We ensure that the query algorithm avoids reading the same corrupted element twice by ensuring that any element is read at most once. The exact layout of the tree and the details of the search operation are as follows.

Structure: Let S be the set of elements contained in the dictionary and let N denote the size of S. The dictionary is a B^c-ary search tree T built on $\frac{N}{\delta}$ leaves. The elements of S are distributed to the leaves in faithful order such that each leaf contains δ elements. Each leaf is represented by a *guiding element* which is smaller than the smallest uncorrupted element in the leaf and larger than the largest uncorrupted element in the preceding leaf. The top tree is built using these guiding elements. The tree is stored in a breadth-first left-to-right layout on disk, such that no pointers are required.

Each internal node u in T stores three types of elements; guiding elements, minimum elements, and maximum elements, stored consecutively on disk. The guiding elements are stored in $\lceil (2\delta + 1)/B^{1-c} \rceil$ identical blocks. Each block contains B^{1-c} copies of each of the B^c guiding elements in sorted order such that the first B^{1-c} elements are copies of the smallest guiding element. This means that each guiding element is stored $2\delta + 1$ times and can be retrieved reliably. The minimum elements are $2\delta + 1$ copies of the guiding element for the leftmost leaf in the subtree defined by u, stored consecutively in $\lceil \frac{2\delta+1}{B} \rceil$ blocks. Similarly the maximum elements are $2\delta + 1$ copies of the guiding element for the leaf following the rightmost leaf in the subtree defined by u, stored consecutively in $\lceil \frac{2\delta+1}{B} \rceil$ blocks. Additionally, minimum and maximum elements are stored with each leaf. The minimum are 4δ copies of the guiding element representing the leaf, stored consecutively in $\frac{4\delta}{B}$ blocks, and the maximum elements are 4δ copies of the guiding element representing the subsequent leaf, stored consecutively in $\frac{4\delta}{B}$ blocks. These are used to verify that the algorithm found the only leaf that may store an uncorrupted element matching the search element.

Query: A query operation for an element q, uses an index k that indicates how many chunks of B^{1-c} elements the algorithm has discarded during the search, initially $k = 0$. Intuitively, a chunk is discarded if the algorithm detects that $\Omega(B^{1-c})$ of its elements are corrupted. The query operation traverses the tree top-down, storing in safe memory the index k, and $O(1)$ extra variables required to traverse the tree using the knowledge of its layout on disk. In an internal node u, the algorithm starts by checking whether u is on the correct path in the tree using the copies of the minimum and maximum elements stored in u. This is done by scanning B^{1-c} of the $2\delta + 1$ copies of the minimum element starting with the kB^{1-c}'th copy, counting how many of these that are larger than q. If $B^{1-c}/2$ or more copies of the minimum element are larger than q the block is discarded by incrementing k and the search is restarted (backtracked) at node v, where $v = u$ if u is root of the tree and the parent of u otherwise. The maximum elements are checked similarly. If the algorithm backtracks, k is increased ensuring that the same element is never read more than once.

If the checks succeed the k'th block storing copies of the B^c guiding elements of u is scanned from left to right. The majority value of each of the B^{1-c} copies of each guiding element is extracted in sorted order using the majority algorithm [18] and compared to q, until a retrieved guiding element larger than q is found or the entire block is read. The traversal then continues to the corresponding child. If any invocation of the majority algorithm fails to select a value, or two fetched guiding elements are out of order, the block is discarded as above by increasing k and backtracking the search to the parent node.

Upon reaching a leaf, the algorithm verifies whether the search found the correct leaf. This is achieved by running a variant of the verification procedure designed for the same purpose in [11]. Counters c_l and c_r, which are initially 1, are stored in safe memory. Then the copies of the minimum and maximum element are scanned in chunks of B^{1-c} elements, starting from the $2kB^{1-c}$'th element. If the majority of elements in a chunk of B^{1-c} copies of the minimum element are smaller than the search element, c_l is increased by 1. Otherwise, c_l is decreased and k increased by one. The copies of maximum elements are treated similarly. Note that every decrement of c_l or c_r signals that at least $\frac{B^{1-c}}{2}$ corruptions have been found. Thus, c_l represents the number of chunks scanned that has not yet been contradicted, where the majority of copies indicates that the search element is in the current leaf or in leafs to the right. Similar for c_r. If $\min\{c_l, c_r\}$ reaches 0, we backtrack to the parent of the leaf as above. If $\min\{c_l, c_r\}\frac{B^{1-c}}{2}$ gets larger than $\delta - k(\frac{B^{1-c}}{2}) + 1$ the verification succeeds. The algorithm finishes by scanning the δ elements stored in the leaf, returning whether it finds q or not.

Lemma 2. *The data structure is a linear space resilient dictionary supporting queries in $O\left(\frac{1}{c}\log_B N + \frac{\alpha}{B^{1-c}} + \frac{\delta}{B}\right)$ I/Os, for any $1/\log B \leq c \leq 1$.*

The correctness portion of the proof is similar to the proof for the optimal binary search algorithm in [11]. The complexity analysis uses the observation that if a search is guided in the wrong direction, the majority of the B^{1-c} copies of a guiding element in the relevant block are corrupted and each additional step requires $\frac{B^{1-c}}{2}$ additional corruptions in order to pass the check against the minimum and maximum elements. Details will appear in the full paper.

To obtain optimal time bounds for the dictionary, we use the resilient binary search algorithm of [11] on each block, instead of scanning it. If more than $\frac{B^{1-c}}{2}$ corruptions are discovered during the search of a block, it is discarded as above. Otherwise, $\frac{B^{1-c}}{2}$ supporting elements are found on both sides of an element, and the algorithm continues to the corresponding child as before. This reduces the time used per node to $O(\log B + B^{1-c})$. Verification takes $O(\delta)$ time in total.

Lemma 3. *For any* $\frac{1}{\log B} \leq c \leq 1$, *queries use* $O((B^{1-c} + \log B)(\frac{1}{c} \log_B N + \frac{\alpha}{B^{1-c}}) + \delta)$ *time.*

Corollary 3. *If* $\delta > B \log N$, *queries use* $O(\frac{\delta}{B})$ *I/Os and* $O(\delta)$ *time.*

Corollary 4. *If* $\delta < \log N$, *queries use* $O(\log_B N)$ *I/Os and* $O(\log N)$ *time.*

Corollary 5. *If* $\log N \leq \delta \leq B \log N$ *for any* $\frac{1}{\log B} \leq c \leq 1$, *queries use* $O(\frac{1}{c} \log_B N + \frac{\alpha}{B^{1-c}} + \frac{\delta}{B})$ *I/Os and* $O(\log n + \delta)$ *time.*

The corollaries follow from Lemma 2 and 3 by setting $c = \frac{1}{\log B}$, $c = 1 - \frac{\log \log B}{\log B}$, and $c \in [\frac{1}{\log B}, 1 - \frac{\log \log B}{\log B}]$ such that $\frac{1}{c} \log_B N = \frac{\delta}{B^{1-c}}$ respectively.

By adapting the techniques of [11,19] and the static dictionary presented above we obtain a dynamic dictionary. Details will appear in the full paper.

Theorem 3. *There is a deterministic dynamic resilient dictionary supporting searches and updates in* $O(\frac{1}{c} \log_B N + \frac{\alpha}{B^{1-c}} + \frac{\delta}{B})$ *I/Os and* $O(\log N + \delta)$ *time, worst-case and amortized respectively with c in the range* $\frac{1}{\log B} \leq c \leq 1$.

5 Priority Queue and Sorting

In this section we present a resilient multi-way merging algorithm and use it to design a resilient sorting algorithm and priority queue. First we show how to merge γ faithfully ordered lists of total size x when $\gamma \leq \min\{\frac{M}{B}, \frac{M}{\delta}\}$.

Multi-way Merging: Initially, the algorithm constructs a perfectly balanced binary tree, T, in memory on top of the γ buffers being merged. Each edge of the binary tree is equipped with a buffer of size $5\delta + 1$. Each internal node $u \in T$ stores the state of a running instance of the *PurifyingMerge*, a resilient binary merging algorithm from [10] that works in rounds. In each round $O(\delta)$ elements from both input buffers are read and the next δ elements in the faithful order are output. If corrupted elements are found, these are moved to a fail buffer and the round is restarted. The algorithm merges elements from the buffers on u's left child edge and right child edge into the buffer of u's parent edge. The states and sizes of all buffers are stored as reliable variables. The entire tree including all buffers and state variables are stored in internal memory, along with one block from each of the γ input streams and one block for the output stream of the root. Instead of storing a fail buffer for each instance of *PurifyingMerge*, a global shared fail buffer F is stored containing all detected corrupted elements.

Let b_l and b_r be the buffers on the edges to the left and right child respectively and let b denote the buffer on the edge from u to its parent. If u is the root, b is the output buffer. The elements are merged using the *fill* operation, which operates on u, as follows. First, it checks whether b_l and b_r contain at least $4\delta + 1$ elements, and if not they are filled recursively. Then, the stored instance of the *PurifyingMerge* algorithm is resumed by running a round of the algorithm outputting the next δ elements to its output stream. The multi-way merging algorithm is initiated by invoking *fill* on the root of T which runs until all elements have been output. Then, the elements moved to F during the *fill* are merged into the output using *NaiveSort* and *UnbalancedMerge* as in [10].

Lemma 4. *Merging $\gamma = \min\{\frac{M}{B}, \frac{M}{\delta}\}$ buffers of total size $x \geq M$ using $O(x/B)$ I/Os and $O(x \log \gamma + \alpha\delta)$ time.*

Proof. The correctness follows from Lemma 1 in [10]. The size of T is $O(\gamma(\delta + B)) = O(\min\{\frac{M}{B}, \frac{M}{\delta}\}(\delta + B)) = O(M)$. We use γ I/Os to load the first block in each leaf of T and $O(x/B)$ I/Os for reading the entire input and writing the output. The final merge with F takes $O(x/B)$ I/Os. Since T fits completely in memory we perform no other I/Os.

Merging two buffers of total size n using *PurifyingMerge* takes $O(n + \alpha\delta)$ time where α is the number of detected corruptions in the input buffers. Since detected corruptions are moved to the global fail buffer each corruption is only charged once. Each element passes through $\log \gamma$ nodes of T and the final merge using *NaiveSort* and *UnbalancedMerge* takes $O(x + \alpha\delta)$ time.   □

Sorting: Assuming $\delta \leq M^\varepsilon$ for $0 \leq \varepsilon < 1$, we can use the multi-way merging algorithm to implement the standard external memory $M^{1-\varepsilon}$-way mergesort from [8] matching the optimal external memory sorting bound for constant ε.

Theorem 4. *Our resilient sorting algorithm uses $O(\frac{1}{1-\varepsilon}Sort(N))$ I/Os and $O(N \log N + \alpha\delta)$ time assuming $\delta \leq M^\varepsilon$.*

Priority Queue: Our comparison based resilient priority queue is optimal with respect to both time and I/O performance assuming that $\delta \leq M^\varepsilon$ for $0 \leq \varepsilon < 1$. An optimal I/O-efficient priority queue uses $\Theta(1/B \log_{M/B}(N/M))$ I/Os amortized per operation [8]. An $\Omega(\log N + \delta)$ time lower bound for comparison based resilient priority queues was proved in [14]. A resilient priority queue as defined in [14] maintains a set of elements under the operations *insert* and *delete-min*, where *insert* adds an element and a *delete-min* deletes and returns the minimum uncorrupted element or a corrupted one.

Our priority queue is based on an amortized version of the worst-case optimal external memory priority queue of [20] using our new resilient multi-way merging algorithm to move elements between disk and internal memory. Details will appear in the full paper.

Theorem 5. *There is a linear space resilient priority queue supporting* insert *and* delete-min *in amortized $O(\frac{1}{1-\varepsilon}(1/B) \log_{M/B}(N/M))$ I/Os and $O(\log N + \delta)$ time assuming $\delta \leq M^\varepsilon$ where $0 \leq \varepsilon < 1$.*

References

1. Tezzaron Semiconductor: Soft errors in electronic memory - a white paper (2004), http://www.tezzaron.com/about/papers/papers.html
2. van de Goor, A.J.: Testing Semiconductor Memories: Theory and Practice. ComTex Publishing, Gouda, The Netherlands (1998) ISBN 90-804276-1-6
3. Huang, K.H., Abraham, J.A.: Algorithm-based fault tolerance for matrix operations. IEEE Transactions on Computers 33, 518–528 (1984)
4. Rela, M.Z., Madeira, H., Silva, J.G.: Experimental evaluation of the fail-silent behaviour in programs with consistency checks. In: Proc. 26th Annual International Symposium on Fault-Tolerant Computing, pp. 394–403 (1996)
5. Yau, S.S., Chen, F.C.: An approach to concurrent control flow checking. IEEE Transactions on Software Engineering SE-6(2), 126–137 (1980)
6. Pradhan, D.K.: Fault-tolerant computer system design. Prentice-Hall, Inc., Englewood Cliffs (1996)
7. Finocchi, I., Italiano, G.F.: Sorting and searching in the presence of memory faults (without redundancy). In: Proc. 36th Annual ACM Symposium on Theory of Computing, pp. 101–110. ACM Press, New York (2004)
8. Aggarwal, A., Vitter, J.S.: The input/output complexity of sorting and related problems. Commun. ACM 31, 1116–1127 (1988)
9. Finocchi, I., Grandoni, F., Italiano, G.F.: Designing reliable algorithms in unreliable memories. Computer Science Review 1(2), 77–87 (2007)
10. Finocchi, I., Grandoni, F., Italiano, G.F.: Optimal resilient sorting and searching in the presence of memory faults. In: Bugliesi, M., Preneel, B., Sassone, V., Wegener, I. (eds.) ICALP 2006. LNCS, vol. 4051, pp. 286–298. Springer, Heidelberg (2006)
11. Brodal, G.S., Fagerberg, R., Finocchi, I., Grandoni, F., Italiano, G., Jørgensen, A.G., Moruz, G., Mølhave, T.: Optimal resilient dynamic dictionaries. In: Azar, Y., Erlebach, T. (eds.) ESA 2006. LNCS, vol. 4168, pp. 347–358. Springer, Heidelberg (2006)
12. Petrillo, U.F., Finocchi, I., Italiano, G.F.: The price of resiliency: a case study on sorting with memory faults. In: Proc. 14th Annual European Symposium on Algorithms, pp. 768–779 (2006)
13. Finocchi, I., Grandoni, F., Italiano, G.F.: Resilient search trees. In: Proc. 18th ACM-SIAM Symposium on Discrete Algorithms, pp. 547–554 (2007)
14. Jørgensen, A.G., Moruz, G., Mølhave, T.: Priority queues resilient to memory faults. In: Proc. 10th International Workshop on Algorithms and Data Structures, pp. 127–138 (2007)
15. Arge, L.: External memory data structures. In: Abello, J., Pardalos, P.M., Resende, M.G.C. (eds.) Handbook of Massive Data Sets, pp. 313–358. Kluwer Academic Publishers, Dordrecht (2002)
16. Vitter, J.S.: Algorithms and data structures for external memory. Foundations and Trends in Theoretical Computer Science 2(4), 305–474 (2008)
17. Bayer, R., McCreight, E.: Organization and maintenance of large ordered indexes. Acta Informatica 1, 173–189 (1972)
18. Boyer, R.S., Moore, J.S.: MJRTY: A fast majority vote algorithm. In: Automated Reasoning: Essays in Honor of Woody Bledsoe, pp. 105–118 (1991)
19. Brodal, G.S., Fagerberg, R., Jacob, R.: Cache-oblivious search trees via binary trees of small height. In: Proc. 13th Annual ACM-SIAM Symposium on Discrete Algorithms, pp. 39–48 (2002)
20. Brodal, G.S., Katajainen, J.: Worst-case efficient external-memory priority queues. In: Arnborg, S. (ed.) SWAT 1998. LNCS, vol. 1432, pp. 107–118. Springer, Heidelberg (1998)

Inspecting a Set of Strips Optimally

Tom Kamphans[1] and Elmar Langetepe[2]

[1] Braunschweig University of Technology, Computer Science, Algorithms Group,
38106 Braunschweig, Germany
[2] University of Bonn, Institute of Computer Science I, 53117 Bonn, Germany

Abstract. We consider a set of axis-parallel nonintersecting strips in the plane. An observer starts to the left of all strips and ends to the right, thus visiting all strips in the given order. A strip is *inspected* as long as the observer is inside the strip. How should the observer move to inspect all strips? We use the path length outside a strip as a quality measure which should be minimized. Therefore, we would like to find a directed path that minimizes the maximal measure among all strips. We present an optimal algorithm designed according to the structural properties of the optimal solution.

Keywords: Computational geometry, motion planning, watchman routes, optimal inspection path, optimal algorithm.

1 Introduction

In the last decades, routes from which an agent can see every point in a given environment have drawn a lot of attention. For example, the optimal offline exploration path in a simple polygon (the *shortest watchman route*) was first considered by Chin and Ntafos [4] for the special case of orthogonal polygons. Some work has been done on shortest watchman routes, until Dror et al. [7] presented an algorithm for the more general problem of visiting a sequence of intersecting polygons under the presence of fences. Similar problems include the shortest watchman *path* with different start and end points [3] or routes with additional constraints such as zookeeper routes [5]. The corresponding online task was considered, for example, by Deng et al. [6] for orthogonal simple polygons, by Hoffmann et al. [8, 9] for general simple polygons, and by Icking et al. [10] for grid polygons. See also the surveys by Mitchell [11] or Icking et al. [9].

Usually, the objective is to find a short path or route (i.e., a closed path); either the shortest possible route (the optimum) or an approximation. In this paper, we focus on another criterion for routes: We want to minimize the time[1] in which a certain area of the environment is *not* seen. Imagine a guard in an art gallery whose objective is to be as vigilant as possible and to minimize the time an object is unguarded. Thus, for a given *inspection route* the route's performance wrt. a single object is given by the maximal time interval where

---

[1] We assume that the agent travels with constant speed; thus, we use *time* and *path length* synonymously.

F. Dehne et al. (Eds.): WADS 2009, LNCS 5664, pp. 423–434, 2009.
© Springer-Verlag Berlin Heidelberg 2009

the object is unguarded. The task is to find a path that minimizes the maximal unguarded time interval among all objects.

Mark Overmars [12] introduced this problem by posing the question whether the shortest watchman route inside a simple polygon is the best inspection route. In this setting the set of objects is given by the set of *all* points inside the polygon. The shortest watchman route *inspects* some of these points only in a single moment. Therefore the conjecture is that the performance of the optimal inspection route for the polygon is given by the length of the shortest watchman route. This question is still open. In this paper, we restrict ourselves to a simple type of environments—parallel strips in the plane. More complicated environments are the subject of ongoing research. We present an optimal algorithm that solves the problem for L_1- and L_2-metric and gives some insight into the structural property of optimal inspection paths.

The paper is organized as follows. In Section 2, we present notational conventions and define an objective function which has to be minimized. Then, in Section 3 we first prove some structural properties of an optimal solution for the Euclidean case. At the end we present an efficient algorithm. The ideas can be adapted to the L_1-case which is mentioned in Section 4. Finally, we summarize the results and discuss future work, see Section 5.

2   Preliminaries

Let $\{S_1, \ldots, S_n\}$ be a set of nonintersecting vertical strips, $S = (s_x, s_y)$ be a start point to the left of all strips, and $T = (t_x, t_y)$ be an end point to the right of all strips. W.l.o.g. we can assume that S is below T (i.e., $s_y \leq t_y$). Strip S_i has width w_i.

An inspection path, P, from S to T visits the strips successively from left to right, see Fig. 1. For a given path P let P_i denote the part of P within strip S_i. Let $|P_i|$ denote the corresponding path length, and $\text{last}(P)$ the last segment of P (i.e., from S_n to T).

While P visits S_i, the strip is entirely visible. The performance of P for a single strip S_i therefore is given by $\text{Perf}(P, P_i) := |P| - |P_i|$. The performance of

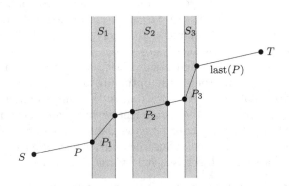

Fig. 1. Visiting three strips in a given order

the path P for all strips is given by the worst performance achieved for a single strip. That is $\text{Perf}(P) := \max_i \text{Perf}(P, P_i)$. Finally, the task is to find among all inspection paths the path that gives the best performance for the given situation; that is, an inspection path with minimal performance:

$$\text{Perf} := \min_P \max_i \text{Perf}(P, P_i) \ .$$

This problem belongs to the class of LP-type problems [14], but the basis could have size n as we will see later. Therefore, we solve the problem directly. It may also be seen as a *Time and Space* Problem (see, e.g., [2, 1]).

3 The Euclidean Case

In this section, we first collect some properties of the optimal solution and then design an efficient algorithm.

3.1 Structural Properties

We can assume that the optimal inspection path is a polygonal chain with straight line segments inside and between the strips. There can be no arcs or kinks inside the strips or outside the strips (in the *free space*). The inspection path enters a strip and leaves a strip and the straight line between these points does not influence the performance of the corresponding strip but optimizes the length of the corresponding subpath. Analogously, between two strips the path leaves a strip and enters another strip and the straight line between these point optimizes the length of the corresponding subpath.

Let us further assume that we have an inspection path as depicted in Fig. 2(i). The first simple observation is that we can rearrange the set of strips in any nonintersecting order and combine the elements of the given path adequately by shifting the segments horizontally without changing the path length; see Fig. 2(ii).

Now, it is easy to see that an optimal solution has the same slope between two successive strips. We simply rearrange the strips such that they stick together and start from the X-coordinate of the start point. The last part of the solution should have no kinks as mentioned earlier.

Lemma 1. *The optimal solution is a polygonal chain without kinks between the strips or inside the strips. The path has the same slope between all strips.*

In the following, we assume that the strips are ordered by widths $w_1 \leq w_2 \leq \cdots \leq w_n$, starting at the X-coordinate, s_x, of the start point and lie side by side (i.e., without overlaps or gaps), see Fig. 2(ii). For $t_y = s_y$ the optimal path is simply the horizontal connection between S and T. Thus, we assume $t_y > s_y$ the following.

Now, we show some structural properties of an optimal solution. The optimal solution visits some strips with the same value $d = |P_i|$ until it finally moves directly to the end point, see Fig. 3 for an example.

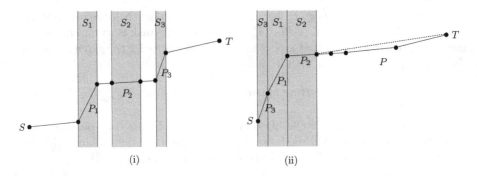

Fig. 2. Rearranging strips and path yields the same objective value. Thus, the optimal solution has the same slope between all strips and we can assume that the strips are ordered by widths.

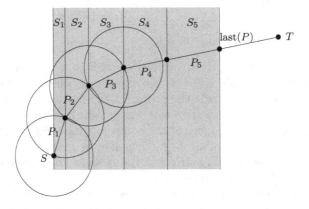

Fig. 3. The structure of an optimal solution: The first three strips are visited with the same value $|P_i| = d$, every other strip with $|P_i| > d$

Lemma 2. *In a setting as described above, the optimal path P visits the first $k \leq n$ strips with the same distance d and then moves directly to the end point. That is, for $i = 1, \ldots, k$ we have $|P_i| = d$ and for $i = k+1, \ldots, n$ we have $|P_i| > d$. The path is monotonically increasing and convex with respect to the segment ST.*

Proof. Let P denote an optimal path for n strips. First, we show that the path is monotonically increasing. There is at least one segment, P_i, with positive slope, because $t_y > s_y$. Let us assume—for contradiction—that there is also a segment, P_j, with negative slope. We rearrange the strips and the path such that P_i immediately succeeds P_j, see Fig. 4(i). Now, we can move the common point of P_j and P_i upwards. Both segments decrease and the performance of P improves. Thus, there is no segment with negative slope and P is monotonically increasing.

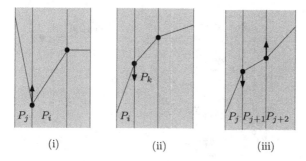

(i)   (ii)   (iii)

Fig. 4. Global optimization by local changes: The connection point can be moved (i) upwards, (ii) downwards or (iii) upwards or downwards. In any case the solution can be improved by local changes.

The performance of P is given by $|P_k| := \min_j |P_j|$. Let k be the greatest index such that P_k is responsible for the performance. By contradiction, we show that for $i < k$ there is no P_i with $|P_i| > |P_k|$. So let us assume that $|P_i| > |P_k|$ holds for $i < k$. We can again rearrange the strips in such a way that P_k immediately succeeds P_i. From $w_i \le w_k$ we conclude that the path $P_i P_k$ makes a right turn. Because $|P_i| > |P_k|$ holds, we can globally optimize the solution by moving the connection point downwards, see Fig. 4(ii). Although $|P| - |P_i|$ increases, the total path length decreases. Thus, $|P_i| = |P_k|$ for $i = 1, \ldots, k-1$. For $i = k+1, \ldots, n$ we have $|P_i| > |P_k|$ by assumption.

Now, we show that there is no kink in the path $P_{k+1} P_{k+2} \cdots P_n$. As $|P_j| > |P_k|$ for $j > k$ we can globally optimize the solution by moving a kink point downwards or upwards, see Fig. 4(iii). The path length decreases. Therefore, $P_{k+1} P_{k+2} \cdots P_n$ is a straight line segment.

Altogether, for $i = k+1, \ldots, n$ we have $|P_i| > d$. For $i = 1, \ldots, k$ we have $|P_i| = d$ and the part $P_{k+1} P_{k+2} \cdots P_n \text{last}(P)$ is a straight line segment. Path P is monotonically increasing. The first part of P makes only right turns as already seen. The last part is a line segment. The concatenation of P_k and $P_{k+1} P_{k+2} \cdots P_n$ also makes a right turn; otherwise, we can again improve the performance. Because $|P_{k+1}| > |P_k|$ we can move the connection point upwards, which decreases the path length and improves the performance of strip S_k and S_{k+1}. Altogether, P is convex with respect to ST.   □

One might think that with the result of Lemma 2 there is an easy way to find a solution by application of Snell's law, which describes how light bends when traveling from one medium to the next. In the formulas below an application or extension of Snell's law seems to be difficult to achieve.

In the following, we show that we can compute the optimal solution incrementally; that is, we successively add new strips and consider the corresponding optimal solutions.

Let $S_1, S_2, \ldots, S_n$ be a set of strips and let S and T be fixed. Let P^i denote the optimal solution for the first i strips. For increasing i the parameter k of Lemma 2 is strictly increasing until it remains fixed:

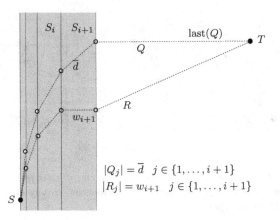

$$|Q_j| = \bar{d} \quad j \in \{1, \dots, i+1\}$$
$$|R_j| = w_{i+1} \quad j \in \{1, \dots, i+1\}$$

Fig. 5. An optimal solution for $k = i + 1$ strips can be found between the extremes R and Q. $|R_j| = w_{i+1}$ holds and $\bar{d}$ fulfills $|Q_j| = \bar{d}$ with horizontal last(Q).

Lemma 3. *For P^i either the index k (see Lemma 2) is equal to i or P^i is already given by P^{i-1}. If P^i is identical to P^{i-1}, also P^j is identical to P^{i-1} for $j = i + 1, \dots, n$.*

We postpone the complete proof of Lemma 3 and first show how to adapt a solution P^i to a solution P^{i+1}. Let us assume that we have a solution P^i and that k in the sense of Lemma 2 is equal to i. That is, the first i strips are visited with the same distance d. If $|P^i_{i+1}| < d$ holds, the solution P^i is not optimal for $i + 1$ strips. Therefore we want to adapt P^i. Lemma 3 states that it migth be useful to search for a solution with identical path length in $k = i + 1$ strips.

We now show that this solution can be computed efficiently. The task is to compute the path P^{i+1} between two extreme solutions, R and Q, as follows: Let R be the path with path lengths $|R_j| = w_{i+1}$ for $j = 1, \dots, i + 1$ and let Q be the path with $|Q_j| = \bar{d}$ for $j = 1, \dots, i + 1$, where the last segment, last(Q), is horizontal. Starting from $d = w_{i+1}$, let $P^{i+1}(d)$ denote the unique path that starts with $|P^{i+1}_j| = d$ for $j = 1, \dots, i + 1$ and ends with a straight line segment, last$(P^{i+1}(d))$. The path R always exists, we can construct it by starting from the first strip. The path Q exists only if the path R does not exceed the Y-coordinate t_y of T. In this case, the strip S_{i+1} will have no influence on the optimal solution in the sense of Lemma 2, an optimal solution will directly pass through S_{i+1} and the following strips.

So let us assume that R and Q exist and that we would like to compute the performance of $P^{i+1}(d)$ starting at $d = w_{i+1}$. The performance of $P^{i+1}(d)$ is given by the function

$$f_{i+1}(d) := d \cdot i + |\text{last}(P^{i+1}(d))|.$$

Note that we can express $|\text{last}(P^j(d))|$ in terms of d: For convenience, let y_j be the vertical height of $P^{i+1}_j(d)$ (i.e., $d^2 = w_j^2 + y_j^2$), and X be the horizontal distance from the last strip to T. With $T = (t_x, t_y)$ and $S = (0,0)$ we have $X := t_x - \sum_{j=1}^{i+1} w_j$. Now, we have $|\text{last}(P^{i+1}(d))| = \sqrt{X^2 + \left(t_y - \sum_{j=1}^{i+1} y_j\right)^2}$. With $y_j := \sqrt{d^2 - w_j^2}$ we get

$$f_{i+1}(d) = d \cdot i + \sqrt{X^2 + \left(t_y - \sum_{j=1}^{i+1} \sqrt{d^2 - w_j^2}\right)^2}. \tag{1}$$

By simple analysis, we can show that $f_{i+1}(d)$ has a unique minimum in d while increasing d from w_j until $\text{last}(P^{i+1}(d))$ gets horizontal:

Lemma 4. *The function $f_{i+1}(d)$ (Eq. 1) has exactly one minimum for $d \in [w_{i+1}, \bar{d}]$, where $\bar{d}$ is the solution of $t_y - \sum_{j=1}^{i+1} \sqrt{\bar{d}^2 - w_j^2} = 0$ (i.e., the last segment is horizontal).*

Proof. Let us consider the first derivative of $f_{i+1}(d)$ in d, which is given by

$$f'_{i+1}(d) = i - \frac{t_y - \sum_{j=1}^{i+1} \sqrt{d^2 - w_j^2}}{\sqrt{X^2 + \left(t_y - \sum_{j=1}^{i+1} \sqrt{d^2 - w_j^2}\right)^2}} \cdot \sum_{j=1}^{i+1} \frac{d}{\sqrt{d^2 - w_j^2}}, \tag{2}$$

where $X := t_x - \sum_{j=1}^{i+1} w_j$. The first summand, i, of Eq. 2 is constant. The second summand of Eq. 2 is given by

$$h_{i+1}(d) := \frac{t_y - \sum_{j=1}^{i+1} \sqrt{d^2 - w_j^2}}{\sqrt{X^2 + \left(t_y - \sum_{j=1}^{i+1} \sqrt{d^2 - w_j^2}\right)^2}} \cdot \sum_{j=1}^{i+1} \frac{d}{\sqrt{d^2 - w_j^2}}.$$

If $h_{i+1}(d)$ is strictly monotone in d, there will be at most one solution for $f'_{i+1}(d) = 0$. Note that $h_{i+1}(d)$ is always positive. Let us consider the derivative of $h_{i+1}(d)$ which is $g'_{i+1}(d)l_{i+1}(d) + g_{i+1}(d)l'_{i+1}(d)$ for

$$g_{i+1}(d) = \frac{t_y - \sum_{j=1}^{i+1} \sqrt{d^2 - w_j^2}}{\sqrt{X^2 + \left(t_y - \sum_{j=1}^{i+1} \sqrt{d^2 - w_j^2}\right)^2}} \quad \text{and} \quad l_{i+1}(d) := \sum_{j=1}^{i+1} \frac{d}{\sqrt{d^2 - w_j^2}}.$$

It is clear that $l_{i+1}(d) > 0$ and $g_{i+1}(d) > 0$ holds until $\text{last}(P^{i+1}(d))$ is horizontal. It remains to show that $g'_{i+1}(d)$ and $l'_{i+1}(d)$ both have negative sign. This means that $h_{i+1}(d)$ is strictly decreasing and $i - h_{i+1}(d)$ changes from positive to negative only once. Thus, $f_{i+1}(d)$ has a unique minimum.

By simple derivation we have

$$l'_{i+1}(d) = - \sum_{j=1}^{i+1} \frac{w_j^2}{(d^2 - w_j^2)\sqrt{d^2 - w_j^2}} \quad \text{and}$$

$$g'_{i+1}(d) = -\sum_{j=1}^{i+1} \frac{d}{\sqrt{d^2 - w_j^2}} \cdot \frac{X^2}{\left(X^2 + \left(t_y - \sum_{j=1}^{i+1}\sqrt{d^2 - w_j^2}\right)^2\right)^{\frac{3}{2}}}$$

which both have negative sign in the given interval for d. Altogether, the statement follows.   □

Now, it is easy to successively compute the minimum of $f_{i+1}(d)$ for $i = 0, \dots, n-1$. For example, we can apply efficient numerical methods for getting a solution of $f'_{i+1}(d) = 0$, especially because of the strictly decreasing behaviour of f'_{i+1}. In the following, we assume that we can compute this minimum in time proportional to the number of terms of the given functions, i.e., the minimum of $f_{i+1}(d)$ is computed in $O(i)$. This assumption is well justified. We can choose an appropriate starting interval and numerical methods will achieve a good convergence rate, for details see Schwarz [13].

Note that Eq. 2 contains the sum of cosines of the bending angles in the strips times the cosine of the bending angle in the last segment. The parameter d has to be adjusted and it changes all angles simultaneously. That is, there is a global criterion involved and, thus, it seems hard to find a simple ratio that resembles the refractive index in Snell's law.

Using Lemma 4 we can now prove Lemma 3:

Proof of Lemma 3. Let us assume that we have a solution P^i computed for i strips and let $|P_k^i| = d$, where k denotes the index in the sense of Lemma 2. We use this solution for the first $i+1$ strips. If $|P_{i+1}^i| \geq d$ holds, the given solution is also optimal for $i + 1$ strips because the overall performance remains the same: The last segment, last(P^i), of P^i is a line segment with positive slope. Further, $w_j \geq w_{j-1}$ holds. Thus, P^i is the overall optimum; that is, we can apply P^i to all n strips and $|P_j^i| \geq d$ holds for $j = i+1, i+2, \dots, n$.

This means: If we have found a solution P^i with $|P_{i+1}^i| \geq |P_k^i|$ where k denotes the index from Lemma 2, then we are done for all strips.

It remains to show that the index k is strictly increasing until it is finally fixed. From the consideration above we already conclude that there is only one strip, for which k does not increase. Namely, if k does not increase from i to $i+1$ we have $k < i+1$ and $d = |P_k^{i+1}| < |P_j^{i+1}|$ for $j = k+1, \dots, n$. Thus, P^{i+1} is the overall optimum and k is fixed.

Finally, we show that indeed k can never decrease (i.e., fall back to some $k < i$) if strip S_{i+1} is added. Let us assume from $\ell = 1$ to $\ell = i$ we have always a solution P^ℓ for ℓ strips and $k = \ell$ for every P^ℓ. Let us further assume that for $i + 1$ strips the solution P^{i+1} comes along with $k = j < i$. We compare the two solutions P^{i+1} (with $k = j < i$ and $|P_j^{i+1}| =: d_{i+1}$) and P^j (with $k = j$ and $|P_j^j| =: d_j$), see Fig. 6.

As P^j is optimal for j strips but not for $j+1$ strips, we have $|P_{j+1}^j| < d_j$. On the other hand, P^{i+1} is optimal for $i + 1$ strips. Thus, $|P_{j+1}^{i+1}| > d_{i+1}$ holds; see Lemma 2.

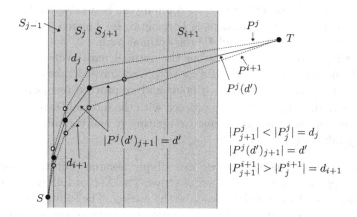

Fig. 6. Between P^{i+1} and P^j there has to be a solution with $P(d)$ which is better than P^{i+1}

Now for a parameter d consider a monotone path $P^j(d)$ that starts from S, has equal path length d in the first j strips and then moves toward T. While d increases, the slope of the last segment strictly decreases. Therefore, the path length $|P^j(d)_{j+1}|$ is strictly decreasing in d. This means that $d_j > d_{i+1}$ and P^j runs above P^{i+1}. The path $P^j(d)$ changes continuously, therefore in $[d_{i+1}, d_j]$ there has to be a value d' such that $|P^j(d')_{j+1}|$ is equal to d', see Fig. 6. The path $P^j(d')$ runs between P^{i+1} and P^j. Obviously, $d' < P^j(d')_l$ holds for $l = j+1, \ldots, n$.

We show that $P(d')$ is better than P^{i+1}. This is a direct consequence of Lemma 4. The value of $P^j(d)$ strictly decreases from $d = d_{i+1}$ to the unique minimum $d = d_j$ and d' is in between. Altogether, P^{i+1} is not optimal which is a contradiction. □

The result of Lemma 3 now suggests a method for computing the optimal path efficiently. Starting from $j = 1$ we compute an optimal path for the first j strips. Let P^j denote this path. If $|P_{j+1}^j| > |P_j^j|$ holds we are done. Otherwise, we have to compute P^{j+1} for $j+1$ strips and so on.

3.2 Algorithm and Its Analysis

Theorem 5. *For a set of n axis-aligned strips the optimal inspection path can be computed in $O(n \log n)$ time and linear space.*

Proof. First, we sort the strips by width which takes $O(n \log n)$ time. Then we apply binary search. That is, in a first step we compute a solution with respect to $j = \lfloor \frac{n}{2} \rfloor$ strips.

We compute the path R with $|R_l| = w_j$ for $l = 1, \ldots, j$. If R does already exceeds the Y-coordinate t_y of T, the optimal path should directly pass through S_j and all successive strips as mentioned above. Therefore we proceed with the interval $[1, j]$ in this case. Otherwise, we compute the best value for $f_j(d)$ starting from $w_j = d$ until the last segment is horizontal, see Lemma 4. Let d_j denote the optimal value for j strips and P^j the optimal path.

Now, we have to determine whether the optimal path visits $i \leq j$ or $i > j$ strips with the same distance d_i. If $d_j > |P^j_{j+1}|$, we have to take into account at least the strip S_{j+1}. Therefore, $i > j$ holds and we proceed recursively with the interval $[j+1, n]$. If $d_j \leq |P^j_{j+1}|$ we proceed with the interval $[1, j]$. Therefore we will find the optimum in $\log n$ steps. Computing R and the minimum of $f_j(d)$ for index j takes $O(j)$ time.   □

For a lower bound construction we can simply assume that the input of an algorithm is given by an unsorted set of strips. The X-coordinates of the strip's left boundaries and their widths describe the setting. The solution is given by a polygonal chain from left to right representing the order of the left boundaries. Thus, sorting a set of n elements can be reduced to the given problem.

Theorem 6. *For a set of n axis-aligned unsorted strips the optimal inspection path is computed in $\Theta(n \log n)$ time and $\Theta(n)$ space.*

4   The L_1-Case

Fortunately, if we measure the distance by the L_1 metric the structural properties are equivalent. Computing the optimal path becomes much easier.

Note that the path segments between the strips have to be horizontal. We have to distribute the vertical distance from S to T among a subset of the strips. Again we sort the strips by their widths and rearrange the scenario. Fig. 7 shows an example of an optimal L_1-path after rearrangement.

Theorem 7. *The optimal L_1-path P visits the first $k \leq n$ strips with the same L_1-distance d and then moves horizontally to the end point T. For $i = 1, \ldots, k$ we*

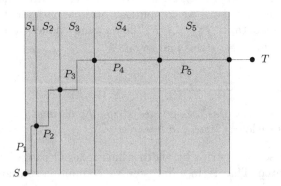

Fig. 7. The structure of an optimal solution in the L_1-case after rearrangement. The first three strips are visited with the same value d, for all other strips $|P_i| > d$ holds. The path is horizontal between strips.

have $|P_i| = d$ and for $i = k+1, \ldots, n$ we have $w_i > d$. Additionally, $\sum_{i=1}^{k} |P_i| = t_y$ holds. If the number of strips, n, increases, the index k increases until it remains fixed. The optimal path can be computed in $\Theta(n \log n)$ time and $\Theta(n)$ space.

An algorithm for the L_1 problem is given as follows. First, we sort the strips by their widths. Then starting from $i = 1$ we distribute $t_y + \sum_{j=1}^{i} w_j$ among i strips. For an optimal path, P^i, for i strips we have $|P_j^i| = \frac{1}{i}(t_y + \sum_{j=1}^{i} w_j)$ for $j = 1, \ldots, i$. If $|P_i^i| < w_{i+1}$ this path is also optimal for $i+1$ (and n) strips. For $|P_i^i| > w_{i+1}$ we distribute $t_y + \sum_{j=1}^{i+1} w_j$ among $i+1$ strips; that is, $|P_j^{i+1}| = \frac{1}{i+1}(t_y + \sum_{j=1}^{i+1} w_j)$ for $j = 1, \ldots, i+1$. We can compute the sum of the weights successively. For a single step only a constant number of operations is necessary.

Altogether, if the strips are given by ordered widths, the algorithm runs in $\Theta(n)$ time and space.

5   Conclusion and Future Work

We presented an optimal algorithm that computes the shortest inspection path for a set of axis-aligned strips which has to be visited in a given order.

The performance of a path P for a single strip S_i is given by the time where the strip is not inspected, i.e. $|P| - |P_i|$. The maximum value $|P| - |P_i|$ among all strips gives the performance of the inspection path. In turn, we compute a path P with minimal performance among all paths in optimal time and space. The approach works for L_1- and L_2-metric.

The structural properties of the solution show that a set of strips with increasing widths determines the solution, the remaining strips are of greater widths and they will be simply passed. This shows that the problem is of LP-Type [14]. The set of strips is the set H and $w(G)$ gives the performance of the optimal solution for a subset $G \subseteq H$. Obviously, w is *monotone*, that is, if we add more strips the performance of the solution cannot decrease. On the other hand *monotonicity* holds. If two subsets $F \subseteq G \subseteq H$ have the same performance and adding an additional strip $h \in H$ does not change the performance of F, the strip h can also not change the performance of G if added. Unfortunately, all strips might determine the solution. Thus the basis of the problem is not a single constant and it was worth computing a solution directly.

One might think that a relative performance is more intuitive. That is, for an inspection path P, $\frac{|P|}{|P_i|}$ defines the performance for a single strip. But for a single strip it is then optimal to make a large detour inside the strip. This might also hold for more than one strip. So this measure can be considered to be counterintuitive.

But there are other extensions which might be interesting to consider. One could consider axis-parallel rectangles instead of full strips. Or the objects might be of arbitrary type and even not ordered by Y-coordinate. Furthermore, obstacles could intersect.

On the other hand, one might consider dynamical versions of the problem. Consider a set of axis-aligned rectangles which separately move inside the given strips in a specified direction. Compute an inspection path that takes the movement of the rectangles into account.

Finally, the question of Mark Overmars mentioned in the introduction is still open. We are searching for a roundtrip so that the maximal time interval where a point is not seen should be minimized. The main problem is whether subpolygons induced by reflex vertices have to be visited in an order along the boundary in the optimal inspection route, see also Dror et al. [7]. Note, that the corresponding subpolygons might be visited more than once in an optimal inspection route.

References

[1] Benkert, M., Djordjevic, B., Gudmundsson, J., Wolle, T.: Finding popular places. In: Tokuyama, T. (ed.) ISAAC 2007. LNCS, vol. 4835, pp. 776–787. Springer, Heidelberg (2007)

[2] Berger, F., Grüne, A., Klein, R.: How many lions can one man avoid? Technical Report 006, Department of Computer Science I, University of Bonn (2007), http://web.informatik.uni-bonn.de/I/publications/bgk-hmlcm-07.pdf

[3] Carlsson, S., Jonsson, H.: Computing a shortest watchman path in a simple polygon in polynomial time. In: Sack, J.-R., Akl, S.G., Dehne, F., Santoro, N. (eds.) WADS 1995. LNCS, vol. 955, pp. 122–134. Springer, Heidelberg (1995)

[4] Chin, W., Ntafos, S.: Optimum watchman routes. In: Proc. 2nd Annu. ACM Sympos. Comput. Geom, pp. 24–33 (1986)

[5] Chin, W.-P., Ntafos, S.: The zookeeper problem. Inform. Sci. 63, 245–259 (1992)

[6] Deng, X., Kameda, T., Papadimitriou, C.: How to learn an unknown environment I: The rectilinear case. J. ACM 45(2), 215–245 (1998)

[7] Dror, M., Efrat, A., Lubiw, A., Mitchell, J.S.B.: Touring a sequence of polygons. In: Proc. 35th Annu. ACM Sympos. Theory Comput, pp. 473–482 (2003)

[8] Hoffmann, F., Icking, C., Klein, R., Kriegel, K.: The polygon exploration problem. SIAM J. Comput. 31, 577–600 (2001)

[9] Icking, C., Kamphans, T., Klein, R., Langetepe, E.: On the competitive complexity of navigation tasks. In: Bunke, H., Christensen, H.I., Hager, G.D., Klein, R. (eds.) Sensor Based Intelligent Robots. LNCS, vol. 2238, pp. 245–258. Springer, Berlin (2002)

[10] Icking, C., Kamphans, T., Klein, R., Langetepe, E.: Exploring simple grid polygons. In: Wang, L. (ed.) COCOON 2005. LNCS, vol. 3595, pp. 524–533. Springer, Heidelberg (2005)

[11] Mitchell, J.S.B.: Geometric shortest paths and network optimization. In: Sack, J.-R., Urrutia, J. (eds.) Handbook of Computational Geometry, pp. 633–701. Elsevier Science Publishers B.V. North-Holland, Amsterdam (2000)

[12] Overmars, M.: Personel communication

[13] Schwarz, H.R.: Numerical Analysis: A Comprehensive Introduction. B. G. Teubner, Stuttgart (1989)

[14] Sharir, M., Welzl, E.: A combinatorial bound for linear programming and related problems. In: Proc. 9th Sympos. Theoret. Aspects Comput. Sci. LNCS, vol. 577, pp. 569–579. Springer, Heidelberg (1992)

A Pseudopolynomial Algorithm
for Alexandrov's Theorem

Daniel Kane[1,*], Gregory N. Price[2,**], and Erik D. Demaine[2,***]

[1] Department of Mathematics, Harvard University,
1 Oxford Street, Cambridge MA 02139, USA
dankane@math.harvard.edu
[2] MIT Computer Science and Artificial Intelligence Laboratory,
32 Vassar Street, Cambridge, MA 02139, USA
{price,edemaine}@mit.edu

Abstract. Alexandrov's Theorem states that every metric with the global topology and local geometry required of a convex polyhedron is in fact the intrinsic metric of some convex polyhedron. Recent work by Bobenko and Izmestiev describes a differential equation whose solution is the polyhedron corresponding to a given metric. We describe an algorithm based on this differential equation to compute the polyhedron to arbitrary precision given the metric, and prove a pseudopolynomial bound on its running time.

1 Introduction

Consider the intrinsic metric induced on the surface M of a convex body in $\mathbb{R}^3$. Clearly M under this metric is homeomorphic to a sphere, and locally convex in the sense that a circle of radius r has circumference at most $2\pi r$.

In 1949, Alexandrov and Pogorelov [1] proved that these two necessary conditions are actually sufficient: every metric space M that is homeomorphic to a 2-sphere and locally convex can be embedded as the surface of a convex body in $\mathbb{R}^3$. Because Alexandrov and Pogorelov's proof is not constructive, their work opened the question of how to produce the embedding given a concrete M.

To enable computation we require that M be a polyhedral metric space, locally isometric to $\mathbb{R}^2$ at all but n points (vertices). Now the theorem is that every polyhedral metric, a complex of triangles with the topology of a sphere and positive curvature at each vertex, can be embedded as an actual convex polyhedron in $\mathbb{R}^3$. This case of the Alexandrov-Pogorelov theorem was proven by Alexandrov in 1941 [1], also nonconstructively. Further, Cauchy showed in 1813 [3] that such an embedding must be unique. All the essential geometry of the general case is preserved in the polyhedral case, because every metric satisfying the general hypothesis can be polyhedrally approximated.

---

* Partially supported by an NDSEG Fellowship.
** Partially supported by an NSF Graduate Research Fellowship.
*** Partially supported by NSF CAREER award CCF-0347776.

F. Dehne et al. (Eds.): WADS 2009, LNCS 5664, pp. 435–446, 2009.
© Springer-Verlag Berlin Heidelberg 2009

Algorithms for Alexandrov's Theorem are motivated by the problem of folding a polygon of paper into precisely the surface of a convex polyhedron. There are efficient algorithms to find one or all gluings of a given polygon's boundary to itself so that the resulting metric satisfies Alexandrov's conditions [4,9]. But this work leaves open how to find the actual 3D polyhedra that can be folded from the polygon of paper.

In 1996, Sabitov [12,11,13,5] showed how to enumerate all the isometric maps $M \to \mathbb{R}^3$ for a polyhedral metric M, so that one could carry out this enumeration and identify the one map that gives a convex polyhedron. In 2005, Fedorchuk and Pak [6] showed an exponential upper bound on the number of such maps. An exponential lower bound is easy to find, so this algorithm takes time exponential in n and is therefore unsatisfactory.

Recent work by Bobenko and Izmestiev [2] produced a new proof of Alexandrov's Theorem, describing a certain ordinary differential equation (ODE) and initial conditions whose solution contains sufficient information to construct the embedding by elementary geometry. This work was accompanied by a computer implementation of the ODE [14], which empirically produces accurate approximations of embeddings of metrics on which it is tested.

In this work, we describe an algorithm based on the Bobenko-Izmestiev ODE, and prove a pseudopolynomial bound on its running time. Specifically, call an embedding of M ε-accurate if the metric is distorted by at most a factor $1 + \varepsilon$, and ε-convex if each dihedral angle is at most $\pi + \varepsilon$. For concreteness, M may be represented by a list of triangles with side lengths and the names of adjacent triangles. Then we show the following theorem:

Theorem 1. *Given a polyhedral metric M with n vertices, ratio S between the largest and smallest distance between vertices, and defect (discrete Gaussian curvature) between ε_1 and $2\pi - \varepsilon_8$ at each vertex, an ε_6-accurate ε_9-convex embedding of M can be found in time $O\left(n^{913/2} S^{831} / (\varepsilon^{121} \varepsilon_1^{445} \varepsilon_8^{616})\right)$ where $\varepsilon = \min(\varepsilon_6/nS, \varepsilon_9 \varepsilon_1^2 / nS^6)$.*

The exponents in the time bound of Theorem 1 are remarkably large. Thankfully, no evidence suggests our algorithm actually takes as long to run as the bound allows. On the contrary, our analysis relies on bounding approximately a dozen geometric quantities, and to keep the analysis tractable we use the simplest bound whenever available. The algorithm's actual performance is governed by the actual values of these quantities, and therefore by whatever sharper bounds could be proven by a stingier analysis.

To describe our approach, consider an embedding of the metric M as a convex polyhedron in $\mathbb{R}^3$, and choose an arbitrary origin O in the surface's interior. Then it is not hard to see that the n distances $r_i = \overline{Ov_i}$ from the origin to the vertices v_i, together with M and the combinatorial data describing which polygons on M are faces of the polyhedron, suffice to reconstruct the embedding: the tetrahedron formed by O and each triangle is rigid in $\mathbb{R}^3$, and we have no choice in how to glue them to each other. In Lemma 1 below, we show that in fact the radii alone suffice to reconstruct the embedding, to do so efficiently, and to do so even with radii of finite precision.

Therefore in order to compute the unique embedding of M that Alexandrov's Theorem guarantees exists, we compute a set of radii $r = \{r_i\}_i$ and derive a triangulation T. The exact radii satisfy three conditions:

1. the radii r determine nondegenerate tetrahedra from O to each face of T;
2. with these tetrahedra, the dihedral angles at each exterior edge total at most π; and
3. with these tetrahedra, the dihedral angles about each radius sum to 2π.

In our computation, we begin with a set of large initial radii $r_i = R$ satisfying Conditions 1 and 2, and write $\kappa = \{\kappa_i\}_i$ for the differences by which Condition 3 fails about each radius. We then iteratively adjust the radii to bring κ near zero and satisfy Condition 3 approximately, maintaining Conditions 1 and 2 throughout.

The computation takes the following form. We describe the Jacobian $\left(\frac{\partial \kappa_i}{\partial r_j}\right)_{ij}$, showing that it can be efficiently computed and that its inverse is pseudopoly-nomially bounded. We show further that the Hessian $\left(\frac{\partial \kappa_i}{\partial r_j \partial r_k}\right)_{ijk}$ is also pseu-dopolynomially bounded. It follows that a change in r in the direction of smaller κ as described by the Jacobian, with some step size only pseudopolynomially small, makes progress in reducing $|\kappa|$. The step size can be chosen online by doubling and halving, so it follows that we can take steps of the appropriate size, pseudopolynomial in number, and obtain an r that zeroes κ to the desired precision in pseudopolynomial total time. Theorem 1 follows.

The construction of [2] is an ODE in the same n variables r_i, with a similar starting point and with the derivative of r driven similarly by a desired path for κ. Their proof differs in that it need only show existence, not a bound, for the Jacobian's inverse, in order to invoke the inverse function theorem. Similarly, while we must show a pseudopolynomial lower bound (Lemma 11) on the alti-tudes of the tetrahedra during our computation, the prior work shows only that these altitudes remain positive. In general our computation requires that the known open conditions—this quantity is positive, that map is nondegenerate—be replaced by stronger compact conditions—this quantity is lower-bounded, that map's inverse is bounded. We model our proofs of these strengthenings on the proofs in [2] of the simpler open conditions, and we directly employ several other results from that paper where possible.

The remainder of this paper supplies the details of the proof of Theorem 1. We give background in Section 2, and detail the main argument in Section 3. We bound the Jacobian in Section 4 and the Hessian in Section 5. Finally, some lemmas are deferred to Section 6 for clarity.

2 Background and Notation

In this section we define our major geometric objects and give the basic facts about them. We also define some parameters describing our central object that we will need to keep bounded throughout the computation.

2.1   Geometric Notions

Central to our argument are two dual classes of geometric structures introduced by Bobenko and Izmestiev in [2] under the names of "generalized convex polytope" and "generalized convex polyhedron". Because in other usages the distinction between "polyhedron" and "polytope" is that a polyhedron is a three-dimensional polytope, and because both of these objects are three-dimensional, we will refer to these objects as "generalized convex polyhedra" and "generalized convex dual polyhedra" respectively to avoid confusion.

First, we define the objects that our main theorem is about.

Definition 1. *A metric M homeomorphic to the sphere is a* polyhedral metric *if each $x \in M$ has an open neighborhood isometric either to a subset of $\mathbb{R}^2$ or to a cone of angle less than 2π with x mapped to the apex, and if only finitely many x, called the* vertices $V(M) = \{v_i\}_i$ *of M, fall into the latter case.*

The defect δ_i *at a vertex $v_i \in V(M)$ is the difference between 2π and the total angle at the vertex, which is positive by the definition of a vertex.*

An embedding *of M is a piecewise linear map $f : M \to \mathbb{R}^3$. An embedding f is ε-accurate if it distorts the metric M by at most $1 + \varepsilon$, and ε-convex if $f(M)$ is a polyhedron and each dihedral angle in $f(M)$ is at most $\pi + \varepsilon$.*

Definition 2. *In a tetrahedron $ABCD$, write $\angle CABD$ for the dihedral angle along edge AB.*

Definition 3. *A* triangulation *of a polyhedral metric M is a decomposition into Euclidean triangles whose vertex set is $V(M)$. Its vertices are denoted by $V(T) = V(M)$, its edges by $E(T)$, and its faces by $F(T)$.*

A radius assignment *on a polyhedral metric M is a map $r : V(M) \to \mathbb{R}_+$. For brevity we write r_i for $r(v_i)$.*

A generalized convex polyhedron *is a gluing of metric tetrahedra with a common apex O. The generalized convex polyhedron $P = (M, T, r)$ is determined by the polyhedral metric M and triangulation T giving its bases and the radius assignment r for the side lengths.*

Write $\kappa_i \triangleq 2\pi - \sum_{jk} \angle v_j O v_i v_k$ for the curvature *about Ov_i, and $\phi_{ij} \triangleq \angle v_i O v_j$ for the angle between vertices v_i, v_j seen from the apex.*

Our algorithm, following the construction in [2], will choose a radius assignment for the M in question and iteratively adjust it until the associated generalized convex polyhedron P fits nearly isometrically in $\mathbb{R}^3$. The resulting radii will give an ε-accurate ε-convex embedding of M into $\mathbb{R}^3$.

In the argument we will require several geometric objects related to generalized convex polyhedra.

Definition 4. *A* Euclidean simplicial complex *is a metric space on a simplicial complex where the metric restricted to each cell is Euclidean.*

A generalized convex polygon *is a Euclidean simplicial 2-complex homeomorphic to a disk, where all triangles have a common vertex V, the total angle at V is no more than 2π, and the total angle at each other vertex is no more than π.*

*Given a generalized convex polyhedron $P = (M, T, r)$, the corresponding gener-
alized convex dual polyhedron $D(P)$ is a certain Euclidean simplicial 3-complex.
Let O be a vertex called the* apex, A_i *a vertex with $OA_i = h_i \overset{\triangle}{=} 1/r_i$ for each i.*

*For each edge $v_i v_j \in E(T)$ bounding triangles $v_i v_j v_k$ and $v_j v_i v_l$, construct two
simplices $OA_i A_{jil} A_{ijk}$, $OA_j A_{ijk} A_{jil}$ in $D(P)$ as follows. Embed the two tetrahedra $Ov_i v_j v_k$, $Ov_j v_i v_l$ in $\mathbb{R}^3$. For each $i' \in \{i, j, k, l\}$, place $A_{i'}$ along ray $Ov_{i'}$
at distance $h_{i'}$, and draw a perpendicular plane $P_{i'}$ through the ray at $A_{i'}$. Let
A_{ijk}, A_{jil} be the intersection of the planes P_i, P_j, P_k and P_j, P_i, P_l respectively.*

*Now identify the vertices $A_{ijk}, A_{jki}, A_{kij}$ for each triangle $v_i v_j v_k \in F(T)$ to
produce the Euclidean simplicial 3-complex $D(P)$. Since the six simplices produced about each of these vertices A_{ijk} are all defined by the same three planes
P_i, P_j, P_k with the same relative configuration in $\mathbb{R}^3$, the total dihedral angle
about each OA_{ijk} is 2π. On the other hand, the total dihedral angle about OA_i
is $2\pi - \kappa_i$, and the face about A_i is a generalized convex polygon of defect κ_i.*

The Jacobian bound in Section 4 makes use of certain multilinear forms described
in [2] and in the full paper [8].

Definition 5. *The* dual volume $\mathrm{vol}(h)$ *is the volume of the generalized convex
dual polyhedron $D(P)$, a cubic form in the dual altitudes h. The* mixed volume
$\mathrm{vol}(\cdot, \cdot, \cdot)$ *is the associated symmetric trilinear form.*

*Let E_i be the area of the face around A_i in $D(P)$, a quadratic form in the
altitudes within this face. The ith* mixed area $E_i(\cdot, \cdot)$ *is the associated symmetric
bilinear form.*

*Let π_i be the linear map $\pi_i(h)_j \overset{\triangle}{=} \frac{h_j - h_i \cos \phi_{ij}}{\sin \phi_{ij}}$. so that $\pi_i(h) = g(i)$. Then
define $F_i(a, b) \overset{\triangle}{=} E_i(\pi_i(a), \pi_i(b))$ so that $F_i(h, h)$ is the area of face i.*

By elementary geometry $\mathrm{vol}(h, h, h) = \frac{1}{3} \sum_i h_i F_i(h, h)$, so that by a simple computation $\mathrm{vol}(a, b, c) = \frac{1}{3} \sum_i a_i F_i(b, c)$.

2.2   Weighted Delaunay Triangulations

The triangulations we require at each step of the computation are the weighted
Delaunay triangulations used in the construction of [2]. We give a simpler definition inspired by Definition 14 of [7].

Definition 6. *The* power $\pi_v(p)$ *of a point p against a vertex v in a polyhedral
metric M with a radius assignment r is $pv^2 - r(v)^2$.*

The center $C(v_i v_j v_k)$ *of a triangle $v_i v_j v_k \in T(M)$ when embedded in $\mathbb{R}^2$ is the
unique point p such that $\pi_{v_i}(p) = \pi_{v_j}(p) = \pi_{v_k}(p)$, which exists by the radical
axis theorem from classical geometry. The quantity $\pi_{v_i}(p) = \pi(v_i v_j v_k)$ is the*
power *of the triangle.*

A triangulation T of a polyhedral metric M with radius assignment r is locally
convex *at edge $v_i v_j$ with neighboring triangles $v_i v_j v_k, v_j v_i v_l$ if $\pi_{v_l}(C(v_i v_j v_k)) >
\pi_{v_l}(v_k)$ and $\pi_{v_k}(C(v_j v_i v_l)) > \pi_{v_k}(v_l)$ when $v_i v_j v_k, v_j v_i v_l$ are embedded together
in $\mathbb{R}^2$.*

A weighted Delaunay triangulation *for a radius assignment r on a polyhedral metric M is a triangulation T that is locally convex at every edge.*

A weighted Delaunay triangulation can be computed in time $O(n^2 \log n)$ by a simple modification of the *continuous Dijkstra algorithm* of [10]. The original analysis of this algorithm assumes that each edge of the input triangulation is a shortest path. In the full paper [8] we show that the same algorithm works in time $O(S\varepsilon_8^{-1} n^2 \log n)$ in the general case. Therefore we perform the general computation once at the outset, and use the resulting triangulation as the basis for subsequent runs of the continuous Dijkstra algorithm in time $O(n^2 \log n)$ each.

The radius assignment r and triangulation T admit a tetrahedron $Ov_iv_jv_k$ just if the power of $v_iv_jv_k$ is negative, and the squared altitude of O in this tetrahedron is $-\pi(v_iv_jv_k)$. The edge v_iv_j is convex when the two neighboring tetrahedra are embedded in $\mathbb{R}^3$ just if it is locally convex in the triangulation as in Definition 6. A weighted Delaunay triangulation with negative powers therefore gives a valid generalized convex polyhedron if the curvatures κ_i are positive. For each new radius assignment r in the computation of Section 3 we therefore compute the weighted Delaunay triangulation and proceed with the resulting generalized convex polyhedron, in which Lemma 11 guarantees a positive altitude and the choices in the computation guarantee positive curvatures.

2.3   Notation for Bounds

Definition 7. *Let the following bounds be observed:*

1. *n is the number of vertices on M.*
2. *$\varepsilon_1 \overset{\Delta}{=} \min_i \delta_i$ is the minimum defect.*
3. *$\varepsilon_2 \overset{\Delta}{=} \min_i(\delta_i - \kappa_i)$ is the minimum defect-curvature gap.*
4. *$\varepsilon_3 \overset{\Delta}{=} \min_{ij \in E(T)} \phi_{ij}$ is the minimum angle between radii.*
5. *$\varepsilon_4 \overset{\Delta}{=} \max_i \kappa_i$ is the maximum curvature.*
6. *$\varepsilon_5 \overset{\Delta}{=} \min_{v_iv_jv_k \in F(T)} \angle v_iv_jv_k$ is the smallest angle in the triangulation. Observe that obtuse angles are also bounded: $\angle v_iv_jv_k < \pi - \angle v_jv_iv_k \leq \pi - \varepsilon_5$.*
7. *ε_6 is used for the desired accuracy in embedding M.*
8. *$\varepsilon_7 \overset{\Delta}{=} (\max_i \frac{\kappa_i}{\delta_i})/(\min_i \frac{\kappa_i}{\delta_i}) - 1$ is the extent to which the ratio among the κ_i varies from that among the δ_i. We will keep $\varepsilon_7 < \varepsilon_8/4\pi$ throughout.*
9. *$\varepsilon_8 \overset{\Delta}{=} \min_i(2\pi - \delta_i)$ is the minimum angle around a vertex.*
10. *ε_9 is used for the desired approximation to convexity in embedding M.*
11. *D is the diameter of M.*
12. *L is the maximum length of any edge in the input triangulation.*
13. *ℓ is the shortest distance v_iv_j between vertices.*
14. *$S \overset{\Delta}{=} \max(D, L)/\ell$ is the maximum ratio of distances.*
15. *$d_0 \overset{\Delta}{=} \min_{p \in M} Op$ is the minimum height of the apex off of any point on M.*
16. *$d_1 \overset{\Delta}{=} \min_{v_iv_j \in E(T)} d(O, v_iv_j)$ is the minimum distance to any edge of T.*

17. $d_2 \triangleq \min_i r_i$ is the minimum distance from the apex to any vertex of M.

18. $H \triangleq 1/d_0$; the name is justified by $h_i = 1/r_i \leq 1/d_0$.

19. $R \triangleq \max_i r_i$, so $1/H \leq r_i \leq R$ for all i.

20. $T \triangleq HR$ is the maximum ratio of radii.

Of these bounds, $n, \varepsilon_1, \varepsilon_8$, and S are fundamental to the given metric M or the form in which it is presented as input, and D, L, and ℓ are dimensionful parameters of the same metric input. The values ε_6 and ε_9 define the objective to be achieved, and our computation will drive ε_4 toward zero while maintaining ε_2 large and ε_7 small. In Section 6 we bound the remaining parameters $\varepsilon_3, \varepsilon_5, R, d_0, d_1$, and d_2 in terms of these.

Definition 8. *Let* $\mathbf{J}$ *denote the Jacobian* $\left(\frac{\partial \kappa_i}{\partial r_j}\right)_{ij}$, *and* $\mathbf{H}$ *the Hessian* $\left(\frac{\partial \kappa_i}{\partial r_j \partial r_k}\right)_{ijk}$.

3   Main Theorem

In this section, we prove our main theorem using the results proved in the remaining sections. The algorithm of Theorem 1 obtains an approximate embedding of the polyhedral metric M in $\mathbb{R}^3$. Its main subroutine is described by the following theorem:

Theorem 2. *Given a polyhedral metric M with n vertices, ratio S (the spread) between the diameter and the smallest distance between vertices, and defect at least ε_1 and at most $2\pi - \varepsilon_8$ at each vertex, a radius assignment r for M with maximum curvature at most ε can be found in time $O\left(n^{913/2} S^{831} / (\varepsilon^{121} \varepsilon_1^{445} \varepsilon_8^{616})\right)$.*

Proof. Let a *good* assignment be a radius assignment r that satisfies two bounds: $\varepsilon_7 < \varepsilon_8/4\pi$ so that Lemmas 9–11 apply and r therefore by the discussion in Subsection 2.2 produces a valid generalized convex polyhedron for M, and $\varepsilon_2 = \Omega(\varepsilon_1^2 \varepsilon_8^3 / n^2 S^2)$ on which our other bounds rely. By Lemma 6, there exists a good assignment r^0. We will iteratively adjust r^0 through a sequence r^t of good assignments to arrive at an assignment r^N with maximum curvature $\varepsilon_4^N < \varepsilon$ as required. At each step we recompute T as a weighted Delaunay triangulation according to Subsection 2.2.

Given a good assignment $r = r^n$, we will compute another good assignment $r' = r^{n+1}$ with $\varepsilon_4 - \varepsilon_4' = \Omega\left(\varepsilon_1^{445} \varepsilon_4^{121} \varepsilon_8^{616} / (n^{907/2} S^{831})\right)$. It follows that from r^0 we can arrive at a satisfactory r^N with $N = O\left((n^{907/2} S^{831}) / (\varepsilon^{121} \varepsilon_1^{445} \varepsilon_8^{616})\right)$.

To do this, let $\mathbf{J}$ be the Jacobian $\left(\frac{\partial \kappa_i}{\partial r_j}\right)_{ij}$ and $\mathbf{H}$ the Hessian $\left(\frac{\partial \kappa_i}{\partial r_j \partial r_k}\right)_{ijk}$, evaluated at r. The goodness conditions and the objective are all in terms of κ, so we choose a desired new curvature vector κ^* in κ-space and apply the inverse Jacobian to get a new radius assignment $r' = r + \mathbf{J}^{-1}(\kappa^* - \kappa)$ in r-space. The actual new curvature vector κ' differs from κ^* by an error at most $\frac{1}{2}|\mathbf{H}||r' - r|^2 \leq \left(\frac{1}{2}|\mathbf{H}||\mathbf{J}^{-1}|^2\right)|\kappa^* - \kappa|^2$, quadratic in the desired change in curvatures with a coefficient

$$C \triangleq \frac{1}{2}|\mathbf{H}||\mathbf{J}^{-1}|^2 = O\left(\frac{n^{3/2}S^{14}}{\varepsilon_5^3} \frac{R^{23}}{D^{14}d_0^3d_1^8}\left(\frac{n^{7/2}T^2}{\varepsilon_2\varepsilon_3^3\varepsilon_4}R\right)^2\right) = O\left(\frac{n^{905/2}S^{831}}{\varepsilon_1^{443}\varepsilon_4^{121}\varepsilon_8^{616}}\right)$$

by Theorems 3 and 4 and Lemmas 7, 6, 11, and 8.

Therefore pick a step size p, and choose κ^* according to $\kappa_i^* - \kappa_i = -p\kappa_i - p\left(\kappa_i - \delta_i \min_j \frac{\kappa_j}{\delta_j}\right)$. The first term diminishes all the curvatures together to reduce ε_4, and the second rebalances them to keep the ratios $\frac{\kappa_j}{\delta_j}$ nearly equal so that ε_7 remains small. In the full paper [8] we show that the resulting actual curvatures κ' make r' a good assignment and put $\varepsilon_4' \leq \varepsilon_4 - p\varepsilon_4/2$, so long as

$$p \leq \varepsilon_1^2/64\pi^2 n\varepsilon_4 C. \qquad (1)$$

This produces a good radius assignment r' in which ε_4 has declined by at least

$$\frac{p\varepsilon_4}{2} = \frac{\varepsilon_1^2}{128\pi^2 nC} = \Omega\left(\frac{\varepsilon_1^{445}\varepsilon_4^{121}\varepsilon_8^{616}}{n^{907/2}S^{831}}\right)$$

as required.

As a simplification, we need not compute p exactly according to (1). Rather, we choose the step size p^t at each step, trying first p^{t-1} (with p^0 an arbitrary constant) and computing the actual curvature error $|\kappa' - \kappa^*|$. If the error exceeds its maximum acceptable value $p\varepsilon_1^2\varepsilon_4/16\pi^2$ then we halve p^t and try step t again, and if it falls below half this value then we double p^t for the next round. Since we double at most once per step and halve at most once per doubling plus a logarithmic number of times to reach an acceptable p, this doubling and halving costs only a constant factor. Even more important than the resulting simplification of the algorithm, this technique holds out the hope of actual performance exceeding the proven bounds.

Now each of the N iterations of the computation go as follows. Compute the weighted Delaunay triangulation T^t for r^t in time $O(n^2 \log n)$ as described in Subsection 2.2. Compute the Jacobian $\mathbf{J}^t$ in time $O(n^2)$ using formulas (14, 15) in [2]. Choose a step size p^t, possibly adjusting it, as discussed above. Finally, take the resulting r' as r^{t+1} and continue. The computation of κ^* to check p^t runs in linear time, and that of r' in time $O(n^\omega)$ where $\omega < 3$ is the time exponent of matrix multiplication. Each iteration therefore costs time $O(n^3)$, and the whole computation costs time $O(n^3 N)$ as claimed. □

Now with our radius assignment r for M and the resulting generalized convex polyhedron P with curvatures all near zero, it remains to approximately embed P and therefore M in $\mathbb{R}^3$. To begin, we observe that this is easy to do given exact values for r and in a model with exact computation: after triangulating, P is made up of rigid tetrahedra and we embed one tetrahedron arbitrarily, then embed each neighboring tetrahedron in turn.

In a realistic model, we compute only with bounded precision, and in any case Theorem 2 gives us only curvatures near zero, not equal to zero. Lemma 1 produces an embedding in this case, settling for less than exact isometry and exact convexity.

Lemma 1. *There is an algorithm that, given a radius assignment r for which the corresponding curvatures κ_i are all less than $\varepsilon = O\left(\min(\varepsilon_6/nS, \varepsilon_9\varepsilon_1^2/nS^6)\right)$ for some constant factor, produces explicitly by vertex coordinates in time $O(n^2\log n)$ an ε_6-accurate ε_9-convex embedding of M.*

Proof (sketch). As in the exact case, triangulate M, embed one tetrahedron arbitrarily, and then embed its neighbors successively. The positive curvature will force gaps between the tetrahedra. Then replace the several copies of each vertex by their centroid, so that the tetrahedra are distorted but leave no gaps. This is the desired embedding. The proofs of ε_6-accuracy and ε_9-convexity are straightforward and left to the full paper [8].

A weighted Delaunay triangulation takes time $O(n^2\log n)$ as discussed in Subsection 2.2, and the remaining steps take time $O(n)$.   □

We now have all the pieces to prove our main theorem.

Proof (Theorem 1). Let $\varepsilon \triangleq O\left(\min(\varepsilon_6/nS, \varepsilon_9\varepsilon_1^2/nS^6)\right)$, and apply the algorithm of Theorem 2 to obtain in time $O\left(n^{913/2}S^{831}/(\varepsilon^{121}\varepsilon_1^{445}\varepsilon_8^{616})\right)$ a radius assignment r for M with maximum curvature $\varepsilon_4 \leq \varepsilon$.

Now apply the algorithm of Lemma 1 to obtain in time $O(n^2\log n)$ the desired embedding and complete the computation.   □

4   Bounding the Jacobian

Theorem 3. *The Jacobian $\mathbf{J} = \left(\frac{\partial\kappa_i}{\partial r_j}\right)_{ij}$ has inverse pseudopolynomially bounded by $|\mathbf{J}^{-1}| = O\left(\frac{n^{7/2}T^2}{\varepsilon_2\varepsilon_3^3\varepsilon_4}R\right)$.*

Proof. Our argument parallels that of Corollary 2 in [2], which concludes that the same Jacobian is nondegenerate. Theorem 4 of [2] shows that this Jacobian equals the Hessian of the volume of the dual $D(P)$. The meat of the corollary's proof is in Theorem 5 of [2], which begins by equating this Hessian to the bilinear form $6\operatorname{vol}(h,\cdot,\cdot)$ derived from the mixed volume we defined in Definition 5. So we have to bound the inverse of this bilinear form.

To do this it suffices to show that the form $\operatorname{vol}(h,x,\cdot)$ has norm at least $\Omega\left(\frac{\varepsilon_2\varepsilon_3^3\varepsilon_4}{n^{7/2}T^2}\frac{|x|}{R}\right)$ for all vectors x. Equivalently, suppose some x has $|\operatorname{vol}(h,x,z)| \leq |z|$ for all z; we show $|x| = O\left(\frac{n^{7/2}T^2}{\varepsilon_2\varepsilon_3^3\varepsilon_4}R\right)$.

To do this we follow the proof in Theorem 5 of [2] that the same form $\operatorname{vol}(h,x,\cdot)$ is nonzero for x nonzero. Throughout the argument we work in terms of the dual $D(P)$.

Recall that for each i, $\pi_i x$ is defined as the vector $\{x_{ij}\}_j$. It suffices to show that for all i

$$|\pi_i x|_2^2 = O\left(\frac{n^3T^3}{\varepsilon_2^2\varepsilon_3\varepsilon_4}R^2 + \frac{n^2T^2}{\varepsilon_2\varepsilon_3\varepsilon_4}R|x|_1\right)$$

since then by Lemma 2

$$|x|_2^2 \leq \frac{4n}{\varepsilon_3^2}\max_i|\pi_i x|_2^2 = O\left(\frac{n^4T^3}{\varepsilon_2^2\varepsilon_3^3\varepsilon_4}R^2 + \frac{n^3T^2}{\varepsilon_2\varepsilon_3^3\varepsilon_4}R|x|_1\right),$$

and since $|x|_1 \leq \sqrt{n}|x|_2$ and $X^2 \leq a + bX$ implies $X \leq \sqrt{a} + b$, $|x|_2 = O\left(\frac{n^{7/2}T^2}{\varepsilon_2\varepsilon_3^3\varepsilon_4}R\right)$. Therefore fix an arbitrary i, let $g = \pi_i h$ and $y = \pi_i x$, and we proceed to bound $|y|_2$.

We break the space on which E_i acts into the 1-dimensional positive eigenspace of E_i and its $(k-1)$-dimensional negative eigenspace, since by Lemma 3.4 of [2] the signature of E_i is $(1, k-1)$, where k is the number of neighbors of v_i. Write λ_+ for the positive eigenvalue and $-E_i^-$ for the restriction to the negative eigenspace so that E_i^- is positive definite, and decompose $g = g_+ + g_-$, $y = y_+ + y_-$ by projection into these subspaces. Then we have

$$G \triangleq E_i(g,g) = \lambda_+ g_+^2 - E_i^-(g_-,g_-) \triangleq \lambda_+ g_+^2 - G_-$$
$$E_i(g,y) = \lambda_+ g_+ y_+ - E_i^-(g_-,y_-)$$
$$Y \triangleq E_i(y,y) = \lambda_+ y_+^2 - E_i^-(y_-,y_-) \triangleq \lambda_+ y_+^2 - Y_-$$

and our task is to obtain an upper bound on $Y_- = E_i^-(y_-,y_-)$, which will translate through our bound on the eigenvalues of E_i away from zero into the desired bound on $|y|$.

We begin by obtaining bounds on $|E_i(g,y)|$, G_-, G, and Y. Since $|z| \geq |\text{vol}(h,x,z)|$ for all z and $\text{vol}(h,x,z) = \sum_j z_j F_j(h,x)$, we have $|E_i(g,y)| = |F_i(h,x)| \leq 1$. Further, $\det\begin{pmatrix} E_i(g,g) & E_i(y,g) \\ E_i(g,y) & E_i(y,y) \end{pmatrix} < 0$ because E_i has signature $(1,1)$ restricted to the (y,g) plane, so by Lemma 3 $Y = E_i(y,y) < \frac{R^2}{\varepsilon_2}$.

Now by further calculation and the use of Lemma 4, the theorem follows; the details are left to the full paper [8] for brevity. □

Three small lemmas used above follow from the geometry of spherical polygons and of generalized convex dual polyhedra. Their proofs are left to the full paper [8] for brevity.

Lemma 2. $|x|^2 \leq (4n/\varepsilon_3^2) \max_i |\pi_i x|^2$.

Lemma 3. $F_i(h,h) > \varepsilon_2/R^2$.

Lemma 4. The inverse of the form E_i is bounded by $|E_i^{-1}| = O(n/\varepsilon_4)$.

5   Bounding the Hessian

In order to control the error in each step of our computation, we need to keep the Jacobian $\mathbf{J}$ along the whole step close to the value it started at, on which the step was based. To do this we bound the Hessian $\mathbf{H}$ when the triangulation is fixed, and we show that the Jacobian does not change discontinuously when changing radii force a new triangulation.

Each curvature κ_i is of the form $2\pi - \sum_{j,k:v_iv_jv_k \in T} \angle v_j O v_i v_k$, so in analyzing its derivatives we focus on the dihedral angles $\angle v_j O v_i v_k$. When the tetrahedron $O v_i v_j v_k$ is embedded in $\mathbb{R}^3$, the angle $\angle v_j O v_i v_k$ is determined by elementary geometry as a smooth function of the distances among O, v_i, v_j, v_k. For a given triangulation T this makes κ a smooth function of r. Our first lemma shows that no error is introduced at the transitions where the triangulation $T(r)$ changes.

Lemma 5. *The Jacobian* $\mathbf{J} = \left(\frac{\partial \kappa_i}{\partial r_j}\right)_{ij}$ *is continuous at the boundary between radii corresponding to one triangulation and to another.*

Proof (sketch). The proof, which can be found in the full paper [8], uses elementary geometry to compare the figures determined by two triangulations near a radius assignment on their boundary.   □

It now remains to control the change in $\mathbf{J}$ as r changes within any particular triangulation, which we do by bounding the Hessian.

Theorem 4. *The Hessian* $\mathbf{H} = \left(\frac{\partial \kappa_i}{\partial r_j \partial r_k}\right)_{ijk}$ *is bounded in norm by* $O\left(n^{5/2} S^{14} R^{23} / (\varepsilon_5^3 d_0^3 d_1^8 D^{14})\right)$.

Proof. By direct computation and computer algebra. See the full paper [8] for the details.   □

6   Intermediate Bounds

Here we bound miscellaneous parameters in the computation in terms of the fundamental parameters $n, S, \varepsilon_1, \varepsilon_8$ and the computation-driving parameter ε_4.

Lemma 6. *Given a polyhedral metric space M, there exists a radius assignment r with curvature skew $\varepsilon_7 < \varepsilon_8/4\pi$, maximum radius $R = O(nD/\varepsilon_1\varepsilon_8)$, and minimum defect-curvature gap $\varepsilon_2 = \Omega(\varepsilon_1^2 \varepsilon_8^3 / n^2 S^2)$.*

Proof (sketch). Take $r_i = R$ for all i, with R sufficiently large. Then each κ_i is nearly equal to δ_i, so that ε_7 is small. For the quantitative bounds and a complete proof, see the full paper [8].   □

Two bounds on angles can be proven by elementary geometry; details are left to the full paper [8] for brevity.

Lemma 7. $\varepsilon_3 > \ell d_1 / R^2$.

Lemma 8. $\varepsilon_5 > \varepsilon_2 / 6S$.

Finally we bound O away from the surface M. The bounds are effective versions of Lemmas 4.8, 4.6, and 4.5 respectively of [2], and the proofs, left for brevity to the full paper [8], are similar but more involved.

Recall that d_2 is the minimum distance from O to any vertex of M, d_1 is the minimum distance to any edge of T, and d is the minimum distance from O to any point of M.

Lemma 9. $d_2 = \Omega\left(D\varepsilon_1\varepsilon_4\varepsilon_5^2\varepsilon_8 / (nS^4)\right)$.

Lemma 10. $d_1 = \Omega\left(D\varepsilon_1^2\varepsilon_4^4\varepsilon_5^6\varepsilon_8^2 / (n^2 S^{10})\right)$.

Lemma 11. $d_0 = \Omega\left(D\varepsilon_1^4\varepsilon_4^9\varepsilon_5^{12}\varepsilon_8^4 / (n^4 S^{22})\right)$.

Acknowledgments. We thank Jeff Erickson and Joseph Mitchell for helpful discussions about shortest paths on non-shortest-path triangulations.

References

1. Alexandrov, A.D.: Convex Polyhedra. Springer, Berlin (2005)
2. Bobenko, A.I., Izmestiev, I.: Alexandrov's theorem, weighted Delaunay triangulations, and mixed volumes. Annales de l'Institut Fourier (2006) arXiv:math.DG/0609447
3. Cauchy, A.L.: Sur les polygones et les polyèdres, seconde mémoire. J. École Polytechnique, XVIe Cahier, Tome IX, 113–148 (1813); OEuvres Complètes, IIe Sèrie, Paris, vol. 1, pp. 26–38 (1905)
4. Demaine, E.D., Demaine, M.L., Lubiw, A., O'Rourke, J.: Enumerating foldings and unfoldings between polygons and polytopes. Graphs Comb. 18, 93–104 (2002)
5. Demaine, E.D., O'Rourke, J.: Geometric Folding Algorithms. Cambridge University Press, Cambridge (2007)
6. Fedorchuk, M., Pak, I.: Rigidity and polynomial invariants of convex polytopes. Duke Math. J. 129, 371–404 (2005)
7. Glickenstein, D.: Geometric triangulations and discrete Laplacians on manifolds (2005) arXiv:math/0508188v1
8. Kane, D., Price, G.N., Demaine, E.D.: A pseudopolynomial algorithm for Alexandrov's theorem (2008) arXiv:0812.5030
9. Lubiw, A., O'Rourke, J.: When can a polygon fold to a polytope? Technical Report 048, Department of Computer Science, Smith College (1996); presented at Am. Math. Soc. Conf., October 5 (1996)
10. Mitchell, J.S.B., Mount, D.M., Papadimitriou, C.H.: The discrete geodesic problem. SIAM J. Comput. 16, 647–668 (1987)
11. Sabitov, I.: The volume of a polyhedron as a function of its metric. Fundam. Prikl. Mat. 2(4), 1235–1246 (1996)
12. Sabitov, I.: The volume of a polyhedron as a function of its metric and algorithmical solution of the main problems in the metric theory of polyhedra. In: International School-Seminar Devoted to the N. V. Efimov's Memory, pp. 64–65 (1996)
13. Sabitov, I.: The volume as a metric invariant of polyhedra. Discrete Comput. Geom. 20, 405–425 (1998)
14. Sechelmann, S.: Alexandrov polyhedron editor (2006),
   http://www.math.tu-berlin.de/geometrie/ps/software.shtml#AlexandrovPolyhedron

A Scheme for Computing Minimum Covers within Simple Regions

Matthew J. Katz and Gila Morgenstern*

Department of Computer Science, Ben-Gurion University of the Negev
Beer-Sheva 84105, Israel
{matya,gilamor}@cs.bgu.ac.il

Abstract. Let X be a simple region (e.g., a simple polygon), and let Q be a set of points in X. Let O be a convex object, such as a disk, a square, or an equilateral triangle. We present a scheme for computing a minimum cover of Q with respect to X, consisting of homothetic copies of O. In particular, a minimum disk cover of Q with respect to X, can be computed in polynomial time.

1 Introduction

Let X be a simple region (e.g., a simple n-gon), and let Q be a set of m points in X. A *disk cover of Q with respect to X* is a set $\mathcal{D}$ of disks (of variable radii), such that the union of the disks of $\mathcal{D}$ covers (i.e., contains) Q and is contained in X. In other words, (i) each disk $D \in \mathcal{D}$ is contained in X, and (ii) each point $q \in Q$, lies in a disk $D \in \mathcal{D}$. A *minimum disk cover of Q with respect to X* is a disk cover of Q with respect to X of minimum cardinality. In this paper, we study the problem of computing such a cover.

The problem of computing a minimum disk cover of a point set Q with respect to a simple region X arises, e.g., in the following setting. X represents a secured area, and each point of Q represents a client of a radio network. One must place the smallest possible number of transmitters, such that each client is served by at least one of the transmitters (i.e., is within the transmission range of at least one of the transmitters), and any point outside the area, is outside the range of each of the transmitters.

Geometric covering problems have been studied extensively. These problems are instances induced by geometric settings of the well-known set cover problem. Most of them are known to be NP-hard. Below, we mention several geometric covering problems that are related to the problems studied in this paper. In the unit disk cover problem, the goal is to cover a given set of points with the smallest possible number of unit disks. A polynomial-time approximation scheme for this problem was given by Hochbaum and Maas [12]. In the discrete version of this problem, the covering unit disks must come from a given set of unit disks. This version is apparently more difficult that the non-discrete version, and only constant-factor approximation algorithms are known; see [2,3,4,22].

---

* Partially supported by the Lynn and William Frankel Center for Computer Sciences.

F. Dehne et al. (Eds.): WADS 2009, LNCS 5664, pp. 447–458, 2009.

The problem of covering a set Q of points by a single disk that is contained in X (if possible), arises as a sub-problem in the study of our minimum disk cover problem. Hurtado et al. [13] studied the related problem of computing a minimum enclosing disk of a given set of m points, whose center must lie in a given convex n-gon; they presented an $O(m + n)$ time algorithm for this problem. The 2-center problem with obstacles was studied by Halperin et al. [11]. In this problem, the goal is to find two congruent disks of smallest radius whose union covers a given set of m points and whose centers lie outside a given set of disjoint simple polygons with a total of n edges. They presented a randomized $O(n \log^2(mn) + mn \log^2 m \log(mn))$ expected time algorithm for this problem. The analogous 1-center problem was studied by Halperin and Linhart [10], who presented an $O((m + n) \log(mn))$ time algorithm for this problem.

In the context of wireless networks, one often wants to minimize the sum of the radii (alternatively, the radii to some power $\alpha > 1$) of the covering disks. Alt et al. [1] gave exact and approximation algorithms for this problem, where the centers of the covering disks must lie on a given line.

Although the problem of computing a minimum disk cover with respect to, e.g., a simple polygon seems quite natural, we are not aware of any previous work on this problem. In Sect. 2, we describe an algorithm that computes such a cover in time polynomial in n and m. Our solution uses the "perfect graph approach." In the perfect graph approach, first, a graph G corresponding to the input scene is defined. Next, the following two theorems are proven: (i) there is a one to one correspondence between a minimum cover of the desired kind (e.g., disk cover or cover by visibility polygons) and a minimum clique cover of G, and (ii) G is perfect. Note that the second claim is crucial, since, in general, minimum clique cover is NP-complete, but is polynomial for chordal and perfect graphs [7,8,9]. The perfect graph approach was used in the solution of several art-gallery problems, under restricted models of visibility; see, e.g., [16,17,20,21,23].

In Sect. 3, we extend the results of Sect. 2 to any fixed convex shape. More precisely, we describe a scheme for computing (in time polynomial in n and m) a minimum cover of Q with respect to a simple region X, where the covering objects are homothetic copies of a given convex object of constant description complexity. This implies, for example, that a minimum cover consisting of axis-parallel squares or equilateral triangles with a horizontal edge can be computed in polynomial time.

2   Minimum Disk Cover

Given a simple polygon X and a set Q of points in X, we define the graph $G_o = \langle V, E \rangle$, such that $V = Q$, and
$E = \{(q_1, q_2) \in V \times V : \text{there exists a disk } D \subseteq X \text{ s.t. } q_1, q_2 \in D\}$. Note that if there exists a disk $D \subseteq X$ covering both q_1 and q_2, then there also exists such a disk for which q_1 and q_2 are on its boundary.

The following lemma is trivially true.

Lemma 1. *Let x, y, z be three points in the plane, and let D be the disk defined by these points (i.e., $x, y, z \in \partial D$). Then, for any disk D' such that $x, y \in \partial D'$, either D' covers z, or D' covers the arc $\widehat{xy}$ of D (that does not include z).*

Lemma 2. *Let q_1, q_2, q_3 be three points in Q, such that for each pair of them q_i, q_j, $i < j$, there exists a disk $D_{ij} \subseteq X$ with $q_i, q_j \in \partial D_{ij}$. Then, there exists a disk $D \subseteq X$, such that $q_1, q_2, q_3 \in D$.*

Proof. If one of the three disks D_{12}, D_{13}, D_{23} covers all three points, then let D be this disk and we are done. Otherwise, let D be the disk defined by q_1, q_2, q_3, and consider its three arcs $\widehat{q_1 q_2}, \widehat{q_1 q_3}, \widehat{q_2 q_3}$. By Lemma 1, since D_{ij} does not cover the third point, it must cover $\widehat{q_i q_j}$. Thus, $\partial D \subseteq D_{12} \cup D_{13} \cup D_{23}$, and since X is simple, it follows that $D \subseteq X$.

Lemma 2 above states that any three points corresponding to a triangle of G_o, can be covered by a single disk that is contained in X. Lemma 3 below states that this is also true for cliques of size greater than 3. We use Helly's Theorem in order to prove it, but before then we prove a simple claim.

Let c, q be two points in the plane. We denote by $D_c(q)$ the disk centered at c with q on its boundary. We denote by $\mathcal{DC}(q)$ the set of centers of all disks that are contained in X and cover q. Notice that $D_c(q)$ is contained in all disks centered at c and covering q, thus $\mathcal{DC}(q) = \{c : D_c(q) \subseteq X\}$.

Claim. Let $q \in Q$. Then, $\mathcal{DC}(q)$ is convex.

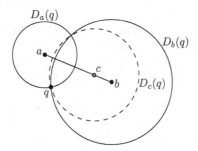

Fig. 1. $D_c(q)$ is contained in $D_a(q) \cup D_b(q)$.

Proof. Let $a, b \in \mathcal{DC}(q)$, and let c be any point on the line segment between a and b. We show that $c \in \mathcal{DC}(q)$. By definition, $D_a(q), D_b(q) \subseteq X$. Consider the disk $D_c(q)$. Clearly, $D_c(q) \subseteq D_a(q) \cup D_b(q) \subseteq X$ (see Fig. 1), and therefore $c \in \mathcal{DC}(q)$.

Lemma 3. *Let C be a subset of Q corresponding to a clique of G_o. Then, there exists a disk $D \subseteq X$, such that $C \subseteq D$.*

Proof. If $|C| = 2$ or $|C| = 3$, then the statement is true by definition or by Lemma 2, respectively. Assume therefore that $|C| \geq 4$. Consider the collection of all convex sets $\mathcal{DC}(q)$, where $q \in C$. We claim that the intersection of any three sets of this collection is nonempty, and therefore, by Helly's Theorem, there exists a point $c \in \bigcap_{q \in C} \mathcal{DC}(q)$. Now, let $q \in C$ be the farthest from c among all points of C, then the disk $D_c(q)$ is contained in X and contains C.

Indeed, let $x, y, z \in C$. By Lemma 2, there exists a disk $D \subseteq X$ such that $x, y, z \in D$, and, by definition, D's center belongs to each of the sets $\mathcal{DC}(x), \mathcal{DC}(y), \mathcal{DC}(z)$.

Theorem 1. *A minimum clique cover of G_o corresponds to a minimum disk cover of Q with respect to X.*

Proof. Let $\mathcal{D}$ be a disk cover of Q with respect to X. Then, clearly, for each $D \in \mathcal{D}$, the subset $Q \cap D$ corresponds to a clique of G_o.

Now, let $\mathcal{C}$ be a clique cover of G_o, and let $C \in \mathcal{C}$. By Lemma 3, the subset of Q corresponding to C can be covered by a single disk $D \subseteq X$.

Theorem 1 guarantees that finding a minimum clique cover of G_o is sufficient. Our next goal is to prove Theorem 2 below that states that G_o is chordal, implying that a minimum clique cover of G_o can be found in time linear in the size of G_o; see [7,8]. We shall need the following two lemmas, where Lemma 4 is a more general version of Lemma 1.

Lemma 4. *Let a, b, c, d be four consecutive points on a line l, and let D be a disk whose boundary passes through the middle points b, c. Then, any disk whose boundary passes through a, d covers at least one of the arcs $\overset{\frown}{bc}$ and $\overset{\frown}{cb}$ of D.*

Proof. Let D' be a disk whose boundary passes through the points a, d. If $D \subseteq D'$, then we are done. Otherwise, the boundaries of D and D' must cross each other twice, and, since $\overline{bc} \subseteq \overline{ad} \subset D'$, these crossing points lie on the same side of l. Therefore, D' covers the arc of D on the other side of l.

Lemma 5. *Let $s_a = \overline{a_1 a_2}$ and $s_b = \overline{b_1 b_2}$ be two segments such that $s_a \cap s_b \neq \emptyset$, and let D_a and D_b be disks, such that $a_1, a_2 \in \partial D_a$ and $b_1, b_2 \in \partial D_b$. Then, at least one of the following is true: $a_1 \in D_b$, $a_2 \in D_b$, $b_1 \in D_a$, $b_2 \in D_a$.*

Proof. If one of the four endpoints, e.g., a_1, lies on the other segment, then, $a_1 \in s_b \subseteq D_b$ and we are done. Otherwise, put $x = s_a \cap s_b$. Clearly, $x \in D_a \cap D_b$, since $s_a \subseteq D_a$ and $s_b \subseteq D_b$. Also, since x is not an endpoint of one of the segments, we know that x is not on D_a's boundary or D_b's boundary. Therefore, ∂D_a and ∂D_b cross each other twice, and let l be the line through these two crossing points.

Assume w.l.o.g. that l is vertical, and that D_a's center is to the left of D_b's center. Assume, e.g., that x lies to the right of l. Then, at least one of s_a's endpoints lies on the arc of D_a that is covered by D_b (see Fig. 2); thus, there exists $i \in \{1, 2\}$ such that $a_i \in D_b$.

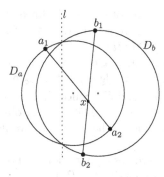

Fig. 2. $a_2 \in \partial D_a \cap D_b$, thus D_b covers the points b_1, b_2, a_2

Theorem 2. G_o *is chordal.*

Proof. Let $\langle 0, 1, 2, 3, \ldots, k-1 \rangle$ be a cycle of G_o of length at least 4, and let H be the subset of Q corresponding to this cycle. There exist disks D_0, D_1, ..., D_{k-1}, such that $D_i \subseteq X$ and $i, i+1 \pmod{k} \in \partial D_i$, for $i = 0, 1, \ldots, k-1$. If there exists $0 \leq i \leq k-1$ for which D_i covers three or more points of H, then the cycle contains a chord and we are done.

Otherwise, let π be the closed path that is obtained by connecting the point i with the point $i+1 \pmod{k}$, for $i = 0, 1, \ldots, k-1$. We claim that π must be simple. Otherwise, there are two cases: (i) There exists $i \in H$ that lies in the interior of an edge $[j, j+1]$ of π (where $i \neq j, j+1$), but then $i \in D_j$ and D_j covers three points of H. (ii) There exist two edges $e_i = [i, i+1]$ and $e_j = [j, j+1]$ of π that cross each other, where $i, i+1, j, j+1$ are all distinct, and by Lemma 5 we have that either D_i or D_j covers at least three of the points $i, i+1, j, j+1$.

Consequently, π defines a simple polygon $P_\pi \subseteq X$ (see Fig. 3(a)). Consider the Delaunay triangulation $\mathcal{DT}(H)$ of H. One of the well-known properties of $\mathcal{DT}(H)$ is that (i, j) is a Delaunay edge if and only if there exists a disk D, such

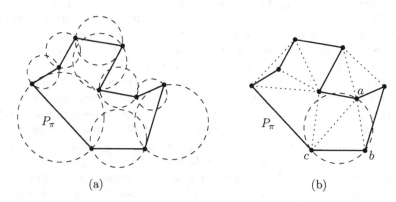

(a)   (b)

Fig. 3. (a) The simple polygon P_π. Each edge of P_π is an edge of $\mathcal{DT}(H)$. (b) $\mathcal{DT}(H)$, the triangle $\triangle_{abc}$ and its circumcircle.

that $i, j \in \partial D$ and $l \notin D$, for any other $l \in H$. Since we are assuming that none of the disks $D_0, D_1, \ldots, D_{k-1}$ covers a third point of H, we conclude that each edge of P_π is also an edge of $\mathcal{DT}(H)$.

Now, consider any Delaunay triangle $\triangle_{abc} \subseteq P_\pi$. (Such a triangle exists since all edges of P_π are also Delaunay edges.) Let D_{abc} be $\triangle_{abc}$'s circumcircle, and consider its three arcs $\widehat{a\,b}$, $\widehat{b\,c}$, and $\widehat{c\,a}$. Since $\triangle_{abc}$ is a Delaunay triangle, D_{abc} does not cover any other point of H. Therefore, if one of these arcs crosses an edge of P_π, it must cross it twice (otherwise D_{abc} would contain a vertex of P_π other than a, b, c).

The points at which ∂D_{abc} crosses P_π partition the boundary of D_{abc} into m arcs (see Fig. 3(b)). An arc that is contained in P_π is clearly also contained in X. And, an arc that is contained in the exterior of P_π has both its endpoints on the same edge of P_π, and, by Lemma 4, such an arc is covered by the disk corresponding to that edge and therefore is also contained in X. We showed that ∂D_{abc} is contained in X and therefore, since X is simple, $D_{abc} \subseteq X$, and a chord exists.

Theorem 3. *Let X be a simple n-gon, and let $Q \subset X$ be a set of m points. Then, a minimum disk cover of Q with respect to X can be computed in time $O(nm^2)$.*

Proof. We need to bound the time needed to construct $G_\circ$, and the time needed to compute a disk corresponding to a clique of $G_\circ$. Since $G_\circ$ is chordal, a minimum clique cover of $G_\circ$ can be computed in time $O(|V|+|E|) = O(m^2)$; see [7,8].

We shall use the medial axis of X, which can be computed in time $O(n)$; see [6]. Consider first the task of constructing $G_\circ$. For each pair of points in Q, we need to determine whether there exists a disk contained in X that covers both points. Let $q_1, q_2 \in Q$. By the observation just above Lemma 1, it is enough to determine whether there exists a disk $D \subseteq X$, such that $q_1, q_2 \in \partial D$. In other words, it is enough to determine whether there exists a point c on bisector(q_1, q_2), such that the disk centered at c and whose boundary passes through q_1 and q_2 is contained in X. This can be done in linear (in n) time as follows. The edges of the medial axis of X partition bisector$(q_1, q_2) \cap X$ into $O(n)$ segments. For each of these segments, one can determine in constant time whether there exists a point c on it, such that the distance between c and the points q_1, q_2 is not greater than the distance between c and the edge of X that is closest to the segment. Thus, $G_\circ$ can be constructed in time $O(nm^2)$.

Consider now the task of computing a disk corresponding to a clique of $G_\circ$. Let Q' be a subset of Q corresponding to a clique of $G_\circ$. By Lemma 3, there exists a disk $D \subseteq X$, such that $Q' \subset D$. We now describe how to compute such a disk in time $O(n|Q'|)$. Notice that we may require that D's boundary pass through at least two points of Q'. In other words, we may require that D's center lie on an edge of the farthest site Voronoi diagram of Q'. Thus, we compute this diagram, whose size is $O(|Q'|)$. Each edge of this diagram, is partitioned into $O(n)$ segments by the boundary of X and by the edges of the medial axis of X. Let s be one of the $O(n|Q'|)$ resulting segments. Since s is contained in an edge of the farthest site Voronoi diagram, there exist two points $q_1, q_2 \in Q'$, such that, for any point c on s, $\text{dist}(c, q_1) = \text{dist}(c, q_2) = \max_{q \in Q'} \text{dist}(c, q)$. Moreover, since s is contained in a single cell of the medial axis of X, there exists an edge e_s of X, such that, for any point c on s, $\text{dist}(c, e_s) = \min\{\text{dist}(c, e) : e \text{ an edge of } X\}$. Thus, one can determine in constant time whether there exists a point c on s, such that the disk centered at c of radius $\text{dist}(c, q_1) = \text{dist}(c, q_2)$ is contained in X. We conclude that a disk corresponding to Q' can be computed in time $O(n|Q'|)$, and therefore a disk cover corresponding to a given clique cover of $G_\circ$ can be computed in time $O(nm)$.

3 Minimum O-Cover

Let X and Q be as in Sect. 2. In this section, we show that the main result of Sect. 2 can be generalized to any convex shape of constant description complexity. More precisely, we show that for any convex object O (of constant description complexity), one can compute, in polynomial time, a minimum cover of Q, using only positive homothetic copies of O that are contained in X. (Recall that a homothetic copy O_h of O is a scaled and translated copy of O.)

Given such a convex object O, we define the graph $G_* = \langle V, E \rangle$, such that $V = Q$, and $E = \{(q_1, q_2) \in V \times V : \text{there exists } O_h \subseteq X \text{ s.t. } q_1, q_2 \in O_h\}$. As in Sect. 2, we need to prove that (i) a minimum clique cover of G_* corresponds to a minimum cover of Q with respect to X, using homothetic copies of O, and (ii) G_* is chordal.

It is tempting to try to prove Lemma 7 below (the subset of points corresponding to a clique of G_* can be covered by a single homothetic copy of O), in a similar way to the proof of Lemma 3; namely, by applying Helly's Theorem. However, this will fail, since, in general, the sets $\mathcal{DC}(q)$ are not convex; Fig. 4 shows that the set of all centers of (axis-parallel) squares that are contained in X and cover q, is not convex. Our proof of Lemma 7 is based on the pseudo-disks property of the set of all homothetic copies of O. That is, for any pair of homothetic copies of O, their boundaries are either disjoint, or intersect at exactly two points, or a few other degenerate cases. This proof also applies to the case where O is a disk, so it is an alternative proof for Lemma 3.

Before proving Lemma 7, we consider (in Lemma 6 below) the case where a set C of points cannot be covered by a homothet of O having three points of C on its boundary; see Fig. 5. Note that there always exists a homothet of O covering C with two points on its boundary (assuming, of course, that $|C| \geq 2$).

Lemma 6. *Let C be a set of points in the plane, and let O_h be a homothet of O covering C and maximizing the number of points of C lying on its boundary. If $|C \cap \partial O_h| = 2$, then any homothet O'_h of O whose boundary passes through the points of $C \cap \partial O_h$, covers C.*

Proof. Put $C \cap \partial O_h = \{p_1, p_2\}$, and assume that there exists a homothet O'_h of O, such that $p_1, p_2 \in \partial O'_h$ but $C \not\subseteq O'_h$. Put $C' = C \setminus O'_h = \{q_1, q_2, \ldots, q_k\}$, and assume, e.g., that O_h covers the portion of $\partial O'_h$ between p_1 and p_2, moving

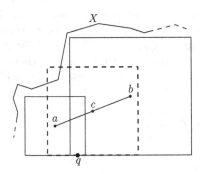

Fig. 4. There exist two squares centered at a and b, respectively, that cover q and are contained in X. However, any square centered at $c \in \overline{ab}$ and covers q, is not contained in X.

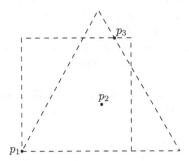

Fig. 5. There exist no axis-parallel square or equilateral triangle with a horizontal edge whose boundary passes through p_1, p_2, p_3

clockwise from p_1 to p_2. We show that there exists a homothet of O that covers C and has three points of C on its boundary, a contradiction.

For each $1 \leq i \leq k$, we have that $q_i \in O_h$ but $q_i \notin O'_h$. Also O_h and O'_h are homothets whose boundaries intersect at p_1 and p_2, thus, by continuity, there exists, for each $1 \leq i \leq k$, a homothet O^i_h such that $p_1, p_2, q_i \in \partial O^i_h$ (i.e., the points p_1, p_2, q_i are "non-collinear"). Consider the set of homothets $\{O^1_h, O^2_h, \ldots, O^k_h\}$ (see Fig. 6). This is a set of pseudo-disks, and the boundaries of each pair of them intersect at p_1, p_2. For each $1 \leq i < j \leq k$, we have that either O^i_h covers q_j, or O^j_h covers q_i. Moreover, this relation is transitive; that is, if O^i_h covers q_j, then it also covers all points of C' that are covered by O^j_h. Thus, there exists $1 \leq i \leq k$ for which $O^i_h \supseteq C'$, and assume, e.g., that $i = 1$.

We claim that $C \subseteq O^1_h$. Indeed, on the one hand, we have that since $q_1 \in O_h$, O^1_h covers the counterclockwise portion of O_h from p_1 to p_2, on the other hand, since $q_1 \notin O'_h$, O^1_h covers the clockwise portion of $\partial O'_h$ from p_1 to p_2. We conclude that O^1_h covers $O_h \cap O'_h$, and, as $C \setminus C' \subseteq O_h \cap O'_h$, we are done.

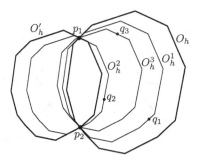

Fig. 6. For each point $q_i \in C'$, there exists a homothet O_h^i, such that $p_1, p_2, q_i \in \partial O_h^i$

Lemma 7. *Let C be a subset of Q corresponding to a clique of G_*. Then, there exists a homothet O_h of O, such that $C \subseteq O_h \subseteq X$.*

Proof. The proof is by induction on $|C|$. If $|C| = 2$, then the theorem is true by definition. Assume $|C| \geq 3$, and let O_h be a homothet of O that covers C and maximizes the number of points of C lying on its boundary. By Lemma 6, if $|C \cap \partial O_h| = 2$, then any homothet of O whose boundary passes through the two points of $C \cap \partial O_h$, covers C. In particular, the one that is associated with these two points by G_*'s construction.

Assume therefore that $|C \cap \partial O_h| \geq 3$, and put $C \cap \partial O_h = \{p_1, p_2, \ldots, p_k\}$. Let $C_i = C \setminus \{p_i\}$, for $i = 1, \ldots, k$. By the induction's hypothesis, there exists a homothet O_h^i of O, such that $C_i \subseteq O_h^i \subseteq X$, for $i = 1, \ldots, k$. Denote by $\widehat{i}$ the portion of ∂O_h that is between the two points on ∂O_h that are adjacent to p_i (and includes p_i). Now, if for some $1 \leq i \leq k$, $\widehat{i} \subseteq O_h^i$, then O_h^i covers p_i as well, and we are done. Otherwise, for each $1 \leq i \leq k$, we have that O_h^i covers the complement of $\widehat{i}$ (including the endpoints of $\widehat{i}$). Thus, $\partial O_h \subseteq \bigcup_{1 \leq i \leq k} O_h^i$, and, as X is simple, $O_h \subseteq X$.

Consider now the chordality proof. The proof of Theorem 2 above relies on two well-known properties of the Delaunay triangulation (the dual graph of the Voronoi diagram): Let P be a set of points in the plane, and let $\mathcal{DT}(P)$ be its Delaunay triangulation. Then, (i) $\overline{pq}$ is an edge of $\mathcal{DT}(P)$ if and only if there exists a closed disk C, such that $p, q \in \partial C$ and C does not cover any other point of P, and (ii) $\triangle_{pqr}$ is a triangle of $\mathcal{DT}(P)$ if and only if the closed disk defined by p, q, r does not cover any other point of P. Usually, the Voronoi diagram is defined under the Euclidean metric (as above), but it can be defined under any L_p-metric, for $1 \leq p \leq \infty$; see [14,18,19]. The properties stated above hold for any $1 \leq p \leq \infty$, just replace disk by L_p-disk (e.g., diamond for $p = 1$ and axis-parallel square for $p = \infty$).

Chew and Drysdale [5] studied Voronoi diagrams under convex distance functions, and showed that similar properties hold. Let O be a convex object and let O' be the reflection of O about its center. Then, the O'-Delaunay triangulation $\mathcal{DT}_{O'}(P)$ of P has the following properties: (i) $\overline{pq}$ is an edge of $\mathcal{DT}_{O'}(P)$ if and

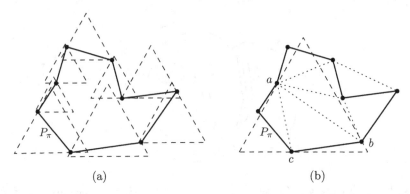

(a)   (b)

Fig. 7. Analogous to Fig. 3, where O is an equilateral triangle with a horizontal edge

only if there exists a homothetic copy O_h of O, such that p, q lie on its boundary and O_h does not cover any other point of P, and (ii) $\triangle_{pqr}$ is a triangle of $\mathcal{DT}_{O'}(P)$ if and only if there exists a homothetic copy O_h of O, such that p, q, r lie on its boundary and O_h does not cover any other point of P.

Thus, one can prove that G_* is chordal by adapting the proof of Theorem 2. Disks are replaced by homothetic copies of O, and the standard Delaunay triangulation is replaced by the O'-Delaunay triangulation. (Figure 7 is analogous to Fig. 3 above, where O is an equilateral triangle with a horizontal edge.) As for the corresponding versions of Lemmas 4 and 5, note that the existing proofs already hold for any set of pseudo-disks, and, in particular, for the set of homothetic copies of O. We conclude that

Theorem 4. G_* *is chordal.*

It remains to show how G_* is constructed, and how a homothetic copy of O corresponding to a clique of G_* is found. In principle, we adapt the algorithms described in the proof of Theorem 3, using appropriate generalizations of the medial-axis and of the farthest site Voronoi diagram.

Theorem 5. *Let X be a simple n-gon, and let $Q \subset X$ be a set of m points. Let O be a convex object of constant description complexity. Then, a minimum cover of Q with respect to X, using homothetic copies of O, can be computed in time polynomial in n and m.*

4   Additional Remarks

We conclude with two remarks.

Remark 1. Since MAX-CLIQUE (i.e., find a clique of G_* of maximum size) is also solvable in time $O(|V|+|E|)$ for chordal graphs [7,8], we obtain the following corollary. A homothetic copy O_h of O that is contained in X and covers the largest possible number of points of Q, can be found in polynomial time. (In the case of L_2-disks, the bound implied by Theorem 3 is $O(nm^2)$ or $O(nm + m^2)$ in case G_o is already given).

Remark 2. We assumed that X is a simple polygon, but, in fact, it could be defined by any simple closed curve. Moreover, our results also hold if X is a set of rays or semi-infinite curves. For example, if X is a set of possibly intersecting rays, then we wish to cover Q with the smallest possible number of disks that are not intersected by rays in X.

References

1. Alt, H., Arkin, E.M., Brönnimann, H., Erickson, J., Fekete, S.P., Knauer, C., Lenchner, J., Mitchell, J.S.B., Whittlesey, K.: Minimum-cost coverage of point sets by disks. In: Proc. 22nd Sympos. Comput. Geom., pp. 449–458 (2006)
2. Brönnimann, H., Goodrich, M.T.: Almost optimal set covers in finite VC-dimension. Discrete & Comput. Geom. 14, 463–479 (1995)
3. Calinescu, G., Mandoiu, I.I., Wan, P.-J., Zelikovsky, A.: Selecting forwarding neighbors in wireless ad hoc networks. MONET 9(2), 101–111 (2004)
4. Carmi, P., Katz, M.J., Lev-Tov, N.: Covering points by unit disks of fixed location. In: Proc. 18th Int. Sympos. on Algorithms and Computation, pp. 644–655 (2007)
5. Chew, L.P., Drysdale, R.L.: Voronoi diagrams based on convex distance functions. In: Proc. 1st Sympos. Comput. Geom., pp. 235–244 (1985)
6. Chin, F., Snoeyink, J., Wang, C.A.: Finding the medial axis of a simple polygon in linear time. Discrete & Comput. Geom. 21(3), 405–420 (1999)
7. Gavril, F.: Algorithms for minimum coloring, maximum clique, minimum covering by cliques, and maximum independent set of a chordal graph. SIAM J. on Computing 1, 180–187 (1972)
8. Golumbic, M.C.: Algorithmic Graph Theory and Perfect Graphs, 1st edn. Academic Press, New York (1980); 2nd edn. Annals of Discrete Mathematics 57, Elsevier, Amsterdam (2004)
9. Grötschel, M., Lovász, L., Schrijver, A.: Polynomial algorithms for perfect graphs. In: Berge, C., Chvátal, V. (eds.) Topics on perfect graphs. Annals of Discrete Mathematics, vol. 21, pp. 325–356. North-Holland, Amsterdam (1984)
10. Halperin, D., Linhart, C.: The minimum enclosing disk with obstacles (manuscript) (1999)
11. Halperin, D., Sharir, M., Goldberg, K.: The 2-center problem with obstacles. J. Algorithms 42, 109–134 (2002)
12. Hochbaum, D.S., Maas, W.: Approximation schemes for covering and packing problems in image processing and VLSI. J. ACM 32, 130–136 (1985)
13. Hurtado, F., Sacristán, V., Toussaint, G.: Some constrained minimax and maximin location problems. Studies in Locational Analysis 15, 17–35 (2000)
14. Hwang, F.K.: An $O(n \log n)$ algorithm for rectilinear minimal spanning trees. J. ACM 26(2), 177–182 (1979)
15. Johnson, D.S.: The NP-completeness column: An ongoing guide. J. Algorithms 3, 182–195 (1982)
16. Katz, M.J., Morgenstern, G.: Guarding orthogonal art galleries with sliding cameras. In: Proc. 25th European Workshop on Comput. Geom., pp. 159–162 (2009)
17. Katz, M.J., Roisman, G.S.: On guarding the vertices of rectilinear domains. Comput. Geom. Theory Appl. 39(3), 219–228 (2008)
18. Lee, D.T.: Two-dimensional Voronoi diagrams in the L_p-metric. J. ACM 27(4), 604–618 (1980)

458   M.J. Katz and G. Morgenstern

19. Lee, D.T., Wong, C.K.: Voronoi diagrams in L_1 (L_∞) Metrics with 2-dimensional storage applications. SIAM J. on Computing 9(1), 200–211 (1980)
20. Motwani, R., Raghunathan, A., Saran, H.: Covering orthogonal polygons with star polygons: The perfect graph approach. J. of Computer and System Sciences 40, 19–48 (1990)
21. Motwani, R., Raghunathan, A., Saran, H.: Perfect graphs and orthogonally convex covers. SIAM J. on Discrete Math. 2(3), 371–392 (1989)
22. Narayanappa, S., Vojtechovsky, P.: An improved approximation factor for the unit disk covering problem. In: Proc. 18th Canadian Conf. Comput. Geom., pp. 15–18 (2006)
23. Worman, C., Keil, J.M.: Polygon decomposition and the orthogonal art gallery problem. Int. J. Comput. Geom. Appl. 17(2), 105–138 (2007)

Better Approximation Algorithms for the Maximum Internal Spanning Tree Problem

Martin Knauer and Joachim Spoerhase

Lehrstuhl für Informatik I, Universtität Würzburg Am Hubland,
97074 Würzburg, Germany
spoerhase@informatik.uni-wuerzburg.de

Abstract. We examine the problem of determining a spanning tree of a given graph such that the number of internal nodes is maximum. The best approximation algorithm known so far for this problem is due to Prieto and Sloper and has a ratio of 2. For graphs without pendant nodes, Salamon has lowered this factor to $\frac{7}{4}$ by means of local search. However, the approximative behaviour of his algorithm on general graphs has remained open. In this paper we show that a simplified and faster version of Salamon's algorithm yields a $\frac{5}{3}$-approximation even on general graphs. In addition to this, we investigate a node weighted variant of the problem for which Salamon achieved a ratio of $2 \cdot \Delta(G) - 3$. Modifying Salamon's approach we obtain a factor of $3 + \epsilon$ for any $\epsilon > 0$. We complement our results with worst case instances showing that our bounds are tight.

1 Introduction and Preliminaries

The MAXIMUM INTERNAL SPANNING TREE problem (MAXIST) consists in finding a spanning tree of a given graph such that the number of internal nodes is maximized. MAXIST is a natural optimization version of the HAMILTONIAN PATH problem and can be motivated by the design of cost-efficient communication networks [1]. It is closely related to the MINIMUM LEAF SPANNING TREE problem (MINLST) which asks for a spanning tree with a minimum number of leaves. Although MINLST and MAXIST lead to the same constructive problem they have different approximability properties.

Lu and Ravi [2] introduced MINLST and showed that it admits no constant factor approximation algorithm unless some NP-hard problem can be solved in deterministic quasi-polynomial time. Flandrin et al. [3] investigated conditions for the existence of spanning trees with few leaves. Interestingly, the aforementioned non-approximability result for MINLST does *not* carry over to MAXIST since the latter has a different objective function. Indeed, Prieto and Sloper [4] gave an efficient 2-approximation algorithm for MAXIST based on local search. Later, Salamon and Wiener [1] developed a simple linear time algorithm, which is basically a slight modification of depth first search, and showed that it yields a 2-approximation, too. Moreover, they obtained that MAXIST is approximable within factors $\frac{3}{2}$ and $\frac{6}{5}$ on claw-free and cubic graphs, respectively. The fixed parameter complexity of MAXIST was studied by Prieto and Sloper [4,5]. They

F. Dehne et al. (Eds.): WADS 2009, LNCS 5664, pp. 459–470, 2009.

gave an $O(k^2)$ kernel which proves that the problem is fixed parameter tractable. Here, k denotes the number of internal nodes. Fernau et al. [6] presented exact exponential time algorithms. In particular they gave $O^*(3^n)$ and $O^*(1.8916^n)$ algorithms for solving MAXIST on general graphs and graphs of maximum degree three, respectively. Here, the O^*-notation disregards polynomial factors.

The existence of efficient approximation algorithms with a ratio better than 2 has been an open problem so far [1]. Recently some progress was made towards answering this question: Salamon [7] suggested a local search based algorithm for MAXIST and proved by means of linear programming techniques that it terminates with a $\frac{7}{4}$-approximation on graphs without pendant nodes (nodes of degree 1). Although this algorithm applies also to arbitrary graphs, its approximative behaviour has remained open for the general case.

Using a different technique, we will show in this paper that Salamon's algorithm reaches a $\frac{5}{3}$-approximation even on *general graphs*. Thereby, we surpass the approximation ratio of 2 which is often a critical bound for combinatorial optimization problems. Moreover, it turns out that a substantially smaller neighborhood structure in the local search is sufficient to guarantee the approximation ratio. This leads to a better running time.

Additionally, Salamon [7] investigated a node weighted version of the problem (MAXWIST) for which he provided a local search algorithm with an approximation ratio of $2 \cdot \Delta(G) - 3$ where $\Delta(G)$ denotes the maximum degree of G. We will also tackle this case: By extending the neighborhood of Salamon's algorithm we are able to approximate MAXWIST within a factor of $3 + \epsilon$ for any $\epsilon > 0$.

The paper is organized as follows: In Section 2 we will present our algorithm LOSTLIGHT which constitutes a substantial simplification of the algorithm LOST proposed by Salamon. Section 3 is devoted to the analysis of the approximation ratio of our algorithm. Here we will prove the factor of $\frac{5}{3}$ which is the heart of this paper. The node weighted version of MAXIST will then be examined in Section 4. We will complete our results with suitable worst case instances in Section 5 demonstrating that the analyses of our algorithms are best possible.

Preliminaries. Let $G = (V, E)$ be an undirected, simple graph and G' be a subgraph of G. Then we denote by $E(G')$ the set of edges of G'. We say that v is a G'-*neighbor* of u, if u and v are adjacent in G'. If v is not a G'-neighbor of u then u and v are called G'-*independent*. By $d_{G'}(u)$ we denote the *degree* of u in G'. Let T be a subtree of G. Then any node u with degree $d_T(u) = 1$ is called a T-*leaf* or *leaf of* T. All other nodes of T are referred to as T-*internal nodes* or alternatively *internal nodes of* T. The set of leaves and the set of internal nodes of T are denoted by $L(T)$ and $I(T)$, respectively. By $P_T(u, v)$ we denote the unique path between nodes u, v in T and by $u^{\rightarrow v}$ the unique T-neighbor of u on $P_T(u, v)$. A T-*branching* is a node u in T with degree $d_T(u) \geq 3$. If T is not a path and l is a leaf of T then $b(l)$ is the unique branching which is closest to l in T. Moreover, l is is said to be x-*supported*, if node x does not lie on path $P_T(l, b(l))$ and x is G-adjacent to l. The *length* of a path is the number of its edges.

If $S' \subseteq S$ is a subset of an arbitrary set S weighted by a function $c \colon S \to \mathbb{Q}$ then let us use the notation $c(S') := \sum_{s \in S'} c(S')$.

Problem Definition. An instance of the MAXIMUM INTERNAL SPANNING TREE problem (MAXIST) is a connected, undirected graph $G = (V, E)$. The goal is to find a spanning tree T of G such that $|I(T)|$ is maximum among all spanning trees of G.

An instance of the MAXIMUM WEIGHTED INTERNAL SPANNING TREE problem (MAXWIST) is a connected, undirected graph $G = (V, E)$ whose nodes are weighted by a function $c \colon V \to \mathbb{Q}^+$. The goal is to find a spanning tree T of G such that $c(I(T))$ is maximum among all spanning trees of G.

2 The Algorithm

Salamon [7] suggests a local search algorithm, named LOST (locally optimal spanning tree), for approximating MAXIST. This algorithm maintains a spanning tree T which is initialized with a depth-first search tree of the input graph $G = (V, E)$. Thereafter, it performs a sequence of local improvement operations, taken from a fixed set of so called *rules*. Each of these rules consists of a precondition and an action which replaces edges of T with non-tree edges of G. The algorithm terminates when no more rule is applicable.

Our algorithm LOSTLIGHT applies the same framework but it employs only five of the 14 rules used by LOST. This is reflected in a running time reduced by a linear factor. In terms of the worst case ratio, we do not lose anything when dropping the additional rules of LOST, as we shall see in Section 5.

Let us adopt the following technical conventions: From now on, the notations $u^{\to v}$ and $b(l)$ will always be interpreted with respect to the spanning tree T maintained by LOSTLIGHT. This is particularly important to avoid confusions, since later we will also deal with trees different from T such as the global optimum T^*. Moreover, we will assume that T is not already a path which ensures that $b(l)$ is defined for every leaf of T.

LOSTLIGHT uses the following five rules (confer Figure 1):

Rule 1. If there are two G-adjacent leaves l_1, l_2 in T then add edge (l_1, l_2) and remove $(b(l_1), b(l_1)^{\to l_1})$.

Rule 2. If there is an x-supported leaf l in T such that $x^{\to l}$ is a branching in T then add (l, x) and remove $(x, x^{\to l})$.

Rule 3. If there are two T-leaves l_1, l_2 and a node x such that $x^{\to l_1}$ has degree two in T and edges $(l_1, x), (l_2, x^{\to l_1})$ are in $E - E(T)$ then add (l_1, x) remove $(x, x^{\to l_1})$. Afterwards apply Rule 1 to l_2 and the new leaf $x^{\to l_1}$.

Rule 4. If there is an x-supported leaf l in T such that $b(l)^{\to x}$ is a branching then add (l, x) and remove $(b(l), b(l)^{\to x})$.

Rule 5. If there is an x-supported leaf l_1 and a leaf l_2 such that $d_T(b(l_1)^{\to x}) = 2$ and $(l_2, b(l_1)^{\to x}) \in E - E(T)$ then add (l_1, x) and remove $(b(l_1), b(l_1)^{\to x})$. Afterwards apply Rule 1 to l_2 and the new leaf $b(l_1)^{\to x}$.

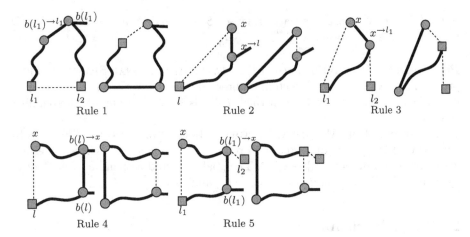

Fig. 1. The rules of LOSTLIGHT. Squares represent T-leaves. Thick and dashed straight lines mark edges of T and non-tree edges, respectively. Wiggly lines depict paths in T. Note that Rule 3 and 5 first generate constellations with T-adjacent leaves which allows for the application of Rule 1 afterwards.

It is easy to see that each of the Rules 1–5 increases the number of internal nodes in T. Thus the algorithm performs at most $|V|$ iterations. Each of the rules, in turn, can be carried out in $O(|V|^2)$ as shown in [7]. We obtain:

Lemma 1. *Algorithm* LOSTLIGHT *runs in time* $O(|V|^3)$.   □

By comparison, algorithm LOST needs time $O(|V|^4)$. This is basically because some of its additional rules do not discard leaves which may lead to overall $\Omega(|V|^2)$ iterations.

Since each of the above rules increases the number of internal nodes they correspond to improvements of the objective function of MaxIST. Thus LOSTLIGHT may be viewed as an implementation of the well known hill climbing algorithm where the neighborhood of T consists of all spanning trees obtained by executing any applicable rule on T.

3   The Analysis

In this section we will analyse the quality of spanning trees generated by LOST-LIGHT. We will show that each spanning tree T, which is locally optimal with respect to the Rules 1–5, is a $\frac{5}{3}$-approximation. More formally, if T^* is globally optimal then $|I(T^*)| \leq \frac{5}{3}|I(T)|$ holds.

3.1   Covered and Uncovered Leaves

Let T^* be an optimal solution for MaxIST and T an output of LOSTLIGHT. Let further r be an arbitrary internal node of T and consider T^* as rooted at r.

A T-leaf l is said to be *covered*, if there is a T-leaf l' such that l' is a descendant of l in T^*. Otherwise, l is called *uncovered*. We denote the set of covered and the set of uncovered leaves by C and U, respectively.

Consider an uncovered leaf l and the subtree T_l^* of T^* hanging from l. Since l is uncovered, T_l^* cannot contain any T-leaf except for l itself. Therefore, if l' is another uncovered leaf then T_l^* and $T_{l'}^*$ are disjoint. Since any subtree hanging from some uncovered leaf contains at least one T^*-leaf we conclude that $|U| \leq |L(T^*)|$. Hence

$$|I(T^*)| = (|I(T)| + |L(T)|) - |L(T^*)|$$
$$\leq |I(T)| + |L(T)| - |U|$$
$$= |I(T)| + |C|.$$

Dividing both hands by $|I(T)|$ we obtain the following central inequality:

$$\frac{|I(T^*)|}{|I(T)|} \leq 1 + \frac{|C|}{|I(T)|}. \tag{1}$$

Inequality (1) says that in order to obtain a good approximation we have to bound $|I(T)|$ from below in terms of $|C|$. Indeed, as a result of our analysis, we will show that $|I(T)| \geq \frac{3}{2}|C|$ which yields the desired approximation ratio of $\frac{5}{3}$.

The central idea of our proof consists in identifying for each covered leaf l a set $I(l)$ of distinctive internal nodes referred to as *companions of l*. For technical reasons we will divide the companions of l into sets $I_p(l)$ and $I_a(l)$ of so called *path companions* and *auxiliary companions*. Showing that the total number of companions is at least $\frac{3}{2}|C|$ we obtain our factor of $\frac{5}{3}$ by Inequality (1). The main difficulty will be that the companion sets $I(\cdot)$ need not be disjoint. To keep the overall set of companions from collapsing we will bound the *frequency* of single companions in the family of companion sets.

3.2   Path Companions

Our first step in identifying distinctive companions for any covered leaf l consists in assigning to l a unique subpath $P(l)$ of T^*. We will then single out certain internal nodes from $P(l)$ forming the set $I_p(l)$ of path companions of l. The path $P(l)$ will also play a crucial role in the definition of auxiliary companions in the next section.

Consider an arbitrary covered leaf l. We pick a T-leaf, denoted by $c(l)$, such that $c(l)$ is a descendant of l in T^* and there are no other T-leaves on path $P_{T^*}(l, c(l))$ except for l and $c(l)$. We set $P(l) := P_{T^*}(l, c(l))$. Clearly there may be more than one T-leaf satisfying the conditions imposed on $c(l)$ but we need only one of them. For an illustration of this path construction confer Figure 2.

Since Rule 1 is not applicable in T we can state that any pair of distinct T-leaves is G-independent. Therfore, it is impossible that some $P(l)$ consists of only one single edge. Moreover, it is easy to see that the union F of the paths $P(\cdot)$ constitutes a subforest of T^* having no F-branchings. We condense the foregoing considerations in the following lemma:

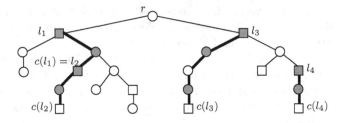

Fig. 2. Illustration of T^* with T-leaves represented by squares. The grey squares $l_1, \ldots, l_4$ are the covered leaves. The paths $P(l_i)$ are marked by thick lines. Grey circles are path companions.

Lemma 2. *Let l and l' be distinct covered leaves. Then $P(l)$ has always length at least 2. Moreover, if $P(l)$ and $P(l')$ share common nodes then either $l = c(l')$ or $l' = c(l)$.* □

We are now ready to define our path companions (confer also Figure 2):

Definition 1 (Path companions). *Let l be a covered leaf. The* upper path companion *of l is the T^*-neighbor of l on path $P(l)$. The T^*-neighbor of $c(l)$ on $P(l)$ is called* lower path companion *of l. The set $I_p(l)$ of path companions of l contains the upper and the lower path companion of l.*

Observe that in the above definition the upper and the lower path companion of a covered leaf may coincide, i. e., a leaf may have only one path companion. As a byproduct of Lemma 2 we obtain that T is a 2-approximation: Since no $I_p(\cdot)$ is empty and all those sets are pairwisely disjoint we conclude that there are at least $|C|$ companions and so $|I(T^*)| \leq 2 \cdot |I(T)|$ by Inequality (1).

3.3   Auxiliary Companions

In the sequel we are going to improve the above ratio of 2 by identifying for each covered leaf l a set $I_a(l)$ of additional companions which we call *auxiliary companions* of l. The definition of auxiliary companions is inspired by the specific structure of our local optimum T reflected in the following two simple observations:

Observation 1. *Let l be a T-leaf and $(l, x) \in E - E(T)$. Then $x^{\rightarrow l}$ has degree two in T and is G-independent from any leaf in $L(T) - l$.*

Proof. The claim follows immediately since Rules 2 and 3 are not applicable.   □

Observation 2. *Let l be an x-supported T-leaf then $b(l)^{\rightarrow x}$ has degree two in T and is G-independent from any leaf of T.*

Proof. The claim follows immediately since Rules 2,4, and 5 are not applicable.
□

Both of the above properties predict the existence of certain degree-two nodes which we will now associate with covered leaves in an (almost) unique way:

Definition 2 (Auxiliary companions). *Let l be a covered leaf. Then the set $I_a(l)$ of auxiliary companions of l consists of the following internal nodes:*

(i) *Nodes of the form $x^{\to l}$ if x is the* upper *path companion of l but not T-adjacent to l.*

(ii) *Nodes of the form $x^{\to c(l)}$ if x is the* only *path companion of l but not T-adjacent to $c(l)$.*

(iii) *Nodes of the form $b(l)^{\to x}$ if $b(l)$ is the the* upper *path companion of l, and l is x-supported.*

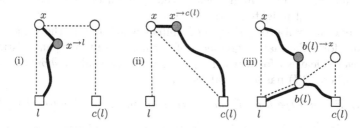

Fig. 3. Illustration of T with auxiliary companions represented by grey nodes. Thick and dashed lines mark edges of T and T^*, respectively. Wiggly lines depict paths of T.

At first glance Definition 2 might appear somewhat confusing. But a closer look reveals that all auxiliary companions share the following nice properties which guarantee (confer next section) that they never occur in more than two sets $I_a(\cdot)$.

Lemma 3. *Let l be a covered leaf and $u \in I_a(l)$ be an auxiliary companion of l. Then u satisfies the following conditions:*

(i) *u is not a path companion of l,*

(ii) *u is G-adjacent to at most one T-leaf, and this is always in $\{l, c(l)\}$,*

(iii) *u has degree $d_T(u) = 2$,*

(iv) *u is T-adjacent to the upper path companion of l.*

Proof. Statements (ii) and (iii) are immediate consequences of Observations 1 and 2. Statement (iv) follows immediately from the construction of auxiliary companions. In order to show statement (i) assume that u is simultaneously an auxiliary and a path companion of l. Consider first that u complies with Definition 2 (i), i.e., u equals $x^{\to l}$ where x is the upper path companion of l. Since u cannot equal x it must be the lower path companion of l and therefore be T^*-adjacent to $c(l)$. As l is not T-adjacent to x this contradicts Observation 1. Obviously u cannot comply with Definition 2 (ii) as in this case u is T-adjacent and therefore not equal to the only path companion of l. Also the remaining case that u complies with Definition 2 (iii) is easily ruled out with the help of Observation 2 since u is T^*-adjacent to some leaf of T. □

3.4   Counting the Companions

In this section we will show that there are at least $\frac{3}{2}|C|$ companions which implies the desired approximation factor of $\frac{5}{3}$ for LOSTLIGHT.

To this end consider the family $\mathcal{I} := \{\, I_a(l), I_p(l) \mid l \in C \,\}$ of all companion sets $I_p(\cdot)$ and $I_a(\cdot)$. Our goal is to count the number of all companions $I_c := \bigcup \mathcal{I}$. Of course it is not sufficient to consider the sum $\sum_{J \in \mathcal{I}} |J|$ of cardinalities of companions sets since these sets need not be disjoint. In order to overcome these difficulties, we introduce for each companion $u \in I_c$ the *frequency* freq$(u) := |\{J \in \mathcal{I} \mid J \ni u\}|$ of u within the family $\mathcal{I}$ and a weight $w(u) := \frac{1}{\mathrm{freq}(u)}$. This weight is extended to covered leaves l by setting $w(l) := w(I_p(l) \cup I_a(l))$. The weight $w(l)$ may be regarded as the *contribution* of the companions of l to the total number $|I_c|$ of companions. Indeed, we can express $|I_c|$ as the sum $w(C) = \sum_{l \in C} w(l)$ of weighted covered leaves.

In the sequel we will prove a series of lemmas which bound the values freq$(\cdot)$ and $w(\cdot)$ and form the main ingredients of our final result. We shall distinguish between covered leaves l for which $P(l)$ has length 2, so called *short leaves*, and *long leaves* where $P(l)$ has length at least 3.

Lemma 4. *For any (path or auxiliary) companion u, we have* freq$(u) \le 2$.

Proof. Consider first the case u is a path companion of some covered leaf l. Since the sets $I_p(\cdot)$ are pairewisely disjoint u has no further occurrences as a path companion in $\mathcal{I}$. So assume that u is an auxiliary companion for covered leaves l', l''. We claim that $l' = l''$: It follows from Lemma 3 (i) that $l', l'' \ne l$. Let us assume further that u is the *upper* path companion of l. Then u is T^*-adjacent to l and hence $c(l') = l = c(l'')$ by Lemma 3 (ii). Thus $l' = l''$ by Lemma 2. If u is the *lower* path companion of l then u is T^*-adjacent to $c(l)$ and hence $l' = c(l) = l''$.

Now consider the case that u is not a path companion. Moreover, let us assume that u is an auxiliary companion of some covered leaf l. Then u is T-adjacent to some path companion of l. Since u has degree $d_T(u) = 2$ and the path companion sets $I_p(\cdot)$ are pairwisely disjoint we deduce that there are at most two such covered leaves.   □

The following lemma is an immediate consequence:

Lemma 5. *For any long leaf l, we have $w(l) \ge 1$.*   □

Lemma 6. *For any short leaf l with path companion u, we have* freq$(u) = 1$ *and $w(l) \ge \frac{3}{2}$.*

Proof. Since u is G-adjacent to two T-leaves u cannot be an auxiliary companion and therefore freq$(u) = 1$.

We claim that l has at least one auxiliary companion. To this end we assume first that $P(l)$ is a subpath of T. Since T is not a path u must be a branching. This implies $u = b(l)$. Let x be the father of l in T^*. Then l is not T-adjacent to x as l is a T-leaf with T-neighbor $u \ne x$. Hence $u^{\to x}$ is an auxiliary companion of l and therefore $w(l) \ge \frac{3}{2}$.

Consider the case that $P(l)$ is not a subpath of T. Then u is T-independent from some $l' \in \{l, c(l)\}$. But this implies that $u \to l'$ is an auxiliary companion of l and therefore $w(l) \geq \frac{3}{2}$. □

Let us denote by $C_{=\alpha}$ and $C_{\geq\alpha}$ the sets of covered leaves l with $w(l) = \alpha$ and $w(l) \geq \alpha$, respectively. Lemma 4 shows that the weights of covered leaves are always multiples of $\frac{1}{2}$. Since we would like to have $w(C) \geq \frac{3}{2}|C|$, leaves of weight $w(l) = 1$ could be problematic. But fortunately we can show that those light covered leaves are compensated by heavy ones:

Lemma 7. *We have* $|C_{=1}| \leq |C_{\geq 2}|$.

Proof. Let l be a covered leaf with weight $w(l) = 1$. According to Lemma 6 it must be a long leaf. Since l has two path companions with weight at least $\frac{1}{2}$ there can be no auxiliary companion of l. Consider the upper path companion u of l. If u was not T-adjacent to l then l would have an auxiliary companion due to Definition 2 (i). We conclude that l is T-adjacent to u. Moreover, $w(u) = \frac{1}{2}$ by premise. This implies that u is auxiliary companion of some covered leaf $l' \neq l$ and thus $l = c(l')$ by Lemma 3 (ii). Since u is T-adjacent to leaf l we can rule out that u complies with Definition 2 (iii). Let x be the upper path companion of l'. Then either $u = x \to l'$ or $u = x \to c(l')$ according to Definition 2 (i) and (ii). Since $x \neq l$ only the latter case may occur which implies that l' is a short leaf.

We will now show that $w(l') \geq 2$ which completes the proof since l' is uniquely determined by l through the relation $c(l') = l$. According to Lemma 6 leaf l' has a path companion u' with weight $w(u') = 1$. Since l' has also the auxiliary companion u with weight $w(u) \geq \frac{1}{2}$ it suffices to identify a second auxiliary companion for l'. To this end assume first that l' is T-adjacent to x (confer Figure 4 (i)). Then x must be a branching in T, for otherwise, $c(l)$ would be disconnected from T. Let v be the father of l' in T. Then $(l', v) \notin E(T)$. Hence $x \to v \neq u$ is an additional auxiliary companion of l'. Let us finally assume that l' is not T-adjacent to x (confer Figure 4 (ii)). Then $x \to l' \neq u$ is an additional auxiliary companion of l' and therefore $w(l') \geq 2$. □

From Lemmas 4–7 we infer immediately:

$$|I(T)| \overset{\text{Lm. 4}}{\geq} w(C_{=1}) + w(C_{=\frac{3}{2}}) + w(C_{\geq 2})$$

$$\geq |C_{=1}| + \frac{3}{2}|C_{=\frac{3}{2}}| + 2|C_{\geq 2}|$$

$$= \left(|C_{=1}| + \frac{1}{2}|C_{\geq 2}|\right) + \frac{3}{2}|C_{=\frac{3}{2}}| + \frac{3}{2}|C_{\geq 2}|$$

$$\overset{\text{Lm. 7}}{\geq} \frac{3}{2}|C_{=1}| + \frac{3}{2}|C_{=\frac{3}{2}}| + \frac{3}{2}|C_{\geq 2}|$$

$$= \frac{3}{2}|C|.$$

And finally by Inequality (1):

Theorem 1. LOSTLIGHT *is a* $\frac{5}{3}$-*approximation algorithm for the* MAXIMUM INTERNAL SPANNING TREE *problem with running time* $O(|V|^3)$. □

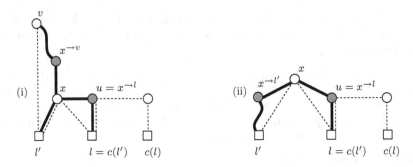

Fig. 4. The illustration shows the two cases for an upper path companion u of a long leaf where $w(u) = \frac{1}{2}$. Thick and dashed *straight* lines mark the edges of T and T^*, respectively. Wiggly lines depict paths in T.

4   The Weighted Case

In this section we investigate the MAXIMUM WEIGHTED INTERNAL SPANNING TREE problem (MAXWIST). Salamon [7] developed a local search algorithm, called WLOST, which runs in polynomial time and gives a $(2 \cdot \Delta(G) - 3)$-approximation for MAXWIST. We will present the first constant-factor approximation algorithm for this problem. Our result is based on a new pseudo-polynomial local search algorithm WLOSTADVANCED which yields a 3-approximation for MAXWIST. This result can then be extended to an efficient $(3 + \epsilon)$-approximation scheme.

Our algorithm WLOSTADVANCED maintains a spanning tree T. Let l be a T-leaf, u its unique T-neighbor and (l, x) a non-tree edge for some node x. We introduce the following six rules:

Rule W1. If $x^{\to l}$ is a branching then add edge (l, x) and remove $(x, x^{\to l})$.

Rule W2. If $x^{\to l}$ has degree 2 and $c(x^{\to l}) < c(l)$ then add (l, x) and remove $(x, x^{\to l})$.

Rule W3. If u and $u^{\to x}$ are branchings then add (l, x) and remove $(u, u^{\to x})$.

Rule W4. If u is a branching, $u^{\to x}$ has degree 2 and $c(u^{\to x}) < c(l)$ then add (l, x) and remove $(u, u^{\to x})$.

Rule W5. If $u^{\to x}$ is a branching, u has degree 2 and $c(u) < c(l)$ then add (l, x) and remove $(u, u^{\to x})$.

Rule W6. If u and $u^{\to x}$ have degree 2 and $c(u) + c(u^{\to x}) < c(l)$ then add (l, x) and remove $(u, u^{\to x})$.

Again we start with a DFS-tree of the input graph and apply Rules W1–W6 until we reach a local optimum. Similar to LOSTLIGHT all rules of WLOSTAD-VANCED improve the objective function $c(I(T))$. Moreover, all rules can be tested and performed in $O(n^2)$. Multiplying all node weights with the smallest common denominator we may assume that they are integral. Hence we obtain a pseudo-polynomial running time $O(n^2 \cdot c(T))$.

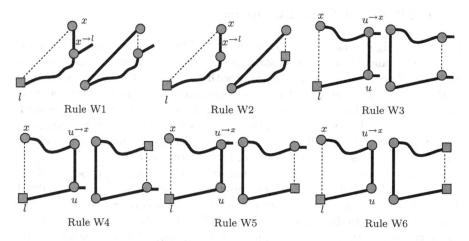

Fig. 5. The rules of WLOSTADVANCED. Thick and dashed lines mark T-edges and non-tree edges, respectively. Wiggly lines depict paths of T.

We are able to show that WLOSTadvanced terminates with a 3-approximation. Using recent findings [8] on local search we can extend this to the subsequent approximability result. The proofs use ideas similar to those of the unweighted case. Due to space limitations we omit them here.

Theorem 2. MaxWIST *can be approximated within a factor of* $3 + \epsilon$ *for any* $\epsilon > 0$ *in polynomial time.* $\square$

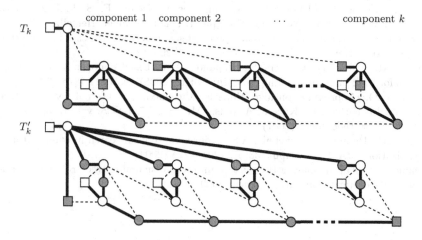

Fig. 6. Two spanning trees of the same graph G_k. The spanning tree T_k is locally optimal with respect to LOSTLIGHT and WLOSTADVANCED. The tree T_k' achieves a deviation close to the upper bounds. In the weighted case the grey nodes are heavy in comparison to the white ones.

5   Tight Worst Case Instances

For the unweighted case consider graph G_k in Figure 6 consisting of k components. Let T and T'_k be the two spanning trees depicted. It is easy to verify that T_k is a DFS-tree which is locally optimal with respect to LOSTLIGHT since none of Rules 1–5 is applicable. T_k has $3k + 3$ internal nodes. On the other hand the spanning tree T'_k has $5k + 1$ internal nodes. Increasing k we come arbitrarily close to factor $\frac{5}{3}$. It is worth noting that T_k is even locally optimal for Salamon's algorithm LOST. Thus that algorithm does not perform better in the worst case although it involves far more rules than LOSTLIGHT does.

For the weighted case consider the same graph where each component contains three grey nodes weighted with a large number Ω. White nodes are unit weighted. The spanning tree T_k is locally optimal with respect to WLOSTADVANCED (and also with respect to Salamon's algorithm WLOST). The approximation factor converges to 3 with increasing Ω and k. We conclude that our analyses of the algorithms LOSTLIGHT and WLOSTADVANCED are best possible.

References

1. Salamon, G., Wiener, G.: On finding spanning trees with few leaves. Information Processing Letters 105, 164–169 (2008)
2. Lu, H., Ravi, R.: The power of local optimization: approximation algorithms for maximum-leaf spanning tree. Technical report, Department of Computer Science, Brown University (1996)
3. Flandrin, E., Kaiser, T., Kuzel, R., Li, H., Ryjcek, Z.: Neighborhood unions and extremal spanning trees. Discrete Mathematics 308(12), 2343–2350 (2008)
4. Prieto, E., Sloper, C.: Either/or: Using vertex cover structure in designing fpt-algorithms - the case of k-internal spanning tree. In: Dehne, F., Sack, J.-R., Smid, M. (eds.) WADS 2003. LNCS, vol. 2748, pp. 474–483. Springer, Heidelberg (2003)
5. Prieto, E., Sloper, C.: Reducing to independent set structure – the case of k-internal spanning tree. Nord. J. Comput. 12(3), 308–318 (2005)
6. Fernau, H., Raible, D., Gaspers, S., Stepanov, A.A.: Exact exponential time algorithms for max internal spanning tree. In: CoRR (2008) abs/0811.1875
7. Salamon, G.: Approximation algorithms for the maximum internal spanning tree problem. In: Kučera, L., Kučera, A. (eds.) MFCS 2007. LNCS, vol. 4708, pp. 90–102. Springer, Heidelberg (2007)
8. Orlin, J.B., Punnen, A.P., Schulz, A.S.: Approximate local search in combinatorial optimization. SIAM J. Comput. 33(5), 1201–1214 (2004)
9. Salamon, G.: Approximating the maximum internal spanning tree problem. To appear in Theoretical Computer Science (2009)

Two Approximation Algorithms for ATSP with Strengthened Triangle Inequality*

Łukasz Kowalik and Marcin Mucha

Institute of Informatics, University of Warsaw, Poland
{kowalik,mucha}@mimuw.edu.pl

Abstract. In this paper, we study the asymmetric traveling salesman problem (ATSP) with strengthened triangle inequality, i.e. for some $\gamma \in [\frac{1}{2}, 1)$ the edge weights satisfy $w(u, v) \leq \gamma(w(u, x) + w(x, v))$ for all distinct vertices u, v, x.

We present two approximation algorithms for this problem. The first one is very simple and has approximation ratio $\frac{1}{2(1-\gamma)}$, which is better than all previous results for all $\gamma \in (\frac{1}{2}, 1)$. The second algorithm is more involved but it also gives a much better approximation ratio: $\frac{2-\gamma}{3(1-\gamma)} + O(\frac{1}{n})$ when $\gamma > \gamma_0$, and $\frac{1}{2}(1 + \gamma)^2 + \epsilon$ for any $\epsilon > 0$ when $\gamma \leq \gamma_0$, where $\gamma_0 \approx 0.7003$.

1 Introduction

The Traveling Salesman Problem is one of the most researched NP-hard problems. In its classical version, given a set of vertices V and a symmetric weight function $w : V^2 \to \mathbb{R}_{\geq 0}$ one has to find a Hamiltonian cycle of minimum weight. Asymmetric Traveling Salesman Problem (ATSP) is a natural generalization where the weight function w does not need to be symmetric. Both TSP and ATSP without additional assumptions do not allow for any reasonable polynomial-time approximation algorithm, i.e. they are NPO-complete problems. A natural assumption, ofter appearing in applications, is the triangle inequality, i.e. $w(u, v) \leq w(u, x) + w(x, v)$, for all distinct vertices u, v, x. With this assumption, TSP has a 3/2-approximation by the well-known algorithm of Christofides [5]. On the other hand, no constant-factor polynomial time algorithm is known for ATSP with triangle inequality and the best algorithm up to date, due to Feige and Singh [6], has approximation ratio $\frac{2}{3} \log_2 n$.

1.1 γ-Parameterized Triangle Inequality and Previous Results

The TSP and ATSP problems have also been studied under the γ-parameterized triangle inequality, i.e. for some constant γ, and for all distinct vertices u, v, and x,

$$w(u, v) \leq \gamma(w(u, x) + w(x, v)).$$

---

* This research is partially supported by a grant from the Polish Ministry of Science and Higher Education, project N206 005 32/0807.

F. Dehne et al. (Eds.): WADS 2009, LNCS 5664, pp. 471–482, 2009.

It can be easily seen that only values $\gamma \geq \frac{1}{2}$ make sense, since otherwise all edge weights would need to be 0, see e.g. [3]. In the first work on ATSP problem with the γ-parameterized triangle inequality, Chandran and Ram [3] showed a $\gamma/(1-\gamma)$-approximation algorithm for any $\gamma \in [\frac{1}{2}, 1)$. Note that for any fixed γ the approximation ratio is bounded. Next, Bläser [1] announced an algorithm with approximation ratio $1/(1 - \frac{1}{2}(\gamma + \gamma^3))$, which is an improvement for $\gamma \in [0.66, 1)$. Later, it was extended by Bläser, Manthey and Sgall [2] to $(1 + \gamma)/(2 - \gamma - \gamma^3)$ which otperforms the earlier algorithms for $\gamma \in [0.55, 1)$. Finally, there is a very recent algorithm of Zhang, Li and Li [11] which is better than the previous methods for $\gamma \in [0.59, 0.72]$.

In their work Chandran and Ram [3] were also interested in bounding the ratio $\mathrm{ATSP}(G)/\mathrm{AP}(G)$ where $\mathrm{ATSP}(G)$ and $\mathrm{AP}(G)$ are the minimum weight of a Hamiltonian cycle in graph G and the minimum weight of a cycle cover in graph G, respectively. Analysis of their algorithm implies that $\mathrm{ATSP}(G)/\mathrm{AP}(G) \leq \gamma/(1 - \gamma)$. On the other hand they show an infinite family of graphs for which $\mathrm{ATSP}(G)/\mathrm{AP}(G) = \frac{1}{2(1-\gamma)}$.

1.2 Our Results

In Section 2 we describe a very simple algorithm, using methods similar to those used by Kostochka and Serdyukov [8] for the max-ATSP with triangle inequality. We show that its approximation ratio is $\frac{1}{2(1-\gamma)}$, which is better than all previous results for any $\gamma \in (\frac{1}{2}, 1)$. This result implies that $\mathrm{ATSP}(G)/\mathrm{AP}(G) \leq \frac{1}{2(1-\gamma)}$ for any graph G, which is tight.

In Section 3 we present an even more efficent method, which outperforms our first algorithm for $\gamma \geq 0.619$. There is a constant $\gamma_0 \approx 0.7003$, such that for $\gamma \in (\gamma_0, 1)$ the second method gives the approximation ratio of $\frac{2-\gamma}{3(1-\gamma)} + O(\frac{1}{n})$, while for $\gamma \in [\frac{1}{2}, \gamma_0)$ it can achieve approximation ratio of $\frac{1}{2}(1 + \gamma)^2 + \epsilon$ for any $\epsilon > 0$ (see Figure 1 for comparison with previous results).

We show that for $\gamma \in (\gamma_0, 1)$ our approximation factor is essentially optimal w.r.t. the relaxation used.

1.3 Notation

Throughtout the paper, V is the vertex set of the input complete graph and $w : V^2 \to \mathbb{R}_{\geq 0}$ is a weight function which satisfies the γ-parameterized triangle inequality and such that $w(v, v) = 0$ for any $v \in V$. The vertex sets of all the graphs and multigraphs in the paper are subsets of V and w naturally induces weights on their edges.

For any (multi)set of edges S we define $w(S) = \sum_{(x,y) \in S} w(x, y)$. We will also write $w(S)$ when S is a (multi)graph, a cycle, a walk etc. always meaning the corresponding (multi)set of edges.

We will say that a directed graph G is *connected* when the underlying undirected graph is connected. Similarly, a *connected component* of G is a inclusion-wise maximal subgraph of G that is connected.

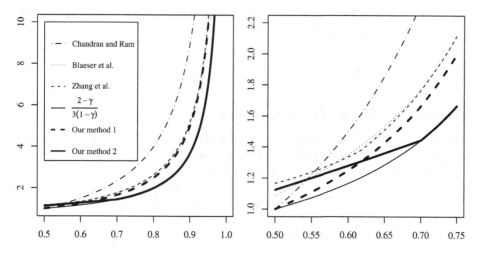

Fig. 1. Comparision of our methods with previous results

2   A Simple $\frac{1}{2(1-\gamma)}$-Approximation Algorithm

In this section we give a very simple algorithm that achieves an approximation ratio of $\frac{1}{2(1-\gamma)}$. Our algorithm starts, as is typical when approximating variants of TSP, by finding a minimum weight cycle cover of the input graph. Such a cover can be found in polynomial time. The cycles are then broken and patched to form a Hamiltonian cycle. The technique we use to guarantee that the patching edges have low weight is very similar to the one introduced by Kostochka and Serdyukov [8] for a variant of max-TSP. Interestingly, this technique has not so far been used to solve min-TSP problems.

2.1   A Randomized Algorithm

We begin by showing a randomized version of our algorithm since it is more natural this way.

Throughout this paper we will use the following lemma.

Lemma 1. *Let $W = (v_1, v_2, \ldots, v_k, v_1)$ be a closed walk and let x be a vertex not visited by W. Then,*

$$\sum_{j=1}^{k} w(v_j, x) \le \frac{\gamma}{1-\gamma} w(W).$$

Proof. From the γ-parametrized triangle inequality it follows that

$$w(v_j, x) \le \gamma \left(w(v_j, v_{j+1}) + w(v_{j+1}, x) \right)$$

for $j = 1, \ldots, k$ (we let $v_{k+1} = v_1$ to avoid having to consider it separately). By summing the above inequality over all $j = 1, \ldots, k$, we get

$$\sum_{j=1}^{k} w(v_j, x) \leq \gamma \left(w(W) + \sum_{j=1}^{k} w(v_j, x) \right),$$

which easily implies the claim of the lemma. $\qquad \square$

Theorem 1. *Let C be a directed cycle of length l, and let P be a path created by randomly removing a single edge from C. Also let u be the last vertex on P, and let v be a vertex not contained in C. Then*

$$E[w(P) + w(uv)] \leq \frac{l - 1 - (l - 2)\gamma}{l(1 - \gamma)} w(C),$$

Proof. Let $C = v_1 v_2 \ldots v_l v_1$. We have

$$E[w(P) + w(uv)] = w(C) - \frac{w(C)}{l} + \frac{\sum_{i=1}^{l} w(v_i v)}{l}. \qquad (1)$$

From Lemma 1 we get

$$E[w(P) + w(uv)] \leq \left(1 - \frac{1}{l} + \frac{\gamma}{l(1 - \gamma)} \right) w(C) = \left(\frac{l - 1 - (l - 2)\gamma}{l(1 - \gamma)} \right) w(C). \qquad (2)$$

$\qquad \square$

Since $\frac{l-1-(l-2)\gamma}{l(1-\gamma)} = 1 + \frac{2\gamma-1}{l(1-\gamma)}$ is a monotonically decreasing function of l, and $l \geq 2$ for all cycles, it follows that

Corollary 1. *With C, P, u and v as in Theorem 1, we have*

$$E[w(P) + w(uv)] \leq \frac{1}{2(1 - \gamma)} w(C).$$

We are now ready to present our basic algorithm.

Theorem 2. *Let $\mathcal{C} = \{C_1, \ldots, C_k\}$ be a directed cycle cover of graph G. Then a Hamiltonian cycle H in G with expected weight $E[w(H)] \leq \frac{1}{2(1-\gamma)} w(\mathcal{C})$ can be found in polynomial time.*

Proof. Remove a random edge from each cycle C_i thus turning it into a directed path P_i. Connect the last vertex of P_i with the first vertex of P_{i+1} for $i = 1, \ldots, k - 1$ and connect the last vertex of P_k with the first vertex of P_1. Let H be the resulting Hamiltonian cycle. We claim that $E[w(H)] \leq \frac{1}{2(1-\gamma)} w(\mathcal{C})$.

Consider any cycle in $\mathcal{C}$, say C_i, and assume that we have already removed a random edge from all the other cycles, and in particular from C_{i+1}. Let us now randomly break C_i (creating P_i) and let e_i be the edge connecting the last vertex of P_i with the first vertex of P_{i+1}. It follows from Corollary 1 that

$$E[w(P_i) + w(e_i)] \leq \frac{1}{2(1 - \gamma)} w(C_i).$$

This holds for cycle C_i regardless of the random choices made for the other cycles, so it also holds when all cycles are broken randomly. Summing this inequality over all cycles proves the theorem.   □

Since the minimum weight of a cycle cover lowerbounds the minimum weight of a Hamiltonian cycle we get the following.

Corollary 2. *For any* $\gamma \in [\frac{1}{2}, 1)$ *there is a randomized polynomial time algorithm for ATSP with* γ*-parameterized triangle inequality, which has expected approximation ratio of* $\frac{1}{2(1-\gamma)}$.

2.2   A Deterministic Algorithm

The algorithm presented in the previous section can be derandomized using a generic conditional expected value approach. The resulting algorithm is particularly simple, but we defer its explicit description to the journal version due to space limitations.

Here, we give only a deterministic version of Theorem 1 and Corollary 1, as they will be used in a further section.

Corollary 3. *Let C be a directed cycle of length l and let v be a vertex not contained in C. Then, an edge can be removed from C so that the resulting directed path P satisfies*

$$w(P) + w(uv) \leq \frac{l - 1 - (l - 2)\gamma}{l(1 - \gamma)} w(C),$$

where u is the last vertex of P. In particular,

$$w(P) + w(uv) \leq \frac{1}{2(1 - \gamma)} w(C).$$

Proof. Remove an edge that gives the lowest value of $w(P) + w(uv)$. The first bound follows easily from Theorem 1. It implies the second one since $l \geq 2$.   □

Derandomizing the algorithm of Theorem 2 is a bit harder (we skip it in this conference version).

Theorem 3. *Let $\mathcal{C} = \{C_1, \ldots, C_k\}$ be a directed cycle cover of graph G. Then, a Hamiltonian cycle H in G with weight $w(H) \leq \frac{1}{2(1-\gamma)} w(\mathcal{C})$ can be found in polynomial time.*

Remark 1. Note that the above analysis implies that $\mathrm{ATSP}(G)/\mathrm{AP}(G) \leq \frac{1}{2(1-\gamma)}$ for any graph G, which is tight by the result of Chandran and Ram [3]. In particular the approximation ratio of our algorithm is optimal w.r.t. minimum weight cycle cover relaxation.

3   Improved Approximation Using Double Cycle Covers

In the algorithm from Section 2 the 2-cycles are the obvious bottleneck, i.e. if we were able to find the minimum weight cycle cover with no 2-cycles, the algorithm would have a significantly better approximation ratio of $\frac{2-\gamma}{3(1-\gamma)}$ (see Fig 1). Unfortunately, finding such a cover is APX-hard even when w has exactly two values as shown by Manthey [10], and when these values are 1 and 2γ the γ-parameterized triangle inequality holds. A similar phenomenon occurs for some other min-ATSP and max-ATSP variants. Nevertheless, Kaplan, Lewenstein, Shafrir and Sviridenko [7] obtained a major progress for three ATSP variants and motivated a series of other improvements (see [6,4,9]) by proving the following.

Theorem 4 (Kaplan et al. [7]). *Let G be a directed weighted graph. One can find in polynomial time a pair of cycle covers $\mathcal{C}_1$, $\mathcal{C}_2$ such that (i) $\mathcal{C}_1$ and $\mathcal{C}_2$ share no 2-cycles, (ii) total weight $w(\mathcal{C}_1) + w(\mathcal{C}_2)$ of the two covers is at most $2OPT$, where OPT is the weight of the minimum weight Hamiltonian cycle in G.*

In this section we will show that the double cycle covers can be also used for ATSP with γ-parameterized triangle inequality. The resulting algoritm has approximation ratio of $\frac{2-\gamma}{3(1-\gamma)} + O(\frac{1}{n})$ for $\gamma \in (\gamma_0, 1)$, $\gamma_0 \approx 0.7003$.

The sketch of our algorithm is as follows. We begin by finding a pair of cycle covers $\mathcal{C}_1$, $\mathcal{C}_2$ described in Theorem 4. The union of these covers corresponds to a 2-regular directed graph M. We then replace the connected components of M by paths which eventually form a Hamiltonian cycle. We present two methods of doing it.

In the first method, each connected component Q of M is replaced by a low-weight cycle which contains all vertices of Q. These cycles are then joined to form a single path using the method from Section 2, Corollary 3. This approach is efficient enough provided that the components are *big* i.e. they have size at least $f(\gamma)$, for certain function f, to be defined later.

For small components we use a different method. Using another deep result of Kaplan et al. we show that a component Q can be replaced by a collection of paths of total weight $\frac{2-\gamma}{3(1-\gamma)} w(Q)$, provided that we have $|V(Q)|$ previously constructed paths to work with. Guaranteeing that is a technical detail, and it increases the final approximation ratio by $O(\frac{1}{n})$.

In the next two subsections we show how to deal with small and big connected components of M. Then we give a more detailed decription of the complete approximation algorithm.

3.1   Dealing with Small Components

A directed graph which is a union of vertex-disjoint paths will be called a *path graph*. The following theorem is due to Kaplan et al. [7] (Theorem 5.1).

Theorem 5. *Let G be a directed 2-regular multigraph that contains neither two copies of the same 2-cycle, nor two copies, oppositely oriented, of the same 3-cycle. Then G is a union of three path graphs, and such a decomposition can be found in polynomial time.*

We will need the following corollary from Lemma 1.

Corollary 4. *Let Q be a 2-regular subgraph in a directed multigraph G and let x be a vertex not in $V(Q)$. Then,*

$$\sum_{v \in V(Q)} w(v, x) \leq \frac{\gamma}{2(1 - \gamma)} w(Q).$$

Proof. We apply Lemma 1 with W being an Eulerian cycle in Q. Since each vertex of Q is visited by W exactly twice, the claimed bound follows. □

Lemma 2. *Let Q be a connected component in a directed multigraph G such that Q is 2-regular and has t vertices, $t \geq 3$. Let X be a set of k vertices of G, disjoint with $V(Q)$, $k \geq t$. Then one can find in polynomial time a set of k vertex-disjoint paths $\mathcal{P}$ such that*

(i) paths in $\mathcal{P}$ end in vertices of X,
(ii) $V(\mathcal{P}) = V(Q) \cup X$, and
(iii) $w(\mathcal{P}) \leq \frac{2-\gamma}{6(1-\gamma)} w(Q)$.

Proof. We begin by decomposing Q into three path graphs $\mathcal{P}_1$, $\mathcal{P}_2$ and $\mathcal{P}_3$ using Theorem 5. This is possible unless Q is a union of two oppositely oriented 3-cycles, in which case we replace the heavier of these cycles by another copy of the lighter one and then apply Theorem 5. We assume that for each $i \in \{1, 2, 3\}$, $V(\mathcal{P}_i) = V(Q)$. If that is not the case, we add the missing vertices (treated as paths of length 0).

Consider any vertex v in Q. Since Q is 2-regular, $\text{outdeg}_{\mathcal{P}_1}(v) + \text{outdeg}_{\mathcal{P}_2}(v) + \text{outdeg}_{\mathcal{P}_3}(v) = 2$. Hence $\text{outdeg}_{\mathcal{P}_i}(v) = 0$ for exactly one $i \in \{1, 2, 3\}$. It follows that for every vertex v of Q there is exactly one path in $\mathcal{P}_1$, $\mathcal{P}_2$ and $\mathcal{P}_3$ that ends in v. In particular, graphs $\mathcal{P}_i$ contain exactly t paths in total. Now, we are going to choose a distinct vertex x of X for each such path, and connect its end v to x by a new edge. In order to guarantee that the total weight of the added edges is small, we consider t ways of assigning vertices to paths and choose the best one.

More precisely, for each $i = 0, \ldots, t - 1$ we construct three path graphs $\mathcal{P}_{i,1}$, $\mathcal{P}_{i,2}$, and $\mathcal{P}_{i,3}$, each with the vertex set $V(Q) \cup X$. Let $\{x_0, x_1, \ldots, x_{t-1}\}$ be a set of t vertices of X (arbitrarily selected) and let $V(Q) = \{v_0, \ldots, v_{t-1}\}$. For any $i = 0, \ldots, t-1, q = 1, 2, 3$, the graph $\mathcal{P}_{i,q}$ is obtained from $\mathcal{P}_q$ by extending each of its paths with an edge — if a path ends in v_j we extend it with $(v_j, x_{(j+i) \bmod t})$.

Our algorithm returns the lightest among the path graphs $\mathcal{P}_{i,q}$. Denote it by $\mathcal{P}$. Let $A = \{(v_j, x_{(j+i) \bmod t}) : i, j = 0, \ldots, t - 1\}$ be the set of edges added to the paths of $\mathcal{P}_1$, $\mathcal{P}_2$ and $\mathcal{P}_3$. Then,

$$w(\mathcal{P}) \leq \frac{\sum_{i=0}^{t-1} \sum_{q=1}^{3} w(\mathcal{P}_{i,q})}{3t} = \frac{t \cdot w(Q) + w(A)}{3t}. \tag{3}$$

It suffices to bound $w(A)$. We proceed as follows.

$$w(A) = \sum_{j=0}^{t-1}\sum_{i=0}^{t-1} w(v_j, x_{(j+i) \bmod t}) = \sum_{j=0}^{t-1}\sum_{\ell=0}^{t-1} w(v_j, x_\ell) = \sum_{\ell=0}^{t-1}\sum_{j=0}^{t-1} w(v_j, x_\ell) \le$$

$$\le \sum_{\ell=0}^{t-1} \frac{\gamma}{2(1-\gamma)} w(Q) = \frac{t\gamma}{2(1-\gamma)} w(Q), \tag{4}$$

where the ineaquality follows from Corollary 4. After plugging (4) to (3) we get the claimed bound on $w(\mathcal{P})$.   □

3.2   Dealing with Big Components

An easy proof of the following lemma is deferred to a journal version because of space limitations.

Lemma 3. *For $k \ge 3$, let $W = (x_1, x_2, \ldots, x_k)$ be a walk such that each vertex appears at most twice in W. Then,*

$$w(x_1 x_k) \le \gamma w(x_1 x_2) + \gamma^2 \sum_{i=2}^{k-2} w(x_i x_{i+1}) + \gamma w(x_{k-1} x_k).$$

Lemma 4. *Let Q be a connected 2-regular directed multigraph. Then there is a randomized polynomial time algorithm which finds in Q a Hamiltonian cycle of expected weight at most $\frac{1}{4}(1+\gamma)^2 w(Q)$.*

Proof. Let t be the number of vertices in Q. Our algorithm begins by finding an Eulerian cycle $\mathcal{E} = (v_0, v_1, \ldots, v_{2t-1}, v_0)$ in Q. For each pair of indices i, j such that $v_i = v_j$, our algorithm chooses one index uniformly at random, and the chosen index will be called a *stop point*. Observe that every vertex of Q appears in $\mathcal{E}$ precisely twice, so for each vertex x there is exactly one stop point i such that $v_i = x$. Let $i_0, i_1, \ldots, i_{t-1}$ be the sequence of stop points such that $i_j < i_{j+1}$ for $j = 0, \ldots, t-2$. Assume w.l.o.g. that $i_0 = 0$. Finally, the algorithm returns the Hamiltonian cycle $C = v_{i_0}, v_{i_1}, \ldots, v_{i_{t-1}}$.

Now we are going to bound $E[w(C)]$. Consider arbitrary choice of stop points. In what follows, indices at v_q and i_p are modulo $2t$ and t, respectively. We have

$$w(C) = \sum_{p=0}^{t-1} w(v_{i_p} v_{i_{p+1}}).$$

Using Lemma 3 we can bound the values of $w(v_{i_p} v_{i_{p+1}})$ for $p = 0, \ldots, t-1$. Then we get

$$w(C) \le \sum_{q=0}^{2t-1} \text{contrib}(v_q v_{q+1}) w(v_q v_{q+1}),$$

where for any q such that $i_p \le q < i_{p+1}$

$$\text{contrib}(v_q v_{q+1}) = \begin{cases} 1 & \text{when both } q \text{ and } q+1 \text{ are stop points ,} \\ \gamma & \text{when exactly one of } q, q+1 \text{ is a stop point,} \\ \gamma^2 & \text{when neither } q \text{ nor } q+1 \text{ is a stop point.} \end{cases}$$

It follows that

$$E[\text{contrib}(v_q v_{q+1})] \le \frac{1}{4} + \frac{1}{2}\gamma + \frac{1}{4}\gamma^2 = \frac{1}{4}(1+\gamma)^2. \tag{5}$$

The lemma follows by the linearity of expectation. $\qquad\square$

Note that the algorithm in Lemma 4 can be easily derandomized using the method of conditional expectation.

Lemma 5. *Let Q be as before. Then there is a deterministic polynomial-time algorithm which finds in Q a Hamiltonian cycle of weight at most $\frac{1}{4}(1+\gamma)^2 w(Q)$.*
$\qquad\square$

Lemma 5 and Corollary 3 immediately imply the following.

Corollary 5. *Let Q be a connected component in a directed multigraph G such that Q is 2-regular and has t vertices, $t \ge 3$. Let x be a vertex of G which does not belong to $V(Q)$. Then one can find in polynomial time a path P such that*

(i) path P ends in x,
(ii) $V(P) = V(Q) \cup \{x\}$, and
(iii) $w(P) \le \frac{1}{4}(1+\gamma)^2 \cdot \frac{t-1-(t-2)\gamma}{t(1-\gamma)} w(Q)$.
$\qquad\square$

3.3   The Complete Algorithm

Now we are going to combine the ingredients developed in the two previous sections in a complete approximation algorithm. Let K be an integer constant whose value will be determined later, $K \ge 1$. (K depends on the constant γ only). Intuitively, K is the maximum size of what we call a small component.

Let G be the input complete graph. Our algorithm begins by computing for each vertex $v \in V$ the value of $D(v) = \max_{x \in V} w(v, x)$, i.e. the maximum weight of an edge in G that leaves v. Let X_0 be a set of K vertices of V with smallest values of $D(v)$. The algorithm finds in $G[V \setminus X_0]$ the two cycle covers described in Theorem 4. Let M be the corresponding 2-regular multigraph.

Our algorithm builds a collection $\mathcal{H}$ of K vertex-disjoint directed paths which end in the vertices of X_0. Initially we put $\mathcal{H} = X_0$ (regarded as a collection of paths of length 0). For each connected component Q of M the algorithm extends the paths in $\mathcal{H}$ using the paths returned by the algorithm from Lemma 2 if Q is of size at most K or from Corollary 5 otherwise. We use the first vertices of the paths in $\mathcal{H}$ as the set X in Lemma 2, and x in Corollary 5 is the first vertex of any of these paths. Note that after each such extension $\mathcal{H}$ is still a collection of K vertex-disjoint paths ending in X_0.

After processing all the connected components of M, $\mathcal{H}$ is a collection of vertex-disjoint paths incident to all vertices of V. The algorithm finishes by patching $\mathcal{H}$ arbitrarily to a Hamiltonian cycle H.

It is clear that our algorithm returns a Hamiltonian cycle. Now we are going to bound its weight. Let $\mathcal{H}$ be the collection of K paths just before patching them to a Hamiltonian cycle. Let M_{small} (resp. M_{big}) be the union of the components of M that have size at most K (resp. at least $K+1$). By Lemma 2 and Corollary 5 we get (recall that $\frac{t-1-(t-2)\gamma}{t(1-\gamma)}$ is a decreasing function of t)

$$w(\mathcal{H}) \leq \frac{2-\gamma}{6(1-\gamma)} w(M_{\text{small}}) + \frac{1}{4}(1+\gamma)^2 \frac{K-(K-1)\gamma}{(K+1)(1-\gamma)} w(M_{\text{big}}).$$

Hence,

$$w(\mathcal{H}) \leq \max\left\{ \frac{2-\gamma}{6(1-\gamma)}, \frac{1}{4}(1+\gamma)^2 \frac{K-(K-1)\gamma}{(K+1)(1-\gamma)} \right\} w(M). \tag{6}$$

Now consider a minimum weight Hamiltonian cycle H^* in G and let OPT $= w(H^*)$. Let $H^*_{V \setminus X_0}$ be a minimum weight Hamiltonian cycle in $G[V \setminus X_0]$. Let C be a Hamiltonian cycle in $G[V \setminus X_0]$ which visitis vertices of $V \setminus X_0$ in the order they appear in H^*. Then,

$$w(H^*_{V \setminus X_0}) \leq w(C) \leq w(H^*) = \text{OPT}, \tag{7}$$

where the second inequality follows from the γ-parameterized triangle inequality. Since $w(M) \leq 2w(H^*_{V \setminus X_0})$ by Theorem 4, we get

$$w(\mathcal{H}) \leq \max\left\{ \frac{2-\gamma}{3(1-\gamma)}, \frac{1}{2}(1+\gamma)^2 \frac{K-(K-1)\gamma}{(K+1)(1-\gamma)} \right\} \text{OPT}. \tag{8}$$

Now it suffices to bound the weight of the edges added during the patching phase. We use the following lemma (proof deferred to the journal version).

Lemma 6. *For any fixed $\gamma \in [\frac{1}{2}, 1)$,*

$$\sum_{x \in X_0} D(x) = O(\tfrac{K}{n}\text{OPT})$$

Note that in our patching phase we use only edges leaving vertices of X_0. Hence using the above lemma we bound their weight by $O(\frac{K}{n}\text{OPT})$. This, together with (8) gives the following theorem.

Theorem 6. *For any integer $K \geq 1$ there is a polynomial time algorithm which finds a Hamiltonian cycle of weight at most*

$$\left[\max\left\{ \frac{2-\gamma}{3(1-\gamma)}, \frac{1}{2}(1+\gamma)^2 \frac{K-(K-1)\gamma}{(K+1)(1-\gamma)} \right\} + O(\tfrac{K}{n}) \right] \text{OPT}.$$

Corollary 6. *There is a constant γ_0, $\gamma_0 \approx 0.7003$ such that*

(i) *For any $\gamma \in (\gamma_0, 1)$ there is a polynomial time algorithm for ATSP with γ-parameterized triangle inequality, which has approximation ratio of $\frac{2-\gamma}{3(1-\gamma)} + O(\frac{1}{n})$.*

(ii) *For any $\gamma \in [\frac{1}{2}, \gamma_0]$, and for any $\epsilon > 0$ there is a polynomial time algorithm for ATSP with γ-parameterized triangle inequality, which has approximation ratio of $\frac{1}{2}(1+\gamma)^2 + \epsilon$.*

Proof. One can easily check that the inequality in the variable x

$$\frac{1}{2}(1+\gamma)^2 \frac{x - (x-1)\gamma}{(x+1)(1-\gamma)} \leq \frac{2-\gamma}{3(1-\gamma)}$$

has the set of solutions of the form $x \geq f(\gamma)$ for some function f, if γ satisfies the inequality

$$\frac{2-\gamma}{3} - \frac{1}{2}(1-\gamma)(1+\gamma)^2 > 0, \tag{9}$$

and has no solutions otherwise. Moreover, the set of solutions of (9) that belong to $[\frac{1}{2}, 1)$ is of the form $\gamma \in (\gamma_0, 1)$, where γ_0 is an irrational number, $\gamma_0 \approx 0.7003$. It follows that for any $\gamma \in (\gamma_0, 1)$ we can put $K = \lceil f(\gamma) \rceil$ in Theorem 6 and we obtain an algorithm with approximation ratio of $\frac{2-\gamma}{3(1-\gamma)} + O(\frac{1}{n})$.

For the second claim we know that whenever $\gamma \in [\frac{1}{2}, \gamma_0]$, the algortihm from Thorem 6 has an approximation ratio of $\frac{1}{2}(1+\gamma)^2 \frac{K-(K-1)\gamma}{(K+1)(1-\gamma)}$. It is easy to check that $\frac{1}{2}(1+\gamma)^2 + \epsilon$ upperbounds this for

$$K \geq \frac{\frac{1}{2}(1+\gamma)^2(2\gamma - 1)}{(1-\gamma)\epsilon} - 1.$$

$\square$

3.4   Tightness

As we mentioned earlier, the algorithm from Section 2 is optimal with respect to the cycle cover relaxation. Recall that for $\gamma \in (\gamma_0, 1)$ the algorithm from Section 2 has approximation ratio of $\frac{2-\gamma}{3(1-\gamma)} + O(\frac{1}{n})$. In what follows we show that this ratio is nearly optimal w.r.t. the double cycle cover relaxation of Kaplan et al. It is also nearly optimal w.r.t. the minimum cycle cover with no 2-cycles. Let $DC(G)$ be the half of the minimum weight of a pair of cycle covers of G described in Theorem 4 and let $AP_3(G)$ be the minimum weight of a cycle cover of G with no 2-cycles. The proof of the following theorem is deferred to the journal version.

Theorem 7. *For every $\gamma \in [\frac{1}{2}, 1)$, there exists an infinite family of graphs $\mathcal{G}$ such that for every $G \in \mathcal{G}$,*

$$\frac{ATSP(G)}{DC(G)} = \frac{ATSP(G)}{AP_3(G)} = \frac{2-\gamma}{3(1-\gamma)}.$$

It is an interesting open problem whether the approximation ratio of $\frac{2-\gamma}{3(1-\gamma)}$ can be achieved for all $\gamma \in [\frac{1}{2}, 1)$.

References

1. Bläser, M.: An improved approximation algorithm for the asymmetric tsp with strengthened triangle inequality. In: Baeten, J.C.M., Lenstra, J.K., Parrow, J., Woeginger, G.J. (eds.) ICALP 2003. LNCS, vol. 2719, pp. 157–163. Springer, Heidelberg (2003)
2. Bläser, M., Manthey, B., Sgall, J.: An improved approximation algorithm for the asymmetric tsp with strengthened triangle inequality. J. Discrete Algorithms 4(4), 623–632 (2006)
3. Chandran, L.S., Ram, L.S.: On the relationship between atsp and the cycle cover problem. Theor. Comput. Sci. 370(1-3), 218–228 (2007)
4. Chen, Z.-Z., Nagoya, T.: Improved approximation algorithms for metric max TSP. In: Brodal, G.S., Leonardi, S. (eds.) ESA 2005. LNCS, vol. 3669, pp. 179–190. Springer, Heidelberg (2005)
5. Christofides, N.: Worst-case analysis of a new heuristic for the travelling salesman problem. Technical Report 388, Graduate School of Industrial Administration, CMU (1976)
6. Feige, U., Singh, M.: Improved approximation ratios for traveling salesperson tours and paths in directed graphs. In: Charikar, M., Jansen, K., Reingold, O., Rolim, J.D.P. (eds.) RANDOM 2007 and APPROX 2007. LNCS, vol. 4627, pp. 104–118. Springer, Heidelberg (2007)
7. Kaplan, H., Lewenstein, M., Shafrir, N., Sviridenko, M.: Approximation algorithms for asymmetric TSP by decomposing directed regular multigraphs. J. ACM 52(4), 602–626 (2005)
8. Kostochka, A.V., Serdyukov, A.I.: Polynomial algorithms with the estimates 3/4 and 5/6 for the traveling salesman problem of the maximum (in Russian). Upravlyaemye Sistemy 26, 55–59 (1985)
9. Kowalik, L., Mucha, M.: 35/44-approximation for asymmetric maximum TSP with triangle inequality. In: Dehne, F., Sack, J.-R., Zeh, N. (eds.) WADS 2007. LNCS, vol. 4619, pp. 589–600. Springer, Heidelberg (2007)
10. Manthey, B.: On approximating restricted cycle covers. SIAM J. Comput. 38(1), 181–206 (2008)
11. Zhang, T., Li, W., Li, J.: An improved approximation algorithm for the atsp with parameterized triangle inequality. J. Algorithms (in press) (2008)

Streaming Embeddings with Slack*

Christiane Lammersen[1], Anastasios Sidiropoulos[2], and Christian Sohler[1]

[1] Department of Computer Science,
TU Dortmund, 44221 Dortmund, Germany
[2] Department of Computer Science,
University of Toronto, Ontario M5S 3G4

Abstract. We study the problem of computing low-distortion embeddings in the streaming model. We present streaming algorithms that, given an n-point metric space M, compute an embedding of M into an n-point metric space M' that preserves a $(1-\sigma)$-fraction of the distances with small distortion (σ is called the *slack*). Our algorithms use space polylogarithmic in n and the spread of the metric. Within such space limitations, it is impossible to store the embedding explicitly. We bypass this obstacle by computing a compact representation of M', without storing the actual bijection from M into M'.

1 Introduction

Over the last few years, computation on large data sets has become an important algorithmic paradigm with many practical applications. Examples of such data sets include the web graph, internet traffic logs, click-streams, and genome data. In many scenarios, the data is only available in the form of a stream of objects. From an application point of view, we are typically interested in a small summary rather than the whole raw data. For example, when analyzing internet traffic logs, we do not want to track every single packet, but we want to have a high-level view of the data to detect anomalies like denial of service attacks, spreading viruses, etc., or we simply want to be able to give trustworthy statistics about network load, packets exchanged with other internet service providers, etc.

One approach to data stream analysis is the development of streaming algorithms, i.e. algorithms that process data sequentially while using only a small amount of memory compared to the size of the data set. Most of these algorithms can be assigned to two categories: they either maintain specific statistics of the stream, such as the number of distinct elements or the frequency moments, or they try to approximate the raw data by a small summary, for example, by maintaining a histogram or a clustering of the observed data.

In this paper, we give a general approach to the second type of problems. We consider the case where the data stream is a sequence of points from a metric space $M = (X, D)$. Our goal is to compute a compact representation of M that captures the pairwise distances of M well and uses only sublinear space. Unfortunately, unless M is very simple (e.g., X is a multiset with many duplicates),

---

* Partially supported by DFG grant So 514/1-2.

F. Dehne et al. (Eds.): WADS 2009, LNCS 5664, pp. 483–494, 2009.

one cannot find a sublinear space representation of M, such that all pairwise distances are preserved. It is not even possible to guarantee that all pairwise distances are preserved up to any fixed factor. Simply said, it is unavoidable that we loose some distances in the sense that they can get arbitrarily distorted. The total loss of information is quantified via the notion of low-distortion embeddings with slack. A (non-contracting) embedding from a metric space $M = (X, D)$ into a metric space $M' = (X', D')$ is a mapping $\phi : X \rightarrow X'$, such that, for every $x, y \in X$, we have $D(x, y) \leq D'(\phi(x), \phi(y))$. Such an embedding has distortion c if, for any $x, y \in X$, $D'(\phi(x), \phi(y)) \leq c \cdot D(x, y)$. An embedding with distortion c and slack σ is a mapping that satisfies $D(x, y) \leq D'(\phi(x), \phi(y)) \leq c \cdot D(x, y)$ for all but a σ-fraction of pairs $x, y \in X$.

Our Results. We initiate the study of streaming algorithms for metric embeddings with slack. We obtain the following results:

- There is a streaming algorithm which for any $\varepsilon, \sigma > 0$, given a dynamic data stream of insert and delete operations of points from a (possibly high-dimensional) Euclidean space, maintains a point set P' in a constant-dimensional Euclidean space, such that the current input point set P embeds into P' with distortion $1 + \varepsilon$ and slack σ.
- We use an embedding technique to show that there is a streaming algorithm that maintains a $(1 + \varepsilon)$-approximation of the Max-Cut problem for a dynamic data stream of high-dimensional, Euclidean points.
- There is a streaming algorithm which for any $\varepsilon, \sigma > 0$, given a stream of points from an n-point metric space $M = (X, D)$ with bounded doubling dimension, computes an implicit representation of an n-point metric space $M' = (X', D)$, such that M embeds into M' with distortion $1 + \varepsilon$ and slack σ.
- There is a streaming algorithm which for any $\sigma > 0$, given a stream of points from a general n-point metric space $M = (X, D)$, computes an implicit representation of an n-point metric space $M' = (X', D')$, such that M embeds into M' with $\mathcal{O}(1)$ distortion and slack σ.
- Any algorithm that computes with constant probability, for every n-point metric space $M = (X, D)$, an (implicit or explicit) representation of another metric space $M' = (X', D')$, such that M embeds into M' with distortion less than 2 and slack at most $1/5$, requires space $\Omega(n/\log n + \log \log \Delta)$, where Δ is the ratio between the smallest and largest distance in M.

Our Techniques. The construction of the compact representation for Euclidean metrics is based on a certain quadtree partition of the input space into a few cells and an elaborate refinement of this partition, where each cell is further subdivided into a few subcells. The set of points within each subcell is replaced by a single weighted point, whose weight is determined by the number of points it replaces. We show that we can use a random sampling technique to estimate the number of points in every cell. The resulting streaming algorithm works well for low-dimensional data. In case that the input points are high-dimensional, we first use a dimension reduction to map the points to a constant-dimensional,

Euclidean space with distortion $1 + \varepsilon$ and slack σ. Afterwards, we apply our streaming algorithm on the projected data.

We develop a different approach in order to obtain a constant distortion for general metrics. In this case, our embedding result is based on the existence of certain subsets of points called *edge-dense nets*. Intuitively, an edge-dense net N of a metric space $M = (X, D)$ has the property that, for a $(1-\sigma)$-fraction of pairs of points $x, y \in X$, the distance between N and both x and y is small compared to $D(x, y)$. The existence of such nets follows from results on embeddings with beacons by Kleinberg et al. [12]. After some modifications, this allows us to compute the embedding with a single-pass streaming algorithm. Our algorithm resembles the construction of spanners with slack of Chan et al. [5].

Related Work. Kleinberg et al. [12] introduced the notion of embeddings with slack. Among other results, they obtained, for any $\sigma > 0$, a beacon-based embedding of general metric spaces with distortion $\mathcal{O}(1)$ and slack σ. For metrics with bounded doubling dimension, they improved the distortion to $1 + \varepsilon$, for any $\varepsilon > 0$. Abraham et al. [1] extended these results to arbitrary metric spaces and for embeddings under any ℓ_p norm, $p \geq 1$. In [2], the authors considered embeddings with low distortion and slack for arbitrary metric spaces that additionally guarantee constant average distortion. Metric approximation with slack has also been investigated in the setting of graph spanners. Chan et al. [5] showed that, for any weighted graph G and any $\varepsilon > 0$, there exists a spanner of G with linear number of edges achieving stretch $\mathcal{O}(\log(1/\varepsilon))$ and slack ε. The authors also gave a spanner construction, which we take as a starting point for our embedding of general metric spaces. In order to transform this construction to the streaming model, we use a technique that has been earlier applied by Czumaj and Sohler [6] to achieve 2-pass streaming algorithms for clustering problems. Embeddings of point sets into trees (via a quadtree partitioning) have been used by Indyk [10] to obtain approximation algorithms for several geometric problems. Also, Frahling and Sohler [9] applied a similar quadtree partitioning to get streaming algorithms for different clustering problems. A somewhat similar partitioning technique is used for embedding Euclidean metric spaces in this paper.

2   Preliminaries

An n-point metric space M is a pair (X, D), where X is a set of n points and D is a symmetric, non-negative mapping $D : X \times X \to \mathbb{R}$ that satisfies the triangle inequality. Throughout the paper, we assume that the minimum pairwise distance in M is at least 1 and the maximum pairwise distance at most Δ. Furthermore, we assume that each coordinate of a point can be represented by using one memory cell and, given two point labels, we can compute the distance between the associated points in a constant number of time units. These assumptions are commonly made in computational geometry. Thus, unless otherwise stated, we measure the running time of our algorithms in time units und the space requirement in memory cells.

We will consider three kinds of metric spaces, namely general metrics spaces, where the only restrictions on the distance function D are as given above, Euclidean spaces, where D is determined by the L_2-norm, and doubling metric spaces. A metric $M = (X, D)$ is called doubling metric if, for any constant dimension λ, each ball with any radius r centered at any point in X can be covered by 2^λ balls each of radius $r/2$ and centered at a point in X. Given n points of such metric spaces, our streaming algorithms compute compact representations that uses only sublinear space. Note that we assume that the parameter n is known in advance by our algorithms. Besides the space requirement, the quality of a representation is measured by the quantity of the distortion and the slack.

Definition 1. *Given $c \geq 1$ and $\sigma > 0$, an embedding $\phi : X \to X'$ from a finite metric space $M = (X, D)$ into a target metric space $M' = (X', D')$ has distortion c and slack σ if $D(x, y) \leq D'(\phi(x), \phi(y)) \leq c \cdot D(x, y)$ is true for all but a σ-fraction of pairs $x, y \in X$.*

Our streaming algorithms for general and doubling metric spaces work in the insertion-only model, whereas the ones for Euclidean metric spaces work in the dynamic geometric data stream model. In the first model, the input is a sequence of insert operations of points. Algorithms are only allowed to perform one sequential scan over the data and to use local memory that is polylogarithmic in the size of the input stream. In our scenario, the available local memory space is $\log^{\mathcal{O}(1)}(n + \Delta)$, where n is the length of the input stream and Δ is the spread of the input points. In the latter model, the input is a sequence of m INSERT and DELETE operations of points from a discrete, Euclidean space $\{1, \ldots, \Delta\}^d$ [10]. We assume that the stream is consistent, i.e., no point is removed that is not present in the current point set, and no point is added twice. We use n as an upper bound on the size of the current point set. Obviously, we have $n = \mathcal{O}(\Delta^d)$. Algorithms are only allowed to perform one sequential scan over the input stream and to use local memory space that is polylogarithmic in m, n, and Δ.

3   Embedding Euclidean Metrics

In this section, we consider compact representations of Euclidean metric spaces. Let $M = (P, D)$ be any given n-point Euclidean metric space. Then, the goal is to find an embedding $\phi : M \to M'$ with low distortion and low slack, such that M' requires only sublinear space. One useful concept is the notion of well-separated pair decomposition (WSPD) [4].

A WSPD of size t allows all pairwise distances to be compactly summarized by t distances. Formally speaking, for any constant $\varepsilon > 0$, two point sets A and B are ε-well-separated if $\max\{\mathrm{diam}(A), \mathrm{diam}(B)\} \leq \varepsilon \cdot D(A, B)$, where $D(A, B)$ is the minimum distance from any point in A to any point in B and $\mathrm{diam}(A)$ and $\mathrm{diam}(B)$ is the diameter of A and B, respectively. Based on this definition, an ε-WSPD for M of size t is a collection of ε-well-separated pairs of subsets $\mathcal{P} = \{(A_1, B_1), \ldots, (A_t, B_t)\}$ with $A_j, B_j \subseteq P$ for $1 \leq j \leq t$, such that every pair of points $(a, b) \in P \times P$, $a \neq b$, lies in $A_j \times B_j$ or $B_j \times A_j$ for exactly one index

j, $1 \leq j \leq t$. The usefulness is now that if we store instead of each pair (A_j, B_j) a pair of representative points $(R(A_j), R(B_j))$, such that the maximum distance from a point in A_j (B_j) to $R(A_j)$ ($R(B_j)$) is $\text{diam}(A_j)/2$ ($\text{diam}(B_j)/2$), then $D(R(A_j), R(B_j))$ is a $(1 + \varepsilon)$-approximation for all the distances between pairs of points from $A_j \times B_j$.

In general, one assumes that the size of an ε-WSPD is linear in n. Since we restrict the space requirement of the representation of M to be sublinear in n and there does not exist an ε-WSPD for any metric M and any constant ε that has sublinear size, we introduce the notion of slack WSPD.

Definition 2. *For any n-point metric $M = (P, D)$, let $\mathcal{P}$ be a collection of pairs of subsets $\{(A_1, B_1), \dots, (A_t, B_t)\}$, where $A_j, B_j \subseteq P$ for $1 \leq j \leq t$, and let I_ε be the subset of indices, such that, for all $i \in I_\varepsilon$, (A_i, B_i) is ε-well-separated. $\mathcal{P}$ is called an ε-WSPD with slack σ for M, if every pair of points $(a, b) \in P \times P$, $a \neq b$, lies in $A_j \times B_j$ or $B_j \times A_j$ for exactly one index $j \in \{1, \dots, t\}$ and $\sum_{i \in I_\varepsilon} |A_i| \cdot |B_i| \geq (1 - \sigma) \cdot n^2$.*

3.1   Slack WSPD in Low Dimensions

In order to construct an ε-WSPD with slack σ for any n-point Euclidean space $M = (P, D)$ with constant dimension d, we impose $\log_2(\Delta) + 1$ nested squared grids over P denoted by $G(0), G(1), \dots, G(\log_2(\Delta))$. The side length of each cell in grid $G(i)$ is 2^i. We say that the grid cells in $G(i)$ are in level i.

In this subsection, we describe the construction without considering streaming. Our algorithm consists of three phases. In the first phase, we compute a partitioning of the space based on the heavy cells in the grids (see Definition 3). Then, it follows a refinement phase, where each cell of the space partitioning is further subdivided into cubelets. In the last phase, we determine a so-called *representative* for each cubelet and compute an ε-WSPD with slack σ from the set of representatives.

Definition 3. *We call a grid cell heavy if it contains at least $h(\sigma) \cdot n$ points of P, where $h(\sigma) := \sigma/2^d$ is a function dependent on σ. A grid cell that is not heavy is light.*

Now, we give a detailed description of the three phases. In the first phase, we build a quadtree partitioning of the point space. We start our construction with the coarsest grid $G(\log_2(\Delta))$. We identify every heavy cell in $G(\log_2(\Delta))$, i.e. cells containing at least $h(\sigma) \cdot n$ points. Then, we subdivide every heavy cell C into 2^d equal sized subcells. These subcells are contained in grid $G(\log_2(\Delta) - 1)$. We call C the parent cell of these subcells. If none of the subcells is heavy, we stop our process. Otherwise, the algorithm recursively subdivides every heavy cell, such that, at the end of the first phase, we have only light cells in our space partitioning. The refinement phase consists of three steps. The first refinement is that we build a so-called balanced or restricted quadtree of the quadtree that we obtained so far, i.e., the side length of each cell is allowed to differ from the side lengths of all neighboring cells by a factor of at most 2 [3]. That means that we further subdivide every leaf cell C of the quadtree which has a neighboring cell whose side length is less than half of the side length of C. We say that two cells are neighbors if they share some part of the boundary. In the second step, every leaf cell of the balanced quadtree is subdivided into ℓ^d equal sized cubes, where $\ell = 6\sqrt{d}$. Finally, we subdivide every cube into $\ell^*(\varepsilon)^d$ equal sized cubelets, where $\ell^*(\varepsilon) = 2\sqrt{d}/\varepsilon$ is a function dependent on ε. Note that we could have merged the second and third refinement step into one step, but the definition of cubes makes the analysis easier. For each cubelet C, we replace all the points inside of C by one representative, which is set in the center of C and weighted by the number of replaced points. The collection of all representative pairs is our ε-WSPD with slack σ for M.

Analysis. Let $L(i)$ be the subset of all the leaf cells of the quadtree whose side length is 2^i, i.e. leaf cells in level i, and let $\mathcal{L}(i)$ be the set of leaf cells of the balanced quadtree whose side length is 2^i. Furthermore, we define $\mathcal{L}^*(i)$ to be the set of all the cubes contained in a cell in $\mathcal{L}(i)$. Finally, we denote the set of heavy cells in level i that do not have a heavy subcell by $H(i)$. Notice that the parent cell of any cell in $L(i)$ is in $H(i+1)$.

The following lemmas give evidence that the collection of all representative pairs is an ε-WSPD with slack σ for M. Moreover, we show that, for each point in P, the distances to its σn closest neighbors in P can be arbitrarily distorted, but the distances to all other points in P are $(1+\varepsilon)$-preserved.

Lemma 1. *If each cube in the set $\bigcup_{i=0}^{\log_2(\Delta)} \mathcal{L}^*(i)$ is divided into $\ell^*(\varepsilon)^d$ equal sized cubelets, where $\ell^*(\varepsilon) = 2\sqrt{d}/\varepsilon$, then any two cubelets, which are not contained in the same cube or in neighboring cubes, are ε-well-separated.*

Proof. Let C_1 and C_2 be any two cubelets, which are not contained in the same cube or in neighboring cubes. Furthermore, let C_1 be in any level i and C_2 be in any level j. We consider the two cases $j \in \{i, i+1\}$ and $j \geq i+2$.

We start with the case $j \geq i+2$. Due to the balanced quadtree partitioning, the side lengths of neighboring cells in $\bigcup_{i=0}^{\log_2(\Delta)} \mathcal{L}(i)$ differ at most by a factor of 2. Hence, the distance between any cell in $\mathcal{L}(i)$ and any cell in $\mathcal{L}(j)$ with $j \geq i+2$ is at least $\sum_{k=i+1}^{j-1} 2^k = 2^j - 2^{i+1} \geq 2^{j-1}$. Since C_1 is contained in a cell in $\mathcal{L}(i)$ and C_2 is contained in a cell in $\mathcal{L}(j)$, the distance between C_1 and C_2 is at least 2^{j-1}. Since the diagonal of the bigger cubelet C_2 is $\sqrt{d} \cdot 2^j/(\ell \cdot \ell^*(\varepsilon)) \leq \varepsilon \cdot 2^{j-1}$, the two cubelets C_1 and C_2 are ε-well-separated.

In the case $j \in \{i, i+1\}$, the distance between C_1 and C_2 is at least $2^i/\ell$. Since the diagonal of C_2 is at most $\varepsilon \cdot 2^i/\ell$, C_1 and C_2 are ε-well-separated. $\square$

Lemma 2. *Let $h(\sigma) = \sigma/2^d$ and let p_1 and p_2 be any two points in P. If the cubelet that contains p_1 and the cubelet that contains p_2 are not ε-well-separated, then p_2 belongs to the σn closest points of p_1.*

Proof. At first, we bound the maximum distance $D(p_1, p_2)$ between p_1 and p_2. Let $C_1^* \in \mathcal{L}^*(i)$ be the cube that contains p_1 and C_2^* be the cube that contains p_2. Due to Lemma 1, C_1^* and C_2^* must be neighbors. Since we use a balanced quadtree partitioning, the side lengths of C_1^* and C_2^* differ at most by a factor of 2. Since $C_1^* \in \mathcal{L}^*(i)$, the side lengths of C_1^* is $2^i/\ell$ with $\ell = 6\sqrt{d}$. We have to consider the cases (i) $C_2^* \in \mathcal{L}^*(i)$, (ii) $C_2^* \in \mathcal{L}^*(i+1)$, and (iii) $C_2^* \in \mathcal{L}^*(i-1)$. In case (i), the maximum distance between p_1 and p_2 is at most $\sqrt{d} \cdot (2^i/\ell + 2^i/\ell) \leq 2^{i-1} - 2^i/\ell$. Since the cube C_1^* is contained in a cell in $\mathcal{L}(i)$ and we use a balanced quadtree partitioning, the side lengths of all neighboring cells of the cell containing C_1^* is at least 2^{i-1}. Thus, the ball with center p_1 and radius $2^{i-1} - 2^i/\ell \geq D(p_1, p_2)$ is covered by at most 2^d cells in $\bigcup_{k=0}^{\log_2(\Delta)} \mathcal{L}(k)$. In case (ii), the maximum distance between p_1 and p_2 is at most $\sqrt{d} \cdot (2^i/\ell + 2^{i+1}/\ell) \leq 2^i - 2^i/\ell$. Furthermore, since C_2^* is contained in a cell in $\mathcal{L}(i+1)$, the side lengths of all common neighbors of the cells containing C_1^* and C_2^* is at least 2^i. Thus, the ball with center p_1 and radius $2^i - 2^i/\ell \geq D(p_1, p_2)$ can be covered by at most 2^d cells in $\bigcup_{k=0}^{\log_2(\Delta)} \mathcal{L}(k)$. Case (iii) is symmetric to case (ii). As a result, in all cases we have to count the number of points in 2^d cells in $\bigcup_{k=0}^{\log_2(\Delta)} \mathcal{L}(k)$. Since the cells in $\bigcup_{k=0}^{\log_2(\Delta)} \mathcal{L}(k)$ are light cells, each one contains at most $h(\sigma) \cdot |P|$ points. It follows that the number of points, whose distance from p_1 is at most $D(p_1, p_2)$, is at most $\sigma \cdot |P|$. $\square$

The complexity of our algorithm is given as follows.

Lemma 3. *The space partitioning consists of $\mathcal{O}(\frac{2^{\mathcal{O}(d)} \cdot d^d \cdot \log(\Delta)}{\varepsilon^d \sigma})$ cubelets and can be computed in $\mathcal{O}(n \cdot \log(\Delta) + \frac{2^{\mathcal{O}(d)} \cdot \log^2(\Delta)}{\sigma})$ time and $\mathcal{O}(n + \frac{2^{\mathcal{O}(d)} \cdot d^d \cdot \log(\Delta)}{\varepsilon^d \sigma})$ space.*

Due to our construction, the ε-WSPD with slack σ for M can also be seen as an embedding for M with distortion $1+\varepsilon$ and slack σ. Let $R(p)$ be the representative of a point $p \in P$, then the embedding is given by $R : P \to P'$.

Theorem 1. *Given any $\varepsilon, \sigma > 0$, there exists an algorithm that computes for n points P from a Euclidean space $\mathbb{R}^d$ a point set $P' \subset \mathbb{R}^d$ of size $\mathcal{O}(\frac{2^{\mathcal{O}(d)} \cdot d^d \cdot \log(\Delta)}{\varepsilon^d \sigma})$, such that P embeds into P' with distortion $1 + \varepsilon$ and slack σ. The algorithm requires $\mathcal{O}(n \cdot \log(\Delta) + \frac{2^{\mathcal{O}(d)} \cdot \log^2(\Delta)}{\sigma})$ time and $\mathcal{O}(n + \frac{2^{\mathcal{O}(d)} \cdot d^d \cdot \log(\Delta)}{\varepsilon^d \sigma})$ space.*

3.2   Streaming Algorithm for Slack WSPD

In this subsection, we explain how to compute a compact representation of a Euclidean metric $M = (P, D)$ given as a dynamic geometric data stream. The idea is simply to maintain a random sample of the current point set and to apply the algorithm described in Subsection 3.1 on the sample set. This is done as follows.

We read the items of the input stream one by one. Each time, we decide whether we use the associated point for further computations or not. For that purpose, we use the technique given in [8] to maintain a sample set of the current point set P with size $s = \Theta(N \log(n)/\sigma^3)$, where $N = \mathcal{O}(2^{\mathcal{O}(d)} d^d \log(\Delta)/(\varepsilon^d \sigma))$ is an upper bound on the number of cubelets in the space partition (cf. Lemma 3). We denote this sample set by S. After the sample step, we build the balanced quadtree partitioning for S and perform the refinement into equal sized cubelets as described in Subsection 3.1. For each cubelet C that contains at least $\lceil \ln(n)/\sigma^2 \rceil$ sample points, we replace the points in C by one representative. This point is set to the center of C and weighted by $\lceil |C| \cdot n/s \rceil$, where $|C|$ denotes the number of replaced points. To avoid that the total weight of the representatives differs from n, we sum up all weights and increase or decrease the weight of an arbitrary representative by the required amount. Finally, the collection of all representative pairs is our ε-WSPD with slack σ for M.

Analysis. The technique described in [8] allows us to maintain a sample set, such that every sample point is chosen nearly uniformly at random from P.

Lemma 4 (Frahling et al. [8]). *Let $\delta > 0$ be an error probability parameter. Given a sequence of* INSERT *and* DELETE *operations of points from the discrete, Euclidean space $\{1, \ldots, \Delta\}^d$, there is a data structure that with probability $1 - \delta$ returns s points $q_0, \ldots, q_{s-1}$ from the current point set $P = \{p_0, \ldots, p_{n-1}\}$, such that $\mathbf{Pr}\left[q_i = p_j\right] = \frac{1}{n} \pm \frac{\delta}{\Delta^d}$ for every $j \in \{0, \ldots, n-1\}$. The algorithm has an update time of $\mathcal{O}((s + \log(1/\delta)) \cdot d \cdot \log(\Delta))$ and needs a memory space of $\mathcal{O}((s + \log(1/\delta)) \cdot d^2 \cdot \log^2(\Delta/\delta))$.*

Due to the fact that we use a sample set to estimate the number of points in a cubelet, we make an error which increases the slack. The following result can be obtained by applying Chernoff Bounds.

Lemma 5. *Let N be the number of cubelets in the space partitioning. For a large enough constant c, the number of points in cubelets that contain at most $\frac{\sigma n}{cN}$ points from P is at most σn. The number of points in cubelets that contain at least $\frac{\sigma n}{cN}$ points from P can be σ-approximated by S with probability $1 - 1/n$.*

We summarize our results in the following theorem. The time and memory space requirement of our algorithm is dominated by the time and memory requirement needed by the sampling data structure.

Theorem 2. *Let $\varepsilon > 0$, $\sigma > 0$, and $\delta > 0$. Given a stream of* INSERT *and* DELETE *operations of points from a discrete, Euclidean space $\{1, \ldots, \Delta\}^d$, there is a streaming algorithm that maintains with probability $1 - \delta$, for the current point set P of size n, a point set $P' \subset \mathbb{R}^d$ of size $\mathcal{O}(\frac{2^{\mathcal{O}(d)} \cdot d^d \cdot \log(\Delta)}{\varepsilon^d \sigma})$, such that P embeds into P' with distortion $1 + \varepsilon$ and slack σ. The algorithm has an update time of $\mathcal{O}((\frac{2^{\mathcal{O}(d)} \cdot d^d \cdot \log(n) \cdot \log(\Delta)}{\varepsilon^d \sigma^4} + \log(1/\delta)) \cdot d \cdot \log(\Delta))$ and needs a memory space of $\mathcal{O}((\frac{2^{\mathcal{O}(d)} \cdot d^d \cdot \log(n) \cdot \log(\Delta)}{\varepsilon^d \sigma^4} + \log(1/\delta)) \cdot d^2 \cdot \log^2(\Delta/\delta))$.*

3.3   Slack WSPD in High Dimensions

If the points in P have a high dimension, we first use the Johnson-Lindenstrauss embedding [11] with $d(\varepsilon, \sigma) = \Theta(1/(\varepsilon^2 \sigma))$ dimensions to get an embedding into a constant-dimensional space with distortion $1 + \varepsilon$ and slack σ. Afterwards, we apply the techniques described in Subsections 3.1 and 3.2 on the constant-dimensional point set. The resulting representation is an 3ε-WSPD with slack 2σ. We get the following result.

Theorem 3. *Let P be a set of n points in $\mathbb{R}^d$, $\varepsilon, \sigma > 0$, and $d(\varepsilon, \sigma) = 2/(\varepsilon^2\sigma\delta)$ be a function dependent on ε and σ. Then there exists an embedding $\pi : P \to \mathbb{R}^{d(\varepsilon,\sigma)}$, such that $(1 - \varepsilon) \cdot D(p,q) \leq D(\pi(p), \pi(q)) \leq (1 + \varepsilon) \cdot D(p,q)$ is true for at least $(1 - \sigma) \cdot n^2$ pairs $p, q \in P$ with probability at least $1 - \delta$.*

3.4   Max-Cut in High Dimensions

In this subsection, we show how to embed a set of high-dimensional, Euclidean points into a constant-dimensional, Euclidean space, such that the sum of the pairwise distances is $(1 + \varepsilon)$-preserved. Afterwards, we apply our result to the Max-Cut problem.

Let $\phi : P \to \mathbb{R}^{d(\varepsilon)}$ be the Johnson-Lindenstrauss embedding, where each point is mapped into a Euclidean space with $d(\varepsilon) = \Theta(1/(\varepsilon^2\delta^2))$ dimensions. By using similar techniques as for the proof of Theorem 3, we can show that, for a pair of points $p, q \in P$, the expected value of $|D(\phi(p), \phi(q)) - D(p,q)|$ is $\delta\varepsilon \cdot D(p,q)$ and $|D(\phi(p), \phi(q)) - D(p,q)|$ is concentrated about its expected value with probability $1 - \delta$. These facts imply the following lemma.

Lemma 6. *Let P be a set of n points in $\mathbb{R}^d$, $0 < \varepsilon < 1$, and $d(\varepsilon) = 50/(\varepsilon^2\delta^2)$ be a function dependent on ε. Then there exists an embedding $\phi : P \to \mathbb{R}^{d(\varepsilon)}$, such that*

$$(1 - \varepsilon) \sum_{p,q \in P^2} D(p,q) \leq \sum_{p,q \in P^2} D(\phi(p), \phi(q)) \leq (1 + \varepsilon) \sum_{p,q \in P^2} D(p,q)$$

is true with probability at least $1 - \delta$.

Given any point set P, the embedding described above is useful for all geometric problems that satisfy the following four properties:

(i) The cost of an optimal solution $\text{cost}(P)$ for P is a function, whose set of input parameters is a subset of all pairwise distances of P.

(ii) The cost of an optimal solution $\text{cost}(P)$ for P is at least $\sum_{p,q \in P^2} 1/c \cdot D(p,q)$, where $c \geq 1$ is any small constant.

(iii) If the distance $D(p,q)$ between any two points $p, q \in P$ is increased or decreased by any value $\alpha > 0$, then the cost of an optimal solution $\text{cost}(P)$ for P is increased or decreased by at most $\mathcal{O}(\alpha)$.

(iv) The complexity of all known $(1 + \varepsilon)$-approximation algorithms depends exponentially on the dimension of P.

To handle these problems efficiently, we just embed the input points first and afterwards apply any $(1 + \varepsilon)$-approximation algorithm on the embedded points.

One suitable problem is the Max-Cut problem in the dynamic data stream model. Here, the goal is to find a partition of a point set P into two subsets C_1 and C_2, such that the sum $\sum_{p,q \in C_1 \times C_2} D(p,q)$ of inter-cluster distances is maximized. Obviously, the Max-Cut problem satisfies properties (i) and (iii). Furthermore, the authors of [9] showed that property (ii) is fulfilled for $c = 4$. By combining the embedding of Lemma 6 with the approximation algorithm presented in [9], we obtain the following result.

Theorem 4. *Let $\varepsilon > 0$ be a precision parameter. Given a stream of m* INSERT *and* DELETE *operations of points from a high-dimensional, discrete, Euclidean space $\{1, \ldots, \Delta\}^d$, there is a streaming algorithm that maintains with probability at least $5/8$, for the current point set P of size n, a data structure of size $\mathcal{O}(\log^7(\Delta mn/\varepsilon)/\varepsilon^{\mathcal{O}(1/\varepsilon^2)})$ from which an implicit solution for the max-cut problem can be extracted in $\mathrm{poly}((1/\varepsilon)^{1/\varepsilon^2}, \log(\Delta), \log(n), \log(m))$ time. An update requires $\mathcal{O}(\log^3(\Delta nm/\varepsilon) + d/\varepsilon^2)$ time and $\mathcal{O}(\log(d)/\varepsilon^2)$ space.*

4   Embedding Doubling Metrics

Our approach for Euclidean spaces can be tranfered to doubling metrics. The idea is just to replace the nested squared grids in Subsection 3.1 by a hierarchical cut decomposition. Due to space limitations, we only summarize our results.

Theorem 5. *Let $\varepsilon > 0$ and $\sigma > 0$. Given a stream of points from an n-point metric $M = (X, D)$ with bounded doubling dimension λ, there exists a streaming algorithm that maintains with probability $1 - 1/n$ a metric space $M' = (X', D)$ with $|X'| = \mathcal{O}(\frac{2^{\mathcal{O}(\lambda)} \cdot \log(\Delta)}{\varepsilon^\lambda \sigma})$, such that X embeds into X' with distortion $1 + \varepsilon$ and slack σ. The algorithm requires $\mathcal{O}(\frac{2^{\mathcal{O}(\lambda)} \cdot \log(n) \cdot \log(\Delta)}{\varepsilon^\lambda \cdot \sigma^4})$ memory space and has a constant update time. The set X' can be extracted in $\mathcal{O}(\frac{2^{\mathcal{O}(\lambda)} \cdot \log^2(n) \cdot \log^3(\Delta)}{\varepsilon^{2\lambda} \cdot \sigma^8})$ time.*

5   Embedding General Metrics

In this section, we give a streaming algorithm for embedding a general metric space M into a metric space $M' = (X', D')$ with constant distortion and slack σ in the insertion-only model. The algorithm relies on techniques developed in [5] and [6].

Let $p = 2m \log_2(n) \log_2(\Delta)/(\sigma^2 n)$, where m is a constant depending on σ. We sample each point in the stream with probability p. Let $S = s_1, \ldots, s_k$ be the set of sampled points. We maintain the distance between each pair of points in S. Moreover, for each $i \in [k]$, we maintain $\log_2(\Delta)$ counters $c_{i,1}, \ldots, c_{i,\log_2(\Delta)}$, which are initially set to 0. For each point $x \notin S$, we compute its nearest neighbor $\tau(x) = s_i$ in S, at the point of time when x appears in the stream, and we increment $c_{i,j}$, where $j = \lceil \log_2(D(x, s_i)) \rceil$.

By storing the distances between points in S and the counters $c_{i,j}$, we implicitly store the following metric space M'. The metric M' is the shortest-path metric of a graph G with vertex set X. For each pair of points $s_i, s_j \in S$, the graph G contains an edge $\{s_i, s_j\}$ of length $D(s_i, s_j)$. For each point $x \notin S$, the graph G contains an edge $\{x, \tau(x)\}$, of length $2^{\lceil \log_2(D(x, \tau(x))) \rceil}$. Let φ be the resulting embedding. Note that we do not store the mapping $\varphi : X \to X'$, since this would require $\Omega(n)$ space.

Analysis. In order to prove that the embedding has constant distortion and slack σ, we first show that M contains some small edge-dense net.

Definition 4 ((σ, γ)-Edge-Dense Net). *For a metric space $M = (X, D)$, a subset $N \subset X$ is a (σ, γ)-edge-dense net if, for at least a $(1 - \sigma)$-fraction of pairs $x, y \in X$, there exist $b_x, b_y \in N$, such that $\max\{D(x, b_x), D(y, b_y)\} \leq \gamma \cdot D(x, y)$.*

Lemma 7 (Kleinberg et al. [12]). *For any metric space $M = (X, D)$ and for any slack parameter $\sigma > 0$, there exists $N \subset X$ with $|N| = C(\sigma)$, where $C(\sigma)$ is a constant depending on σ, such that, for at least a $(1 - \sigma)$-fraction of pairs $x, y \in X$, we have $\min_{b \in N}\{D(x, b), D(y, b)\} \leq D(x, y)$.*

The next lemma follows immediately from Lemma 7 and the triangle inequality.

Lemma 8. *For a metric space $M = (X, D)$ and for any $\sigma > 0$, there exists a $(\sigma, 2)$-edge-dense net $N \subset X$ with $|N| = C(\sigma)$, where $C(\sigma)$ is a constant depending on σ.*

Now, let $N = \{y_1, \ldots, y_m\}$ be a $(\sigma, 2)$-edge-dense net for the input metric M and $\sigma \in (0, 1)$. For each $\ell \in [m]$, let X_ℓ be the set of points in X, for which the nearest neighbor in N is y_ℓ (breaking ties arbitrarily). Also, for each $j \in [\log_2(\Delta)]$, let $X_{\ell,j} = \{x \in X_\ell : D(x, y_\ell) \in (2^{j-1}, 2^j]\}$. We say that $X_{\ell,j}$ is *good*, if after $\sigma|X_{\ell,j}|$ points from $X_{\ell,j}$ have appeared in the stream, the set S contains at least one point from $X_{\ell,j}$. Since any $X_{\ell,j}$ with $|X_{\ell,j}| \geq \frac{n\sigma}{m\log_2(\Delta)}$ is bad with probability $(1 - p)^{\sigma|X_{\ell,j}|} < 1/n^2$, the next lemma follows by a Union bound over all ℓ, j.

Lemma 9. *With probability at least $1 - 1/n$, for each $\ell \in [m]$ and $j \in [\log_2(\Delta)]$ with $|X_{\ell,j}| \geq \frac{n\sigma}{m\log_2(\Delta)}$, $X_{\ell,j}$ is good.*

Lemma 10. *With probability at least $1 - 1/n$, for at least a $(1 - 3\sigma)$-fraction of pairs $x, y \in X$, we have $D(x, y) \leq D'(\varphi(x), \varphi(y)) \leq 19 \cdot D(x, y)$.*

Proof. The total number of points that are contained in sets $X_{\ell,j}$ with $|X_{\ell,j}| \leq \frac{\sigma n}{m\log_2(\Delta)}$ is at most σn. Now, by Lemmas 7, 8, and 9 and since N is a $(\sigma, 2)$-edge-dense net, it follows with probability at least $1 - 1/n$ that, for at least a $(1 - 3\sigma)$-fraction of pairs $x, y \in X$, there exist $b_x, b_y \in N$, and $x', y' \in S$, such that:

- x' and y' appear in the stream before x and y, respectively,
- $D(x, b_x) \leq D(x, y)$ or $D(y, b_y) \leq D(x, y)$,
- $\max\{D(x, b_x), D(y, b_y)\} \leq 2 \cdot D(x, y)$,
- $D(x', b_x) \leq 2 \cdot D(x, b_x)$, and $D(y', b_y) \leq 2 \cdot D(y, b_y)$.

Due to our construction and the triangle inequality, for a pair x, y, we get

$$D'(\varphi(x), \varphi(y)) = D'(\varphi(x), \varphi(\tau(x))) + D'(\varphi(\tau(x)), \varphi(\tau(y))) + D'(\varphi(\tau(y)), \varphi(y))$$
$$\leq D(x, b_x) + D(b_x, x') + D(x', y') + D(y', b_y) + D(b_y, y) .$$

By using the facts above, we can upper bound the last sum by $19 \cdot D(x, y)$. $\square$

Combining Lemmas 8 and 10, we obtain the following result.

Theorem 6. *Given any $\sigma > 0$ and any n-point metric space M, there exists a streaming algorithm that computes with probability at least $1 - 1/n$ an n-point metric space M', such that M embeds into M' with distortion $\mathcal{O}(1)$ and slack σ. The space requirement of the algorithm is $\mathcal{O}(\frac{C(\sigma)\cdot\log^2(n)\cdot\log^2(\Delta)}{\sigma^2})$, where $C(\sigma)$ is a constant depending on σ.*

6 Lower Bounds

We derive the following lower bound on the space requirement. The proof is based on the pigeonhole principle.

Theorem 7. *Let $M = (X, D)$ be an arbitrary n-point metric space and let $\sigma \leq 1/5$. Then any algorithm that computes for every input metric space M with positive probability an (implicit or explicit) representation of another metric space $M' = (X', D')$, such that M embeds into M' with distortion less than 2 and slack σ, requires $\Omega(n/\log n + \log\log\Delta)$ bits of memory, where Δ is the spread of the metric M.*

References

1. Abraham, I., Bartal, Y., Chan, T.-H., Dhamdhere, K., Gupta, A., Kleinberg, J., Neiman, O., Slivkins, A.: Metric Embeddings with Relaxed Guarantees. In: Proc. 46th IEEE Sympos. Found. Comput. Sci., pp. 83–100 (2005)
2. Abraham, I., Bartal, Y., Neiman, O.: Advances in metric embedding theory. In: Proc. 38th ACM Sympos. Theory Comput., pp. 271–286 (2006)
3. de Berg, M., van Kreveld, M., Overmars, M., Schwarzkopf, O.: Computational Geometry: Algorithms and Applications. Springer, Heidelberg (2000)
4. Callahan, P.B., Kosaraju, S.R.: A decomposition of multidimensional point sets with applications to k-nearest neighbors and n-body potential fields. Journal of the ACM 42(1), 67–90 (1995)
5. Chan, T.-H.H., Dinitz, M., Gupta, A.: Spanners with slack. In: Azar, Y., Erlebach, T. (eds.) ESA 2006. LNCS, vol. 4168, pp. 196–207. Springer, Heidelberg (2006)
6. Czumaj, A., Sohler, C.: Small Space Representations for Metric Min-Sum k-Clustering and their Applications. In: Thomas, W., Weil, P. (eds.) STACS 2007. LNCS, vol. 4393, pp. 536–548. Springer, Heidelberg (2007)
7. Fakcharoenphol, J., Rao, S., Talwar, K.: A Tight Bound on Approximating Arbitrary Metrics by Tree Metrics. In: Proc. ACM Sympos. Theory Comput. (2003)
8. Frahling, G., Indyk, P., Sohler, C.: Sampling in dynamic data streams and applications. In: Proc. 21st ACM Sympos. Comput. Geom., pp. 142–149 (2005)
9. Frahling, G., Sohler, C.: Coresets in dynamic geometric data streams. In: Proc. 37th ACM Sympos. Theory Comput., pp. 209–217 (2005)
10. Indyk, P.: Algorithms for Dynamic Geometric Problems over Data Streams. In: Proc. 36th ACM Sympos. Theory Comput., pp. 373–380 (2004)
11. Johnson, W., Lindenstrauss, J.: Extensions of Lipschitz mappings into a Hilbert space. In: Conference in Modern Analysis and Probability (1982)
12. Kleinberg, J., Slivkins, A., Wexler, T.: Triangulation and Embedding using Small Sets of Beacons. In: Proc. 45th IEEE Sympos. Found. Comput. Sci. (2004)

Computing the Implicit Voronoi Diagram in Triple Precision

David L. Millman and Jack Snoeyink

Department of Computer Science, University of North Carolina - Chapel Hill,
Box 3175, Brooks Computer Science Building, Chapel Hill, NC, 27599-3175, USA
{dave,snoeyink}@cs.unc.edu
http://cs.unc.edu/~dave

Abstract. In a paper that considered arithmetic precision as a limited resource in the design and analysis of algorithms, Liotta, Preparata and Tamassia defined an "implicit Voronoi diagram" supporting logarithmic-time proximity queries using predicates of twice the precision of the input and query coordinates. They reported, however, that computing this diagram uses five times the input precision. We define a reduced-precision Voronoi diagram that similarly supports proximity queries, and describe a randomized incremental construction using only three times the input precision. The expected construction time is $O(n(\log n + \log \mu))$, where μ is the length of the longest Voronoi edge; we can construct the implicit Voronoi from the reduced-precision Voronoi in linear time.

Keywords: Voronoi diagram, Low-degree primitives, Randomized algorithm, Robust computation.

1   Introduction

Geometric algorithms that have been proved correct may still fail due to numerical errors that occur because geometric predicates and constructions require higher precision than is readily available. For example, computing the topological structure of the Voronoi diagram of n sites requires four times the input precision, Voronoi vertices of sites with integer coordinates have rational coordinates of triple precision over double, and testing whether a query point is above or below a segment joining two Voronoi vertices requires six times input precision. Liotta, Preparata, and Tamassia [1] derived a structure from the Voronoi diagram that supports logarithmic-time proximity queries for points on a grid using only two times the input precision. Unfortunately, they report that computing their diagram requires five times the input precision.

We introduce a structure that similarly supports proximity queries, but is computed incrementally using at most triple precision in $O(n(\log n + \log \mu))$ expected time, where μ is the length of the longest Voronoi edge. From our structure it is easy to obtain the structure of Liotta *et al.* in linear time.

Computing Voronoi diagrams is a well studied problem and many optimal algorithms have been proposed [2, 3, 4, 5]. Most are designed for a RealRAM or other computational model in which coordinate computations may be carried out to arbitrary precision, allowing the computer to work with exact Euclidean geometry.

F. Dehne et al. (Eds.): WADS 2009, LNCS 5664, pp. 495–506, 2009.
© Springer-Verlag Berlin Heidelberg 2009

There are four popular ways to handle the numerical precision issues that arise when geometric algorithms are implemented on computers with limited precision: rounding, exact geometric computation, topological consistency, and degree-driven algorithm design. *Rounding* to machine precision is simplest, and results in fast execution, but calculations with incorrect values may cause algorithms to have unexpected behavior or even fail. The *exact geometric computation* paradigm encapsulates the numerical computations in geometric and combinatorial predicates and constructions that are guaranteed to produce correct decisions. These predicates can be built into libraries, such as CORE [6, 7], CGAL [8] and LEDA [9], for reuse by many algorithms. These libraries support various techniques for implementing correct predicates, including *arbitrary precision* in software, which is slow but always correct, *arithmetic filters* [10, 11, 12], which use precomputed error bounds for machine arithmetic so that arbitrary precision is needed only when machine precision is insufficient, and *adaptive predicates* [13], which evaluate only to the precision needed to guarantee a correct solution. Sugihara and Iri [14] suggest that *topological consistency* is more important than geometric correctness – that inaccurate values of coordinates can be tolerated provided that the data structures satisfy topological invariants needed by algorithms for correct operation. For example, a Voronoi diagram algorithm may be allowed to round vertex coordinates so that the embedding becomes non-planar, but as the graph itself is connected and planar, a graph-based traversal will at least terminate. Topological consistency produces correct results when the numerical computation gives correct predicate decisions, and at least avoids catastrophic failure when one or more predicate decisions are incorrect. *Degree-driven algorithm design* considers arithmetic precision as a limited resource that should be optimized along with running time and memory. Input is assumed to be single precision; often restricted to an integer grid for convenient analysis. Liotta, Preparata, and Tamassia [1] named this technique in their work on point location, in which they suggested polynomial degree to capture the complexity of predicates. The technique has also been applied to computing segment intersections [15, 16, 17]. Our work described here falls into degree-driven algorithm design.

2   Geometric Preliminaries and Related Work

The Voronoi diagram is well known in computational geometry, but because we will be concerned with the precision of input, we start with a restricted definition and remind the reader of some properties. Assume that we are given a set of n *sites*, $S = \{s_1, s_2, \ldots, s_n\}$, with each $s_i = (x_i, y_i)$ having single precision coordinates and all of them lying in a region of interest in the plane that can be described with a $O(1)$ coordinate values. (The easiest assumption is that S lay in a bounded rectangle in the integer grid.) The distance metric is Euclidean.

The Voronoi diagram, VoD(S), is the partition of our bounded rectangle into maximally connected regions with the same set of closest sites. The partition includes *Voronoi regions* closest to one site, *Voronoi edges* equidistant to two closest sites, and *Voronoi vertices* equidistant to three or more closest sites. The Voronoi cell $\mathrm{VR}_S(s_i)$ is the closure of the region of s_i:

$$\mathrm{VR}_S(s_i) = \{x \in \mathbb{R}^2 \mid \|x - s_i\| \leq \|x - s_j\|, \ \forall s_j \in S\}.$$

Note that we will suppress the S and write $\mathrm{VR}(s_i)$ when S is clear from the context. Let b_{ij} denote the locus of points equidistant to s_i and s_j, the perpendicular bisector of the segment $\overline{s_i s_j}$. Note that as we are interested in a particular region, we can clip bisectors and Voronoi edges to finite segments.

Aurenhammer surveys [18] properties of the Voronoi diagram. In particular, we use the following:

- Bisectors are straight lines.
- A Voronoi edge on the boundary of cells $\mathrm{VR}(s_i)$ and $\mathrm{VR}(s_j)$ lies on the bisector of s_i and s_j.
- A Voronoi cell is the intersection of closed half planes; this implies that Voronoi cells are convex.
- A site s_i is contained in its cell, $s_i \in \mathrm{VR}(s_i)$.

The following properties of the Voronoi diagram are also known, but we state them in a form that will be helpful for our constructions later in the paper.

Lemma 1. *The order in which Voronoi cells intersect a line ℓ is the same as the order of the corresponding sites orthogonal projection onto ℓ.*

For our incremental construction we will need to decide if a new cell intersects a horizontal or a vertical segment. A corollary of Lemma 1 will give us a convenient way of making this decision.

Corollary 2. *Without loss of generality, consider a horizontal segment σ and the set of sites S whose Voronoi cells intersect σ. Let q be a new site, and let s_i, s_j be the sites of S whose projections onto σ form the smallest interval containing the projection of q; site s_i or s_j can be taken as infinite if no finite interval exists. To determine if $\mathrm{VR}_S(q)$ intersects σ, it suffices to test if an endpoint of σ or the intersection of $\sigma \cap b_{ij}$ is closer to q than to both s_i and s_j.*

Proof. Assume that q is above σ and that we would like to determine if $\mathrm{VR}_S(q)$ appears below σ. Let $x_i < x_j$, and $c_i, c_j \in \sigma$ be points in the cells of s_i and s_j respectively. Consider the point q' that has the same x coordinate as q, but is raised to infinity. Now, lower q' continuously, computing the Voronoi diagram of $\sigma \cup \{q'\}$ until the cell of q' intersects σ, and let $c_{q'}$ be this intersection point. Lemma 1 tells us that $c_i < c_{q'} < c_j$. In addition, $c_{q'}$ must be on b_{ij}, otherwise $c_{q'}$ would be in the middle of the cell of s_i or s_j causing the cell to be non-convex. In fact, $c_{q'}$ is the point equidistant to s_i, q' and s_j. Now, if q' is above q then the point equidistant to s_i, q and s_j must be below $c_{q'}$ so the cell of q must intersect σ. Alternatively, if q' is below q then the point equidistant to s_i, q and s_j is above $c_{q'}$ so the cell of q is completely above σ and therefore, their intersection is empty. ∎

Next we show that the intersection of a Voronoi diagram and a convex region is a collection of trees.

Lemma 3. *Given a set of sites S and a convex region R containing no sites of S; the edges and vertices of the $\text{VoD}(S)$ in R form a forest.*

Proof. If $R \cap \text{VoD}(S)$ contained a cycle then R would contain a Voronoi cell, and therefore a site. Since R contains no sites, we conclude that $R \cap \text{VoD}(S)$ does not contain a cycle. ∎

The Voronoi diagram can be used to determine the closest site to a query point q if we build a point location structure on top of it. The trapezoid method of point location [2] builds a logarithmic-depth directed acyclic graph (DAG) with two types of nodes: x-nodes evaluate whether q is left or right of a vertical line through some vertex v by comparing x coordinates, and y-nodes evaluate whether q is above or below the line through some edge e.

As suggested in the introduction, Voronoi vertices are rational polynomials in the input coordinates of degree three over degree two. So to compare x coordinates with a single precision input, it would suffice to clear fractions and evaluate the sign of a degree three polynomial using triple precision computation. (Triple precision is also necessary; the polynomial for this predicate is irreducible of degree three.)

If edge e were defined by two arbitrary Voronoi vertices, then y-node test would require degree six, but since it is enough to test Voronoi edges, which lie on bisectors, we can compare squared distances to pairs using double precision. Liotta et al. [1] further observed that when the query points are on a grid, an x-node can store coordinates of v rounded to half grid points, which reduces the x-node evaluation to single precision. Thus, they defined their *implicit Voronoi diagram*, which stores the topology of $\text{VoD}(S)$ as a point location DAG, and for each edge stores the pair of sites defining the bisector, and for each vertex the Voronoi vertex rounded to a half-integer grid. This has the anomaly that the stored vertices do not lie on the stored edges. Nevertheless, point location with a grid point q as input will report the containing cell correctly, and in logarithmic time.

Unfortunately, the only method that Liotta et al. [1] suggest to build their implied Voronoi diagram is to build the true Voronoi diagram and round, which they report is a degree five computation. We will reduce this to degree three.

Voronoi diagrams on a pixel grid have been considered in both graphics and image processing. In graphics, Hoff *et al.* [19] used the GPU to render an image of the Voronoi diagram and to recover an approximate Voronoi diagram from the screen buffer. This method generalizes easily to sites that are segments, curves, or areas. The work does not consider precision or accuracy guarantees but discusses errors created from multi-precision distance computations rounded to machine precision.

In image processing, the distance and nearest neighbor transforms are two operations that can be viewed as querying only at pixels for the distance or name of the nearest feature pixel. Breu *et al.* [20] developed a linear-time algorithm for these transforms by computing the Euclidean Voronoi diagram and querying. Here, linear is in the number of pixels in the grid; the number of sites may be proportional. They avoid extra logarithmic factors by using the locality of the grid in point location and in divide and conquer construction. Their algorithm assumes a RealRAM and uses at least four times input precision; a divide and conquer version of our algorithm would reduce the precision of computing distance and neighbor transforms without sacrificing their worst-case running time.

3 Predicates and Constructions

The traditional measures in the theory of algorithms are asymptotic time and space, usually described up to a multiplicative constant by big-O notation. Liotta *et al.* [1, 15] suggest that we can analyze the arithmetic precision required by combinatorial and geometric algorithms, up to an additive constant, by expressing predicates and constructions as rational polynomial functions of the input variables and looking only at the polynomial degree.

We assume that input coordinates are single variables, which are degree-one polynomials. The degree of a monomial is the sum of the degrees of its variables, and the degree of a polynomial is the maximum degree of its monomials. The degree of a predicate is the maximum degree of its polynomials, and the degree of an algorithm is the maximum degree of its predicates.

Bisector Side Predicate: To clarify, we illustrate this concept with a *bisectorSide* predicate that determines whether a query point q is closer to site p or site r by comparing squared distances:

(a1) Evaluate $(q_x - p_x)^2 + (q_y - p_y)^2 \lesseqqgtr (q_x - r_x)^2 + (q_y - r_y)^2$.

(a2) The result $<$ implies that p is closer, $>$ implies r is closer, and $=$ implies q is on the bisector of p and r.

Since this computation can be performed by evaluating the sign of a degree 2 polynomial, it suffices to use double precision plus a couple of bits for possible carries. In the rest of this section we briefly define three other predicates or constructions that operate on bisectors.

Stabbing Order Predicate: Given two bisectors b_{12}, and b_{34}, defined by input sites, and a vertical grid line ℓ that both bisectors intersect, the *stabbingOrder* predicate determines if the intersection $b_{12} \cap \ell$ is above, below or at the same point as the intersection $b_{34} \cap \ell$.

Lemma 4. *We can determine the stabbingOrder of two bisectors on a grid line using degree three computation in constant time.*

Proof omitted for extended abstract.

Bisector-Segment Intersection Predicate: Given bisector b_{12} and a non-vertical segment $\sigma \subset b_{34}$ with left and right endpoints on horizontal gridlines, ℓ_l and ℓ_r, the *bisectorSegmentIntersection* predicate determines if σ intersects b_{12}.

Lemma 5. *We can determine if a bisector intersects a segment whose end points lay on gridlines with degree three computation in constant time.*

Proof. The stabbingOrder of b_{12} and b_{34} on ℓ_l and ℓ_r is different if and only if we have found an intersection; two stabbingOrder tests suffice. ∎

Bisector Intersection Construction: Given a bisector b_{12} that intersects a non-vertical segment σ as defined for Lemma 5, the *bisectorIntersection* construction identifies the grid cell containing the intersection of b_{12} and σ.

Lemma 6. *We can identify the grid cell containing the intersection of a bisector and a segment whose end points lay on gridlines with degree three computation in time proportional to log of the length of the segment.*

Proof. We can do a binary search on the segment for the grid cell containing the intersection using the degree three stabbingOrder predicate.   ∎

4   The Reduced-Precision Voronoi Diagram

Given a set of n sites $S = \{s_1, s_2, \ldots, s_n\}$, whose coordinates are b-bit integers, we define a reduced-precision Voronoi diagram that is intermediate between the true Voronoi diagram VoD(S) and the implicit diagram of Liotta *et al.* [1]. Because we use predicates of at most degree three, we cannot know exactly how bisectors intersect inside of a grid cell. The predicates of the previous section, however, do provide full information at the grid cell borders. This low level of information inside of a grid cell gives us a "fuzzy" picture of Voronoi vertices that we contract to *rp-vertices*. Since we do however know precise information at the grid boundaries we maintain *rp-edges* that keep the same edge ordering as Voronoi edges entering the grid cell. In this way we keep enough control of the Voronoi vertices to perform constructions efficiently; in contrast, the implicit diagram rounds Voronoi vertices off their defining edges.

Let us consider the integer grid $\mathcal{G}$ as a partition into *grid cells* of the form $[i, i+1) \times [j, j+1)$ for integers i, j. The *rp-Voronoi* $\widehat{V}(S)$ is the graph with *rp-vertices* and *rp-edges* defined by contracting every edge of the Voronoi diagram VoD(S) that lies entirely inside some grid cell.

Figure 1(a) depicts and example grid cell $G \in \mathcal{G}$ and shows the intersection $G \cap \text{VoD}(S)$, which by Lemma 3 is a forest. In the graph structure of the rp-Voronoi, $\widehat{V}(S)$, we therefore contract each tree of the forest to an rp-vertex. Edges that leave the grid cell are preserved as rp-edges. Notice that the planar embedding of the Voronoi VoD(S) gives a natural planar embedding of $\widehat{V}(S)$ in which the ordering of edges entering a grid cell is preserved as the ordering of edges around the rp-vertex. We find it useful to depict these rp-vertices as the convex hulls of the intersections of rp-edges, as in Figure 1(b). Although we never actually compute these convex hulls they bound the locations where the tree of VoD(S) can lie.

Each rp-vertex v maintains the grid cell G_v containing v, and an list of its incident rp-edges in counter-clockwise order by their entry to G_v. Each rp-edge e stores the generator sites s_1 and s_2 of the corresponding Voronoi diagram edge, and pointers to its location in the lists of its two rp-vertices. Standard data structures, like the doubly-connected edge list [2], allow us to maintain the order in the planar subdivision represented by $\widehat{V}(S)$.

Finally we define the *boundary of the rp-region* of s to be the alternating sequence of rp-edges storing site s and the connecting rp-vertices that form a

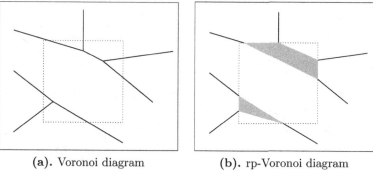

(a). Voronoi diagram   (b). rp-Voronoi diagram

Fig. 1. The Voronoi vertices in a grid cell on the left contract to two rp-vertices, depicted as gray convex polygons on the right. The ordering of the edges entering the grid cell is maintained in both diagrams.

cycle. We call the *rp-region* all the points enclosed in this cycle, and the *rp-cell* the union of the rp-region with its boundary.

Observation 7. *The number of rp-vertices and rp-edges in the rp-Voronoi diagram of S is less than or equal to the number of Voronoi vertices and Voronoi edges in the Voronoi diagram of S respectively.*

As we will show in Section 5.3, we can retrieve the implicit Voronoi diagram once we have constructed the rp-Voronoi.

5   Constructing the Reduced-Precision Voronoi Diagram

Next we describe how to construct the rp-Voronoi, analyze the expected time and space, describe how to use the rp-Voronoi for point location and show how to convert the rp-Voronoi to the implicit Voronoi diagram.

We create the rp-Voronoi by a randomized incremental construction [2] that parallels Sugihara and Iri's method [14]: inserting a new site by "carving" out the new cell from the previous diagram. Inserting a new site invalidates a subgraph of the Voronoi diagram, referred to as the *conflict region*. Sugihara and Iri made the observation that the conflict region is a tree, and that by walking the tree we identify the invalid sub-graph.

Specifically, their method constructs a Voronoi diagram of the first $k-1$ sites and then inserts site s_k. To start carving, the site s_i closest to s_k is identified and the bisector b_{ik} is traced until it enters the neighboring Voronoi cell, VR(s_j). The bisector b_{jk} is then traced, and the process continues until it returns back to VR(s_i). The tracing process requires the identification of the next bisector intersection with a Voronoi edge. Sugihara and Iri do this by walking around the cell of s_j on the side of the new site s_k until the next intersection is found (see Figure 2a).

Our method does the same computation, but since we restrict ourselves to degree three, it is too costly to compute and compare coordinates of bisector interesections.

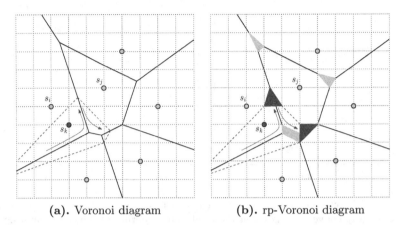

(a). Voronoi diagram   (b). rp-Voronoi diagram

Fig. 2. The cell for the new dark gray site s_k is "carved" out of the diagram of light gray sites. The traced bisectors are emphasized with dotted lines, and the tree walk is shown with gray arrows.

5.1   Incremental Construction

We initialize the rp-Voronoi diagram with two sites s_1 and s_2, and use their bisector b_{12} to split the initial region of interest (the grid) by using the binary search of bisectorIntersection.

Now, assume that we have already constructed the rp-Voronoi of $k-1$ sites S_{k-1} and that we would like to insert site s_k. The *rp-Voronoi Update Procedure* takes as input the rp-Voronoi of S_{k-1} and a new input site s_k and returns the rp-Voronoi of $S_{k-1} \cup \{s_k\}$.

rp-Voronoi Update Procedure: We sketch the procedure in this paragraph and then fill in the details in the remainder of the section. We first locate the site $s_i \in S_{k-1}$ closest to s_k, and proceed in two steps. We find the subgraph T that consists of the set of rp-vertices and rp-edges that are no longer part of the rp-Voronoi of $S_{k-1} \cup \{s_k\}$. In the Voronoi diagram, the conflict region is a tree and the sum of all conflict region sizes is linear in expectation. In the rp-Voronoi we walk a subset T of the edges of this tree, and their vertices; once we have identified this subset, we maintain our data structure in time proportional to its size, which is therefore also linear (see Figure 2b).

To identify T we start by tracing out the s_i, s_k bisector b_{ik}. We walk around the boundary of the region of s_i until we find the grid cell G containing the intersection of b_{ik} and the boundary of the rp-region of s_i. As in Sugihara and Iri's algorithm, we would like to pick the next bisector to trace, thus, allowing us to continue our tree walk. To *pick the next bisector* for the rp-Voronoi with limited precision there are two cases: one simple and the other interesting.

In the simple case, b_{ik} intersects the rp-edge e that stores sites s_i and s_j. This intersection is determined by applying the bisectorIntersection construction. We switch to the s_k, s_j bisector and continue building T by walking around the boundary of the region of s_j on the side of s_k.

In the more interesting case, b_{ik} intersects an rp-vertex v. This intersection is determined by first checking if b_{ik} passes through the grid cell containing v. If it does, we compare the stabbingOrder of b_{ik} and the two rp-edges incident on v of the rp-cell of s_k to determine if b_{ik} intersects v.

Let G_N, G_E, G_S and G_W be the north, east, south and west grid walls, respectively, of G, and without loss of generality, assume that we have entered G from the south, with s_i below the b_{ik} bisector (see Figure 3).

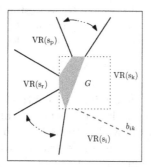

Fig. 3. We enter grid cell G from the south while walking the tree of the new site s_k along the s_i, s_k bisector b_{ik} in dashed gray. The projections of sites s_p and s_r onto the west grid line are directly above and below the projection of s_k onto the west grid line respectively.

Since Voronoi cells are convex, the new cell of s_k can intersect each of the grid cell boundary walls at most twice. This gives us four cases for how the traced bisectors of the new Voronoi cell enter and exit a grid wall G_x of G (see Figure 4). Traced bisectors of the new Voronoi cell,

(c1) do not exit through G_x.
(c2) exit through G_x and do not return to G.
(c3) exit through G_x and return through G_x.
(c4) exit through G_x and return through a different grid wall G_y.

First, we determine if the Voronoi cell $\mathrm{VR}(s_i)$ pokes out of the G_W grid wall. We find the two sites s_p and s_r whose Voronoi cells intersect G_W and whose y coordinate is directly above and below the y coordinate of s_k, respectively. We then determine if $\mathrm{VR}(s_i)$ pokes out by applying Corollary 2.

If $\mathrm{VR}(s_k)$ does not poke out of G_W (case 1) we repeat the process with G_N, followed by G_E. If $\mathrm{VR}(s_k)$ does poke out then there are some points in the $\mathrm{VR}(s_r)$ that are now in the $\mathrm{VR}(s_k)$. Since no site has an empty cell there must be a Voronoi edge $e \in \mathrm{VR}(s_k)$ that is a subset of the s_r, s_k bisector. We trace b_{rk}, following the tree, towards b_{ik} until we return back to G. Now, we have identified the next bisector to trace for our tree walk. In addition, we just walked backwards through a subtree of T, and we continue the procedure by tracing b_{rk} in the opposite direction as before.

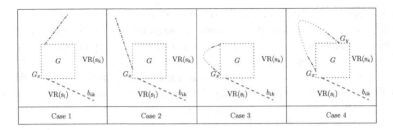

Fig. 4. Grid cell G with grid walls G_x and G_y as west and north walls respectively. A bisector enters G through the south wall by tracing the s_i, s_k bisector b_{ik}, in dashed gray. In alternating dashed and dotted gray are the four cases per wall for bisector tracing.

The other three cases are determined by continuing the walk. Case 3 corresponds to the walk returning back to the grid cell through the same cell wall it exited. Case 4 occurs when the walk returns back to the grid cell, but through a different grid wall. This allows us to determine cases 3 and 4 that can cause multiple rp-vertices to occur in one grid cell.

We continue this process until we have completed the cycle, identified T and the new rp-vertices and rp-edges. We update the rp-Voronoi of S_{k-1} to get the rp-Voronoi of $S_{k-1} \cup \{s_k\}$.

Note that if a Voronoi vertex v is outside our region of interest we do not need to identify the grid cell containing v since it will not be used for proximity queries. However, to continue with our tree walk we can apply Corollary 2 similar to the case where a bisector intersects an rp-vertex.

5.2   Analysis

Point location is accomplished by the standard method of maintaining the construction history [4] allowing for point location in expected $O(\log n)$ time. To achieve a degree two algorithm we use grid cells in x-nodes and bisectorSide for y-nodes, much like the structure in [1]. The incremental construction described above relies on bisectorSide and bisectorIntersection operations, which are of degree two and three respectively, as shown in Section 3.

As explained by Observation 7, the rp-Voronoi and Voronoi diagram have the same combinatorial complexity. The update procedure creates at most as many rp-vertices as Voronoi vertices. As shown by [2, 4] the number of Voronoi vertices created is expected linear throughout the algorithm. Furthermore, the tree walk touches only edges that are modified, and the number of modified edges is constant in expectation [4].

However, we must pay two additional charges in each update. First there is an extra $O(\log \mu)$ charge for finding bisector segment intersections. Secondly, we must pay an $O(\log n)$ to find the upper and lower neighbors when a bisector intersects an rp-vertex. So we have shown the expected time to insert a new site into the reduced-precision Voronoi diagram of size n is $O(\log n + \log \mu)$ where μ

is the length of the longest Voronoi edge, and that this insertion can be done with degree three predicates. We conclude:

Theorem 8. *We can construct a reduced-precision Voronoi diagram of n sites with a degree three algorithm in expected $O(n(\log n + \log \mu))$ time where μ is the length of the longest Voronoi edge.*

5.3   Reduced-Precision Voronoi to Implicit Voronoi

Next we describe how to convert the reduced-precision Voronoi to the implicit Voronoi of [1], described in Section 4. An rp-vertex corresponds to a tree T, of Voronoi vertices and Voronoi edges. Some of the Voronoi vertices of T may be on grid lines and the implicit Voronoi would assign these vertices integer coordinates. To create the implicit Voronoi we must separate these vertices from an rp-vertex.

Theorem 9. *We can convert the reduced-precision Voronoi diagram to the implicit Voronoi diagram in $O(n)$ time with degree three computations.*

Proof omitted for extended abstract.

6   Conclusion and Open problems

To our knowledge, this is the first construction of a proximity query structure in less than degree four and sub-quadratic time, that allows for logarithmic query times. In addition, we believe that this is the first construction of the implicit Voronoi diagram without fully computing the Voronoi diagram. Is there a reasonable algorithm for creating a proximity query structure with only degree two predicates? We believe that there is, but perhaps at the cost of an additional logarithmic factor in space and time.

There is more to discover with respect to restricted predicates for computing Voronoi diagrams. It is easy to generalize the ideas presented in this paper to the power diagram, but further investigation is necessary to understand how these methods effect the complexity of the power diagram, as well as other generalized Voronoi diagrams with linear bisectors. The basis for the method of bisector intersection relies on linearity. What if the diagram's bisectors are more complicated, such as hyperbolic arcs, as in the Voronoi diagram of disks?

One final observation is the algebraic complexity of standard predicates, such as orientation and inSphere, increases with dimension. Distance based predicates, on the other hand, maintain the same algebraic complexity regardless of the dimension. Our last question is can we compute some form of a Voronoi diagram with reduced precision in any dimension?

References

[1] Liotta, G., Preparata, F.P., Tamassia, R.: Robust proximity queries: An illustration of degree-driven algorithm design. SIAM J. Comput. 28(3), 864–889 (1999)

[2] de Berg, M., Cheong, O., van Kreveld, M., Overmars, M.: Computational Geometry: Algorithms and Applications, 3rd edn. Springer, New York (2008)

[3] Fortune, S.: A sweepline algorithm for Voronoi diagrams. Algorithmica 2, 153–174 (1987)

[4] Guibas, L.J., Knuth, D.E., Sharir, M.: Randomized incremental construction of Delaunay and Voronoi diagrams. Algorithmica 7, 381–413 (1992)

[5] Shamos, M.I., Hoey, D.: Closest-point problems. In: SFCS 1975: Proceedings of the 16th Annual Symposium on Foundations of Computer Science, Washington, DC, USA, pp. 151–162. IEEE Computer Society, Los Alamitos (1975)

[6] Karamcheti, V., Li, C., Pechtchanski, I., Yap, C.: A core library for robust numeric and geometric computation. In: SCG 1999: Proceedings of the fifteenth annual symposium on Computational geometry, pp. 351–359 (1999)

[7] Yap, C.K.: Towards exact geometric computation. Comput. Geom. Theory Appl. 7(1-2), 3–23 (1997)

[8] CGAL, Computational Geometry Algorithms Library, http://www.cgal.org

[9] Burnikel, C., Könemann, J., Mehlhorn, K., Näher, S., Schirra, S., Uhrig, C.: Exact geometric computation in LEDA. In: SCG 1995: Proceedings of the eleventh annual symposium on Computational geometry, pp. 418–419 (1995)

[10] Devillers, O., Preparata, F.P.: A probabilistic analysis of the power of arithmetic filters. Discrete and Computational Geometry 20(4), 523–547 (1998)

[11] Devillers, O., Preparata, F.P.: Further results on arithmetic filters for geometric predicates. Comput. Geom. Theory Appl. 13(2), 141–148 (1999)

[12] Fortune, S., Van Wyk, C.J.: Efficient exact arithmetic for computational geometry. In: SCG 1995: Proceedings of the eleventh annual symposium on Computational geometry, pp. 163–172 (1993)

[13] Shewchuk, J.R.: Adaptive Precision Floating-Point Arithmetic and Fast Robust Geometric Predicates. Discrete & Computational Geometry 18(3), 305–363 (1997)

[14] Sugihara, K., Iri, M.: Construction of the Voronoi diagram for 'one million' generators in single-precision arithmetic. Proceedings of the IEEE 80(9), 1471–1484 (1992)

[15] Boissonnat, J.D., Preparata, F.P.: Robust plane sweep for intersecting segments. SIAM J. Comput. 29(5), 1401–1421 (2000)

[16] Boissonnat, J.D., Snoeyink, J.: Efficient algorithms for line and curve segment intersection using restricted predicates. In: SCG 1999: Proceedings of the fifteenth annual symposium on Computational geometry, pp. 370–379 (1999)

[17] Chan, T.M.: Reporting curve segment intersections using restricted predicates. Comput. Geom. Theory Appl. 16(4), 245–256 (2000)

[18] Aurenhammer, F.: Voronoi diagrams a survey of a fundamental geometric data structure. ACM Comput. Surv. 23(3), 345–405 (1991)

[19] Hoff, K.E., Keyser, J., Lin, M., Manocha, D., Culver, T.: Fast computation of generalized voronoi diagrams using graphics hardware. In: SIGGRAPH 1999: Proceedings of the 26th annual conference on Computer graphics and interactive techniques, pp. 277–286 (1999)

[20] Breu, H., Gil, J., Kirkpatrick, D., Werman, M.: Linear time Euclidean distance transform algorithms. IEEE Transactions on Pattern Analysis and Machine Intelligence 17, 529–533 (1995)

Efficient Approximation of Combinatorial Problems by Moderately Exponential Algorithms

Nicolas Bourgeois, Bruno Escoffier, and Vangelis Th. Paschos

LAMSADE, CNRS FRE 3234 and Université Paris-Dauphine, France
{bourgeois,escoffier,paschos}@lamsade.dauphine.fr

Abstract. We design approximation algorithms for several **NP**-hard combinatorial problems achieving ratios that cannot be achieved in polynomial time (unless a very unlikely complexity conjecture is confirmed) with worst-case complexity much lower (though super-polynomial) than that of an exact computation. We study in particular MAX INDEPENDENT SET, MIN VERTEX COVER and MIN SET COVER and then extend our results to MAX CLIQUE, MAX BIPARTITE SUBGRAPH and MAX SET PACKING.

1   Introduction

The two most known paradigms for solving **NP**-hard problems are either the exact computation, or the polynomial approximation. Both of them are very active areas in theoretical computer science and combinatorial optimization.

Dealing with the former, very active research has been recently conducted on the development of optimal algorithms with non-trivial worst-case complexity. As an example, let us consider MAX INDEPENDENT SET. It can be optimally solved with complexity $O^*(2^n)$ (where $O^*(\cdot)$ is as $O(\cdot)$ ignoring polynomial factors), where n is the order of G (i.e, the cardinality of V) by a trivial algorithm consisting of exhaustively examining all the subsets in 2^V and by taking the largest among them that forms an independent set. Hence, an interesting question is if we can compute a maximum independent set with complexity $O^*(\gamma^n)$, for $\gamma < 2$. More about such issues for several combinatorial problems can be found in the seminal paper by [1]. Recently, this area has gained renewed interest by the computer science community. This is partly due to numerous pessimistic results in polynomial approximation, but also due to the fantastic increase of the computational power of modern computers. On the other hand, dealing with polynomial approximation, very intensive research since the beginnings of 70's has lead to numerous results exhibiting possibilities but also limits to the approximability of **NP**-hard problems. Such limits are expressed as statements that a given problem cannot be approximated within a certain approximation level (for instance, within a constant approximation ratio) unless a very unlikely complexity condition (e.g., **P = NP**) holds. Reference works about this field are the books by [2,3,4].

This paper combines ideas from exact computation and polynomial approximation in order to devise approximation algorithms for some **NP**-hard problems that achieve approximation ratios that cannot be achieved in polynomial

F. Dehne et al. (Eds.): WADS 2009, LNCS 5664, pp. 507–518, 2009.
© Springer-Verlag Berlin Heidelberg 2009

time, with a worst-case complexity that is significantly lower (though super-polynomial) than the complexity of an exact computation.

Since the beginning of 90's, and using the celebrated PCP theorem ([5]), numerous natural hard optimization problems are proved to admit more or less pessimistic inapproximability results. For instance, MAX INDEPENDENT SET is inapproximable within approximation ratio better than $n^{\epsilon-1}$, unless $\mathbf{P} = \mathbf{NP}$ ([6]). Similar results, known as *inapproximability* or *negative results*, have been provided for numerous other paradigmatic optimization problems, as MIN COLORING, etc. Such inapproximability results exhibit large gaps between what it is possible to do in polynomial time and what becomes possible in exponential time. Hence, for the case of MAX INDEPENDENT SET, for example, a natural question is how much time takes the computation of an r-approximate solution, for $r \in [n^{\epsilon-1}, 1[$? Of course, we have a lower bound to this time (any polynomial to the size of the instance) and also an upper bound (the running time of exact computation). But *can we devise, for some ratio r, a r-approximate algorithm with an improved running time located somewhere between these bounds? Is this possible for any ratio r, i.e., can we specify a global relationship between running time and approximation ratio?*.

Here we try to bring some answers to these questions. The issue we follow has been also marginally handled by [7] for minimum coloring. Also, a similar approach has been simultaneously and independently developed by [8]. Moderately exponential approximation has been also handled by [9,10,11], though in a different setting and with different objectives oriented towards development of fixed-parameter algorithms. Finally, a different but very interesting kind of trade-off between exact computation and polynomial approximation is settled by [12].

In what follows, in Section 2, we give approximation results of a broad class of maximization graph-problems whose solutions are subgraphs of the input-graph that satisfy some non-trivial hereditary property[1]. In Section 3 we develop efficient non-polynomial approximation algorithms for another paradigmatic problem in combinatorial optimization, the MIN VERTEX COVER. In Section 4, we propose randomized approaches that improve complexity results obtained in Sections 2 (in particular for the case of MAX INDEPENDENT SET) and 3. In Section 5, we consider specific classes of graphs where MAX INDEPENDENT SET is polynomially approximable. We show there how approximation algorithms for MAX INDEPENDENT SET can be improved in order to guarantee any approximation ratio with low exponential complexity. In Section 6, we handle MIN SET COVER by devising an approximation algorithm based upon "pruning the search tree", a very well known technique in exact computation. This algorithm allows, for instance, to compute a 7-approximate solution in time $O^*(1.0007^d)$, where, denoting by m the size of the set-system $\mathcal{S}$ and by n the size of the ground set C of the MIN SET COVER-instance, $d = m + n$. Finally, in Section 7, we present

---

[1] A graph G is said to satisfy a hereditary property π if every subgraph of G satisfies π whenever G satisfies π. Furthermore, π is non-trivial if it is satisfied for infinitely many graphs and it is false for infinitely many graphs.

approximation results for other combinatorial problems linked to MAX INDEPEN-
DENT SET by simple approximation-preserving reductions. In particular, for one
of the problems handled in this section that is MAX CLIQUE, we also produce a
parameterized complexity result that is interesting per se.

Given a graph $G(V,E)$, we denote by n the size of V, by $\alpha(G)$ the size of a
maximum independent set of G and by $\tau(G)$ the size of a minimum vertex cover
of G. Also, we denote by $\Delta(G)$ the maximum degree of G. Given a subset V'
of V, $G[V']$ denotes the subgraph of G induced by V'. Sometimes, for a graph G,
we denote by $V(G)$ its vertex-set. Finally, for shortness, we omit proofs of many
of the results claimed as well as definition of the problems handled. Definitions
of the problems handled can be found in [2], while omitted proofs can be found
in [13,14].

2   Maximum Induced Subgraph Problems with Property π

We handle in this section a large class of graph-problems that is defined as
follows: given a graph $G(V,E)$ and some hereditary property π, find a subset
$V' \subseteq V$, of maximum size, such that the subgraph of G induced by V' satisfies the property π. For a fixed property π we denote by MAX HEREDITARY-π
the particular **NPO** problem resulting when considering π. For instance, if π
is "independent set", then MAX HEREDITARY-"independent set" is exactly MAX
INDEPENDENT SET.

The idea of the method proposed consists in splitting the instance into several
subinstances (of much smaller size) and in solving the problem on these subinstances using an exact algorithm. The ratio obtained is directly related to the
size of the subinstances, hence to the global running time of the algorithm.

Proposition 1. *Fix a hereditary property π and assume that there exists an
exact algorithm A for* MAX HEREDITARY-π *with worst-case complexity $O^*(\gamma^n)$
for some $\gamma \in \mathbb{R}$, where n is the order of the input-graph, for* MAX HEREDITARY-
π. *Then for any $\rho \in \mathbb{Q}$, $\rho \leqslant 1$, there exists a ρ-approximation algorithm for* MAX
HEREDITARY-π *that runs in time $O^*(\gamma^{\rho n})$.*

Proof. Consider a graph G of order n and fix a rational $\rho \leqslant 1$. Since $\rho \in \mathbb{Q}$, it
can be written as $\rho = p/q$, $p,q \in \mathbb{N}$, $p \leqslant q$. Run the following algorithm, denoted
by SPLIT and called with parameters G and ρ:

1. arbitrarily partition G into q induced subgraphs $G_1, \ldots, G_q$ of order (except
   eventually for G_q) n/q and build the q subgraphs $G'_1, \ldots G'_q$ that are unions
   of p consecutive subgraphs $G_{i+1}, \ldots, G_{i+p}$, $i = 1, \ldots, q$ (where of course
   $G_{q+1} = G_1$);
2. optimally solve MAX HEREDITARY-π in every G'_i, $i = 1, \ldots, q$ and output the
   best of the solutions computed.

Denote by S the solution returned by Algorithm SPLIT and fix an optimal so-
lution S^* of G. Then, $|S| \geqslant (p/q)\operatorname{opt}(G) = \rho\operatorname{opt}(G)$. Indeed, let $S_i^* = S^* \cap G_i$.

Then, by heredity, $|S^*_{i+1}| + |S^*_{i+2}| + \ldots + |S^*_{i+p}| \leqslant \mathrm{opt}(G'_i) \leqslant |S|$. Summing up for $i = 1, 2, \ldots, q$, we get: $p|S^*| = p \sum_{i=1}^q |S^*_i| \leqslant q|S|$, that proves the approximation ratio claimed.

It is easy to see that the above algorithm involves q executions of A (the exact algorithm for MAX HEREDITARY-π) on graphs of order roughly $pn/q = \rho n$. Hence, its complexity is of $O^*(\gamma^{\rho n})$, that proves the running time claimed and the proposition. □

Note that, hereditary properties as "independent set", "clique", "planar graph", "bipartite graph", etc., perfectly fit Proposition 1.

Denote by IS the instantiation of the algorithm above to MAX INDEPENDENT SET and assume that it is parameterized by two parameters: the input-graph G and the ratio ρ to be achieved. To the best of our knowledge, the best γ known for MAX INDEPENDENT SET is 1.18 due to [15]. In Table 1 at the end of Section 4.1, time-performance of Algorithm IS is shown for some values of ρ.

From Proposition 1 we can note the following two interesting facts: (i) the algorithm of Proposition 1 can be implemented to use *polynomial space provided that the exact algorithms used do so*; (ii) any improvement to the basis γ of the exponential for the running time of the exact algorithm for MAX HEREDITARY-π is immediately transferred to Proposition 1.

3   MIN VERTEX COVER

MIN VERTEX COVER is approximable within approximation ratio 2 and one of the most known open problems in polynomial approximation is either to improve this ratio, or to prove that such an improvement is impossible until a strong unlikely complexity condition (e.g., $\mathbf{P} = \mathbf{NP}$) holds. A recent result by [16] gives a strong evidence that the latter alternative might be true.

On the other hand, from an exact computation point of view, the well-known complementarity relation between MIN VERTEX COVER and MAX INDEPENDENT SET has as immediate corollary that an optimal vertex cover can be determined in time $O^*(\gamma^n)$. Furthermore, the following parameterized complexity result is proved by [17].

Theorem 1. *([17]) There exists a δ such that, for $k \leqslant n$, it takes time $O^*(\delta^k)$ to determine if a graph G contains a vertex cover of size k or not and, if yes, to compute it. The best δ actually known is 1.2852.*

In what follows, we denote by OPT_VC the algorithm claimed in Theorem 1 and we assume that it is called with parameters G and k.

Thanks to the seminal result by [18] characterizing the polyhedron of MAX INDEPENDENT SET (or, equivalently, of MIN VERTEX COVER), we can assume that the input graph G verifies $\alpha(G) \leqslant n/2$ (via a polytime preprocessing of the initial instance). Under this hypothesis, the following lemma holds.

Lemma 1. *Let G be a graph where $\alpha(G) \leqslant n/2$. If S is a ρ-approximate independent set of G, then $V \setminus S$ is a $(2 - \rho)$-approximate vertex cover of G.*

Proposition 1 and Lemma 1 immediately derive the following result.

Proposition 2. *For any* $\rho \leqslant 1$, MIN VERTEX COVER *is approximable within ratio* $2 - \rho$ *in time* $O^*(\gamma^{\rho n})$.

In other words, any approximation ratio $r \in [1, 2[$ for MIN VERTEX COVER can be attained in $O^*(\gamma^{(2-r)n})$ by an algorithm, denoted by VC1, that simply takes as solution the set $V \setminus S$, where S is the independent set computed by Algorithm IS.

Proposition 2 can be improved by showing that one can get approximation ratio ρ, for every $\rho > 1$, in time smaller than $O^*(\gamma^{(2-\rho)n})$. For this, we propose a method (Algorithm VC2) based upon a tradeoff between the exact algorithm in time $O^*(\gamma^n)$ for MAX INDEPENDENT SET and the fixed-parameter algorithm in time $O^*(\delta^k)$ for MIN VERTEX COVER. The idea is the following:

- If $\alpha(G)$ is small then the result of Proposition 2 can be further improved. More precisely, if, say, $\alpha(G) \leqslant \lambda n$, then $C_0 = V \setminus \text{IS}(G, \mu)$ where $\mu = \rho - ((\rho - 1)/\lambda)$ is a ρ approximate vertex cover. This is obtained in running time $O^*(\gamma^{\mu n})$.
- If $\alpha(G)$ is large, then $\tau(G)$ is small and the parameterized algorithm OPT_VC is efficient. More precisely, we split the instance in subinstances of size $(2 - \rho)n$, and find on these subinstances an optimum vertex cover of size at most $(2 - \rho)(1 - \lambda)n$ (if it exists). This can be done in time $O^*\left(\delta^{(2-\rho)(1-\lambda)n}\right)$.

Choosing λ such that $\gamma^{(\rho-(\rho-1/\lambda))} = \delta^{(1-\lambda)(2-\rho)}$ (the first increases, while the second decreases with λ), the following theorem holds.

Theorem 2. *For any* $\rho > 1$, MIN VERTEX COVER *can be solved approximately within ratio* ρ *and with running time* $O^*(\gamma^{(\rho-(\rho-1)/\lambda)n})$, *where* λ *is such that* $\gamma^{(\rho-(\rho-1)/\lambda)n} = \delta^{(2-\rho)(1-\lambda)n}$.

Note that this improvement cannot be transferred to MAX INDEPENDENT SET, since it is based upon Lemma 1 that does not work in both directions.

Also, using the same technique as in Proposition 1 (up to the facts that the problem is not hereditary and the algorithm upon which it is based is parameterized) we can extend the result of Theorem 1 by proving that, *for every graph G and for any* $r \in \mathbb{Q}$, *if there exists a solution for minimal vertex cover in G whose size is less than k, it is possible to determine with complexity* $O^*(\delta^{rk})$ *a* $2 - r$-*approximation of it.* The fixed-parameter approximation algorithm (denoted by VC3) doing this consists of splitting the instance and applying OPT_VC on the subinstances.

More about the running times of Algorithms VC2 and VC3 can be found in Table 2 (Section 4.2). Throughout the paper, when such running times are given, they are obtained with $\gamma = 1.18$ and (for simplicity) $\delta = 1.28$ (instead of 1.2852).

4   Randomized Algorithms

We give in this section randomized algorithms for MAX INDEPENDENT SET and MIN VERTEX COVER that, with probability $1 - \exp\{-cn\}$, for some constant c, turn to efficient approximation algorithms with running-time lower (though exponential) than the one of the deterministic algorithms seen in Sections 2 and 3.

4.1 MAX INDEPENDENT SET

In the deterministic algorithm for MAX INDEPENDENT SET seen previously, we split the instance into subinstances of size rn to get a r-approximation algorithm. Here, we show that by splitting into subinstances of smaller size βn, with $\beta < r$, we can achieve the same ratio by iterating the splitting a very large (exponential) number of times. The tradeoff between the size of the subinstances and the number of times that we iterate splitting to get the ratio, is given in Theorem 3.

Theorem 3. *For any $\rho < 1$ and for any β, $\rho/2 \leqslant \beta \leqslant \rho$, it is possible to find an independent set that is, with probability $1 - \exp\{-cn\}$ (for some constant c), a ρ-approximation for* MAX INDEPENDENT SET, *with running time $O^*(K_n \gamma^{\beta n})$, where $K_n = nC_n^{n/2}/(C_{\beta n}^{\rho n/2} C_{n-\beta n}^{(1-\rho)n/2})$.*

Proof (Sketch). As mentioned previously, we can assume $\alpha(G)/n \leqslant 1/2$. Fix a maximum independent set S^* of G and consider a subgraph B of G (for simplicity, denote also by B the vertex-set of B), whose size is $\beta n \geqslant \rho n/2 \geqslant \rho\alpha(G)$. The probability that B contains $\rho\alpha(G)$ vertices from S^* is given by the following formula: $p_{\beta,\alpha} = \Pr[|S^* \cap B| = \rho\alpha(G)] = C_{\beta n}^{\rho\alpha(G)} C_{n-\beta n}^{(1-\rho)\alpha(G)}/C_n^{\alpha(G)}$.

If we take at random K_n such different subgraphs B_i, the probability that $|S^* \cap B_i|$ is never greater than $\rho\alpha(G)$ is bounded above by $\Pr[|S^* \cap B_i| < \rho\alpha(G), \forall i \leqslant K_n] \leqslant \exp\{-np_{\beta,\alpha}/p_{\beta,n/2}\}$.

We now study $p_{\beta,\alpha}$ to show that the the previous probability is bounded by $\exp\{-cn\}$. Fix $\lambda = \alpha(G)/n$. From Stirling's formula we get $p_{\beta,\alpha} = \theta(q_{\beta,\alpha}/\sqrt{n})$, where:

$$q_{\beta,\alpha} = \left(\frac{\beta^\beta(1-\beta)^{1-\beta}\lambda^\lambda(1-\lambda)^{1-\lambda}}{(\beta - \rho\lambda)^{\beta-\rho\lambda}(\rho\lambda)^{\rho\lambda}((1-\rho)\lambda)^{(1-\rho)\lambda}(1-\beta-(1-\rho)\lambda)^{1-\beta-(1-\rho)\lambda}} \right)^n$$

Function $f(\rho, \lambda, \beta) = -\log(q_{\beta,\lambda n})/n$. grows with λ, while $q_{\beta,\lambda n}$ decreases with λ. Thus, the minimum for $q_{\beta,\alpha}$ is reached for $\lambda = 1/2$, and $p_{\beta,\alpha}/p_{\beta,n/2} > c$, for some constant c. This fact derives that $\Pr[\max_{i \leqslant K_n} |S^* \cap B_i| \geqslant rm^*] \geqslant 1 - \exp\{-cn\}$.

Run now a straightforward algorithm (denoted by RIS1 and called with parameters G and ρ), that is the randomized counterpart of Algorithm IS of Section 2. It calls an exact algorithm OPT_IS on all the B_i's (i from 1 to K_n) and returns the largest independent set found. It is easy to see that the running time of Algorithm RIS1 is $O^*(K_n \gamma^{\beta n})$, while the probability that it returns a ρ-approximate solution is $1 - \exp\{-cn\}$. □

Algorithm RIS1 can be improved in two different ways, leading to Algorithms RIS2 and RIS3, respectively.

The way the first improvement is obtained (Algorithm RIS2) is somehow analogous to that of Theorem 2. The basic idea is to show that, informally, the smaller the independent set, the higher the probability of finding a good approximation by splitting. In other words, when $\alpha(G)$ is small, a smaller running time suffices to get the same approximation ratio with high probability, i.e., Algorithm RIS1 is more efficient. On the other hand, when $\alpha(G)$ is large, the

Table 1. Running times of Algorithms IS, RIS1, RIS2 and RIS3 with $\gamma = 1.18$

Ratio	0.1	0.2	0.3	0.4	0.5	0.6	0.7	0.8	0.9
IS	1.017^n	1.034^n	1.051^n	1.068^n	1.086^n	1.104^n	1.123^n	1.142^n	1.161^n
RIS1	1.016^n	1.032^n	1.048^n	1.065^n	1.083^n	1.101^n	1.119^n	1.139^n	1.159^n
RIS2	1.015^n	1.031^n	1.047^n	1.063^n	1.080^n	1.098^n	1.117^n	1.136^n	1.157^n
RIS3	1.013^n	1.027^n	1.042^n	1.057^n	1.075^n	1.093^n	1.115^n	1.139^n	1.169^n

fixed-parameter Algorithm OPT_VC runs fast. Then, Algorithm RIS2 combines these two algorithms.

The second improvement follows a different approach, based upon an exhaustive lookup of all the candidate values for $\alpha(G)$, and using an exact algorithm for MIN VERTEX COVER rather than for MAX INDEPENDENT SET on the subinstances considered in the splitting. Informally, the underlying idea for this approach (leading to Algorithm RIS3) is that randomization allows to split the input graph into "small" subgraphs, on which a fixed-parameter algorithm can be efficiently used to reach both a good overall running time and any a priori fixed approximation ratio. Then, Algorithm RIS3 consists of running Algorithm OPT_VC on subgraphs of size $\beta n < rn$ taken at random and for a sufficient number of times, where β is optimally determined as a function of $\alpha(G)$. Of course, as we can see in Table 1, this algorithm is especially interesting when we seek small approximation ratios (subgraphs are very small in this case).

It is important to notice that these two improvements are strongly related to randomization. In particular, the same ideas cannot be used to improve the deterministic Algorithm IS.

Comparisons on running times of Algorithm IS, RIS1, RIS2 and RIS3, are presented in Table 1, for different values of the approximation ratio. For ratios smaller than 0.75, Algorithm RIS3 hits the other ones, while for ratios greater than 0.75, it is dominated by Algorithm RIS2 (and even by IS).

4.2 MIN VERTEX COVER

Obviously, Lemma 1 still holds for randomized algorithms. Hence, the complements of the solutions provided by Algorithms RIS1, RIS2 and RIS3 are vertex covers for G achieving ratios $2 - \rho$ with probability $1 - \exp\{-cn\}$. In what follows, we propose randomized efficient approximation algorithms for MIN VERTEX COVER with running times better than those got in Section 4.1. Underlying ideas are similar to the previous ones but, taking into account once again Lemma 1, a more involved analysis leads to better results.

We can first simply mix the randomization technique of Algorithm RIS1 and the fixed-parameter approximate Algorithm VC3 to get the following Algorithm RVC1: set $C_1 = \text{VC3}(G, 2 - r)$, $C_2 = V \setminus \text{RIS1}(G, r)$ and output $C = \text{argmin}\{|C_1|, |C_2|\}$.

We further improve the above result by another algorithm denoted by RVC2, devised in the same spirit as Algorithm RIS3: we repeatedly (for a sufficient

Table 2. Running times of Algorithms VC1, VC2, RVC1 and RVC2 with $\gamma = 1.18$ and $\delta = 1.28$

Ratio	1.9	1.8	1.7	1.6	1.5	1.4	1.3	1.2	1.1
VC1	1.017^n	1.034^n	1.051^n	1.068^n	1.086^n	1.104^n	1.123^n	1.142^n	1.161^n
VC2	1.013^n	1.026^n	1.039^n	1.054^n	1.069^n	1.086^n	1.104^n	1.124^n	1.148^n
RVC1	1.013^n	1.026^n	1.039^n	1.053^n	1.068^n	1.085^n	1.102^n	1.122^n	1.146^n
RVC2	1.010^n	1.021^n	1.032^n	1.043^n	1.056^n	1.069^n	1.083^n	1.099^n	1.127^n

number of times) apply the fixed-parameter Algorithm OPT_VC on subinstances of size $\beta n < rn$ taken at random. In this way OPT_VC runs fast and guarantees a good approximation ratio.

In Table 2, the running times of Algorithms VC1, VC2, RVC1 and RVC2 are shown for some ratio's values.

5   Approximation of MAX INDEPENDENT SET in Particular Classes of Graphs

In this section we consider particular classes of MAX INDEPENDENT SET-instances admitting polynomial approximation algorithms achieving some ratio ρ. For instance, a notable example of such a class is the class of bounded-degree graphs. For these graphs, denoting by $\Delta(G)$ the bound on the degrees, MAX INDEPENDENT SET can be polynomially approximated within ratio $\rho = 5/(\Delta(G) + 3)$ ([19]).

Let $\mathcal{C}$ be a class of graphs where MAX INDEPENDENT SET is approximable in polynomial time within approximation ratio ρ by an algorithm called APIS in what follows. The graph-splitting technique used previously can be efficiently applied to get interesting tradeoffs between running times and approximation ratios (greater than ρ) by means of the following algorithm, denoted by EIS1 (where $\Gamma(H)$ denotes the set of neighbors of H in $V \setminus H$): 1. apply step 1 of Algorithm SPLIT and let V_i' be the vertex set of G_i', $i = 1, \ldots, q$; 2. for any V_i' and any $H \subseteq V_i'$, if H is independent, then $S = H \cup \text{APIS}(G[V \setminus (H \cup \Gamma(H))])$; 3. output the best among S's computed at step 2.

Proposition 3. *For any rational $r < 1$, it is possible to compute, for any graph $G \in \mathcal{C}$, a $(r + (1 - r)\rho)$-approximation of MAX INDEPENDENT SET, with running time $O^*(2^{rn})$.*

Let us note that Algorithm EIS1 is interesting only if its ratio is better than that of IS that has the same running time. For instance, for graphs with maximum degree 3, this means that $r + (1 - r)\rho \leq 0.870$.

Result in Proposition 3 can be further improved. Roughly speaking, if S^* is not "uniformly" distributed over the G_i''s, then the ratio improves. If S^* is not "uniformly" distributed then, informally, generating only "small" subsets of G_i''s is sufficient in this case.

Proposition 4. *For any $r < 1$ and with a suitable choice of $\beta < 1$ it is possible to compute, in any graph $G \in \mathcal{C}$, a $(r + (1 - r)\rho)$-approximation of* MAX INDEPENDENT SET, *with running time $O^*(2^{\beta r n})$.*

6 MIN SET COVER

Pruning the search tree is one of the most classical techniques to get exact algorithms with non trivial exponential complexity. Here, we show that this technique can be adapted to get approximation algorithms realizing interesting tradeoffs between time complexity and approximation. The algorithm is based upon two "speedups" with respect to an exact search tree algorithm. First, since we only seek an approximate solution, the algorithm may, when branching, make some "errors" by being "less careful" than an exact one. For instance, if one wants a 2-approximate solution, each time the algorithm branches on a set, in the case it puts it in the cover it can add another set (obtaining recursively a ratio 2): taking this additional set reduces the size of the remaining problem, thus reducing the complexity of the pruning algorithm. The second improvement consists of stopping the development of the tree before the end: if at some point the remaining instance is polynomially approximable within the desired ratio, there is no need to continue branching.

Considering these two ways of better pruning the search tree, we propose Algorithm SC1, parameterized by the ratio q we want to guarantee: let p be the largest integer such that $\mathcal{H}(p) - 1/2 \leqslant q$, where $\mathcal{H}$ is the harmonic number sequence. Then SC1 repeats the following four steps until C is covered: *1.* if there exists an item of C that belongs to a single subset $S \in \mathcal{S}$, then add S to the solution; *2.* if there exist two sets S, R in $\mathcal{S}$ such that S is included into R, then remove S without branching; *3.* if all the residual subsets have cardinality at most p, then run the algorithm by [20] in order to compute a q-approximation of the optimal solution in the surviving instance; *4.* determine q sets $S_1, \ldots, S_q$ from $\mathcal{S}$ such that $\cup_{i \leqslant q} S_i$ has maximum cardinality and perform the the following branching step: either add every S_i to the solution (and remove $\cup_{i \leqslant q} S_i$ from C), or remove all of them.

Note that the corresponding best known exact algorithms for MIN SET COVER that we are aware of have complexity $O^*(1.23^d)$ ([21]), $O^*(2^n)$ ([22]) and $O^*(2^m)$ (brute force algorithm).

Theorem 4. *For any integer $q \geqslant 1$, Algorithm SC1 computes with running time $O^*(\alpha^d)$ a q-approximation of* MIN SET COVER, *where α is the solution of equation $x^{q(2+p)} - x^{q(1+p)} - 1 = 0$ and p is the largest integer such that $\mathcal{H}(p) - 1/2 \leqslant q$.*

Remark that, since at least q sets are removed in each branch, SC1 computes a q-approximation of MIN SET COVER in time $O^*(2^{m/q})$. Measuring complexity with m instead of d may be useful if m is "small", for instance if it is smaller than n. Table 3 gives complexity of Algorithm SC1 for some values of the ratio q.

Table 3. Complexity of SC1 for some values of q

Ratio q	1	2	3	4	5	6	7	8
Time(d)	1.380^d	1.110^d	1.038^d	1.014^d	1.005^d	1.002^d	1.0007^d	1.0003^d
Time(m)	2^m	1.414^m	1.260^m	1.189^m	1.149^m	1.123^m	1.104^m	1.091^m

7   MAX SET PACKING, MAX BIPARTITE SUBGRAPH and MAX CLIQUE

We show in this section that for MAX SET PACKING, MAX BIPARTITE SUBGRAPH and MAX CLIQUE, any of the results of the previous sections identically apply with parameters γ' and δ' that depend on those of MAX INDEPENDENT SET and MIN VERTEX COVER, respectively. Note that, to the best of our knowledge, all these problems have not been autonomously studied under the exponential time hypothesis.

Let us first handle the case of MAX SET PACKING. An instance $I(\mathcal{S}, C)$ of MAX SET PACKING with $\mathcal{S} = \{S_1, \ldots, S_m\}$, $S_i \subseteq C$, $i = 1, \ldots, m$, and $C = \{c_1, \ldots, c_n\}$, can be transformed into an instance $G(V, E)$ of MAX INDEPENDENT SET by adding a vertex for each S_i and by linking two vertices if and only if the corresponding sets have non-empty intersection. This classical reduction makes that *for* MAX SET PACKING *parameters γ and δ are the same as for* MAX INDEPENDENT SET *and* MIN VERTEX COVER, *respectively. The exponent for* MAX SET PACKING *is the cardinality m of the set-family $\mathcal{S}$*.

Consider now the reduction from MAX BIPARTITE SUBGRAPH to MAX INDEPENDENT SET given in [23]. According to this reduction, an instance of size n for MAX BIPARTITE SUBGRAPH transforms into an instance of size $2n$ for MAX INDEPENDENT SET. So, *for* MAX BIPARTITE SUBGRAPH, *parameters γ and δ of* MAX INDEPENDENT SET *and* MIN VERTEX COVER, *are transformed into γ^2 and δ^2 respectively. The exponent for* MAX BIPARTITE SUBGRAPH *is the size n of the input-graph.* In other words, considering $\gamma = 1.18$ and $\delta = 1.28$, the corresponding bases for MAX BIPARTITE SUBGRAPH become 1.39 and 1.64, respectively.

We conclude the paper by handling another famous combinatorial optimization problem that is MAX CLIQUE. It is very well known that an independent set in a graph G becomes a clique of the same size in the complement $\bar{G}$ of G. So, results of previous sections for independent set trivially apply to MAX CLIQUE. In what follows, we improve these results replacing exponent n, the order of the input graph G for MAX CLIQUE, by $\Delta(G)$, the maximum degree of G.

Consider the following reduction from MAX CLIQUE to MAX INDEPENDENT SET. Let $G(V, E)$ be the input graph of MAX CLIQUE, $V = \{v_1, \ldots, v_n\}$ and, for $i = 1, \ldots, n$, denote by $\Gamma(v_i)$ the neighbors of v_i. Build the n graphs $G_i = G[\{v_i\} \cup \Gamma(v_i)]$. Since in any clique of G any vertex is a neighbor of any other vertex of a clique, any clique is subset of the neighborhood of each of its vertices. So, a maximum clique is a subset of some graph G_i. For every G_i, build its complement $\bar{G}_i$ and solve MAX INDEPENDENT SET in $\bar{G}_i$. Let S_i, $i = 1, \ldots, n$, the independent sets so computed. These sets are cliques of G_i. Then, take the largest of these sets as solutions. Obviously, if an exact algorithm for MAX INDEPENDENT SET is used, then the largest among the sets S_i is a maximum clique in G. By taking into account that the order of any of the graphs G_i is bounded above by $\Delta(G)+1$, we immediately deduce that computing a maximum clique in a graph G takes time $O^*(n\gamma^{\Delta(G)+1}) = O^*(\gamma^{\Delta(G)})$, where γ is the basis of the exponential of MAX INDEPENDENT SET.

Discussion just above derives, at very first, the following parameterized complexity result for the exact computation of MAX CLIQUE, interesting per se.

Theorem 5. MAX CLIQUE *can be exactly solved in time* $O^*(\gamma^{\Delta(G)})$, *where* $\Delta(G)$ *is the maximum degree of the input graph and* γ *is the basis of the exponential of* MAX INDEPENDENT SET *(currently 1.18).*

Also, any of the results dealing with MAX INDEPENDENT SET seen in the previous sections, identically applies to MAX CLIQUE also. So, the following theorem holds and concludes this section.

Theorem 6. *For the efficient approximation of* MAX CLIQUE, *parameters* γ *and* δ *are the same as* MAX INDEPENDENT SET *and* MIN VERTEX COVER, *respectively. The exponent for* MAX CLIQUE *is the maximum degree* $\Delta(G)$ *of the input-graph.*

References

1. Woeginger, G.J.: Exact algorithms for NP-hard problems: a survey. In: Jünger, M., Reinelt, G., Rinaldi, G. (eds.) Combinatorial Optimization - Eureka, You Shrink! LNCS, vol. 2570, pp. 185–207. Springer, Heidelberg (2003)
2. Ausiello, G., Crescenzi, P., Gambosi, G., Kann, V., Marchetti-Spaccamela, A., Protasi, M.: Complexity and approximation. In: Combinatorial optimization problems and their approximability properties. Springer, Berlin (1999)
3. Hochbaum, D.S. (ed.): Approximation algorithms for NP-hard problems. PWS, Boston (1997)
4. Vazirani, V.: Approximation algorithms. Springer, Berlin (2001)
5. Arora, S., Lund, C., Motwani, R., Sudan, M., Szegedy, M.: Proof verification and intractability of approximation problems. J. Assoc. Comput. Mach. 45, 501–555 (1998)
6. Zuckerman, D.: Linear degree extractors and the inapproximability of max clique and chromatic number. In: Proc. STOC 2006, pp. 681–690 (2006)
7. Björklund, A., Husfeldt, T.: Inclusion-exclusion algorithms for counting set partitions. In: Proc. FOCS 2006, pp. 575–582 (2006)
8. Cygan, M., Kowalik, L., Pilipczuk, M., Wykurz, M.: Exponential-time approximation of hard problems. arXiv:0810.4934v1 (2008),
   http://arxiv.org/abs/0810.4934v1
9. Cai, L., Huang, X.: Fixed-parameter approximation: conceptual framework and approximability results. In: Bodlaender, H.L., Langston, M.A. (eds.) IWPEC 2006. LNCS, vol. 4169, pp. 96–108. Springer, Heidelberg (2006)
10. Chen, Y., Grohe, M., Grüber, M.: On parameterized approximability. In: Bodlaender, H.L., Langston, M.A. (eds.) IWPEC 2006. LNCS, vol. 4169, pp. 109–120. Springer, Heidelberg (2006)

11. Downey, R.G., Fellows, M.R., McCartin, C.: Parameterized approximation problems. In: Bodlaender, H.L., Langston, M.A. (eds.) IWPEC 2006. LNCS, vol. 4169, pp. 121–129. Springer, Heidelberg (2006)

12. Vassilevska, V., Williams, R., Woo, S.: Confronting hardness using a hybrid approach. In: Proc. Symposium on Discrete Algorithms, SODA 2006, pp. 1–10 (2006)

13. Bourgeois, N., Escoffier, B., Paschos, V.T.: Efficient approximation by "low-complexity" exponential algorithms. Cahier du LAMSADE 271, LAMSADE, Université Paris-Dauphine (2007),
    http://www.lamsade.dauphine.fr/cahiers/PDF/cahierLamsade271.pdf

14. Bourgeois, N., Escoffier, B., Paschos, V.T.: Efficient approximation of min set cover by "low-complexity" exponential algorithms. Cahier du LAMSADE 278, LAMSADE, Université Paris-Dauphine (2008),
    http://www.lamsade.dauphine.fr/cahiers/PDF/cahierLamsade278.pdf

15. Robson, J.M.: Finding a maximum independent set in time $O(2^{n/4})$. Technical Report 1251-01, LaBRI, Université de Bordeaux I (2001)

16. Khot, S., Regev, O.: Vertex cover might be hard to approximate to within $2 - \epsilon$. In: Proc. Annual Conference on Computational Complexity, CCC 2003, pp. 379–386 (2003)

17. Chen, J., Kanj, I., Jia, W.: Vertex cover: further observations and further improvements. J. Algorithms 41, 280–301 (2001)

18. Nemhauser, G.L., Trotter, L.E.: Vertex packings: structural properties and algorithms. Math. Programming 8, 232–248 (1975)

19. Berman, P., Fujito, T.: On the approximation properties of independent set problem in degree 3 graphs. In: Sack, J.-R., Akl, S.G., Dehne, F., Santoro, N. (eds.) WADS 1995. LNCS, vol. 955, pp. 449–460. Springer, Heidelberg (1995)

20. Duh, R., Fürer, M.: Approximation of k-set cover by semi-local optimization. In: Proc. STOC 1997, pp. 256–265 (1997)

21. van Rooij, J., Bodlaender, H.: Design by measure and conquer, a faster exact algorithm for dominating set. In: Albers, S., Weil, P. (eds.) Proc. International Symposium on Theoretical Aspects of Computer Science, STACS 2008, pp. 657–668 (2008)

22. Björklund, A., Husfeldt, T., Koivisto, M.: Set partitioning via inclusion-exclusion. SIAM J. Comput. (To appear in the special issue dedicated to selected papers from FOCS 2006)

23. Simon, H.U.: On approximate solutions for combinatorial optimization problems. SIAM J. Disc. Math. 3, 294–310 (1990)

Integer Programming:
Optimization and Evaluation Are Equivalent

James B. Orlin[1], Abraham P. Punnen[2], and Andreas S. Schulz[1]

[1] Massachusetts Institute of Technology, Cambridge, MA
[2] Simon Fraser University, Surrey, BC

Abstract. We show that if one can find the optimal value of an integer linear programming problem in polynomial time, then one can find an optimal solution in polynomial time. We also present a proper generalization to (general) integer programs and to local search problems of the well-known result that optimization and augmentation are equivalent for 0/1-integer programs. Among other things, our results imply that PLS-complete problems cannot have "near-exact" neighborhoods, unless PLS = P.

1 Introduction

The following question arises naturally in the study of optimization problems (see, e.g., [12, Chap. 15.2]): Is computing the value of an optimal solution as hard as actually finding an optimal solution? Crescenzi and Silvestri [4] initiated the formal study of the relative complexity of evaluating the optimal cost of an optimization problem versus constructing an optimal solution, and they provided sufficient and necessary conditions for the existence of optimization problems for which obtaining an optimal solution is harder than computing the optimal value. Ausiello et al. [2] and Johnson [8] pointed out that evaluation is actually as hard as finding an optimal solution for all optimization problems whose associated decision problems are NP-complete. Schulz [16] studied the relative complexity of several problems related to 0/1-integer programming,[1] including augmentation, optimization and evaluation, all of which are polynomial-time equivalent. In this paper, we prove that evaluation and optimization are polynomial-time equivalent for all integer linear programming problems. That is, given a matrix $A \in \mathbb{Z}^{m \times n}$ and a vector $b \in \mathbb{Z}^m$, a polynomial-time algorithm for finding the optimal value of $\min\{cx : Ax \geq b, x \in \mathbb{Z}_+^n\}$, for any $c \in \mathbb{Z}^n$, implies the existence of such an algorithm for finding an optimal solution, $\arg\min\{cx : Ax \geq b, x \in \mathbb{Z}_+^n\}$. In fact, our result is slightly stronger than this. As long as we are given bounds on the values that individual variables may attain, the matrix A and the vector b need not be known explicitly. An evaluation oracle, which accepts as input

---

[1] In a 0/1-integer programming problem all variables can have values 0 or 1 only.

F. Dehne et al. (Eds.): WADS 2009, LNCS 5664, pp. 519–529, 2009.

any objective function vector c and returns the optimal objective function value, suffices. Our proof is constructive.

The proof itself gives rise to a new problem, related to questions typically brought up in postoptimality analysis of optimization problems, which we call the "unit increment problem:" Given an optimal solution x^0 with respect to an objective function vector c, find an optimal solution for $c + e_j$, where e_j is the j-th unit vector.[2] We show that an integer linear program can be solved in polynomial time if and only if its unit increment problem can be solved in polynomial time. For 0/1-integer programs, we prove that the unit increment problem is polynomial-time equivalent to the augmentation problem.[3] Hence, we have a proper generalization (to general integer programs) of a result by Grötschel and Lovász [6] and Schulz et al. [17], who showed that optimization and augmentation are polynomial-time equivalent for 0/1-integer programs. A relaxation to the augmentation problem, the ϵ-augmentation problem, can be defined as follows: Given an objective function vector c and a feasible solution x, find a feasible solution with better objective function value, or assert that x is ϵ-optimal. Here, $\epsilon > 0$, and a solution x is ϵ-optimal if $cx \leq (1 + \epsilon)cx'$ for all feasible solutions x'. The corresponding unit increment problem, the ϵ-unit increment problem, is defined as follows: Given an index j and an ϵ-optimal solution with respect to an objective function vector c, find an ϵ-optimal solution for $c + e_j$. We show that an ϵ-optimal solution can be obtained in polynomial time if and only if the ϵ-unit increment problem can be solved in polynomial time. Moreover, we show that for 0-1 integer programs, the ϵ-augmentation problem and the ϵ-unit increment problem are polynomial-time equivalent as well.

The concepts of unit increment and augmentation extend naturally to local search, with interesting implications. For an integer programming or combinatorial optimization problem with a neighborhood function N, the local augmentation problem, given a feasible solution x and an objective function vector c, asks for a solution in the neighborhood $N(x)$ of x of better objective function value, if one exists. The local unit increment problem is defined similarly: Given an index j and a locally optimal solution x with respect to c, find a locally optimal solution for $c + e_j$. We show that for a given neighborhood function, a locally optimal solution can be computed in polynomial time if and only if the local unit increment problem can be solved in polynomial time. However, in contrast to the cases of global optimization and ϵ-optimization, for 0/1-integer programs, the local unit increment problem and the local augmentation problem are not known to be equivalent. In fact, it follows from our results that if polynomial solvability of the local augmentation problem implies the polynomial solvability

---

[2] In this part of the paper we assume, for convenience, that all objective function coefficients are nonnegative. Most of our results hold true in general, if the unit increment problem is extended to finding optimal solutions for $c \pm e_j$.

[3] The augmentation problem is defined as follows: Given a feasible solution and an objective function vector, find a feasible solution of better objective function value, if one exists.

of the local unit increment problem, then all PLS-complete[4] problems can be solved in polynomial time.

A neighborhood function is said to be "exact" if every locally optimal solution is guaranteed to be globally optimal. A neighborhood function is said to be "near exact" if the objective function value of any locally optimal solution is no worse than that of all but a polynomial number of feasible solutions. Near-exact neighborhoods are related to the domination number of local search heuristics [7]. We show that, for 0/1-integer programs, polynomial solvability of the local augmentation problem implies polynomial solvability of the local optimization problem whenever the corresponding neighborhood is near exact. This implies that no PLS-complete problem can possess a near-exact neighborhood, unless PLS = P.

The rest of the paper is organized as follows. In Sect. 2 we establish that optimization and evaluation are polynomial-time equivalent for integer programming problems. In Sect. 3, we show that the unit increment problem and the optimization problem are polynomial-time equivalent. We also give a direct proof that, for 0/1-integer programming problems, augmentation and unit increment are polynomial-time equivalent. In Sect. 4 we extend these results to ϵ-optimization, ϵ-augmentation, and ϵ-unit increment. Section 5 contains our results on local search; in particular, we show that even for 0/1-integer programs, a local unit increment oracle is stronger than a local augmentation oracle, unless PLS = P.

2   Evaluation versus Optimization

It is well known (see, e.g. [15, Chap. 17.1]) that if an integer program has a finite optimum, then it has an optimal solution of size (i.e., encoding length) polynomially bounded by the size of the input. Hence, instead of considering $\min\{cx : Ax \geq b, x \in \mathbb{Z}_+^n\}$, we may restrict ourselves to solving $\min\{cx : Ax \geq b, x \leq u, x \in \mathbb{Z}_+^n\}$, for a vector $u \in \mathbb{Z}_+^n$ whose encoding length is polynomial in that of A, b, and c. From now on, we therefore consider a family $\mathcal{F}$ of integer programming problems that is described as follows. For each instance of the family we are given a vector $u \in \mathbb{Z}_+^n$ such that the set $X \subseteq \mathbb{Z}^n$ of feasible solutions is contained in $\{0, 1, \ldots, u_1\} \times \{0, 1, \ldots, u_2\} \times \cdots \times \{0, 1, \ldots, u_n\}$. We are also given an evaluation oracle that contains the only additional information that we have on X.[5] (In particular, we do not explicitly need to know a matrix A and a vector b such that $X = \{x \in \mathbb{Z}_+^n : Ax \geq b, x \leq u\}$.) For input vector $c \in \mathbb{Z}^n$,

---

[4] The complexity classes PLS and PLS-complete were introduced by Johnson et al. [9] to capture the difficulty of finding local optima. Prominent PLS-complete problems include the max-cut problem with the flip neighborhood and the graph partitioning problem with the swap neighborhood [14], the traveling salesman problem with the k-exchange neighborhood (for sufficiently large, but constant k) [10], and the problem of finding pure-strategy Nash equilibria in congestion games [5].

[5] We may assume, without loss of generality, that there exists a feasible solution, i.e., $X \neq \emptyset$. Otherwise both oracles, evaluation and optimization, would have to detect infeasibility.

the oracle returns the optimal objective function value of $\min\{cx : x \in X\}$. The following is our main result.

Theorem 1. *Given a family $\mathcal{F}$ of integer programming problems described by an evaluation oracle, there is an oracle-polynomial time algorithm for solving the optimization problem.*

Proof. Let $\min\{cx : x \in X\}$ be the optimization problem to be solved, given by an evaluation oracle and a vector $u \in \mathbb{Z}_+^n$ such that $X \subseteq \{0 \le x \le u\}$. The main idea is as follows. Among all optimal solutions, let x' be the one that is lexicographically minimal. We perturb c in such a way that x' remains optimal for the perturbed vector c' and is also optimal for the objective function vectors $c' + e_j$, for all $j = 1, 2, \ldots, n$. With $n + 1$ calls of the evaluation oracle we can then recover x' via $x'_j = (c' + e_j)x' - c'x'$, for $j = 1, 2, \ldots, n$. If the size of c' is sufficiently small, this yields an oracle-polynomial time algorithm.

Here are the details. Let $U := \max\{u_j : j = 1, 2, \ldots, n\} + 1$. We define c' as follows:

$$c'_j := U^{2n+1} c_j + U^{2(n-j)+1}, \quad \text{for } j = 1, 2, \ldots, n.$$

Note that the encoding length of c' is indeed polynomial in that of c and u. We first show that, (i), every solution x^* that is optimal for c' is also optimal for c. In fact, for any $x \in X$, we obtain that

$$cx^* \le \frac{c'x^*}{U^{2n+1}} \le \frac{c'x}{U^{2n+1}} = cx + \frac{\sum_{j=1}^{n} U^{2(n-j)+1} x_j}{U^{2n+1}} < cx + 1.$$

Together with the integrality of c, x^*, and x, this implies that $cx^* \le cx$, proving (i). We now show that, (ii), if x is an optimal solution for c that is different from x', then $c'(x' - x) \le -U$. Let i be the first index for which $x'_i < x_i$. Then,

$$c'(x' - x) = U^{2(n-i)+1}(x'_i - x_i) + \sum_{j=i+1}^{n} U^{2(n-j)+1}(x'_j - x_j)$$

$$\le -U^{2(n-i)+1} + \sum_{j=i+1}^{n} U^{2(n-j)+2}$$

$$\le -U.$$

It remains to show that x' is optimal for $c' + e_j$, for an arbitrary, but fixed index $j \in \{1, 2, \ldots, n\}$. So, let x be some feasible solution different from x'. We distinguish two cases. If x is optimal for c, then, with the help of (ii), we get

$$(c' + e_j)(x' - x) = c'(x' - x) + (x'_j - x_j) \le -U + U = 0.$$

If x is not optimal for c, we have

$$c'(x' - x) = U^{2n+1} c(x' - x) + \sum_{j=1}^{n} U^{2(n-j)+1}(x'_j - x_j)$$

$$\le -U^{2n+1} + \sum_{j=1}^{n} U^{2(n-j)+2} \le -U.$$

Hence, $(c' + e_j)(x' - x) \leq 0$. Thus, x' is optimal for c' and $c' + e_j$, and $x'_j = (c' + e_j)x' - c'x'$. In particular, the j-th component of x' can be computed by two calls of the evaluation oracle. □

3 The Unit Increment Problem and Global Optimization

In this section we assume that all objective function vectors are nonnegative, for convenience. All results can be extended in a straightforward way to arbitrary objective function vectors. If c is an objective function vector, let $C := \max\{c_j : j = 1, 2, \ldots, n\}$ and $\alpha := 1 + \lceil \log_2 C \rceil$. Then each c_j can be represented as a binary number using α bits. Let $b^j = (b^j_1, b^j_2, \ldots, b^j_\alpha)$ with $b^j_i \in \{0, 1\}$ be this representation. Moreover, let c^k_j be the number represented by the k leading bits of b^j. That is, $c^k_j := \sum_{i=1}^k 2^{k-i} b^j_i$. Thus $c^1_j \in \{0, 1\}$, $c^\alpha_j = c_j$, and $c^{k+1}_j = 2c^k_j + b^j_{k+1}$ for all $k = 1, 2, \ldots, \alpha - 1$ and $j = 1, 2, \ldots, n$.

Let $\min\{cx : x \in X\}$ with $X \subseteq \mathbb{Z}^n_+$ be an instance of the optimization problem. We assume that an oracle UNIT-INC is available which with input j, $c + e_j$, and an optimal solution x^0 with respect to c, computes an optimal solution x^* for $\min\{(c + e_j)x : x \in X\}$. We consider the following algorithm.

Algorithm UI

begin
    let x^0 be any feasible solution
    set $c^*_j := 0$ for $j = 1$ to n
    for $k = 1$ to α **do**
        for $j = 1$ to n **do** $c^*_j := 2c^*_j$
        $S := \{j : b^j_k = 1\}$
        while $S \neq \emptyset$ **do**
            choose $j \in S$
            $S := S \setminus \{j\}$
            $c^*_j := c^*_j + 1$
            If $x^0_j > 0$ **then**
                call UNIT-INC(c^*, x^0, x^*, j)
                $x^0 := x^*$
            endif
        endwhile
    endfor
    output x^0
end

Theorem 2. *Let a family of optimization problems with linear objective functions be given by a unit-increment oracle. Then algorithm UI computes an optimal solution in oracle-polynomial time.*

Proof. Note that at the end of the k-th iteration of the main loop we have $c^* = c^k$. Assume that at the beginning of the k-th iteration of the main loop, x^0

is an optimal solution to $\min\{c^{k-1}x : x \in X\}$. (For convenience, we let $c^0 = 0$.) Then x^0 continues to be an optimal solution if we change the objective function to $2c^{k-1}$. Thus at the end of the while loop, the oracle UNIT-INC guarantees that x^0 is an optimal solution to $\min\{c^k x : x \in X\}$. The correctness of the algorithm follows by induction over k.   □

Hence, if one can find a feasible solution in polynomial time and if one can solve the unit increment problem in polynomial time, then one can determine an optimal solution in polynomial time. For the assignment problem on a bipartite graph on n nodes and m edges, the general algorithm described above terminates in $O(nm \log C)$ time. This is because the unit increment problem for this special case can be solved in $O(m)$ time. Although there are special-purpose algorithms with better worst case bounds to solve the assignment problem, it is interesting to note that the general algorithm UI achieves a good time bound.

Algorithm UI is a generalization of the bit scaling algorithm studied extensively in the network flow literature (see, e.g., [1]) and in the context of 0/1-integer programming (see, e.g., [16]). The new feature here is the use of the unit increment oracle. This allows us to compare the computational complexity of optimization problems and unit increment problems and also provides a framework for our study of ϵ-optimization and local optimization. Theorem 2 establishes that an optimization problem with linear objective function can be solved in polynomial time if and only if the corresponding unit increment problem can be solved in polynomial time. Alternatively, if the optimization problem is NP-hard, then the corresponding unit increment problem is also NP-hard. Thus the additional information available for the unit increment problem (i.e., an optimal solution for the original objective function) is not of much help for NP-hard problems. This provides additional evidence that postoptimality analysis is typically hard for NP-hard problems (see, e.g., [3,13,18] for related results).

We now examine the relationship between the unit increment problem and the augmentation problem.

Lemma 3. *Let x^0 be an optimal solution to $\min\{cx : x \in X\}$, and let j be a given index, $1 \le j \le n$. If $x \in X$ is a feasible solution, then $(c+e_j)(x^0-x) \le x_j^0$.*

Proof. Since x is a feasible solution to $\min\{cx : x \in X\}$, $cx^0 \le cx$. Thus $(c + e_j)x^0 = cx^0 + x_j^0 \le cx + x_j^0 \le (c + e_j)x + x_j^0$.   □

The next theorem is, in principle, a consequence of the before-mentioned equivalence between augmentation and optimization for 0/1-integer programs, and Theorem 2. However, the following proof provides a Karp reduction from the unit increment problem to the augmentation problem.

Theorem 4. *For 0/1-integer programs, the unit increment problem and the augmentation problem are polynomial-time equivalent.*

Proof. Assume that $X \subseteq \{0,1\}^n$ is given by an augmentation oracle. Consider the instance $\min\{cx : x \in X\}$ and its unit increment version $\min\{(c+e_j)x : x \in$

$X\}$ together with respective optimal solutions x^0 (given) and x^* (unknown). By Lemma 3,

$$(c + e_j)(x^0 - x^*) \leq 1. \tag{1}$$

Since $c \in \mathbb{Z}^n$, one application of the augmentation oracle starting with x^0 either declares that x^0 is optimal for $\min\{(c + e_j)x : x \in X\}$ or finds an improving solution which must be optimal for $\min\{(c + e_j)x : x \in X\}$ in view of (1). The other direction is implied by Theorem 2. $\square$

Using Lemma 3 we have the following result for general integer programs. Let $\min\{cx : x \in X\}$ be an instance and x^0 be an optimal solution. Let $\min\{(c+e_j)x : x \in X\}$ be the corresponding j-th unit increment instance.

Theorem 5. *Given $x^0 \in \arg\min\{cx : x \in X\}$ and an augmentation oracle, $\min\{(c + e_j)x : x \in X\}$ can be solved by $O(x_j^0)$ calls of the augmentation oracle.*

Theorems 2 and 5 show that for an integer linear program for which $x \leq u$ for all $x \in X$ and where the components of u are bounded above by a polynomial of the remaining input data, the optimization problem can be solved in polynomial time whenever the augmentation problem can be solved in polynomial time.

4   The Unit Increment Problem and ϵ-Optimization

In this section we explore the complexity of finding near-optimal solutions if ϵ-augmentation or ϵ-unit increment oracles are available. We need the assumption that $c_j \geq 0$ for all $j = 1, 2, \ldots, n$. We also fix $\epsilon > 0$. Let $\mathrm{UI}(\epsilon)$ denote the variation of the algorithm UI where the oracle UNIT-INC is replaced by ϵ-UNIT-INC which takes as input $c + e_j$, an ϵ-optimal solution x^0 of $\min\{cx : x \in X\}$, an index j, and computes an ϵ-optimal solution x^* of $\min\{(c + e_j)x : x \in X\}$. Using arguments similar to that in the proof of Theorem 2 one can show the following result.

Theorem 6. *Given an ϵ-unit increment oracle and an initial feasible solution, algorithm UI(ϵ) computes an ϵ-optimal solution to $\min\{cx : x \in X\}$ in oracle-polynomial time.*

Thus if a feasible solution can be computed in polynomial time and the ϵ-unit increment problem can be solved in polynomial time, then an ϵ-optimal solution can be obtained in polynomial time. Alternatively, an optimization problem is not approximable if and only if the corresponding unit increment problem is not approximable. This result is interesting in several ways. For example, even if we have an ϵ-optimal solution to the traveling salesman problem, if one of the edge weights is increased by one, then getting an ϵ-optimal solution is still NP-hard. Also, there exists a (fully) polynomial-time approximation scheme for an optimization problem if and only if there is a (fully) polynomial-time approximation scheme for the unit increment problem.

Interestingly, we can show that the ϵ-augmentation problem and the ϵ-unit increment problem are equivalent for 0/1-integer programs.

Theorem 7. *For 0/1-integer programs, the ε-augmentation problem and the ε-unit increment problem are polynomial-time equivalent.*

Proof. Assume first that an ε-augmentation oracle and an ε-optimal solution x^1 to $\min\{cx : x \in X\}$ are given. Consider an instance $\min\{cx : x \in X\}$ with $X \subseteq \{0,1\}^n$ and its j-th unit increment instance $\min\{(c + e_j)x : x \in X\}$. Let x^0 and x^* be (unknown) optimal solutions of the former problem and the latter problem, respectively.

If the ε-augmentation oracle declares that x^1 is an ε-optimal solution to $\min\{(c + e_j)x : x \in X\}$, we are done. Thus suppose that starting with the solution x^1 to $\min\{(c + e_j)x : x \in X\}$ the ε-augmentation oracle produces an improved solution, say x^2. We will show that x^2 is an ε-approximate solution to $\min\{(c + e_j)x : x \in X\}$. By Lemma 3 we have

$$(c + e_j)x^* = cx^0 \text{ or } (c + e_j)x^* = cx^0 + 1. \tag{2}$$

Since x^1 is ε-optimal for $\min\{cx : x \in X\}$,

$$\frac{cx^1 - cx^0}{cx^0} \leq \epsilon.$$

Case 1: $x_j^1 = 1$. In this case $(c + e_j)x^1 = cx^1 + 1$. Since x^2 is an improved solution for $\min\{(c + e_j)x : x \in X\}$ obtained from x^1, $(c + e_j)x^2 < (c + e_j)x^1$ and hence

$$(c + e_j)x^2 \leq cx^1.$$

From (2) we have $(c + e_j)x^* = cx^0$ or $(c + e_j)x^* = cx^0 + 1$. If $(c + e_j)x^* = cx^0$, then

$$\frac{(c + e_j)x^2 - (c + e_j)x^*}{(c + e_j)x^*} \leq \frac{cx^1 - cx^0}{cx^0} \leq \epsilon.$$

If $(c + e_j)x^* = cx^0 + 1$, then

$$\frac{(c + e_j)x^2 - (c + e_j)x^*}{(c + e_j)x^*} \leq \frac{cx^1 - cx^0 - 1}{cx^0 + 1} \leq \frac{cx^1 - cx^0}{cx^0} \leq \epsilon.$$

Case 2: $x_j^1 = 0$. In this case $(c + e_j)x^1 = cx^1$. We will show that x^1 is an ε-optimal solution to $\min\{(c + e_j)x : x \in X\}$. If $(c + e_j)x^* = cx^0$, then

$$\frac{(c + e_j)x^1 - (c + e_j)x^*}{(c + e_j)x^*} = \frac{cx^1 - cx^0}{cx^0} \leq \epsilon.$$

If $(c + e_j)x^* = cx^0 + 1$ then,

$$\frac{(c + e_j)x^1 - (c + e_j)x^*}{(c + e_j)x^*} = \frac{cx^1 - cx^0 - 1}{cx^0 + 1} \leq \frac{cx^1 - cx^0}{cx^0} \leq \epsilon.$$

Thus if the ε-augmentation oracle does not declare x^1 as ε-optimal, the improved solution x^2 is guaranteed to be ε-optimal for $\min\{(c + e_j)x : x \in X\}$.

The converse of the theorem follows from Theorem 6. □

One consequence of the above theorem is that a $0/1$-integer program has a (fully) polynomial-time approximation scheme if and only if the corresponding augmentation problem has a (fully) polynomial-time approximation scheme. The same result was obtained by Orlin et al. using different arguments [11].

5   The Unit Increment Problem and Local Optimization

In this section we consider the complexity of computing a locally optimal solution with respect to a given neighborhood function N. Recall that the *local augmentation problem* has as input a feasible solution x and an objective function vector c, and it outputs a solution $y \in N(x)$ with $cy < cx$, unless x is already a local optimum. The *local* unit increment problem accepts as input an index j and a locally optimal solution x^0 with respect to c, and it returns a locally optimal solution x^* with respect to $c + e_j$.

As in the case of (global) optimization and ϵ-optimization, we first observe that if a feasible solution can be obtained in polynomial time and the local unit increment problem can be solved in polynomial time, then a local optimum can be computed in polynomial time. To establish this, we simply modify algorithm UI by replacing the unit increment oracle, UNIT-INC, with a local unit increment oracle, LOCAL-UNIT-INC. We call the resulting algorithm LUI.

Theorem 8. *Given a* LOCAL-UNIT-INC *oracle, algorithm LUI computes a locally optimal solution in oracle-polynomial time.*

The proof of Theorem 8 is similar to that of Theorem 2. Theorem 8 establishes that the complexity of finding a local optimum is captured by that of the local unit increment problem.

Unlike the case of optimization and ϵ-optimization, we are not able to establish the equivalence of the local unit increment problem and the local augmentation problem for $0/1$-integer programs. In fact, if they are equivalent, then, by Theorem 8, there is a polynomial-time algorithm for finding a local optimum for any problem in PLS, including PLS-complete problems. In other words, this would imply PLS = P. However, the two problems are equivalent if the neighborhood is exact. This follows from Theorem 4. Interestingly, we can show that this is also true for near-exact neighborhoods. Recall from the introduction that a neighborhood is called near exact if the objective function value of any local optimum is worse than that of at most a polynomial number of other feasible solutions.

Theorem 9. *For $0/1$-integer programs with near-exact neighborhoods, a polynomial-time algorithm for local augmentation implies a polynomial-time algorithm for the local unit increment problem.*

Proof. Let x^0 be a locally optimal solution with respect to the near-exact neighborhood N and the objective function vector c. As usual, X denotes the set of feasible solutions.

Since N is near exact, there exists $X^* \subseteq X$ such that $cx^0 \leq cx$ for all $x \in X^*$ and $|X \setminus X^*| \leq f(|I|)$ for some polynomial f. Here, $|I|$ is the size of the input.

By Lemma 3, $(c+e_j)(x^0 - x) \leq 1$ for all $x \in X^*$. Thus in one augmentation step, starting from x^0, we get a solution that is no worse than any solution in X^* w.r.t $(c + e_j)$. This solution may or may not be a local optimum with respect to N. But outside X^* there are only $f(|I|)$ solutions and hence the local augmentation oracle cannot be called more than $f(|I|)$ additional times before reaching a local optimum.   □

Corollary 10. *If there exists a near-exact neighborhood for a PLS-complete optimization problem with linear objective function, then there is a polynomial-time algorithm that finds a local optimum for all problems in PLS. That is, PLS = P.*

However, near-exact neighborhoods are unlikely to exist, at least for the TSP [7].

Acknowledgements

This work was supported in part by ONR grant N00014-01208-1-0029.

References

1. Ahuja, R.K., Magnanti, T.L., Orlin, J.B.: Network Flows: Theory, Algorithms, and Applications. Prentice-Hall, Englewood Cliffs (1993)
2. Ausiello, G., Crescenzi, P., Gambosi, G., Kann, V., Marchetti-Spaccamela, A., Protasi, M.: Complexity and Approximation. Springer, Heidelberg (1999)
3. Chakravarti, N., Wagelmans, A.P.M.: Calculation of stability radii for combinatorial optimization problems. Operations Research Letters 23, 1–7 (1998)
4. Crescenzi, P., Silvestri, R.: Relative complexity of evaluating the optimum cost and constructing the optimum for maximization problems. Information Processing Letters 33, 221–226 (1990)
5. Fabrikant, A., Papadimitriou, C.H., Talwar, K.: The complexity of pure Nash equilibria. In: Proceedings of the 36th Annual ACM Symposium on Theory of Computing, Chicago, IL, pp. 604–612 (2004)
6. Grötschel, M., Lovász, L.: Combinatorial optimization. In: Graham, R.L., Grötschel, M., Lovász, L. (eds.) Handbook of Combinatorics, ch. 28, vol. 2, pp. 1541–1597. Elsevier, Amsterdam (1995)
7. Gutin, G., Yeo, A., Zverovitch, A.: Exponential neighborhoods and domination analysis for the TSP. In: Gutin, G., Punnen, A.P. (eds.) The Traveling Salesman Problem and Its Variations, ch. 6, pp. 223–256. Kluwer, Dordrecht (2002)
8. Johnson, D.S.: The NP-completeness column: Finding needles in haystacks. ACM Transactions on Algorithms 3 (2007)
9. Johnson, D.S., Papadimitriou, C.H., Yannakakis, M.: How easy is local search? Journal of Computer and System Sciences 37, 79–100 (1988)
10. Krentel, M.W., Structure in locally optimal solutions, in Proceedings of the 30th Annual Symposium on Foundations of Computer Science, Research Triangle Park, NC, 1989, 216–221.
11. Orlin, J.B., Punnen, A.P., Schulz, A.S.: Approximate local search in combinatorial optimization. SIAM Journal on Computing 33, 1201–1214 (2004)
12. Papadimitriou, C.H., Steiglitz, K.: Combinatorial Optimization: Algorithms and Complexity. Prentice-Hall, Englewood Cliffs (1982)

13. Ramaswamy, R., Chakravarti, N.: Complexity of determining exact tolerances for min-sum and min-max combinatorial optimization problems, Working Paper WPS-247/95, Indian Institute of Management, Calcutta, India (1995)
14. Schäffer, A.A., Yannakakis, M.: Simple local search problems that are hard to solve. SIAM Journal on Computing 20, 56–87 (1991)
15. Schrijver, A.: Theory of Linear and Integer Programming. Wiley, Chichester (1986)
16. Schulz, A.S.: On the relative complexity of 15 problems related to 0/1-integer programming. In: Cook, W.J., Lovász, L., Vygen, J. (eds.) Research Trends in Combinatorial Optimization, ch. 19, pp. 399–428. Springer, Berlin (2009)
17. Schulz, A.S., Weismantel, R., Ziegler, G.M.: 0/1-integer programming: Optimization and augmentation are equivalent. In: Spirakis, P.G. (ed.) ESA 1995. LNCS, vol. 979, pp. 473–483. Springer, Heidelberg (1995)
18. van Hoesel, S., Wagelmans, A.P.M.: On the complexity of postoptimality analysis of 0/1 programs. Discrete Applied Mathematics 91, 251–263 (1999)

Resolving Loads with Positive Interior Stresses

Günter Rote[1] and André Schulz[2,*]

[1] Institut für Informatik, Freie Universität Berlin, Germany
rote@inf.fu-berlin.de
[2] Department of Computer Science, Smith College, USA
aschulz@email.smith.edu

Abstract. We consider the pair $(\mathbf{p}_i, \mathbf{f}_i)$ as a force with two-dimensional direction vector $\mathbf{f}_i$ applied at the point $\mathbf{p}_i$ in the plane. For a given set of forces we ask for a non-crossing geometric graph on the points $\mathbf{p}_i$ that has the following property: There exists a weight assignment to the edges of the graph, such that for every $\mathbf{p}_i$ the sum of the weighted edges (seen as vectors) around $\mathbf{p}_i$ yields $-\mathbf{f}_i$. As additional constraint we restrict ourselves to weights that are non-negative on every edge that is not on the convex hull of the point set. We show that (under a generic assumption) for any reasonable set of forces there is exactly one pointed pseudo-triangulation that fulfils the desired properties. Our results will be obtained by linear programming duality over the PPT-polytope. For the case where the forces appear only at convex hull vertices we show that the pseudo-triangulation that resolves the load can be computed as weighted Delaunay triangulation. Our observations lead to a new characterization of pointed pseudo-triangulations, structures that have been proven to be extremely useful in the design and analysis of efficient geometric algorithms.

As an application, we discuss how to compute the maximal locally convex function for a polygon whose corners lie on its convex hull.

1 Introduction

Let $P = \{\mathbf{p}_1, \ldots, \mathbf{p}_n\}$ be a set of distinct points in the plane in general position and let $F = \{\mathbf{f}_1, \ldots, \mathbf{f}_n\}$ denote a set of two-dimensional vectors. We think of the pair $(\mathbf{p}_i, \mathbf{f}_i)$ as a force in direction $\mathbf{f}_i$ that is applied at the point $\mathbf{p}_i$. The set of pairs $L = \{(\mathbf{p}_1, \mathbf{f}_1), \ldots, (\mathbf{p}_n, \mathbf{f}_n)\}$ is called a *load*. The objects we study in this paper are geometric graphs $G = (P, E)$ with point set P and edge set E. An edge of E that is not part of the convex hull of P is considered as *interior* edge.

A *stress* is a (symmetric) assignment of scalars to the edges E. Throughout the paper we denote a stress of G with $\omega \colon P \to \mathbb{R}$. We say that a graph *resolves a load L with stress ω*, if

$$\forall \mathbf{p}_i : \quad \sum_{(i,j) \in E} \omega_{ij}(\mathbf{p}_i - \mathbf{p}_j) = -\mathbf{f}_i. \tag{1}$$

* Funded by the German Science Foundation (DFG).

F. Dehne et al. (Eds.): WADS 2009, LNCS 5664, pp. 530–541, 2009.

If there exists some stress ω for which G resolves the load L we say that G resolves L. Furthermore, if there exists a stress ω, positive or zero on every interior edge, we say that G resolves L with *positive interior* stress. Stressed graphs have a physical interpretation. Their edges can be considered as a system of springs. Due to Hooke's law the force induced by a spring is proportional to its length. Hence the values ω_{ij} are the spring constants from this point of view. A negative spring constant models a rubber band—thus by considering only positive interior stresses we restrict ourselves to (expansive) springs.

In this paper we study the problem how to find a graph that resolves a given load with positive interior stress. Figure 1(a) shows a small introductory example of a problem instance. A possible solution with the corresponding stress is depicted in Figure 1(b).

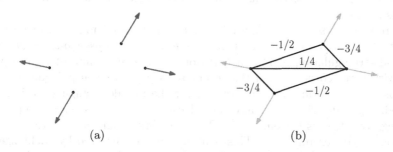

<table>
</table>

(a)   (b)

Fig. 1. A small example

Not all loads can be resolved by a graph. In particular, a load has to contribute neither a linear momentum (that is $\sum_i \mathbf{f}_i = \mathbf{0}$) nor an angular momentum (that is $\sum_i \langle \mathbf{f}_i, \mathbf{p}_i^{\perp} \rangle = 0$)[1]. To see this, observe the following: Every geometric graph with fixed stress can resolve exactly one load. If the graph consists of a single edge (i, j) it can resolve the load given by $\mathbf{f}_i = \omega_{ij}(\mathbf{p}_j - \mathbf{p}_i)$ and $\mathbf{f}_j = \omega_{ij}(\mathbf{p}_i - \mathbf{p}_j)$. We notice that $\mathbf{f}_i$ and $\mathbf{f}_j$ sum up to $\mathbf{0}$ and that $\langle \mathbf{f}_i, \mathbf{p}_i^{\perp} \rangle + \langle \mathbf{f}_j, \mathbf{p}_j^{\perp} \rangle = 0$ holds. A load resolved by a stressed graph is the composition of these "atomic" forces induced by single edges. Since the total linear (angular) momentum is the sum of the linear (angular) momenta of the atomic forces we cannot resolve a load with non-vanishing linear or angular momentum. In the following we consider only loads without linear and angular momentum. If we want to emphasize the absence of a momentum we use the term *moment-free* for loads.

The load given by $\mathbf{f}_i = \mathbf{0}$ for all $\mathbf{p}_i$ is called *zero load*. A stress that resolves this load on G is called *equilibrium stress* for G. Graphs with equilibrium stresses have several nice properties. An old results that goes back to James Clerk Maxwell states that the equilibrium stresses of a planar graph are in one-to-one correspondence with its spatial liftings [16]. Furthermore, the sign of the stress on an edge indicates the curvature along this edge in the lifting. This observation is known as the Maxwell-Cremona correspondence.

---

[1] The vector $\mathbf{p}^{\perp} := (y, -x)^T$ denotes the rotation of $\mathbf{p} = (x, y)^T$ by 90 degrees around the origin.

Considering positive interior stresses only is also motivated by the construction of a discrete Laplace-Beltrami operator [25]. There exist different versions of the discrete Laplace-Beltrami operator, but none can guarantee all properties of the continuous equivalent [25]. In particular, a discrete Laplace-Beltrami operator for a triangulated surface mesh should be modeled by an equilibrium stress (with every vertex incident to at least one non-zero stress) whenever the surface lies in the plane. On the other hand the edge weights should be non-negative to guarantee the *maximum principle*, which is a natural property for the classical Laplace-Beltrami operator and which should also hold for the discrete version [5]. As mentioned in [25], the existence of non-regular triangulations makes it impossible to construct a "perfect" discrete Laplacian. However, it is open how to fix this inconvenience by local adjustments to the original mesh. A better understanding how positive equilibrium stresses behave could lead to a solution for this problem.

We restrict ourselves to a special class of graphs that might resolve a given load with a positive interior stress. These are the *pointed pseudo-triangulations*. A pseudo-triangulation of P is a partition of the convex hull of P into polygons with three corners[2] such that every $\mathbf{p}_i$ is part of some polygon. If every point is incident to an angle greater than π the pseudo-triangulation is called pointed. A pseudo-triangulations can resolve any moment-free load with positive and negative stresses (Streinu [24]). Pseudo-triangulations are related to maximal locally convex functions. This relationship was observed by Aichholzer et al. [2] and was further extended by Aurenhammer and Krasser [3]. Similar to triangulations pointed pseudo-triangulations appear as geometric data structures and they are used to prove the correctness and efficiency of algorithms. They find applications in ray shooting [6], motion planning [4], and art gallery type problems [23]. This list is far from complete, for a comprehensive discussion on pseudo-triangulations we direct the reader to the survey by Rote, Santos and Streinu [20].

The reasons why we focus on pointed pseudo-triangulations are the following. Since they can resolve loads with general stresses, they seem to be powerful enough. On the other hand, every vertex which is not pointed is in some sense "over-constrained" to resolve the force (with positive stresses) at this point, because the effect of at least one stressed edge can be expressed by adjusting the stresses on two other edges. Moreover, pseudo-triangulations are non-crossing geometric graphs and therefore easy to understand for the viewer.

Results: As our main result we show that there exists for every moment-free load L a pointed pseudo-triangulation that can resolve L with positive interior stress. Moreover, up to degenerate situations, this pointed pseudo-triangulation is unique. We extend these results to a constrained version of the original problem. This means, that even if we restrict a certain set of edges to appear in the solution, we can find a pointed pseudo-triangulation that contains this set and resolves L with positive stress on the unconstrained interior edges. This

---

[2] A corner is a vertex of a polygon with interior angle smaller than π.

is true for any constraints that allow the completion to some pointed pseudo-triangulation. As in the original setting the obtained solution is unique for almost every load. The constrained and the unconstrained problem can be solved by linear programming. For the special situation when all forces appear only at convex hull vertices we provide an algorithm that computes the load resolving pointed pseudo-triangulation without linear programming.

As application we show how we can compute pointed pseudo-triangulations that refer to maximal locally convex functions, for polygons whose corners lie on its convex hull. Our approach combines the load-resolving method with the Maxwell-Cremona correspondence.

2   Resolving Loads with Pointed Pseudo-triangulations

2.1   General Solution

Let us start with some preliminary observations about pointed pseudotriangulations. In the following we use pointed pseudo-triangulations of point sets and (later in Section 3) also of simple polygons.

Definition 1. *A* pointed pseudo-triangulation *of a point set P is the partition of the convex hull of P into polygons with three corners, such that every vertex in P is incident to an angle greater than π.*

A pointed pseudo-triangulation *of a polygon $\mathcal{P}$ is the partition of $\mathcal{P}$ into polygons with three corners, such that every vertex of $\mathcal{P}$ is incident to an angle greater than π.*

There exists a high-dimensional polytope whose corners correspond to the pointed pseudo-triangulations a point set can have [19]. This polytope is called PPT-polytope and it is based on the fact that every pointed pseudo-triangulation has an expansive infinitesimal motion, if one removes a convex hull edge (see Streinu [24]). The infinitesimal velocities $\mathbf{v}_1, \ldots, \mathbf{v}_n$ (each $\mathbf{v}_i$ is a two-dimensional vector) act as unknowns in the description of the polytope:

$$\langle \mathbf{v}_i - \mathbf{v}_j, \mathbf{p}_i - \mathbf{p}_j \rangle \geq \langle \mathbf{p}_i, \mathbf{p}_j^{\perp} \rangle^2 \quad \forall i, j \leq n,$$
$$\langle \mathbf{v}_i - \mathbf{v}_j, \mathbf{p}_i - \mathbf{p}_j \rangle = \langle \mathbf{p}_i, \mathbf{p}_j^{\perp} \rangle^2 \quad \text{for } (i,j) \in \text{conv}(P), \tag{2}$$
$$\sum_{i=1}^{n} \mathbf{v}_i = \mathbf{0}, \tag{3}$$
$$\sum_{i=1}^{n} \langle \mathbf{v}_i, \mathbf{p}_i^{\perp} \rangle = 0. \tag{4}$$

The PPT-polytope is a simple polytope with dimension $2n - 3$. Hence, in each of its vertices $2n - 3$ inequalities are tight (including the equations of the convex hull edges). The pointed pseudo-triangulation that is associated with a specific vertex of the PPT-polytope is given by the edges induced by its tight inequalities.

We study the minimization of the function

$$\sum_{i=1}^{n} \langle \mathbf{v}_i, -\mathbf{f}_i \rangle \tag{5}$$

over the PPT-polytope given by (2–4) (in the following considered as primal program). As we will see later our choice of the objective function (5) leads to a solution that is capable to resolve the load L.

The constraints of the corresponding dual program have the following form:

$$1 \leq i \leq n : \quad \sum_{j=1}^{n} u_{ij}(\mathbf{p}_i - \mathbf{p}_j) + \mathbf{t} + r\mathbf{p}_i^{\perp} = -\mathbf{f}_i. \tag{6}$$

The variables $\mathbf{t}$ and r correspond to the equations (3) and (4). For every possible interior edge we obtain by LP duality the dual constraint $u_{ij} \geq 0$. By complementary slackness we deduce that if a constraint is not tight in the primal solution (there is no edge defined by this inequality), then the corresponding dual variable u_{ij} is zero in the dual solution. Thus, a (non-zero) u_{ij} appears only on the edges of the primal solution.

We observe that the dual variables $\mathbf{t}$ and r that come from the conditions (3) and (4) are the only difference between (6) and (1). Fortunately, we can show that under our assumptions the variables $\mathbf{t}$ and r can only be zero.

Lemma 1. *If the load in the primal program is moment-free we have for the dual variables $\mathbf{t} = \mathbf{0}$ and $r = 0$.*

Proof. Equation (6) denotes n restrictions of the dual program. Adding up all these equations cancels the u_{ij} variables and gives

$$n\mathbf{t} + r\left(\sum_{i=1}^{n} \mathbf{p}_i\right)^{\perp} = -\sum_{i=1}^{n} \mathbf{f}_i. \tag{7}$$

Because the objective function is moment-free the last equation equals zero. Now, we take the scalar product of both sides of equation (6) with $\mathbf{p}_i^{\perp}$. This gives n equations of the form

$$-\sum_{j=1}^{n} u_{ij}\langle \mathbf{p}_i^{\perp}, \mathbf{p}_j \rangle + \langle \mathbf{p}_i^{\perp}, \mathbf{t} \rangle + r\|\mathbf{p}_i\|^2 = -\langle \mathbf{p}_i^{\perp}, \mathbf{f}_i \rangle. \tag{8}$$

If we sum up all these equations, the u_{ij} variables cancel, since $\langle \mathbf{p}_i^{\perp}, \mathbf{p}_j \rangle = \langle \mathbf{p}_i^{\perp\perp}, \mathbf{p}_j^{\perp} \rangle = \langle -\mathbf{p}_i, \mathbf{p}_j^{\perp} \rangle = -\langle \mathbf{p}_j^{\perp}, \mathbf{p}_i \rangle$. We obtain

$$\left\langle \left(\sum_{i=1}^{n} \mathbf{p}_i\right)^{\perp}, \mathbf{t} \right\rangle + r\sum_{i=1}^{n} \|\mathbf{p}_i\|^2 = -\sum_{i=1}^{n} \langle \mathbf{p}_i^{\perp}, \mathbf{f}_i \rangle. \tag{9}$$

The u_{ij}s cancel, since $\langle \mathbf{p}_i^{\perp}, \mathbf{p}_j \rangle = \langle \mathbf{p}_i^{\perp\perp}, \mathbf{p}_j^{\perp} \rangle = \langle -\mathbf{p}_i, \mathbf{p}_j^{\perp} \rangle = -\langle \mathbf{p}_j^{\perp}, \mathbf{p}_i \rangle$. Again this equation is zero because the objective function is moment-free.

The variables $\mathbf{t}$ and r can be computed by solving a homogeneous linear equation system given by the three equations from (7) and (9). It remains to show that $\mathbf{t} = \mathbf{0}$ and $r = 0$ is the only solution of this system. We can rephrase (7) to express $\mathbf{t}$ as

$$\mathbf{t} = -\frac{r}{n}\left(\sum_{i=1}^{n} \mathbf{p}_i\right)^{\perp}. \tag{10}$$

If $r = 0$ then $\mathbf{t} = \mathbf{0}$ and we get the trivial solution. Therefore, let us assume that r is nonzero. We plug (10) into equation (9) and obtain

$$-\frac{r}{n}\left\langle\left(\sum_{i=1}^{n}\mathbf{p}_i\right)^{\perp},\left(\sum_{i=1}^{n}\mathbf{p}_i\right)^{\perp}\right\rangle + r\sum_{i=1}^{n}\|\mathbf{p}_i\|^2 = 0.$$

Further simplifications give

$$\sum_{i=1}^{n}\|\mathbf{p}_i\|^2 - \frac{1}{n}\left\|\sum_{i=1}^{n}\mathbf{p}_i\right\|^2 = 0. \tag{11}$$

Let $\bar{\mathbf{p}} := \frac{1}{n}\sum_{i=1}^{n}\mathbf{p}_i$ denote the center of gravity of P. We deduce

$$\|\mathbf{p}_i\|^2 = \|\bar{\mathbf{p}} + \mathbf{p}_i - \bar{\mathbf{p}}\|^2 = \|\bar{\mathbf{p}}\|^2 + 2\langle\bar{\mathbf{p}}, \mathbf{p}_i - \bar{\mathbf{p}}\rangle + \|\mathbf{p}_i - \bar{\mathbf{p}}\|^2.$$

Plugging this equivalence into equation (11) leads to

$$\sum_{i=1}^{n}(\|\bar{\mathbf{p}}\|^2 + 2\langle\bar{\mathbf{p}}, \mathbf{p}_i - \bar{\mathbf{p}}\rangle + \|\mathbf{p}_i - \bar{\mathbf{p}}\|^2) - n\|\bar{\mathbf{p}}\|^2 = n\|\bar{\mathbf{p}}\|^2 + \sum_{i=1}^{n}\|\mathbf{p}_i - \bar{\mathbf{p}}\|^2 - n\|\bar{\mathbf{p}}\|^2,$$

$$= \sum_{i=1}^{n}\|\mathbf{p}_i - \bar{\mathbf{p}}\|^2.$$

We observe that the last expression is strictly positive as long as not all $\mathbf{p}_i$ are the same, which is not allowed in our case. Hence, $\mathbf{t} = \mathbf{0}$ and $r = 0$ is the only solution of the homogeneous system and the lemma follows.   □

As consequence of Lemma 1 the dual variables u_{ij} define a stress that resolves L and is positive on every interior edge. Hence, the solution of the primal program computes a graph with the desired properties. Notice that almost every objective function has a unique solution, and thus there is for almost every load exactly one pointed pseudo-triangulation that resolves it with positive interior stress.

Theorem 1 (Main Theorem). *For every moment-free load L there exists a pointed pseudo-triangulation that resolves L with positive interior stress. Up to degenerate situations this pointed pseudo-triangulation is unique and it is the solution of the linear program (2–5).*

Algorithmically, the computation of the desired pointed pseudo-triangulation boils down to solving a linear program with $2n$ variables, whose length is in $O(n^2)$. Various methods and tools are applicable to solve LP programs. The interior point method of Karmarkar [13] runs in $O(n^{3.5}K)$, where K is the number of input bits.

2.2   The Constrained Problem

We are looking now for a pointed pseudo-triangulation that resolves a load L with positive interior stress and that contains a prescribed set of edges E_c. Notice that a set of non-crossing edges that leaves an angle greater than π at every vertex can always be completed to a pointed pseudo-triangulation [24]. Let us assume that E_c allows the completion to a pointed pseudo-triangulation on P.

The solution in this constrained setting can be computed with the same method we used for the general case. Again, we use a linear program to compute the pointed pseudo-triangulation. But this time we optimize only over a facet of the PPT-polytope. This facet can be obtained by turning all inequalities of (2) that refer to edges in E_c into equations:

$$\langle \mathbf{v}_i - \mathbf{v}_j, \mathbf{p}_i - \mathbf{p}_j \rangle = \langle \mathbf{p}_i, \mathbf{p}_j^\perp \rangle^2, \quad \text{for } (i,j) \in E_c. \tag{12}$$

The facet of the PPT-polytope is simple and, more important, it is not empty.

Forcing edges to appear in the graph has the following consequences for our LP approach: The inequalities of E_c are now equations. Thus, we have no information about the sign of the corresponding dual variables u_{ij} anymore. On the other hand the dual restrictions (6) are not affected. We have still $u_{ij} \geq 0$ for every u_{ij} that does not refer to an edge in E_c. Notice that Lemma 1 can be applied in this constrained setting without modifications. As a consequence we can deduce:

Theorem 2. *Let $L = P \times F$ be a moment free load and E_c be a set of edges that allows the completion to a pointed pseudo-triangulation on P. There exists a pointed pseudo-triangulation that contains E_c and resolves L with a stress that is positive on every interior edge that is not in E_c. Up to degenerate situations this pointed pseudo-triangulation is unique and it is the solution of the linear program given by (2–5) and (12).*

2.3   No Interior Forces

Let us assume for this section that $\mathbf{f}_i = \mathbf{0}$ for every vertex that is not on the convex hull of P. Under this assumption, we can find the pseudo-triangulation that resolves L by geometric methods, without solving a linear program. If we have no edge constraints, all interior points can simply be ignored: in this setting, we are looking for a triangulation that resolves forces whose points of application are in convex position.

Let us assume that the facial structure of a planar graph G is given by some combinatorial embedding (a planar map), and in addition, the vertices of G are drawn in the plane and the edges are realized as straight lines (a geometric graph). Note that the drawing in the plane can have crossings, and is not necessarily related to the facial structure. A height assignment $h: P \rightarrow \mathbb{R}$ is a *lifting* of a graph if all vertices that belong to a face lie on a common plane in $\mathbb{R}^3$ when giving $\mathbf{p}_i$ the additional coordinate $z_i = h(\mathbf{p}_i)$. The relation between liftings of a planar graph that is drawn as a geometric graph in the plane and its equilibrium stresses is expressed by the Maxwell-Cremona correspondence [16].

Theorem 3 (Maxwell-Cremona correspondence). *For a planar 2-connected graph G drawn in the plane, with a designated face $\hat{f}$, there is a one-to-one correspondence between*

1. *the liftings of G with $\hat{f}$ in the xy-plane, and*
2. *the equilibrium stresses on G.*

When we apply this theorem in this paper, the lifted surface is usually a polyhedral surface that consists of an "upper surface" and a "lower surface" that are glued together at their common boundary. Each part, when individually projected to the plane, will yield a planar drawing without crossings, but the overlay of the two parts will in general have crossings. It is possible that the lower and the upper surface intersect each other; this is no problem. When G, or a part of G, is drawn without crossings and the faces of G are the faces of this drawing, an edge with a positive stress lifts to a convex edge and an edge with a negative stress a concave edge.

The complete proof of the Maxwell-Cremona correspondence is due to Whiteley [26] A more constructive proof is due to Richter-Gebert [18]. The Maxwell-Cremona correspondence finds application in polygon unfolding [9] and grid embeddings of 3-polytopes [17]. Once we have a equilibrium stress, the computation of the lifting is easy and can be computed face by face, starting with $\hat{f}$. For detailed rules how to compute the lifting we refer to Richter-Gebert's book [18].

We first discuss the unconstrained load resolving problem. We look for a graph that resolves $L_C := \{(\mathbf{p}_i, \mathbf{f}_i) \mid \mathbf{p}_i$ lies on the convex hull$\}$. But this time negative stresses on interior edges are allowed. It is known that every triangulation can resolve L_C since it is a so called Laman graph and hence statically rigid [11]. Let us pick an arbitrary triangulation T of the convex hull. We compute the stress that resolves L_C on T and multiply all stresses by -1. This gives a stressed graph that "produces" the load L_C. The stressed triangulation T combined with the triangulation T' that resolves L_C with positive interior stresses yields a planar graph with equilibrium stress. Thus, by Maxwell-Cremona there is a lifting of this composition. From this lifting we know the heights of the lifted points, because every point lies on a face of T and we can compute the lifting partially for the faces of T, when we choose $\hat{f}$ as face of T. Since the interior edges of T' are restricted to have a positive stress their curvature in the lifting of $T \cup T'$ yields a convex bending. Therefore, the combinatorial structure of T' coincides with the weighted Delaunay triangulation of conv(P). The Delaunay-weights are given by $\|p_i\|^2 - z_i$ for every convex hull vertex $\mathbf{p}_i$. The weighted Delaunay triangulation for convex point sets can be computed in linear time [1].

The stresses of T can be computed with help of an ear decomposition of T. Let $\mathbf{p}_c$ be the corner of an ear of T. The force $(\mathbf{p}_c, \mathbf{f}_c)$ can be canceled by the two boundary edges of $\mathbf{p}_c$ by a unique stress which can be easily computed. We can update the forces assigned to the neighbors of $\mathbf{p}_c$ by subtracting the vector induced by the (newly) stressed edges incident to $\mathbf{p}_c$. Now we can eliminate the ear and continue the ear decomposition in this fashion until we reduced T to a triangle. Notice that by updating the neighboring forces we deal with a moment-free load in every step of the ear decomposition. Thus for the final triangle we have three moment-free forces which can be canceled. The stresses of the final triangle can be computed by a small linear system. Since the ear decomposition clearly runs in linear time we can compute the triangulation that resolves L_C with positive interior stresses in linear time.

For the constrained setting we can use the same ideas. More precisely, we fix again an arbitrary triangulation T and compute again a stress that resolves L on T. The heights of the associated lifting can be computed by the Maxwell-Cremona correspondence. The pointed pseudo-triangulation that resolves L_C is characterized by the polyhedral surface that is convex on every line segment inside $\mathrm{conv}(P)$ that doesn't cross a constraint edge. Aichholzer et al. [2] give an algorithm (not based on linear programming) that computes this polyhedral surface. Unfortunately, no practical bounds for the running time of this algorithm are known.

3   Computing Optimal Pointed Pseudo-triangulations of Polygons

Let $\mathcal{P}$ be a polygon with vertex set $P = \{\mathbf{p}_1, \ldots, \mathbf{p}_n\}$ and let P_c denote the set of corners of $\mathcal{P}$. Furthermore, let $h\colon P_c \to \mathbb{R}$ be a height assignment for the corners of $\mathcal{P}$. We study the maximal function $f^*\colon \mathcal{P} \to \mathbb{R}$ that is convex on every line inside $\mathcal{P}$ and that fulfills $h(\mathbf{p}_i) = f^*(\mathbf{p}_i)$ for all corners $\mathbf{p}_i \in P_c$. It was proved by Aichholzer et al. [2] that f^* describes a piecewise linear surface, whose non-linearities project down to a pointed pseudo-triangulation $\mathcal{PT}_h$. The framework of [2] is more general and covers also polygons with additional points inside. It can also be used to define optimal non-pointed pseudo-triangulations.

We can compute $\mathcal{PT}_h$ (or f^*) by (i) picking an arbitrary pseudo-triangulation of $\mathcal{P}$ and then (ii) applying a sequence of local adjustments which are called *flips*. A flip is a transformation of a pseudo-triangulation that exchanges, inserts or removes a single edge and produces a new pseudo-triangulation. There exists a criterion which tells us whether a flip brings us "closer" to $\mathcal{PT}_h$ or not. Thus, we can compute $\mathcal{PT}_h$ by a sequence of such (improving) flips. As a result of [2, Optimality Theorem] we know that the sequence is finite and terminates at $\mathcal{PT}_h$. For a polygon, an improving sequence can have super-polynomial length [22] but we can always find a short sequence of $O(n^2)$ flips [2, Lemma 7.3]. During the flipping process we must keep track of the heights on all vertices of $P \setminus P_C$. Only this information allows us to decide if a flip is improving. It was noticed in [12, page 55] that the recomputation of the heights can be expressed by a linear equation system that is based on a planar structure. As mentioned in [7], the Planar Separator Theorem provides a solution in $O(M(\sqrt{n}))$ time in this case, where $M(n)$ is the upper bound for multiplying two $n \times n$ matrices [14,15]. The current record for $M(n)$ is $O(n^{2.325})$, which is due Coppersmith and Winograd [10]. Thus, a flip can be carried out in $O(n^{1.163})$ time, and the whole algorithm takes $O(n^{3.163})$ time.

We give an alternative algorithm how to compute $\mathcal{PT}_h$ for polygons whose corners lie on its convex hull. Our approach uses a completely new technique and is based on the observations of the Maxwell-Cremona correspondence, introduced in the previous section. Here, in contrast to the previous section, we use a linear programming approach to solve a geometric problem.

We first give a high-level description of our method. We construct a polyhedron that consists of two shells. Roughly speaking, the upper shell is the (unknown) surface given by f^*—the lower shell is a cone with apex at the origin that spans the lifted corners of the polygon. To glue the two shells together, we have to extend f^* to the set $\text{conv}(P) \setminus P$ (the pockets of $\mathcal{P}$). Due to Maxwell-Cremona, the vertical projection of the polyhedron has an equilibrium stress, with positive stressed edges on the interior edges of $\mathcal{P}$. Each shell alone produces a moment-free load that shows up on the convex hull vertices. Since we know the height and the position of the convex hull vertices we know the geometric shape of the lower shell. Hence, we can compute all of its (interior) stresses and therefore the induced moment-free load. The graph that resolves this load with positive interior stresses gives $\mathcal{PT}_h$ with triangulated pockets. Of course we have to enforce the boundary edges of $\mathcal{P}$ to appear in our solution, which can be done by solving the constrained problem as discussed in Section 2.2.

We continue with the detailed construction. Let $[i, j, k, l]$ denote the signed volume of the tetrahedron spanned by the lifted vertices $\mathbf{p}_i, \mathbf{p}_j, \mathbf{p}_k, \mathbf{p}_l$ and let $[i, j, k]$ be the signed area of the triangle spanned by the (plane) vertices $\mathbf{p}_i, \mathbf{p}_j, \mathbf{p}_k$. We introduce a new vertex $\mathbf{p}_0 = (0, 0)^T$ as the apex of the lower shell. As observed in [8,22], the corresponding stresses on every edge connecting $\mathbf{p}_0$ to a corner $\mathbf{p}_i$ can be expressed as

$$\omega_{0i} := \frac{[0, h, i, j]}{[0, h, i][0, i, j]},$$

where $\mathbf{p}_h$ is the left neighbor, and $\mathbf{p}_j$ is the right neighbor of $\mathbf{p}_i$ on the convex hull of P. By construction, the stressed edges incident to $\mathbf{p}_0$ sum up to $\mathbf{0}$. For every corner $\mathbf{p}_i$ we obtain a vector $\mathbf{f}_i := \omega_{0i}(\mathbf{p}_i - \mathbf{p}_0) = \omega_{0i}\mathbf{p}_i$. We use the boundary edges of $\mathcal{P}$ as constraints and compute the pointed pseudo-triangulation that resolves the forces $\mathbf{f}_i$.

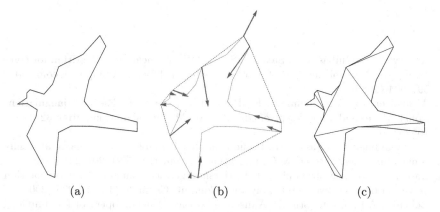

(a)   (b)   (c)

Fig. 2. A polygon (a), the induced forces that will yield the lifting to the paraboloid (b), and the corresponding pointed Delaunay pseudo-triangulation (c)

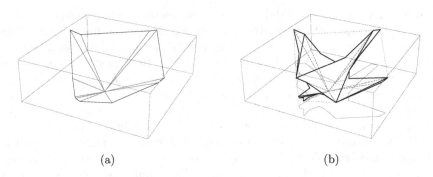

(a)   (b)

Fig. 3. Construction of a polyhedron whose upper shell gives f^*

We conclude with an example of our method. Figure 2(a) shows a polygon whose corners lie on its convex hull. As height assignment we choose the paraboloid lifting $h(\mathbf{p}_i) := \|\mathbf{p}_i\|^2$—this lifting gives the pointed Delaunay pseudo-triangulation of $\mathcal{P}$ (see [21]). The induced lower shell is shown in Figure 3(a) and its induced forces are shown in Figure 2(b). The surface that "fits" into the lower shell and that fulfills all the requirements is depicted in Figure 3(b). It gives the pointed pseudo-triangulation $\mathcal{PT}_h$ of Figure 2(c).

Our method extends the result presented in [21] because it allows the computations for a wider class of polygons. The question how to solve the general case with the load resolving method is still open but not out of reach. One has to find a way how to fix the values of the stresses of the edges appearing at the pockets of $\mathcal{P}$ to specify the heights of the corners that are not part of conv(P). Our algorithm is slower than the execution of the improving flip sequence of [2]. On the other hand our method allows a very simple implementation with the help of an LP-solver.

References

1. Aggarwal, A., Guibas, L.J., Saxe, J., Shor, P.W.: A linear-time algorithm for computing the Voronoi diagram of a convex polygon. Discrete Comput. Geom. 4(6), 591–604 (1989)
2. Aichholzer, O., Aurenhammer, F., Brass, P., Krasser, H.: Pseudo-triangulations from surfaces and a novel type of edge flip. SIAM Journal on Computing 32, 1621–1653 (2003)
3. Aurenhammer, F., Krasser, H.: Pseudo-simplicial complexes from maximal locally convex functions. Discrete & Computational Geometry 35(2), 201–221 (2006)
4. Basch, J., Erickson, J., Guibas, L.J., Hershberger, J., Zhang, L.: Kinetic collision detection between two simple polygons. Comput. Geom. 27(3), 211–235 (2004)
5. Bobenko, A.I., Springborn, B.: A discrete Laplace-Beltrami operator for simplicial surfaces. Discrete & Computational Geometry 38(4), 740–756 (2007)
6. Chazelle, B., Edelsbrunner, H., Grigni, M., Guibas, L.J., Hershberger, J., Sharir, M., Snoeyink, J.: Ray shooting in polygons using geodesic triangulations. Algorithmica 12, 54–68 (1994)

7. Chrobak, M., Goodrich, M.T., Tamassia, R.: Convex drawings of graphs in two and three dimensions (preliminary version). In: Proc. 12th Ann. Symposium on Computational Geometry, pp. 319–328 (1996)

8. Colin de Verdière, É., Pocchiola, M., Vegter, G.: Tutte's barycenter method applied to isotopies. Comput. Geom. 26(1), 81–97 (2003)

9. Connelly, R., Demaine, E.D., Rote, G.: Straightening polygonal arcs and convexifying polygonal cycles. Discr. Comput. Geometry 30, 205–239 (2003)

10. Coppersmith, D., Winograd, S.: Matrix multiplication via arithmetic progressions. J. Symb. Comput. 9(3), 251–280 (1990)

11. Föppl, A.: Theorie des Fachwerks. Verlag Arthur Felix (1880)

12. Haas, R., Orden, D., Rote, G., Santos, F., Servatius, B., Servatius, H., Souvaine, D.L., Streinu, I., Whiteley, W.: Planar minimally rigid graphs and pseudo-triangulations. Comput. Geom. 31(1-2), 31–61 (2005)

13. Karmarkar, N.: A new polynomial-time algorithm for linear programming. Combinatorica 4(4), 373–396 (1984)

14. Lipton, R.J., Rose, D., Tarjan, R.: Generalized nested dissection. SIAM J. Numer. Anal. 16(2), 346–358 (1979)

15. Lipton, R.J., Tarjan, R.E.: Applications of a planar separator theorem. SIAM J. Comput. 9(3), 615–627 (1980)

16. Maxwell, J.C.: On reciprocal figures and diagrams of forces. Phil. Mag. Ser. 27, 250–261 (1864)

17. Ribó Mor, A., Rote, G., Schulz, A.: Embedding 3-polytopes on a small grid. In: Erickson, J. (ed.) Proc. 23rd Symposium on Computational Geometry, pp. 112–118. ACM Press, New York (2007)

18. Richter-Gebert, J.: Realization Spaces of Polytopes. Lecture Notes in Mathematics, vol. 1643. Springer, Heidelberg (1996)

19. Rote, G., Santos, F., Streinu, I.: Expansive motions and the polytope of pointed pseudo-triangulations. In: Discrete and Computational Geometry–The Goodman-Pollack Festschrift, vol. 25, pp. 699–736. Springer, Heidelberg (2003)

20. Rote, G., Santos, F., Streinu, I.: Pseudo-triangulations — a survey. In: Surveys on Discrete and Computational Geometry—Twenty Years Later. Contemporary Mathematics, vol. 453, pp. 343–410 (2008)

21. Rote, G., Schulz, A.: A pointed Delaunay pseudo-triangulation of a simple polygon. In: Proceedings of the 21st European Workshop on Computational Geometry, Eindhoven, pp. 77–80 (2005)

22. Schulz, A.: Lifting planar graphs to realize integral 3-polytopes and topics in pseudo-triangulations. PhD thesis, Freie Universität Berlin (2008)

23. Speckmann, B., Tóth, C.D.: Allocating vertex pi-guards in simple polygons via pseudo-triangulations. Discrete & Computational Geometry, 33(2):345–364 (2005)

24. Streinu, I.: Pseudo-triangulations, rigidity and motion planning. Discrete & Computational Geometry, 34(4):587–635 (2005)

25. Wardetzky, M., Mathur, S., Kälberer, F., Grinspun, E.: Discrete Laplace operators: no free lunch. In: SGP 2007: Proceedings of the Fifth Eurographics Symposium on Geometry Processing, pp. 33–37 (2007)

26. Whiteley, W.: Motion and stresses of projected polyhedra. Structural Topology 7, 13–38 (1982)

On Making Directed Graphs Transitive

Mathias Weller*, Christian Komusiewicz**, Rolf Niedermeier,
and Johannes Uhlmann***

Institut für Informatik, Friedrich-Schiller-Universität Jena
Ernst-Abbe-Platz 2, D-07743 Jena, Germany
{mathias.weller,c.komus,rolf.niedermeier,johannes.uhlmann}@uni-jena.de

Abstract. We present the first thorough theoretical analysis of the
TRANSITIVITY EDITING problem on digraphs. Herein, the task is to perform a minimum number of arc insertions or deletions in order to make
a given digraph transitive. This problem has recently been identified as
important for the detection of hierarchical structure in molecular characteristics of disease. Mixing up TRANSITIVITY EDITING with the companion problems on undirected graphs, it has been erroneously claimed
to be NP-hard. We correct this error by presenting a first proof of NP-
hardness, which also extends to the restricted cases where the input
digraph is acyclic or has maximum degree four. Moreover, we improve
previous fixed-parameter algorithms, now achieving a running time of
$O(2.57^k + n^3)$ for an n-vertex digraph if k arc modifications are sufficient
to make it transitive. In particular, providing an $O(k^2)$-vertex problem
kernel, we positively answer an open question from the literature. In case
of digraphs with maximum degree d, an $O(k \cdot d)$-vertex problem kernel
can be shown. We also demonstrate that if the input digraph contains no
"diamond structure", then one can always find an optimal solution that
exclusively performs arc deletions. Most of our results (including NP-
hardness) can be transferred to the TRANSITIVITY DELETION problem,
where only arc deletions are allowed.

1 Introduction

To make a directed graph (digraph for short) transitive by a minimum number
of arc modifications has recently been identified to have important applications
in detecting hierarchical structure in molecular characteristics of disease [3,11].
A digraph $D = (V, A)$ is called *transitive* if $(u, v) \in A$ and $(v, w) \in A$ implies
$(u, w) \in A$ (also cf. [1, Section 4.3]). Thus, the central problem TRANSITIV-
ITY EDITING studied here asks, given a digraph and an integer $k \geq 0$, to find
a set of at most k arcs to insert or delete in order to make the resulting digraph transitive. We provide a first thorough theoretical study of TRANSITIVITY
EDITING, complementing previous work that focused on heuristics, integer linear programming, and simple fixed-parameter algorithms [3,11]. We also study

---

* Partially supported by the DFG, research project DARE, GU 1023/1.
** Supported by a PhD fellowship of the Carl-Zeiss-Stiftung.
*** Supported by the DFG, research project PABI, NI 369/7.

F. Dehne et al. (Eds.): WADS 2009, LNCS 5664, pp. 542–553, 2009.
© Springer-Verlag Berlin Heidelberg 2009

the special case when only arc deletions (TRANSITIVITY DELETION) are allowed and restricted classes of digraphs (acyclic and bounded-degree). Note that the corresponding problem TRANSITIVITY COMPLETION (where only arc insertions are allowed) is nothing but the well-studied problem of computing the transitive closure of a digraph; this is clearly solvable in polynomial time [13].

Previous work. TRANSITIVITY EDITING can be seen as the "directed counterpart" of the so far much better studied problem CLUSTER EDITING on undirected graphs (see [2,5,7,8,15]). Indeed, both problems are also referred to as TRANSITIVE APPROXIMATION problem on directed and undirected graphs, respectively. Unfortunately, this is perhaps a reason why TRANSITIVITY EDITING has erroneously been claimed to be NP-hard [11,3] by referring to work that only considers problems on undirected graphs, including CLUSTER EDITING. On the positive side, however, the close correspondence between CLUSTER EDITING and TRANSITIVITY EDITING helped Böcker et al. [3] to transfer their previous results for CLUSTER EDITING [2] to TRANSITIVITY EDITING, delivering the currently fastest implementations that exactly solve TRANSITIVITY EDITING (by means of integer linear programming and fixed-parameter algorithms). In particular, their computational experiments demonstrate that their exact algorithms are by far more efficient in practice than the previously used purely heuristic approach by Jacob et al. [11].

Our contributions. We eventually prove the so far only claimed NP-hardness[1] of TRANSITIVITY EDITING, also extending this result to TRANSITIVITY DELETION. Moreover, we show that both problems remain NP-hard when restricted to acyclic digraphs or digraphs with maximum vertex degree four (more precisely, indegree two and outdegree two). To this end, we also make the helpful combinatorial observation that if a digraph does not contain a so-called "diamond structure", then there is an optimal solution for TRANSITIVITY EDITING that only deletes arcs. This observation is also useful for developing more efficient fixed-parameter algorithms than the ones presented in previous work. First, we provide a polynomial-time data reduction that yields an $O(k^2)$-vertex problem kernel for TRANSITIVITY EDITING and TRANSITIVITY DELETION. This answers an open question of Böcker et al. [3]. In the special case of digraphs with maximum vertex degree d, we can actually prove an $O(k \cdot d)$-vertex kernel. Finally, exploiting the aforementioned observation on diamond-freeness, we develop an improved search tree for TRANSITIVITY EDITING. That is, whereas the fixed-parameter algorithm of Böcker et al. [3] runs in $O(3^k \cdot n^3)$ time on n-vertex digraphs, our new algorithm runs in $O(2.57^k + n^3)$ time (note that in our algorithm the cubic term n^3 has become additive instead of multiplicative due to our kernelization result). Finally, we mention that TRANSITIVITY DELETION can be solved in $O(2^k + n^3)$ time. To conclude, note that Gutin and Yeo [9] asked in their recent survey about parameterized problems on digraphs for extending the so far small list of fixed-parameter tractability results for NP-hard problems on *digraphs*—we hope that our work makes a useful addition to this list. Due to the lack of space, several details are deferred to a full version of this article.

---

[1] Indeed, all corresponding decision problems are NP-complete.

2   Preliminaries and a Structural Result

Our algorithmic results are in the context of fixed-parameter algorithms. Parameterized complexity is a two-dimensional framework for studying the computational complexity of problems [4,6,14]. One dimension is the input size n (as in classical complexity theory), and the other one is the *parameter* k (usually a positive integer). A problem is called *fixed-parameter tractable* (fpt) if it can be solved in $f(k) \cdot n^{O(1)}$ time, where f is a computable function only depending on k. This means that when solving a combinatorial problem that is fpt, the combinatorial explosion can be confined to the parameter. A core tool in the development of fixed-parameter algorithms is polynomial-time preprocessing by *data reduction*. Here, the goal is for a given problem instance x with parameter k to transform it into a new instance x' with parameter $k' \leq k$ such that the size of x' is upper-bounded by some function only depending on k and the instance (x, k) is a yes-instance iff (x', k') is a yes-instance. The reduced instance, which must be computable in polynomial time, is called a *problem kernel*, and the whole process is called *reduction to a problem kernel* or simply *kernelization*.

A *directed graph* or *digraph* is a pair $D = (V, A)$ with $A \subseteq V \times V$. The set V contains the *vertices* of the digraph, while A contains the *arcs*. Throughout this work, let $n := |V|$. If $V' \subseteq V$, then $D[V'] := (V', A \cap (V' \times V'))$ denotes the subgraph of D that is *induced* by V'. Furthermore, we write $D - u$ for $D[V \setminus \{u\}]$. The *symmetric difference* of two sets of arcs A and A' is $A \Delta A' := (A \cup A') \setminus (A \cap A')$. In this work, we only consider simple digraphs, that is digraphs without self-loops and double arcs. For any $u \in V$, $\operatorname{pred}_A(u) := \{v \in V \mid (v, u) \in A\}$ denotes the set of *predecessors* of u with respect to A, while $\operatorname{succ}_A(u) := \{v \in V \mid (u, v) \in A\}$ denotes its *successors*. The vertices in $\operatorname{pred}_A(u) \cup \operatorname{succ}_A(u)$ are said to be *adjacent to* u.

A digraph $D = (V, A)$ is called *transitive* if

$$\forall_{u,v,w \in V} \ ((u, v) \in A \land (v, w) \in A) \Rightarrow (u, w) \in A.$$

In other words, D is transitive if A is a transitive relation on $(V \times V)$. The central problem of this work (formulated as decision problem, but our algorithms can also solve the corresponding minimization problem) is defined as follows.

> TRANSITIVITY EDITING:
> **Input**: A digraph $D = (V, A)$ and an integer $k \geq 0$.
> **Question**: Does there exist a digraph $D' = (V, A')$ that is transitive and $|A \Delta A'| \leq k$?

Analogously, TRANSITIVITY DELETION is defined via only allowing arc deletions.

To derive our results, we make use of the fact that transitive digraphs can be characterized by "forbidden P_3s". Slightly abusing notation, in our setting, the P_3s of a digraph are all vertex triples (u, v, w), such that $(u, v) \in A$, $(v, w) \in A$, and $(u, w) \notin A$. We say that the P_3 (u, v, w) *contains* the arcs (u, v) and (v, w) and the vertices u, v, and w. As also noted by Böcker et al. [3], transitive digraphs can be characterized as the digraphs without P_3s, that is, a digraph is transitive iff it does not contain a P_3.

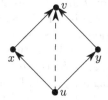

	u	v	x	y
u	-	0	1	1
v	*	-	*	*
x	*	1	-	*
y	*	1	*	-

Fig. 1. The diamond structure and its adjacency matrix. In order to meet the definition, the solid arcs must be present and the dashed arc must be absent. All other arcs may or may not be present. In the adjacency matrix, for each vertex, the endpoints of its outgoing arcs are determined by its row. Stars represent wildcards, that is, these entries do not matter for the definition.

A central tool for our combinatorial studies is based on the consideration of "diamonds". The absence of diamonds in a given digraph simplifies the TRANSITIVITY EDITING problem. This helps us in proving NP-hardness and in our algorithmic results. A *diamond* in a digraph $D = (V, A)$ is a triple $(u, \{x, y\}, v)$, where $u, x, y, v \in V$, $(u, v) \notin A$, and $(u, z), (z, v) \in A$ for $z \in \{x, y\}$ (see Fig. 1).[2] If D does not contain a diamond, then it is said to be *diamond-free*.

A set $S \subseteq V \times V$ is called a *solution set* of TRANSITIVITY EDITING for the digraph (V, A) if $(V, A \triangle S)$ is transitive. A solution set S is *optimal* if there is no solution set S' with $|S'| < |S|$. For each solution set S we consider its two-partition $S = S_{DEL} \uplus S_{INS}$, where S_{DEL} denotes the set of arc deletions and S_{INS} denotes the set of arc insertions. The following lemma shows that the property of being diamond-free is preserved by deleting the arcs of a solution set.

Lemma 1. *Let $D = (V, A)$ be a diamond-free digraph and let S be a solution set for D. Then $D_{DEL} := (V, A \triangle S_{DEL})$ is diamond-free.*

The following important result shows that in order to solve TRANSITIVITY EDITING on diamond-free digraphs, it is optimal to only perform arc deletions.

Lemma 2. *Let (D, k) with $D = (V, A)$ be a diamond-free input instance of TRANSITIVITY EDITING. Then, there is an optimal solution set S for D that inserts no arc, that is, $S = S_{DEL}$.*

Proof. Let S' be any optimal solution set for D. By Lemma 1, we can apply all arc deletions of a given solution set without destroying diamond-freeness. Hence, we assume the solution set S' to only consist of arc insertions. We now construct S from S':

$$S := \{(a, b) \mid \exists_{c \in V} (a, c) \in S' \wedge (a, b) \in A \wedge (b, c) \in A\}.$$

Since D is diamond-free, for each vertex pair (a, c), there is at most one b meeting the criteria $(a, b) \in A$ and $(b, c) \in A$. Hence, for each inserted arc (a, c) in S', there is at most one arc (a, b) in S and hence $|S| \leq |S'|$.

---

[2] Note that this is not a common definition and should not be mixed-up for instance with diamonds in undirected graphs.

Let $D' := (V, A')$ with $A' := A \setminus S$. We now show that S is a solution set for D by proving that D' is transitive: Assume that there is a P_3 $p = (x, y, z)$ in D'. Since $S \subseteq A$ (that is, S contains only arc deletions), we know that $(x, y) \in A$ and $(y, z) \in A$ and, since S' is a solution set for D, we know that p is not a P_3 in $(V, A \Delta S')$, implying either $(x, z) \in S'$ or $(x, z) \in S$. However, $(x, z) \notin S'$, because otherwise $(x, y) \in S$, contradicting p being a P_3 in D'. Hence, $(x, z) \in A$ and $(x, z) \in S$. By definition of S, this implies that there is a vertex $v \in V$ with $(z, v) \in A$ and $(x, v) \in S'$. Also, $(y, v) \notin A$, since, otherwise, (x, z, v) and (x, y, v) would form a diamond in D. Hence, $q = (y, z, v)$ is a P_3 in D. As p, also q cannot be a P_3 in $(V, A \Delta S')$. However, S' does only contain insert operations, which implies $(y, v) \in S'$. Since $(y, z) \in A$ and $(z, v) \in A$, this implies $(y, z) \in S$, contradicting p being a P_3 in D'. $\square$

3   NP-Hardness Results

In this section, we prove the NP-hardness of TRANSITIVITY EDITING and TRANSITIVITY DELETION in degree-four digraphs and in acyclic digraphs. Both results are derived by a reduction from POSITIVE-NOT-ALL-EQUAL-3SAT, which is an NP-complete variant of 3SAT [12].

POSITIVE-NOT-ALL-EQUAL-3SAT (PNAE-3SAT):

Input: A Boolean formula φ in n variables $x_0, \ldots, x_{n-1}$ which is a conjunction of m clauses C_i, $0 \leq i < m$, each consisting of three positive literals.

Question: Is there a truth assignment to all n variables such that for each clause C_i exactly one or two of its variables are assigned true, that is, for no clause the truth values of its variables are all equal?

First, we show that PNAE-3SAT can be reduced to TRANSITIVITY EDITING in degree-four digraphs. To this end, we construct an input instance of TRANSITIVITY EDITING from a given input instance of PNAE-3SAT in polynomial time as follows. For each of the n Boolean variables, we construct a *variable cycle*, that is, a directed cycle of length $8m$, with m being the number of clauses in the given formula φ. More specifically, for each variable x_i, the corresponding variable cycle consists of the vertices $V_i := \{i_0, \ldots, i_{8m-1}\}$. The vertices in V_i are connected into a cycle by adding the arcs $A_i := \{\{i_p, i_{p+1}\} \mid 0 \leq p \leq 8m-1\}$ (for the ease of presentation, let $i_{8m} = i_0$). The collection of all variable cycles is then referred to by (V, A) with $V := \bigcup_{i=0}^{n-1} V_i$ and $A := \bigcup_{i=0}^{n-1} A_i$. In the following, we refer to the arcs $(i_0, i_1), (i_2, i_3), \ldots, (i_{8m-2}, i_{8m-1})$ as *even arcs* and to all other arcs in the variable cycle as *odd arcs*.

Moreover, for each clause $C_j = \{x_p, x_q, x_r\}$ in φ with $0 \leq j < m$, we construct a *clause cycle*, that is, a directed length-three cycle between the variable cycles of its three variables consisting of the arcs $A'_j := \{(p_{8j}, q_{8j}), (q_{8j}, r_{8j}), (r_{8j}, p_{8j})\}$. See Fig. 2(a) for an illustration. This completes the construction.

The set of all arcs in the clause cycles is denoted by $A' := \bigcup_{j=0}^{m-1} A'_j$. Note that two vertices of a variable cycle contained in different clause cycles have distance

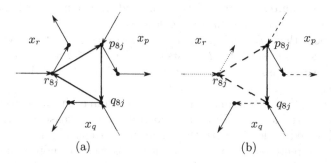

Fig. 2. (a): The clause cycle of clause $C_i = \{x_p, x_q, x_r\}$ connecting the corresponding variable cycles. Bold arcs are in A'. (b): All P_3s containing an arc of the clause cycle can be destroyed by deleting two arcs if in the variable cycle of x_p all odd arcs are deleted and in the variable cycle of x_q all even arcs are deleted. Dashed lines indicate deleted arcs and exactly one of the two dotted arcs incident to r_{8j} is deleted.

at least 8, which makes it easy to see that the constructed digraph is diamond-free. Finally, let $D := (V, A \cup A')$ denote the resulting digraph and $k := 2m + 4mn$. Observe that D is diamond-free and has maximum degree four.

Theorem 1. TRANSITIVITY EDITING *is NP-complete, even if the maximum degree is bounded by four (indegree two and outdegree two).*

Proof. Obviously, one can verify in polynomial time whether a digraph is transitive. This implies that TRANSITIVITY EDITING is in NP. We now show that it is NP-hard by reducing from PNAE-3SAT. Let $D = (V, A \cup A')$ be a digraph constructed as described above from a given instance φ of PNAE-3SAT. We show that (D, k) with $k := 2m + 4mn$ is a yes-instance for TRANSITIVITY EDITING iff there is a satisfying assignment to the variables of φ.

"$\Leftarrow$": Suppose that there is a satisfying assignment β to the variables of a PNAE-3SAT input instance φ. Then, we can construct a transitive digraph by modifying D in the following way: First, for each variable x_i, we remove all odd arcs of its variable cycle if $\beta(x_i) = $ true, and all even arcs if $\beta(x_i) = $ false. All in all, we remove $4m$ arcs for each of the n variable cycles, which is a total of $4mn$ arc deletions. Note that all remaining P_3s contain at least one arc of a clause cycle.

To destroy these P_3s, for each clause $C_j = \{x_p, x_q, x_r\}$, $0 \leq j < m$, the clause cycle is modified in the following way: Since β is a satisfying assignment for the PNAE-3SAT instance, we can assume without loss of generality that $\beta(x_p) = $ true and $\beta(x_q) = $ false. Hence, we have deleted all odd arcs in the variable cycle of x_p and all even arcs in the variable cycle of x_q, that is, the arcs (p_{8j-1}, p_{8j}) and (q_{8j}, q_{8j+1}) are deleted and the arcs (p_{8j}, p_{8j+1}) and (q_{8j-1}, q_{8j}) are not deleted. Moreover, observe that deleting the arcs (q_{8j}, r_{8j}) and (r_{8j}, p_{8j}) of the clause cycle makes p_{8j} a source and q_{8j} a sink. Hence, all P_3s containing an arc of the clause cycle of C_j are destroyed. See Fig. 2(b) for an illustration. For all clauses, this requires $2m$ arc deletions in total. In summary, it is possible to make D transitive with $2m + 4mn$ arc deletions.

"$\Rightarrow$": Suppose that $(D, 2m + 4mn)$ is a yes-instance of TRANSITIVITY EDIT-ING. Hence, a solution set S for D exists such that $|S| \leq 2m + 4mn$. Since D is diamond-free we can assume, by Lemma 2, that $S \subseteq A \cup A'$. Let $\hat{A} := (A \cup A') \setminus S$ and $\hat{D} := (V, \hat{A})$.

Next, we show that S contains exactly two arcs from each clause cycle and $4m$ arcs from each variable cycle. First, note that one needs at least two arc deletions to make a directed cycle of length three transitive. Hence, turning all m clause cycles transitive requires at least $2m$ arc deletions. Second, note that making a variable cycle (which has length $8m$) transitive requires at least $4m$ arc deletions since it contains $4m$ arc-disjoint P_3s. This implies that S contains exactly two arcs from each clause cycle and $4m$ arcs from each variable cycle (note that the variable and clause cycles are arc-disjoint). Moreover, observe that, to make a variable cycle transitive by deleting $4m$ arcs, either all $4m$ even or all $4m$ odd arcs must be deleted (since it is clearly optimal to delete every second arc).

Consider a clause $C_j = \{x_p, x_q, x_r\}$, $0 \leq j < m$. We show that for one of the three corresponding variable cycles all even arcs and for another all odd arcs are deleted, and, as a consequence, the assignment β with $\beta(x_i) := $ true if all odd arcs of the corresponding variable cycle are deleted and $\beta(x_i) := $ false, otherwise, is satisfying. Assume towards a contradiction that there exists a clause $C_j = \{x_p, x_q, x_r\}$, $0 \leq j < m$, such that for all three variables x_p, x_q, and x_r all even (odd) arcs of the variable cycles are deleted. Recall that for each clause cycle all but one arc are deleted. Without loss of generality, let (p_{8j}, q_{8j}) be this arc, that is, $(p_{8j}, q_{8j}) \in \hat{A}$. If all even arcs are deleted in the variable cycles, then the odd arc $(p_{8j-1}, p_{8j}) \in \hat{A}$ and $(p_{8j-1}, p_{8j}, q_{8j})$ is a P_3 in $\hat{D}$. Otherwise, if all odd arcs are deleted, then the even arc $(q_{8j}, q_{8j+1}) \in \hat{A}$ and $(p_{8j}, q_{8j}, q_{8j+1})$ is a P_3 in $\hat{D}$. Both cases contradict the fact that S is a solution. $\square$

In the above proof, we never employ arc insertions. This implies that TRANSITIVITY DELETION is also NP-complete.

Corollary 1. TRANSITIVITY DELETION *is NP-complete, even if the maximum degree is bounded by four (indegree two and outdegree two).*

The undirected "sister" problem CLUSTER EDITING becomes polynomial-time solvable when the input is a tree, that is, acyclic. It is thus natural to study the complexity of TRANSITIVITY EDITING on acyclic digraphs. Somewhat surprisingly, we find that TRANSITIVITY EDITING remains NP-hard for acyclic digraphs, unlike for example DISJOINT PATHS [16] which is NP-hard in general but polynomial-time solvable on acyclic digraphs. However, we have to give up the bounded degree constraint.

To show the NP-hardness, we reduce again from PNAE-3SAT. The technical effort, however, significantly increases. The trickiness of the proof lies in incorporating an "information feedback" between the variable gadgets while using only acyclic variable and clause gadgets.

Theorem 2. TRANSITIVITY EDITING *and* TRANSITIVITY DELETION *are NP-complete, even when restricted to acyclic digraphs.*

4   Fixed-Parameter Tractability Results

In this section, we complement the NP-hardness results of the previous section with encouraging algorithmic results. Note that Böcker et al. [3] observed that "most graphs derived from real-world applications are almost transitive". Consequently, as Böcker et al., we study how the parameter k (denoting the number of arc modifications) influences the computational complexity. We deliver improved fixed-parameter tractability results; in particular, we positively answer Böcker et al.'s [3] question for the existence of a polynomial-size problem kernel. Thus, in what follows, we first develop kernelization results, and then we present an improved search tree strategy, altogether yielding the so far fastest fixed-parameter algorithms for TRANSITIVITY EDITING.

First, observe that TRANSITIVITY EDITING is fixed-parameter tractable with respect to the parameter k: The task is simply to destroy all P_3s in a given digraph. Clearly, there are exactly three possibilities to destroy a P_3, either by deleting one of the two arcs or by inserting the "missing" one. This yields a search tree of size $O(3^k)$ (cf. [3]), which indeed can be used to enumerate *all* solutions of size at most k because it exhaustively tries all possibilities to destroy P_3s.

Kernelization. In the following, we describe a kernelization for TRANSITIVITY EDITING. We show a kernel consisting of $O(k^2)$ vertices for the general problem and a kernel of $O(k)$ vertices for digraphs with bounded degree. In the latter case, already the following data reduction rule suffices.

Rule 1. *Let $(D = (V, A), k)$ be an input instance of* TRANSITIVITY EDITING. *If there is a vertex $u \in V$ that does not take part in any P_3 in D, then remove u and all arcs that are incident to it.*

Lemma 3. *Rule 1 is correct and can be exhaustively applied in $O(n^3)$ time.*

Proof. To prove the correctness, we construct a sequence of arc modifications that form an optimal solution set. Then, we will prove that, if at some point in this sequence a vertex u does not take part in any P_3, then u does not take part in any P_3 at any later point in the sequence. Thus, removing u never changes the set of P_3s to be destroyed.

Let (D, k) with $D = (V, A)$ denote the given input instance and let S denote an optimal solution set for D with $s := |S| \leq k$ and $D' := (V, A \triangle S)$. Let Q be the straightforward search-tree algorithm that searches a P_3 in the digraph and destroys it by branching into all three possibilities of inserting or deleting an arc. Clearly, Q returns a shortest sequence of digraphs $(D = D_0, D_1, \ldots, D_s = D')$ with $D_i := (V, A_i)$ and a sequence of arc modifications $F_1, \ldots, F_s$ with $F_i := A_{i-1} \triangle A_i$ for each $1 \leq i \leq s$. We prove the following: For each $i \geq 1$, if a vertex $u \in V$ does not take part in any P_3 in D_{i-1}, then it does not take part in any P_3 in D_i. Hence, by induction, if u does not take part in any P_3 in D_0, then there is no $j > 0$ such that u takes part in a P_3 in D_j. Thus, D and $D - u$ yield the same sequence of arc modifications $F_1, \ldots, F_s$ and thus $(D, k) \in$ Transitivity Editing $\Leftrightarrow (D - u, k) \in$ Transitivity Editing.

In the following, we show the contraposition of the claim: For each $i \geq 1$, if a vertex $u \in V$ takes part in a P_3 p in D_i, then it takes part in a P_3 q in D_{i-1}. Let $F_i = \{(a, b)\}$. Since $\mathcal{Q}$ only inserts or deletes (a, b) to destroy a P_3, we know that there is a P_3 r in D_{i-1} that contains both a and b. Hence, if $u = a$ or $u = b$, then $q = r$ and thus u takes part in q. Otherwise, we consider the following cases.

Case 1: (a, b) is inserted.

Clearly, there is a vertex $v \in V$ such that (a, v, b) is a P_3 in D_{i-1}; hence, if $u = v$, then $q = (a, u, b)$. Furthermore, if $p \neq (a, b, u)$ and $p \neq (u, a, b)$, then $q = p$. Otherwise, without loss of generality, assume that $p = (a, b, u)$. Obviously, $(a, u) \notin A_i$. Since $(a, u) \notin F_i$, we know that $(a, u) \notin A_{i-1}$. If $(v, u) \in A_{i-1}$, then $q = (a, v, u)$, otherwise $q = (v, b, u)$.

Case 2: (a, b) is deleted.

Clearly, there is a vertex $v \in V$ such that either (a, b, v) or (v, a, b) is a P_3 in D_{i-1}; hence, if $u = v$, then $q = (a, b, u)$ or $q = (u, a, b)$. If $u \neq v$, without loss of generality assume that (a, b, v) is a P_3 in D_{i-1}. Furthermore, if $p \neq (a, u, b)$, then $q = p$. If $p = (a, u, b)$, then if $(u, v) \in A_{i-1}$, then $q = (a, u, v)$; otherwise, $q = (u, b, v)$.

Finally, the running time can be seen as follows. We enumerate all P_3s in $O(n^3)$ time, thereby labeling all vertices that are part of a P_3. Afterwards, we remove all unlabeled vertices. □

Surprisingly, this data reduction rule is sufficient to show a linear-size problem kernel if the maximum degree of the given digraph is constant.

Theorem 3. Transitivity Editing *restricted to digraphs with maximum degree* d *admits a problem kernel containing at most* $2k \cdot (d + 1)$ *vertices.*

Proof. Let $D = (V, A)$ be a digraph that is reduced with respect to Rule 1 and let S be a solution set for D with $|S| \leq k$. We show that $|V| \leq 2k(d+1)$. Consider the two-partition of V into $Y := \{v \in V \mid \exists_{u \in V} (u, v) \in S \vee (v, u) \in S\}$ and $X := V \backslash Y$. Since $|S| \leq k$, we have $|Y| \leq 2k$. Note that, since D is reduced with respect to Rule 1, every $x \in X$ is contained in a P_3 q. It is clear that the other two vertices of q are in Y and thus every $x \in X$ is adjacent to at least one vertex in Y. However, each vertex in Y has at most d neighbors and thus $|X| \leq d|Y|$, implying $|V| = |X| + |Y| \leq 2k + d2k = 2k(d + 1)$. □

The above data reduction also works for Transitivity Deletion:

Corollary 2. Transitivity Deletion *restricted to digraphs with maximum degree* d *admits a problem kernel containing at most* $2k \cdot (d + 1)$ *vertices.*

Next, we prove an $O(k^2)$-vertex kernel for general digraphs. The following data reduction rule roughly follows an idea for Cluster Editing [7]: If there is some vertex pair (a, b) such that not modifying (a, b) results in a solution size of at least $k + 1$, then every solution of size at most k must contain (a, b).

Rule 2. *Let $(D = (V, A), k)$ be an input instance of* TRANSITIVITY EDITING.

1. *Let $(u, v) \in (V \times V) \backslash A$ and $Z := \mathrm{succ}_A(u) \cap \mathrm{pred}_A(v)$. If $|Z| > k$, then insert (u, v) into A and decrease k by one.*
2. *Let $(u, v) \in A$, $Z_u := \mathrm{pred}_A(u) \setminus \mathrm{pred}_A(v)$ and $Z_v := \mathrm{succ}_A(v) \setminus \mathrm{succ}_A(u)$. If $|Z_u| + |Z_v| > k$, then delete (u, v) from A and decrease k by one.*

Lemma 4. *Let (D, k) be an input instance of* TRANSITIVITY EDITING. *Then, Rule 2 causes an arc modification iff it destroys more than k P_3s in D.*

Lemma 4 is decisive for proving the correctness of Rule 2.

Lemma 5. *Rule 2 is correct and can be exhaustively applied in $O(n^3)$ time.*

We now show that the exhaustive application of both rules leads to a problem kernel of $O(k^2)$ vertices.

Theorem 4. TRANSITIVITY EDITING *admits a problem kernel containing at most $k(k + 2)$ vertices.*

Proof. Assume that there is a digraph $D = (V, A)$ with $|V| > k(k + 2)$, D is reduced with respect to Rules 1 and 2, and it is possible to make D transitive by applying at most k arc modifications. Let $D' = (V, A')$ denote a transitive digraph obtained by the application of k arc modifications and let $S := A \Delta A'$ denote the corresponding solution set. Consider a two-partition (X, Y) of V, where $Y := \{v \in V \mid \exists_{u \in V} (u, v) \in S \vee (v, u) \in S\}$ and $X := V \backslash Y$. Note that all vertices in X are adjacent to at least one vertex in Y because D is reduced with respect to Rule 1. Also note that in order to destroy a P_3 p in D, the solution set S must contain an arc incident to two of the vertices of p, hence for each P_3 p in D at most one of the vertices of p is in X.

Since we assume that D can be made transitive with at most k arc modifications, we know that $|S| \leq k$ and consequently $|Y| \leq 2k$. Clearly, $|V| = |X| + |Y|$, hence the assumption that $|V| > k(k + 2)$ implies $|X| > k^2$. With the above observation, it follows that there are more than k^2 P_3s in D.

For each $(a, b) \in S$, let $Z_{(a,b)} := \{p \mid \text{modifying } (a, b) \text{ destroys the } P_3 \ p \text{ in } D\}$. Since there are more than k^2 P_3s in D, but $|S| \leq k$, we know that there is an $(a, b) \in S$ with $|Z_{(a,b)}| > k$, a contradiction to Lemma 4. $\square$

The above data reduction works also for TRANSITIVITY DELETION:

Corollary 3. TRANSITIVITY DELETION *admits a problem kernel containing at most $k(k + 2)$ vertices.*

Search Tree Algorithm. As mentioned before, a straightforward algorithm that finds an optimal solution set for a given digraph branches on each P_3 (u, v, w) in the digraph, trying to destroy it by either deletion of (u, v), deletion of (v, w), or insertion of (u, w). This directly gives a search tree algorithm solving TRANSITIVITY EDITING on an n-vertex digraph in $O(3^k \cdot n^3)$ time (cf. [3]). Note that, to solve TRANSITIVITY DELETION, the search only needs to branch into two cases, yielding an algorithm running in $O(2^k \cdot n^3)$ time. Indeed, using more clever data structures, these running times can be improved to $O(3^k \cdot n \log n + n^3)$ and $O(2^k \cdot n \log n + n^3)$, respectively. Using the so-called interleaving technique [14] together with the polynomial-size problem kernel results, however, one actually can achieve running times $O(3^k + n^3)$ and $O(2^k + n^3)$, respectively.

In the following, we shrink the search tree size for TRANSITIVITY EDITING from 3^k to 2.57^k by applying our combinatorial result on diamond-freeness.

Theorem 5. TRANSITIVITY EDITING *and* TRANSITIVITY DELETION *can be solved in* $O(2.57^k + n^3)$ *and* $O(2^k + n^3)$ *time, respectively.*

Proof. Recall from Lemma 2 that in diamond-free digraphs we only need to consider arc deletions. This helps us to improve the branching strategy. The modified algorithm employs the following search structure. Upon finding a diamond $(u, \{x, y\}, v)$ in the given digraph $D = (V, A)$, the algorithm recursively asks whether

1. $(V, A \setminus \{(u, x), (u, y)\})$ can be made transitive with $\leq k - 2$ operations,
2. $(V, A \setminus \{(u, x), (y, v)\})$ can be made transitive with $\leq k - 2$ operations,
3. $(V, A \setminus \{(x, v), (u, y)\})$ can be made transitive with $\leq k - 2$ operations,
4. $(V, A \setminus \{(x, v), (y, v)\})$ can be made transitive with $\leq k - 2$ operations, or
5. $(V, A \cup \{(u, v)\})$ can be made transitive with $\leq k - 1$ operations.

Thus, the search branches into five cases and the recurrence for the corresponding search tree size reads as $T_k = 1 + 4 \cdot T_{k-2} + T_{k-1}$, where $T_0 = T_1 = 1$. Resolving this recurrence yields $O(2.57^k)$ for the search tree size under the assumption that the branching is always performed in this way. The correctness of this branching is easy to check. If there are no diamonds in the input graph, then the straightforward search tree for TRANSITIVITY DELETION is used to solve the problem, which runs in $O(2^k \cdot n^3)$ time. The correctness of the overall search tree algorithm easily follows.

Applying the interleaving technique [14], and making use of the polynomial-size problem kernels from Theorem 3 results in the running times $O(2.57^k + n^3)$ for TRANSITIVITY EDITING and $O(2^k + n^3)$ for TRANSITIVITY DELETION.   □

5   Conclusion

Two immediate theoretical challenges (of significant practical relevance) arising from our work are to find out whether there is an $O(k)$-vertex problem kernel for TRANSITIVITY EDITING in the case of general digraphs (see [5,8] for corresponding results in the case of undirected graphs, that is, CLUSTER EDITING) or to investigate whether *linear-time* polynomial size kernelization (so far the kernelization takes cubic time in the number of vertices) is possible (see [15] for corresponding results in case of CLUSTER EDITING). Finally, note that we focused on arc modifications to make a given digraph transitive—it might be of similar interest to start an investigation of the TRANSITIVITY VERTEX DELETION problem, where the graph shall be made transitive by as few *vertex deletions* as possible (see [10] for corresponding results in the case of undirected graphs, that is, CLUSTER VERTEX DELETION). Finally, from a more general point of view, there seems to be a rich field of studying further modification problems on digraphs. For instance, the concept of quasi-transitivity is of considerable interest in the theory of directed graphs (cf. [1]), hence one might start investigations on problems such as QUASI-TRANSITIVITY EDITING.

References

1. Bang-Jensen, J., Gutin, G.: Digraphs: Theory, Algorithms and Applications. Springer, Heidelberg (2002)
2. Böcker, S., Briesemeister, S., Klau, G.W.: Exact algorithms for cluster editing: Evaluation and experiments. Algorithmica (to appear, 2009)
3. Böcker, S., Briesemeister, S., Klau, G.W.: On optimal comparability editing with applications to molecular diagnostics. BMC Bioinformatics 10(suppl. 1), S61 (2009); Proc. 7th APBC
4. Downey, R.G., Fellows, M.R.: Parameterized Complexity. Springer, Heidelberg (1999)
5. Fellows, M.R., Langston, M.A., Rosamond, F.A., Shaw, P.: Efficient parameterized preprocessing for Cluster Editing. In: Csuhaj-Varjú, E., Ésik, Z. (eds.) FCT 2007. LNCS, vol. 4639, pp. 312–321. Springer, Heidelberg (2007)
6. Flum, J., Grohe, M.: Parameterized Complexity Theory. Springer, Heidelberg (2006)
7. Gramm, J., Guo, J., Hüffner, F., Niedermeier, R.: Graph-modeled data clustering: Exact algorithms for clique generation. Theory Comput. Syst. 38(4), 373–392 (2005)
8. Guo, J.: A more effective linear kernelization for cluster editing. Theor. Comput. Sci. 410(8-10), 718–726 (2009)
9. Gutin, G., Yeo, A.: Some parameterized problems on digraphs. Comput. J. 51(3), 363–371 (2008)
10. Hüffner, F., Komusiewicz, C., Moser, H., Niedermeier, R.: Fixed-parameter algorithms for cluster vertex deletion. Theory Comput. Syst. (to appear, 2009)
11. Jacob, J., Jentsch, M., Kostka, D., Bentink, S., Spang, R.: Detecting hierarchical structure in molecular characteristics of disease using transitive approximations of directed graphs. Bioinformatics 24(7), 995–1001 (2008)
12. Kratochvíl, J., Tuza, Z.: On the complexity of bicoloring clique hypergraphs of graphs. J. Algorithms 45(1), 40–54 (2002)
13. Munro, J.I.: Efficient determination of the transitive closure of a directed graph. Inf. Process. Lett. 1(2), 56–58 (1971)
14. Niedermeier, R.: Invitation to Fixed-Parameter Algorithms. Oxford University Press, Oxford (2006)
15. Protti, F., da Silva, M.D., Szwarcfiter, J.L.: Applying modular decomposition to parameterized cluster editing problems. Theory Comput. Syst. 44(1), 91–104 (2009)
16. Yang, B., Zheng, S.Q., Lu, E.: Finding two disjoint paths in a network with minsum-minmin objective function. In: Proc. International Conference on Foundations of Computer Science (FCS 2007), pp. 356–361. CSREA Press (2007)

Bit-Parallel Tree Pattern Matching Algorithms for Unordered Labeled Trees*

Hiroaki Yamamoto[1] and Daichi Takenouchi[2]

[1] Department of Information Engineering, Shinshu University,
4-17-1 Wakasato, Nagano-shi, 380-8553 Japan
yamamoto@cs.shinshu-u.ac.jp
[2] NTT Advanced Technology Corporation

Abstract. The following tree pattern matching problem is considered: Given two unordered labeled trees P and T, find all occurrences of P in T. Here P and T are called a *pattern tree* and a *target tree*, respectively. We first introduce a new problem called *the pseudo-tree pattern matching problem*. Then we show two efficient bit-parallel algorithms for the pseudo-tree pattern matching problem. One runs in $O(L_P \cdot n \cdot l \cdot \lceil \frac{h}{W} \rceil)$ time and $O(n \cdot l \cdot \lceil \frac{h}{W} \rceil)$ space, and another one runs in $O((L_P \cdot n + h \cdot 2^l) \cdot \lceil \frac{h \cdot l}{W} \rceil)$ time and $O((n + h \cdot 2^l) \cdot \lceil \frac{h \cdot l}{W} \rceil)$ space, where n is the number of nodes in T, h and l are the height of P and the number of leaves of P, respectively, and W is the length of a computer-word. The parameter L_P, called a *recursive level of P*, is defined to be the number of occurrences of the same label on a path from the root to a leaf. Hence we have $L_P \leq h$. Finally, we give an algorithm to extract all occurrences from pseud-occurrences in $O(n \cdot L_P \cdot l^{3/2})$ time and $O(n \cdot L_P \cdot l)$ space.

1 Introduction

In recent years, XML has been recognized as a common data format for data storages and exchanging data over the Internet, and has been widely spread. The tree pattern matching problem is a central part of XML query problems. In addition, this problem has a number of applications in the fields of computer science. Therefore, many researches have been done on developing an efficient tree pattern matching algorithm. The tree pattern matching problem is as follows: Given two labeled trees P and T, find all occurrences of P in T. Here P and T are called a pattern tree and a target tree, respectively. For this problem, ordered trees and unordered trees have been considered. An ordered tree is a tree such that the left-to-right order among siblings is significant. Hence, the order must usually be preserved in the tree pattern matching problem. On the other hand, an unordered tree is a tree such that any order among siblings is not defined, and hence the order is not significant in the tree pattern matching problem.

* This research was supported by the Ministry of Education, Sports, Culture, Science and Technology, Grant-in-Aid for Scientific Research (C).

F. Dehne et al. (Eds.): WADS 2009, LNCS 5664, pp. 554–565, 2009.

For ordered trees, the tree pattern matching problem under the matching condition preserving parent-child relationship and the position of a child has been studied. The obvious algorithm runs in $O(n \cdot m)$ time, where n and m are the number of nodes of a target tree and a pattern tree, respectively. Hoffman and Donnell [7] proposed several algorithms. Dubiner, Galil and Magen [5] improved $O(n \cdot m)$ time and presented an $O(n \cdot \sqrt{m} \cdot polylog(m))$ time algorithm. Cole and Hariharan [3,4] presented $O(n \log^2 m)$ time tree pattern matching algorithm by introducing a subset matching problem. Chauve [2] consider a more general matching condition, and has given an $O(n \cdot l)$ time algorithm, where l is the number of leaves of a pattern tree.

Several researches on unordered trees have also been done. Kilpeläinen and Mannila [9] studied the tree inclusion problem, which can be regarded as the unordered tree pattern matching problem with an ancestor-descendant relationship. They presented $O(n \cdot m)$ time algorithm for ordered trees and showed NP-completeness for unordered trees. Shamir and Tsur [11] gave an $O(n \cdot \frac{m^{3/2}}{\log m})$ time algorithm to solve the subtree isomorphism problem, in which unrooted and unlabeled trees are considered. This algorithm can solve the unordered tree pattern matching problem with a parent-child relationship. Furthermore several researches on XML query problems have been done (for example, see [6,12,13]) because the order among siblings is not significant in many practical applications for querying XML.

In this paper, we are concerned with a tree pattern matching problem on unordered labeled trees. We here introduce two new notions of *a pseudo-tree pattern matching problem* and *the recursive level of a labeled tree*. A pseudo-tree pattern matching problem is defined by allowing a many-to-one mapping from nodes of P to nodes of T. Note that tree pattern matching problems are normally defined based on a one-to-one mapping. Then the pseudo-tree pattern matching problem is to find out all pseud-occurrences of P in T. Götz, Koch and Martens [6] have studied on the tree homeomorphism problem for searching XML data. This problem can be regarded as a pseudo-tree pattern matching problem with ancestor-descendant relationship. They gave an $O(n \cdot m \cdot h)$ time algorithm. *The recursive level of a labeled tree* is defined to be the maximum number of occurrences of the same label over a path from the root to a leaf. In XML applications, a labeled tree with the recursive level 1, called a *non-recursive labeled tree*, is well studied (for example see [6,12]). We present two efficient bit-parallel algorithms for solving the pseudo-tree pattern matching problem as follows. Here h and l are the height and the number of leaves of P, respectively, and W is the length of a computer-word, and L_P is the recursive level of P. Our algorithms make use of the Shift-OR technique which has been developed on the string matching problem [1].

- One algorithm runs in $O(L_P \cdot n \cdot l \cdot \lceil \frac{h}{W} \rceil)$ time and $O(n \cdot l \cdot \lceil \frac{h}{W} \rceil)$ space.
- Another one runs in $O((L_P \cdot n + h \cdot 2^l) \cdot \lceil \frac{h \cdot l}{W} \rceil)$ time and $O((n + h \cdot 2^l) \cdot \lceil \frac{h \cdot l}{W} \rceil)$ space. This algorithm consists of two parts *a preprocessing part* and *a matching part*. The preprocessing part, which generates bit-masks from a pattern tree P, takes $O(h \cdot 2^l \cdot \lceil \frac{h \cdot l}{W} \rceil)$ time and the matching part takes $O(L_P \cdot n \cdot \lceil \frac{h \cdot l}{W} \rceil)$ time.

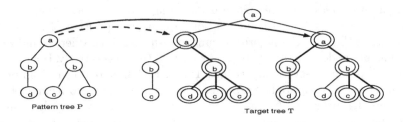

Fig. 1. A pseud-occurrence and an occurrence. The dotted arrow indicates a pseud-occurrence and the solid arrow indicates an occurrence (an exact occurrence).

In general, W is defined as $W = O(\log n)$ on conventional computing models. Hence, if h is at most $\log n$, then the first algorithm runs in $O(L_P \cdot n \cdot l)$ time, and if $h \cdot l$ is at most $\log n$, then the second algorithm runs in $O(L_P \cdot n)$ time. This time, if $L_P = O(1)$, then the second algorithm solves the pseudo-tree pattern matching problem in $O(n)$ time. Thus our algorithms run faster for pattern trees with small size.

Finally we give an algorithm to extract occurrences from pseud-occurrences for the tree pattern matching problem. If there are not any nodes with the same label among siblings in P, then a pseud-occurrence of P is identical to an occurrence of P, and hence the bit-parallel algorithms for the pseudo-tree pattern matching problem solve the tree pattern matching problem. If there are nodes with the same label among siblings in P, then a pseud-occurrence does not always become an occurrence. For this case, we can show an algorithm to obtain all occurrences of P from pseud-occurrences using an algorithm finding a maximum matching on bipartite graphs. Our algorithm runs in $O(n \cdot L_P \cdot l^{3/2})$ time and $O(n \cdot L_P \cdot l)$ space.

2   Tree Pattern Matching Problem and Related Definitions

Let Σ be an alphabet. Then we concentrate on a labeled tree such that each node of the tree is labeled by a symbol of Σ. Let T a labeled tree. For any node v of T, the children of node v are *siblings* of each other. If the order among siblings is significant, then the tree is said to be *ordered*; otherwise it is said to be *unordered*. The height of T is defined as follows. The depth of the root is defined to be 1. For any node v of T, the depth of v is defined to be the depth of the parent plus 1. Then the height of T is defined to be the maximum depth over all nodes of T. We introduce a notion of a *recursive level of T*. For any node v of T, *the recursive level of v* is defined to be the number of occurrences of the same symbol as v over the path from the root to v. The recursive level of T is defined to be the maximum recursive level over all nodes of T. In addition, for any $\sigma \in \Sigma$, we define *the recursive level of σ* to be the maximum recursive level over all nodes with label σ. We use two orders when traversing over the nodes of T. One is *preorder*, which is recursively defined as follows: First the root of T is visited. Let $T_1, \ldots, T_t$ be subtrees rooted by children of the root in the left-to-right order. Then each T_j is visited in the order from T_1 to T_t. Another one is *postorder*, in which the leftmost leaf is first visited, and then each node is visited after having visited all the children of the node.

Let P and T be unordered labeled trees, which are called a *pattern tree* and a *target tree*, respectively. We first define a notion of pseud-occurrence of P in T and a pseudo-tree pattern matching problem.

Definition 1 (a pseud-occurrence). *We say that P nearly matches T at a node d of T if there is a mapping ϕ from nodes of P into nodes of T such that*

1. *the root of P is mapped to d,*
2. *for any node u of P, there is a node $\phi(u)$ of T such that the label of u is equal to the label of $\phi(u)$,*
3. *for any nodes u, v of P, u is the parent of v if and only if $\phi(u)$ is the parent of $\phi(v)$.*

We say that d is a pseud-occurrence of P in T.

We give an example of a pseud-occurrence in Fig.1. Note that two nodes with label b of P are mapped to one node of T, that is, the right-hand child of a node a. Thus, in the definition of a pseud-occurrence, a mapping ϕ is allowed to be many-to-one. *The pseudo-tree pattern matching problem* is to find out all pseud-occurrences of P in T. Next we define a tree pattern matching problem, which is defined based on a one-to-one mapping.

Definition 2 (an occurrence). *We say that P matches T at a node d of T if there is a one-to-one mapping ϕ from nodes of P into nodes of T such that*

1. *it satisfies three conditions of the pseud-occurrence,*
2. *for any nodes u, v of P, if $u \neq v$, then $\phi(u) \neq \phi(v)$.*

We say that d is an occurrence (or an exact occurrence) of P in T.

We give an example of an occurrence in Fig.1. Note that the mapping ϕ is required to be a one-to-one mapping. *The tree pattern matching problem* is to find out all occurrences of P in T. It is clear from the definitions that a pseud-occurrence implies an occurrence, but the reverse does not always hold. In this paper, we first discuss the pseudo-tree pattern matching problem and then discuss the tree pattern matching problem.

3 Algorithms for the Pseudo-tree Pattern Matching Problem

In this section, we give a bit-parallel algorithm to find all pseud-occurrences of a pattern tree P in a target T. We make use of a Shift-OR technique on a string matching problem, which was developed by Baeza-Yates and Gonnet [1].

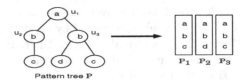

Fig. 2. Decomposition of a pattern tree P into path patterns P_1, P_2 and P_3

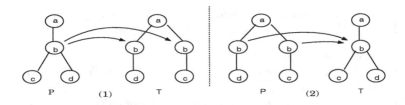

Fig. 3. Two problems in BasicTreeMatch

Let P_i be the string consisting of labels on a path from the root to a leaf in P. Then we call P_i a *path pattern* and denote by $|P_i|$ the length of path pattern P_i. We decompose P into path patterns for the Shift-OR technique. If P has l leaves, then P is decomposed into l path patterns $P_1, \ldots, P_l$. We say a node v of P appears on a path pattern P_i when the label of v appears on P_i. Fig. 2 illustrates an example of path patterns in which P is decomposed into three path patterns $P_1 = abc$, $P_2 = abd$ and $P_3 = abc$. We say that a path pattern occurs at a node d of T if the path pattern becomes just a prefix of the string consisting of labels on the path from node d to a leaf.

3.1 Bit-Masks for Path Patterns

To make use of the Shift-OR technique, we generate a bit-mask $B[P_i, \sigma]$ of h bits for every path pattern $P_i = p_1^i \cdots p_{|P_i|}^i$ and symbol σ, where h is the height of P. The value of $B[P_i, \sigma]$ is defined to be a bit sequence $b_1 \cdots b_h$ ($b_i \in \{0, 1\}$) such that for $j \leq |P_i|$, $b_j = 0$ if and only if $p_j^i = \sigma$, and for $|P_i| + 1 \leq j \leq h$, $b_j = 0$. For instance, bit-masks for three path patterns P_1, P_2, P_3 in Fig. 2 are defined as follows: $B[P_1, a] = 011$, $B[P_2, a] = 011$, $B[P_3, a] = 011$, $B[P_1, b] = 101$, $B[P_2, b] = 101$, $B[P_3, b] = 101$, $B[P_1, c] = 110$, $B[P_2, c] = 111$, $B[P_3, c] = 110$, $B[P_1, d] = 111$, $B[P_2, d] = 110$, $B[P_3, d] = 111$.

3.2 A Basic Algorithm

In this section, we give a simple algorithm *BasicTreeMatch*(P, T) using the Shift-OR technique, which is given in Fig. 4. This algorithm finds all nodes in a target tree T at which all path patterns of a pattern tree P occurs. We use an array $M[P_i, d]$ of bit-sequences, called a *matching state*, whose element is an h-bit

---

Algorithm BasicTreeMatch(P, T)

Step 0. /* Initialization */
    1. set bit-masks $B[P_i, c]$,
    2. for all path patterns P_i, set the initial matching state $IM[P_i]$ to $1^{|P_i|}0^{h-|P_i|}$.

Visit each node d of T in postorder and do the following for d.

Step 1. /* This step computes the matching state $M[P_i, d]$ of d from matching states of the children.
    Let $d_1, \ldots, d_t$ be children of d. If d is a leaf, then we use $IM[P_i]$ instead. */
    For all path pattern P_i,
    1. if d a leaf, then $M[P_i, d] := IM[P_i]$;
        otherwise $M[P_i, d] := M[P_i, d_1] \& \cdots \& M[P_i, d_t]$,
    2. $M[P_i, d] := (M[P_i, d] << 1) \mid B[P_i, \sigma]$.
Step 2. /* This step determines whether all path patterns occur. */
    1. $Temp := M[P_1, d] \mid \cdots \mid M[P_l, d]$,
    2. the first bit of $Temp$, that is, the bit corresponding to the root of P, is 0, then return d.

Fig. 4. The algorithm *BasicTreeMatch*

sequence $b_1 \cdots b_h$. While traversing over nodes of T in postorder, we compute $M[P_i, d]$ for each node d of T. Let $P_i = p_1^i \cdots p_{|P_i|}^i$. This time, for any node d of T, $M[P_i, d] = b_1 \cdots b_h$ satisfies that for any $j \leq |P_i|$, $b_j = 0$ if and only if $p_j^i \cdots p_{|P_i|}^i$ occurs at node d. $M[P_i, d]$ is initially set to $1^{|P_i|}0^{h-|P_i|}$. The operation $M[P_i, d] << 1$ used in the algorithm denotes a shift operation which shifts a bit-sequence in $M[P_i, d]$ one bit to the left and sets the rightmost bit to 0. In addition, the operator "$\mid$" denotes a bitwise OR, and the operator "$\&$" denotes a bitwise AND. *BasicTreeMatch* decomposes a pattern tree P into path patterns $P_1, \ldots, P_l$, and searches for a node of T at which all path patterns occur by traversing over T in postorder. The following proposition holds.

Proposition 1. *Let d be a node of T returned by* BasicTreeMatch. *Then all path patterns of P occur at the node d.*

The algorithm *BasicTreeMatch* completely cannot solve the tree pattern matching problem. That is, there are the following two problems: (1) one is that one node of P may be mapped to two or more nodes of T (see (1) in Fig.3); (2) another one is that two or more nodes of P may be mapped to one node of T (see (2) in Fig.3). *BasicTreeMatch* regards these cases as a match. The first case can be solved by introducing a notion of synchronization in a matching stage. Hence the pseudo-tree pattern matching problem can be solved. We will show this in the rest of this section. The second case will be discussed in Section 5.

3.3   A Pseudo-tree Pattern Matching Algorithm

BasicTreeMatch regards two cases in Fig.3 as a match. In this section, we give a bit-parallel algorithm which does not regard the case (1) as a match. In (1) of Fig.3, a node b of P is mapped to two nodes with b of T. Thus two path patterns bd and bc are separated in T, and hence this does not satisfy the condition of the pseud-occurrence. Therefore, by solving the case (1), we can solve the pseudo-tree pattern matching problem. Our algorithm checks whether bd and bc occur

---

Algorithm PseudoTreeMatch(P, T)

Step 0. /* Initialization */
1. set bit-masks $B[P_i, c]$,
2. for all path patterns P_i, set the initial matching state $IM[P_i]$ to $1^{|P_i|}0^{h-|P_i|}$,
3. for all path patterns P_i and all nodes u of P, set a synchronization bit-mask $Syn[P_i, u]$,
4. for all nodes u of P, set $\mathcal{P}_u = \{P_{i_1}, \ldots, P_{i_e}\}$ such that a path pattern P_{i_j} is in $\mathcal{P}_u$ if and only if u appears on P_{i_j}.

Visit each node d of T in postorder and compute the matching state $M[P_1, d], \ldots, M[P_l, d]$ for d. Here the label of d is σ.

Step 1. /* This step computes the matching state of d from matching states of the children. Let $d_1, \ldots, d_t$ be children of d. If d is a leaf, then we use $IM[P_i]$ instead. */
For all path pattern P_i,
1. if d a leaf, then $M[P_i, d] := IM[P_i]$;
   otherwise $M[P_i, d] := M[P_i, d_1] \& \cdots \& M[P_i, d_t]$,
2. $M[P_i, d] := (M[P_i, d] << 1) \mid B[P_i, \sigma]$.

Step 2. /* Synchronization between path patterns */
For $lev = 1, \ldots, L_\sigma$, /* L_σ denotes the recursive level of symbol σ.*/
   for all nodes u other than the root of P such that it has σ and the recursive level lev
1. for $P_i \in \mathcal{P}_u$, $SYN[P_i] := M[P_i, d] \& Syn[P_i, u]$,
2. $SynMask := SYN[P_{i_1}] \mid \cdots \mid SYN[P_{i_e}]$, where $\mathcal{P}_u = \{P_{i_1}, \ldots, P_{i_e}\}$,
3. for $P_i \in \mathcal{P}_u$, $M[P_i, d] := M[P_i, d] \mid SynMask$.

Step 3. /* This step determines whether a pseud-occurrence occurs.*/
1. $Temp := M[P_1, d] \mid \cdots \mid M[P_l, d]$,
2. the first bit of $Temp$, that is, the bit corresponding to the root of P, is 0, then return d as a pseud-occurrence.

Fig. 5. The algorithm *PseudoTreeMatch*

---

at the same node in T, and if they occur at the same node, then the algorithm regards them as a match; otherwise does not so.

For this purpose, we introduce a new bit-mask called a *synchronization bit-mask* for any node and path pattern of P. Let u be any node of P with the height h. Then, for any path pattern P_i, a synchronization mask $Syn[P_i, u] = b_1 \cdots b_h$ is defined as follows: for any $1 \leq j \leq h$, if the node corresponding to b_j is just u, then $b_j = 1$; otherwise $b_j = 0$. Thus, in $Syn[P_1, u], \ldots, Syn[P_l, u]$, only the bits corresponding to the node u are set to 1; the other bits are set to 0. For instance, we show synchronization bit-masks for the pattern tree given in Fig. 2. For nodes u_1, u_2 and u_3, we have the following: $Syn[P_1, u_1] = 100$, $Syn[P_2, u_1] = 100$, $Syn[P_3, u_1] = 100$, and $Syn[P_1, u_2] = 010$, $Syn[P_2, u_2] = 000$, $Syn[P_3, u_2] = 000$, and $Syn[P_1, u_3] = 000$, $Syn[P_2, u_3] = 010$, $Syn[P_3, u_3] = 010$.

The algorithm *PseudoTreeMatch* in given Fig. 5 is constructed by adding the synchronization stage of Step 2 to *BasicTreeMatch*; it can find out all pseud-occurrences of P in T. From Proposition 1, we know that *BasicTreeMatch* finds out all nodes in T at which all path patterns occur. We explain how Step 2 works. Let u be a node of P and the depth is j. Then, for any P_i and any node d of T, the j-th bit b_j of $M[P_i, d]$ corresponds to u. Step 2 checks out the bit b_j of $M[P_i, d]$ for all $P_i \in \mathcal{P}_u$, and if the bit b_j of at least one $M[P_i, d]$ is 1, then the bits b_j of all $M[P_i, d]$ are set to 1. To do this, we first check out the value of b_j in (a) of Step 2 using $Syn[P_i, u]$. If b_j of $M[P_i, d]$ is 1, then the j-th bit of $SYN[P_i]$ becomes 1. Hence if there is at least one $SYN[P_i]$ such that the j-th

M[P$_1$,d]	M[P$_2$,d]	. . . . .	M[P$_l$,d]

Fig. 6. A packed matching state $M[d]$

B[a]	011	011	011
B[b]	101	101	101
B[c]	111	110	110
B[d]	110	111	111

Fig. 7. Packed bit-masks of Fig.2

bit is 1, then the j-th bit of *SynMask* becomes 1 in (b) of Step 2. Finally, in this case, for all $P_i \in \mathcal{P}_u$, the j-bit b_j of $M[P_i, d]$ is set to 1 in (c) of Step 2. We have the following theorem.

Theorem 1. *The algorithm* PseudoTreeMatch *finds all pseud-occurrences of* P *in* $O(L_P \cdot n \cdot l \cdot \lceil \frac{h}{W} \rceil)$ *time and* $O(n \cdot l \cdot \lceil \frac{h}{W} \rceil)$ *space, where* n *is the number of nodes of* T, L_P, h, *and* l *are the recursive level, the height, and the number of leaves of* P, *respectively, and* W *is the length of a computer-word.*

3.4   Improving the Algorithm by Packing Bit-Sequences

Let n be the number of nodes of T, and let l and h be the number of leaves and height of P, respectively. The algorithm *PseudoTreeMatch* is checking a matching on each path pattern. Therefore it requires at least $n \times l$ time because there are l path patterns. In this section, we improve this matching process by packing path patterns into computer-words. This allows us to carry out matching processes on path patterns simultaneously. We give the improved algorithm *FastPseudoTreeMatch* in Fig. 8. In the algorithm, we pack matching states $M[P_1, d], \ldots, M[P_l, d]$ into one word $M[d]$ as in Fig. 6 (if the packed bit-sequence is long, then multiple words are used). We denote by $(M[P_1, d], \ldots, M[P_l, d])$ such a packed bit-sequence $M[d]$. Similarly, we also pack $B[P_1, \sigma], \ldots, B[P_l, \sigma]$ into $B[\sigma] = (B[P_1, \sigma], \ldots, B[P_l, \sigma])$ for each symbol σ as in Fig. 7. By these packing, we can simultaneously compute the matching state of each node in Step 1. Let $h_i = |P_i|$. We here use an h_i-bit sequence for bit-sequences such as $M[P_i, d]$ and $B[P_i, \sigma]$ to make as compact a packed bit-sequence as possible. In addition, we would like to carry out the synchronization task of Step 2 in *PseudoTreeMatch* simultaneously. To do this, we pack synchronization bit-masks into *PSyn*[σ, *lev*] as follows, where σ is a symbol and *lev* is a recursive level. We classify nodes of P into subsets $N_{(\sigma, lev)}$ such that $N_{(\sigma, lev)}$ consists of all nodes which are labeled by the symbol σ and have the recursive level *lev*. Let $N_{(\sigma, lev)} = \{u_1, \ldots, u_s\}$ for any symbol σ and any recursive level *lev*. Then

---

Algorithm FastPseudoTreeMatch(P, T)

Step 0. /* Initialization */
1. compute bit-masks $B[P_i, \sigma]$, and set $B[c] := (B[P_1, \sigma], \ldots, B[P_l, \sigma])$ for each symbol σ in P,
2. for all path patterns P_i, set the initial matching state $IM[P_i]$ to 1^{h_i}, and then set the packed initial matching state $IM := (IM[P_1], \ldots, IM[P_l])$,
3. for all path patterns P_i and all nodes u of P, compute a synchronization bit-mask $Syn[P_i, u]$,
4. for all symbols σ and recursive level lev, set the packed synchronization bit-mask $PSyn[\sigma, lev] = (PSyn_1, \ldots, PSyn_l)$, where if $u_j \in N_{(\sigma, lev)}$ appears on P_i, then $PSyn_i = Syn[P_i, u_j]$; otherwise $PSyn_i = 0^{h_i}$.
5. $SetPSynMask(P)$, /* set $PSynMask[SYN, lev]$ for at most $h2^l$ distinct values of SYN and a recursive level lev, */
6. set $ZMask := (1^{h_1-1}0, \ldots, 1^{h_l-1}0)$ and $AccCheck := (01^{h_1-1}, \ldots, 01^{h_l-1})$.

Visit each node d of T in postorder and compute the matching state of d as follows.

Step 1. /* Computing the matching state of d from the matching states of the children. Let $d_1, \ldots, d_t$ be children of d. The label of d is σ. */
1. If d is a leaf, then $M[d] := IM$; otherwise $M[d] := M[d_1]\& \cdots \&M[d_t]$,
2. $M[d] := ((M[d] << 1) \& ZMask) | B[\sigma]$.
Step 2. /* Synchronization between path patterns */
  For $lev = 1, \ldots, L_\sigma$ do /* L_σ denotes the recursive level of σ. */
1. $SYN := M[d] \& PSyn[\sigma, lev]$,
2. $M[d] := M[d] | PSynMask[SYN, lev]$.
Step 3. /* This step determines whether a pseudo-match occurs. */
1. $Acc := M[d] | AccCheck$,
2. if $Acc = AccCheck$, then return d as a pseud-occurrence.

Fig. 8. The algorithm *FastPseudoTreeMatch*

we define $PSyn[\sigma, lev] = (PSyn_1, \ldots, PSyn_l)$, where if a node u_j $(1 \leq j \leq s)$ appears on P_i, then $PSyn_i = Syn[P_i, u_j]$; otherwise $PSyn_i = 0^{h_i}$.

We introduce an array $PSynMask[SYN, lev]$ of bit-sequences to reflect the result of a synchronization to a matching state, where $SYN = (SYN_1, \ldots, SYN_l)$ and each SYN_i corresponds to $SYN[P_i]$ in *PseudoTreeMatch*. Let us define $\mathcal{P}_u$ to be the set of path patterns P_i such that node u appears on P_i. Then, for any nodes $u, v \in N_{(\sigma, lev)}$, we have $\mathcal{P}_u \cap \mathcal{P}_v = \emptyset$. Hence we can represent $SYN[P_i]$ for all nodes in $N_{(\sigma, lev)}$ by SYN. The algorithm *FastPseudoTreeMatch* computes SYN in one step, while *PseudoTreeMatch* compute $SYN[P_i]$ for each node of $N_{(\sigma, lev)}$. The value of $PSynMask[SYN, lev]$ is defined to be $(PSynMask_1, \ldots, PSynMask_l)$, where each $PSynMask_i$ corresponds to $SynMask$ computed for $\mathcal{P}_u$ corresponding to $u \in N_{(\sigma, lev)}$. Hence we can update a matching state $M[d]$ using $PSynMask[SYN, lev]$ in the same way as Step 2 of *PseudoTreeMatch*. *FastPseudoTreeMatch* carries out this task in one step at 2 of Step 2. We compute $PSynMask[SYN, lev]$ in Step 0 by the procedure $SetPSynMask(P)$ in given Fig. 9. In $SetPSynMask(P)$, $\mathcal{K}_{e_j^i}$ is defined to be a bit sequence $(K_1, \ldots, K_l)$ such that only the $d(u_i)$-th bit from the leftmost bit of $K_{e_j^i}$ is 1 and all other bits are 0, where $\mathcal{P}_{u_i} = \{P_{e_1^i}, \ldots, P_{e_{t_i}^i}\}$ for any $u_i \in N_{(\sigma, lev)}$ and $d(u_i)$ is the depth of u_i. Furthermore we make use of two special bit-masks, $ZMask$ and $AccCheck$. $ZMask$ is set to $(1^{h_1-1}0, \ldots, 1^{h_l-1}0)$ and is used for clearing the rightmost bit of each matching state in a packed bit-sequence. We need such a clearing

Procedure *SetPSynMask*(P)
For all symbols σ which occurs in P, do the following:

Step 1. /* set $SubMask[SYN, lev]$. */
   1.   for $lev = 1, \ldots, L_P$ do
   2.   for $u_i \in \{u_1, \ldots, u_s\}$ ($= N_{(\sigma, lev)}$) other than the root of P, do
   3. $TMask := (TMask_1, \ldots, TMask_l)$, where $TMask_k = Syn[P_k, u_i]$,
   4.   for $SYN = \mathcal{K}_{e_1^i}, \ldots, \mathcal{K}_{e_{t_i}^i}$ do
   5. $SubMask[SYN, lev] := TMask$,
   6.   end-for
   7.   end-for
   8. end-for
Step 2. /* compute $PSynMask[SYN, lev]$. */
   1. for $lev = 1, \ldots, L_P$ do
   2. $IX := 0$, $Val[0] := 0$ and $PSynMask[0, lev] := (0^{h_1}, \ldots, 0^{h_l})$,
   3.   for $k_1 = \mathcal{K}_{e_1^1}, \ldots, \mathcal{K}_{e_{t_1}^1}, \ldots, \mathcal{K}_{e_1^s}, \ldots, \mathcal{K}_{e_{t_s}^s}$ do
   4. $t := IX$,
   5.   for $k_2 := 0, \ldots, t$ do
   6. $PSynMask[k_1 + Val[k_2], lev] := PSynMask[Val[k_2], lev] \mid SubMask[k_1, lev]$,
   7. $IX := IX + 1$,
   8. $Val[IX] := k_1 + Val[k_2]$,
   9.   end-for
  10.   end-for
  11. end-for

Fig. 9. The procedure *SetPSynMask*

Algorithm ExactTreeMatch(P, T)

Step 1. Do *PseudoTreeMatch*(P, T) or *FastPseudoTreeMatch*(P, T).
Step 2. For all pseud-occurrences d of P in T do
    if $CheckMatch(\{v_P\}, d)$ returns $\{v_P\}$, then return d as an exact occurrence, where v_P is the
    root of P.

Fig. 10. The algorithm *ExactTreeMatch*

process because a shift operation sets the rightmost bit to 0. *AccCheck* is set to $(01^{h_1 - 1}, \ldots, 01^{h_l - 1})$ and is used for checking whether or not a pseud-occurrence occurs. We have the following theorem.

Theorem 2. *FastPseudoTreeMatch can find out all pseud-occurrences of P in T in* $O((L_P \cdot n + h \cdot 2^l) \cdot \lceil \frac{h \cdot l}{W} \rceil)$ *time and* $O((n + h \cdot 2^l) \cdot \lceil \frac{h \cdot l}{W} \rceil)$ *space.*

Let m be the number of nodes in P. Then we have $h \cdot 2^l \leq 2^m$. Hence if $m = \log n$, then *FastPseudoTreeMatch* runs in $O(L_P \cdot n \cdot \lceil \frac{h \cdot l}{W} \rceil)$ time and space. Furthermore, if $h \cdot l = O(W)$ and $L_P = O(1)$, then it runs in $O(n)$ time.

4 A Tree Pattern Matching Algorithm

In the previous section, we have given an algorithm to find all pseud-occurrences of P in T. A pseud-occurrence allows a mapping to map multiple nodes of P to one node of T, but the condition of an occurrence of P does not. Therefore, to find out all occurrences of P, we must check whether a pseud-occurrences satisfies the condition of an occurrence or not.

Function CheckMatch(G, d)

Step 1. $Match := \emptyset$,
Step 2. for all $g \in G$ do
    1. if g is a leaf of P, then add g to $Match$;
    2. otherwise do the following:
       (a) classify children $d_1, \ldots, d_t$ of d into groups $D_{a_1}, \ldots, D_{a_r}$ such that D_{a_j} $(1 \leq j \leq r)$ consists of all nodes with label a_j, and similarly classify children $g_1, \ldots, g_{t'}$ of g into groups $F_{a_1}, \ldots, F_{a_r}$ in the same way.
       (b) for all groups $D_{a_j} = D_{a_1}, \ldots, D_{a_r}$ do
          i. $R := \emptyset$,
          ii. for all nodes $d_k \in D_{a_j}$ do
             A. compute the set G_k^j consisting of all nodes g' of P such that g' is a child of g and d_k becomes a pseud-occurrence of the subtree rooted by node g' using the matching state $M[d_k]$ of d_k.
             B. $F_k^j := CheckMatch(G_k^j, d_k)$,
             C. add all pairs (g', d_k) with $g' \in F_k^j$ to R,
          iii. if $ExistMap(F_{a_j}, D_{a_j}, R) = false$, then go to next node of G.
       (c) add g to $Match$,
Step 3. return $Match$.

Fig. 11. The function *CheckMatch*

4.1 A Special Case

Here let us consider a special case; for any node v of a pattern tree P, all children of v have distinct labels. If P is the case, then pseud-occurrences become occurrences of P. Hence, we have the following theorem.

Theorem 3. *Let P be a pattern tree such that for any node of P, labels of any two children of the node are distinct. Then the algorithms* PseudoTreeMatch *and* FastPseudoTreeMatch *can find out all occurrences of P in T.*

4.2 A General Case

Let us consider a general case, that is, there are siblings in P such that they have the same label. The most difficult problem is that two or more nodes of P may be mapped to one node of T. We extract occurrences from pseud-occurrences by checking whether or not there is a one-to-one mapping. As in [11], the algorithm is designed using an algorithm finding a maximum matching on bipartite graphs. The algorithm *ExactTreeMatch* in Fig.10 checks whether a pseud-occurrence of P in T is an occurrence of P using the function *CheckMatch* given in Fig.11.

Given a node d in T and a subset G of nodes of P such that d becomes a pseud-occurrence of the subtree rooted by node $g \in G$, the function *CheckMatch*(G, d) returns the subset $Match$ of G such that d becomes an occurrence. *CheckMatch* recursively checks whether a pseud-occurrence satisfies a one-to-one mapping in (b) of Step 2. The function $ExistMap(F_{a_j}, D_{a_j}, R)$ returns *true* if there is a subset R' of R such that (1) for any $g \in F_{a_j}$, there is $d \in D_{a_j}$ with $(g, d) \in R'$, and (2) for any $(g_1, d_1), (g_2, d_2) \in R'$, if $g_1 \neq g_2$, then $d_1 \neq d_2$; otherwise returns *false*. We can view (F_{a_j}, D_{a_j}, R) as a bipartite graph having the vertex set $F_{a_j} \cup D_{a_j}$ and the edge set R. This time, $ExistMap(F_{a_j}, D_{a_j}, R)$ can be implemented using an algorithm for a maximum matching on bipartite graphs. For a pseud-occurrence d of P in T, let T_d be the subtree of T such that the root is d and the other nodes consists of all descendants of d which are at most h away from d. Then we define N_p to be $\sum_d |T_d|$, where d takes all pseud-occurrences of P and $|T_d|$ denotes the number of nodes in T_d. Then if we use the algorithm by Hopcroft and Karp [8], we have the following theorem. Here note that since we have $N_p \leq n \cdot L_P$ and $L_p \leq h$, the algorithm runs in $O(n \cdot h \cdot l^{3/2})$ time and $O(n \cdot h \cdot l)$ space in the worst case.

Theorem 4. *The algorithm* ExactTreeMatch *can find out all occurrences of P in T in $O(N_P \cdot l^{3/2})$ time and $O(N_P \cdot l)$ space plus the complexity of* PseudoTreeMatch *or* FastPseudoTreeMatch.

Acknowledgments. We are grateful to anonymous referees for many variable comments, which helped to improve algorithms and the presentation.

References

1. Baeza-Yates, R., Gonnet, G.H.: A New Approach to Text Searching. Communications of the ACM 35(10), 74–82 (1992)
2. Chauve, C.: Tree pattern matching with a more general notion of occurrence of the pattern. Information Processing Letters 82, 197–201 (2001)
3. Cole, R., Hariharan, R.: Verifying Candidate Matches in Sparse and Wildcard Matching. In: Proc. of the 34th ACM STOC, pp. 592–601 (2002)
4. Cole, R., Hariharan, R.: Tree Pattern Matching to Subset Matching in Linear Time. SIAM J. Comput. 32(4), 1056–1066 (2003)
5. Dubiner, M., Galil, Z., Magen, E.: Faster Tree Pattern Matching. Journal of the ACM 41(2), 205–213 (1994)
6. Götz, M., Koch, C., Martens, W.: Efficient Algorithms for the Tree Homeomorphism Problem. In: Arenas, M., Schwartzbach, M.I. (eds.) DBPL 2007. LNCS, vol. 4797, pp. 17–31. Springer, Heidelberg (2007)
7. Hoffman, C.M., O'Donnell, M.J.: Pattern matching in trees. Journal of the ACM 29(1), 68–95 (1982)
8. Hopcroft, J.E., Karp, R.M.: An $n^{5/2}$ Algorithm for Maximum Matchings in Bipartite Graphs. SIAM J. Comput. 2(4), 225–231 (1973)
9. Kilpeläinen, P., Mannila, H.: Ordered and Unordered Tree Inclusion. SIAM J. Comput. 24(2), 340–356 (1995)
10. Leeuweno, J.V.: Graph Algorithms. In: Leeuwen, J.V. (ed.) Handbook of Theoretical Computer Science, Elsevier Science Pub., Amsterdam (1990)
11. Shamir, R., Tsur, D.: Faster Subtree Isomorphism. J. of Algorithms 33, 267–280 (1999)
12. Yao, J.T., Zhang, M.: A Fast Tree Pattern Matching Algorithm for XML Query. In: Proc. of the WI 2004, pp. 235–241 (2004)
13. Zezula, P., Mandreoli, F., Martoglia, R.: Tree Signatures and Unordered XML Pattern Matching. In: Van Emde Boas, P., Pokorný, J., Bieliková, M., Štuller, J. (eds.) SOFSEM 2004. LNCS, vol. 2932, pp. 122–139. Springer, Heidelberg (2004)

Compact and Low Delay Routing Labeling Scheme for Unit Disk Graphs

Chenyu Yan, Yang Xiang, and Feodor F. Dragan

Algorithmic Research Laboratory, Department of Computer Science
Kent State University, Kent, Ohio, U.S.A.
{cyan,yxiang,dragan}@cs.kent.edu

Abstract. In this paper, we propose a *new compact and low delay routing labeling scheme* for *Unit Disk Graphs (UDGs)* which often model wireless ad hoc networks. We show that one can assign each vertex of an n-vertex UDG G a compact $O(\log^2 n)$-bit label such that, given the label of a source vertex and the label of a destination, it is possible to compute efficiently, based solely on these two labels, a neighbor of the source vertex that heads in the direction of the destination. We prove that this *routing labeling scheme* has a constant *hop route-stretch* (= *hop delay*), i.e., for each two vertices x and y of G, it produces a routing path with $h(x,y)$ hops (edges) such that $h(x,y) \le 3 \cdot d_G(x,y) + 12$, where $d_G(x,y)$ is the hop distance between x and y in G. To the best of our knowledge, this is the first compact routing scheme for UDGs which not only guaranties delivery but has a low hop delay and polylog label size. Furthermore, our routing labeling scheme has a constant length route-stretch.

1   Introduction

A common assumption for wireless ad hoc networks is that all nodes have the same maximum transmission range. By proper scaling, one can model these networks with *Unit Disk Graphs (UDGs)*, which are defined as the intersection graphs of equal sized circles in the plane [3]. In other words, there is an edge between two vertices in an UDG if and only if their Euclidean distance is no more than one.

Communications in networks are performed using *routing schemes*, i.e., mechanisms that can deliver packets of information from any vertex of a network to any other vertex. In most strategies, each vertex v of a graph has full knowledge of its neighborhood and uses a piece of global information available to it about the graph topology – some "sense of direction" to each destination – stored locally at v. Based only on this information and the address of a destination vertex, vertex v needs to decide whether the packet has reached its destination, and if not, to which neighbor of v to forward the packet. The *efficiency* of a routing scheme is measured in terms of its *multiplicative route-stretch* (or *additive route-stretch*), namely, the maximum ratio (or surplus) between the cost (which could be the *hop-count* or the *length* of a route, produced by the scheme for a pair of vertices, and the cost of an optimal route available in graph for that

F. Dehne et al. (Eds.): WADS 2009, LNCS 5664, pp. 566–577, 2009.

pair. Here, the *hop-count* of a route is defined as the number of edges on it and the *length* of a route is defined as the sum of the Euclidean length of its edges. Using different cost functions, for a given graph G and a given routing scheme on G, one can define two different notions of route-stretch: *hop route-stretch* and *length route-stretch*.

The most popular strategy in wireless networks is the *geographic routing* (sometimes called also the *greedy geographic routing*), where each vertex forwards the packet to the neighbor geographically closest to the destination (see survey [12] for this and many other strategies). Each vertex of the network knows its position (e.g., Euclidean coordinates) in the underlying physical space and forwards messages according to the coordinates of the destination and the coordinates of neighbors. Although this greedy method is effective in many cases, packets may get routed to where no neighbor is closer to the destination than the current vertex. Many recovery schemes have been proposed to route around such voids for guaranteed packet delivery as long as a path exists [4,14,16]. These techniques typically exploit planar subgraphs (e.g., Gabriel graph, Relative Neighborhood graph), and packets traverse faces on such graphs using the well-known right-hand rule. Although these techniques guarantee packet delivery, none of them give any guaranties on how the routing path traveled is "close" to an optimal path; the worst-case route-stretch can be linear in the network size.

All earlier papers assumed that vertices are aware of their physical location, an assumption which is often violated in practice for various of reasons (see [7,15,21]). In addition, implementations of recovery schemes are either based on non-rigorous heuristics or on non-trivial planarization procedures. To overcome these shortcomings, recent papers [7,15,21] propose routing algorithms which assign virtual coordinates to vertices in a metric space X and forward messages using geographic routing in X. In [21], the metric space is the Euclidean plane, and virtual coordinates are assigned using a distributed version of Tutte's "rubber band" algorithm for finding convex embeddings of graphs. In [7], the graph is embedded in R^d for some value of d much smaller than the network size, by identifying d beacon vertices and representing each vertex by the vector of distances to those beacons. The distance function on R^d used in [7] is a modification of the ℓ_1 norm. Both [7] and [21] provide substantial experimental support for the efficacy of their proposed embedding techniques – both algorithms are successful in finding a route from the source to the destination more than 95% of the time – but neither of them has a provable guarantee. Unlike embeddings of [7] and [21], the embedding of [15] guarantees that the geographic routing will always be successful in finding a route to the destination, if such a route exists. Algorithm of [15] assigns to each vertex of the network a virtual coordinate in the hyperbolic plane, and performs greedy geographic routing with respect to these virtual coordinates. However, although the experimental results of [15] confirm that the greedy hyperbolic embedding yields routes with low route-stretch when applied to typical unit-disk graphs, the worst-case route-stretch is still linear in the network size.

In this paper, we propose a new compact and low delay routing labeling scheme for Unit Disk Graphs. We show that one can assign each vertex of an n-vertex UDG G a compact $O(\log^2 n)$-bit label such that, given the label of a source vertex and the label of a destination, it is possible to compute efficiently, based solely on these two labels, a neighbor of the source vertex that heads in the direction of the destination. We prove that this *routing labeling scheme* has a constant *hop route-stretch* ($=$ *hop delay*), i.e., for each two vertices x and y of G, it produces a routing path with $h(x, y)$ hops such that $h(x, y) \leq 3 \cdot d_G(x, y) + 12$, where $d_G(x, y)$ is the hop distance between x and y in G. To the best of our knowledge, this is the first compact routing scheme for UDGs which not only guaranties delivery but has a low hop delay and polylog label size. Furthermore, our routing labeling scheme has a constant length route-stretch. Note also that, unlike geographic routing or any other strategies discussed in [4,7,12,14,15,16,21], our routing scheme is *degree-independent*. That is, each current vertex makes routing decision based only on its label and the label of destination, does not involve any labels of neighbors. The label assigned to a vertex in our scheme can be interpreted as its virtual coordinates. To assign those labels to vertices, we need to know only the topology of the input unit disk graph and relative Euclidean lengths of its edges.

To obtain our routing scheme, we establish a novel *balanced separator* theorem for UDGs, which mimics the well-known Lipton and Tarjan's planar balanced shortest paths separator theorem. We prove that, in any n-vertex UDG G, one can find two hop-shortest paths $P(s, x)$ and $P(s, y)$ such that the removal of the 3-hop-neighborhood of these paths (i.e., $N_G^3[P(s, x) \cup P(s, y)]$) from G leaves no connected component with more than $2/3n$ vertices. The famous Lipton and Tarjan's planar balanced separator theorem has two variants (see [19]). One variant (called *planar balanced $\sqrt{n}$-separator theorem*) states that any n-vertex planar graph G has a set S of vertices such that $|S| = O(\sqrt{n})$ and the removal of S from G leaves no connected component with more than $2/3n$ vertices. Another variant (called *planar balanced shortest-paths separator theorem*) states that any n-vertex planar graph G has two shortest paths removal of which from G leaves no connected component with more than $2/3n$ vertices. Although the first variant of the planar balanced separator theorem has an extension to the class of disk graphs (which includes UDGs) (see [1]), the second variant of the theorem proved to be more useful in designing compact routing (and distance) labeling schemes for planar graphs (see [13,22]). To the date, there was not known any extension of the planar balanced shortest-paths separator theorem to unit disk graphs. The paper [11] notes that *"Unfortunately, Thorup's algorithm uses balanced shortest-path separators in planar graphs which do not obviously extend to the unit-disk graphs"* and uses the well-separated pair decomposition to get fast approximate distance computations in UDGs. We do not know how to use the well-separated pair decomposition of an UDG G to design a compact and low delay routing labeling scheme for G. Application of the balanced $\sqrt{\cdot}$-separator theorem of [1] to UDGs can result only in routing (and distance) labeling schemes with labels of size no less than $O(\sqrt{n} \log n)$-bits per vertex. Our separator theorem allows us to get $O(\log^2 n)$-bit labels which is more suitable for the wireless ad hoc and sensor networks where the issues of memory size and power-conservation are critical.

Our new *balanced shortest-paths—3-hop-neighborhood separator* theorem allows us to build, for any n-vertex UDG $G = (V, E)$, a system $\mathcal{T}(G)$ of at most $2 \log_{\frac{3}{2}} n + 2$ spanning trees of G such that, for any two vertices x and y of G, there exists a tree T in $\mathcal{T}(G)$ with $d_T(x, y) \leq 3 \cdot d_G(x, y) + 12$. That is, the distances in any UDG can be approximately represented by the distances in at most $2 \log_{\frac{3}{2}} n + 2$ of its spanning trees. An earlier version of these results has appeared in [24] (see Section 3.4 and pages 124 and 125 of Section 3.5.5). Taking the union of all these spanning trees of G, we obtain a *hop* $(3, 12)$-*spanner* H of G (i.e., a spanning subgraph H of G with $d_H(x, y) \leq 3 \cdot d_G(x, y) + 12$ for any $x, y \in V$) with at most $O(n \log n)$ edges. There is a number of papers describing different types of *length-spanners* and *hop-spanners* for UDGs (see [2,8,10,17,18] and literature cited therein). Many of those spanners have nice properties of being planar or sparse, or having bounded maximum degree or bounded length (or hop) *spanner-stretch*, or having localized construction. Unfortunately, neither of those papers develops or discusses any routing schemes which could translate the constant spanner-stretch bounds into some constant route-stretch bounds.

2 Notions and Notations

Let V be a set of $n = |V|$ nodes on the Euclidean plane and let $G = (V, E)$ be the unit disk graph (UDG) induced by those nodes. Let also $m = |E|$. For each edge (a, b) of G, by (a, b) we denote also the open straightline segment representing it, and by $|ab|$ the Euclidean length of the edge/segment (a, b). For simplicity, in what follows, we will assume that any two edges in G can intersect at no more than one point (i.e., no two intersecting edges are on the same straight line), and no three edges intersect at the same point.

For a path P of G, the *hop-count* of P is defined as the number of edges on P and the *length* of P is defined as the sum of the Euclidean length of its edges. For any two vertices x and y of G, we denote: by $d_G(x, y)$, the *hop-distance* (or simply *distance*) in G between x and y, i.e., the minimum hop-count of any path connecting x and y in G; by $l_G(x, y)$, the *length-distance* in G between x and y, i.e., the minimum length of any path connecting x and y in G.

A graph family Γ is said (see [20]) to have an $l(n)$ bit (s, r)-*approximate distance labeling scheme* if there is a function L labeling the vertices of each n-vertex graph in Γ with distinct labels of up to $l(n)$ bits, and there exists an algorithm/function f, called *distance decoder*, that given two labels $L(v), L(u)$ of two vertices v, u in a graph G from Γ, computes, in time polynomial in the length of the given labels, a value $f(L(v), L(u))$ such that $d_G(v, u) \leq f(L(v), L(u)) \leq s \cdot d_G(v, u) + r$. Note that the algorithm is not given any additional information, other that the two labels, regarding the graph from which the vertices were taken. Similarly, a family Γ of graphs is said (see [20]) to have an $l(n)$ *bit routing labeling scheme* if there exist a function L, labeling the vertices of each

n-vertex graph in Γ with distinct labels of up to $l(n)$ bits, and an efficient algorithm/function, called the *routing decision* or *routing protocol*, that given the label $L(v)$ of a current vertex v and the label $L(u)$ of the destination vertex u (the header of the packet), decides in time polynomial in the length of the given labels and using only those two labels, whether this packet has already reached its destination, and if not, to which neighbor of v to forward the packet.

Let $\mathcal{R}$ be a routing scheme and $R(x, y)$ be a route (path) produced by $\mathcal{R}$ for a pair of vertices x and y in a graph G. We say that $\mathcal{R}$ has: *hop (α, β)-route-stretch* if hop-count of $R(x, y)$ is at most $\alpha \cdot d_G(x, y) + \beta$, for any $x, y \in V$; *length (α, β)-route-stretch* if length of $R(x, y)$ is at most $\alpha \cdot l_G(x, y) + \beta$, for any $x, y \in V$.

Let $H = (V, E')$ be a spanning subgraph of a graph $G = (V, E)$. We say that H is: *hop (α, β)-spanner* of G if $d_H(x, y) \le \alpha \cdot d_G(x, y) + \beta$, for any $x, y \in V$; *length (α, β)-spanner* of G if $l_H(x, y) \le \alpha \cdot l_G(x, y) + \beta$, for any $x, y \in V$.

In Section 6, we will need also the notion of collective tree spanners from [6]. It is said that a graph G admits a system of μ collective tree (α, β)-spanners if there is a system $\mathcal{T}(G)$ of at most μ spanning trees of G such that for any two vertices x, y of G a spanning tree $T \in \mathcal{T}(G)$ exists such that $d_T(x, y) \le \alpha \cdot d_G(x, y) + \beta$.

For a vertex v of G, the kth *neighborhood* of v in G is the set $N_G^k[v] = \{u \in V : d_G(v, u) \le k\}$. For a vertex v of G, the sets $N_G[v] = N_G^1[v]$ and $N_G(v) = N_G[v] \setminus \{v\}$ are called the *neighborhood* and the *open neighborhood* of v, respectively. For a set $S \subseteq V$, by $N_G^k[S] = \bigcup_{v \in S} N_G^k[v]$ we denote the kth *neighborhood* of S in G.

3   Intersection Lemmas

In this section we present few auxiliary lemmas. From the definition of unit disk graphs, we immediately conclude the following (proofs of these lemmas and all other omitted proofs can be found in the journal version of the paper).

Lemma 1. *In an UDG $G = (V, E)$, if edges $(a, b), (c, d) \in E$ intersect, then G must have at least one of $(a, c), (b, d)$ and at least one of $(a, d), (c, b)$ in E.*

Let r be an arbitrary but fixed vertex of an UDG $G = (V, E)$, and $L_0, L_1, \ldots L_q$ be the *layering* of G with respect to r, where $L_i = \{u \in V : d_G(r, u) = i\}$. For G, using this layering, we construct a *layering tree* T_{orig} rooted at r as follows: each vertex $v \in L_i$ ($i \in \{1, \ldots, q\}$) chooses a neighbor u in L_{i-1} such that $|vu|$ is minimum (closest neighbor in L_{i-1}) to be its father in T_{orig} (breaking ties arbitrarily). Let $E(T_{orig})$ be the edge set of T_{orig}. This tree T_{orig} will help us to construct a balanced separator for G. It will be convenient, for each vertex $v \in V$, by $L(v)$ to denote the layer index of v, i.e., $L(v) = d_G(r, v)$. In what follows, we will also adopt the following agreements (unless otherwise is specified). When we refer to any edge (a, b) of T_{orig}, we assume $L(a) = L(b) - 1$. When we refer to any two intersecting edges (a, b) and (c, d) of T_{orig} (in that order), we assume that $L(a) \le L(c)$.

Lemma 2. *In T_{orig}, no two edges (a, b) and (c, d) with $L(a) = L(c)$ and $L(b) = L(d)$ can cross.*

Lemma 3. *Let $(a, b), (c, d)$ be two edges in T_{orig} that intersect. If $L(a) = L(b) - 1$, $L(c) = L(d) - 1$ and $L(a) \leq L(c)$, then $L(a) = L(c) - 1$, $(a, d) \notin E$ and $(b, c) \in E$.*

For an UDG $G = (V, E)$, in what follows, by $G_p = (V_p, E_p)$ we denote the planar graph obtained from G by turning each edge intersection point in G into a vertex in G_p. The vertices of T_{orig} (i.e. vertices of G) will be called *real vertices*, to differentiate them from *imaginary* and *null* points that will be defined later. In the following, we will use the term "element" as a general name for real vertices, imaginary points and null points. For any graph $\mathcal{G}$, we will use $E(\mathcal{G})$ to denote the set of its edges and $V(\mathcal{G})$ to denote the set of its vertices (or elements, if $V(\mathcal{G})$ contains imaginary or null points). Below, we will create an imaginary point (details will be given later) at the point where two edges (a, b) and (c, d) from T_{orig} intersect. Recall that we agreed to assume that $L(a) = L(b) - 1$, $L(c) = L(d) - 1$ and $L(a) \leq L(c)$. By Lemma 3, we know that $L(a) = L(c) - 1$. Now, assuming that the imaginary point is m, we define $a(m) = a$, $b(m) = b$, $c(m) = c$ and $d(m) = d$.

4   Balanced Separator for Restricted UDGs

In this section, we consider a special unit disk graph, a simple-crossing UDG. On this simple case, we demonstrate our idea of construction of a balanced separator. It may help the reader to follow the much more complicated case, where we construct a balanced separator for an arbitrary UDG. We define a *simple-crossing UDG* to be an UDG $G = (V, E)$ with each edge crossing at most one other edge.

In what follows, we will transform tree T_{orig} into a special spanning tree T for the planar graph G_p. Let $T = T_{orig}$ initially. For each two intersecting edges (a, b) and (c, d) of T_{orig} (by Lemma 3, we know $L(a) = L(c) - 1$), we do the following. Create a vertex $m_{a,b,c,d}$ at the point where (a, b) and (c, d) intersect. We call $m_{a,b,c,d}$ an *imaginary point*. Remove edges (a, b), (c, d) from T and add vertex $m_{a,b,c,d}$ and edges $(m_{a,b,c,d}, d)$, $(a, m_{a,b,c,d})$ and $(b, m_{a,b,c,d})$ into T. One can see that all the descendants of b and d in T find their way to the root via a.

There are two other kinds of edge intersections in G: the intersection between a tree-edge and a non-tree-edge and the intersection between two non-tree-edges. We handle them separately. First, assume a tree-edge (u, w) intersects a non-tree-edge (s, t). We create a new vertex, called a *null point*, say o, at the point where (u, w) and (s, t) intersect. We remove edge (u, w) from T and add vertex o and edges (u, o), (o, w) into T. Now assume two non-tree-edges (a, b) and (c, d) intersect. We create a new vertex, called a *null point*, say o, at the point where (a, b) and (c, d) intersect. We add vertex o (as a pendant vertex) and edge (a, o) into T.

It is easy to see that T is a spanning tree for the planar graph G_p. We will need the Lipton and Tarjan's planar separator theorem [19] in the following form.

Theorem 1 (Planar Separator Theorem). [19] *Let G be any planar graph with non-negative vertex weights and W be the total weight of G (which is the sum of the weights of its vertices). Let T be any spanning tree of G rooted at a* *vertex r. Then, there exist two vertices x and y in G such that if one removes from G the tree-paths connecting in T r with x and r with y, then each connected component of the resulting graph has total weight at most 2/3W. Vertices x and y can be found in linear time.*

We can apply Theorem 1 to T and G_p by letting the weight of each real vertex be 1 and the weight of each imaginary or null point be 0 in G_p. Then, there must exist in T two paths $P_1 = P_T(r, x)$ and $P_2 = P_T(r, y)$ such that removal of them from G_p leaves no connected component with more than $2/3n$ real vertices.

Using paths $P_1 = (x_0 = r, x_1, \ldots, x_{k-1}, x_k = x)$ and $P_2 = (y_0 = r, y_1, \ldots, y_{l-1}, y_l = y)$ of G_p (of T), we can create a balanced separator for G as follows. (1) Skip all the null points in P_1 and P_2. (2) Skip every imaginary point in P_i which is collinear with its two neighbors in P_i ($i = 1, 2$). (3) For any imaginary point $m_{a,b,c,d}$ in P_i ($i = 1, 2$) which is not collinear with its two neighbors in P_i (the only possible case is where $L(a) = L(c) - 1$ and imaginary point $m_{a,b,c,d}$ connects a and d in P_i), replace the subpath $(a, m_{a,b,c,d}, d)$ by either (a, c, d) (if $(a, c) \in E$) or (a, b, d) (if $(b, d) \in E$). By Lemma 1, (a, c) or (b, d) is in E. Let P_i' be the resulting path obtained from P_i ($i = 1, 2$). It is easy to check that P_1' and P_2' are shortest paths in G. Here and in what follows, by a shortest path we mean a hop-shortest path. We can also show that the union of $N_G^1[P_1']$ and $N_G^1[P_2']$ is a balanced separator for G, i.e., removal of $N_G^1[P_1'] \cup N_G^1[P_2']$ from G leaves no connected component with more that $2/3n$ vertices. Assume that removal of P_1 and P_2 from $G_p = (V_p, E_p)$ results in removing a set of edges E_p' from E_p, and removal of $N_G^1[P_1']$ and $N_G^1[P_2']$ from $G = (V, E)$ results in removing a set of edges E' from E. It is easy to check that, for any edge $e_p' \in E_p'$ there exists an edge $e' \in E'$ that covers e_p'. The latter implies that the union of $N_G^1[P_1']$ and $N_G^1[P_2']$ is a balanced separator for G. A formal proof of this will be presented in the journal version of the paper.

5   Balanced Separator for Arbitrary UDGs

In an arbitrary unit disk graph $G = (V, E)$, an edge may cross any number of other edges. Our basic strategy for building a balanced separator for G is similar to one we used in the case of a simple-crossing UDG, but details are more complicated. Let $T = T_{orig}$ initially. We will revise T to create a special spanning tree for the planar graph G_p obtained from G. Then, we will apply the Planar Separator Theorem from [19] (Theorem 1 above) to G_p and T to get a balanced separator S for G_p. Finally, we will recover from S the required separator for G.

5.1   Building a Special Spanning Tree T of G_p

In what follows, the edges of the tree T_{orig} will be called *original tree-edges*. By Lemma 3, for any two intersecting original tree-edges (a, b) and (c, d) (for which we assumed that $L(a) = L(b) - 1$, $L(c) = L(d) - 1$ and $L(a) \leq L(c)$), we have $L(a) = L(c) - 1$, $(a, d) \notin E(G)$ and $(b, c) \in E(G)$. We handle this kind of intersections (between original tree-edges) using PROCEDURE 1.

PROCEDURE 1. Handle original tree-edge intersections
Input: A layering tree T_{orig} rooted at r.
Output: A tree T where all original tree-edge intersections resolved.
Method: /* Break ties arbitrarily */

(1) Let $L_i = \{v : L(v) = i\}$ and $T = T_{orig}$;
(2) Let q be the maximum layer number of T;
(3) **FOR** $i = 1$ to q **DO**
(4)   **FOR** each vertex $v_j \in L_i$ **DO**
(5)   **FOR** each vertex $v_k \in L_{i+1}$ adjacent to v_j in T **DO**
(6)   **IF** there is an original tree-edge intersection on (v_j, v_k) such that
               $L(v_j)$ is the SECOND smallest layer index among the layer indices
               of all four end-vertices of the two edges giving the intersection
               THEN DO
(7)   Choose such an original tree-edge intersection closest to v_k and
                 assume it is the intersection between (v_j, v_k) and (x, y) in T
                 and between (v_j, v_k) and (v_p, v_h) in T_{orig} (i.e., $(x, y) \subseteq (v_p, v_h)$);
(8)   Create an imaginary point $m_{j,k,p,h}$ at the point where (v_j, v_k) and
                 (x, y) intersect;
(9)   Update T by removing edges (v_j, v_k) and (x, y), and adding vertex
                 $m_{j,k,p,h}$ and edges $(m_{j,k,p,h}, x), (m_{j,k,p,h}, y), (m_{j,k,p,h}, v_k)$;
(10) **RETURN** T

Lemma 4. *PROCEDURE 1 returns a tree T with all original tree-edge intersections resolved (i.e., edges of T do not cross each other).*

In addition, there are two other kinds of intersections remaining: the intersection between an edge in $E(T)$ (T-edge) and an edge in $E(G) \setminus E(T)$ (non-T-edge), and intersection between two non-T-edges.

First we handle intersections between T-edges and non-T-edges. They are resolved the same way as in Section 4. Here, we rephrase the rule. Assume (u, w) is a T-edge, (s, t) is a non-T-edge. Add a null point, say o, at the point where (u, w) and (s, t) intersect. Remove edge (u, w) from T and add vertex o and edges $(u, o), (o, w)$ into T. After resolving all intersections of this kind, T becomes a subgraph of G_p. Note that it is possible that T does not span yet all elements of $V(G_p)$. *Let name this T as T_{sub}.*

Now, we deal with intersections between two non-T_{sub}-edges. This is more complicated than it was in Section 4 for restricted UDGs. We will grow T_{sub} to a spanning tree T_{span} for G_p (extension T_{span} of T_{sub} will cover all elements of $V(G_p)$). We use a procedure similar to one of building a shortest path tree from a set of vertices. We assign to each vertex in T_{sub} a weight according to the following formula. In formula, if v is an imaginary point or a null point, we assume v is at the intersection between edges (a, b) and (c, d) of G.

$$weight(v) = \begin{cases} 0, & \text{if } v \text{ is a real vertex;} \\ min\{|av|, |bv|, |cv|, |dv|\}, & \text{if } v \text{ is an imaginary or a null point.} \end{cases}$$

To build our spanning tree for G_p, we use PROCEDURE 2. At the beginning, for any $v \in V(G_p) \setminus V(T_{sub})$, $distance[v] = \infty$ and father of v is undefined.

PROCEDURE 2. Build a spanning tree for G_p from T_{sub}

Input: A tree $T = T_{sub}$;
Output: A tree T_{span} as a spanning tree for G_p.
Method: /* Break ties arbitrarily */

(1)   **FOR** each i in $V(T)$ **DO**
(2)   **FOR** each neighbor $j \in V(G_p)\backslash V(T)$ of i **DO**
(3) $tmp := weight[i] + |ij|$;
(4)   **IF** $tmp < distance[j]$ **DO**
(5) $distance[j] := tmp$;
(6) $father[j] := i$;
(7) $Q := V(G_p)\backslash V(T)$;
(8)   **WHILE** Q is not empty **DO**
(9) $u :=$node in Q with smallest distance$[\cdot]$;
(10)   remove u from Q and add u into T;
(11)   **FOR** each neighbor $v \in Q$ of u **DO**
(12) $tmp := distance[u] + |uv|$;
(13)   **IF** $tmp < distance[v]$ **DO**
(14) $distance[v] := tmp$;
(15) $father[v] := u$;
(16) **RETURN** $T_{span} := T$.

It is easy to check that T_{span} is a spanning tree of the planar graph G_p.

5.2   Finding a Balanced 2×Shortest-Paths—3-Hop-Neighborhood Separator for G

Now we can apply Theorem 1 to G_p and T_{span} by letting the weight of each real vertex be 1 and the weight of each imaginary or null point be 0, and get a balanced separator S of G_p. Assume that S is the union of paths $P_1 = P_{T_{span}}(r, x)$ and $P_2 = P_{T_{span}}(r, y)$. There are three kinds of elements on P_1 and P_2: real vertices, imaginary points and null points. Generally, each imaginary point or null point is adjacent to at most four elements in G_p, and each element in P_1 or P_2 has the previous element and the next element, except for the root r (it has only the next element) and elements x and y (they have only the previous element). Let u be the last real or imaginary point in P_1 (or P_2). We name all null points after u in P_1 (or P_2) as the *tail null points*. For any element in P_1 or P_2, there are two possible relations between itself, its previous element and its next element: the element, its previous element and its next element are on the same line, which means its previous element and its next element are on the same edge of G (according to our general assumption that no two edges of G are on the same line); the element, its previous element and its next element are not on the same line, which means its previous element and itself are on one edge of G, and its next element and itself are on another edge of G.

Using paths $P_1 = (x_0 = r, x_1, \ldots, x_{k-1}, x_k = x)$ and $P_2 = (y_0 = r, y_1, \ldots, y_{l-1}, y_l = y)$ of G_p (of T_{span}), We will find the corresponding balanced separator for G using the following steps. (1) We skip all null points in P_1 and P_2. Let the resulting paths be P_1' and P_2', respectively. (2) We skip in P_1' and P_2' each imaginary point whose previous element and next element are on the same edge of T_{orig}. For example, let (x_f, x_i, x_j) be a fragment of path P_1' or P_2', where x_i is an imaginary point and $\{x_f, x_i, x_j\}$ are collinear, then (x_f, x_i, x_j) will be replaced with (x_f, x_j). Let the resulting paths be P_1'' and P_2'', respectively. (3) Replace each remaining imaginary point m in P_1'' and P_2'' with two vertices: $b(m)$ followed by $c(m)$ (see end of Section 3 for these notations). For example, let (x_f, x_i, x_j) be a fragment of path P_1'' or P_2'', where x_i is an imaginary point and x_f is closest to the root r among $\{x_f, x_i, x_j\}$. Then, (x_f, x_i, x_j) will be replaced with $(x_f, b(x_i), c(x_i), x_j)$. Let the resulting paths be P_1''' and P_2''', respectively. By Lemma 3, the edge $(b(x_i), c(x_i))$ exists in G. It is easy to check that P_1''' and P_2''' are valid paths in G.

A path P of G is called a 2×shortest path iff for any two vertices x,y in P, $d_P(x,y) \leq 2d_G(x,y)$.

Theorem 2. *P_1''' and P_2''' are 2× shortest paths in G.*

We can show also that the union of $N_G^3[P_1''']$ and $N_G^3[P_2''']$ is a balanced separator for G with 2/3-split, i.e., removal of $N_G^3[P_1'''] \cup N_G^3[P_2''']$ from G leaves no connected component with more than $2/3n$ vertices. Thus, there exist two paths P_1''' and P_2''' in G such that they are 2×shortest paths and the union of $N_G^3[P_1''']$ and $N_G^3[P_2''']$ is a balanced separator for G.

5.3   Finding a Balanced Shortest-Paths—3-Hop-Neighborhood Separator for G

In this section, we will improve the result of Section 5.2. We will show that any UDG G has two shortest paths P_1''' and P_2''' such that the union of $N_G^3[P_1''']$ and $N_G^3[P_2''']$ forms a balanced separator for G. Recall that, by a shortest path we mean a hop-shortest path.

Let P_1, P_2, P_1', P_2', P_1'' and P_2'' be the paths defined in Section 5.2. Analogs of paths P_1''' and P_2''' of Section 5.2 will be obtained from P_1'' and P_2'' in a more careful way (than in Section 5.2). We use PROCEDURE 3 for this.

PROCEDURE 3. Handle imaginary points
Input: Path $P \in \{P_1'', P_2''\}$ (containing still some imaginary points).
Output: Path P as a shortest path of G, with all imaginary points resolved.
Method: /* Break ties arbitrarily. The first vertex in P is the root r, a real vertex.*/

(1) Let $[v_1, \cdots, v_k]$ be the imaginary points in P in the order from r;
(2) **FOR** $i = 1$ to k **DO**
(3)   **IF** vertex $c(v_i)$ is adjacent to $prev_P(v_i)$
         ($c(v_i)$ is always adjacent to $next_P(v_i)$, as it is shown later.)
(4)   Replace v_i with $c(v_i)$ in P;
(5)   **ELSE** (It implies that vertex $b(v_i)$ is adjacent to both $prev_P(v_i)$
         and $next_P(v_i)$, as it is shown later.)
(6)   Replace v_i with $b(v_i)$ in P;
(7) **RETURN** P

We call PROCEDURE 3 for both P_1'' and P_2''. Let the resulting paths be P_1''' and P_2''', respectively. We can show that P_1''' and P_2''' are shortest paths in G. Now, for these paths P_1''' and P_2''', we have.

Theorem 3. *The union of $N_G^3[P_1''']$ and $N_G^3[P_2''']$ is a balanced separator for G with 2/3-split, i.e., removal of $N_G^3[P_1'''] \cup N_G^3[P_2''']$ from G leaves no connected component with more than $2/3n$ vertices.*

6   Application of Balanced Separators for UDGs

In this section, we show how one can use the above balanced separator theorem for UDGs to construct for them collective tree spanners with low stretch and to develop a compact and low delay routing labeling scheme. For this, we combine strategies used in [5,6,13]. The details can be found in the full version of this paper. Here we list only the final results.

Theorem 4. *Any unit disk graph G with n vertices and m edges admits a system $T(G)$ of at most $2\log_{3/2} n + 2$ collective tree $(3, 12)$-spanners, i.e., for any two vertices x and y in G, there exists a spanning tree $T \in T(G)$ with $d_T(x, y) \leq 3d_G(x, y) + 12$. Moreover, such a system $T(G)$ can be constructed in $O((C + m) \log n)$ time, where C is the number of crossings in G.*

Corollary 1. *Any unit disk graph G with n vertices admits a hop $(3, 12)$-spanner with at most $2(n - 1)(\log_{3/2} n + 1)$ edges.*

Theorem 5. *The family of n-vertex unit disk graphs admits an $O(\log^2 n)$ bit $(3, 12)$-approximate distance labeling scheme with $O(\log n)$ time distance decoder.*

Theorem 6. *The family of n-vertex unit disk graphs admits an $O(\log^2 n)$ bit routing labeling scheme. The scheme has hop $(3, 12)$-route-stretch. Once computed by the sender in $O(\log n)$ time, headers never change, and the routing decision is made in constant time per vertex.*

In the journal version of the paper, we show also how to extend this bounded hop route-stretch routing labeling scheme to a routing labeling scheme with bounded length route-stretch.

References

1. Alber, J., Fiala, J.: Geometric separation and exact solutions for the parameterized independent set problem on disk graphs. J. of Algorithms 52, 134–151 (2004)
2. Alzoubi, K., Li, X.-Y., Wang, Y., Wan, P.-J., Frieder, O.: Geometric spanners for wireless ad hoc networks. IEEE Trans. on Par. and Distr. Syst. 14, 408–421 (2003)
3. Clark, B.N., Colbourn, C.J.: Unit Disk Graphs. Discrete Math. 86, 165–177 (1990)
4. Bose, P., Morin, P., Stojmenovic, I., Urrutia, J.: Routing with guaranteed delivery in ad hoc wireless networks. In: 3rd Internat. workshop on discr. algor. and methods for mobile computing and communications, pp. 48–55. ACM Press, New York (1999)

5. Dragan, F.F., Yan, C., Corneil, D.G.: Collective Tree Spanners and Routing in AT-free Related Graphs. J. of Graph Algor. and Applic. 10(2), 97–122 (2006)
6. Dragan, F.F., Yan, C., Lomonosov, I.: Collective tree spanners of graphs. SIAM J. Discrete Math. 20, 241–260 (2006)
7. Fonseca, R., Ratnasamy, S., Zhao, J., Ee, C.T., Culler, D., Shenker, S., Stoica, I.: Beacon vector routing: Scalable point-to-point routing in wireless sensornets. In: 2nd USENIX/ACM Symp. on Netw. Syst. Design and Implement (NSDI 2005) (2005)
8. Fürer, M., Kasiviswanathan, S.P.: Spanners for geometric intersection graphs. In: Dehne, F., Sack, J.-R., Zeh, N. (eds.) WADS 2007. LNCS, vol. 4619, pp. 312–324. Springer, Heidelberg (2007)
9. Fraigniaud, P., Gavoille, C.: Routing in Trees. In: Orejas, F., Spirakis, P.G., van Leeuwen, J. (eds.) ICALP 2001. LNCS, vol. 2076, pp. 757–772. Springer, Heidelberg (2001)
10. Gao, J., Guibas, L.J., Hershberger, J., Zhang, L., Zhu, A.: Geometric spanner for routing in mobile networks. In: 2nd ACM international symposium on mobile ad hoc networking & computing, Long Beach, CA, USA, October 04-05 (2001)
11. Gao, J., Zhang, L.: Well-separated pair decomposition for the unit-disk graph metric and its applications. In: STOC 2003, pp. 483–492 (2003)
12. Giordano, S., Stojmenovic, I.: Position based routing algorithms for ad hoc networks: A taxonomy. In: Ad Hoc Wireless Networking, pp. 103–136. Kluwer, Dordrecht (2004)
13. Gupta, A., Kumar, A., Rastogi, R.: Traveling with a Pez Dispenser (Or, Routing Issues in MPLS). In: FOCS 2001, pp. 148–157 (2001)
14. Karp, B., Kung, H.T.: GPSR: greedy perimeter stateless routing for wireless networks. In: Proceedings of the 6th ACM/IEEE MobiCom, pp. 243–254. ACM Press, New York (2000)
15. Kleinberg, R.: Geographic routing using hyperbolic space. In: INFOCOM 2007, pp. 1902–1909 (2007)
16. Kuhn, F., Wattenhofer, R., Zhang, Y., Zollinger, A.: Geometric ad-hoc routing: of theory and practice. In: PODC 2003, pp. 63–72. ACM Press, New York (2003)
17. Li, X.-Y.: Ad Hoc Wireless Networking. In: Li, X.-Y. (ed.) Applications of Computational Geomety in Wireless Ad Hoc Networks, Kluwer, Dordrecht (2003)
18. Li, X.-Y., Wang, Y.: Geometrical Spanner for Wireless Ad Hoc Networks. In: Handbook of Approx. Algorithms and Metaheuristics. Chapman&Hall/Crc, Boca Raton (2006)
19. Lipton, R.J., Tarjan, R.E.: A Separator Theorem for Planar Graphs. SIAM Journal on Applied Mathematics 36, 177–189 (1979)
20. Peleg, D.: Distributed Computing: A Locality-Sensitive Approach. SIAM Monographs on Discrete Math. Appl. SIAM, Philadelphia (2000)
21. Rao, A., Papadimitriou, C., Shenker, S., Stoica, I.: Geographical routing without location information. In: MobiCom 2003, pp. 96–108 (2003)
22. Thorup, M.: Compact Oracles for Reachability and Approximate Distances in Planar Digraphs. In: FOCS, pp. 242–251 (2001)
23. Thorup, M., Zwick, U.: Compact routing schemes. In: SPAA 2001, pp. 1–10 (2001)
24. Yan, C.: Approximating Distances in Complicated Graphs by Distances in Simple Graphs With Applications. PhD Dissertation, Kent State University (2007), http://www.ohiolink.edu/etd/send-pdf.cgi/Yan%20Chenyu.pdf?kent1184639623

Author Index